Solution Key

Algebra 2
and Trigonometry

Mary P. Dolciani
John A. Graham
Richard A. Swanson
Sidney Sharron

Houghton Mifflin Company • Boston
Atlanta Dallas Geneva, Ill. Palo Alto Princeton Toronto

1989 Impression

Printed in U.S.A.

ISBN: 0-395-45710-6

BCDEFGHIJ-CS-99876543210

CONTENTS

CHAPTER 1 • Review of Essentials

Pages 2–3 • WRITTEN EXERCISES

A **1.** {1, 5, 9, 10, 11, 13, 15, 20, 25} **2.** {4, 5, 6, 7, 8, 10, 15, 20, 25} **3.** {15}
4. {1, 4, 5, 6, 7, 8, 9, 10, 11, 13, 15} **5.** {5, 10} **6.** ∅

B **7.** $X \cap Y = X$, $X \cup Y = Y$

C **8.** $\{a, b, c, d\}$, $\{a, b, c\}$, $\{a, b, d\}$, $\{a, c, d\}$, $\{b, c, d\}$, $\{a, b\}$, $\{a, c\}$, $\{a, d\}$, $\{b, c\}$, $\{b, d\}$, $\{c, d\}$, $\{a\}$, $\{b\}$, $\{c\}$, $\{d\}$, ∅ **9.** 2^n

Page 3 • COMPUTER EXERCISES

1.
```
10 INPUT A, B, E
20 LET C = (B - A)/E
30 PRINT A
40 FOR I = 1 TO E
50 LET A = A + C
60 PRINT A
70 NEXT I
80 END
```
2. 2, 2.4, 2.8, 3.2, 3.6, 4
3. 3, 3.6, 4.3, 4.9, 5.5, 6.2, 6.8, 7.5, 8.1, 8.7, 9.4, 10
4. −1, −0.25, 0.5, 1.25, 2, 2.75, 3.5, 4.25, 5

Page 6 • WRITTEN EXERCISES

A **1.** {3} **2.** {−2, −1, 1, 3} **3.** ∅ **4.** {−2, −1, 1, 2, 3} **5.** {−2, −1, 1, 2, 3}
6. {−2, 2} **7.** {−1, 1, 3} **8.** {−2, −1} **9.** {−2, −1, 1, 2, 3} **10.** ∅

11.
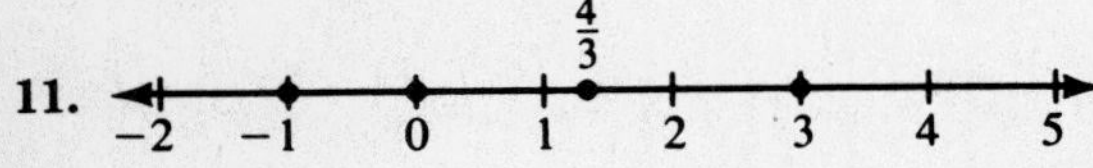

12.
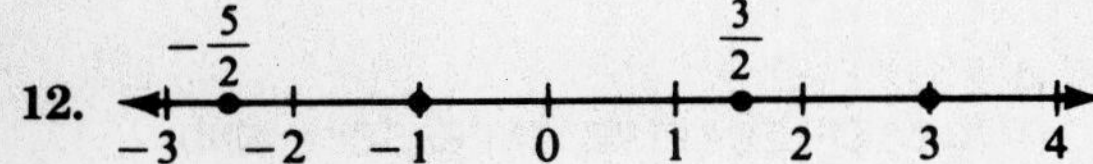

13. −7 −6 −5 −4 −3 −2 −1 0

14. −8 −4 0 4 8 12 16

B **15.** 0 1 2 3 4 5 6 7 8 9 10

16. 0 2 4 6 8 10

17. 0 1 2 3 4 5 6 7 8 9 10

18. 2 3 4 5 6

19. 0 1 2 3 4 5 6 7 8 9 10

20. 2 4 6 8 10

C **21.** 0 1 2 3 4 5 6 7 8 9 10 **22.** 2 3 4 5 6 7 8 9 10 11

Pages 9–10 • WRITTEN EXERCISES

A **1.** 3; Comm. ax. for add. **2.** ab; Assoc. ax. for mult. **3.** y; Symm. prop. of =
4. $3n$; Dist. ax. **5.** -4; Ax. of add. inv. **6.** 0; Iden. ax. for add.
7. (-2); Reflex. prop. of = **8.** $\frac{1}{2r}$; Ax. of mult. inv.
9. $8 + c$; Comm. ax. for mult. **10.** 1; Iden. ax. for mult.
11. p, r; Trans. prop. of =
12. x; Closure ax. for mult. **13.** 11 **14.** 22 **15.** 120 **16.** -840
B **17.** yes; yes **18.** yes; yes; the nth row is identical to the nth column. **19.** 0; yes
20. 1; no
21. $(1 \oplus 2) \oplus 3 = 3 \oplus 3 = 0$; $1 \oplus (2 \oplus 3) = 1 \oplus 1 = 0$; $\therefore (1 \oplus 2) \oplus 3 = 1 \oplus (2 \oplus 3)$
22. $(1 \star 2) \star 3 = 2 \star 3 = 3$; $1 \star (2 \star 3) = 1 \star 3 = 3$; $\therefore (1 \star 2) \star 3 = 1 \star (2 \star 3)$
C **23.** $2 \star (1 \oplus 3) = 2 \star 2 = 1$; $(2 \star 1) \oplus (2 \star 3) = 2 \oplus (2 \star 3) = 2 \oplus 3 = 1$; $\therefore 2 \star (1 \oplus 3) = (2 \star 1) \oplus (2 \star 3)$
24. $3 \star (1 \oplus 2) = 3 \star 3 = 0$; $(3 \star 1) \oplus (3 \star 2) = 3 \oplus 3 \star 2 = 3 \oplus 3 = 0$; $\therefore 3 \star (1 \oplus 2) = (3 \star 1) \oplus (3 \star 2)$
25.

$(x + a)(x + b) = (x + a)x + (x + a)b$	Dist. ax.
$= x^2 + ax + xb + ab$	Dist. ax.
$= x^2 + ax + bx + ab$	Comm. ax. for mult.
$= x^2 + (a + b)x + ab$	Dist. ax.
$(x + a)(x + b) = x^2 + (a + b)x + ab$	Trans. prop. of =

Page 10 • COMPUTER EXERCISES

1.
```
10 INPUT X
20 PRINT "ADDITIVE INVERSE IS "; -X
30 IF X = 0 THEN PRINT "NO MULTIPLICATIVE INVERSE": END
40 PRINT "MULTIPLICATIVE INVERSE IS "; 1/X
50 END
```
2. 5, -0.2 **3.** -0.625, 1.6 **4.** 0, none **5.** -7, 0.142857 **6.** 0.83333, -1.2
7. 3, -0.333333

Page 11 • SELF-TEST 1

1. $3 \in \{y: y < 5 \text{ and } y \in \mathscr{R}\}$ **2.** $\{0\}$ **3.** $\emptyset$

4. 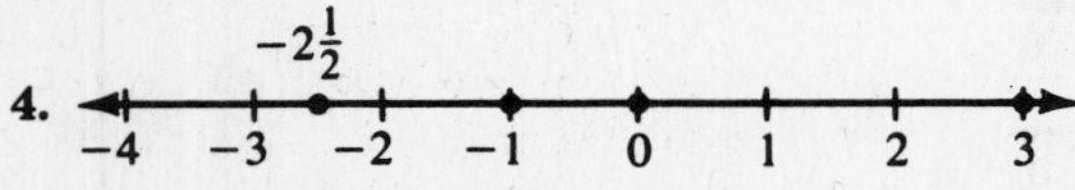

5. −2 −1 0 1 2 3 4 5 6 7 8

6. Identity axiom for addition **7.** Associative axiom for addition
8. Distributive axiom **9.** Axiom of multiplicative inverses **10.** $\{-48\}$

Pages 15–17 • WRITTEN EXERCISES

A **1.** -3 **2.** -15 **3.** y **4.** 0 **5.** 3 **6.** 8 **7.** 4 **8.** 0 **9.** z **10.** $-b$
11. $7c + (-16)$ **12.** $-4 + 2k$ **13.** $3x + (-9)$ **14.** -8 **15.** 2 **16.** $-40 + r$
17. $2w + (-3)$ **18.** 11 **19.** $\{-23\}$ **20.** $\{-18\}$ **21.** $\{-24\}$ **22.** $\{7\}$ **23.** $\{20\}$
24. $\{3\}$

B **25.** 1. Hypothesis 2. Ax. of add. inv. 3. Assoc. ax. for add. 4. Ax. of add. inv. 5. Iden. ax. for add. 6. Trans. prop. of =

26. 1. Hypothesis 2. Ax. of add. inv. 3. Iden. ax. for add. 4. Ax. of add. inv. 5. Assoc. ax. for add. 6. Ax. of add. inv. 7. Iden. ax. for add. 8. Trans. prop. of =

27.

1. $c + a = c + b$	Hypothesis
2. $c + a = a + c$, $c + b = b + c$	Comm. ax. for add.
3. $a + c = b + c$	Subs. prin.
4. $a = b$	Theorem, p. 13

28.

1. $a + (-c) = b + (-c)$	Hypothesis
2. $-c$ is a real number.	Ax. of add. inv.
3. $a = b$	Canc. prop. of add.

29.

1. $b + a = a$	Hypothesis
2. $a = 0 + a$	Iden. ax. for add.
3. $b + a = 0 + a$	Trans. prop. of =
4. $b = 0$	Canc. prop. of add.

30.

1. $a + b = 0$	Hypothesis
2. $0 = a + (-a)$	Ax. of add. inv.
3. $a + b = a + (-a)$	Trans. prop. of =
4. $b = -a$	Canc. prop. of add.

31.

1. $a, b, \in \mathcal{R}$	Hypothesis
2. $-a, -b \in \mathcal{R}$	Ax. of add. inv.
3. $-[(-a) + (-b)] = -(-a) + [-(-b)]$	Prop. of opp. of a sum
4. $-[(-a) + (-b)] = a + b$	Canc. prop. of add. inv.

32.

1. $a = b$ and $c = d$	Hypothesis
2. $a + c = a + c$	Reflex. prop. of =
3. $a + c = b + c$	Subs. prin.
4. $a + c = b + d$	Subs. prin.

33.

1. $a + c = b + d$ and $c = d$	Hypothesis
2. $a + c = b + c$	Subs. prin.
3. $a = b$	Canc. prop. of add.

34.

1. $x + 3 = 7$	Hypothesis
2. $\quad = 7 + 0$	Iden. ax. for add.
3. $\quad = 7 + [(-3) + 3]$	Ax. of add. inv.
4. $\quad = [7 + (-3)] + 3$	Assoc. ax. for add.
5. $\quad = 4 + 3$	Subs. prin.
6. $x + 3 = 4 + 3$	Trans. prop. of =
7. $x = 4$	Canc. prop. of add.

35.

1. $x + (-6) = -1$	Hypothesis
2. $\quad = -1 + 0$	Iden. ax. for add.
3. $\quad = -1 + [6 + (-6)]$	Ax. of add. inv.
4. $\quad = (-1 + 6) + (-6)$	Assoc. ax. for add.
5. $\quad = 5 + (-6)$	Subs. prin.
6. $x + (-6) = 5 + (-6)$	Trans. prop. of =
7. $x = 5$	Canc. prop. of add.

36. 1. $x + 9 = -1$ Hypothesis
2. $= -1 + 0$ Iden. ax. for add.
3. $= -1 + [(-9) + 9]$ Ax. of add. inv.
4. $= [-1 + (-9)] + 9$ Assoc. ax. for add.
5. $= -10 + 9$ Subs. prin.
6. $x + 9 = -10 + 9$ Trans. prop. of =
7. $x = -10$ Canc. prop. of add.

C **37.** 1. $x + b = a$ Hypothesis
2. $= a + 0$ Iden. ax. for add.
3. $= a + (-b + b)$ Ax. of add. inv.
4. $= [a + (-b)] + b$ Assoc. ax. for add.
5. $x + b = [a + (-b)] + b$ Trans. prop. of =
6. $x = a + (-b)$ Canc. prop. of add.

38. 1. $x = a + (-b)$ Hypothesis
2. $x + b = x + b$ Reflex. prop. of =
3. $x + b = [a + (-b)] + b$ Subs. prin.
4. $= a + [(-b) + b]$ Assoc. ax. for add.
5. $= a + 0$ Ax. of add. inv.
6. $= a$ Iden. ax. for add.
7. $x + b = a$ Trans. prop. of =

39. 1. $2x + b = x$ Hypothesis
2. $(x + x) + b = x$ Subs. prin.
3. $x + (x + b) = x$ Assoc. ax. for add.
4. $x + (x + b) = x + 0$ Iden. ax. for add.
5. $x + b = 0$ Canc. prop. of add.
6. $x + b = (-b) + b$ Ax. of add. inv.
7. $x = -b$ Canc. prop. of add.

Pages 20–21 • WRITTEN EXERCISES

A **1.** 420 **2.** -55 **3.** $5xy$ **4.** $24ab$ **5.** -6 **6.** $\frac{19}{6}$ **7.** $30p + (-40)pq$
8. $-2r + (-18)s$ **9.** $4ac + 4bc$ **10.** $-23x + (-xy)$ **11.** $36h + (-132)hk$
12. $-6w + \frac{3}{2}$ **13.** $6cd + \left(-\frac{21}{2}\right)c$ **14.** $15a$ **15.** $-4x + 13y + (-2xy)$
16. $10mn + 5m + 6n$ **17.** $2ab + 13a + (-40b)$ **18.** positive **19.** positive
20. zero **21.** negative **22.** positive **23.** zero

B **24.** 1. Assoc. ax. for mult. 2. Ax. of mult. inv. 3. Iden. ax. for mult.
4. Trans. prop. of =

25. 1. Hypothesis 2. Ax. of mult. inv. 3. Subs. prin. 4. Assoc. ax. for mult.
5. Ax. of mult. inv. 6. Iden. ax. for mult.

26. 1. $ca = cb$; $c \neq 0$ Hypothesis
2. $ca = ac$, $cb = bc$ Comm. ax. for mult.
3. $ac = bc$ Subs. prin.
4. $a = b$ Ex. 25

27. 1. $a \neq 0$; $b \neq 0$ Hypothesis
2. $\frac{1}{ab}, \frac{1}{a}$, and $\frac{1}{b}$ are real numbers. Ax. of mult. inv.
3. $ab\left(\frac{1}{a} \cdot \frac{1}{b}\right) = \left[a\left(\frac{1}{a}\right)\right] \cdot \left[b\left(\frac{1}{b}\right)\right]$ Comm. and assoc. ax. for mult.

4. $= 1 \cdot 1$ — Ax. of mult. inv.
5. $= 1$ — Iden. ax. for mult.
6. $ab\left(\frac{1}{a} \cdot \frac{1}{b}\right) = 1$ — Trans. prop. of =
7. $ab\left(\frac{1}{ab}\right) = 1$ — Ax. of mult. inv.
8. $ab\left(\frac{1}{ab}\right) = ab\left(\frac{1}{a} \cdot \frac{1}{b}\right)$ — Trans. prop. of =
9. $\frac{1}{ab} = \frac{1}{a} \cdot \frac{1}{b}$ — Canc. prop. of mult.

28. 1. $ab = a$, and $a \neq 0$ — Hypothesis
2. $a = a \cdot 1$ — Iden. ax. for mult.
3. $ab = a \cdot 1$ — Trans. prop. of =
4. $b = 1$ — Canc. prop. of mult.

29. 1. $ab = 1$ — Hypothesis
2. $1 = a\left(\frac{1}{a}\right)$ — Ax. of mult. inv.
3. $ab = a\left(\frac{1}{a}\right)$ — Subs. prin.
4. $b = \frac{1}{a}$ — Canc. prop. of mult.

30. 1. $a \neq 0$ — Hypothesis
2. $\frac{1}{a} \cdot a = 1$ — Ax. of mult. inv.
3. $a = \frac{1}{\frac{1}{a}}$ — Ex. 29
4. $\frac{1}{\frac{1}{a}} = a$ — Symm. prop. of =

31. 1. $(-a)b = [(-1)a]b$ — Mult. prop. of -1
2. $= (-1)(ab)$ — Assoc. ax. for mult.
3. $= -ab$ — Mult. prop. of -1
4. $(-a)b = -ab$ — Trans. prop. of =

32. 1. $a(-b) = a[b \cdot (-1)]$ — Mult. prop. of -1
2. $a(-b) = (ab)(-1)$ — Assoc. ax. for mult.
3. $a(-b) = -ab$ — Mult. prop. of -1

33. 1. $(-a)(-b) = [(-1)a][(-1)b]$ — Mult. prop. of -1
2. $= [(-1)(-1)](ab)$ — Comm. and assoc. ax. for mult.
3. $= [-(-1)](ab)$ — Mult. prop. of -1
4. $= 1 \cdot ab$ — Canc. prop. of add. inv.
5. $= ab$ — Iden. ax. for mult.
6. $(-a)(-b) = ab$ — Trans. prop. of =

34. 1. $a = b$, $a \neq 0$, $b \neq 0$ — Hypothesis
2. $a \cdot \frac{1}{a} = 1$, $1 = b \cdot \frac{1}{b}$ — Ax. of mult. inv.
3. $a \cdot \frac{1}{a} = b \cdot \frac{1}{b}$ — Trans. prop. of =

[*Proof continued on next page*]

4. $a \cdot \frac{1}{a} = a \cdot \frac{1}{b}$ Subs. prin.

5. $\frac{1}{a} = \frac{1}{b}$ Canc. prop. of mult.

35. 1. $ab = 0, b \neq 0$ Hypothesis

2. $ab\left(\frac{1}{b}\right) = 0$ Mult. prop. of zero

3. $ab\left(\frac{1}{b}\right) = a\left[b\left(\frac{1}{b}\right)\right]$ Assoc. ax. for mult.

4. $= a \cdot 1$ Ax. of mult. inv.

5. $= a$ Iden. prop. for mult.

6. $ab\left(\frac{1}{b}\right) = a$ Trans. prop. of =

7. $a = 0$ Subs. prin. (in step 2)

C **36.** 1. $xb = a, b \neq 0$ Hypothesis

2. $xb = a \cdot 1$ Iden. ax. for mult.

3. $xb = a\left(\frac{1}{b} \cdot b\right)$ Ax. of mult. inv.

4. $xb = \left(a \cdot \frac{1}{b}\right)b$ Assoc. ax. for mult.

5. $x = a \cdot \frac{1}{b}$ Canc. prop. of mult.

37. 1. $x = a \cdot \frac{1}{b}, b \neq 0$ Hypothesis

2. $bx = b\left(a \cdot \frac{1}{b}\right)$ Subs. prin.

3. $= a\left(b \cdot \frac{1}{b}\right)$ Comm. and assoc. ax. for mult.

4. $= a \cdot 1$ Ax. of mult. inv.

5. $= a$ Iden. ax. for mult.

6. $bx = a$ Trans. prop. of =

38. 1. $ax + b = c, a \neq 0$ Hypothesis

2. $= c + 0$ Iden. ax. for add.

3. $= c + [(-b) + b]$ Ax. of add. inv.

4. $= [c + (-b)] + b$ Assoc. ax. for add.

5. $ax + b = [c + (-b)] + b$ Trans. prop. of =

6. $ax = c + (-b)$ Canc. prop. of add.

7. $xa = c + (-b)$ Comm. prop. for mult.

39. 1. $\frac{1}{x} = a\left(\frac{1}{b}\right)$ Hypothesis

2. $b\left(\frac{1}{x}\right) = a$ Ex. 37

3. $a = b\left(\frac{1}{x}\right)$ Symm. prop. of =

4. $xa = b$ Ex. 37

5. $x = b \cdot \frac{1}{a}$ Ex. 36

40. 1. $\frac{1}{x} = ab$, and a, b, and $x \neq 0$ — Hypothesis

2. $\frac{1}{\frac{1}{x}} = \frac{1}{ab}$ — Ex. 34

3. $x = \frac{1}{\frac{1}{x}}$ — Ex. 30

4. $x = \frac{1}{ab}$ — Subs. prin.

41. 1. $a\left(\frac{1}{x}\right) = b$ — Hypothesis

2. $b = a\left(\frac{1}{x}\right)$ — Symm. prop. of =

3. $xb = a$ — Ex. 37

4. $x = a\left(\frac{1}{b}\right)$ — Ex. 36

Page 22 • COMPUTER EXERCISES

1.
```
10 INPUT X, Y
20 LET C = 0
30 PRINT "PRODUCT IS ";
40 IF X = 0 OR Y = 0 THEN GOTO 100
50 IF X < 0 THEN LET C = C + 1
60 IF Y < 0 THEN LET C = C + 1
70 IF C = 1 THEN PRINT "NEGATIVE": GOTO 110
80 PRINT "POSITIVE"
90 GOTO 110
100 PRINT "ZERO"
110 END
```
2. negative **3.** zero **4.** positive **5.** positive **6.** negative **7.** zero

8.
```
10 INPUT X, Y, Z
20 LET C = 0
30 PRINT "PRODUCT IS ";
40 IF X = 0 OR Y = 0 OR Z = 0 THEN GOTO 110
50 IF X < 0 THEN LET C = C + 1
60 IF Y < 0 THEN LET C = C + 1
70 IF Z < 0 THEN LET C = C + 1
80 IF C = 1 OR C = 3 THEN PRINT "NEGATIVE": GOTO 120
90 PRINT "POSITIVE"
100 GOTO 120
110 PRINT "ZERO"
120 END
```
9. positive **10.** negative **11.** negative **12.** positive **13.** zero **14.** zero

Pages 23–24 • WRITTEN EXERCISES

A **1.** $-2x + 2y$ **2.** $-5a + 5b - 5c$ **3.** $2mp - 2mq - 6m$ **4.** $-7de + 7ef - 35e$
5. $-12r + 60s - 48t$ **6.** $66kn - 594n$ **7.** $18a - 3b + 3ac$ **8.** $-24m + 140mn$

9. $27 + 7w - 6z$ **10.** $28xy - 63x$ **11.** $2f + 14g - 15$ **12.** $6p + 12pq - 60q$
13. 2 **14.** 9 **15.** -20 **16.** -18 **17.** -19 **18.** -192

B **19.** 1. Hypothesis 2. Def. of subt. 3. Subs. prin. 4. Dist. ax.
5. Prop. of opp. in prod. 6. Def. of subt. 7. Trans. prop. of =

20. 1. Hypothesis 2. Ax. of add. inv. 3. Def. of subt. 4. Canc. prop. of add. inv.
5. Subs. prin.

21.

1. $a, b \in \mathscr{R}$	Hypothesis
2. $(a - b) + b = [a + (-b)] + b$	Def. of subt.
3. $= a + [(-b) + b]$	Assoc. ax. for add.
4. $= a + 0$	Ax. of add. inv.
5. $= a$	Iden. ax. for add.
6. $(a - b) + b = a$	Trans. prop. of =

22.

1. $a, b, c \in \mathscr{R}$	Hypothesis
2. $-a(b + c) = -[a(b + c)]$	Prop. of opp. in prod.
3. $= -(ab + ac)$	Dist. prop.
4. $= (-ab) + (-ac)$	Prop. of opp. of sum
5. $= -ab - ac$	Def. of subt.
6. $-a(b + c) = -ab - ac$	Trans. prop. of =

23.

1. $a, b, c \in \mathscr{R}$	Hypothesis
2. $-a(b - c) = -a[b + (-c)]$	Def. of subt.
3. $= -ab - [a(-c)]$	Ex. 22
4. $= -ab - (-ac)$	Prop. of opp. in prod.
5. $= -ab + ac$	Ex. 20
6. $= ac + (-ab)$	Comm. ax. for add.
7. $= ac - ab$	Def. of subt.
8. $-a(b - c) = ac - ab$	Trans. prop. of =

24.

1. $a - c = b - c$	Hypothesis
2. $a + (-c) = b + (-c)$	Def. of subt.
3. $a = b$	Canc. prop. of add.

25.

1. $c - a = c - b$	Hypothesis
2. $c + (-a) = c + (-b)$	Def. of subt.
3. $-a = -b$	Canc. prop. of add.
4. $(-1)a = (-1)b$	Mult. prop. of -1
5. $a = b$	Canc. prop. of mult.

26.

1. $a = b$	Hypothesis
2. $a - c = a - c$	Reflex. prop. of =
3. $a - c = b - c$	Subs. prin.

27.

1. $a = b$	Hypothesis
2. $c - a = c - a$	Reflex. prop. of =
3. $c - a = c - b$	Subs. prin.

28.

1. $a - b = 0$	Hypothesis
2. $= b + (-b)$	Ax. of add. inv.
3. $= b - b$	Def. of subt.
4. $a - b = b - b$	Trans. prop. of =
5. $a = b$	Ex. 24

29.

1. $x - b = a$	Hypothesis
2. $x + (-b) = a$	Def. of subt.
3. $= a + 0$	Iden. ax. for add.
4. $= a + [b + (-b)]$	Ax. of add. inv.
5. $= (a + b) + (-b)$	Assoc. ax. for add.
6. $x + (-b) = (a + b) + (-b)$	Trans. prop. of =
7. $x = a + b$	Canc. prop. of add.

30. 1. $x = a + b$ Hypothesis
2. $a + b = x$ Symm. prop. of =
3. $\quad = x + 0$ Iden. ax. for add.
4. $\quad = x + [(-b) + b]$ Ax. of add. inv.
5. $\quad = [x + (-b)] + b$ Assoc. ax. for add.
6. $\quad = (x - b) + b$ Def. of subt.
7. $a + b = (x - b) + b$ Trans. prop. of =
8. $\quad a = x - b$ Canc. prop. of add.
9. $x - b = a$ Symm. prop. of =

C **31.** 1. $(a - b)(c + d) = (a - b)c + (a - b)d$ Dist. ax.
2. $\quad = [a + (-b)]c + [a + (-b)]d$ Def. of subt.
3. $\quad = [ac + (-b)c] + [ad + (-b)d]$ Dist. ax.
4. $\quad = [ac + (-bc)] + [ad + (-bd)]$ Prop. of opp. in prod.
5. $\quad = ac - bc + ad - bd$ Def. of subt.
6. $(a - b)(c + d) = ac - bc + ad - bd$ Trans. prop. of =

32. 1. $a, b, c, d \in \mathcal{R}$ Hypothesis
2. $(a - b)(c - d) = [a + (-b)][c + (-d)]$ Def. of subt.
3. $\quad = [a + (-b)]c + [a + (-b)](-d)$ Dist. ax.
4. $\quad = ac + (-b)c + a(-d) + (-b)(-d)$ Dist. ax.
5. $\quad = ac - bc - ad + bd$ Prop. of opp. in prod.
6. $(a - b)(c - d) = ac - bc - ad + bd$ Trans. prop. of =

33. 1. $a, b, c, d \in \mathcal{R}$ Hypothesis
2. $a[b - (c + d)] = ab - [a(c + d)]$ Ex. 19
3. $\quad = ab - a(c + d)$ Prop. of opp. in prod.
4. $\quad = ab - ac - ad$ Ex. 22
5. $a[b - (c + d)] = ab - ac - ad$ Trans. prop. of =

34. 1. $a, b, c, d \in \mathcal{R}$ Hypothesis
2. $a[b - (c - d) = ab - [a(c - d)]$ Ex. 19
3. $\quad = ab - (ac - ad)$ Ex. 19
4. $\quad = ab - [ac + (-ad)]$ Def. of subt.
5. $\quad = ab - ac - (-ad)$ Prop. of opp. of sum
6. $\quad = ab - ac + ad$ Prop. of opp. in prod.
7. $a[b - (c - d)] = ab - ac + ad$ Trans. prop. of =

Pages 26–28 • WRITTEN EXERCISES

A **1.** -3 **2.** 6 **3.** $8x - 6$ **4.** $15y$ **5.** $-2k$ **6.** $-\frac{1}{7c}$ **7.** $-62m$ **8.** $-8p$

9. $29 - \frac{64}{n}$ **10.** $-\frac{360ab}{13}$ **11.** $\frac{392r}{15}$ **12.** $-\frac{96}{7v}$ **13. a.** $-2, -2$ **b.** true

14. a. $2, -2$ **b.** false **15. a.** 18, 18 **b.** true **16. a.** $\frac{1}{5}, \frac{1}{5}$ **b.** true

17. a. 14, 11 **b.** false **18. a.** $3, -3$ **b.** false

19. a. $-\frac{11}{3}, \frac{1}{3}$ **b.** false **20. a.** 3, 3 **b.** true

B **21.** 1. Hypothesis 2. Def. of div. 3. Dist. ax. 4. Def. of div. 5. Trans. prop. of =

22. 1. Hypothesis 2. Def. of div. 3. Prop. of opp. in prod. 4. Def. of div. 5. Trans. prop. of =

23. 1. $a \in \mathcal{R}, a \neq 0$ Hypothesis
2. $\frac{a}{a} = a \cdot \frac{1}{a}$ Def. of div.

[*Proof continued on next page*]

	Statement	Reason
	3. $= 1$	Ax. of mult. inv.
	4. $\frac{a}{a} = 1$	Trans. prop. of =
24.	1. $a \in \mathcal{R}, a \neq 0$	Hypothesis
	2. $\frac{-a}{a} = -a \cdot \frac{1}{a}$	Def. of div.
	3. $= -\left(a \cdot \frac{1}{a}\right)$	Prop. of opp. in prod.
	4. $= -1$	Ax. of mult. inv.
	5. $\frac{-a}{a} = -1$	Trans. prop. of =
25.	1. $a, b \in \mathcal{R}, b \neq 0$	Hypothesis
	2. $(ab) \div b = (ab) \cdot \frac{1}{b}$	Def. of div.
	3. $= a \cdot \left(b \cdot \frac{1}{b}\right)$	Assoc. ax. for mult.
	4. $= a \cdot 1$	Ax. of mult. inv.
	5. $= a$	Iden. ax. for mult.
	6. $(ab) \div b = a$	Trans. prop. of =
26.	1. $ax = b, a \neq 0$	Hypothesis
	2. $ax = xa$	Comm. ax. for mult.
	3. $xa = b$	Subs. prin.
	4. $xa \cdot \frac{1}{a} = xa \cdot \frac{1}{a}$	Refl. prop of =
	5. $xa \cdot \frac{1}{a} = b \cdot \frac{1}{a}$	Subs. prin.
	6. $x\left(a \cdot \frac{1}{a}\right) = b \cdot \frac{1}{a}$	Assoc. ax. for mult.
	7. $x \cdot 1 = b \cdot \frac{1}{a}$	Ax. of mult. inv.
	8. $x = b \cdot \frac{1}{a}$	Iden. ax. for mult.
	9. $x = \frac{b}{a}$	Def. of div.
27.	1. $x = \frac{b}{a}, a \neq 0$	Hypothesis
	2. $1 \cdot x = 1\left(\frac{b}{a}\right)$	Iden. ax. for mult.
	3. $\left(\frac{1}{a} \cdot a\right)x = \left(\frac{1}{a} \cdot a\right) \cdot \frac{b}{a}$	Ax. of mult. inv.
	4. $\frac{1}{a}(ax) = \frac{1}{a}\left(a \cdot \frac{b}{a}\right)$	Assoc. ax. for mult.
	5. $ax = a \cdot \frac{b}{a}$	Canc. prop. of mult.
	6. $ax = \frac{b}{a} \cdot a$	Comm. ax. for mult.
	7. $= \left[b\left(\frac{1}{a}\right)\right] \cdot a$	Def. of div.

8. $= b\left[\frac{1}{a} \cdot a\right]$	Assoc. ax. for mult.
9. $= b \cdot 1$	Ax. of mult. inv.
10. $= b$	Iden. ax. for mult.
11. $ax = b$	Trans. prop. of =

28.

1. $a, b, c \in \mathcal{R}$; $b, c \neq 0$	Hypothesis
2. $\frac{ac}{bc} = ac \cdot \frac{1}{bc}$	Def. of div.
3. $= ac\left(\frac{1}{b} \cdot \frac{1}{c}\right)$	Prop. of recip. of a prod.
4. $= \left[a\left(c \cdot \frac{1}{c}\right)\right]\frac{1}{b}$	Comm. and assoc. ax. for mult.
5. $= (a \cdot 1)\frac{1}{b}$	Ax. of mult. inv.
6. $= a \cdot \frac{1}{b}$	Iden. ax. for mult.
7. $= \frac{a}{b}$	Def. of div.
8. $\frac{ac}{bc} = \frac{a}{b}$	Trans. prop. of =

29.

1. $a, b, c, d \in \mathcal{R}$; $b, d \neq 0$	Hypothesis
2. $\frac{a}{b} \cdot \frac{c}{d} = a\left(\frac{1}{b}\right) \cdot c\left(\frac{1}{d}\right)$	Def. of div.
3. $= (ac)\left(\frac{1}{b} \cdot \frac{1}{d}\right)$	Comm. and assoc. ax. for mult.
4. $= ac\left(\frac{1}{bd}\right)$	Prop. of recip. of a prod.
5. $= \frac{ac}{bd}$	Def. of div.
6. $\frac{a}{b} \cdot \frac{c}{d} = \frac{ac}{bd}$	Trans. prop. of =

30.

1. $a, b, c \in \mathcal{R}$; $b \neq 0$	Hypothesis
2. $a + \frac{c}{b} = \frac{ab}{b} + \frac{c}{b}$	Ex. 25
3. $= \frac{ab + c}{b}$	Ex. 21
4. $a + \frac{c}{b} = \frac{ab + c}{b}$	Trans. prop. of =

C 31.

1. $a, b \in \mathcal{R}$; $a, b \neq 0$	Hypothesis
2. $\frac{1}{a} + \frac{1}{b} = \frac{1 \cdot b}{ab} + \frac{1 \cdot a}{ba}$	Ex. 28
3. $= \frac{b}{ab} + \frac{a}{ba}$	Iden. ax. for mult.
4. $= \frac{b}{ab} + \frac{a}{ab}$	Comm. ax. for mult.
5. $= \frac{b + a}{ab}$	Ex. 21
6. $\frac{1}{a} + \frac{1}{b} = \frac{b + a}{ab}$	Trans. prop. of =

32. 1. $a, b, c, d \in \mathcal{R}$; $b, d \neq 0$ — Hypothesis

2. $\frac{a}{b} = \frac{ad}{bd}$; $\frac{c}{d} = \frac{cb}{db}$ — Ex. 28

3. $\frac{cb}{db} = \frac{bc}{bd}$ — Comm. ax. for mult.

4. $\frac{a}{b} + \frac{c}{d} = \frac{ad}{bd} + \frac{bc}{bd}$ — Subs. prin.

5. $\qquad = \frac{ad + bc}{bd}$ — Ex. 21

6. $\frac{a}{b} + \frac{c}{d} = \frac{ad + bc}{bd}$ — Trans. prop. of =

33. 1. $a, b \neq 0$ — Hypothesis

2. $\frac{a}{b} \cdot \frac{b}{a} = \frac{ab}{ba}$ — Ex. 29

3. $\frac{a}{b} \cdot \frac{b}{a} = \frac{ab}{ab}$ — Comm. ax. for mult.

4. $\frac{a}{b} \cdot \frac{b}{a} = 1$ — Ex. 23

34. 1. $a, b \neq 0$ — Hypothesis

2. $\frac{a}{b} \cdot \frac{b}{a} = 1$ — Ex. 33

3. $\frac{b}{a} = \frac{1}{\frac{a}{b}}$ — Uniqueness of mult. inv. (Ex. 29, p. 21)

4. $\frac{1}{\frac{a}{b}} = \frac{b}{a}$ — Symm. prop. of =

35. 1. $b, c, d \neq 0$ — Hypothesis

2. $\frac{\frac{a}{b}}{\frac{c}{d}} = \frac{a}{b} \cdot \frac{1}{\frac{c}{d}}$ — Def. of div.

3. $\qquad = \frac{a}{b} \cdot \frac{d}{c}$ — Ex. 34

4. $\qquad = \frac{ad}{bc}$ — Ex. 29

5. $\frac{\frac{a}{b}}{\frac{c}{d}} = \frac{ad}{bc}$ — Trans. prop. of =

Page 28 • ON THE CALCULATOR

1. 34.795 **2.** 5.25 **3.** 18.23 **4.** −1.88496 **5.** 1810 **6.** −61

Page 29 • SELF-TEST 2

1. 1. $-(a + b) = (-a) + (-b)$ — Prop. of the opp. of a sum

2. $-[(-a) + (-b)] = -[-(a + b)]$ — Subs. prin.

3. $-[(-a) + (-b)] = a + b$ — Canc. prop. of add. inv.

2. 1. $a, b \in \mathcal{R}; a \neq 0, b \neq 0$ Hypothesis

2. $b \cdot \frac{1}{ab} = b\left(\frac{1}{a} \cdot \frac{1}{b}\right)$ Prop. of recip. of prod.

3. $= \left(b \cdot \frac{1}{b}\right)\frac{1}{a}$ Comm. and assoc. ax. for mult.

4. $= 1\left(\frac{1}{a}\right)$ Ax. of mult. inv.

5. $= \frac{1}{a}$ Iden. ax. for mult.

6. $b \cdot \frac{1}{ab} = \frac{1}{a}$ Trans. prop. of =

3. -11 **4.** $(-6n - (-5))(-2 + 6) = (-6n + 5)4 = -24n + 20$
5. $2b(-3) \div 6 + 9 = (-6b \div 6) + 9 = -b + 9$
6. $-(x + 2) - (-3 - x) = -x - 2 + 3 + x = 1$

Pages 30–31 • CHAPTER REVIEW

1. b **2.** a **3.** a **4.** b **5.** d **6.** d **7.** d **8.** d **9.** c **10.** c **11.** a **12.** c
13. b **14.** c **15.** b **16.** a **17.** a

Page 31 • CHAPTER TEST

1. {1, 2, 3} **2.** {0, 1, 2, 3, . . .} **3.** {2, 3}
4. {0, 1, 2, 3}, {0, 1, 2}, {0, 1, 3}, {0, 2, 3}, {1, 2, 3}, {0, 1}, {0, 2}, {0, 3}, {1, 2}, {1, 3}, {2, 3}, {0}, {1}, {2}, {3}, ∅
5. {0} **6.** no solution **7.** {−1, 0, 1, 5, 10} **8.** {1, 10}

9. 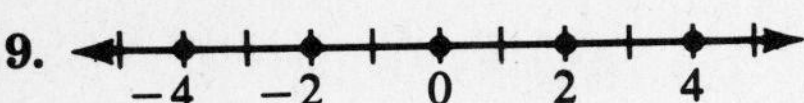

10. 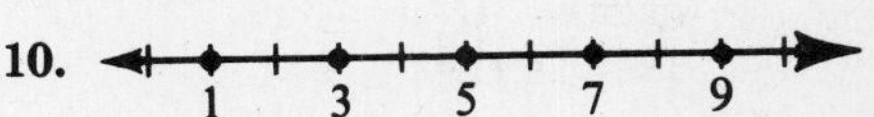

11. Trans. prop. of = (or Subs. prin.)
12. Assoc. ax. for mult. **13.** Dist. ax. **14.** $2x - 7$ **15.** $30 - 7x$
16. $7dr + 14ds$ **17.** $5ab + 6b + 8a$ **18.** $-4mk + 4mh - 32m$ **19.** $10 - 40c$
20. -1 **21.** $-8mn$

Page 32 • APPLICATION

1. a. The current is halved. **b.** The current is halved.
2. $R = \frac{V}{I}; R = \frac{6.0}{0.15} = 40\ \Omega$ **3.** $V = IR; V = (0.15)(36) = 5.4\ \text{V}$

Page 33 • PREPARING FOR COLLEGE ENTRANCE EXAMS

1. D **2.** B **3.** D **4.** A **5.** C **6.** D **7.** C **8.** A

Pages 37–39 • PROGRAMMING IN PASCAL

1. a. `english_vowels : charset;`
b. `english_vowels := ['a','e','i','o','u'];`
c. `punctuation_marks : charset;`
d. `punctuation_marks := ['.', ',', ';', ':', '?', '!'];`
e. `digits : charset;`
f. `digits := ['0'..'9'];`

2.
```
write('B = {');
FOR i := 1 TO 25 DO
    IF i IN set_b
       THEN write(i:3);
writeln('   }');
writeln;
write('C = {');
FOR i := 1 TO 25 DO
    IF i IN set_c
       THEN write(i:3);
writeln('   }');
writeln;
```

3.
```
write('B intersect C = {');
FOR i := 1 TO 25 DO
    IF i IN set_b * set_c
       THEN write(i:3);
writeln('   }');
writeln;
```

4.
```
write('B = ');
display(set_b);
```

5. In the VAR declaration, insert *set_c : nums* and in the main program the following lines must be inserted at the appropriate places.

```
set_c := [ ];
. . .
get_elements(set_c);
. . .
write('C = ');
display(set_c);
```

6.
```
write('A union B = ');
display(set_a + set_b);
write('A intersect B = ');
display(set_a * set_b);
write('A union C = ');
display(set_a + set_c);
write('A intersect C = ');
display(set_a * set_c);
write('B union C = ');
display(set_b + set_c);
write('B intersect C = ');
display(set_b * set_c);
```

Pages 44–45 • WRITTEN EXERCISES

A **1.** $5x^3 + 3x^2 - 4x - 8$ **2.** $-y^3 - 5y^2 + 3y - 5$ **3.** $-7z^3 - 6z^2 + z - 11$
4. $-2n^4 + 6n^3 - 6n^2 - n$ **5.** $2a^3 + 4a^2b - 2ab^2 - b^3$
6. $-11p^3 - 13p^2q + 9pq^2 + 5q^3$ **7.** $-3x^3 + 7x^2 - 10x + 14$
8. $15y^3 - 7y^2 + 3y - 13$ **9.** $-2z^4 - 7z^3 + 6z^2 + 5z + 9$
10. $8n^4 - 6n^3 + 4n^2 - 3n + 6$ **11.** $8a^3 - 12a^2b + b^3$
12. $-p^3 + 13p^2q + 9pq^2 - 19q^3$ **13.** $x - 3$ **14.** $15 - 12y$ **15.** $-4a + 5b$
16. $6cd + 2d^2$ **17.** $-9p^2 - 3$ **18.** $q^2 - 4q - 4$ **19.** $5r^3 + 5r^2 - 9r - 3$ **20.** $5v^2$
21. $-w^2 + 2w$ **22.** $-8k^4 - k^3 + 12$ **23.** $7x^2 - 2xy$ **24.** $-6p^3 - 5p^2q + 5pq^2$
B **25.** $-4c^2 - 2c + 14d$ **26.** $-m^2 - 16mn - 20n^2$ **27.** $16x - 96$ **28.** $-78y - 52$
29. $-5x^4 + 3x^3 - 7x^2 + 2x - 5$ **30.** $-9x^5 + 3x^3 - 4x + 5$

Exs. 31–34: Answers may vary.

C **31.**

1.	$x^2 - (x^2 + 1) = x^2 + [-(x^2 + 1)]$	Def. of subt.
2.	$= x^2 + [(-x^2) + (-1)]$	Prop. of the opp. of a sum
3.	$= [x^2 + (-x^2)] + (-1)$	Assoc. ax. for add.
4.	$= 0 + (-1)$	Ax. of add. inv.
5.	$= -1$	Iden. ax. for add.
6.	$x^2 - (x^2 + 1) = -1$	Trans. prop. of =

32.

1.	$-(x^2 - 2) + x^2 = -[x^2 + (-2)] + x^2$	Def. of subt.
2.	$= [-x^2 + (-(-2))] + x^2$	Prop. of the opp. of a sum
3.	$= (-x^2 + 2) + x^2$	Canc. prop. of add. inv.
4.	$= (-x^2 + x^2) + 2$	Comm., assoc. ax. for add.
5.	$= 0 + 2$	Ax. of add. inv.
6.	$= 2$	Iden. ax. for add.
7.	$-(x^2 - 2) + x^2 = 2$	Trans. prop. of =

33.

1.	$3x^2 - x - (3x^2 + x)$	
	$= 3x^2 + (-x) + [-(3x^2 + x)]$	Def. of subt.
2.	$= 3x^2 + (-x) + [-3x^2 + (-x)]$	Prop. of the opp. of a sum
3.	$= 3x^2 + (-3x^2) + (-x) + (-x)$	Comm. ax. for add.
4.	$= [3x^2 + (-3x^2)] + [(-x) + (-x)]$	Assoc. ax. for add.
5.	$= 0 + [(-x) + (-x)]$	Ax. of add. inv.
6.	$= (-x) + (-x)$	Iden. ax. for add.
7.	$= (-1)x + (-1)x$	Mult. prop. of -1
8.	$= [(-1) + (-1)]x$	Dist. ax.
9.	$= (-2)x$	Subs. prin.
10.	$= -2x$	Prop. of opp. in prod.
11.	$3x^2 - x - (3x^2 + x) = -2x$	Trans. prop. of =

34.

1.	$(x^2 - 8) - (x^2 - 8) = (x^2 - 8) + [-(x^2 - 8)]$	Def. of subt.
2.	$= 0$	Ax. of add. inv.
3.	$(x^2 - 8) - (x^2 - 8) = 0$	Trans. prop. of =

Pages 47–48 • WRITTEN EXERCISES

A **1.** Add 7 to each side; divide each side by 5.
2. Add $4a$ to each side; add 3 to each side.
3. Add 2 to each side; multiply each side by 3.
4. Subtract 7 from each side; multiply each side by (-4).
5. Use the distributive axiom to simplify the left side; combine like terms.
6. Add $4k$ to each side; subtract 14 from each side.
7. Subtract $\frac{1}{2}p$ from each side; add 5 to each side.

8. Use the distributive axiom to simplify the right side; combine like terms.

9. $\{-8\}$ **10.** $\{5\}$ **11.** $\{5\}$ **12.** $\{-3\}$ **13.** $\{0\}$ **14.** $\{8\}$ **15.** $\{20\}$ **16.** $\left\{-\frac{20}{7}\right\}$

17. $\{70\}$ **18.** $\{-12\}$ **19.** $\{-27\}$ **20.** $\{-49\}$ **21.** $\left\{-\frac{12}{5}\right\}$ **22.** $\{-4\}$ **23.** $\{-14\}$

24. $\{-18\}$ **25.** $\{6\}$ **26.** $\{-4\}$ **27.** $\{-3\}$ **28.** $\{1\}$

B **29.** $r = \frac{7b}{3a}$ **30.** $n = \frac{4c}{d}$ **31.** $x = \frac{7h}{3k}$ **32.** $y = \frac{c}{a - b}$ **33.** $v = \frac{2r}{5r - s}$

34. $w = \frac{3}{2d - e}$ **35.** $t = \frac{p - v}{g}$ **36.** $x = \frac{2A}{h} - a$ **37.** $h = \frac{A}{\pi r} - 2r$

38. $p = \frac{7n}{k - 9}$ **39.** $x = \frac{ac}{b}$ **40.** $v = \frac{d}{5d - c}$

C **41.**

1. $a = b$ — Hypothesis
2. $a + c = a + c$ — Refl. prop. of =
3. $a + c = b + c$ — Subs. prin.

42.

1. $a = b$ — Hypothesis
2. $a - c = a - c$ — Refl. prop. of =
3. $a - c = b - c$ — Subs. prin.

43.

1. $a = b$ — Hypothesis
2. $ac = ac$ — Refl. prop. of =
3. $ac = bc$ — Subs. prin.

44.

1. $a = b$ and $c \neq 0$ — Hypothesis
2. $\frac{a}{c} = \frac{a}{c}$ — Refl. prop. of =
3. $\frac{a}{c} = \frac{b}{c}$ — Subs. prin.

45.

1. $ax + b = c;\ a, b, c \in \mathscr{R};\ a \neq 0$ — Hypothesis
2. $ax + b - b = c - b$ — Ex. 42
3. $ax + b + (-b) = c - b$ — Def. of subt.
4. $ax + [b + (-b)] = c - b$ — Assoc. ax. for add.
5. $ax + 0 = c - b$ — Ax. for add. inv.
6. $ax = c - b$ — Iden. ax. for add.
7. $\frac{ax}{a} = \frac{c - b}{a}$ — Ex. 44
8. $ax\left(\frac{1}{a}\right) = \frac{c - b}{a}$ — Def. of div.
9. $x\left[a\left(\frac{1}{a}\right)\right] = \frac{c - b}{a}$ — Comm. and assoc. ax. for mult.
10. $x \cdot 1 = \frac{c - b}{a}$ — Ax. of mult. inv.
11. $x = \frac{c - b}{a}$ — Iden. ax. for mult.

Page 48 • COMPUTER EXERCISES

1.

```
10 INPUT A,B,C,D
20 LET A = A - C
30 IF A = 0 AND B <> D THEN PRINT "NO SOLUTION" : END
35 IF A = 0 AND B = D THEN PRINT "ALL REAL NUMBERS" : END
40 LET D = D - B
```

```
50 PRINT "X = ";D/A
60 END
```

2. -5.2 **3.** No solution **4.** 2.5 **5.** 0 **6.** All real numbers **7.** 1

Pages 52–54 • PROBLEMS

A **1.** Let x = the smallest integer; $x + (x + 1) + (x + 2) = 159$; 52, 53, 54
2. Let x = the charge (in cents) for the first copy; $x + 16(x - 6) = 142$; 14¢
3. Let x = the first integer; $x + 2(x + 1) = 203$; 67, 68
4. Let x = the smallest integer; $x + (x + 1) + (x + 2) = 5x - 11$; 7, 8, 9
5. Let x = the smallest odd integer; $x + (x + 2) + (x + 4) = 2(x + 4) + 13$; 15, 17, 19
6. Let x = the length of the base; $x + 2(2x - 7) = 106$; 24 cm, 41 cm, 41 cm
7. Let x = the measure of the smallest angle; $x + (x + 6) + (x + 12) = 180$; 54°, 60°, 66°
8. Let $DC = x$; $DB = x + 2$, $BC = 2x$; $x + (x + 2) + 2x = 70$; $x = 17$, $BC = 34$ cm
9. Let $DC = AD = x$; $DB = x + 2$, $AB = x + 1$, $BC = 2x$;
$x + x + (x + 2) + (x + 1) + 2x = 45$; $x = 7$; $BC = 14$ cm
10. longer base = 17, shorter base = $17 - 2x$; $A = \frac{1}{2}h(b + b')$; $168 = \frac{1}{2} \cdot 12[17 + (17 - 2x)]$; $x = 3$
11. $m\angle EPG = 90 - x$; $5(90 - x) + x = 180$; $x = 67.5$
12. Let x = no. of gal used for city driving; $20x + 28(9 - x) = 220$; city, 4 gal; highway, 5 gal
13. Let x = no. of lb of extra lean; $3.10x + 2.20(5 - x) = 12.35$; $x = 1.5$ lb

B **14.** Let t = no. of hours on highway; $60t = 36\left(t + \frac{1}{2}\right)$; $t = \frac{3}{4}$ h
15. Let d = round-trip distance; Tues. rate = Wed. rate; $\frac{40 + 3d}{2} = \frac{17 + 2d}{1}$; $40 + 3d = 2(17 + 2d)$; $d = 6$ km
16. Let d = length of bike trail; $1\frac{1}{2} \cdot$ first day's rate = sec. day's rate; $\frac{3}{2}\left(\frac{d + 8}{5}\right) = \frac{d}{3}$; $d = 72$ km
17. Let t = no. of min the newer model operates; $2(160)(1.5 + t) + 200t = 1780$; $t = 2.5$ min; $t + 1.5 = 4.0$ min
18. Let x = length of one leg; $x^2 + 8^2 = (x + 2)^2$; 15 cm, 8 cm, 17 cm
19. Let r = radius of cylinder; $2\pi r \cdot 15 + 2\pi r^2 = 2[\pi(r + 6)^2]$; $r = 12$

C **20.** Let h = no. of Lopez's hits; $\frac{h}{h + 10} = \frac{h + 3}{(h + 3) + 15}$; $h = 6$; batting avg. = 0.375
21. Let x = no. of shares sold for total of \$165; $\frac{165}{x} = \frac{120}{2(30 - x)}$; $x = 22$ shares, $30 - x = 8$ shares

22.

	r	t	d
to city	20	5	100
return trip	r	$\frac{100}{r}$	100
round trip	32	$5 + \frac{100}{r}$	200

$200 = 32\left(5 + \frac{100}{r}\right)$; $r = 80$ km/h

Page 54 • COMPUTER EXERCISES

1.
```
10 INPUT K
20 IF K <= 350 THEN LET C = K * .05876 : GOTO 70
30 LET K = K - 350 : LET C = 350 * .05876
40 IF K <= 400 THEN LET C = C + K * .08252 : GOTO 70
50 LET K = K - 400
60 LET C = C + 400 * .08252 + K * .09064
70 PRINT "COST ";C + 7.42
80 END
```
2. \$42.67 3. \$23.58 4. \$66.61

5.
```
10 INPUT H,T,X,A
20 LET M = INT((A * T + A * X - H) + 1)
30 IF M > X THEN PRINT "IMPOSSIBLE TO ACHIEVE" : END
40 PRINT "NEED AT LEAST ";M;"HITS."
50 END
```
6. 7 7. Impossible to achieve

Page 55 • SELF-TEST 1

1. $y^2 - 2y - 2$ 2. $-6x^2 + 36$ 3. $-a^3 - 5a^2 - 5a + 13$ 4. $5xy + 2x - 2y - 16$
5. $\{1\}$ 6. $\{16\}$ 7. $a = \dfrac{y - 3}{4x}$
8. Let x = amt. each child receives; $10{,}000 = 400 + 3x + (x + x + x)$; $x = 1600$; each child receives \$1700, the adult \$4900.
9. Let x = side of smaller square; $4x + 4(x + 8) = 128$, $x = 12$; 12 cm × 12 cm, 20 cm × 20 cm
10. Let x = first integer; $x + (x + 4) = 3(x + 2) - 8$; 6, 8, 10

Pages 59–60 • WRITTEN EXERCISES

A 1. false; $a = 0$, $b = 1$ 2. true; Addition property of order
3. true; Multiplication property of order 4. true; Transitive property of order
5. false; $a = 1$, $b = 2$, $c = -2$, $d = -1$ 6. false; $k = -16$
7. true; Addition property of order 8. false; $n = 0$

9. $\{y: y < -13\}$

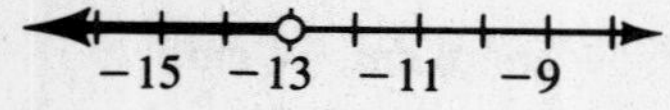

10. $\{x: x < -2\}$

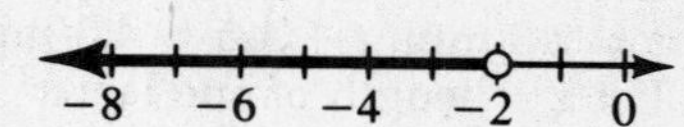

11. $\{a: a > -6\}$

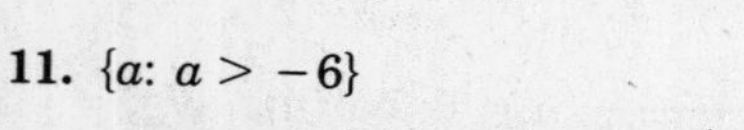

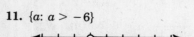

12. $\{k: k > 4\}$

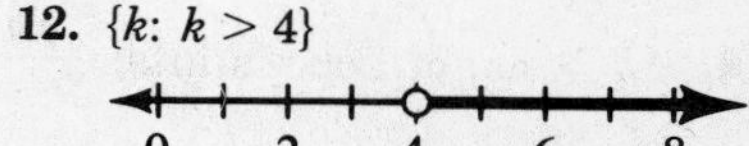

13. $\{c: c < -32\}$

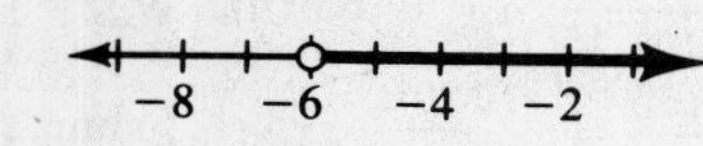

14. $\{p: p < -18\}$

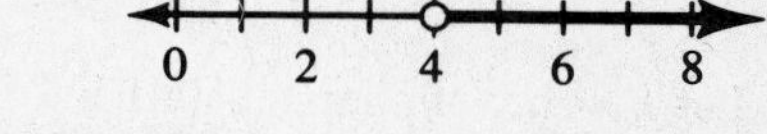

15. $\{r: r < -36\}$

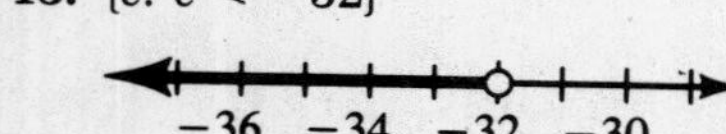

16. $\{n: n < -6\}$

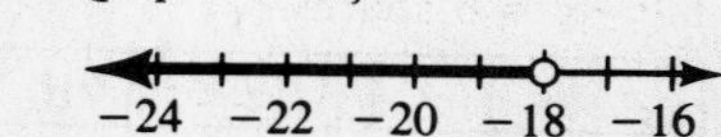

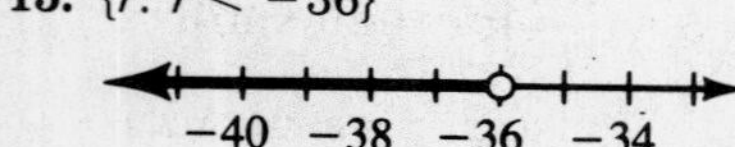

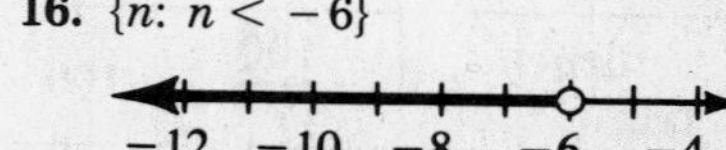

17. $\{h: h > 3\}$

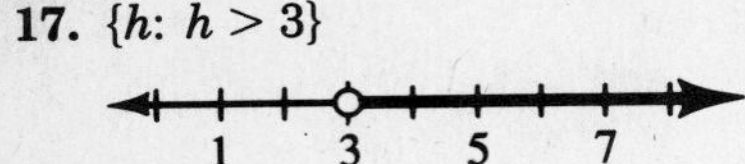

18. $\{m: m > -2\}$

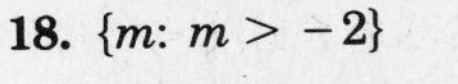

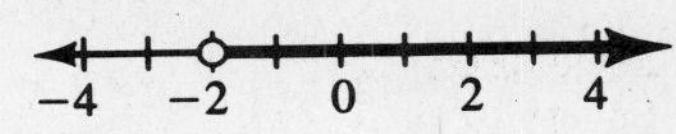

B **19.** $\{d: d > 3\}$

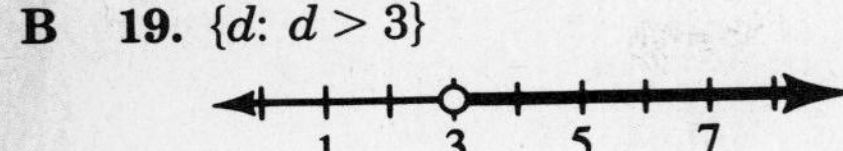

20. $\{z: z < 4\}$

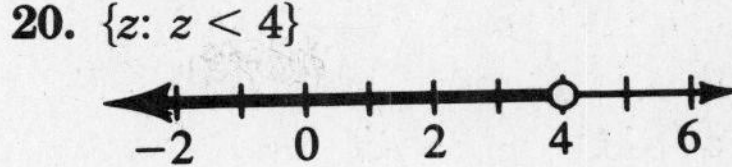

21. $\{v: v > 2\}$

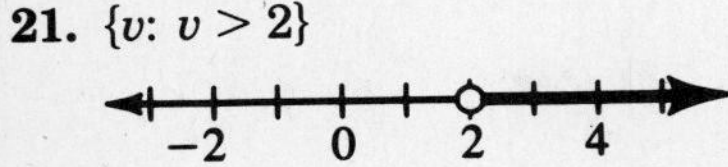

22. $\{w: w < 8\}$

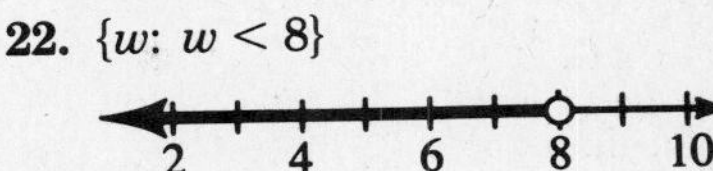

23. $\{x: x < -5\}$

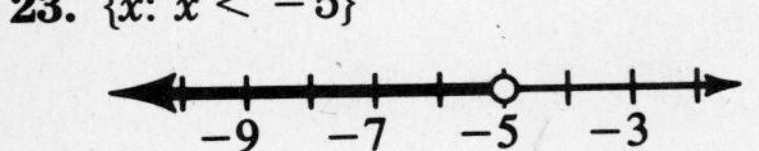

24. $\{x: x > 0\}$

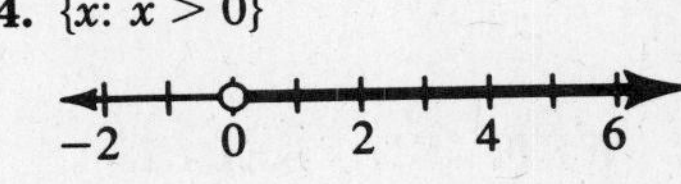

25. 1. Hypothesis 2. Def. of < 3. Add. prop. of = 4. Comm. and assoc. ax. for add. 5. Def. of <

26. 1. Hypothesis 2. Def. of < 3. Subt. prop. of = 4. Subt. prop. and subs. prin. 5. Prop. of the opp. of a sum and assoc. ax. for add. 6. Def. of >

C **27.**

Statement	Reason
1. $a > 0,\ a \in \mathcal{R}$	Hypothesis
2. $a + (-a) > 0 + (-a)$	Add. prop. of order
3. $0 > 0 + -a$	Ax. of add. inv.
4. $0 > -a$	Iden. ax. for add.
5. $-a < 0$	Inequal. rewritten

28.

Statement	Reason
1. $a \in \mathcal{R},\ a < 0$	Hypothesis
2. $a + (-a) < 0 + (-a)$	Add. prop. of order
3. $0 < 0 + (-a)$	Ax. of add. inv.
4. $0 < -a$	Iden. ax. for add.
5. $-a > 0$	Inequal. rewritten

29.

Statement	Reason
1. $a < b,\ c > 0;\ a, b, c \in \mathcal{R}$	Hypothesis
2. For some positive number p, $a + p = b$	Def. of <
3. $(a + p)c = bc$	Mult. prop. of =
4. $ac + pc = bc$	Dist. ax.
5. pc is a positive number	Closure ax. for $\mathcal{R}_+$
6. $ac < bc$	Def. of <

30.

Statement	Reason
1. $a < b;\ c < 0;\ a, b, c \in \mathcal{R}$	Hypothesis
2. For some positive number p, $a + p = b$.	Def. of <
3. $(a + p)(-c) = b(-c)$	Mult. prop. of =
4. $a(-c) + p(-c) = b(-c)$	Dist. ax.
5. $-ac + p(-c) = -bc$	Prop. of opp. in a prod.
6. $ac + (-ac) + p(-c) = ac + (-bc)$	Add. prop. of =
7. $p(-c) = ac + (-bc)$	Ax. of add. inv.
8. $p(-c) + bc = ac + (-bc) + bc$	Add. prop. of =
9. $p(-c) + bc = ac$	Ax. of add. inv.
10. $-c > 0$	Ex. 28
11. $p(-c)$ is positive.	Clos. ax. for $\mathcal{R}_+$
12. $bc < ac$	Def. of <
13. $ac > bc$	Inequal. rewritten

31.

Statement	Reason
1. $a + c > b + c;\ a, b, c \in \mathcal{R}$	Hypothesis
2. $(a + c) + (-c) > (b + c) + (-c)$	Add. prop. of order
3. $a + [c + (-c)] > b + [c + (-c)]$	Assoc. ax. for add.

[*Proof continued on next page*]

	4. $a + 0 > b + 0$	Ax. of add. inv.
	5. $a > b$	Iden. ax. for add.
32.	No; $-1 > -2$, but $(-1)^2 < (-2)^2$.	
33.	1. $a > b$	Hypothesis
	2. $a + c > b + c$	Add. prop. of order
	3. $c > d$	Hypothesis
	4. $b + c > b + d$	Add. prop. of order
	5. $a + c > b + d$	Trans. prop. of order
34.	1. $a, b, c, d \in \mathscr{R}$; $a > b$	Hypothesis
	2. $ac > bc$	Mult. prop. of order
	3. $c > d$	Hypothesis
	4. $bc > bd$	Mult. prop. of order
	5. $ac > bd$	Trans. prop. of order

Page 62 • WRITTEN EXERCISES

A **1.** $\{x: x \le -2\}$

2. $\{r: r \ge -7\}$

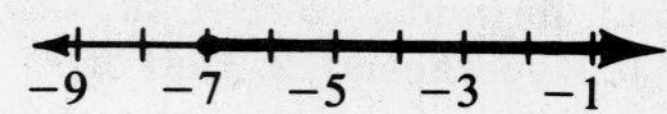

3. $\{a: a \le 9\}$

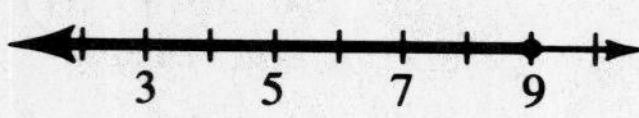

4. $\{y: y \le 9\}$

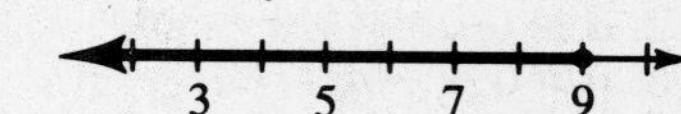

5. $\left\{g: g \le \frac{4}{3}\right\}$

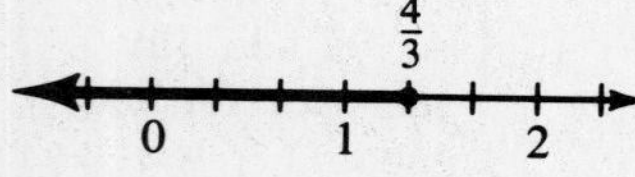

6. $\{p: p \le 6\}$

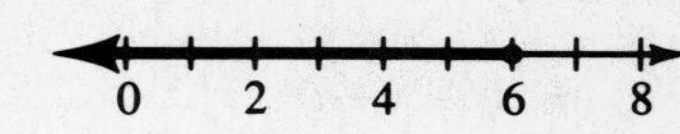

7. $\{k: 4 < k < 5\}$

8. $\{d: -2 < d < 3\}$

9. $\{n: -5 < n < 0\}$

10. $\{m: 5 < m < 9\}$

11. $\{v: -5 \le v \le 10\}$

12. $\{t: -2 \le t \le 6\}$

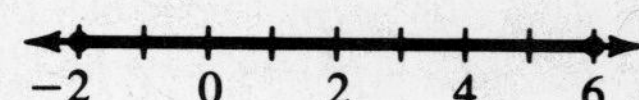

13. $\{u: u \ge 3\}$

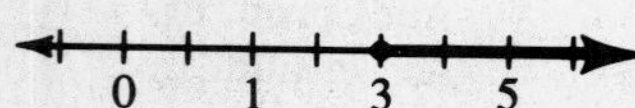

14. $\{w: w \ge 4\}$

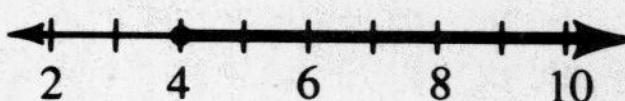

B **15.** $\{z: z \ge 3\}$

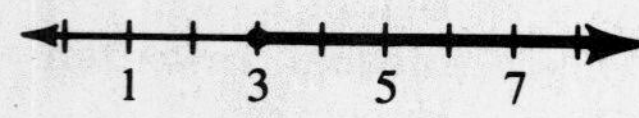

16. $\{w: w \ge 5\}$

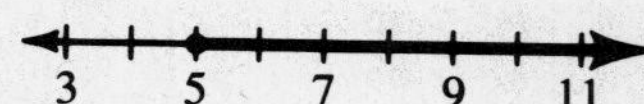

17. $\{x: x \le 4\}$

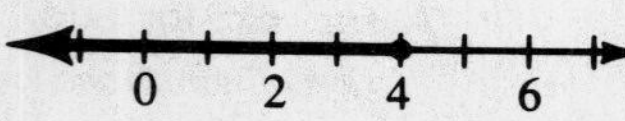

18. $\{y: y \ge -1\}$

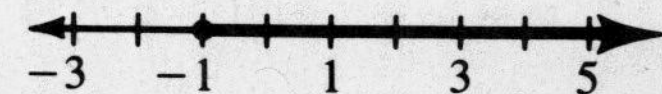

19. $\{b: b < -2\} \cup \{b: b > 4\}$

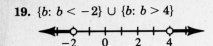

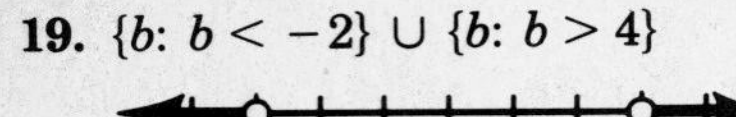

20. $\emptyset$

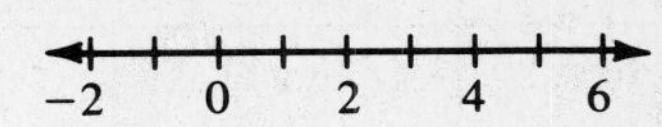

21. $\{t: -6 < t < -3\}$

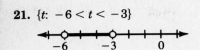

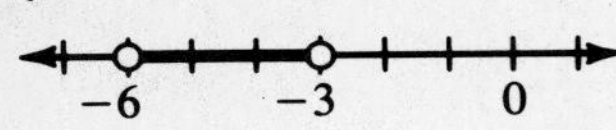

22. $\{d: d > 8\} \cup \{d: d < -6\}$

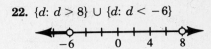

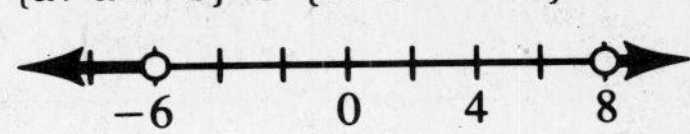

23. $\left\{h: h < \frac{3}{2}\right\} \cup \left\{h: h > 7\right\}$

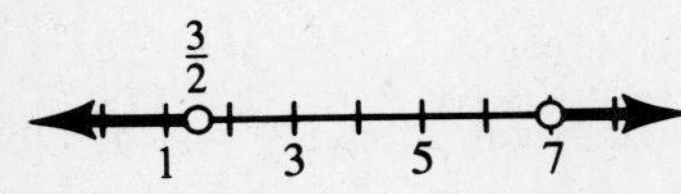

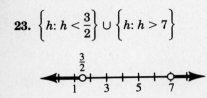

24. $\{r: r > 4\}$

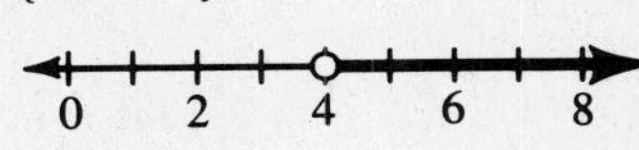

C **25.** $\{v: -2 < v < 4\}$

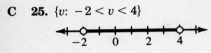

26. $\{x: -1 < x \le 5\}$

27. $\{y: y = -1\}$

28. $\{n: -1 \le n < 2\}$

Pages 63–64 • PROBLEMS

A **1.** Let x = weight of adult portion; $x - 2$ = weight of child's portion; $16x + 15(x - 2) \le 125$; $x \le 5$; 5 oz

2. Let x = length of longer base; $x - 17$ = length of shorter base; $A = \frac{1}{2}h(b + b')$; $\frac{1}{2} \cdot 15[x + (x - 17)] < 165$; $x < 19.5$; 2.5 cm

3. Let x = no. of dividers; $x + 1$ = divisions between dividers on one side; $100 + 50x + 5 \cdot 2(x + 1) \le 770$; $x \le 11$; 11 dividers

4. Let x = no. of studs; $4\left(\frac{3}{2}\right) + 14 + \left(14 + \frac{3}{2}\right)x \ge \frac{233}{2}$; $x \ge 6.2$; 7 additional studs

B **5.** Let h = ht. of cylinder; $2(9\pi) + 2\pi \cdot 3 \cdot h \le 108\pi$; $h \le 15$; $V = 135\pi$ cm^3

6. Let t = no. of hours Jose swims; $2t + 18\left(\frac{1}{3} - t\right) = 2$; $t = \frac{1}{4}$; $\frac{1}{2}$ km

7. Let t = no. of hours the express train travels; $80t + 60\left(t - \frac{1}{4}\right) \ge 195$; $t \ge \frac{3}{2}$; $195 - 120 = 75$ km

8. Let x = no. of ft^2 painted; $x(1.40) + (300 - x)(.60) \le 324.0$; $x \le 180$; 180 ft^2

9. Let x = no. of min first pipe operates; $x - 30$ = no. of min the second pipe operates; $x - 60$ = no. of min the third pipe operates; $\text{min} \cdot \frac{\text{L}}{\text{min}} = \text{L}$; $x \cdot 12 + (x - 30)16 + (x - 60)20 = 1200$; $x = 60$; 60 min

10.

	r	t	d
first part	8	$\frac{1}{2}$	4
second part	15	$\frac{d}{15}$	d
total	12	$\frac{1}{2} + \frac{d}{15}$	$4 + d$

$4 + d = 12\left(\frac{1}{2} + \frac{d}{15}\right)$; $d = 10$; 10 km

C **11.** Let x = no. of balls, let r = radius of one ball; vol. of x balls $= x \cdot \frac{4}{3}\pi r^3$; vol. of can $= \pi r^2(2rx)$; vol. of unoccupied space = vol. of can − vol. of x balls; $(2\pi r^3)x - \left(\frac{4}{3}\pi r^3\right)x \le \frac{4}{3}\pi r^3$; $x \le 2$; 2 balls

12. Let x = length of a side of the larger garden; $\frac{50 - 4x}{4}$ = a side of the smaller garden; $x^2 - (12.5 - x)^2 \le 100$; $x \le 10.25$; 10.25 m

Pages 67–69 • WRITTEN EXERCISES

A **1.** 1. Multiplication property of equality 2. Multiplicative property of zero 3. Axiom of multiplicative inverses 4. $1 > 0$

2. 1. Hypothesis 2. Multiplication property of order 3. Multiplicative property of zero 4. Axiom of multiplicative inverses 5. $1 > 0$

3. 1. Multiplication property of equality 2. Identity axiom for multiplication 3. Definition of division 4. Associative axiom for multiplication 5. Axiom of multiplicative inverses 6. Identity axiom for multiplication 7. $a > b$

4. 1. Hypothesis 2. Multiplicative property of order 3. Identity axiom for multiplication 4. Definition of division 5. Associative axiom for multiplication 6. Axiom of multiplicative inverses 7. Identity axiom for multiplication 8. $a > b$

Exs. 5–18: Answers may vary.

5. Case 1: Assume $\frac{1}{a} = 1$.

1. $a > 1$ and $\frac{1}{a} = 1$	Hypothesis
2. $\frac{1}{a} \cdot a = 1 \cdot a$	Mult. prop. of =
3. $1 = 1 \cdot a$	Ax. of mult. inv.
4. $1 = a$	Iden. ax. for mult.
5. Contradiction	Hypothesis: $a > 1$

Case 2: Assume $\frac{1}{a} > 1$.

1. $a > 1$ and $\frac{1}{a} > 1$	Hypothesis
2. $a \cdot \frac{1}{a} > 1 \cdot a$	Mult. prop. of order
3. $1 > 1 \cdot a$	Ax. of mult. inv.
4. $1 > a$	Iden. ax. for mult.

5. Contradiction	Hypothesis: $a > 1$
Therefore $\frac{1}{a} < 1$.	
6. 1. $1 > \frac{1}{a}$ and $a > 0$	Hypothesis
2. $a \cdot 1 > a \cdot \frac{1}{a}$	Mult. prop. of order
3. $a \cdot 1 > 1$	Ax. of mult. inv.
4. $a > 1$	Iden. ax. for mult.
7. 1. $a > b$	Hypothesis
2. $a + a > a + b$	Add. prop. of order
3. $2a > a + b$	Dist. ax.
4. $2a\left(\frac{1}{2}\right) > (a + b)\frac{1}{2}$	Mult. prop. of order
5. $a\left(2 \cdot \frac{1}{2}\right) > (a + b)\frac{1}{2}$	Comm. and assoc. ax. for mult.
6. $a \cdot 1 > (a + b)\frac{1}{2}$	Ax. of mult. inv.
7. $a > (a + b)\frac{1}{2}$	Iden. ax. for mult.
8. $a > \frac{a + b}{2}$	Def. of div.
9. $\frac{a + b}{2} < a$	Inequal. rewritten
8. 1. $a > b$	Hypothesis
2. $a + b > b + b$	Add. prop. of order
3. $a + b > 2b$	Dist. ax.
4. $(a + b)\frac{1}{2} > 2b\left(\frac{1}{2}\right)$	Mult. prop. of order
5. $\frac{a + b}{2} > 2b\left(\frac{1}{2}\right)$	Def. of div.
6. $\frac{a + b}{2} > b\left(2 \cdot \frac{1}{2}\right)$	Comm. and assoc. ax. for mult.
7. $\frac{a + b}{2} > b \cdot 1$	Ax. of mult. inv.
8. $\frac{a + b}{2} > b$	Iden. ax. for mult.
C **9.** 1. $0 < a < b$	Hypothesis
2. $\frac{1}{a} > 0, \frac{1}{b} > 0$	Theorem, page 65
3. $\frac{1}{a} \cdot \frac{1}{b} > 0$	Clos. ax. for $\mathcal{R}_+$
4. $a\left(\frac{1}{a} \cdot \frac{1}{b}\right) < b\left(\frac{1}{a} \cdot \frac{1}{b}\right)$	Mult. prop. of order
5. $\left(a \cdot \frac{1}{a}\right)\frac{1}{b} < \left(b \cdot \frac{1}{b}\right)\frac{1}{a}$	Comm. and assoc. ax. for mult.
6. $1 \cdot \frac{1}{b} < 1 \cdot \frac{1}{a}$	Ax. of mult. inv.

[*Proof continued on next page*]

	7. $\frac{1}{b} < \frac{1}{a}$	Iden. ax. for mult.
	8. $\frac{1}{a} > \frac{1}{b}$	Inequal. rewritten
10.	1. $a > b > 0$	Hypothesis
	2. $a^2 > ab$	Mult. prop. of order
	3. $ab > b^2$	Mult. prop. of order
	4. $a^2 > b^2$	Trans. prop. of order
11.	Case 1: Assume $a = b$.	
	1. $a = b,\ a^2 > b^2$	Hypothesis
	2. $a \cdot a = a \cdot b;\ a \cdot b = b \cdot b$	Mult. prop. of =
	3. $a^2 = ab;\ ab = b^2$	Subs. prin.
	4. $a^2 = b^2$	Trans. prop. of =
	5. Contradiction	Hypothesis: $a^2 > b^2$
	Case 2: Assume $a < b$.	
	1. $a > 0,\ b > 0,\ a^2 > b^2,\ a < b$	Hypothesis
	2. $a \cdot a < b \cdot a;\ b \cdot a < b \cdot b$	Mult. prop. of order
	3. $a \cdot a < b \cdot b$	Trans. prop. of order
	4. $a^2 < b^2$	Subs. prin.
	5. Contradiction	Hypothesis: $a^2 > b^2$
	Therefore $a > b$.	
12.	1. $a < b < 0$	Hypothesis
	2. $\frac{1}{a} < 0;\ \frac{1}{b} < 0$	Theorem, page 65
	3. $\frac{1}{a} \cdot \frac{1}{b} > 0$	Mult. prop. of order
	4. $a\left(\frac{1}{a} \cdot \frac{1}{b}\right) < b\left(\frac{1}{a} \cdot \frac{1}{b}\right)$	Mult. prop. of order
	5. $\left(a \cdot \frac{1}{a}\right)\left(\frac{1}{b}\right) < \left(b \cdot \frac{1}{b}\right)\left(\frac{1}{a}\right)$	Comm. and assoc. ax. for mult.
	6. $1 \cdot \frac{1}{b} < 1 \cdot \frac{1}{a}$	Ax. of mult. inv.
	7. $\frac{1}{b} < \frac{1}{a}$	Iden. ax. for mult.
	8. $\frac{1}{a} > \frac{1}{b}$	Inequal. rewritten
13.	Case 1: Assume $a = b$.	
	1. $a = b,\ a^2 > b^2$	Hypothesis
	2. $a \cdot a = b \cdot a;\ b \cdot a = b \cdot b$	Mult. prop. of =
	3. $a \cdot a = b \cdot b$	Trans. prop. of =
	4. $a^2 = b^2$	Subs. prin.
	5. Contradiction	Hypothesis: $a^2 > b^2$
	Case 2: Assume $a > b$.	
	1. $a > b,\ a < 0,\ b < 0,\ a^2 > b^2$	Hypothesis
	2. $a \cdot a < b \cdot a;\ b \cdot a < b \cdot b$	Mult. prop. of order
	3. $a \cdot a < b \cdot b$	Trans. prop. of order
	4. $a^2 < b^2$	Subs. prin.
	5. Contradiction	Hypothesis: $a^2 > b^2$
	Therefore $a < b$.	

Statement	Reason
14. 1. $b < a < 0$	Hypothesis
2. $b^2 > ab,\ ab > a^2$	Mult. prop. of order
3. $b^2 > a^2$	Trans. prop. of order
4. $a^2 < b^2$	Inequal. rewritten
15. 1. $a > 0,\ b > 0,\ c > 0,\ d > 0,\ ab < cd,$ $b > d$	Hypothesis
2. $ab > ad$	Mult. prop. of order
3. $ad < ab$	Inequal. rewritten
4. $ad < cd$	Trans. prop. of order
5. $\frac{1}{d} > 0$	Theorem, page 65
6. $ad\left(\frac{1}{d}\right) < cd\left(\frac{1}{d}\right)$	Mult. prop. of order
7. $a\left(d \cdot \frac{1}{d}\right) < c\left(d \cdot \frac{1}{d}\right)$	Assoc. prop. for mult.
8. $a \cdot 1 < c \cdot 1$	Ax. of mult. inv.
9. $a < c$	Iden. ax. for mult.
16. 1. $a, b, c, d < 0;\ ab < cd,\ b < d$	Hypothesis
2. $ab > ad$	Mult. prop. of order
3. $cd > ad$	Trans. prop. of order
4. $\frac{1}{d} < 0$	Theorem, p. 65
5. $cd \cdot \frac{1}{d} < ad \cdot \frac{1}{d}$	Mult. prop. of order
6. $c\left(d \cdot \frac{1}{d}\right) < a\left(d \cdot \frac{1}{d}\right)$	Assoc. ax. for mult.
7. $c \cdot 1 < a \cdot 1$	Ax. of mult. inv.
8. $c < a$	Iden. ax. for mult.
17. 1. $a > b > 0$	Hypothesis
2. $a + a > a + b$	Add. prop. of order
3. $2a > a + b$	Dist. ax.
4. $2ab > (a + b)b$	Mult. prop. of order
5. $a + b > 0$	Clos. ax. for $\mathcal{R}_+$
6. $\frac{1}{a + b} > 0$	Theorem, page 65
7. $2ab\left(\frac{1}{a + b}\right) > [(a + b)b]\left(\frac{1}{a + b}\right)$	Mult. prop. of order
8. $\frac{2ab}{a + b} > [(a + b)b]\left(\frac{1}{a + b}\right)$	Def. of div.
9. $\frac{2ab}{a + b} > b\left[(a + b)\left(\frac{1}{a + b}\right)\right]$	Comm. and assoc. ax. for mult.
10. $\frac{2ab}{a + b} > b \cdot 1$	Ax. of mult. inv.
11. $\frac{2ab}{a + b} > b$	Iden. ax. for mult.
18. 1. $a > b > 0$	Hypothesis
2. $a + b > 2b$	Add. prop. of order and dist. ax.
3. $a(a + b) > a(2b)$	Mult. prop. of order
4. $a + b > 0$	Clos. ax. for $\mathcal{R}_+$

[*Proof continued on next page*]

5. $\frac{1}{a+b} > 0$	Theorem, page 65
6. $[a(a+b)]\left(\frac{1}{a+b}\right) > 2ab\left(\frac{1}{a+b}\right)$	Mult. prop. of order
7. $a \cdot 1 > 2ab\left(\frac{1}{a+b}\right)$	Assoc. ax. for mult.; ax. of mult. inv.
8. $a > \frac{2ab}{a+b}$	Iden. ax. for mult. and def. of div.
9. $\frac{2ab}{a+b} < a$	Inequal. rewritten

Pages 71–72 • WRITTEN EXERCISES

A **1.** 5 **2.** 3 **3.** 12
4. 0 **5.** 7 **6.** 0

7. $\{r: r \geq 4\} \cup \{r: r \leq -4\}$

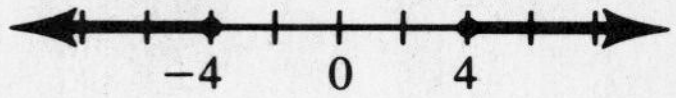

−4 0 4

8. $\left\{t: -\frac{5}{2} < t < \frac{5}{2}\right\}$

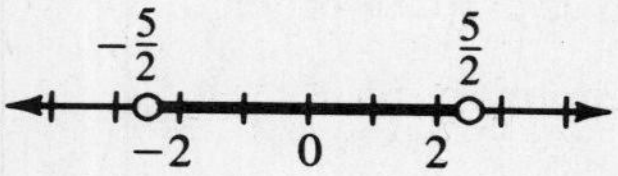

$-\frac{5}{2}$ $\frac{5}{2}$

−2 0 2

9. $\{x: 6 \leq x \leq 8\}$

4 6 8 10

10. $\{y: y > 2\} \cup \{y: y < -4\}$

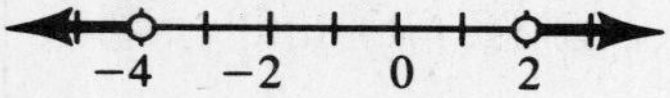

−4 −2 0 2

11. $\{c: c \geq 8\} \cup \{c: c \leq 4\}$

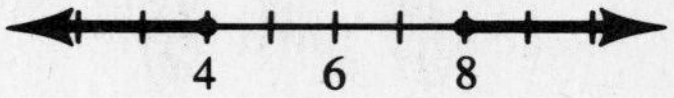

4 6 8

12. $\{d: -7 < d < 1\}$

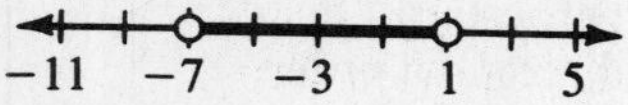

−11 −7 −3 1 5

13. $\{1, 2\}$

0 1 2 3 4

14. $\left\{\frac{2}{3}, 4\right\}$

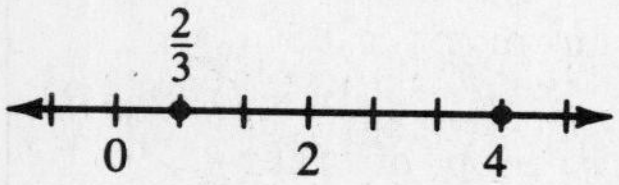

$\frac{2}{3}$

0 2 4

15. $\{n: n \geq -1\} \cup \{n: n \leq -8\}$

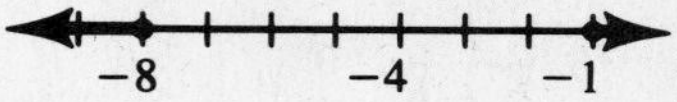

−8 −4 −1

16. $\{r: -10 < r < 6\}$

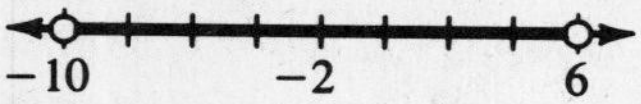

−10 −2 6

17. $\{q: -2 \leq q \leq 10\}$

−2 2 6 10

18. $\{w: 2 < w < 10\}$

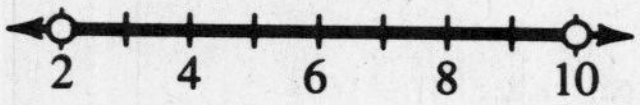

2 4 6 8 10

19. $\left\{m: -1 \leq m \leq \frac{3}{2}\right\}$

$\frac{3}{2}$

−1 0 1 2

B **20.** $\left\{x: x > \frac{5}{2}\right\} \cup \left\{x: x < -\frac{7}{2}\right\}$

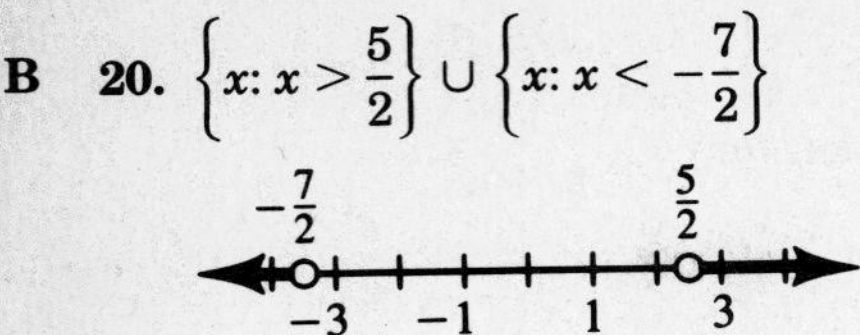

21. $\{u: u \geq 1\} \cup \{u: u \leq -8\}$

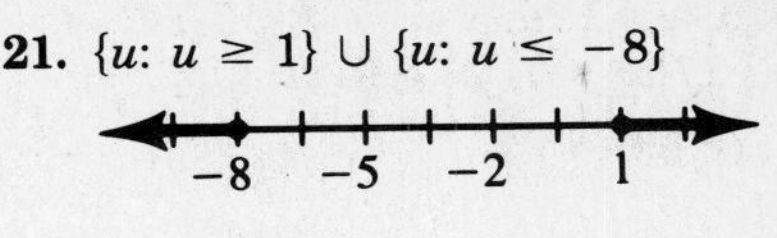

22. $\{v: -1 < v < 4\}$

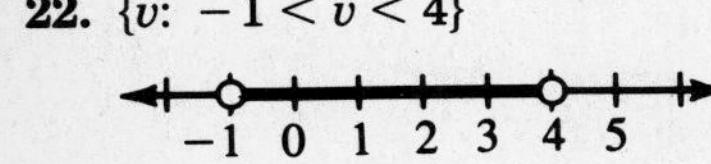

23. $\{a: a > 6\} \cup \{a: a < 3\}$

24. $\left\{c: -\frac{1}{3} \leq c \leq 5\right\}$

25. no; $a = 2$, $b = -1$ **26.** yes **27.** yes **28.** no; $a = 2$, $b = -2$ **29.** yes
30. no; $a = 0$, $b = 1$ **31.** yes **32.** yes **33.** no; $a = 1$, $b = -1$ **34.** yes

35. Case 1: $a \geq 0$

Statement	Reason
1. $\lvert a\rvert = a$	Def. of $\lvert a\rvert$ for $a > 0$
2. $\lvert a\rvert^2 = \lvert a\rvert \cdot \lvert a\rvert$	Def. of x^2
3. $\quad = a \cdot a$	Subs. prin.
4. $\quad = a^2$	Def. of x^2
5. $\lvert a\rvert^2 = a^2$	Trans. prop. of =

Case 2: $a < 0$

Statement	Reason
1. $\lvert a\rvert = -a$	Def. of $\lvert a\rvert$ for $a < 0$
2. $\lvert a\rvert^2 = (-a)(-a)$	Def. of x^2
3. $\quad = a \cdot a$	Prop. of opp. in prod.
4. $\quad = a^2$	Def. of x^2
5. $\lvert a\rvert^2 = a^2$	Trans. prop. of =

36. Case 1: $a = 0$ or $b = 0$

Statement	Reason
1. $ab = 0$	Mult. prop. of 0
2. $\lvert ab\rvert = ab$	Def. of abs. value

Case 2: $a > 0$, $b > 0$

Statement	Reason
1. $ab > 0$	Clos. ax. for $\mathcal{R}_+$
2. $\lvert ab\rvert = ab$	Def. of abs. value

Case 3: $a > 0$, $b < 0$

Statement	Reason
1. $ab < 0$	Mult. prop. of order
2. $\lvert ab\rvert > 0$	Def. of abs. value
3. $ab < \lvert ab\rvert$	Trans. prop. of order

Case 4: $a < 0$, $b < 0$

Statement	Reason
1. $ab > 0$	Clos. ax. for $\mathcal{R}_+$
2. $\lvert ab\rvert = ab$	Def. of abs. value

37.

Statement	Reason
1. $a, b \in \mathcal{R}$	Hypothesis
2. $\lvert a + b\rvert^2 = (a + b)^2$	Ex. 35
3. $\lvert a + b\rvert^2 = a^2 + 2ab + b^2$	Dist. ax.
4. $(\lvert a\rvert + \lvert b\rvert)^2 = \lvert a\rvert^2 + 2\lvert a\rvert \cdot \lvert b\rvert + \lvert b\rvert^2$	Dist. ax.
5. $\lvert a\rvert^2 = a^2$, $\lvert b\rvert^2 = b^2$	Ex. 35
6. $\lvert a\rvert \cdot \lvert b\rvert = \lvert ab\rvert$	Ex. 26
7. $(\lvert a\rvert + \lvert b\rvert)^2 = a^2 + 2\lvert ab\rvert + b^2$	Subs. prin.
8. $ab \leq \lvert ab\rvert$	Ex. 36
9. $2ab \leq 2\lvert ab\rvert$	Mult. prop. of order
10. $a^2 + 2ab + b^2 \leq a^2 + 2\lvert ab\rvert + b^2$	Add. prop. of order
11. $\lvert a + b\rvert^2 \leq (\lvert a\rvert + \lvert b\rvert)^2$	Subs. prin.

38. Case 1: $a \neq 0$, $b \neq 0$

1. $\lvert a\rvert > 0$, $\lvert b\rvert > 0$, $\lvert a + b\rvert > 0$	Def. of abs. value
2. $\lvert a\rvert + \lvert b\rvert > 0$	Clos. ax. for $\mathscr{R}_+$
3. $\lvert a + b\rvert^2 \le (\lvert a\rvert + \lvert b\rvert)^2$	Ex. 37
4. $\lvert a + b\rvert \le \lvert a\rvert + \lvert b\rvert$	Ex. 11, page 68

Case 2: $a = 0$

1. $\lvert a\rvert = 0$	Def. of abs. value
2. $\lvert a + b\rvert = \lvert 0 + b\rvert$	Subs. prin.
3. $\lvert a + b\rvert = \lvert b\rvert$	Iden. ax. for add.
4. $\lvert b\rvert = \lvert b\rvert$	Reflex. prop. of =
5. $\lvert b\rvert \le \lvert b\rvert$	Def. of $\le$
6. $\lvert b\rvert \le 0 + \lvert b\rvert$	Iden. ax. for add.
7. $\lvert a + b\rvert \le \lvert a\rvert + \lvert b\rvert$	Subs. prin.

Case 3: $b = 0$

1. $\lvert b\rvert = 0$	Def. of abs. value
2. $\lvert a + b\rvert = \lvert a + 0\rvert$	Subs. prin.
3. $\lvert a + b\rvert = \lvert a\rvert$	Iden. ax. for add.
4. $\lvert a\rvert = \lvert a\rvert$	Reflex. prop. of =
5. $\lvert a\rvert \le \lvert a\rvert$	Def. of $\le$
6. $\lvert a\rvert \le \lvert a\rvert + 0$	Iden. ax. for add.
7. $\lvert a + b\rvert \le \lvert a\rvert + \lvert b\rvert$	Subs. prin.

Page 72 • COMPUTER EXERCISES

1.
```
10 INPUT A,B,A$
20 IF A$ = ">" THEN 40
30 PRINT -B - A;" < X < ";B - A : END
40 PRINT "X > ";B - A;" OR X < ";-B - A
50 END
```

2. $-2 < x < 12$ **3.** $x < -7$ or $x > -1$ **4.** $x < -17$ or $x > 13$ **5.** $19 < x < 37$

Page 73 • SELF-TEST 2

1. $\{a\colon a > 4\}$

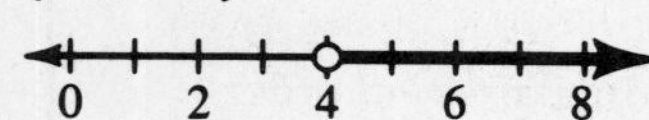

2. $\{x\colon x > -1\}$

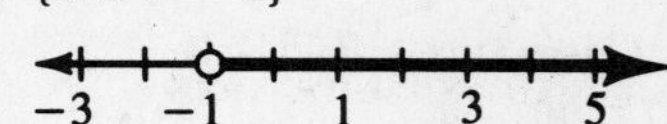

3. $\left\{y\colon y \ge -\frac{23}{2}\right\}$

4. $\left\{b\colon \frac{1}{2} < b \le \frac{3}{2}\right\}$

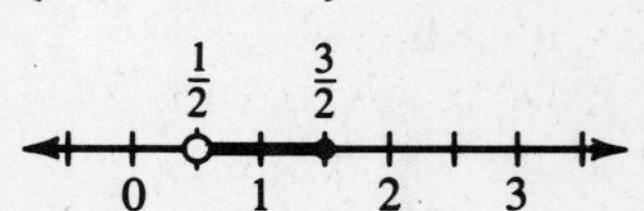

5. Let x = total no. of minutes; $2.05 + (x - 3)(0.35) \le 4.00$; $x \le \frac{60}{7}$; 8 min

6. Case 1: Assume $b = 0$.

1. $a < 0$, $ab > 0$, $b = 0$	Hypothesis
2. $ab = a \cdot 0$	Subs. prin.
3. $ab = 0$	Mult. prop. of 0
4. Contradiction	Hypothesis: $ab > 0$

Case 2: Assume $b > 0$.

1. $a < 0$, $ab > 0$, $b > 0$	Hypothesis
2. $ab < 0 \cdot b$	Mult. prop. of order
3. $ab < 0$	Mult. prop. of 0
4. Contradiction	Hypothesis: $ab > 0$

7. $\{x: -2 \le x \le 12\}$

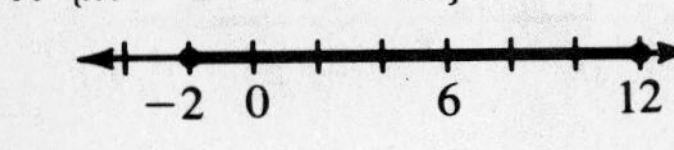

8. $\{y: y < -2\} \cup \left\{y: y > \frac{2}{3}\right\}$

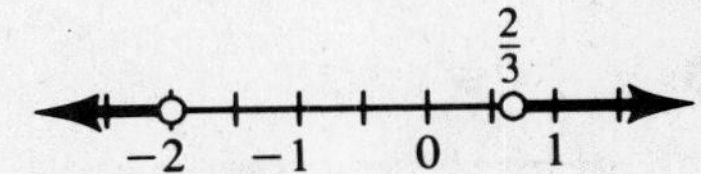

Pages 74–75 • CHAPTER REVIEW

1. c **2.** c **3.** b **4.** b **5.** $x - 1$ = the time traveling when it is not snowing; $75(x - 1) = 60x$; d **6.** b **7.** d **8.** b **9.** b **10.** c **11.** d **12.** b **13.** d **14.** c

Page 76 • CHAPTER TEST

1. $x^2 - 3x + 27$ **2.** $-13d^2 - 15e + 11d$ **3.** $\{4\}$ **4.** $\{-1\}$ **5.** $\frac{3}{2x - 5} = a$

6. $\frac{b}{3m + f} = a$ **7.** Let x = first integer; $x + 2(x + 2) = 3(x + 1) + 1$; $3x + 4 = 3x + 4$; any 3 consecutive integers **8.** Let x = length of base; $x + 2\left(\frac{1}{2}x + 4\right) = 36$; $x = 14$; 14 cm, 11 cm, 11 cm

9. $\{x: x < -10\}$

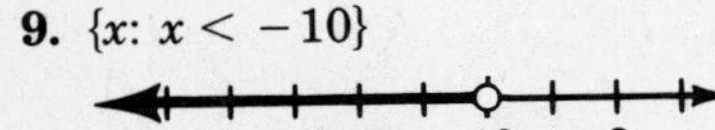

10. $\{x: x > 2\}$

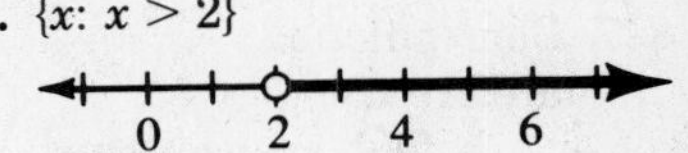

11. $\{n: -1 < n < 5\}$

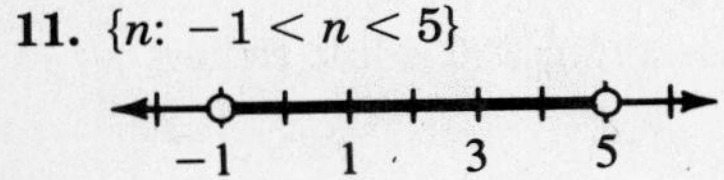

12. $\{q: q < 2\} \cup \{q: q \ge 5\}$

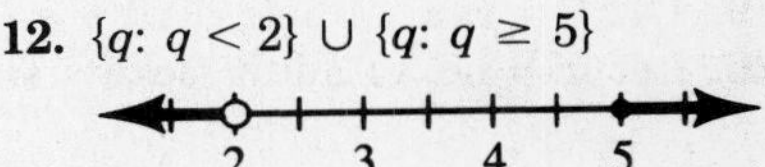

13. Let w = width; $2(w) + 2(2w + 2) \le 70$; $w \le 11$; 11 m

14. Case 1: Assume $a = b$.

1. $a + c > b + c$, $a = b$	Hypothesis
2. $a + c = b + c$	Add. prop. of =
3. Contradiction	Hypothesis: $a + c > b + c$

Case 2: Assume $a < b$.

1. $a + c > b + c$, $a < b$	Hypothesis
2. $a + c < b + c$	Add. prop. of order
3. Contradiction	Hypothesis: $a + c > b + c$

15. $\{x: -2 \le x \le 1\}$

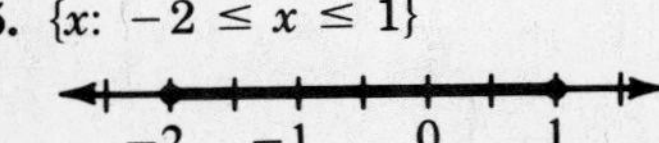

16. $\{v: -2 \le v \le 8\}$

−2 0 2 4 6 8 10

Pages 76–77 • MIXED REVIEW

1. 0 **2.** 10 **3.** $50wz - 80z$ **4.** $7x^2 - 11x + 7$ **5.** $\{-4\}$ **6.** $\{-2\}$ **7.** $\{1\}$

8. $\{5\}$ **9.** $b = \dfrac{2a}{4a - 5}$ **10.** $b = \dfrac{m^2}{t}$ **11.** -16 **12.** 30

13. $\{a: 2 < a \le 7\}$

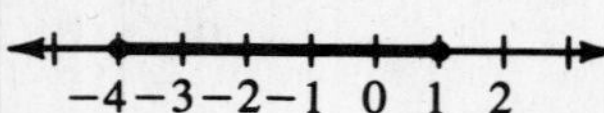

14. $\{b: b > 4\} \cup \{b: b < -1\}$

−1 0 1 2 3 4 5

15. $\{t: -4 \le t \le 1\}$

−4 −3 −2 −1 0 1 2

16. $\{x: x > -6\}$

−8 −6 −4 −2

17. Case 1: Assume $a - b = 0$.

1. $a > b,\ a - b = 0$	Hypothesis
2. $a - b + b = 0 + b$	Add. prop. of =
3. $a + (-b) + b = 0 + b$	Def. of subt.
4. $a + [(-b) + b] = 0 + b$	Assoc. ax. for add.
5. $a + 0 = 0 + b$	Ax. of add. inv.
6. $a = b$	Iden. ax. for add.
7. Contradiction	Hypothesis: $a > b$

Case 2: Assume $a - b < 0$.

1. $a > b,\ a - b < 0$	Hypothesis
2. $a - b + b < 0 + b$	Add. prop. of order
3. $a + (-b) + b < 0 + b$	Def. of subt.
4. $a + [(-b) + b] < 0 + b$	Assoc. ax. for add.
5. $a + 0 < 0 + b$	Ax. of add. inv.
6. $a < b$	Iden. ax. for add.
7. Contradiction	Hypothesis: $a > b$

Therefore $a - b > 0$.

18. Let x = first even integer, $x + (x + 2) + (x + 4) = 4x - 10$; $x = 16$; 16, 18, 20

19. Let r = rate of bus; $90.5 + r \cdot 3 = 795$; $r = 115$; 115 km/h

20. Let a = no. of adult tickets; $3a + 2(400 - a) \ge 1100$; 300 adult tickets

Page 77 • CONTEST PROBLEMS

1. Let x = rate of boat; y = rate of current; C be the point where the man turns around.

	$r\left(\frac{\text{mi}}{\text{h}}\right)$	t (h)	d (mi)
AB	$x - y$	$\frac{1}{x - y}$	1
BC	$x - y$	$\frac{1}{6}$	$\frac{x - y}{6}$
AC	$x + y$		$1 + \frac{x - y}{6}$
hat	y	$\frac{1}{y}$	1

dist. from A to C = dist. from A to B + dist. from B to C; $t = \dfrac{d}{r}$

BC time + AC time = hat time

$\dfrac{1}{6} + \left[\left(1 + \dfrac{x - y}{6}\right) \div x + y\right] = \dfrac{1}{y}$;

$y = 3$; 3 mi/h

2. Let r = rate for the second mile

	r	t	d
first mi	30	$\frac{1}{30}$	1
sec. mi	r	$\frac{1}{r}$	1
total	60	$\frac{1}{30} + \frac{1}{r}$	2

$60\left(\frac{1}{30} + \frac{1}{r}\right) = 2$; $60 = 0$; impossible

Pages 79–81 • PROGRAMMING IN PASCAL

1. a. `s IN [2..17]`
b. `t IN [5..7] + [10..14]`
c. `s IN set_u * set_v`
d. `[t, s] <= set_u`
e. `s IN set_u + set_v`
f. `s IN [2, 4, 6]`
g. `s IN set_v - set_u`
h. `t IN [s - 2..s + 2]`

2. a.
```
PROGRAM abs_val (INPUT, OUTPUT);
VAR
  num: integer;
BEGIN
  write('Enter a number: ');
  readln(num);
  writeln('The absolute value of that number is: ', abs(num));
END.
```

b.
```
PROGRAM abs_val (INPUT, OUTPUT);
VAR
  num : integer;
BEGIN
  REPEAT
    write('Enter a number: ');
    readln(num);
    writeln('Absolute value of that number is: ', abs(num));
  UNTIL num = 0;
END.
```

c. The program is interrupted with an I/O error message.
d. Enter the digits as characters, then use the *ord* function to convert the characters to integers.

e.
```
PROGRAM abs_val (INPUT, OUTPUT);
VAR
  value : string[10];
  num : integer;
  ok : boolean;

(************************************************************)
PROCEDURE convert;
VAR
  i : integer;
BEGIN
  num := 0;
  IF (value[1] = '-') AND (value[2] IN ['0'..'9'])
    THEN BEGIN
           ok := TRUE;
           i := 2;
           REPEAT
             num := 10 * num + ord(value[i]) - ord('0');
             i := i + 1;
           UNTIL (i > 10) OR NOT (value[i] IN ['0'..'9']);
           num := -num;
         END
    ELSE BEGIN
           IF (value[1] IN ['0'..'9'])
             THEN BEGIN
                    ok := TRUE;
                    i := 1;
                    REPEAT
                      num := 10 * num + ord(value[i]) - ord('0');
                      i := i + 1;
                    UNTIL (i > 10) OR NOT (value[i] IN ['0'..'9']);
                  END
             ELSE BEGIN
                    writeln('ILLEGAL INTEGER VALUE');
                    writeln('REENTER A NUMBER');
                  END;
         END;
END;

(************************************************************)
BEGIN
  REPEAT
    value := '          '; {* initialize to 10 blanks *}
    ok := FALSE;
    write('Enter a number: ');
    readln(value);
    convert;
```

```
      IF ok
        THEN writeln('Absolute value of number is ', abs(num));
    UNTIL (num = 0) AND (ok);
  END.
```

3.
```
(**************************************************************)
PROCEDURE instruct;
BEGIN
  writeln('This program will solve an');
  writeln('inequality of the form |ax - b| >= c');
  writeln;
  writeln('The coefficients a, b, and c');
  writeln('are integer values.');
  writeln;
  writeln('The solution is subject to');
  writeln('rounding errors.');
  writeln;
END;

(**************************************************************)
BEGIN {* Main Program *}
  instruct;
  get_coefficients;
  solve_and_show_solution;
END.
```

4. a.
```
write('x < ', left_end_point:3:3);
write(' OR ');
writeln('x > ', right_end_point:3:3);
```

b.
```
writeln(left_end_point:3:3,'<= x <=', right_end_point);
```

5.
```
PROGRAM linear_equation (INPUT, OUTPUT);
VAR
  a, b, c : real;

(**************************************************************)
PROCEDURE instruct;
BEGIN
  writeln('This program will solve any linear equation of');
  writeln('the form ax + b = c for any values of a, b, and');
  writeln('c that are real numbers and such that a is not 0');
END;
```

[*Program continued on next page*]

```
(**************************************************************)
PROCEDURE get_values(VAR a, b, c : real);
BEGIN
  REPEAT
    writeln;
    write('Enter the value of a: ');
    readln(a);
    IF a = 0
      THEN writeln('ILLEGAL VALUE.');
  UNTIL a <> 0;
  write('Enter the value of b: ');
  readln(b);
  write('Enter the value of c: ');
  readln(c);
END;

(**************************************************************)
PROCEDURE solve_and_show(a, b, c : real);
BEGIN
  write('The solution to the equation ',a:1:3, 'x + ',b:1:3 );
  writeln(' = ',c:1:3);
  writeln('is {',((c - b)/a):1:3, '}.');
END;

(**************************************************************)
BEGIN (* main *)
  instruct;
  get_values(a, b, c);
  solve_and_show(a, b, c);
END.
```

Pages 86–87 • WRITTEN EXERCISES

A 1. 37 2. -11 3. -13 4. 59 5. 4 6. 3 7. $\frac{9}{2}$ 8. $\frac{7}{2}$ 9. 37 10. 28

11. 5 12. 1 13. $y = \frac{1}{2}x$; a function 14. $y = 4x$; a function

15. $|y| = \frac{1}{x}$; not a function 16. $y = 2x + 1$; a function 17. $y = \frac{1}{x + 1}$; a function

18. $x = 7$; not a function

B 19. $y = -x^2$; a function 20. $|x| + |y| = 5$; not a function

21. $y = \frac{1}{3}x + 1$; a function 22. $y = x^2 + 1$; a function 23. $-4\frac{1}{2}$ 24. 3

25. $-\frac{43}{9}$ 26. 1 27. $\frac{1}{49}$ 28. $\frac{1}{36}$ 29. $-\frac{117}{25}$ 30. not defined

C 31. a. $f(0 + 2) = f(0) + f(2)$
$f(2) = f(0) + f(2)$
$0 = f(0)$

b. $f(-2 + 2) = f(-2) + f(2)$
$f(0) = f(-2) + f(2)$
$0 = f(-2) + f(2)$
$f(-2) = -f(2)$

32. $f(-0) = -f(0)$
$f(0) = -f(0)$
$f(0) + f(0) = -f(0) + f(0)$
$2f(0) = 0$
$f(0) = 0$

33. a. $f(0 + 2) = f(0) + 5$
$f(2) = 0 + 5 = 5$
$f(2 + 2) = f(2) + 5$
$f(4) = 5 + 5 = 10$
$f(24) = 6f(4) = 6 \cdot 10 = 60$

b. $y = \frac{5}{2}x$

Page 90 • WRITTEN EXERCISES

A 1. a function

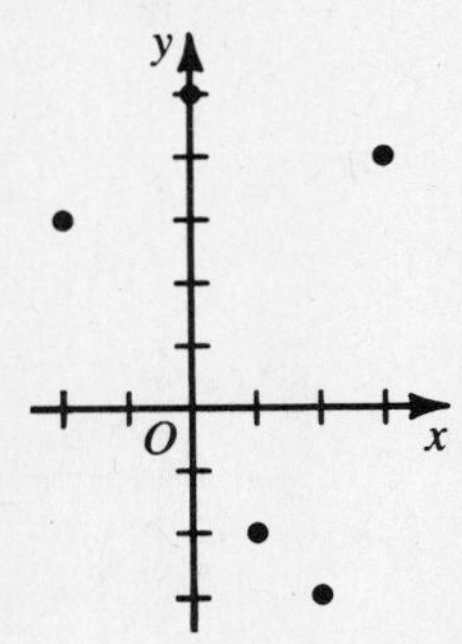

2. not a function

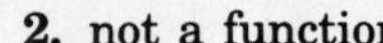

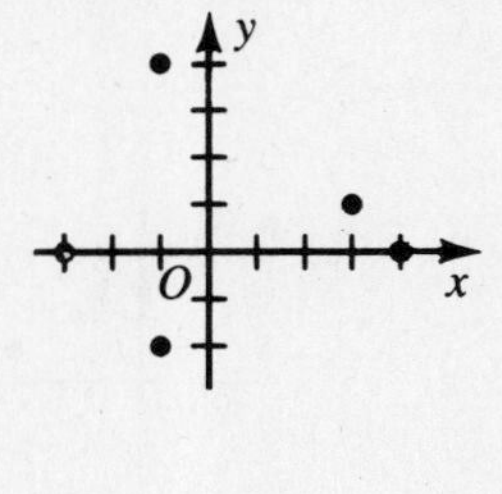

3. a function

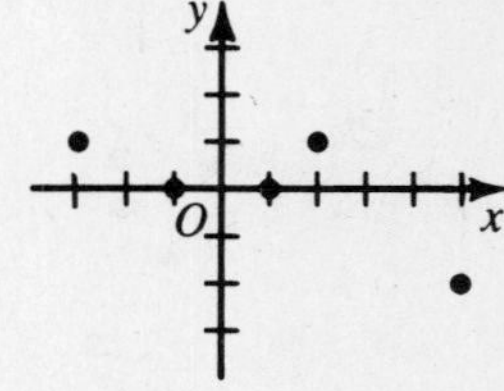

4. a function

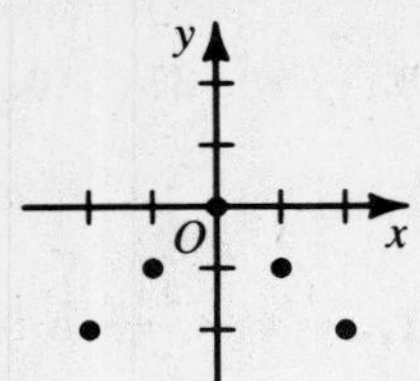

5. not a function

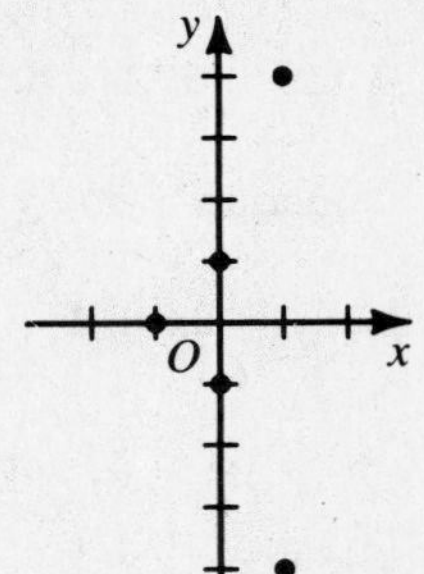

6. a function

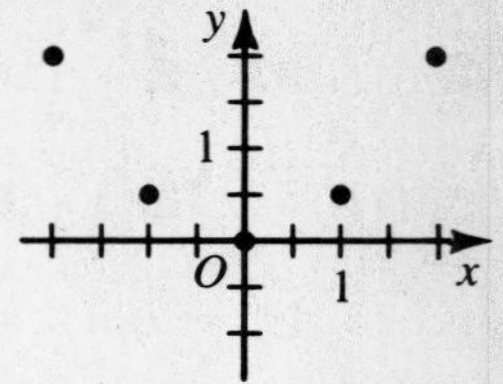

7.

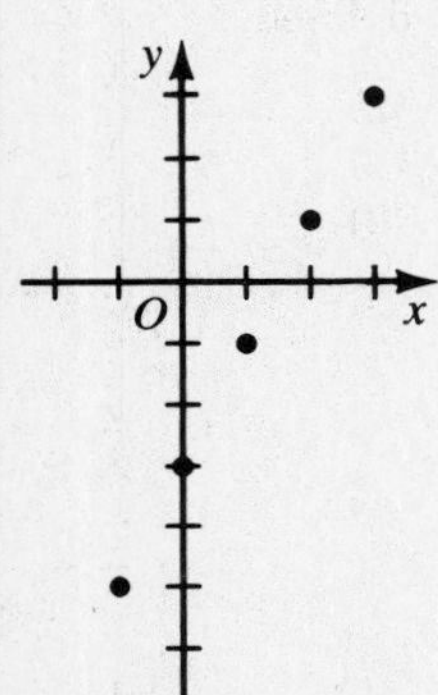

8.

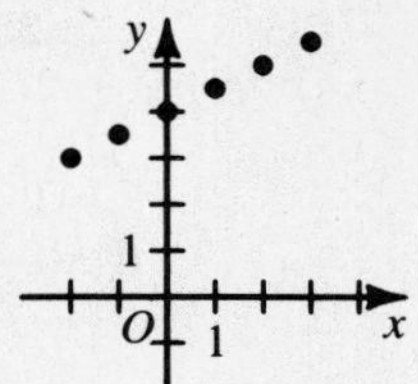

9.

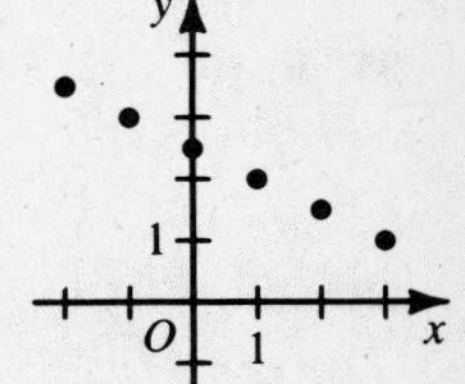

10.

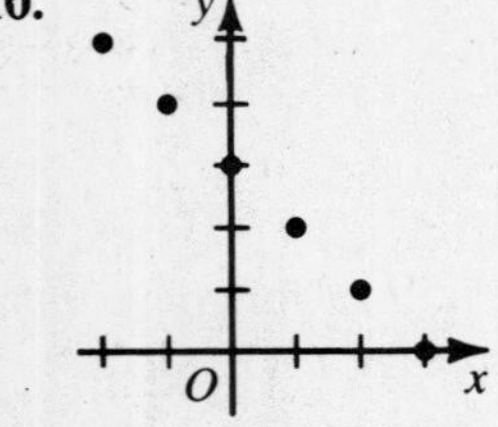

11.

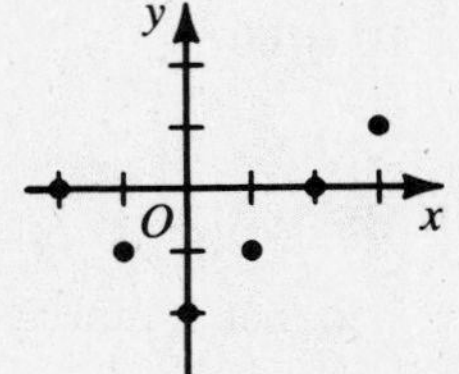

12.

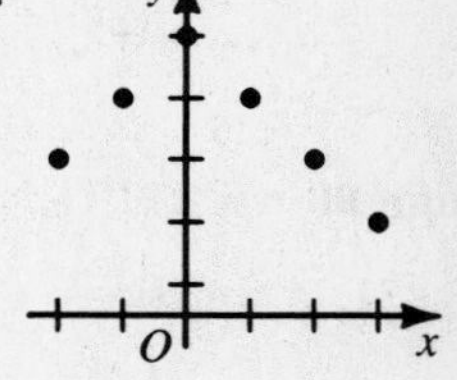

13.

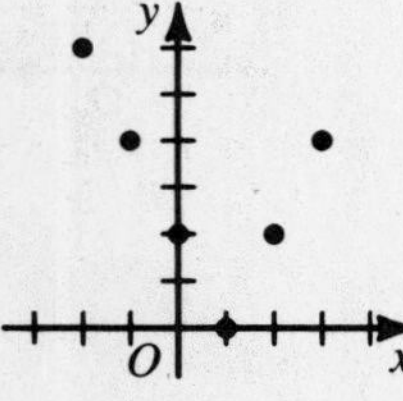

14.

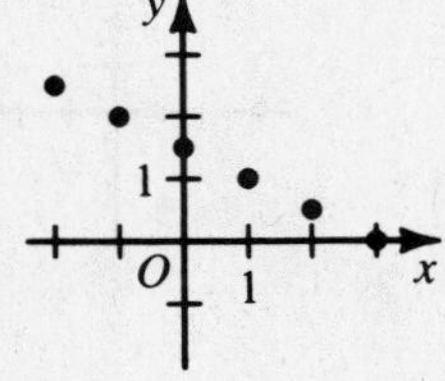

15.

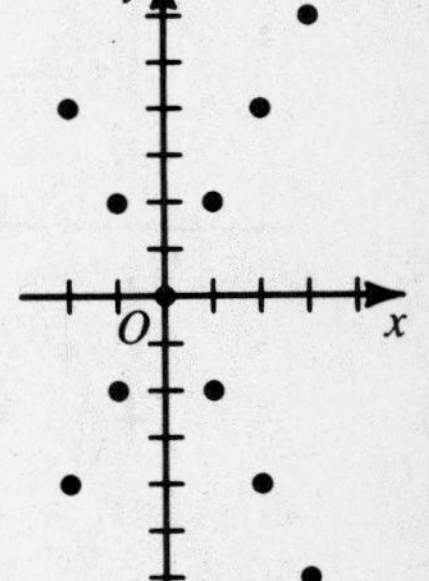

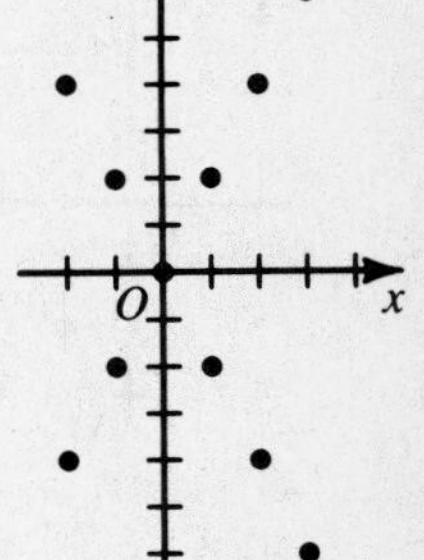

16.

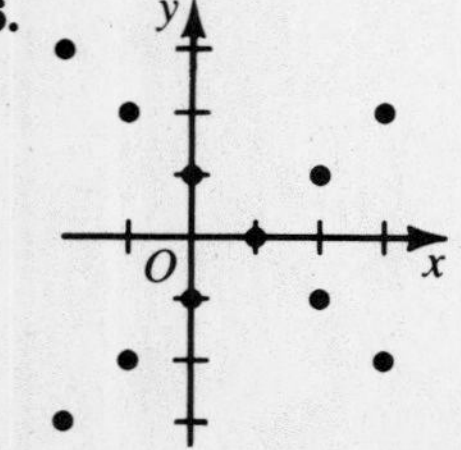

B 17.

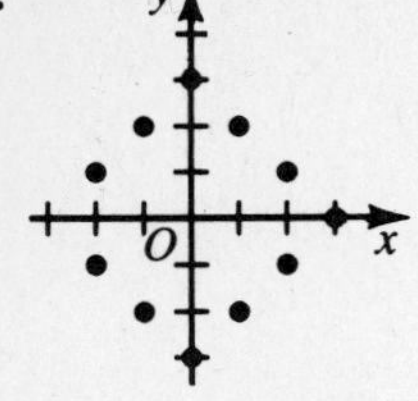

18.

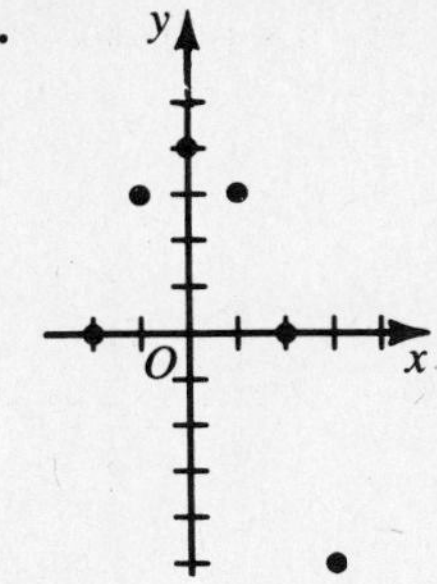

19.

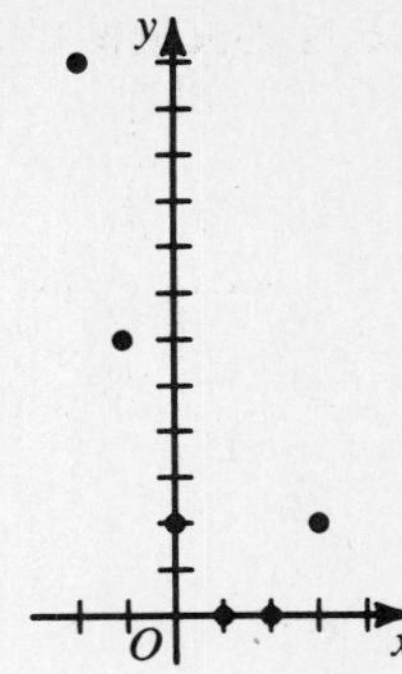

20.

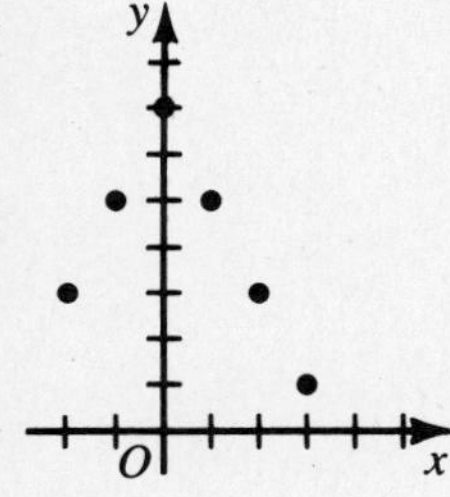

21.

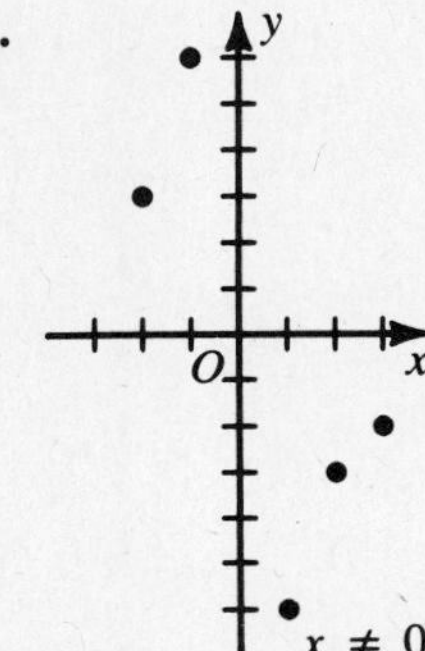

22.

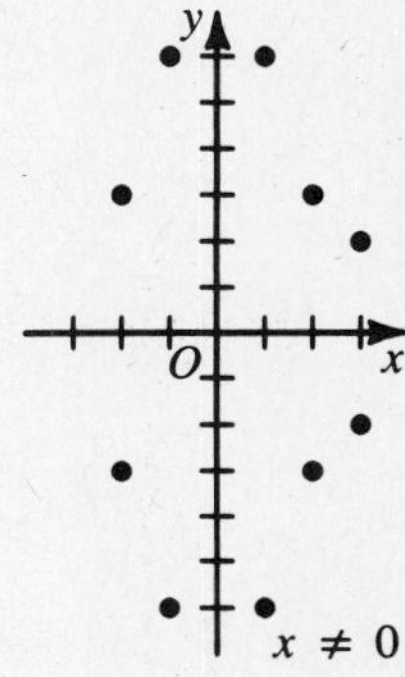

C 23.

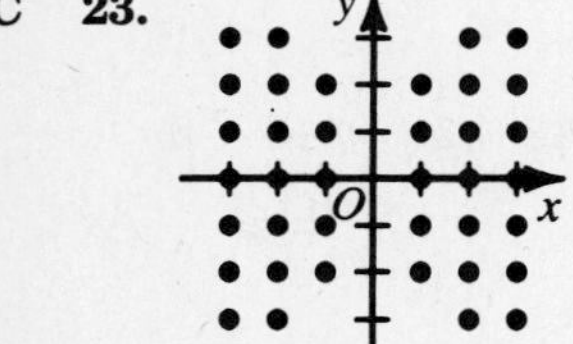

24.

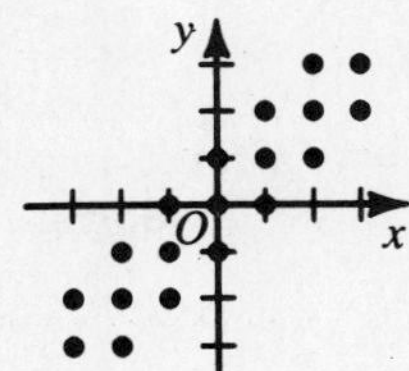

25.

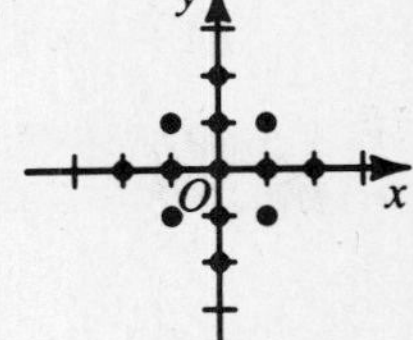

26.

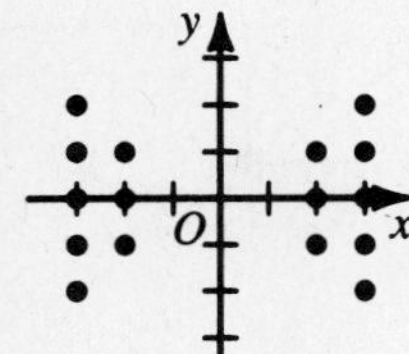

27.

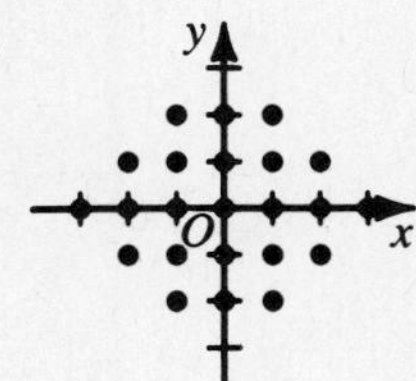

28.

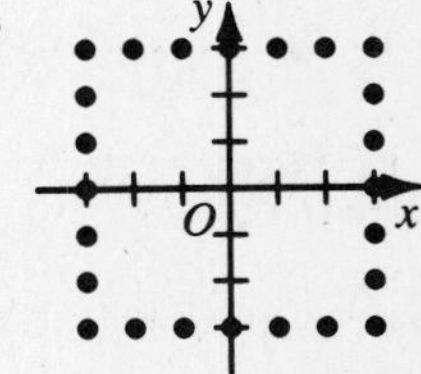

Pages 90–91 · COMPUTER EXERCISES

1.
```
10 INPUT H
20 IF H <> INT(H) THEN LET H = INT(H + 1)
30 IF H <= 12 THEN GOTO 80
40 LET D = H/24
50 IF D <> INT(D) THEN LET D = INT(D + 1)
60 LET C = 6.5 * D
70 GOTO 90
80 LET C = 1.25 + .45 * (H - 1)
```

[*Program continued on next page*]

```
90 PRINT "COST IS $";C
100 END
```

2. $3.95 **3.** $5.30 **4.** $6.50 **5.** $26.00 **6.** $32.50

7.
```
10 INPUT "DAYS OR HOURS ";A$
20 IF A$ = "HOURS" THEN GOTO 70
30 IF A$ <> "DAYS" THEN GOTO 10
40 INPUT DY
50 LET H = DY * 24
60 GOTO 80
70 INPUT H
80 IF H <> INT(H) THEN LET H = INT(H + 1)
90 IF H <= 12 THEN 140
100 LET D = H/24
110 IF D <> INT(D) THEN LET D = INT(D + 1)
120 LET C = 6.5 * D
130 GOTO 150
140 LET C = 1.25 + .45 * (H - 1)
150 PRINT "COST IS $";C
160 END
```

8. $4.40 **9.** $65.00 **10.** $19.50 **11.** $6.20 **12.** $4.40

Page 91 • SELF-TEST 1

1. $y = 2x$; a function **2.** $y = 7$; a function **3.** -6 **4.** 13 **5.** -118 **6.** 11

7. yes, a function

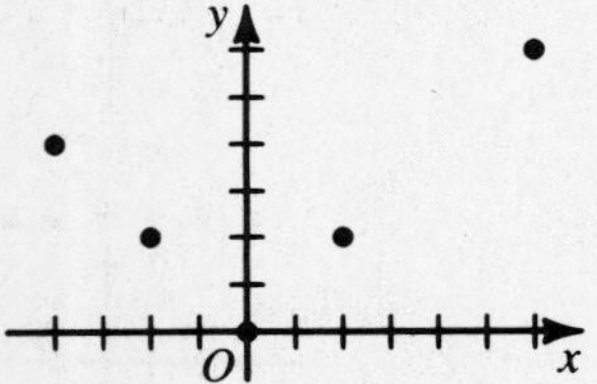

8. yes, a function

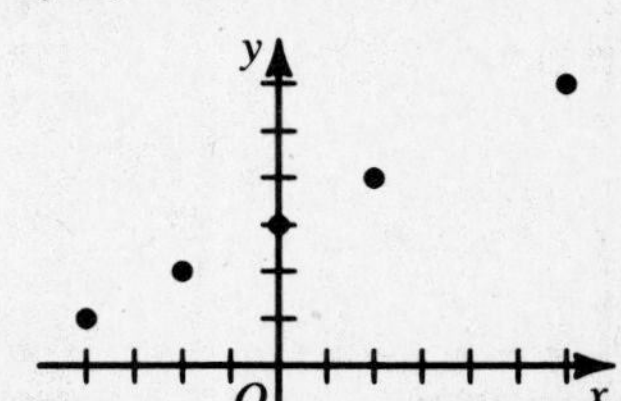

Pages 94–95 • WRITTEN EXERCISES

A **1.**

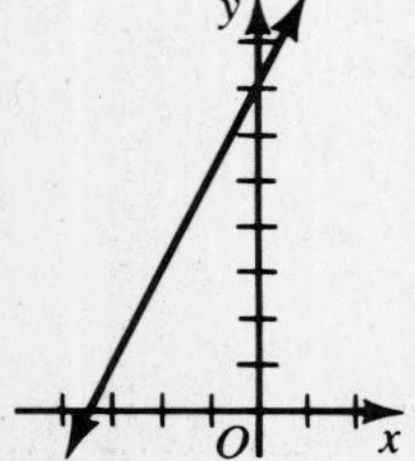

2.

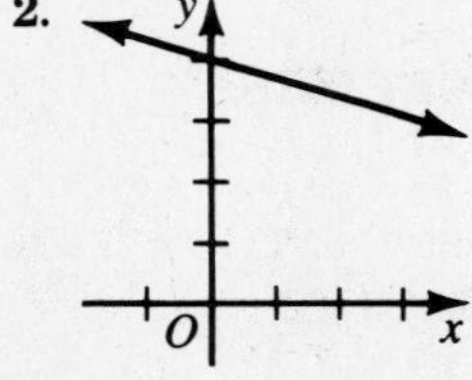

3.

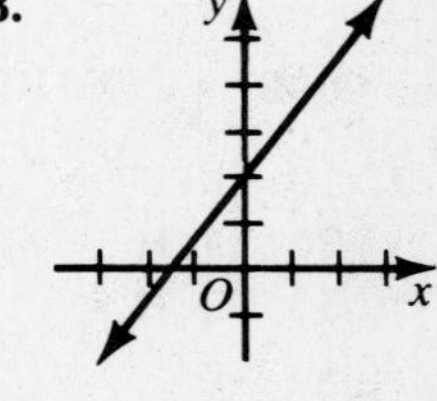

4.

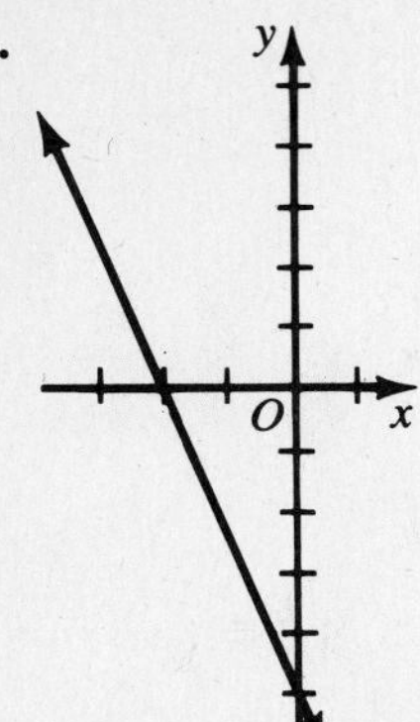

5.

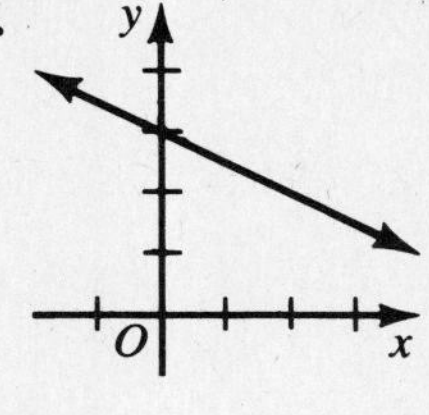

6.

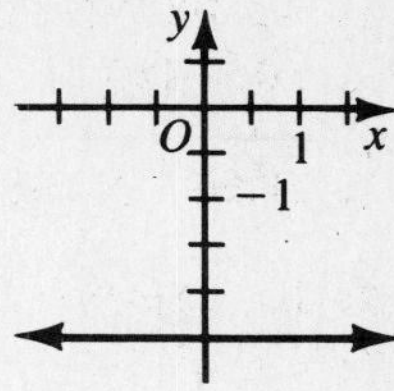

7.

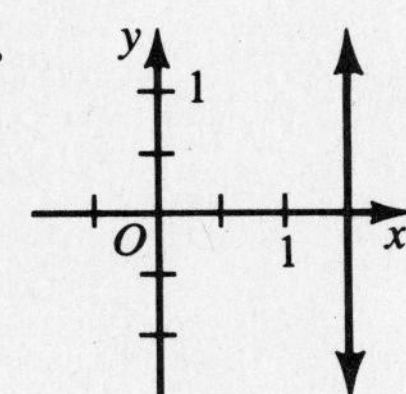

8.

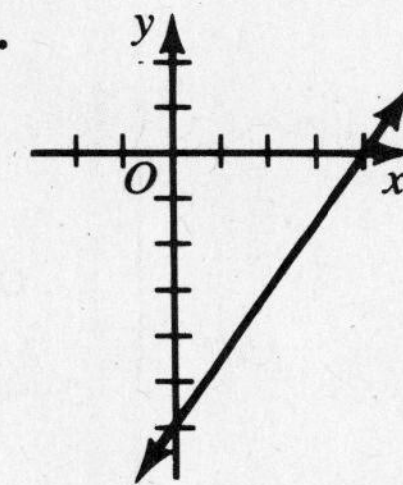

9.

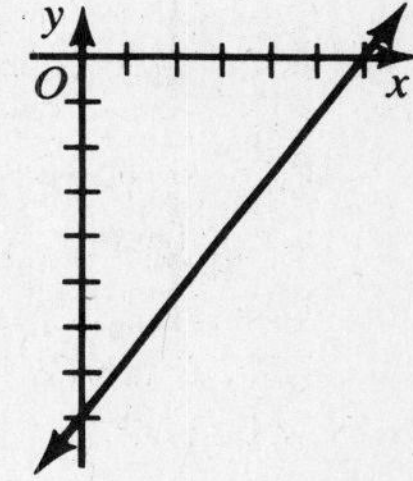

10.

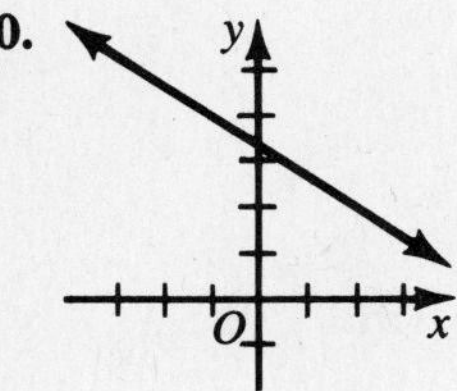

11.

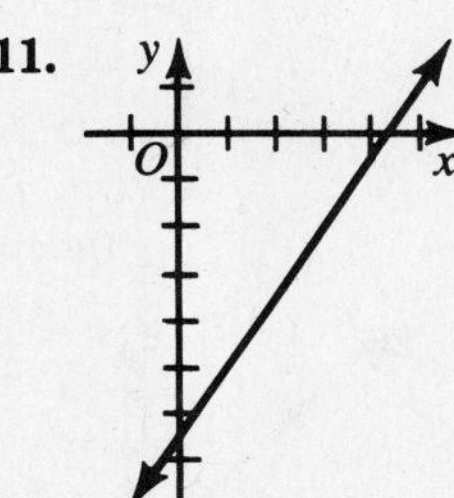

12.

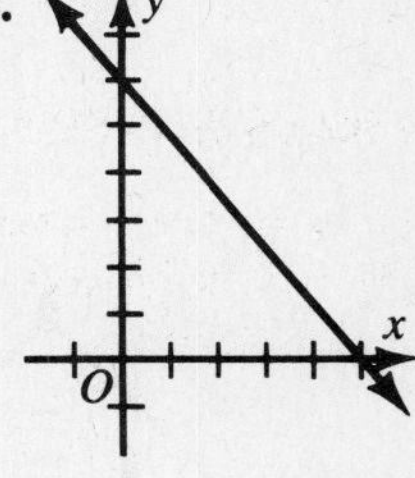

13.

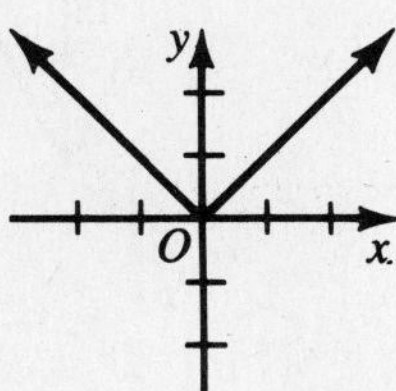

14.

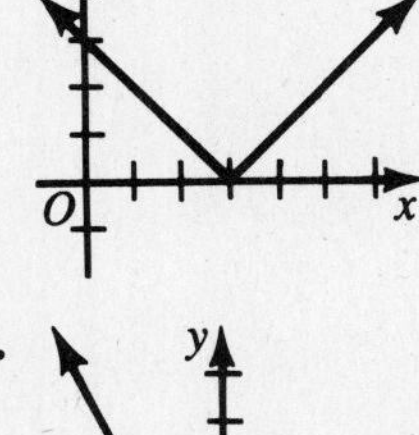

B 15.

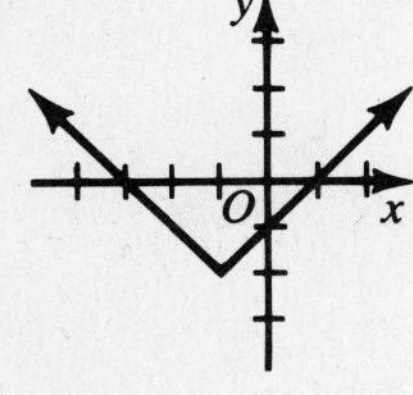

16.

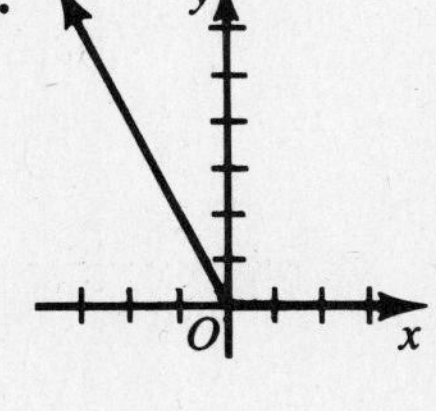

17.

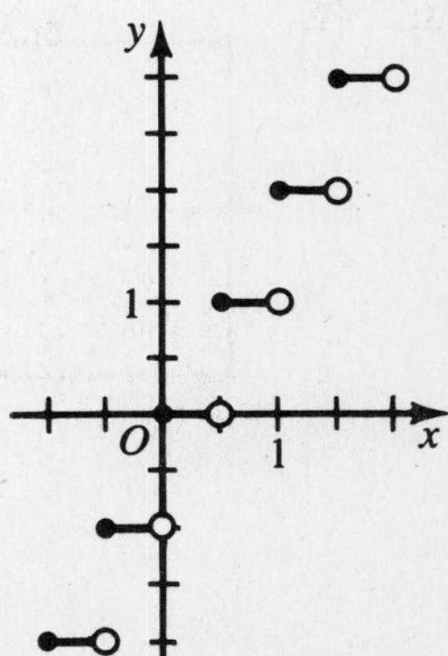

18.

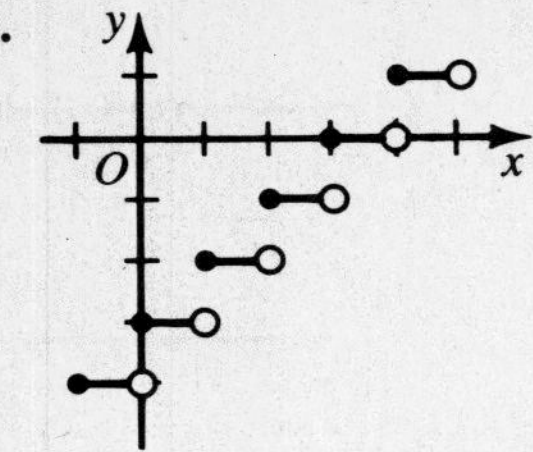

19. $4k + 3 \cdot 9 = -5;\ k = -8$ **20.** $-4 \cdot 3 + 6k = 3(-10);\ k = -3$

21. $2 \cdot 5 - k(-7) = k + 4;\ k = -1$ **22.** $7k + (k - 2)(-3) = 8;\ k = \frac{1}{2}$

23.

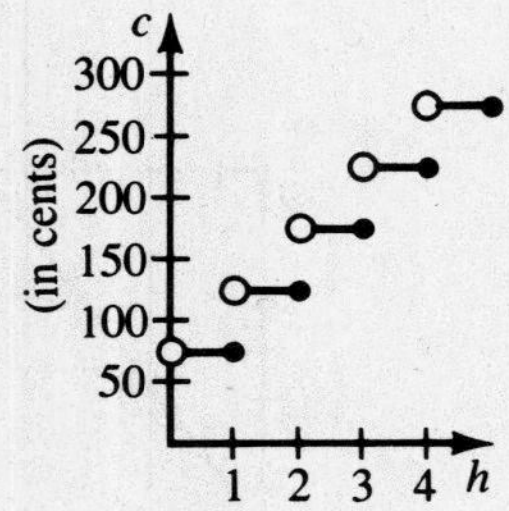

24.

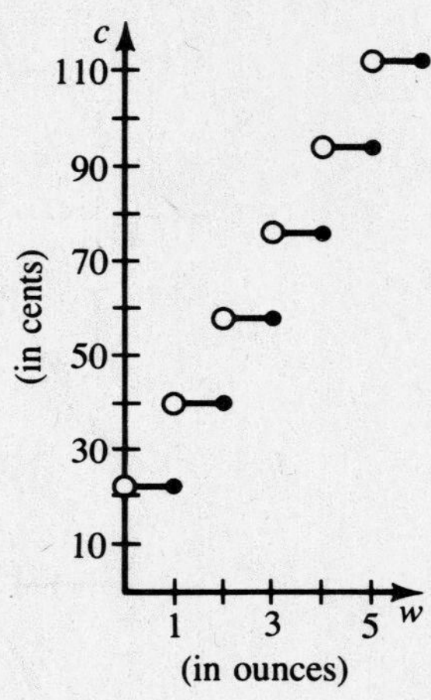

C **25.**

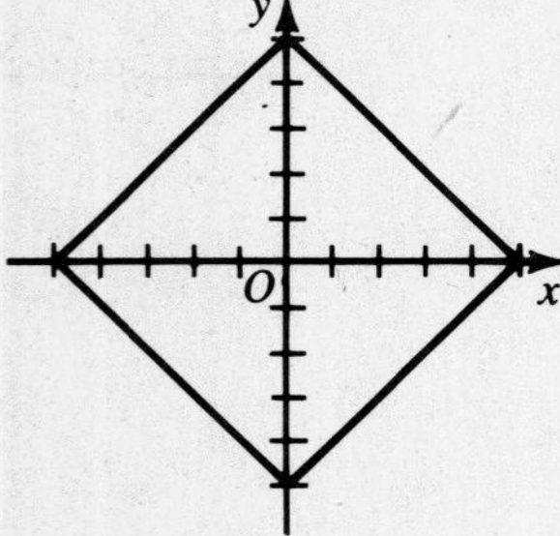

26.

27.

28.

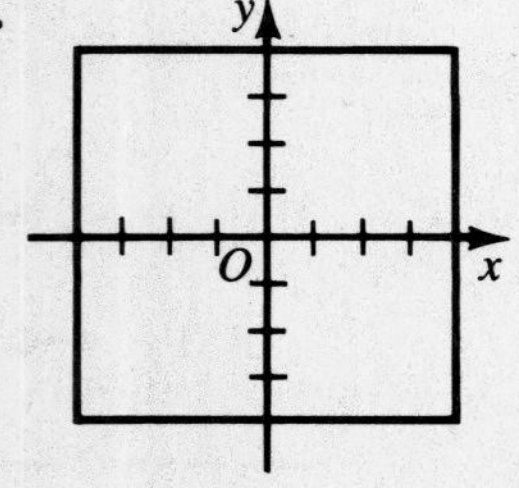

Page 97 • WRITTEN EXERCISES

A 1. 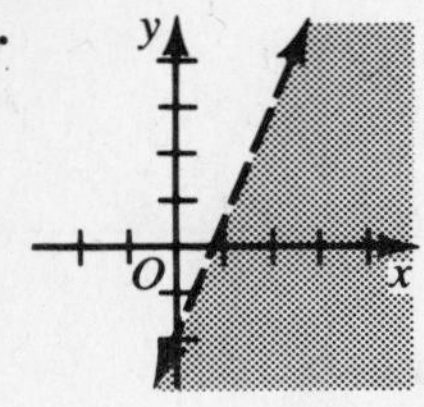

2. 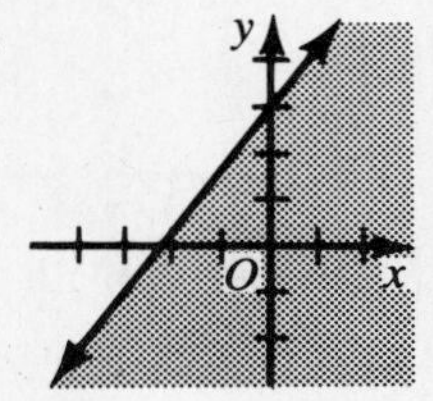

3. 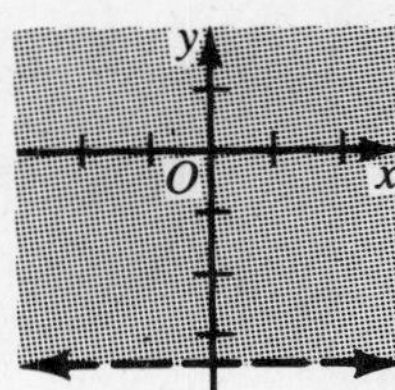

4. 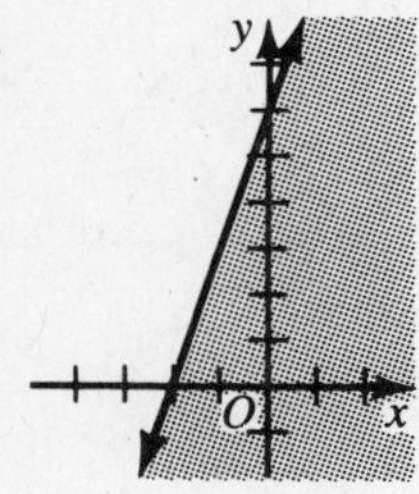

5. 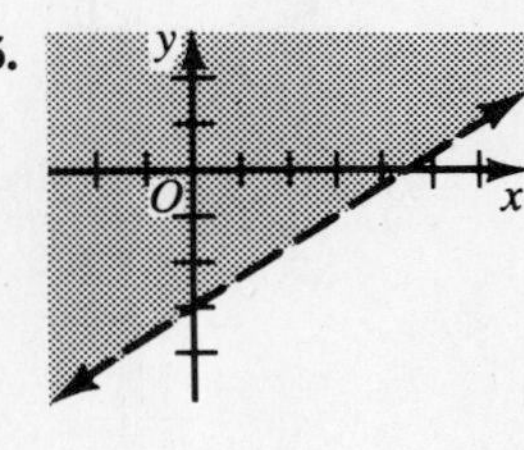

6. 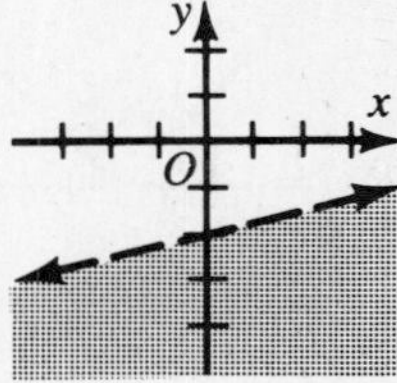

7. 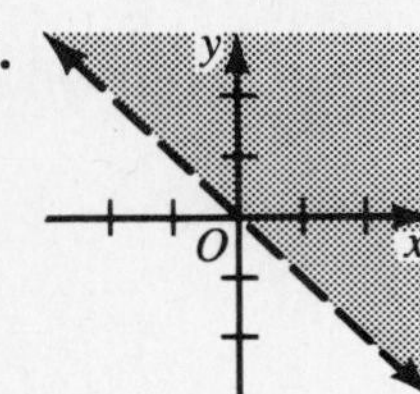

8. 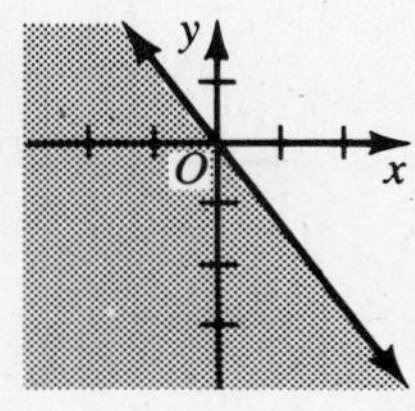

9. 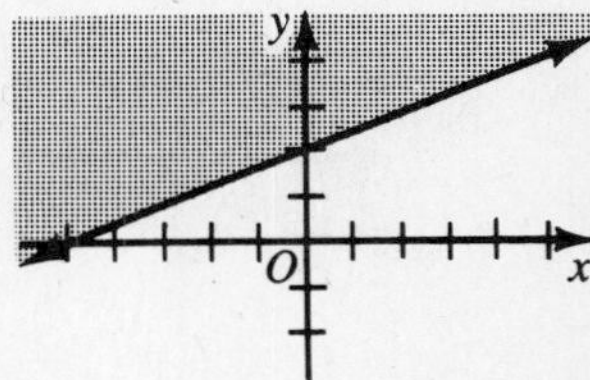

10. 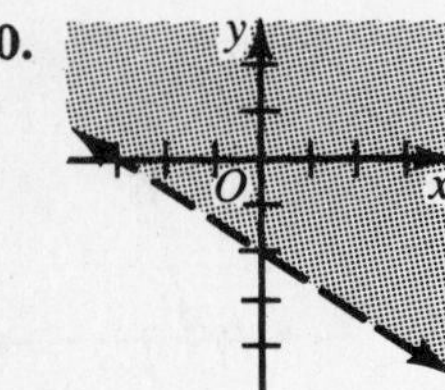

11. 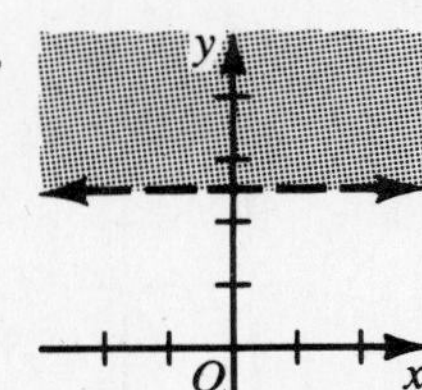

12. 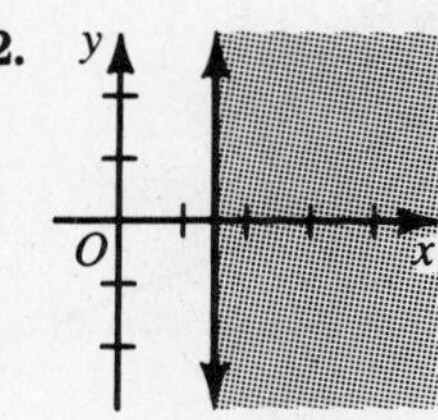

13. 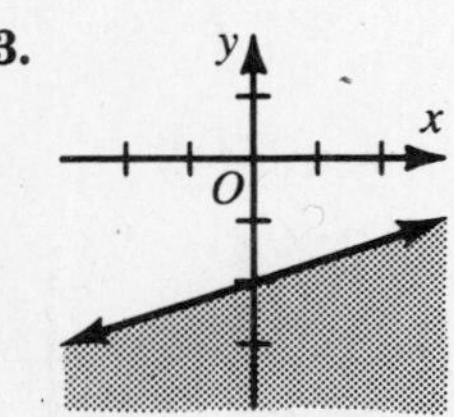

14. 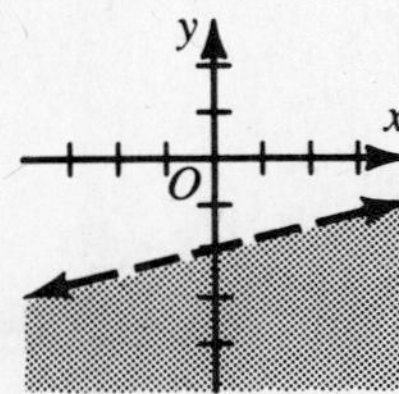

15. 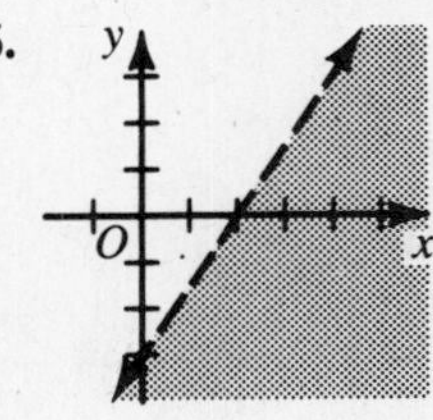

16. 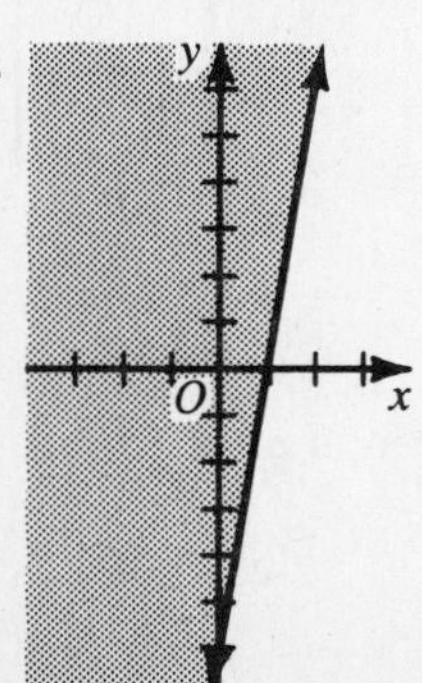

B 17.
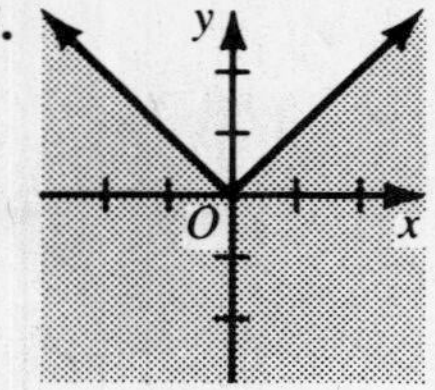
18.
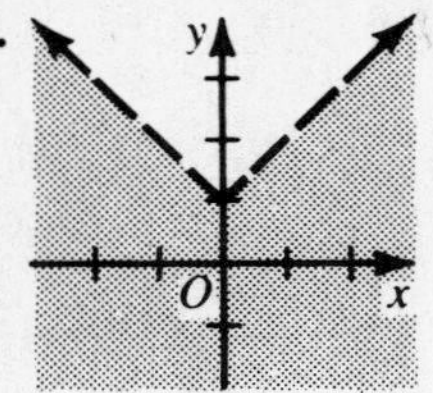
19.
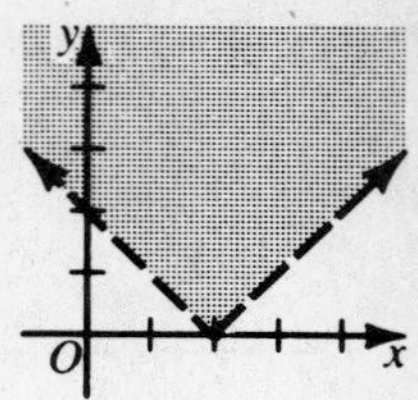
20.
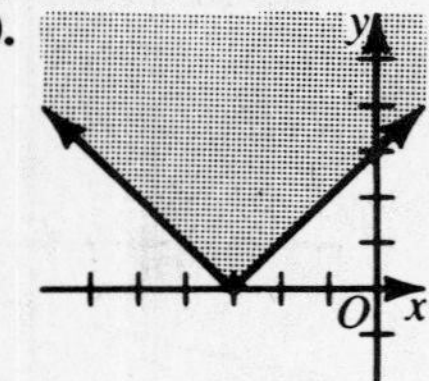
21.
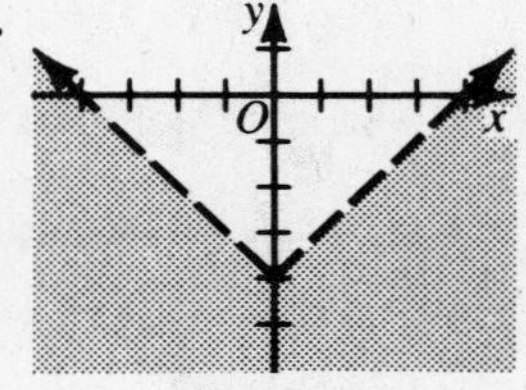
22.
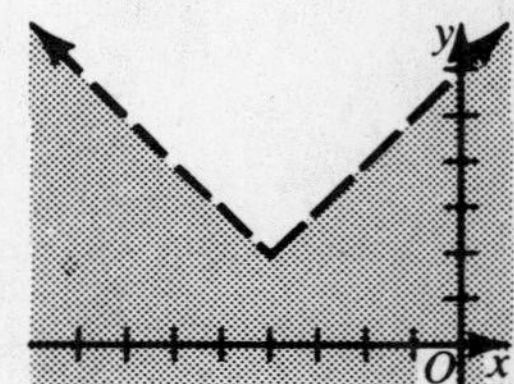

C 23.
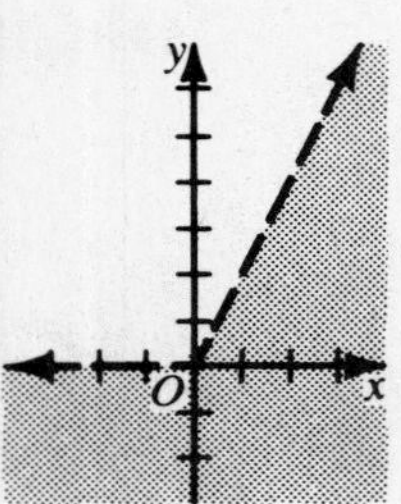
24.
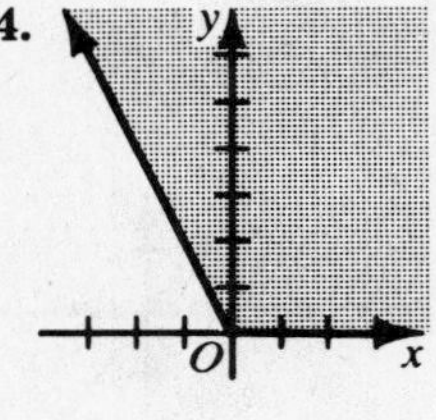

Page 98 • SELF-TEST 2

1.
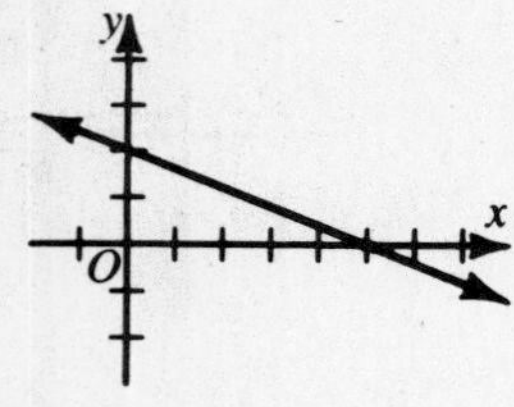
2.
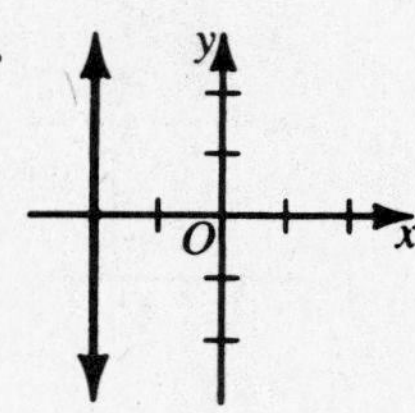
3.
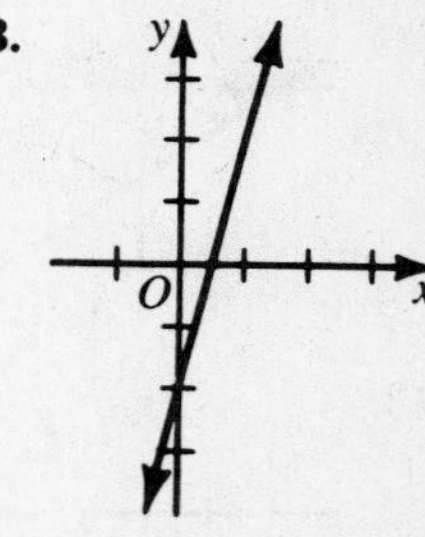
4.
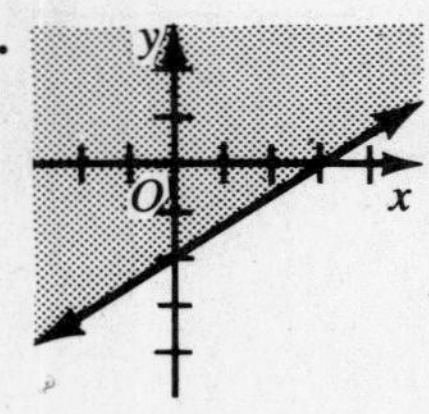
5.
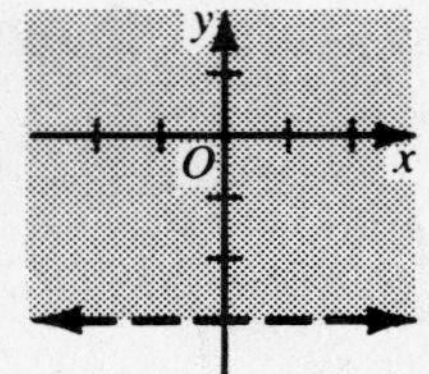
6.
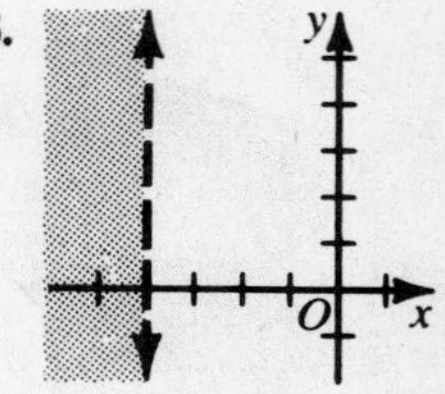

Page 102 • WRITTEN EXERCISES

A **1.** $m = 3$ **2.** $m = -4$ **3.** $m = -\frac{1}{2}$ **4.** $m = 0$ **5.** Slope is undefined. **6.** $m = -5$

7. $m = \frac{4}{3}$

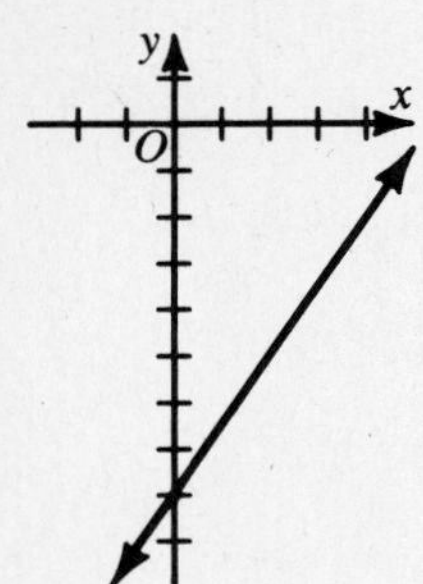

8. $m = -\frac{5}{4}$

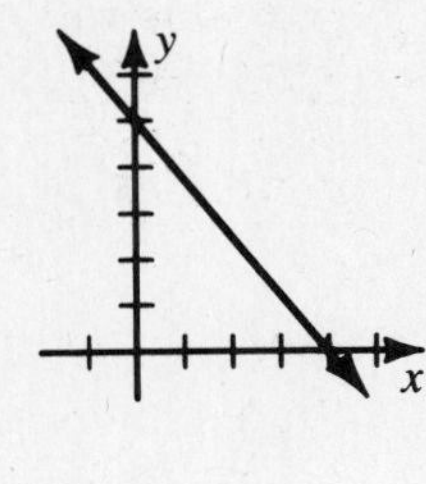

9. $m = \frac{3}{5}$

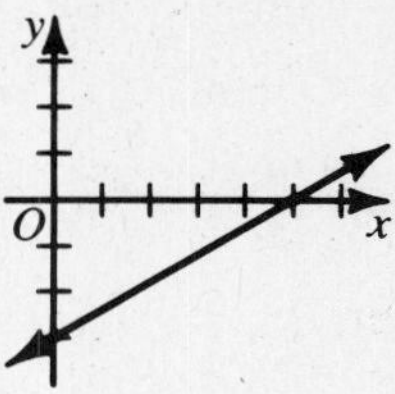

10. $m = \frac{1}{4}$

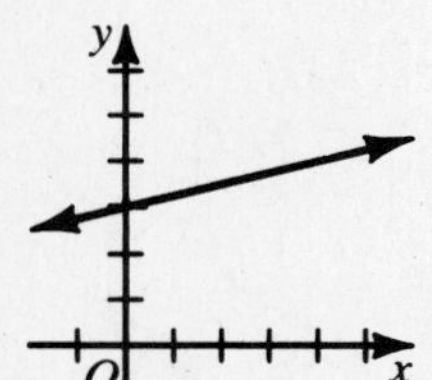

11. $m = -\frac{3}{2}$

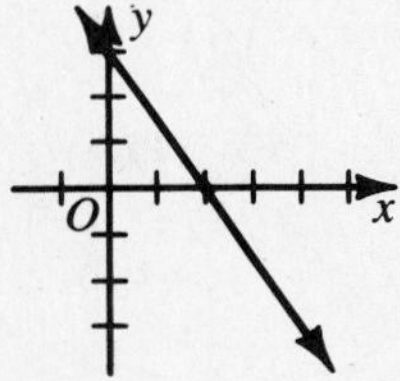

12. $m = \frac{1}{6}$

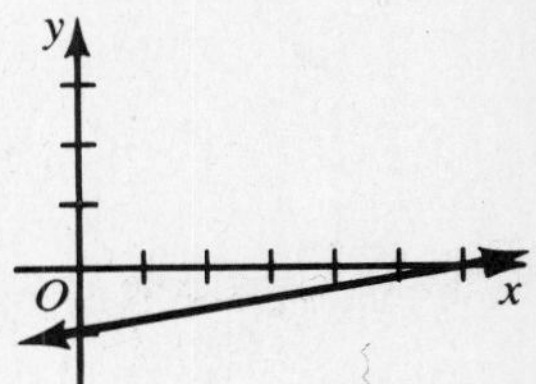

Other coordinates for point Q are possible for Exs. 13-18.

13.

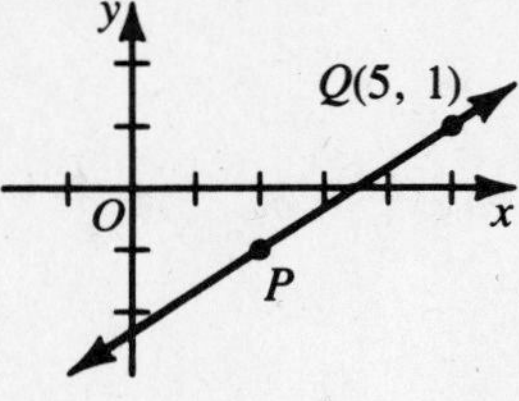

14.

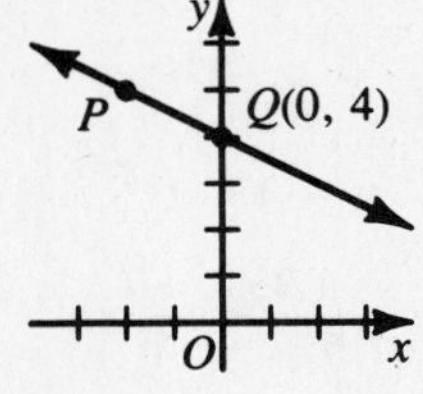

15.

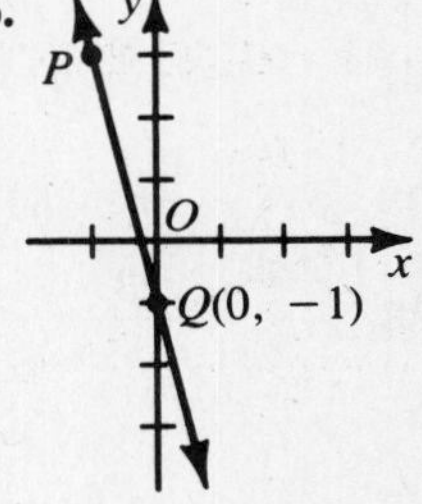

16.

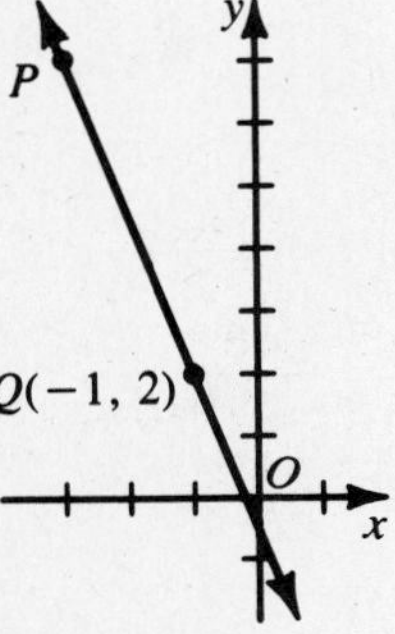

17.

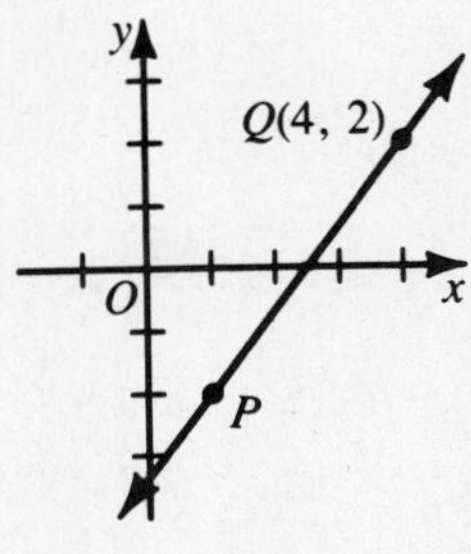

18.

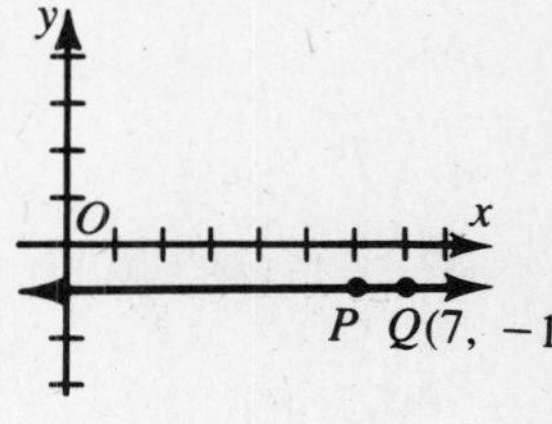

B **19.** $\frac{a - (-3)}{2 - 4} = \frac{1}{4}; a = -\frac{7}{2}$

20. $\frac{-4 - 1}{-2 - a} = \frac{3}{2}; a = \frac{4}{3}$

21. $\frac{-4 - 1}{a - 3a} = -\frac{1}{3}; a = -\frac{15}{2}$

22. $\frac{a - 2 - 5}{1 - (a + 3)} = 4; a = -\frac{1}{5}$

23. $\frac{a - 5}{-a - a} = -2; a = -\frac{5}{3}$

24. $\frac{2a - (-a)}{4 - (-3a)} = \frac{1}{2}; a = \frac{4}{3}$

C **25.** The line determined by (x_1, y_1) and (x_2, y_2) must be the same as the line determined by (x_2, y_2) and (x_3, y_3), which must be the same as the line determined by (x_1, y_1) and (x_3, y_3). This will occur if and only if their slopes are equal. Therefore, $\frac{y_2 - y_1}{x_2 - x_1} = \frac{y_3 - y_2}{x_3 - x_2} = \frac{y_3 - y_1}{x_3 - x_1}$.

26. $m_1 = \frac{(2c - 5) - (-2)}{7 - 3}$, $m_2 = \frac{(c + 4) - 5}{-6 - (-9)}$; $m_1 = m_2$, or $\frac{2c - 3}{4} = \frac{c - 1}{3}$; $c = \frac{5}{2}$

Pages 105–106 • WRITTEN EXERCISES

A **1.** $y = -x + 8$ **2.** $y = 3x + 11$ **3.** $y = \frac{1}{2}x - 5$ **4.** $y = 4$ **5.** $y = \frac{4}{3}x - 5$

6. $y = -2x + 10$ **7.** $y = \frac{3}{2}x - \frac{7}{2}$ **8.** $y = -\frac{2}{3}x + \frac{5}{3}$ **9.** $y = \frac{3}{4}x$ **10.** $y = 2x + 7$

11. $y = -\frac{2}{5}x + 10$ **12.** $y = -\frac{1}{3}x - 12$ **13.** $y = 2x - 5$ **14.** $y = -2$

15. $y = -x + 3$ **16.** $y = -3x + 15$ **17.** $y = \frac{1}{2}x + 2$ **18.** $y = \frac{1}{2}x - 10$

19. $y = -4x + 1$ **20.** $y = 3x + \frac{11}{2}$ **21.** $y = \frac{1}{2}x - \frac{13}{6}$ **22.** $y = 3x + \frac{7}{2}$

23. $y = \frac{2}{3}x + 1$ **24.** $y = -\frac{5}{2}x - 7$

B **25.** $y = -2x - 5$ **26.** $y = \frac{1}{3}x + 9$ **27.** $y = -x + 7$ **28.** $y = \frac{5}{2}x - \frac{9}{2}$

29. $y = -\frac{1}{2}x - 10$ **30.** $y = \frac{2}{5}x + 4$ **31.** $\frac{3}{k} = -\frac{2}{5}$; $k = -\frac{15}{2}$ **32.** $\frac{k}{5} = \frac{9}{2}$; $k = \frac{45}{2}$

33. $-\frac{4}{k} = \frac{3}{k - 1}$; $k = \frac{4}{7}$ **34.** $\frac{k + 2}{5} = \frac{-k}{3}$; $k = -\frac{3}{4}$

C **35.** $y = mx + b$; $m = \frac{k - 3}{8}$, $b = 4$; $3 = \left(\frac{k - 3}{8}\right)(-2) + 4$; $k = 7$

36. $y - y_1 = m(x - x_1)$; $(x, y) = (0, 0)$, $(x_1, y_1) = (k, 0)$;
$m = \frac{2}{k - 2}$; $4 - 0 = \frac{2}{k - 2}(0 - k)$; $k = \frac{4}{3}$

37. $y = mx + b$; $(x, y) = (k, 0)$; $m = \frac{b - 1}{-2}$; $0 = \left(\frac{b - 1}{-2}\right)k + b$; $k = \frac{2b}{b - 1}$

38. $y - y_1 = m(x - x_1)$; $(x, y) = (-1, 3)$, $(x_1, y_1) = (a, 0)$; $m = \frac{3}{-1 - a}$;
$3 = \left(\frac{3}{-1 - a}\right)(-1) + k$; $k = \frac{3a}{a + 1}$

39. $\frac{4 - 2}{k - (k - 1)} = \frac{11 - 4}{3 - k}$; $k = -\frac{1}{2}$

40. $(0, k)$ and $(k, 0)$ lie on the line; $m = \frac{0 - k}{k - 0} = -1$; (c, d) is on the line, so $y - d = (-1)(x - c)$, or $x + y = c + d$.

Page 106 • COMPUTER EXERCISES

1.
```
10 INPUT X1,Y1,X2,Y2
20 LET M = (Y2 - Y1)/(X2 - X1)
30 LET B = Y1 - M * X1
40 PRINT "Y = ";M;"X";
50 IF B < 0 THEN PRINT B:END
60 IF B = 0 THEN END
70 PRINT "+";B
80 END
```
2. $y = 3.4x + 22.2$ **3.** $y = -6.875x + 16.375$ **4.** $y = 1.57143x + 9.57143$
5. $y = 0.8125x + 47.6875$
6.
```
10 INPUT X1,Y1,X2,Y2
12 IF X1 = X2 AND Y1 = Y2 THEN PRINT "SAME POINT":GOTO 10
15 IF X2 - X1 = 0 THEN PRINT "X = ";X1:PRINT "VERTICAL LINE":END
20 LET M = (Y2 - Y1)/(X2 - X1)
30 LET B = Y1 - M * X1
40 PRINT "Y = ";M;"X";
50 IF B < 0 THEN PRINT B:END
60 IF B = 0 THEN 80
70 PRINT "+";B
80 END
```
7. $x = 5$; vertical **8.** $x = -8$; vertical **9.** $x = -7$; vertical **10.** $x = 0$; vertical

Page 106 • SELF-TEST 3

1. $\frac{4}{5}$ **2.** $y = 2x - 4$ **3.** $y = \frac{3}{2}x + \frac{45}{8}$ **4.** $y = 4x - 2$ **5.** $y = -2x + 3$

Page 107 • READING ALGEBRA

1. Yes, half the perimeter is 49 m and the sum of an odd and even integer is an odd integer; no, the sum of two odd integers cannot be an odd integer (49); no, the sum of two even integers cannot be an odd integer (49).
2. Tom's age, 23 yr; sister's age, 20 yr; strategies will vary.
3. Answers will vary.
4. 625; find a pattern for the last three digits using 15^3, 15^4, 15^5, and so on.
5. Sunday; strategies will vary.

Pages 110–111 • WRITTEN EXERCISES

A **1.** $\frac{5}{2}$ **2.** $\frac{16}{3}$ **3.** $\frac{28}{3}$ **4.** $\frac{3}{2}$ **5.** $\frac{28}{5}$ **6.** $\frac{9}{2}$ **7.** $y = -3x + 1$ **8.** $y = 4x - 9$

9. $y = 2x + 5$ **10.** $y = -\frac{2}{3}x + 1$ **11.** $y = -\frac{1}{2}x + 4$ **12.** $y = -\frac{3}{2}x$

13. $y = 4x + 3$ **14.** $y = -3x + 7$ **15.** -1 **16.** $\frac{10}{3}$ **17.** $\frac{7}{6}$ **18.** $-\frac{1}{2}$ **19.** $\frac{5}{6}$

20. $\frac{5}{2}$ **21.** $\frac{45}{2}$ **22.** 23

B **23.** $\frac{-9-5}{4-(-3)} = \frac{y-(-9)}{7-4}$; $y = -15$ **24.** $\frac{1-(-3)}{12-4} = \frac{7-1}{x-12}$; $x = 24$

25. $\frac{-5-5}{-9-6} = \frac{y-5}{1-6}$; $y = \frac{5}{3}$ **26.** $\frac{-9-9}{-4-8} = \frac{-2-9}{x-8}$; $x = \frac{2}{3}$

27. 1. $\frac{y_1}{x_1} = \frac{y_2}{x_2}$; $x_1, y_1, x_2, y_2 \neq 0$ — Hypothesis
2. $x_1 y_2 = x_2 y_1$ — Prod. of means = prod. of extremes
3. $\frac{x_1 y_2}{y_1 y_2} = \frac{x_2 y_1}{y_1 y_2}$ — Div. prop. of =
4. $\frac{x_1}{y_1} = \frac{x_2}{y_2}$ — Ex. 28, page 27

28. 1. $\frac{y_1}{x_1} = \frac{y_2}{x_2}$; $x_1, y_1, x_2, y_2 \neq 0$ — Hypothesis
2. $y_1 x_2 = y_2 x_1$ — Prod. of means = prod. of extremes
3. $\frac{y_1 x_2}{x_2 y_2} = \frac{y_2 x_1}{x_2 y_2}$ — Div. prop. of =
4. $\frac{y_1}{y_2} = \frac{x_1}{x_2}$ — Ex. 28, page 27

29. 1. $\frac{y_1}{x_1} = \frac{y_2}{x_2}$; $x_1, x_2, y_1, y_2 \neq 0$; $x_1 \neq x_2$ — Hypothesis
2. $y_1 = mx_1$ — Def. of direct variation
3. $\frac{mx_1}{x_1} = \frac{y_2}{x_2}$ — Subs. prin.
4. $m = \frac{y_2}{x_2}$ — Ex. 25, page 27
5. $m = \frac{y_1 - y_2}{x_1 - x_2}$ — Def. of slope
6. $\frac{y_1 - y_2}{x_1 - x_2} = \frac{y_2}{x_2}$ — Subs. prin.

C **30.** 1. $g(x) = mx$ — Def. of direct variation
2. $g(r) = mr$ — Subs. prin.
$g(s) = ms$
$g(r + s) = m(r + s)$
3. $g(r + s) = mr + ms$ — Dist. ax.
4. $g(r + s) = g(r) + g(s)$ — Subs. prin.

31. 1. $g(x) = mx + b$; there exist $r, s \in \mathcal{R}$ such that $g(r + s) = g(r) + g(s)$ — Hypothesis
2. $g(r + s) = m(r + s) + b$ — Subs. prin.
3. $g(r + s) = mr + ms + b$ — Dist. ax.
4. $g(r) = mr + b$ — Subs. prin.
$g(s) = ms + b$
$mr + ms + b = mr + b + ms + b$
5. $0 = b$ — Canc. prop.
6. $b = 0$ — Symm. prop. of =

Pages 112–113 • PROBLEMS

A **1.** Let x = total cost for refreshments; $\frac{240}{500} = \frac{300}{x}$; \$625

2. Let x = actual depth of the rock in meters; $\frac{4}{3} = \frac{x}{5}$; 6.6 m

3. Let x = voltage produced by three windings; $\frac{250}{10{,}000} = \frac{3}{x}$; 120 V

4. Let x = amount of coal in kg to produce 90 kW·h of electricity; $\frac{5}{36} = \frac{x}{90}$; 12.5 kg

5. Let x = temperature drop between 11 A.M. and 2 P.M.; $\frac{-10}{4} = \frac{x}{3}$; $x = -7.5$; -9.5°C

6. Let a = fixed fee; let b = fixed hourly rate; $a + 2b = 57$, $a + 4.5b = 97$; \$25

B 7. Let f = rate of snowfall (in./h); let c = rate of clearing (in./h); $9f - 0.5c = 2$, $2 - 0.4c = 0$; 0.5 in./h

8. Let F = Fahrenheit temperature; let C = Celsius temperature; $F = mC + b$; $32 = m \cdot 0 + b$, $b = 32$; $212 = m \cdot 100 + 32$, $m = \frac{9}{5}$; $F = \frac{9}{5} \cdot 35 + 32 = 95$; 95°F

9. Let x = time for light to reach Jupiter from E_1 (in seconds); $x + 1000$ = time for light to reach Jupiter from E_2; $\frac{600}{x} = \frac{900}{x + 1000}$; $x = 2000$; 2000 s

10. Let x = amount invested at 5%; let r = yield of high-yield account; $x(0.05) + (3500 - x)r = 235$, $3735r = 280$; \$1100

Page 113 • SELF-TEST 4

1. $\frac{35}{4}$ 2. 105 km

Pages 114–115 • CHAPTER REVIEW

1. $g(-3) = -1$; $f(-1) = 0$; b 2. d 3. a 4. a 5. b 6. d 7. c

8. $3(-2) + 4p = 10$; $p = 4$; c 9. d 10. b 11. $\frac{3 - a}{5 - (-3)} = \frac{1}{4}$; $a = 1$; d

12. $\frac{7 - 1}{-6 - 3} = \frac{6}{-9} = -\frac{2}{3}$; $y - 7 = -\frac{2}{3}(x + 6)$; $y = \frac{-2}{3}x + 3$; d

13. $m = \frac{3}{4}$; $y - 2 = \frac{3}{4}(x + 4)$; $y = \frac{3}{4}x + 5$; d 14. b 15. c

Pages 115–116 • CHAPTER TEST

1. 20 2. 3 3. $f(2) = 8$ 4. $g(8) = \frac{8}{3}$

5. yes, a function

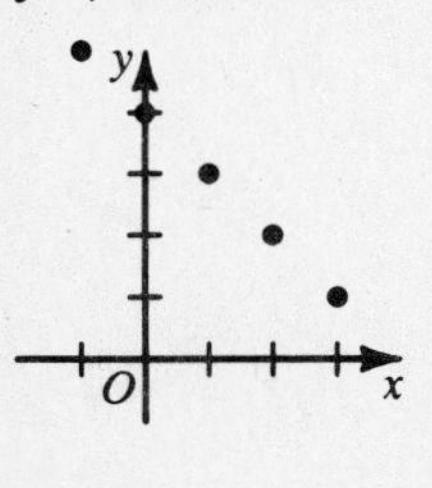

6. no, not a function

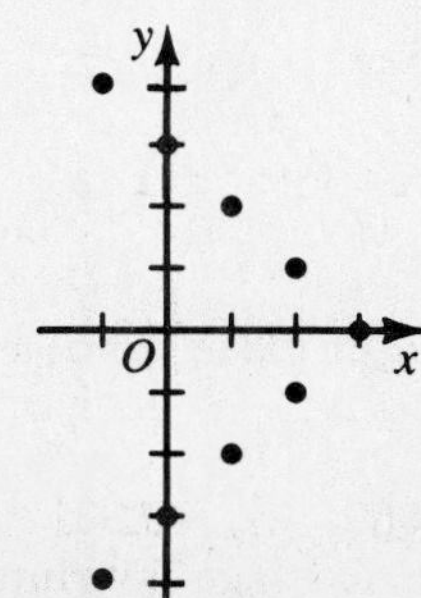

7. yes, a function

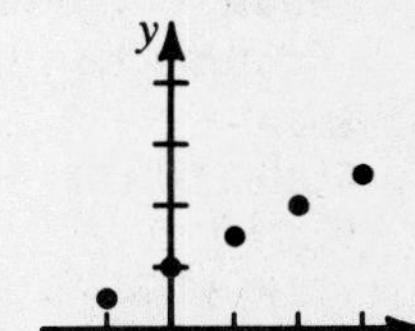

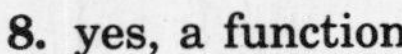

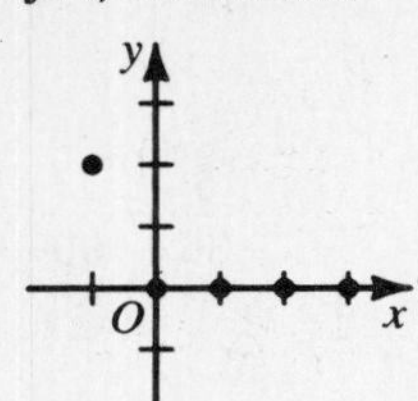

9.

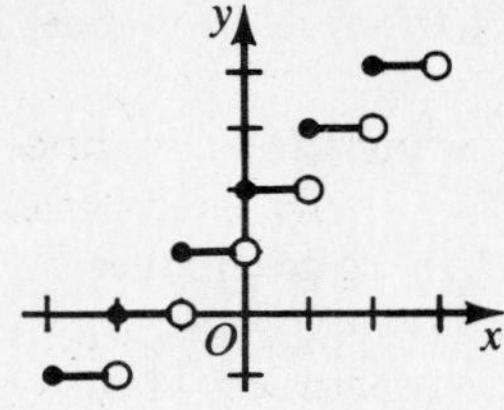

10.

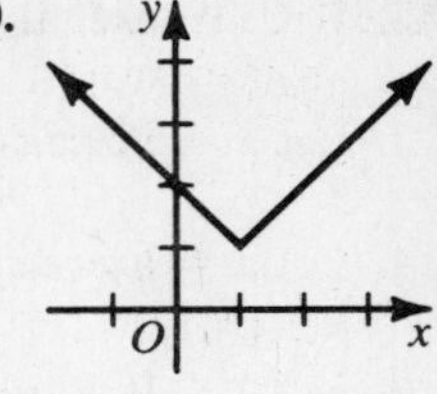

11.

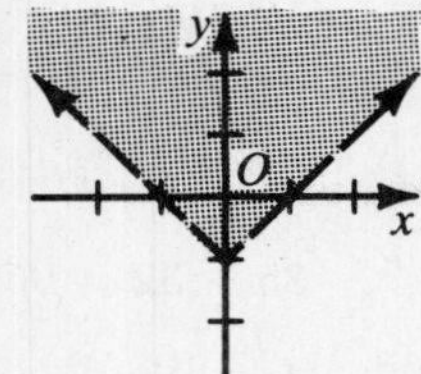

12.

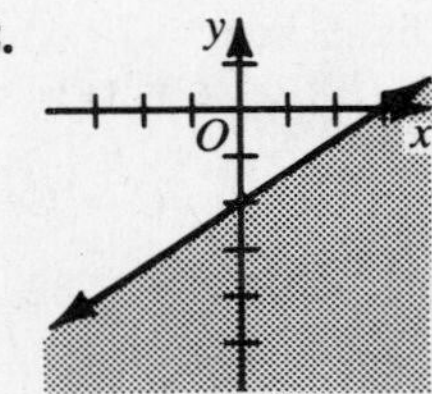

13. -2 **14.** $-\frac{5}{4}$ **15.** $y = -\frac{1}{2}x + \frac{3}{2}$ **16.** $y = \frac{2}{5}x - 3$ **17.** 16 **18.** \$85,000

Page 117 • APPLICATION

1. Substance A **2.** $0.540 = \frac{d}{14.2}$; $d = 7.668$; 7.668 cm

Page 118 • PREPARING FOR COLLEGE ENTRANCE EXAMS

1. D **2.** B **3.** A **4.** C

5. $|x - 3| > 4$; $x - 3 > 4$ or $x - 3 < -4$; $x > 7$ or $x < -1$; E

6. The points (3, 0) and (0, -4) lie on the line; $m = \frac{4}{3}$; $y - 0 = \frac{4}{3}(x - 3)$;
$y = \frac{4}{3}(x - 3)$; E

7. $m = \frac{-1 + 2}{4 - 6} = \frac{1}{-2}$; $-\frac{1}{2} = -\frac{4}{k}$; $k = 8$; E **8.** E

Pages 120–121 • PROGRAMMING IN PASCAL

1.
```
FUNCTION round_to_tenths(num : real) : real;
BEGIN
  num := num/0.1;
  num := round(num);
  round_to_tenths := num * 0.1;
END;
```

2.
```
FUNCTION greatest_integer(x : real) : real;
BEGIN
  IF x >= 0
    THEN greatest_integer := trunc(x)
    ELSE  greatest_integer := -(trunc(abs(x)) + 1);
END;
```

3. **a.** 4.000 **b.** 3.250 **c.** 5.000 **d.** -0.667 **e.** -0.333 **f.** The program is interrupted and an error message is printed.

4.
```
IF x1 = x2
  THEN BEGIN
         writeln('There is no slope.');
         writeln('The line is parallel to the y-axis.');
       END
  ELSE BEGIN
         writeln('The slope is :',slope(x1,y1,x2,y2):3:3);
         IF y1 = y2
          THEN writeln('The line is parallel to the x-axis.');
       END;
```

5.
```
REPEAT
  readln(x1,y1,x2,y2);
  IF (x1 = x2) AND (y1 = y2)
    THEN BEGIN
           writeln;
           writeln('The points are not distinct.');
           writeln('Enter another pair of coordinates.');
           write('Place a space between each of the values:');
         END;
UNTIL (x1 <> x2) OR (y1 <> y2);
```

6.
```
PROGRAM point_slope (INPUT,OUTPUT);
VAR
  x1,y1,x2,y2,m,k :real;

{ ************************************************************* }
FUNCTION slope (x1,y1,x2,y2 :real) :real;

BEGIN
  slope := (y1 - y2)/(x1 - x2);
END;

{ ************************************************************* }
BEGIN    { * main *}
  writeln('This program finds the slope of');
  writeln('the line between two distinct points.');
  writeln;
  writeln('Enter the coordinates of two points.');
  write('Place a space between each of the values:');
  REPEAT
    readln(x1,y1,x2,y2);
    IF (x1 = x2) AND (y1 = y2)
      THEN BEGIN
             writeln;
             writeln('The points are not distinct.');
             writeln('Enter another pair of coordinates.');
             write('Place a space between each of the values:');
           END;
```

[*Program continued on next page*]

```
    UNTIL (x1 <> x2) OR (y1 <> y2);
    writeln;
    IF x1 = x2
      THEN writeln('x = ', x1);
    IF y1 = y2
      THEN writeln('y = ', y1)
      ELSE BEGIN
              m := slope(x1,y1,x2,y2);
              k := y1 - m * x1;
              IF k >= 0
                THEN writeln('y = ', m:3:3,'x + ', k:3:3)
                ELSE writeln('y = ', m:3:3,'x ', k:3:3);
           END;
END.
```

Pages 126–127 • WRITTEN EXERCISES

A **1.** $\{(3, 2)\}$; slopes: $-1, \frac{1}{2}$

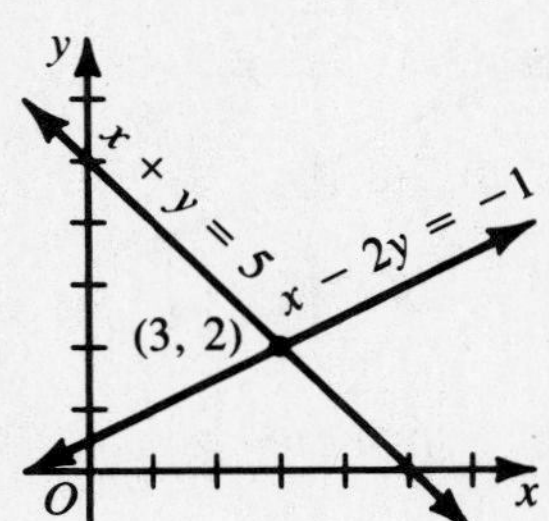

2. $\emptyset$; slopes: $\frac{1}{3}, \frac{1}{3}$; y-intercepts: $1, -1$

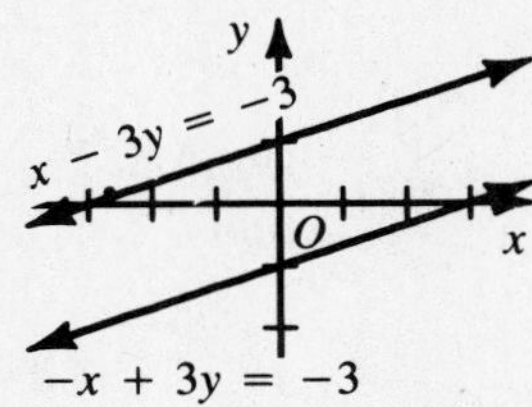

3. $\{(x, y): x + 3y = 6\}$; slopes: $-\frac{1}{3}, -\frac{1}{3}$; y-intercepts: $2, 2$

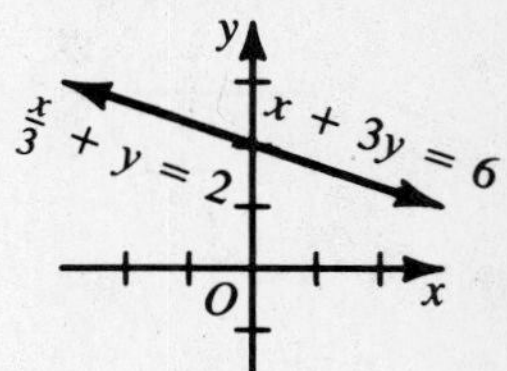

4. $\{(1, -3)\}$; slopes: $-3, 1$

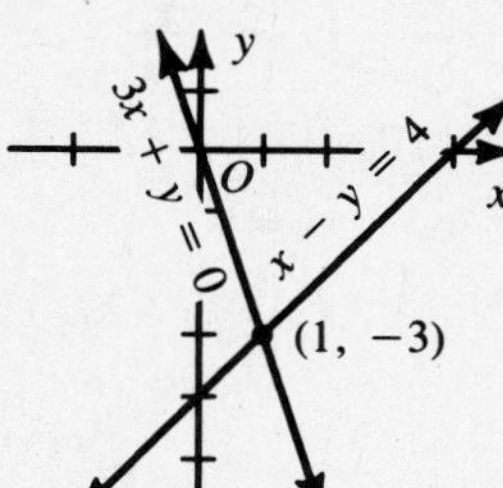

5. $\emptyset$; slopes: $-\frac{1}{2}, -\frac{1}{2}$; y-intercepts: $1, 2$

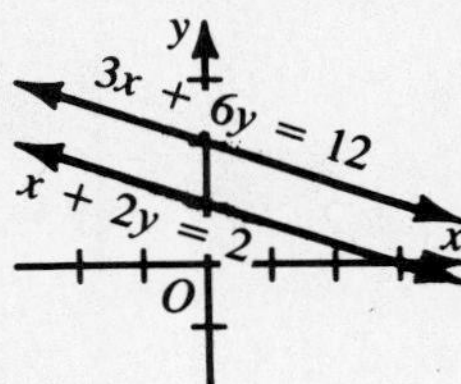

6. $\{(0, 2)\}$; slopes: $-\frac{2}{5}, \frac{2}{5}$

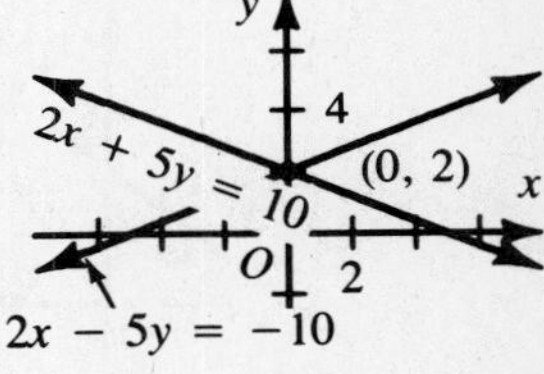

7. $\{(3, -2)\}$; slopes: $-\frac{1}{2}, -3$

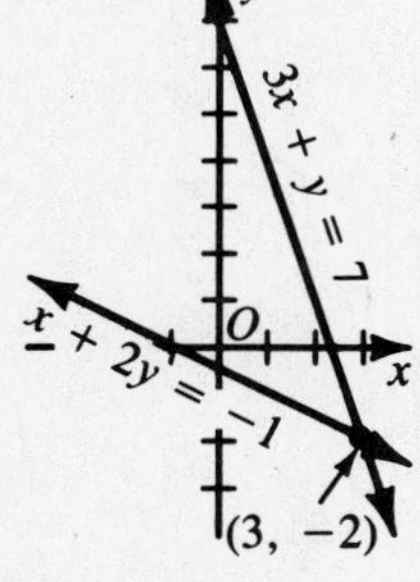

8. $\{(6, 4)\}$; slopes: $\frac{1}{2}, \frac{2}{3}$

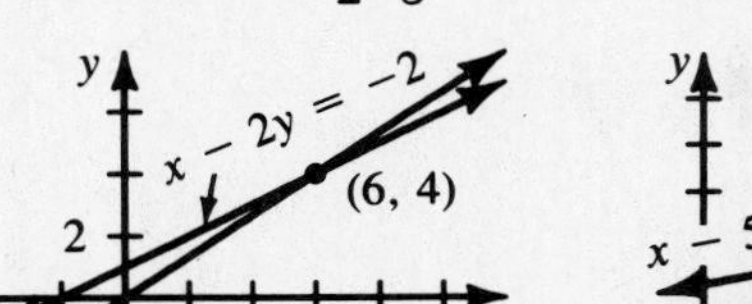

9. $\{(5, 2)\}$; slopes: $-\frac{1}{5}, 2$

10. $\{(x, y): 3x - 4y = 12\}$; slopes: $\frac{3}{4}, \frac{3}{4}$; y-intercepts: $-3, -3$

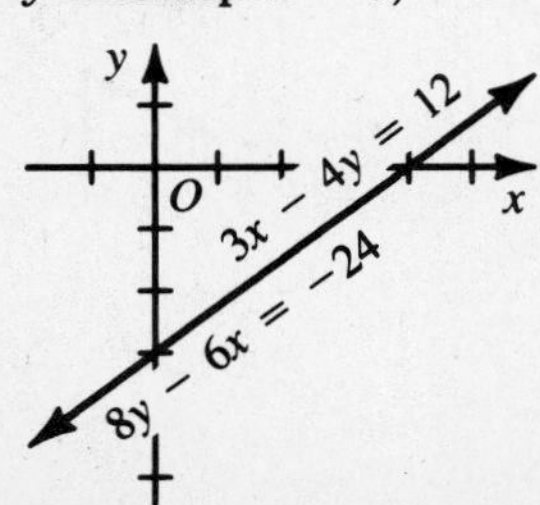

11. $\emptyset$; slopes: $-\frac{1}{6}, -\frac{1}{6}$; y-intercepts: $1, -1$

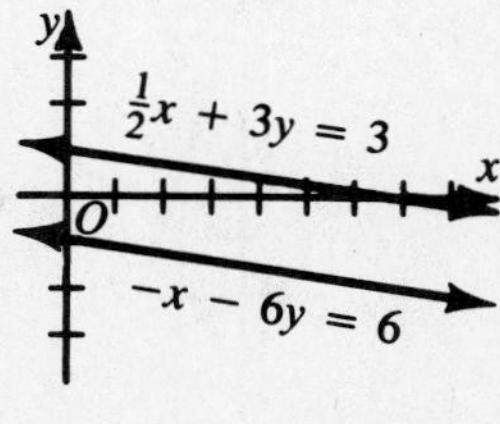

12. $\emptyset$; slopes: $\frac{2}{3}, \frac{2}{3}$; y-intercepts: $1, -2$

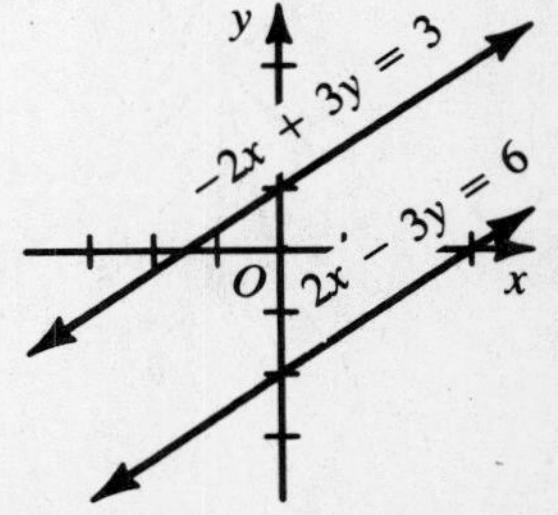

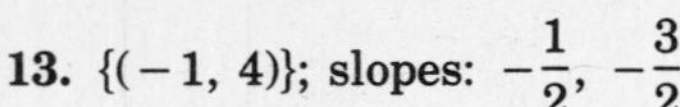

13. $\{(-1, 4)\}$; slopes: $-\frac{1}{2}, -\frac{3}{2}$

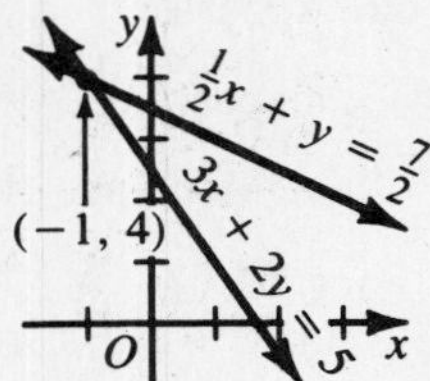

14. $\{(0, -3)\}$; slopes: $-\frac{1}{2}, \frac{3}{2}$

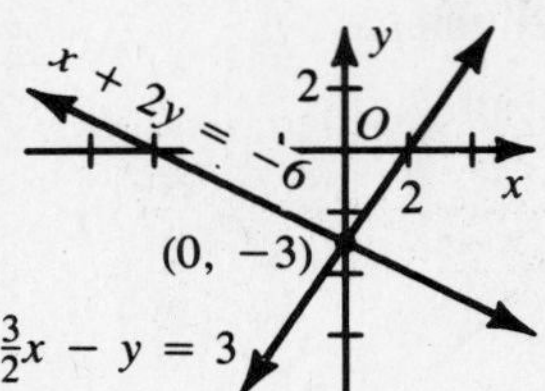

15. $\{(x, y)\colon -3x + 4y = 9\}$;
slopes: $\frac{3}{4}, \frac{3}{4}$;
y-intercepts: $\frac{9}{4}, \frac{9}{4}$

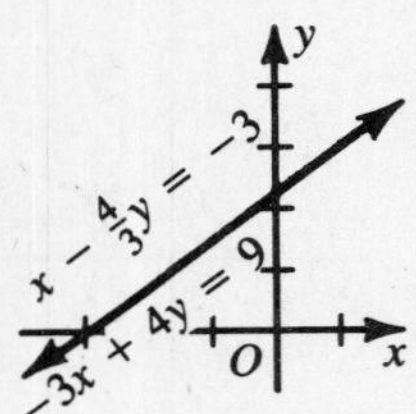

16. $\{(-1, -3)\}$; slopes: $\frac{6}{5}, \frac{1}{3}$

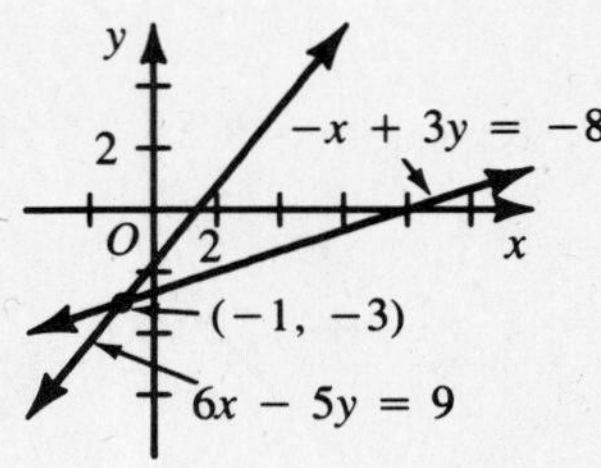

17. $\emptyset$; slopes: $\frac{5}{2}, \frac{5}{2}$;
y-intercepts: 3, -5

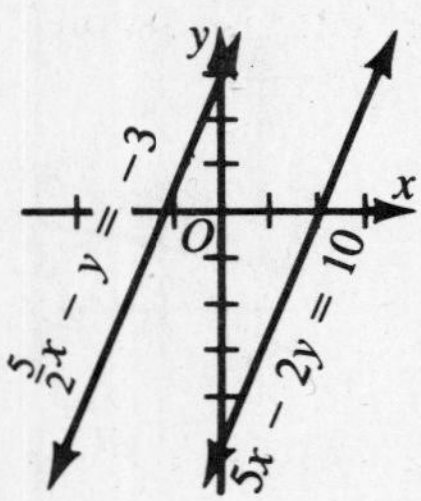

18. $\{(x, y)\colon 4x - 6y = 6\}$;
slopes: $\frac{2}{3}, \frac{2}{3}$;
y-intercepts: $-1, -1$

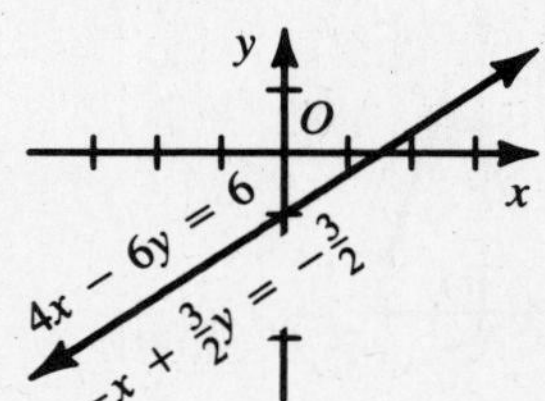

B **19.** $\{(0, 4)\}$; $0 + 4 = 4$;
$0 - 4 = -4$;
$4(0) - 4 = -4$

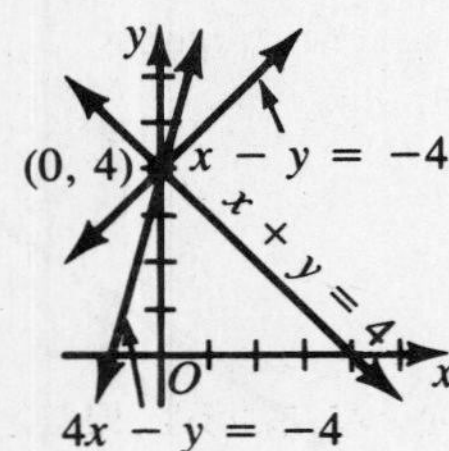

20. $\{(2, -1)\}$; $2 = -2(-1)$;
$3(2) + (-1) = 5$;
$2 - 4(-1) = 6$

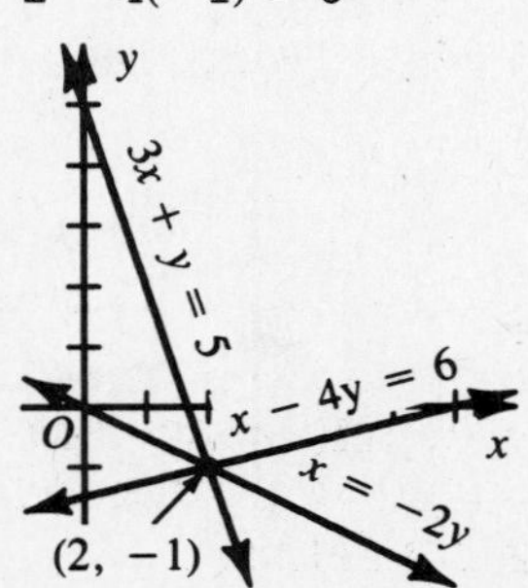

21. $\emptyset$

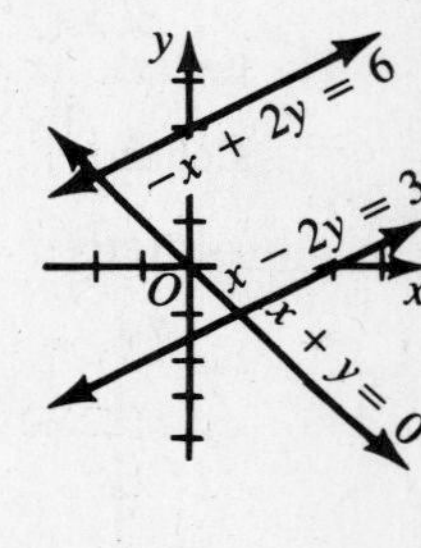

22. ∅

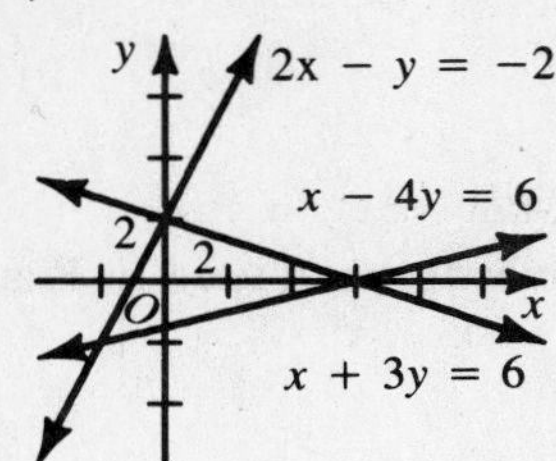

23. {(5, 2)}; 5 − 2 = 3;
5 − 4(2) = −3;
5 + 2 = 7

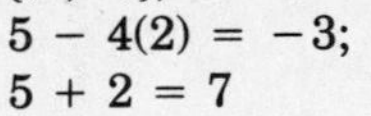

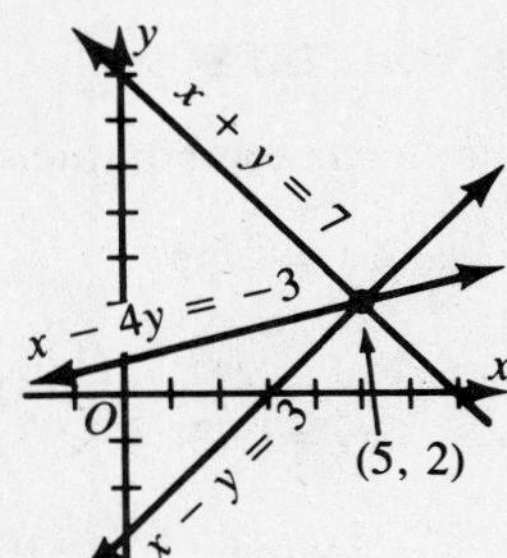

24. ∅

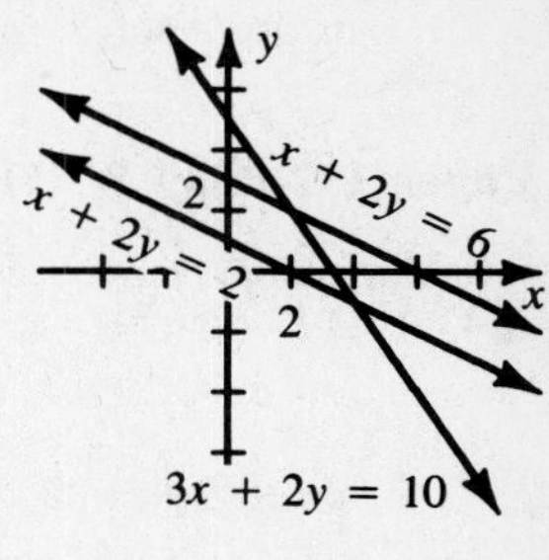

25. {(0, 4)}

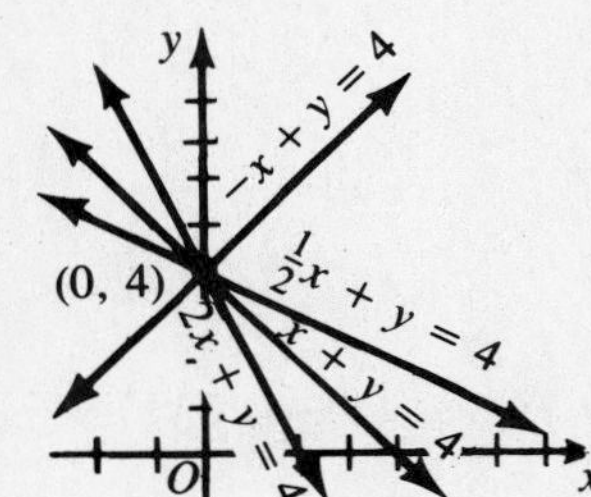

26. {(−3, 0)}

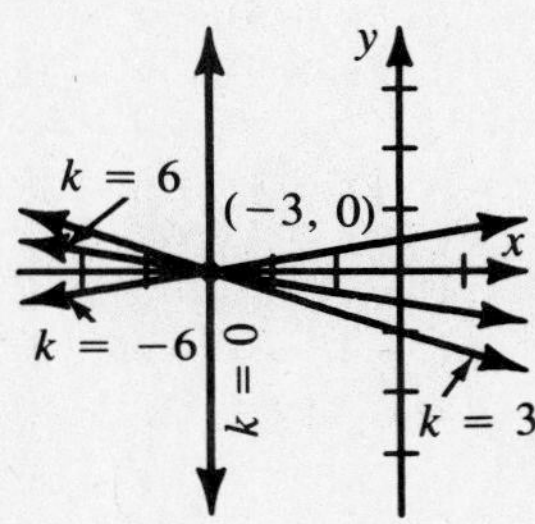

C **27.**

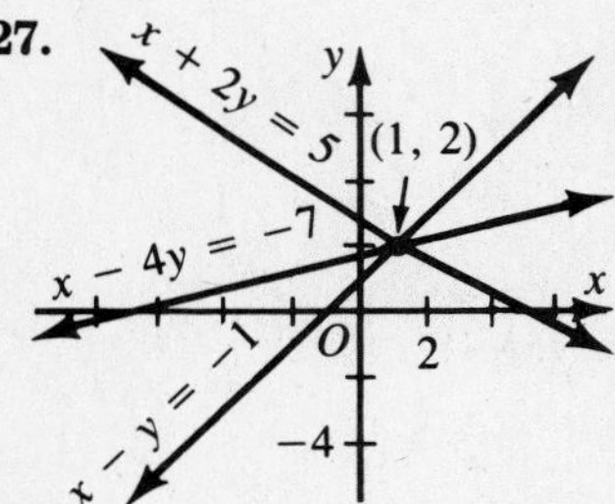

Suppose $r = -1$, $s = 2$. Then $r(x + 2y) + s(x - y) = r(5) + s(-1)$ becomes $-x - 2y + 2x - 2y = -5 - 2$, or $x - 4y = -7$. The apparent solution set is {(1, 2)}.

28. Suppose (a, b) is a solution of the equations $x + 2y = 5$ and $x - y = -1$. Then $a + 2b = 5$ and $a - b = -1$. Substituting (a, b) in $r(x + 2y) + s(x - y) = r(5) + s(-1)$ yields $r(a + 2b) + s(a - b) = r(5) + s(-1)$. Substituting 5 for $a + 2b$ and -1 for $a - b$ yields $r(5) + s(-1) = r(5) + s(-1)$, which is true for all values of r and s. Thus, if (a, b) is a solution of the first two equations, it is also a solution of $r(x + 2y) + s(x - y) = r(5) + s(-1)$.

Page 127 • COMPUTER EXERCISES

1.
```
10  INPUT A, B, C, D, E, F
20  LET B = (-D/A) * B
30  LET C = (-D/A) * C
40  IF B + E = C + F THEN PRINT "INFINITELY MANY SOLUTIONS": END
50  IF ABS(B + E) < 0.0001 THEN PRINT "NO SOLUTION": END
60  PRINT "UNIQUE SOLUTION"
70  END
```

2. infinitely many solutions **3.** unique solution **4.** no solution
5. unique solution

Pages 131–133 • WRITTEN EXERCISES

Variations may occur in the linear combinations and substitutions used in Exs. 1–32.

A **1.** $4x - 3y = -3$ $\quad 4x - 3y = -3$ $\quad -11y = -55$ $\quad x = 3;$
$x + 2y = 13$ $\quad -4x - 8y = -52$ $\quad y = 5;$ $\quad \{(3, 5)\}$

2. $6x + 2y = 22$ $\quad 6x + 2y = 22$ $\quad 12y = -12$ $\quad x = 4;$
$3x - 5y = 17$ $\quad -6x + 10y = -34$ $\quad y = -1;$ $\quad \{(4, -1)\}$

3. $7x - 2y = -2$ $\quad -21x + 6y = 6$ $\quad -16x = 32$ $\quad y = -6;$
$5x - 6y = 26$ $\quad 5x - 6y = 26$ $\quad x = -2;$ $\quad \{(-2, -6)\}$

4. $-3x + 10y = 9$ $\quad -3x + 10y = 9$ $\quad x = -3;$ $\quad \{(-3, 0)\}$
$2x - 5y = -6$ $\quad 4x - 10y = -12$ $\quad y = 0;$

5. $2x - \frac{2}{3}y = 12$ $\quad 4x - \frac{4}{3}y = 24$ $\quad x = 7;$ $\quad \{(7, 3)\}$
$-3x + \frac{4}{3}y = -17$ $\quad -3x + \frac{4}{3}y = -17$ $\quad y = 3;$

6. $\frac{1}{2}x + 2y = 12$ $\quad -4x - 16y = -96$ $\quad -11y = -77$ $\quad x = -4;$
$4x + 5y = 19$ $\quad 4x + 5y = 19$ $\quad y = 7;$ $\quad \{(-4, 7)\}$

7. $3x - 2y = 16$ $\quad 12x - 8y = 64$ $\quad y = -5;$ $\quad \{(2, -5)\}$
$4x - 3y = 23$ $\quad -12x + 9y = -69$ $\quad x = 2;$

8. $8x - 5y = -5$ $\quad 16x - 10y = -10$ $\quad 31x = 0$ $\quad y = 1;$
$3x + 2y = 2$ $\quad 15x + 10y = 10$ $\quad x = 0;$ $\quad \{(0, 1)\}$

9. $3x + 4y = 14$ $\quad 9x + 12y = 42$ $\quad 19x = 76$ $\quad y = \frac{1}{2};$
$5x - 6y = 17$ $\quad 10x - 12y = 34$ $\quad x = 4;$ $\quad \{(4, \frac{1}{2})\}$

10. $5x + 12y = 11$ $\quad 15x + 36y = 33$ $\quad 23x = 69$ $\quad y = -\frac{1}{3};$
$2x - 9y = 9$ $\quad 8x - 36y = 36$ $\quad x = 3;$ $\quad \{(3, -\frac{1}{3})\}$

11. $-\frac{2}{3}x + 3y = 8$ $\quad -\frac{10}{3}x + 15y = 40$ $\quad 7y = 28$ $\quad x = 6;$
$\frac{5}{3}x - 4y = -6$ $\quad \frac{10}{3}x - 8y = -12$ $\quad y = 4;$ $\quad \{(6, 4)\}$

12. $\frac{1}{3}x + 8y = 1$ $\quad x + 24y = 3$ $\quad -7x = 63$ $\quad y = \frac{1}{2};$
$-2x + 6y = 15$ $\quad -8x - 24y = 60$ $\quad x = -9;$ $\quad \{(-9, \frac{1}{2})\}$

13. $6x - 5y = 13$ $\quad y = 47 - 5x$ $\quad 31x - 235 = 13$ $\quad y = 7;$
$5x + y = 47$ $\quad 6x - 5(47 - 5x) = 13$ $\quad x = 8;$ $\quad \{(8, 7)\}$

14. $x - 15y = 3$ $\quad x = 3 + 15y$ $\quad 96y + 21 = -11$ $\quad x = -2;$
$7x - 9y = -11$ $\quad 7(3 + 15y) - 9y = -11$ $\quad y = -\frac{1}{3};$ $\quad \{(-2, -\frac{1}{3})\}$

15. $8x - y = 7$ $\quad y = 8x - 7$ $\quad -18x + 21 = -6$ $\quad y = 5;$
$6x - 3y = -6$ $\quad 6x - 3(8x - 7) = -6$ $\quad x = \frac{3}{2};$ $\quad \{(\frac{3}{2}, 5)\}$

16. $3x - 5y = 14$ $\qquad x = 7y + 18$ $\qquad 16y + 54 = 14$ $\qquad x = \frac{1}{2};$
$-x + 7y = -18$ $\qquad 3(7y + 18) - 5y = 14$ $\qquad y = -\frac{5}{2};$ $\qquad \{(\frac{1}{2}, -\frac{5}{2})\}$

17. $\frac{1}{2}x + 4y = 11$ $\qquad x = 22 - 8y$ $\qquad 66 - 19y = 9$ $\qquad x = -2;$
$3x + 5y = 9$ $\qquad 3(22 - 8y) + 5y = 9$ $\qquad y = 3;$ $\qquad \{(-2, 3)\}$

18. $3x - 2y = 15$ $\qquad y = \frac{1}{2}(3x - 15)$ $\qquad -4x + 45 = -7$ $\qquad y = 12;$
$5x - 6y = -7$ $\qquad 5x - 3(3x - 15) = -7$ $\qquad x = 13;$ $\qquad \{(13, 12)\}$

19. $-5x + y = 20$ $\qquad -15x + 3y = 60$ $\qquad -8x = 48$ $\qquad y = -10;$
$7x - 3y = -12$ $\qquad 7x - 3y = -12$ $\qquad x = -6;$ $\qquad \{(-6, -10)\}$

20. $12x - 7y = 16$ $\qquad -24x + 14y = -32$ $\qquad -4y = 4$ $\qquad x = \frac{3}{4};$
$8x - 6y = 12$ $\qquad 24x - 18y = 36$ $\qquad y = -1;$ $\qquad \{(\frac{3}{4}, -1)\}$

21. $x = \dfrac{5y + 4}{2}$ $\qquad 4\left(\dfrac{5y + 4}{2}\right) - 9y = 10$ $\qquad y = 2;$ $\qquad \{(7, 2)\}$
$x = 7;$
$4x - 9y = 10$ $\qquad y + 8 = 10$

22. $2y = 8 - 3x$ $\qquad 6y = 24 - 9x$ $\qquad -4x + 24 = 40$ $\qquad y = 10;$
$5x + 6y = 40$ $\qquad 5x + 24 - 9x = 40$ $\qquad x = -4;$ $\qquad \{(-4, 10)\}$

23. $\frac{1}{2}x - 5y = 32$ $\qquad -\frac{3}{2}x + 15y = -96$ $\qquad 8y = -51$ $\qquad x = \frac{1}{4};$
$\frac{3}{2}x - 7y = 45$ $\qquad \frac{3}{2}x - 7y = 45$ $\qquad y = -\frac{51}{8};$ $\qquad \{(\frac{1}{4}, -\frac{51}{8})\}$

24. $4x - \frac{1}{3}y = 5$ $\qquad 16x - \frac{4}{3}y = 20$ $\qquad 25x = 25$ $\qquad y = -3;$
$9x + \frac{4}{3}y = 5$ $\qquad 9x + \frac{4}{3}y = 5$ $\qquad x = 1;$ $\qquad \{(1, -3)\}$

B **25.** $4(x - 1) = y - 2$ $\qquad 4x - y = 2$ $\qquad \{(x, y)\colon 4x - y = 2\}$
$12x - 3y = 6$ $\qquad 4x - y = 2$

26. $3(x + y) = 2(y - 3)$ $\qquad 3x + y = -6$ $\qquad 15x + 5y = -30$
$5(x - y) = 2(7x)$ $\qquad -9x - 5y = 0$ $\qquad -9x - 5y = 0$
$6x = -30,\ x = -5;\ y = 9;\ \{(-5, 9)\}$

27. $6(\frac{x}{3} - \frac{y}{2}) = 6(y + 2)$ $\qquad 2x - 9y = 12$ $\qquad 10x - 45y = 60$
$10(\frac{x}{2} - \frac{y}{5}) = 10(x - 3)$ $\qquad -5x - 2y = -30$ $\qquad -10x - 4y = -60$
$-49y = 0,\ y = 0;\ x = 6;\ \{(6, 0)\}$

28. $3x = 7x - 5y$ $\qquad 4x - 5y = 0$ $\qquad -8x + 10y = 0$
$y = 9x - 9y - 20$ $\qquad 9x - 10y = 20$ $\qquad 9x - 10y = 20$
$x = 20;\ y = 16;\ \{(20, 16)\}$

29. $2(x - 1) + 3(y + 1) = 6(0)$ $\qquad 2x + 3y = -1$ $\qquad -4x - 6y = 2$
$4(x - 1) + 15(y + 1) = 6(12)$ $\qquad 4x + 15y = 61$ $\qquad 4x + 15y = 61$
$9y = 63,\ y = 7;\ x = -11;\ \{(-11, 7)\}$

30. $x + 7 - 10y = -15$ $\qquad x - 10y = -22$ $\qquad x - 10y = -22$
$8x + y + 5 = 72$ $\qquad 8x + y = 67$ $\qquad 80x + 10y = 670$
$81x = 648,\ x = 8;\ y = 3;\ \{(8, 3)\}$

31. $\frac{3}{x} + \frac{1}{y} = 1$ $\quad -\frac{15}{x} - \frac{5}{y} = -5$ $\quad -\frac{13}{x} = 13$ $\quad \frac{1}{y} = 4;$

$\frac{2}{x} + \frac{5}{y} = 18$ $\quad \frac{2}{x} + \frac{5}{y} = 18$ $\quad \frac{1}{x} = -1;$ $\quad \left\{\left(-1, \frac{1}{4}\right)\right\}$

32. $\frac{5}{x} - \frac{2}{y} = 5$ $\quad \frac{15}{x} - \frac{6}{y} = 15$ $\quad \frac{7}{x} = 21$ $\quad \frac{1}{y} = 5;$

$\frac{4}{x} - \frac{3}{y} = -3$ $\quad -\frac{8}{x} + \frac{6}{y} = 6$ $\quad \frac{1}{x} = 3;$ $\quad \left\{\left(\frac{1}{3}, \frac{1}{5}\right)\right\}$

C **33.** Let $x = -1$; $5 = a - b$ $\quad 15 = 3a - 3b$ $\quad 36 = 12a$

Let $x = 3$; $21 = 9a + 3b$ $\quad 21 = 9a + 3b$ $\quad a = 3;\ b = -2$

34. Let $x = 3$; $-15 = 9a + 3b$ $\quad -30 = 18a + 6b$ $\quad -3 = 6a$

Let $x = 2$; $-9 = 4a + 2b$ $\quad 27 = -12a - 6b$ $\quad a = -\frac{1}{2};\ b = -\frac{7}{2}$

35.

Statements	Reasons
1. $\left.\begin{array}{l} y = m_1x + b_1 \\ y = m_2x + b_2 \end{array}\right\} m_1 \neq m_2$ $b_1, b_2, m_1, m_2 \in \mathscr{R}$	Hypothesis
2. $0 = m_2x - m_1x + (b_2 - b_1)$ $y = m_1x + b_1$	Trans. 2
3. $-(b_2 - b_1) = m_2x - m_1x$ $y = m_1x + b_1$	Add. prop. of $=$
4. $b_1 - b_2 = m_2x - m_1x$ $y = m_1x + b_1$	Prop. of the opp. of a sum
5. $b_1 - b_2 = (m_2 - m_1)x$ $y = m_1x + b_1$	Dist. ax.
6. $x = \frac{b_1 - b_2}{m_2 - m_1}$ $y = m_1x + b_1$	Div. prop., symm. prop. of $=$
7. $x = \frac{b_1 - b_2}{m_2 - m_1}$ $y = m_1\left(\frac{b_1 - b_2}{m_2 - m_1}\right) + b_1$	Subs. prin.
8. $x = \frac{b_1 - b_2}{m_2 - m_1}$ $y = m_1\left(\frac{b_1 - b_2}{m_2 - m_1}\right) + b_1\left(\frac{m_2 - m_1}{m_2 - m_1}\right)$	Subs. prin.
9. $x = \frac{b_1 - b_2}{m_2 - m_1}$ $y = \frac{m_1(b_1 - b_2) + b_1(m_2 - m_1)}{m_2 - m_1}$	Addition of fractions
10. $x = \frac{b_1 - b_2}{m_2 - m_1}$ $y = \frac{m_1b_1 - m_1b_2 + b_1m_2 - b_1m_1}{m_2 - m_1}$	Dist. ax.
11. $x = \frac{b_1 - b_2}{m_2 - m_1}$ $y = \frac{m_2b_1 - m_1b_2}{m_2 - m_1}$	Subs. prin., $a - a = 0$

36. We have two cases to consider: (1) $b_1 = b_2$ and (2) $b_1 \neq b_2$.

(1) Suppose $b_1 = b_2$, $\begin{cases} y = mx + b_1 \\ y = mx + b_2 \end{cases}$ and $\begin{cases} x = b_1 \\ x = b_2 \end{cases}$. Using the substitution principle, $\begin{cases} y = mx + b_1 \\ y = mx + b_1 \end{cases}$ and $\begin{cases} x = b_1 \\ x = b_1 \end{cases}$. Replacing the second equation with the result when it is subtracted from the first in each case yields $\begin{cases} y = mx + b_1 \\ 0 = 0 \end{cases}$ and $\begin{cases} x = b_1 \\ 0 = 0 \end{cases}$. Since $0 = 0$ is a true statement, the solutions of each system are the solutions of the first equation in each. $y = mx + b_1$ and $x = b_1$ both have an infinite number of solutions.

(2) Suppose $b_1 \neq b_2$, $\begin{cases} y = mx + b_1 \\ y = mx + b_2 \end{cases}$, and $\begin{cases} x = b_1 \\ x = b_2 \end{cases}$. Replacing the second equation with the result when it is subtracted from the first in each case yields $\begin{cases} y = mx + b_1 \\ 0 = b_1 - b_2 \end{cases}$ and $\begin{cases} x = b_1 \\ 0 = b_1 - b_2 \end{cases}$. Since $b_1 \neq b_2$, $b_1 - b_2 \neq 0$ and the second equation in each system is false. Hence there are no solutions.

37.

1. $y = mx + b_1$; $m, b_1, b_2 \in \mathscr{R}$ $x = b_2$	Hypothesis
2. $y = mb_2 + b_1$ $x = b_2$	Trans. 3
3. The solution set is $\{(b_2, mb_2 + b_1)\}$	Def. of sol. set

38. (1) Let $\begin{cases} y = m_1x + b_1 \\ y = m_2x + b_2 \end{cases}$ and $\begin{cases} y = mx + b_1 \\ x = b_2 \end{cases}$ be the two systems in part (1) of the theorem. By hypothesis, $m_1 \neq m_2$. Using Exercises 35 and 37 and the definition of solution set, these systems have the solutions $\left\{\left(\frac{b_1 - b_2}{m_2 - m_1}, \frac{m_2b_1 - m_1b_2}{m_2 - m_1}\right)\right\}$ and $\{(b_2, mb_2 + b_1)\}$, respectively. These are the only solutions in each case, since they are the only ordered pairs satisfying the equivalent set of equations in each case.

(2) Let $\begin{cases} y = mx + b_1 \\ y = mx + b_2 \end{cases}$ and $\begin{cases} x = b_1 \\ x = b_2 \end{cases}$ be the two systems in part (2) of the theorem. $b_1 = b_2$ by hypothesis. Using Exercise 36, the systems have an infinite set of solutions.

(3) Let $\begin{cases} y = mx + b_1 \\ y = mx + b_2 \end{cases}$ and $\begin{cases} x = b_1 \\ x = b_2 \end{cases}$ be the two systems in part (3) of the theorem. $b_1 \neq b_2$ by hypothesis. Using Exercise 36, the systems have no solution.

39. (1) If two lines are parallel, then each line has no slope and different x-intercepts or both lines have the same slope and different y-intercepts.

PROOF *Case 1: One of the lines has no slope.* If the other line has a slope, then the two lines intersect (Ex. 37), which contradicts the hypothesis that they are parallel. If the other line has no slope and the same x-intercept, then the intersection of the two lines is an infinite set (Ex. 36), which again contradicts the hypothesis that the two lines are parallel. Therefore, if one of two parallel lines has no slope, then the other must have no slope and a different x-intercept.

Case 2: Both lines have slopes. If the two lines have different slopes, then they intersect (Ex. 35), which contradicts the hypothesis that they are parallel. If the two lines have the same slope and the same y-intercept, their intersection is an infinite set (Ex. 36), which again contradicts the hypothesis. Therefore if two

parallel lines both have slopes, then they both have the same slope and different y-intercepts.
(2) If each line has no slope and different x-intercepts or both lines have the same slope and different y-intercepts, then the lines are parallel.
PROOF *Case 1:* If two lines have no slope and different x-intercepts, they are of the form $x = b_1$ and $x = b_2$ with $b_1 \neq b_2$. By Ex. 36, there are no solutions (points of intersection) and the lines are parallel.
Case 2: If two lines have the same slope and different y-intercepts, they are of the form $y = mx + b_1$ and $y = mx + b_2$ with $b_1 \neq b_2$. By Ex. 36, there are no solutions (intersections) and the lines are parallel.

Pages 135–136 • WRITTEN EXERCISES

A **1.** $3(4) - 5(-2) = 22$ **2.** $8(7) - (-3)(-2) = 50$ **3.** $9(0) - (-6)(8) = 48$
4. $-8(3) - 4(-6) = 0$ **5.** $63 - k^2$ **6.** $(k - 3)(k - 3) - 5(4) = k^2 - 6k - 11$
7. $k(2k - 10) - 2k(k - 5) = 0$ **8.** $k(3 - 6k) - (2k - 1)(-3k) = 0$

9. $D = \begin{vmatrix} 3 & 7 \\ 2 & 5 \end{vmatrix} = 15 - 14 = 1$; $x = \dfrac{\begin{vmatrix} -2 & 7 \\ 4 & 5 \end{vmatrix}}{1} = -10 - 28 = -38$,

$y = \dfrac{\begin{vmatrix} 3 & -2 \\ 2 & 4 \end{vmatrix}}{1} = 12 + 4 = 16$; $\{(-38, 16)\}$

10. $D = \begin{vmatrix} 3 & 4 \\ -5 & -7 \end{vmatrix} = -21 + 20 = -1$; $x = \dfrac{\begin{vmatrix} 1 & 4 \\ -1 & -7 \end{vmatrix}}{-1} = \dfrac{-7 + 4}{-1} = 3$,

$y = \dfrac{\begin{vmatrix} 3 & 1 \\ -5 & -1 \end{vmatrix}}{-1} = \dfrac{-3 + 5}{-1} = -2$; $\{(3, -2)\}$

11. $D = \begin{vmatrix} 5 & -3 \\ -9 & 5 \end{vmatrix} = 25 - 27 = -2$; $x = \dfrac{\begin{vmatrix} 2 & -3 \\ -6 & 5 \end{vmatrix}}{-2} = \dfrac{10 - 18}{-2} = 4$,

$y = \dfrac{\begin{vmatrix} 5 & 2 \\ -9 & -6 \end{vmatrix}}{-2} = \dfrac{-30 + 18}{-2} = 6$; $\{(4, 6)\}$

12. $D = \begin{vmatrix} 7 & -5 \\ 5 & -4 \end{vmatrix} = -28 + 25 = -3$; $x = \dfrac{\begin{vmatrix} 0 & -5 \\ -3 & -4 \end{vmatrix}}{-3} = \dfrac{0 - 15}{-3} = 5$,

$y = \dfrac{\begin{vmatrix} 7 & 0 \\ 5 & -3 \end{vmatrix}}{-3} = \dfrac{-21 - 0}{-3} = 7$; $\{(5, 7)\}$

13. $D = \begin{vmatrix} -3 & 11 \\ -2 & 10 \end{vmatrix} = -30 + 22 = -8$; $x = \dfrac{\begin{vmatrix} 1 & 11 \\ 2 & 10 \end{vmatrix}}{-8} = \dfrac{10 - 22}{-8} = \dfrac{3}{2}$,

$y = \dfrac{\begin{vmatrix} -3 & 1 \\ -2 & 2 \end{vmatrix}}{-8} = \dfrac{-6 + 2}{-8} = \dfrac{1}{2}$; $\left\{\left(\dfrac{3}{2}, \dfrac{1}{2}\right)\right\}$

14. $D = \begin{vmatrix} 4 & -3 \\ 9 & -5 \end{vmatrix} = -20 + 27 = 7;\ x = \dfrac{\begin{vmatrix} 2 & -3 \\ 1 & -5 \end{vmatrix}}{7} = \dfrac{-10 + 3}{7} = -1,$

$y = \dfrac{\begin{vmatrix} 4 & 2 \\ 9 & 1 \end{vmatrix}}{7} = \dfrac{4 - 18}{7} = -2;\ \{(-1, -2)\}$

15. $D = \begin{vmatrix} 3 & -4 \\ 4 & -7 \end{vmatrix} = -21 + 16 = -5;\ x = \dfrac{\begin{vmatrix} 6 & -4 \\ 8 & -7 \end{vmatrix}}{-5} = \dfrac{-42 + 32}{-5} = 2,$

$y = \dfrac{\begin{vmatrix} 3 & 6 \\ 4 & 8 \end{vmatrix}}{-5} = \dfrac{24 - 24}{-5} = 0;\ \{(2, 0)\}$

16. $D = \begin{vmatrix} 4 & 5 \\ 5 & 6 \end{vmatrix} = 24 - 25 = -1;\ x = \dfrac{\begin{vmatrix} 3 & 5 \\ 2 & 6 \end{vmatrix}}{-1} = \dfrac{18 - 10}{-1} = -8,$

$y = \dfrac{\begin{vmatrix} 4 & 3 \\ 5 & 2 \end{vmatrix}}{-1} = \dfrac{8 - 15}{-1} = 7;\ \{(-8, 7)\}$

B 17. $D = \begin{vmatrix} -7 & 4 \\ 5 & -2 \end{vmatrix} = 14 - 20 = -6;\ x = \dfrac{\begin{vmatrix} -5 & 4 \\ 3 & -2 \end{vmatrix}}{-6} = \dfrac{10 - 12}{-6} = \dfrac{1}{3},$

$y = \dfrac{\begin{vmatrix} -7 & -5 \\ 5 & 3 \end{vmatrix}}{-6} = \dfrac{-21 + 25}{-6} = -\dfrac{2}{3};\ \left\{\left(\dfrac{1}{3}, -\dfrac{2}{3}\right)\right\}$

18. $D = \begin{vmatrix} 2 & -4 \\ 3 & 5 \end{vmatrix} = 10 + 12 = 22;\ x = \dfrac{\begin{vmatrix} -9 & -4 \\ 14 & 5 \end{vmatrix}}{22} = \dfrac{-45 + 56}{22} = \dfrac{1}{2},$

$y = \dfrac{\begin{vmatrix} 2 & -9 \\ 3 & 14 \end{vmatrix}}{22} = \dfrac{28 + 27}{22} = \dfrac{5}{2};\ \left\{\left(\dfrac{1}{2}, \dfrac{5}{2}\right)\right\}$

19. $D = \begin{vmatrix} 11 & 6 \\ -5 & -3 \end{vmatrix} = -33 + 30 = -3;\ x = \dfrac{\begin{vmatrix} -12 & 6 \\ 5 & -3 \end{vmatrix}}{-3} = \dfrac{36 - 30}{-3} = -2,$

$y = \dfrac{\begin{vmatrix} 11 & -12 \\ -5 & 5 \end{vmatrix}}{-3} = \dfrac{55 - 60}{-3} = \dfrac{5}{3};\ \left\{\left(-2, \dfrac{5}{3}\right)\right\}$

20. $D = \begin{vmatrix} 9 & -4 \\ -8 & 3 \end{vmatrix} = 27 - 32 = -5;\ x = \dfrac{\begin{vmatrix} -1 & -4 \\ 7 & 3 \end{vmatrix}}{-5} = \dfrac{-3 + 28}{-5} = -5,$

$y = \dfrac{\begin{vmatrix} 9 & -1 \\ -8 & 7 \end{vmatrix}}{-5} = \dfrac{63 - 8}{-5} = -11;\ \{(-5, -11)\}$

21. $D = \begin{vmatrix} 2 & 5 \\ 8 & -7 \end{vmatrix} = -14 - 40 = -54$; $x = \dfrac{\begin{vmatrix} 3 & 5 \\ -6 & -7 \end{vmatrix}}{-54} = \dfrac{-21 + 30}{-54} = -\dfrac{1}{6}$,

$y = \dfrac{\begin{vmatrix} 2 & 3 \\ 8 & -6 \end{vmatrix}}{-54} = \dfrac{-12 - 24}{-54} = \dfrac{2}{3}$; $\left\{\left(-\dfrac{1}{6}, \dfrac{2}{3}\right)\right\}$

22. $D = \begin{vmatrix} 7 & -6 \\ 4 & -9 \end{vmatrix} = -63 + 24 = -39$; $x = \dfrac{\begin{vmatrix} 4 & -6 \\ -2 & -9 \end{vmatrix}}{-39} = \dfrac{-36 - 12}{-39} = \dfrac{16}{13}$,

$y = \dfrac{\begin{vmatrix} 7 & 4 \\ 4 & -2 \end{vmatrix}}{-39} = \dfrac{-14 - 16}{-39} = \dfrac{10}{13}$; $\left\{\left(\dfrac{16}{13}, \dfrac{10}{13}\right)\right\}$

C **23.** The area of the rectangle is $(a + c)(b + d)$. The area of each of the triangles with sides $\overline{AB}$ and $\overline{CD}$ is $\frac{1}{2}b(a + c)$; of those with sides $\overline{BC}$ and $\overline{AD}$ is $\frac{1}{2}c(b + d)$. The area of the parallelogram is $(a + c)(b + d) - 2 \cdot \frac{1}{2}b(a + c) - 2 \cdot \frac{1}{2}c(b + d) = ad - bc = \begin{vmatrix} a & b \\ c & d \end{vmatrix}$.

24. If the graphs are parallel or coincident, then the slopes are the same, $-\dfrac{a_1}{b_1} = -\dfrac{a_2}{b_2}$, or they are undefined, so that $b_1 = b_2 = 0$. If $-\dfrac{a_1}{b_1} = -\dfrac{a_2}{b_2}$, then $-a_1b_2 = -a_2b_1$ (means and extremes) and $a_1b_2 - a_2b_1 = 0$. If $b_1 = b_2 = 0$, then $a_1b_2 - a_2b_1 = a_1 \cdot 0 - a_2 \cdot 0 = 0$. In any case, $\begin{vmatrix} a_1 & b_1 \\ a_2 & b_2 \end{vmatrix} = a_1b_2 = a_2b_1 = 0$.

25. By hypothesis, $\begin{vmatrix} a_1 & b_1 \\ a_2 & b_2 \end{vmatrix} = a_1b_2 - a_2b_1 = 0$. If $b_1 \neq 0$ and $b_2 \neq 0$, then $\dfrac{a_1}{b_1} - \dfrac{a_2}{b_2} = 0$. Thus, $-\dfrac{a_1}{b_1} = -\dfrac{a_2}{b_2}$, so that the slopes are the same. If $b_1 = 0$, then $a_1b_2 - a_2b_1 = 0$ means that $a_1b_2 = 0$. Therefore, $b_2 = 0$ (if a_1 and b_2 are both zero, there is no line). Similarly, if $b_2 = 0$, then $b_1 = 0$. Therefore, the slopes are the same, or both lines have no slope ($b_1 = b_2 = 0$). Thus, the lines are either parallel or coincident. (See Theorem, p. 125, and Ex. 39, p. 133.)

26. Multiplying the first equation by k ($k \neq 0$) yields the system $\begin{cases} ka_1x + kb_1y = kc_1 \\ a_2x + b_2y = c_2 \end{cases}$.

$D = \begin{vmatrix} ka_1 & kb_1 \\ a_2 & b_2 \end{vmatrix} = k(a_1b_2 - a_2b_1)$; $x = \dfrac{\begin{vmatrix} kc_1 & kb_1 \\ c_2 & b_2 \end{vmatrix}}{k(a_1b_2 - a_2b_1)} = \dfrac{k(c_1b_2 - c_2b_1)}{k(a_1b_2 - a_2b_1)} =$

$\dfrac{c_1b_2 - c_2b_1}{a_1b_2 - a_2b_1}$; $y = \dfrac{\begin{vmatrix} ka_1 & kc_1 \\ a_2 & c_2 \end{vmatrix}}{k(a_1b_2 - a_2b_1)} = \dfrac{k(a_1c_2 - a_2c_1)}{k(a_1b_2 - a_2b_1)} = \dfrac{a_1c_2 - a_2c_1}{a_1b_2 - a_2b_1}$. The solution set is $\left\{\left(\dfrac{c_1b_2 - c_2b_1}{a_1b_2 - a_2b_1}, \dfrac{a_1c_2 - a_2c_1}{a_1b_2 - a_2b_1}\right)\right\}$, which is the solution set of the original system. (See Eqs. (1), (2), and (3) on page 133.)

27. Replacing the first equation by the sum of the equations yields the system $\begin{cases}(a_1 + a_2)x + (b_1 + b_2)y = c_1 + c_2 \\ a_2x + b_2y = c_2\end{cases}$. $D = b_2(a_1 + a_2) - a_2(b_1 + b_2) = a_1b_2 - a_2b_1$; $x = \dfrac{b_2(c_1 + c_2) - c_2(b_1 + b_2)}{a_1b_2 - a_2b_1} = \dfrac{c_1b_2 - c_2b_1}{a_1b_2 - a_2b_1}$; $y = \dfrac{c_2(a_1 + a_2) - a_2(c_1 + c_2)}{a_1b_2 - a_2b_1} = \dfrac{a_1c_2 - a_2c_1}{a_1b_2 - a_2b_1}$. The solution set is $\left\{\left(\dfrac{c_1b_2 - c_2b_1}{a_1b_2 - a_2b_1}, \dfrac{a_1c_2 - a_2c_1}{a_1b_2 - a_2b_1}\right)\right\}$, which is the solution set of the original system. (See Eqs. (1), (2), and (3) on page 133.)

Page 136 • COMPUTER EXERCISES

1.
```
10  INPUT A,B,C,D,E,F
20  LET B = (-D/A) * B
30  LET C = (-D/A) * C
40  IF B + E = C + F THEN PRINT "INFINITELY MANY SOLUTIONS":END
50  IF ABS(B+E) < 0.0001 THEN PRINT "NO SOLUTION":END
60  LET B = B+E
70  LET C = C+F
80  LET Y = C/B
90  LET X = (F - E * Y)/D
100 PRINT "(";X;", ";Y;")"
110 END
```

2. $\{(3.632597, -0.1353591)\}$ **3.** $\{(1.79866, 6.95302)\}$ **4.** $\{(-2, -3)\}$

5. no solution

Pages 142–144 • PROBLEMS

A

1. $c + h = 12$; $25c + 75h = 750$ $\quad$ $c + h = 12$; $-c - 3h = -30$ $\quad$ $h = 9$; $c = 3$; 9 heads of lettuce and 3 cucumbers

2.

	r (km/h)	t (h)	d (km)
upstream	$b - c$	$\frac{1}{2}$	$\frac{1}{2}(b - c)$
downstream	$b + c$	$\frac{1}{3}$	$\frac{1}{3}(b + c)$

$\frac{1}{2}(b - c) = 6$ $\quad$ $b - c = 12$ $\quad$ $b = 15$; speed of boat: 15 km/h

$\frac{1}{3}(b + c) = 6$ $\quad$ $b + c = 18$ $\quad$ $c = 3$; speed of current: 3 km/h

3. $2a + b = 140$; $a = 2b - 20$ $\quad$ $2(2b - 20) + b = 140$; $5b = 180$ $\quad$ $b = 36$; base: 36 cm; $a = 52$; sides: 52 cm each

4. $4a + b = 244$; $a + \frac{3}{2}b = 106$ $\quad$ $12a + 3b = 732$; $-2a - 3b = -212$ $\quad$ $a = 52$; base: 36 cm; $b = 36$; sides: 52 cm each

5. Let x = measure of first angle in degrees; y = measure of second angle in degrees.
$x + y + 35 = 180$ $(2y - 26) + y + 35 = 180$ $y = 57$;
$x = 2y - 26$ $3y + 9 = 180$ $x = 88$; 57°, 88°

6. Let $x = m(\angle A) = m(\angle B) = m(\angle C)$; $y = m(\angle D) = m(\angle E)$.
$3x + 2y = 540$ $3x + 2(x + 30) = 540$ $x = 96$;
$y = x + 30$ $5x + 60 = 540$ $y = 126$
$m(\angle A) = m(\angle B) = m(\angle C) = 96°$; $m(\angle D) = m(\angle E) = 126°$

7. $I_1 + I_2 = 4$ $I_1 = 4 - I_2$ $I_2 = 2.5$ A
$80I_1 = 48I_2$ $80(4 - I_2) = 48I_2$ $I_1 = 1.5$ A

8. Let x = air speed of plane in km/h; y = wind speed in km/h.

	r (km/h)	t (h)	d (km)
with a head wind	$x + y$	2.5	$2.5(x + y)$
with a tail wind	$x - y$	3	$3(x - y)$

$2.5(x + y) = 1440$ $x + y = 576$ $x = 528$; air speed of plane: 528 km/h
$3(x - y) = 1440$ $x - y = 480$

9. Let x = number of \$4.50 tickets sold; y = number of \$2.50 tickets sold.
$x + y = 480$ $5x + 5y = 2400$ $x = 175$; 175 tickets
$4.5x + 2.5y = 1550$ $9x + 5y = 3100$

10. Let x = speed of boat in km/h; y = distance traveled upstream in km.

	r (km/h)	t (h)	d (km)
upstream	$x - 3$	2	$2(x - 3)$
downstream	$x + 3$	$1\frac{1}{3}$	$\frac{4}{3}(x + 3)$

$2(x - 3) = y$ $2x - 6 = y$ $x = 12$; speed of boat: 12 km/h
$\frac{4}{3}(x + 3) = y + 2$ $\frac{4}{3}x + 4 = (2x - 6) + 2$ $y = 18$; distance: 18 km

11. Let $x = HG = GE$; $y = FG = FE$.
$x + x + (x + y) + y = 39$ $3x + 2y = 39$ $x = 7$;
$2(x + y) + 2y = 50$ $2x + 4y = 50$ $y = 9$
$HG = 7$ cm, $DE = HG + GF = 16$ cm, $GF = 9$ cm

12. Let f = number of free throws made; t = number of free throws attempted.
$f = 0.75t$ $0.75t + 1 = 0.65(t + 4)$ $t = 16$;
$f + 1 = 0.65(t + 4)$ $0.10t = 1.6$ $f = 12$
16 free throws attempted; 12 free throws made

B **13.** Let x = number of cars; y = number of trucks.
$x + y = 108$ $x + y = 108$ $y = 24$; 24 trucks
$0.25(2x) + 0.25(5y) = 72$ $x + 2.5y = 144$

14. Let x = width of table in ft; y = length of table in ft.

$x + 2y = 17\frac{1}{2}$ $\quad$ $x + 2y = 17\frac{1}{2}$ $\quad$ $y = 6\frac{1}{2}$;

$2(x + 10) + 2(y + 8) = 58$ $\quad$ $x + y = 11$ $\quad$ $x = 4\frac{1}{2}$; $4\frac{1}{2}$ ft by $6\frac{1}{2}$ ft

15. Let x = time in hours for first worker to run to one end of bridge; y = time in hours for second worker to run to other end of bridge.

	r (km/h)	t (h)	d (km)
first worker	12	x	$12x$
second worker	12	y	$12y$

$y = x + \frac{1}{60}$ $\quad$ $12x + 12\left(x + \frac{1}{60}\right) = 1.2$ $\quad$ $x = \frac{1}{24}$; $\quad$ $12x = 0.5$; 0.5 km

$12x + 12y = 1.2$ $\quad$ $24x = 1$ $\quad$ $y = \frac{7}{120}$; $\quad$ $12y = 0.7$; 0.7 km

16. Let x = rate of each large pipe in L/h; y = rate of each small pipe in L/h.
Recall: rate × time = number of liters.

$2(3x) + 2(2y) = 1000$ $\quad$ $3x + 2y = 500$ $\quad$ $3x + 2y = 500$ $\quad$ $y = 70$;

$\frac{5}{6}(3x) + 5(2y) = 1000$ $\quad$ $x + 4y = 400$ $\quad$ $-3x - 12y = -1200$ $\quad$ $x = 120$

Small pipe: 70 L/h; large pipe: 120 L/h

17. $8A + 3B = 3$ $\quad$ $56A + 21B = 21$ $\quad$ $A = \frac{12}{17}$;

$13A + 7B = 3$ $\quad$ $-39A - 21B = -9$ $\quad$ $B = -\frac{15}{17}$

18. $28 = 16A + 4B$ $\quad$ $7 = 4A + B$ $\quad$ $A = 3$;
$12 = 9A + 3B$ $\quad$ $-4 = -3A - B$ $\quad$ $B = -5$

19. $9 = 1 + A + B$ $\quad$ $-8 = -A - B$ $\quad$ $A = -2$;
$25 = 9 - 3A + B$ $\quad$ $16 = -3A + B$ $\quad$ $B = 10$

20. $6 = 9A + 9 + B$ $\quad$ $3 = -9A - B$ $\quad$ $A = -1$;
$2 = 16A + 12 + B$ $\quad$ $-10 = 16A + B$ $\quad$ $B = 6$

21. $30 = -5 + v_0 + h_0$ $\quad$ $-35 = -v_0 - h_0$ $\quad$ $v_0 = 5$ m/s;
$0 = -45 + 3v_0 + h_0$ $\quad$ $45 = 3v_0 + h_0$ $\quad$ $h_0 = 30$ m

22. $8 = 20b + 400a$ $\quad$ $2 = 5b + 100a$ $\quad$ $b = \frac{1}{2}$;

$0 = 100b + 10000a$ $\quad$ $0 = -b - 100a$ $\quad$ $a = -\frac{1}{200}$

C **23.** Let c = rate for city driving; h = rate for highway driving.

$\frac{120}{c} + \frac{280}{h} = 16$ $\quad$ $-\frac{120}{c} - \frac{280}{h} = -16$ $\quad$ $\frac{320}{c} = 16$ $\quad$ $c = 20$;

$\frac{220}{c} + \frac{140}{h} = 16$ $\quad$ $\frac{440}{c} + \frac{280}{h} = 32$ $\quad$ $\frac{1}{c} = \frac{1}{20}$ $\quad$ $h = 28$; 28 mi/gal

24. 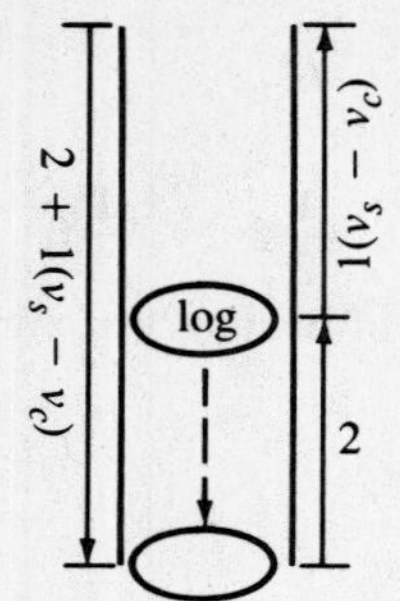

Let v_c = rate of current in km/h;
v_s = rate of rowing in still water in km/h.
See table below for expressions involving v_c and v_s.

$$\frac{2}{v_c} = 1 + \frac{2 + v_s - v_c}{v_s + v_c}$$

$$\frac{2}{v_c} = \frac{2 + 2v_s}{v_s + v_c}$$

$$2(v_s + v_c) = v_c(2 + 2v_s)$$

$$v_s = v_s v_c \quad (v_s \neq 0)$$

$$v_c = 1$$

Rate of current: 1 km/h
Note: The problem is satisfied regardless of the rower's rate.

	d (km)	r (km/h)	t (h)
Rowing to meet log	2		
Rowing from log to turn	$1(v_s - v_c)$	$v_s - v_c$	1
Rowing back to boathouse	$2 + 1(v_s - v_c)$	$v_s + v_c$	$\frac{2 + 1(v_s - v_c)}{v_s + v_c}$
Log's trip to boathouse	2	v_c	$\frac{2}{v_c}$

25. Let n = number of nickels; d = number of dimes; q = number of quarters.
$n + d + q = 27$ $\quad (18 - 2d) + d + (16 - 0.4d) = 27$
$5n + 10d = 90 \quad n = 18 - 2d \quad -1.4d + 34 = 27 \quad n = 8;$
$10d + 25q = 400 \quad q = 16 - 0.4d \quad d = 5; \quad q = 14$
8 nickels, 5 dimes, 14 quarters

Pages 144–145 • SELF-TEST 1

1. {(1, −3)}

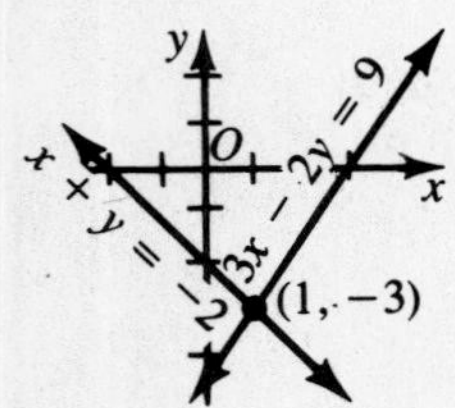

2. slopes: $-2, \frac{3}{2}$; 1 solution

3. slopes: 1, 1; y-intercepts: −3, 4; no solution

4. slopes: $\frac{1}{2}, \frac{1}{2}$; y-intercepts: 2, 2;
infinitely many solutions

5. $3x + 2y = 5 \quad 9x + 6y = 15 \quad 5x = 1 \quad y = \frac{11}{5};$
$2x + 3y = 7 \quad -4x - 6y = -14 \quad x = \frac{1}{5}; \quad \left\{\left(\frac{1}{5}, \frac{11}{5}\right)\right\}$

6. $4x + 2y = 14 \quad y = 3x - 3 \quad 10x - 6 = 14 \quad y = 3;$
$3x - y = 3 \quad 4x + 2(3x - 3) = 14 \quad x = 2; \quad \{(2, 3)\}$

7. $D = \begin{vmatrix} 2 & 1 \\ 3 & 4 \end{vmatrix} = 8 - 3 = 5$; $x = \dfrac{\begin{vmatrix} 8 & 1 \\ 17 & 4 \end{vmatrix}}{5} = \dfrac{32 - 17}{5} = 3$; $y = \dfrac{\begin{vmatrix} 2 & 8 \\ 3 & 17 \end{vmatrix}}{5} = \dfrac{34 - 24}{5} = 2$; $\{(3, 2)\}$

8. Let x = cost/lb of oranges; y = cost/lb of apples.

$3x + 2y = 3.45$ $\quad -6x - 4y = -6.90$ $\quad x = 0.75$; oranges: \$.75/lb

$2x + 4y = 3.90$ $\quad 2x + 4y = 3.90$ $\quad y = 0.60$; apples: \$.60/lb

Pages 146–147 • WRITTEN EXERCISES

A 1.

2.

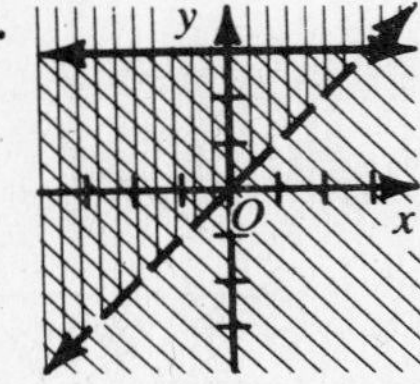

3.

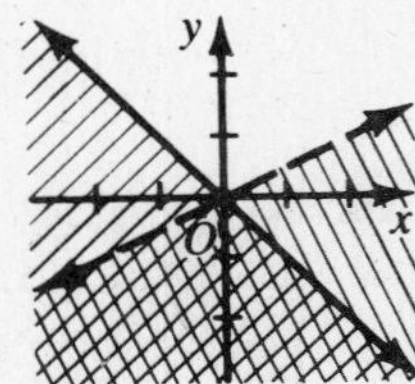

4.

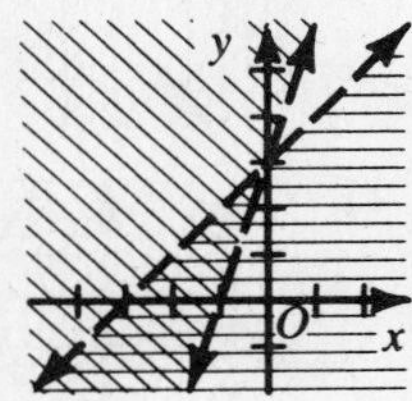

5.

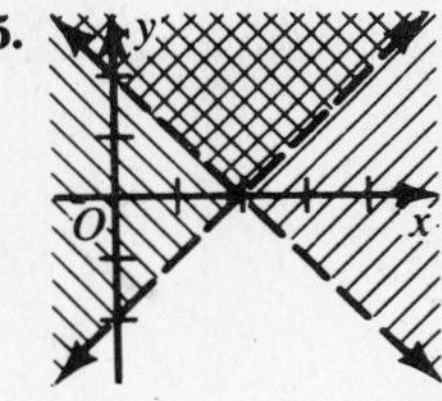

6.

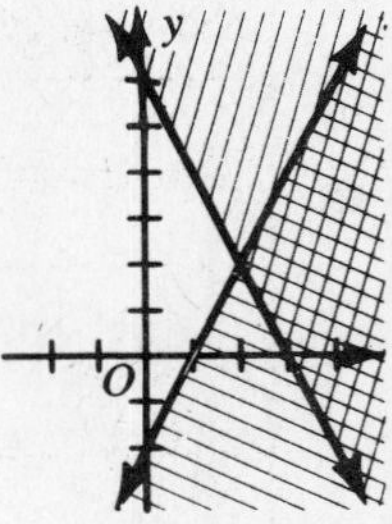

7.

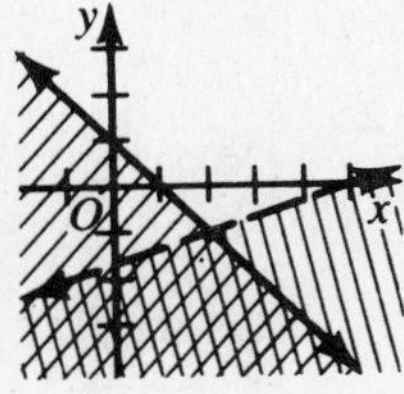

8.

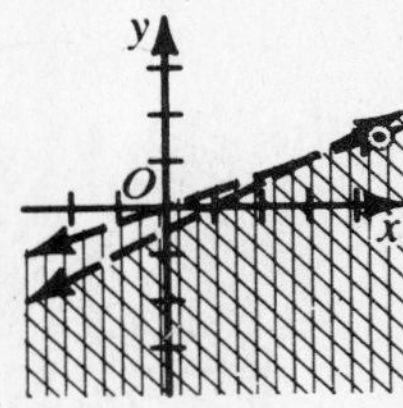

9.

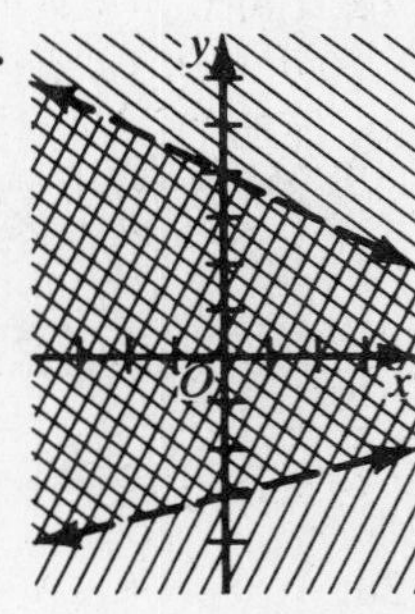

10.

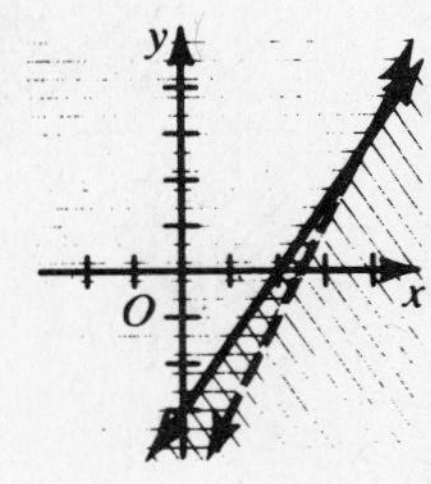

11.

12.

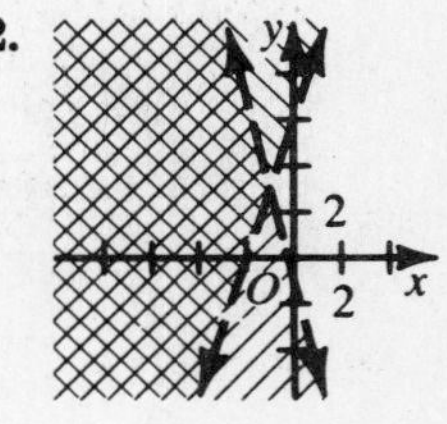

13. 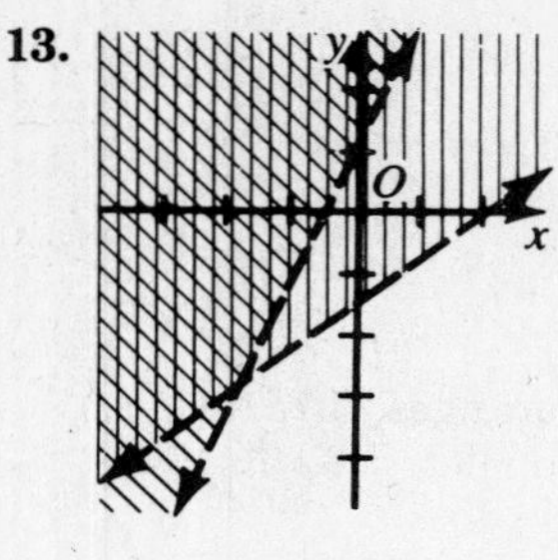

14. 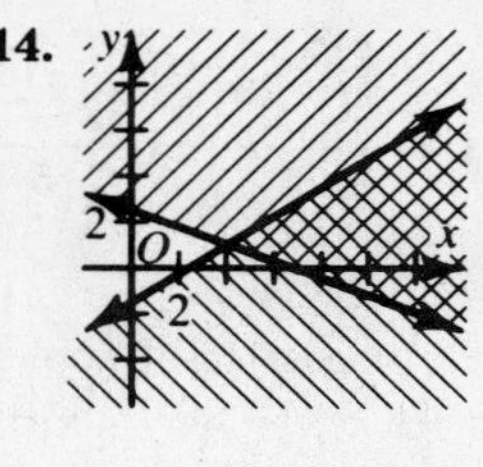

15. 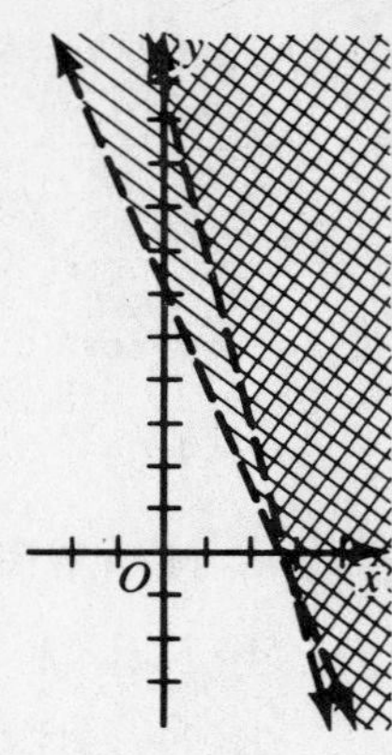

16. 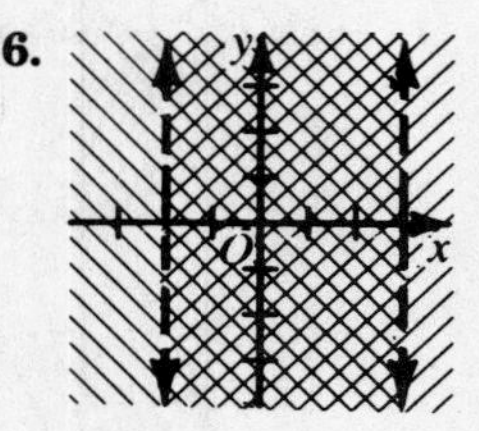

17. 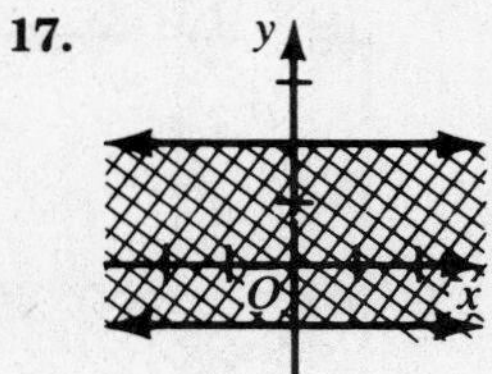

18. 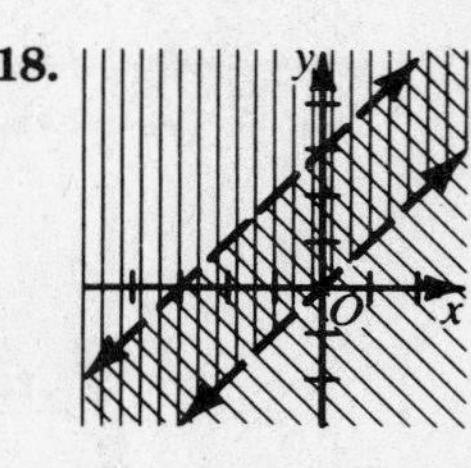

B 19. 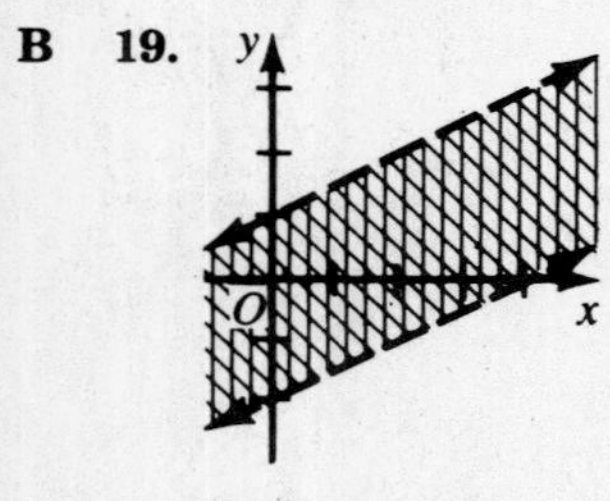

20. 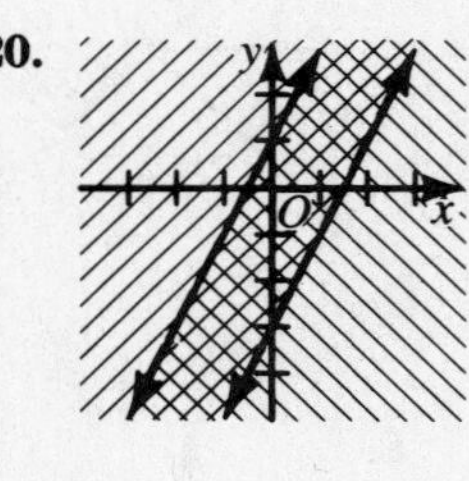

21. 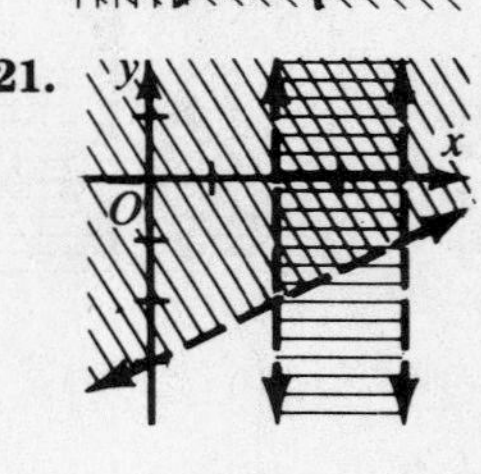

22. 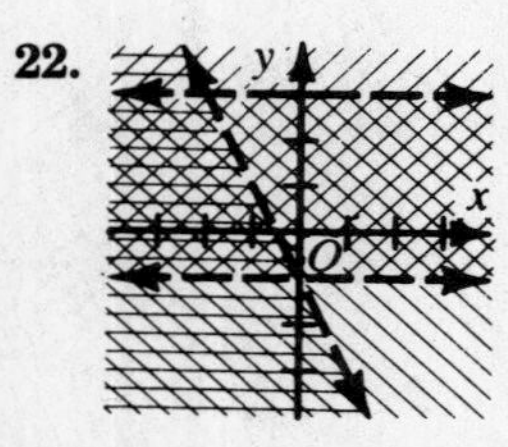

23. 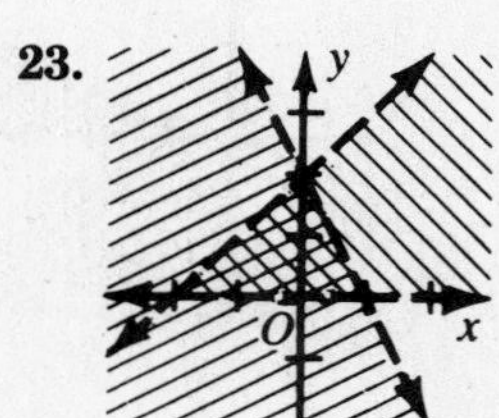

24. 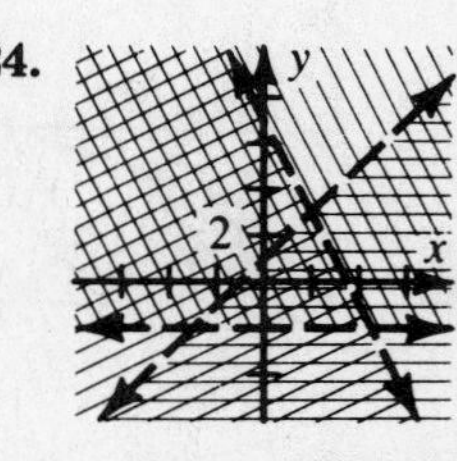

25. 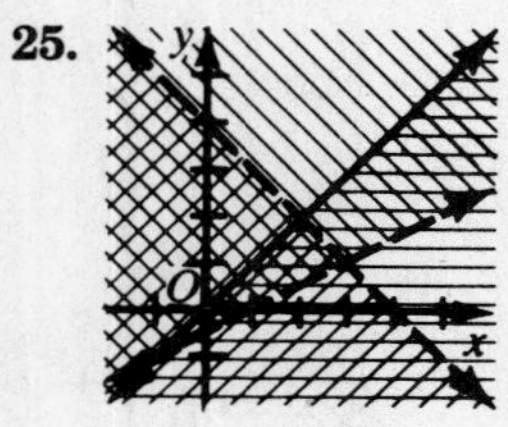

26. 

27. 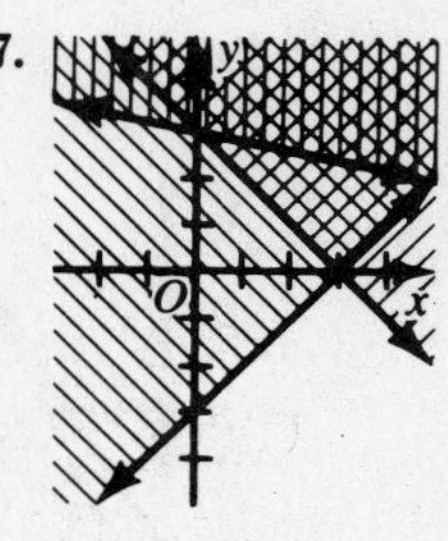

28.

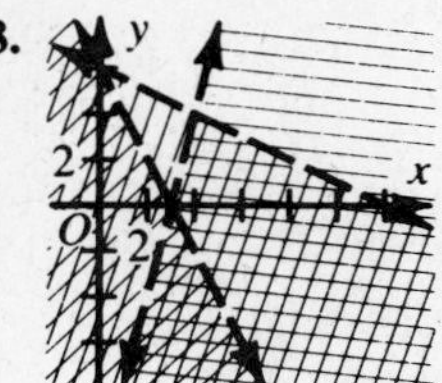

29.

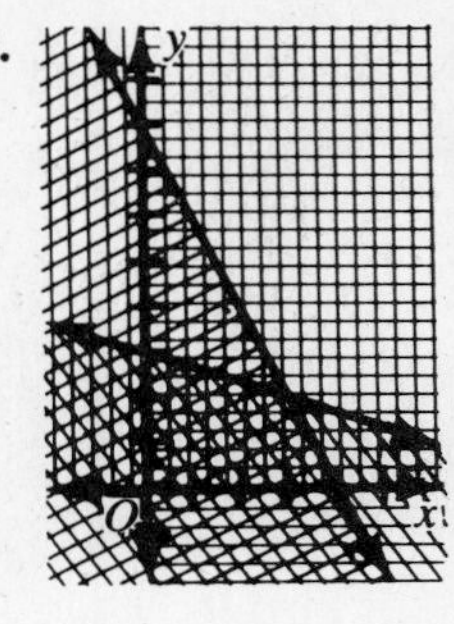

30.

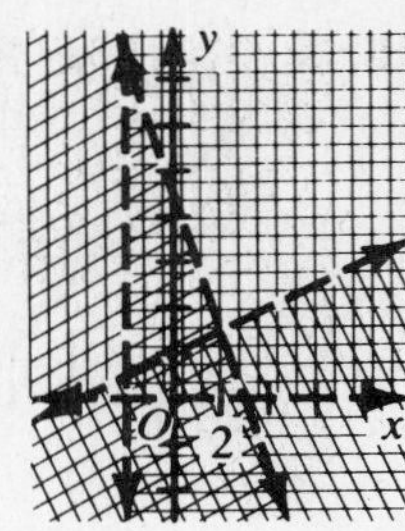

C **31.**

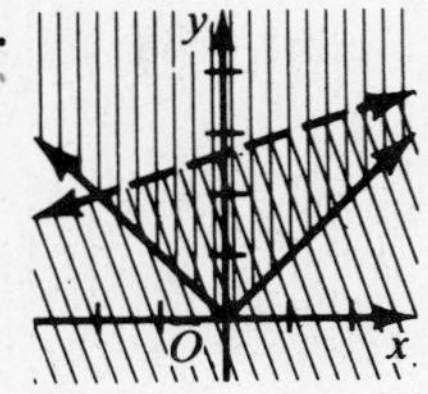

32.

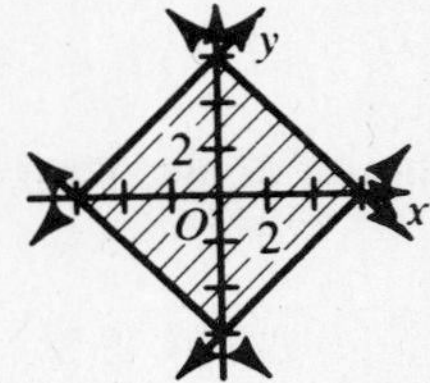

33.

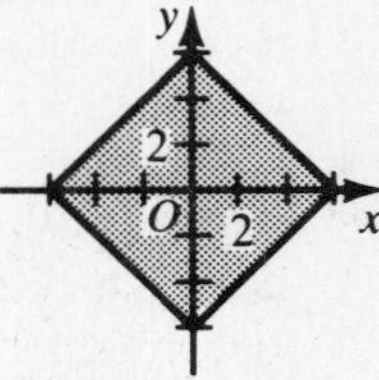

34.

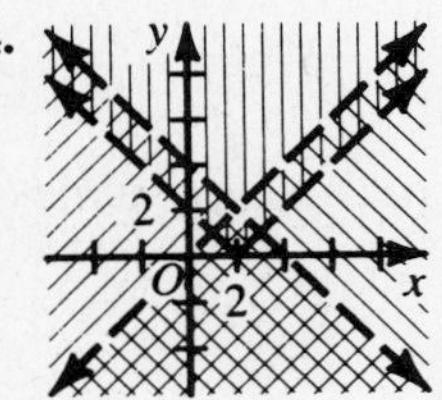

35.

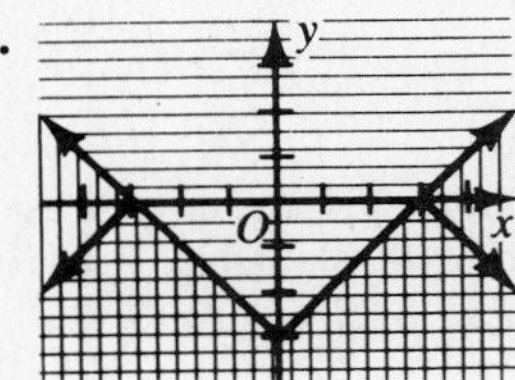

36.

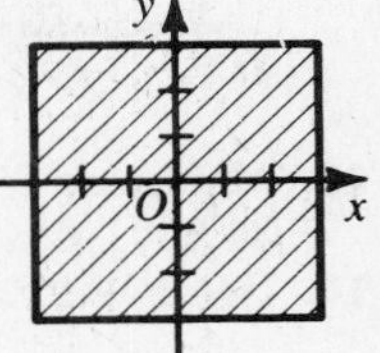

Pages 150–151 • WRITTEN EXERCISES

A **1. a.**

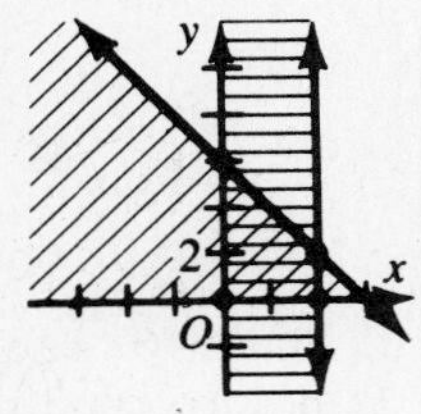

b. (0, 0), (0, 6), (4, 2), (4, 0)
c. 0, 18, 10, 4
d. max.: 18; min.: 0

2. a.

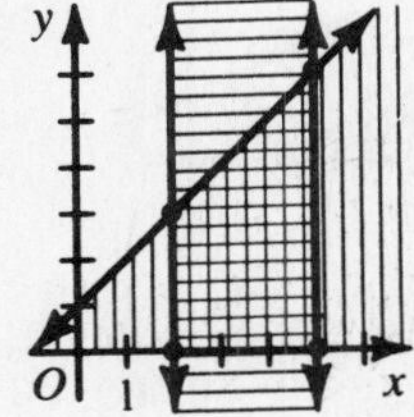
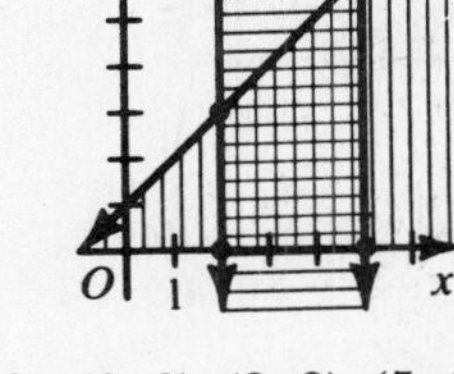

b. (2, 0), (2, 3), (5, 6), (5, 0)
c. 6, 3, 9, 15
d. max.: 15; min.: 3

3. a.

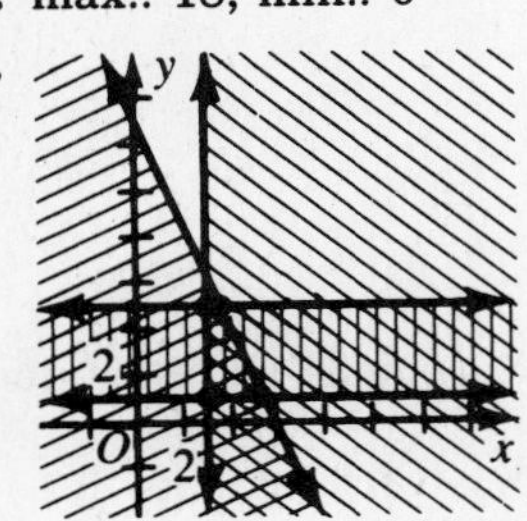

b. (3, 1), (3, 5), (4, 5), (6, 1)
c. −1, 7, 6, −4
d. max.: 7; min.: −4

4. a.

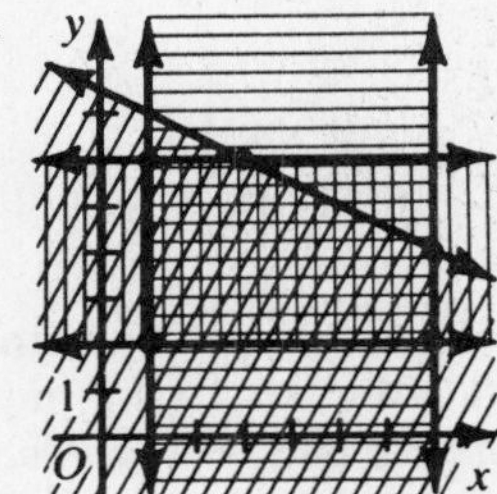

b. (1, 2), (1, 6), (3, 6), (7, 4), (7, 2)
c. 1, −7, 3, 27, 31
d. max.: 31; min.: −7

5. a. 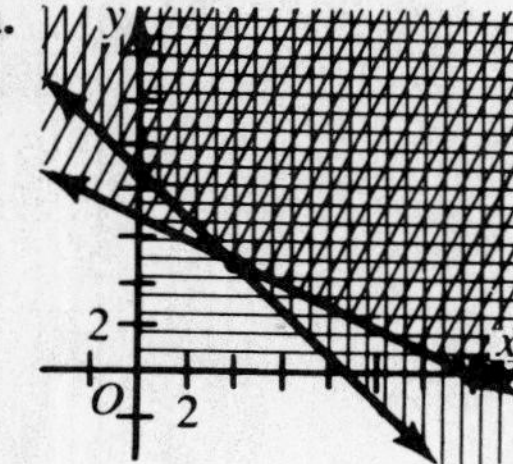

b. (0, 9), (4, 5), (14, 0)
c. 36, 32, 42
d. max.: none; min.: 32

6. a. 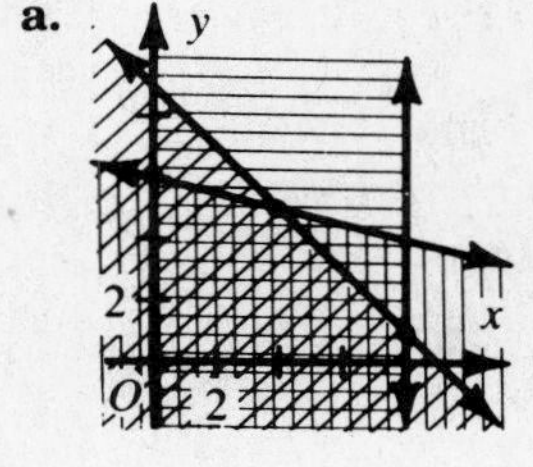

b. (0, 6), (4, 5), (8, 1)
c. 6, 17, 25
d. max.: 25; min.: none

7. $x \geq 0$, $x \leq 12$, $y \geq 0$, $y \leq 8$, $x + 1.5y \leq 18$

8. 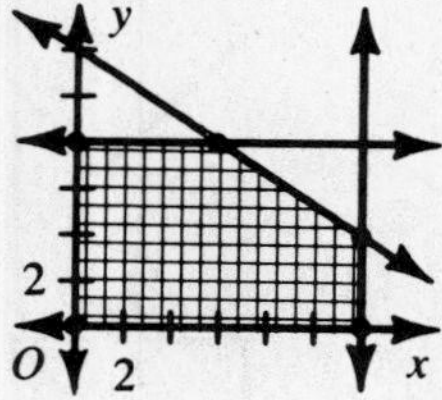

9. (0, 0), (0, 8), (6, 8), (12, 4), (12, 0)

10. Maximize $3000x + 5000y$. Values of expression are: 0; 40,000; 58,000; 56,000; 36,000. Max. value from (6, 8); stove: 6 h, furnace: 8h

11. Use $x + 2y \leq 18$ instead of $x + 1.5y \leq 18$.

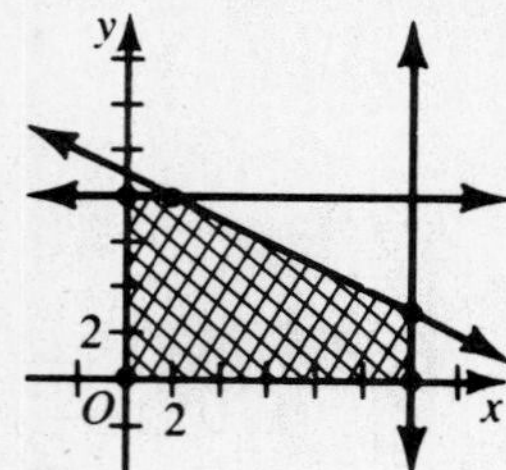

12. Maximize $3000x + 5000y$ using points (0, 0), (0, 8), (2, 8), (12, 3), and (12, 0). Values are: 0; 40,000; 46,000; 51,000; 36,000. Stove: 12 h, furnace: 3 h

13. 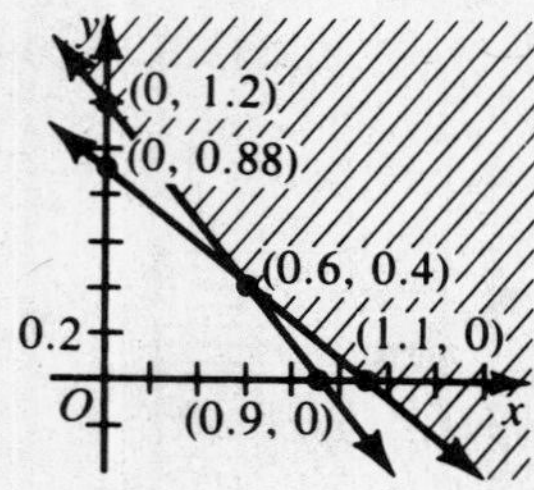

Let x = no. of kg of wheat; y = no. of kg of oats. $80x + 100y \geq 88$, $40x + 30y \geq 36$

B 14. Minimize $30x + 40y$ using points (0, 1.2), (0.6, 0.4), and (1.1, 0). Values are: 48; 34; 33. Wheat: 1.1 kg, no oats

15. Minimize $50x + 40y$. Values are: 48; 46; 55. Wheat: 0.6 kg, oats: 0.4 kg

16. Minimize $70x + 50y$. Values are: 60; 62; 77. No wheat, oats: 1.2 kg

17. Values are: 0; bc; $a + bc$; a. Max.: $a + bc$

C **18.** Values are: $-a + bc$; $-a - bc$; $ac - bc$; $ac + bc$. Min.: $-a - bc$

19. 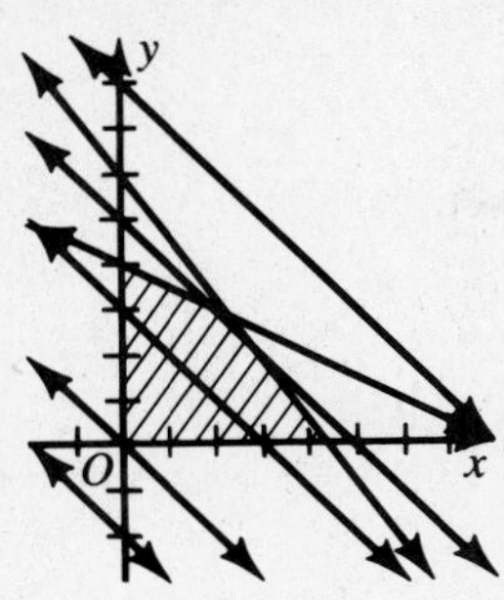

The minimum value of $x + y$ over R is 0 because no line with $p < 0$ will intersect R. The maximum value of $x + y$ over R is 5 because no line with $p > 5$ will intersect R.

Page 151 • COMPUTER EXERCISES

1.
```
10  INPUT A, B
15  LET MN = 999
20  FOR I = 1 TO 5
30  READ X, Y
40  LET G = A * X + B * Y
50  IF G > MX THEN LET MX = G: LET XM = X: LET YM = Y
60  IF G < MN THEN LET MN = G: LET XN = X: LET YN = Y
70  NEXT I
80  PRINT "MAXIMUM VALUE AT (";XM;", ";YM;")"
90  PRINT "MINIMUM VALUE AT (";XN;", ";YN;")"
100  DATA 1, 26, 3, 22, 5, 18, 9, 16, 2, 25
110  END
```

2. max.: (1, 26); min.: (5, 18) **3.** max.: (1, 26); min.: (9, 16)

Page 152 • SELF-TEST 2

1. 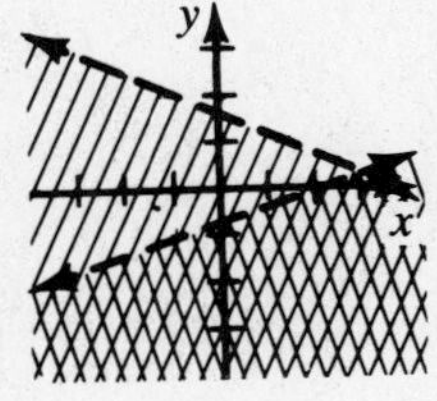

2. 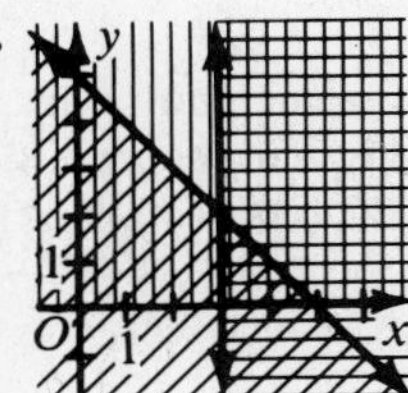

3. 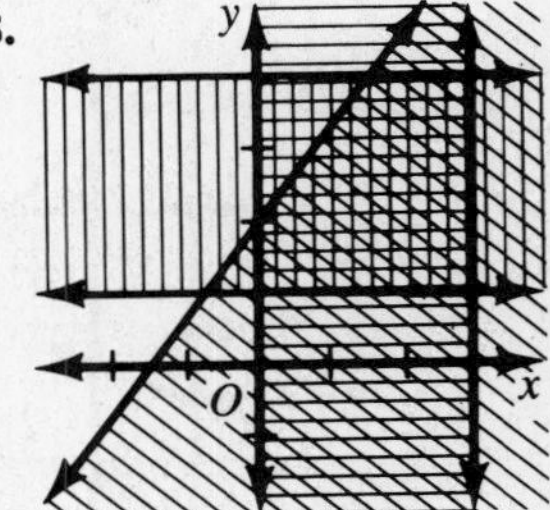

4. Use (2, 1), (2, 2), (5, 5), and (6, 3). Values are: 4; 6; 15; 12. Max.: 15, min.: 4

5. Use (2, 1), (2, 2), (5, 5), and (6, 3). Values are: 7; 8; 20; 21. Max.: 21, min.: 7

6. 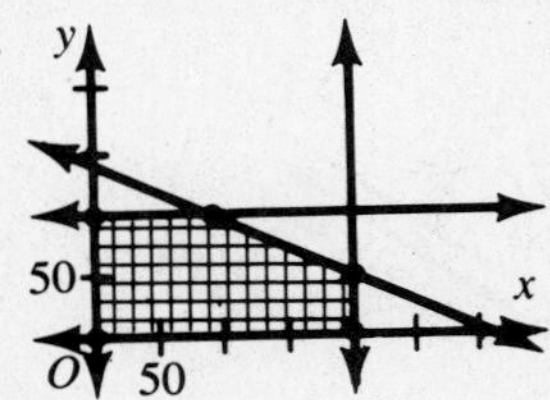

$x \geq 0$, $x \leq 200$, $y \geq 0$, $y \leq 100$, $5x + 11y \leq 1540$
Maximize $3x + 5y$ using points (0, 0), (0, 100), (88, 100), (200, 49), and (200, 0). Values are: 0; 500; 764; 845; 600. Max. profit: $845 from 200 shirts and 49 pairs of slacks

Pages 153–154 • CHAPTER REVIEW

1. slopes: $\frac{3}{4}$, $\frac{3}{4}$; y-intercepts: $-\frac{1}{2}$, 0; no solution; c
2. slopes: $\frac{2}{5}$, $\frac{2}{5}$; y-intercepts: $\frac{1}{5}$, $\frac{1}{5}$; infinite set of solutions; b
3. $\begin{array}{llll} 2x + 4y = 6 & 2x + 4y = 6 & 8x = 24 & y = 0; \\ 3x - 2y = 9 & 6x - 4y = 18 & x = 3; & \{(3, 0)\}; \text{ b} \end{array}$
4. $\begin{array}{llll} 5x + 4y = -1 & y = -2x + 2 & -3x + 8 = -1 & y = -4; \\ 2x + y = 2 & 5x + 4(-2x + 2) = -1 & x = 3; & \{(3, -4)\}; \text{ c} \end{array}$
5. $-6 + 5 = -1$; c
6. $D = \begin{vmatrix} 3 & 1 \\ -5 & -2 \end{vmatrix} = -6 + 5 = -1$; $x = \dfrac{\begin{vmatrix} -5 & 1 \\ 8 & -2 \end{vmatrix}}{-1} = \dfrac{10 - 8}{-1} = -2$;

   $y = \dfrac{\begin{vmatrix} 3 & -5 \\ -5 & 8 \end{vmatrix}}{-1} = \dfrac{24 - 25}{-1} = 1$; $\{(-2, 1)\}$; c
7. $rt = d$; $3(s - c) = 24$; c
8. $\begin{array}{lll} 3(s - c) = 24 & s - c = 8 & s = 11.5 \text{ km/h}; \\ 2(s + c) = 30 & s + c = 15 & c = 3.5 \text{ km/h}; \text{ a} \end{array}$

9.–10. **9.** c **10.** d

11.–12.

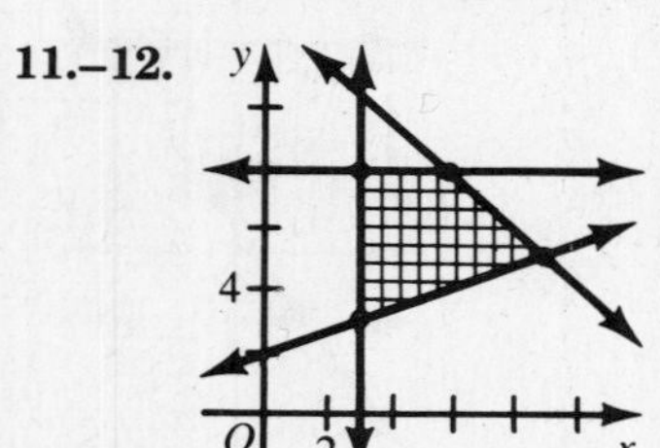

Corner points are (3, 3), (3, 8), (6, 8), and (9, 5).

11. Values are: 3; −2; 4; 13. Min.: −2; b

12. Values are: 21; 36; 48; 51. Max.: 51; c

Pages 154–155 • CHAPTER TEST

1. $\{(-2, 1)\}$

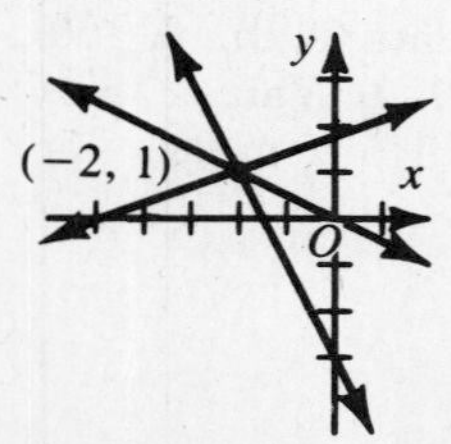

2. $\emptyset$

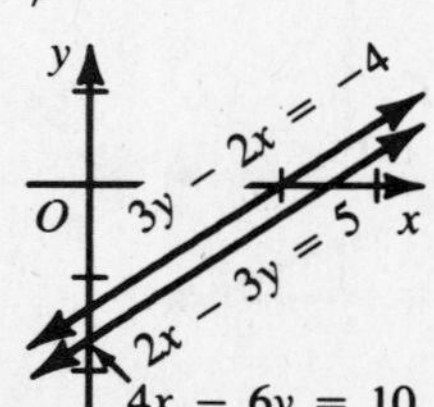

3. $2x - 2y = -6$ $\quad$ $2x - 2y = -6$ $\quad$ $-4x = -16$ $\quad$ $y = 7;$
$-3x + y = -5$ $\quad$ $-6x + 2y = -10$ $\quad$ $x = 4;$ $\quad$ $\{(4, 7)\}$

4. $3(x - 1) = 2(y + 5)$ $\quad$ $3x - 2y = 13$ $\quad$ $-6y = 12$ $\quad$ $x = 3;$
$3x + 4y = 1$ $\quad$ $-3x - 4y = -1$ $\quad$ $y = -2;$ $\quad$ $\{(3, -2)\}$

5. $D = \begin{vmatrix} 4 & -1 \\ -8 & 3 \end{vmatrix} = 12 - 8 = 4;\ x = \dfrac{\begin{vmatrix} 1 & -1 \\ 0 & 3 \end{vmatrix}}{4} = \dfrac{3 - 0}{4} = \dfrac{3}{4};\ y = \dfrac{\begin{vmatrix} 4 & 1 \\ -8 & 0 \end{vmatrix}}{4} =$
$\dfrac{0 + 8}{4} = 2;\ \left\{\left(\dfrac{3}{4}, 2\right)\right\}$

6. $D = \begin{vmatrix} \frac{8}{3} & 6 \\ 4 & 9 \end{vmatrix} = 24 - 24 = 0$; each line has y-intercept $\frac{1}{6}$; system has infinite solution set.

7. $D = \begin{vmatrix} \frac{1}{5} & \frac{2}{3} \\ 1 & 3 \end{vmatrix} = \dfrac{3}{5} - \dfrac{2}{3} = -\dfrac{1}{15};\ x = \dfrac{\begin{vmatrix} 1 & \frac{2}{3} \\ 4 & 3 \end{vmatrix}}{-\frac{1}{15}} = \dfrac{3 - \frac{8}{3}}{-\frac{1}{15}} = -5;\ y = \dfrac{\begin{vmatrix} \frac{1}{5} & 1 \\ 1 & 4 \end{vmatrix}}{-\frac{1}{15}} =$
$\dfrac{\frac{4}{5} - 1}{-\frac{1}{15}} = 3;\ \{(-5, 3)\}$

8. $D = \begin{vmatrix} -\frac{3}{4} & \frac{1}{6} \\ 9 & -2 \end{vmatrix} = \dfrac{3}{2} - \dfrac{3}{2} = 0$; y-intercepts are -2 and -3; ∴ lines are parallel; system is inconsistent.

9. Let x = air speed of plane in km/h; y = speed of wind in km/h.

	r (km/h)	t (h)	d (km)
with a head wind	$x - y$	6	$6(x - y)$
with a tail wind	$x + y$	5	$5(x + y)$

$6(x - y) = 3000$ $\quad$ $x - y = 500$ $\quad$ $x = 550;$ $\quad$ air speed of plane: 550 km/h
$5(x + y) = 3000$ $\quad$ $x + y = 600$ $\quad$ $y = 50;$ $\quad$ speed of wind: 50 km/h

10. Let x = number of hours Ron works; y = number of hours Jim works.
$x + y = 44$ $\quad$ $x = 44 - y$ $\quad$ $198 + 0.75y = 205.5$
$4.5x + 5.25y = 205.5$ $\quad$ $4.5(44 - y) + 5.25y = 205.5$ $\quad$ $y = 10;$
Ron: 34 h, Jim: 10 h $\quad$ $x = 34$

11.

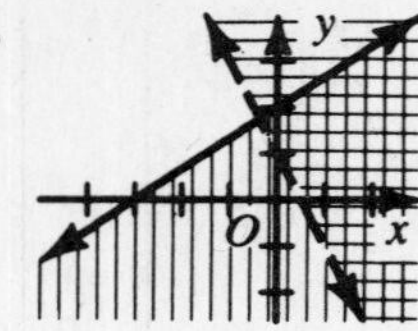

12.

13.

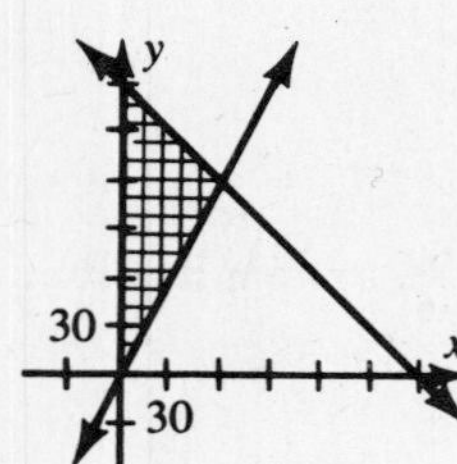

Let x = number of peaches; y = number of apples.
$x \geq 0$, $y \geq 0$, $x + y \leq 180$, $y \geq 2x$
Maximize $8x + 10y$ using points (0, 0), (0, 180), and (60, 120). Values are 0, 1800, and 1680; max.: \$18.00 from 0 peaches and 180 apples

Pages 155–157 • CUMULATIVE REVIEW

1. 120 **2.** $-3x + 15$ **3.** 5 **4.** $6x - 18xy$ **5.** $14z + 2 + 15z + 6 = 29z + 8$
6. $8xy + 2x + 30y$ **7.** $\{0\}$ **8.** $\{-2, 2\}$ **9.** no solution **10.** $\{2\}$
11. $\{-4, -2, 0, 2, 4\}$ **12.** $\{0\}$

13.

1. $\frac{a}{c} = \frac{b}{c}$; $c \neq 0$	Hypothesis
2. $a\left(\frac{1}{c}\right) = b\left(\frac{1}{c}\right)$	Def. of div.
3. $a = b$	Canc. prop. of mult.

14.

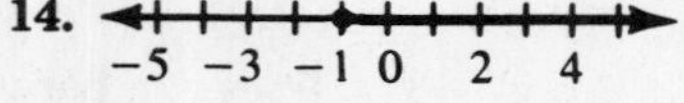

15.

16. $\{2\}$ **17.** $\{-4\}$ **18.** $\{-32\}$ **19.** $\{12\}$ **20.** $\left\{\frac{6}{5}\right\}$

21. $3 - [8x - 2(3 - 4x)] + 16x = 3x$; $3 - 8x + 6 - 8x + 16x = 3x$; $\{3\}$

22. $x = \frac{1}{4}y(a + b)$; $\frac{4x}{y} = a + b$; $b = \frac{4x - ay}{y}$ $(y \neq 0)$

23. $ab - cb = a - c$; $b(a - c) = a - c$; $b = 1$ $(a \neq c)$

24. $\{m: m < -5\}$

−5 0

25. $\{x: 2 < x < 8\}$

0 2 5 8 10

26. $2 - 4r < 10$ and $2 - 4r > -10$; $\{r: -2 < r < 3\}$

27. $t + 4 \geq 2$ or $t + 4 \leq -2$
$\{t: t \geq -2\} \cup \{t: t \leq -6\}$

28. Let x = length of shorter base in cm; $x + 4$ = length of longer base in cm.
$\frac{1}{2}(5)(x + x + 4) = 40$; $5x + 10 = 40$; $x = 6$; 6 cm, 10 cm

29. $\frac{1}{5}(83 + 76 + 92 + 85 + x) \geq 85$; $336 + x \geq 425$; $x \geq 89$; at least 89

30. -14 **31.** $\frac{16}{9}$ **32.** $g(-2) = 4$; $f(4) = 13$ **33.** $f(-2) = -5$; $g(-5) = 25$

34. $y = x - 4$; a function **35.** $x = 4$; not a function

36.

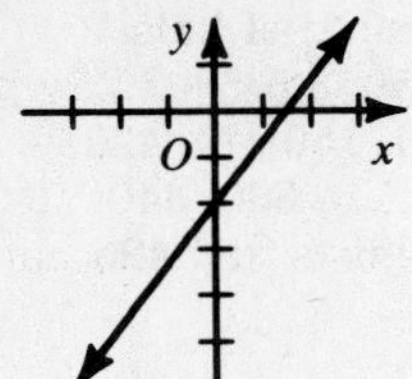

37.

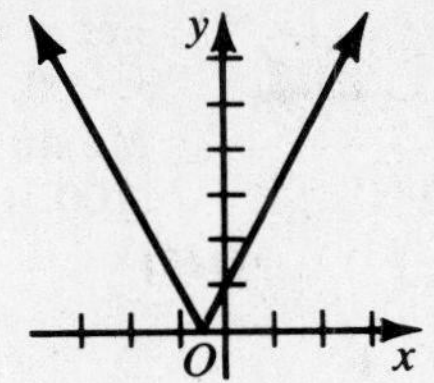

38.

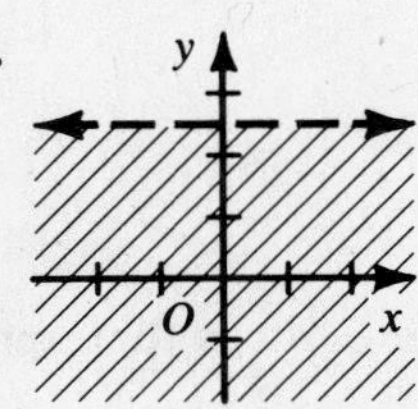

39.

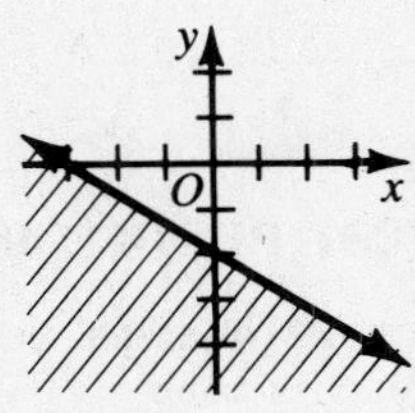

40. $-4(-2) + k(-1) = 6;\ k = 2$ **41.** $5k + 5(-3) = k - 3;\ k = 3$

42. $m = \dfrac{11 - 1}{3 + 2} = 2;\ y - 1 = 2(x + 2);\ y = 2x + 5$

43. $m = -3;\ y + 2 = -3(x - 1);\ y = -3x + 1$

44. Let x = number of gallons of paint needed to cover 4000 ft^2; $\dfrac{600}{1} = \dfrac{4000}{x}$;

$600x = 4000;\ x = 6\dfrac{2}{3};\ 6\dfrac{2}{3}$ gal

45. slopes: $-\dfrac{2}{5}, -\dfrac{3}{2}$; one solution; $2x + 5\left(-\dfrac{3}{2}x - 2\right) = 1;\ -\dfrac{11}{2}x = 11;\ x = -2;$
$y = 1;\ \{(-2, 1)\}$

46. slopes: 5, 5; y-intercepts: -4, 4; no solution

47. slopes: $\dfrac{5}{3}, \dfrac{5}{3}$; y-intercepts: 4, $\dfrac{4}{3}$; no solution

48. slopes: $-\dfrac{9}{4}, \dfrac{6}{5}$; one solution

$$\begin{array}{llll} 9x + 4y = 11 & 45x + 20y = 55 & 69x = 63 & y = \dfrac{16}{23}; \\ -12x + 10y = -4 & 24x - 20y = 8 & x = \dfrac{21}{23}; & \left\{\left(\dfrac{21}{23}, \dfrac{16}{23}\right)\right\} \end{array}$$

49.

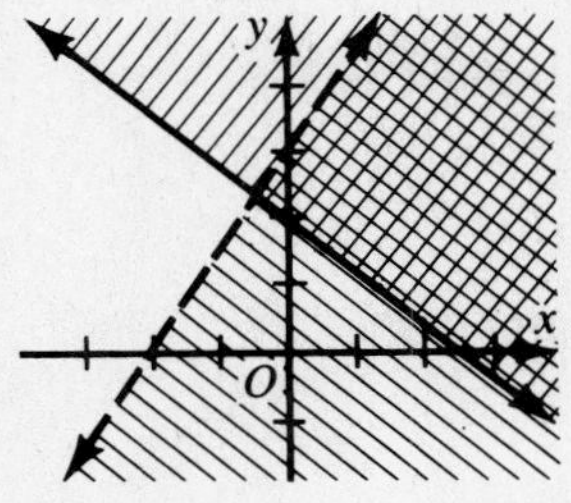

50.

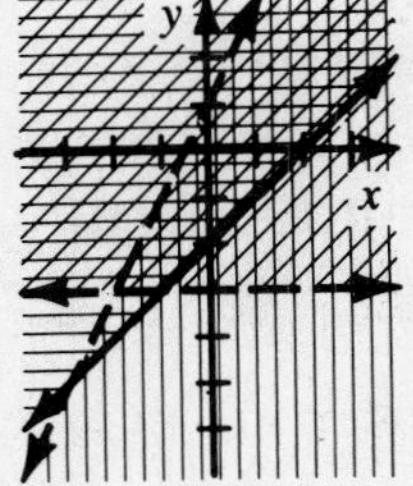

51. Let a = cost of adult ticket; s = cost of student ticket.

$$\begin{array}{lll} a + s = 500 & -a - s = -500 & a = 300; \\ 4.5a + 1.5s = 1650 & 3a + s = 1100 & s = 200; \end{array}$$

300 adult tickets
200 student tickets

52. 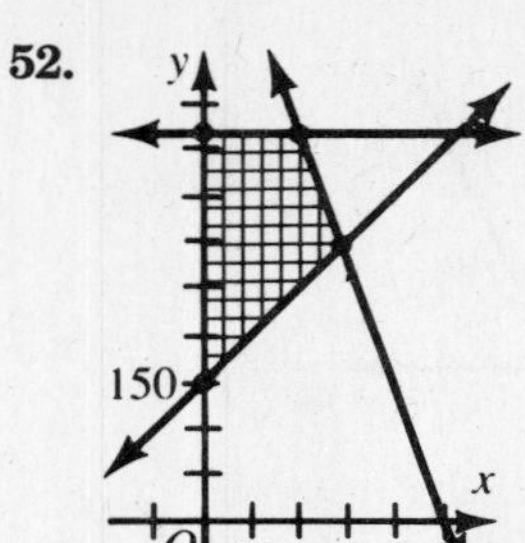

Let x = number of shirts; y = number of hats.
$x \geq 0$, $y \geq 0$, $y \leq 420$, $y \geq x + 150$, $3.5x + 1.25y \leq 875$
Maximize $5x + 2y$ using points (0, 150), (0, 420), (100, 420), and (144, 294). Values are 300, 840, 1340, and 1308. Max.: $1340 from 100 shirts and 420 hats

Page 158 • CONTEST PROBLEMS

1. Let n = number of steps visible at one time; r = number of steps that appear or disappear each second.

$n = 10 + 20r$	$n = 10 + 20r$	$0 = -40 + 50r$	$n = 26$;
$n = 50 - 30r$	$-n = -50 + 30r$	$r = \frac{4}{5}$;	26 steps

2. Let c = cost of the apples ($c \in W$); x = number of apples. Since 2 apples cost $0 and 3 apples cost at least $1:
(1) if x is a multiple of 3, then $x = 3c$.
(2) if x is one more than a multiple of 3, then $x = 3c + 1$.
(3) if x is two more than a multiple of 3, then $x = 3c + 2$.
Solving for the cost c in each case gives:
(1) $c = \frac{1}{3}x$ (2) $c = \frac{1}{3}x - \frac{1}{3}$ (3) $c = \frac{1}{3}x - \frac{2}{3}$
For rule 4 consider the six possible combinations of number types for the values of m and n.
1. Both m and n are multiples of 3.
2. Both m and n are one more than a multiple of 3.
3. Both m and n are two more than a multiple of 3.
4. m is a multiple of 3 and n is one more than a multiple of 3.
5. m is a multiple of 3 and n is two more than a multiple of 3.
6. m is one more than a multiple of 3 and n is two more than a multiple of 3.

For rule 5, since 9999 is a multiple of 3, the cost is $c = \frac{1}{3}x = \frac{1}{3}(9999) = \$3333$.
Since 1982 is two more than a multiple of 3 (i.e., $1982 = 3 \cdot 660 + 2$), then the cost of 1982 apples is $c = \frac{1}{3}x - \frac{2}{3} = \frac{1}{3}(1982) - \frac{2}{3} = 660$; $660

Pages 158–159 • PROGRAMMING IN PASCAL

1.
```
FUNCTION det(s, t, u, v : integer) : integer;

BEGIN
  det := s * v - t * u;
END;
```
2.
```
PROCEDURE get_values (VAR a, b, c : integer);

BEGIN
  writeln('Enter the coefficients and the');
  writeln('constant of the equation.');
```

```
    write('Place a space between each of the values: ');
    readln(a, b, c);
    writeln;
  END;
```

3.
```
PROGRAM cramer2 (INPUT, OUTPUT);

VAR
  a1, b1, c1, a2, b2, c2 : integer;
  x, y : real;
  d, dx, dy : integer;

( ************************************************************ )
FUNCTION det(s, t, u, v : integer) : integer;

BEGIN
  det := s * v - t * u;
END;

( ************************************************************ )
PROCEDURE get_values (VAR a, b, c : integer);

BEGIN
  writeln('Enter the coefficients and the');
  writeln('constant of the equation.');
  write('Place a space between each of the values: ');
  readln(a, b, c);
  writeln;
END;

( ************************************************************ )
BEGIN (*main*)
  get_values(a1, b1, c1);
  get_values(a2, b2, c2);
  dx := det(c1, b1, c2, b2);
  dy := det(a1, c1, a2, c2);
  d := det(a1, b1, a2, b2);
  IF d <> 0
    THEN BEGIN
           x := dx/d;
           y := dy/d;
           writeln('Solution: (', x:3:3, ' , ', y:3:3, ')');
         END
    ELSE writeln('No solution or infinitely many solutions.');
END.
```

4. A REPEAT UNTIL loop must be inserted into the program. See the program fragment in Exercise 5.

5. In the variable declaration section of the program, include:

factor : integer;
answer : char;
solutions : boolean;

Then adjust the main program as shown.

```
BEGIN    (*main*)
  REPEAT
    get_values(a1, b1, c1);
    get_values(a2, b2, c2);
    dx := det(c1, b1, c2, b2);
    dy := det(a1, c1, a2, c2);
    d := det(a1, b1, a2, b2);
    IF d <> 0
      THEN BEGIN
             x := dx/d;
             y := dy/d;
             writeln('Solution: (', x:3:3, ' , ', y:3:3, ')');
           END
      ELSE BEGIN
             IF a1 >= a2
               THEN BEGIN
                      factor := a1 DIV a2;
                      IF factor = c1 DIV c2
                        THEN solutions := TRUE
                        ELSE solutions := FALSE;
                    END
               ELSE BEGIN
                      IF a1 < a2
                        THEN BEGIN
                               factor := a2 DIV a1;
                               IF factor = c2 DIV c1
                                 THEN solutions := TRUE
                                 ELSE solutions := FALSE;
                             END;
                    END;
             IF solutions = TRUE
               THEN writeln('Infinitely many solutions.')
               ELSE writeln('No solutions.');
           END;
    writeln;
    write('Do you want to solve another system <Y/N>: ');
    readln(answer);
    writeln;
  UNTIL answer IN ['N', 'n'];
END.
```

Page 164 • WRITTEN EXERCISES

A 1.

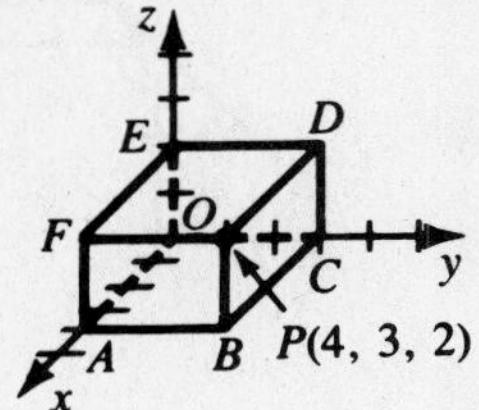

2.

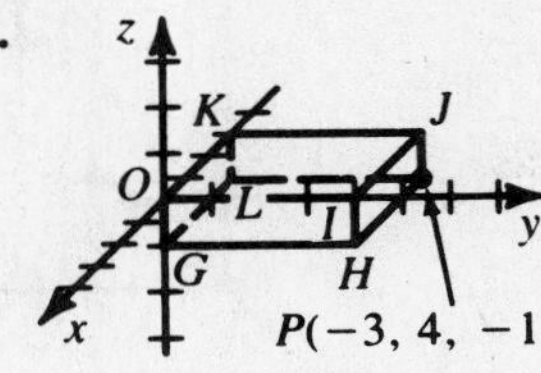

3.

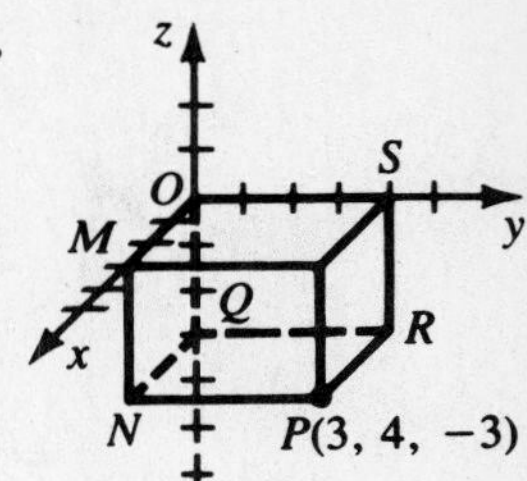

4.

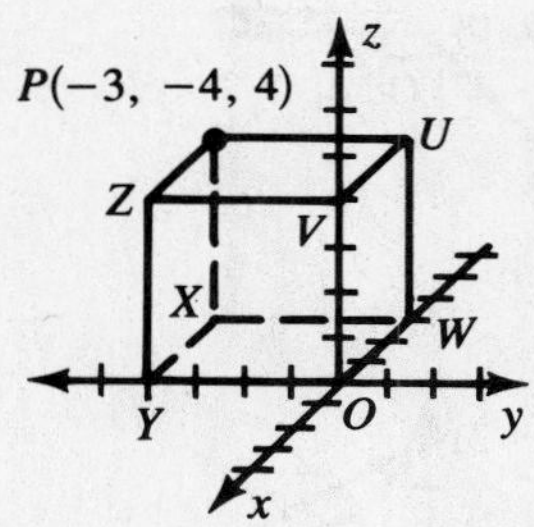

5.

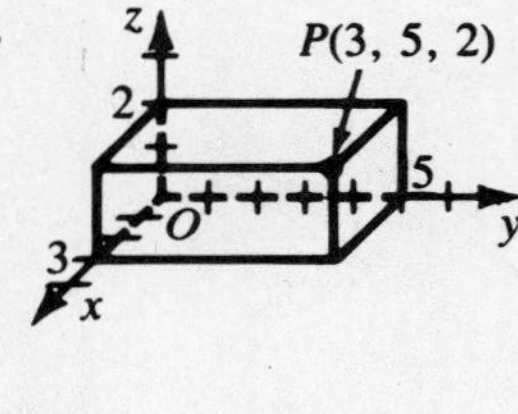

6.

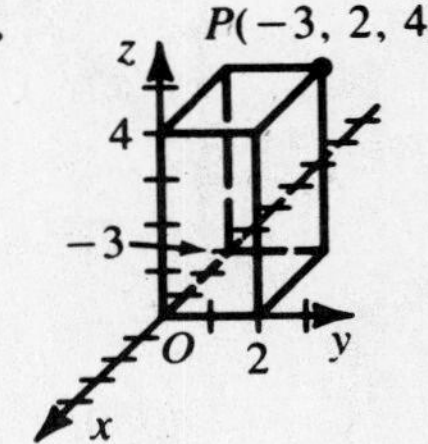

7.

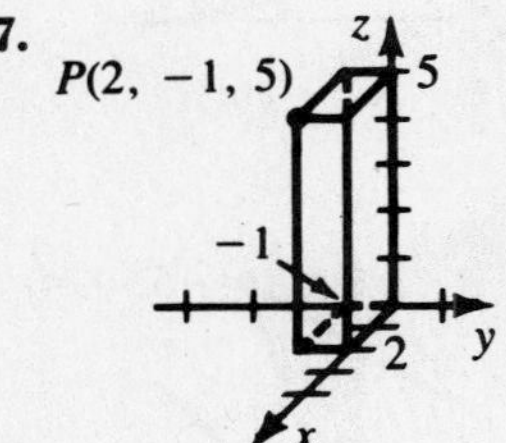

8.

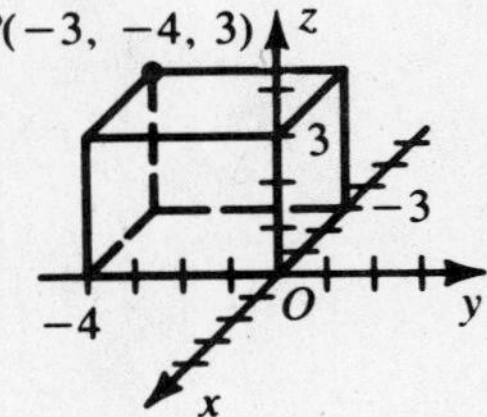

9.

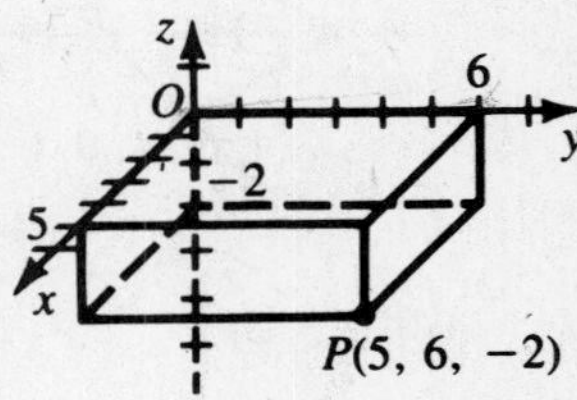

10.

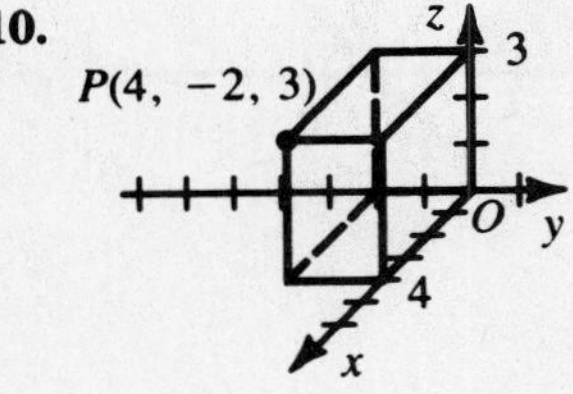

11.

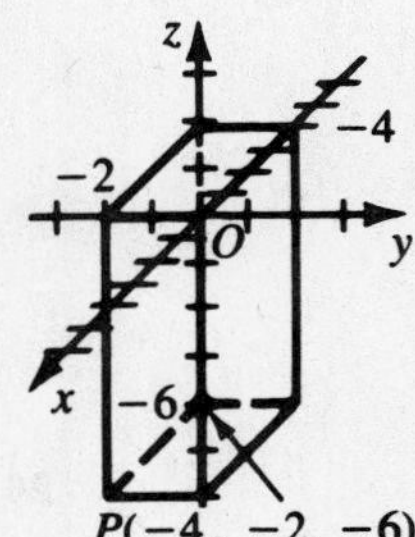

12.

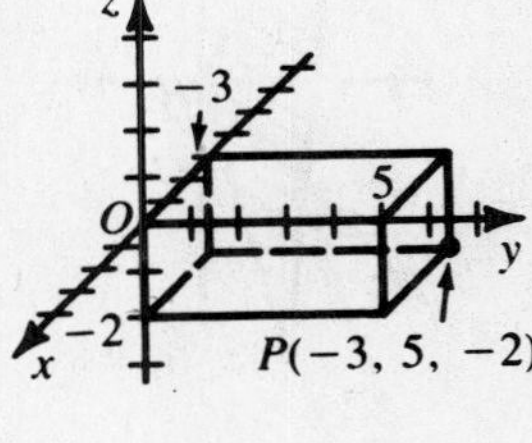

13.

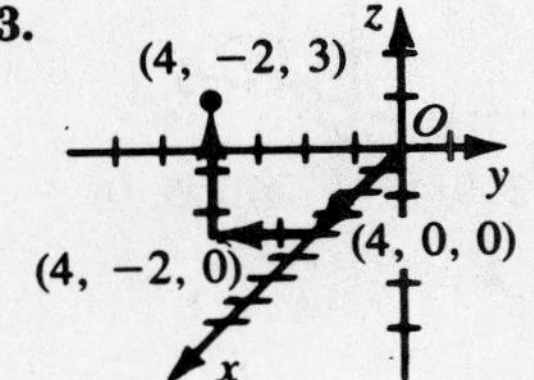

14.

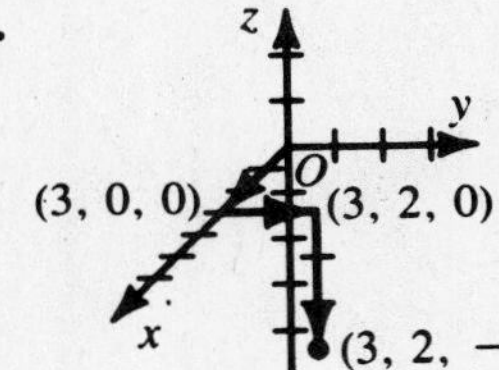

15.

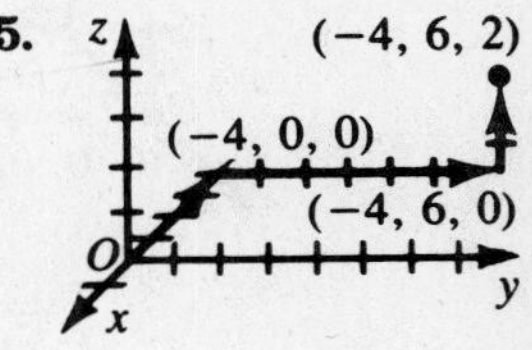

16.

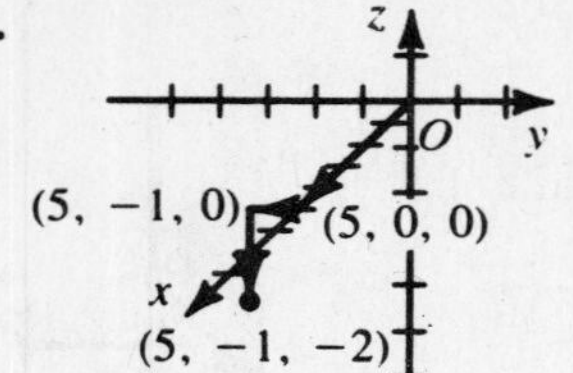

17.

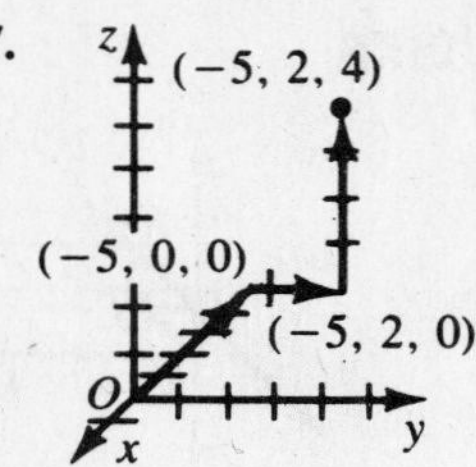

18.

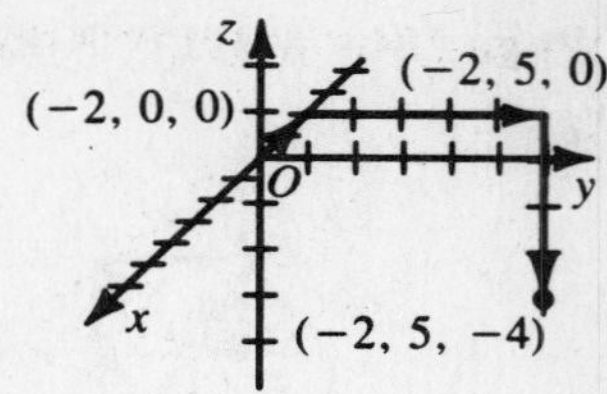

19.

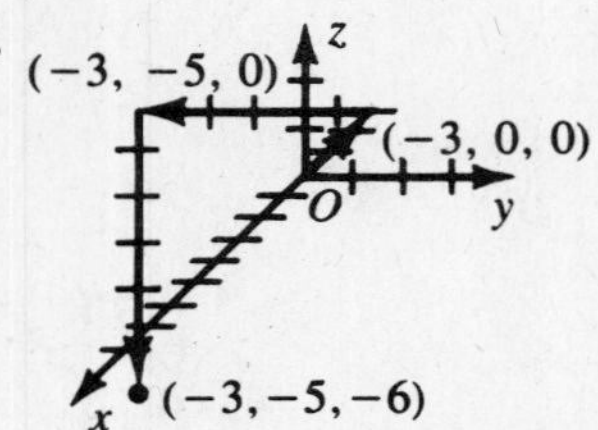

20.

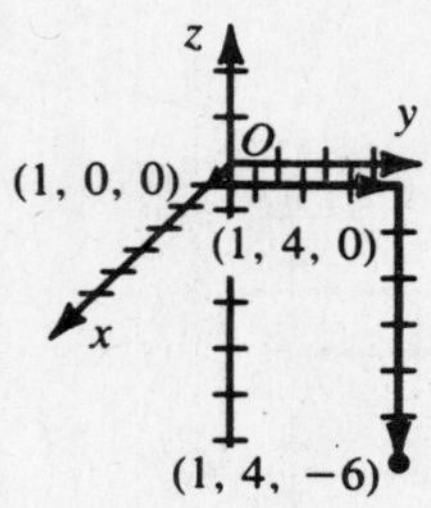

B 21.

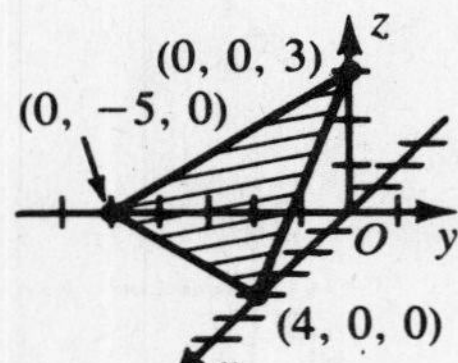

22.

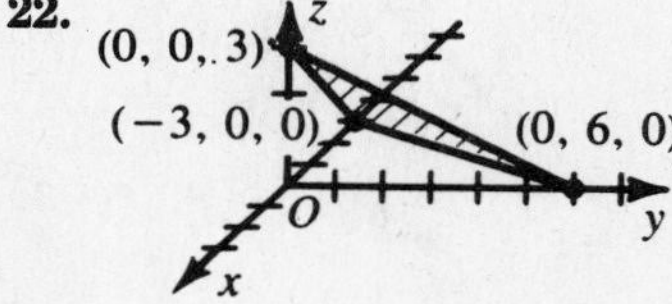

23.

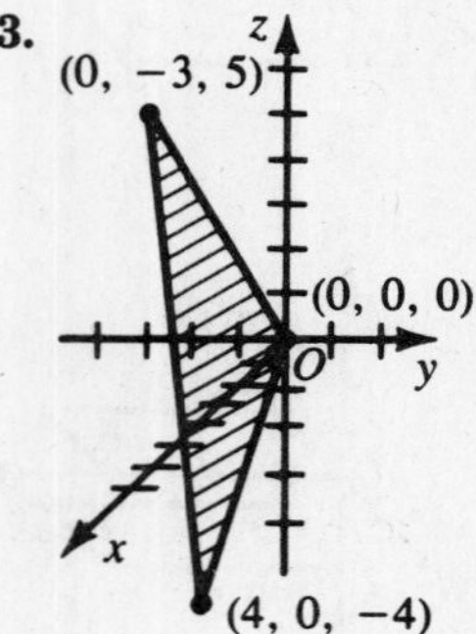

24.

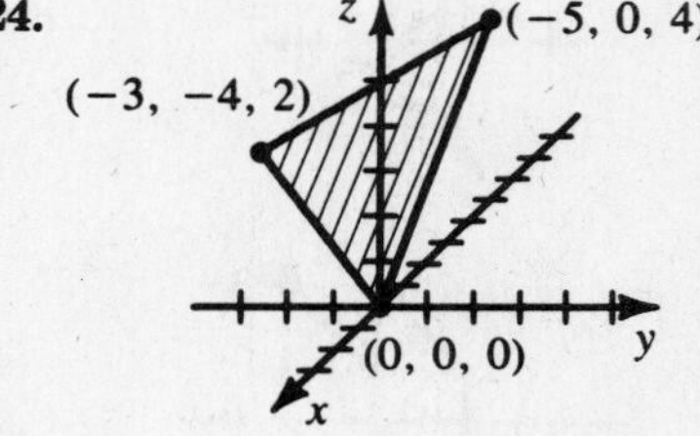

C 25.

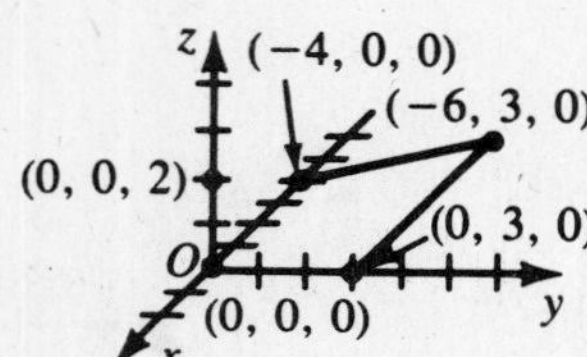

26.

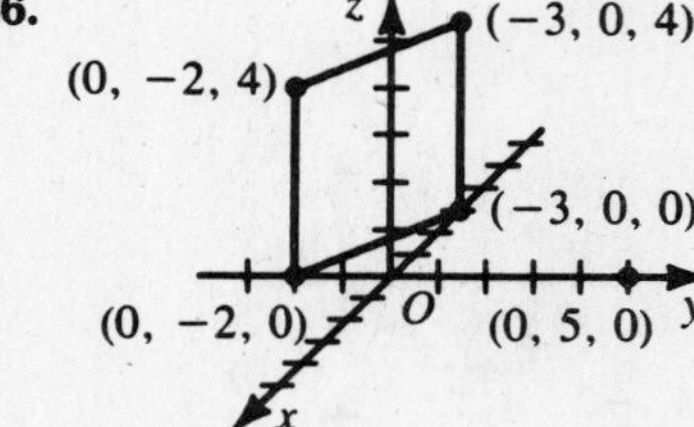

Page 170 • WRITTEN EXERCISES

A **1. a.**

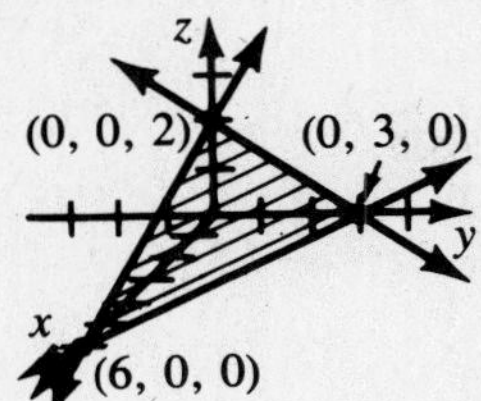

b. $\begin{cases} z = 0 \\ x + 2y = 6 \end{cases}$

$\begin{cases} y = 0 \\ x + 3z = 6 \end{cases}$

$\begin{cases} x = 0 \\ 2y + 3z = 6 \end{cases}$

2. a.

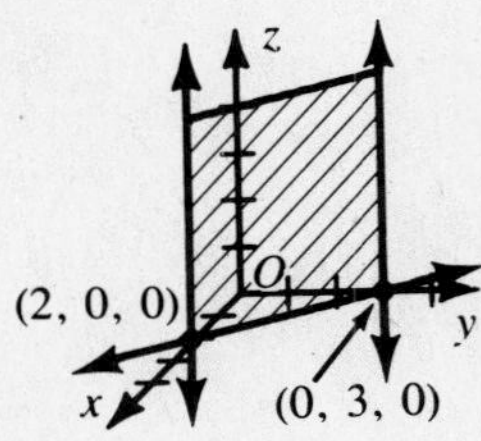

b. $\begin{cases} z = 0 \\ 3x + 2y = 6 \end{cases}$

$\begin{cases} x = 0 \\ y = 3 \end{cases}$

$\begin{cases} y = 0 \\ x = 2 \end{cases}$

The graph is parallel to the z-axis.

3. a.

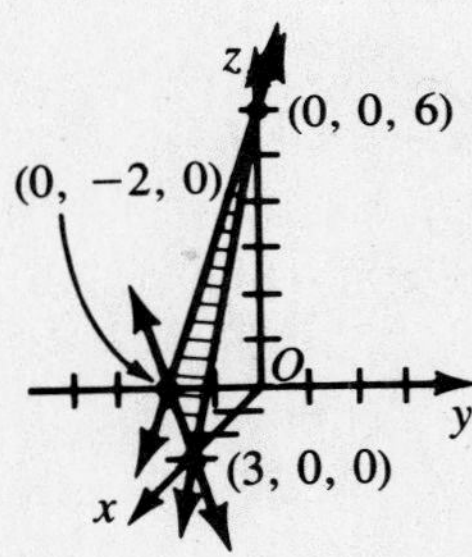

b. $\begin{cases} z = 0 \\ 2x - 3y = 6 \end{cases}$

$\begin{cases} y = 0 \\ 2x + z = 6 \end{cases}$

$\begin{cases} x = 0 \\ -3y + z = 6 \end{cases}$

4. a.

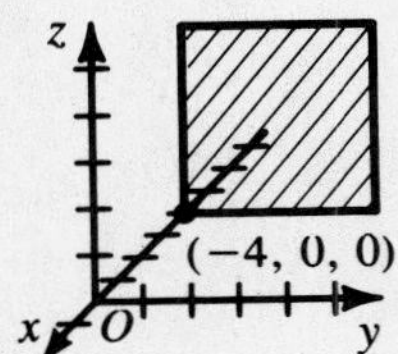

b. $\begin{cases} z = 0 \\ x = -4 \end{cases}$

$\begin{cases} y = 0 \\ x = -4 \end{cases}$

There is no trace in the yz-plane.

5. a.

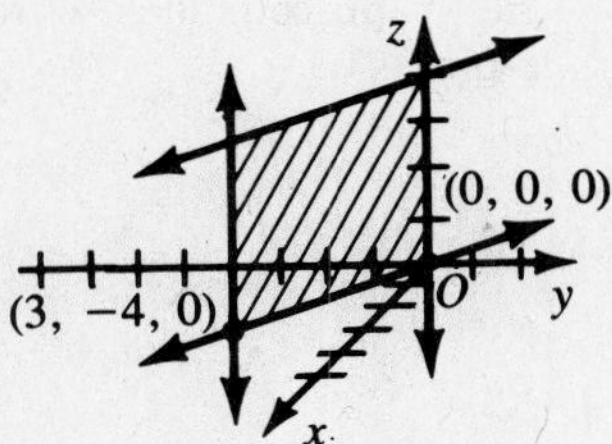

b. $\begin{cases} z = 0 \\ 4x + 3y = 0 \end{cases}$

$\begin{cases} y = 0 \\ x = 0 \end{cases}$

The graph contains the z-axis.

6. a.

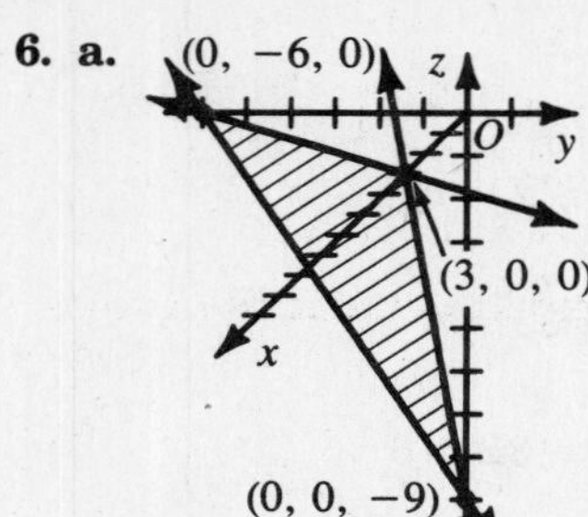

b. $\begin{cases} z = 0 \\ 6x - 3y = 18 \end{cases}$

$\begin{cases} y = 0 \\ 6x - 2z = 18 \end{cases}$

$\begin{cases} x = 0 \\ 3y + 2z = -18 \end{cases}$

7. a.

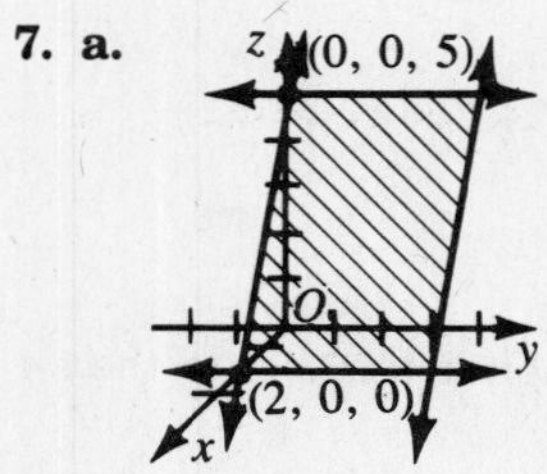

b. $\begin{cases} z = 0 \\ x = 2 \end{cases}$

$\begin{cases} y = 0 \\ 5x + 2z = 10 \end{cases}$

$\begin{cases} x = 0 \\ z = 5 \end{cases}$

8. a.

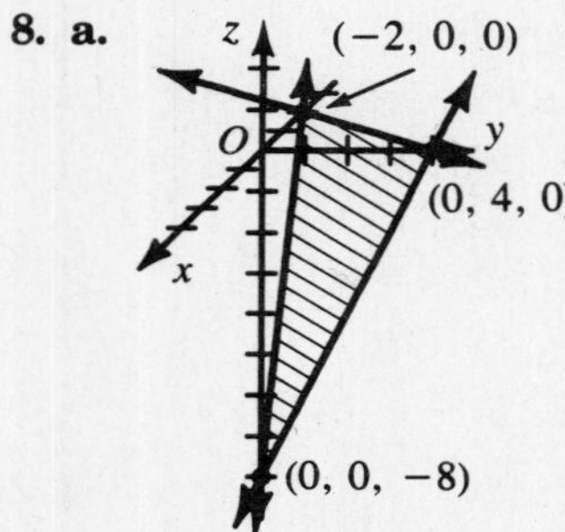

b. $\begin{cases} z = 0 \\ -4x + 2y = 8 \end{cases}$

$\begin{cases} y = 0 \\ 4x + z = -8 \end{cases}$

$\begin{cases} x = 0 \\ 2y - z = 8 \end{cases}$

9. a.

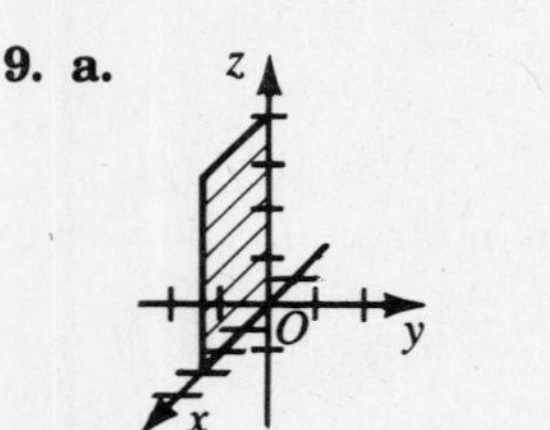

b. $\begin{cases} z = 0 \\ y = 0 \end{cases}$

$\begin{cases} x = 0 \\ y = 0 \end{cases}$

The graph coincides with the *xz*-plane.

10. a.

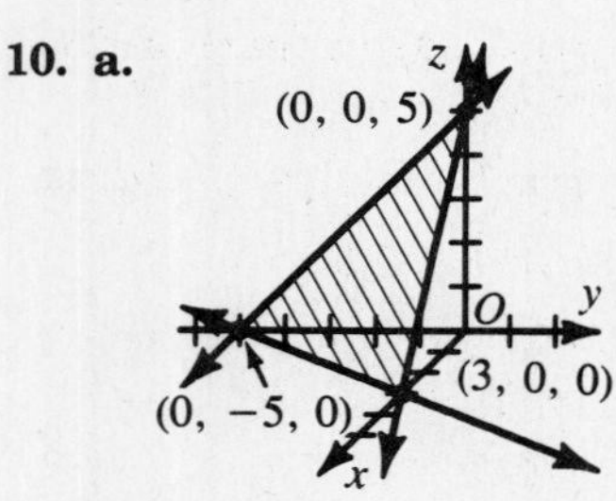

b. $\begin{cases} z = 0 \\ 5x - 3y = 15 \end{cases}$

$\begin{cases} y = 0 \\ 5x + 3z = 15 \end{cases}$

$\begin{cases} x = 0 \\ y - z = -5 \end{cases}$

11. a. 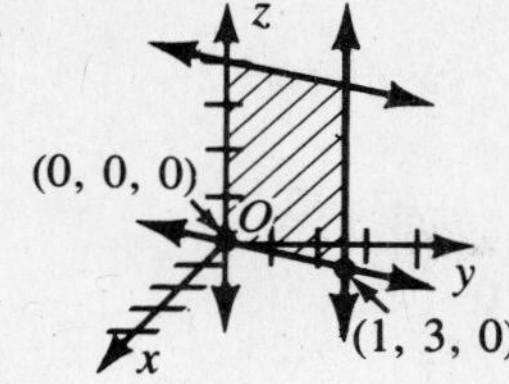

b. $\begin{cases} z = 0 \\ 3x - y = 0 \end{cases}$

$\begin{cases} y = 0 \\ x = 0 \end{cases}$

The graph contains the z-axis.

12. a. 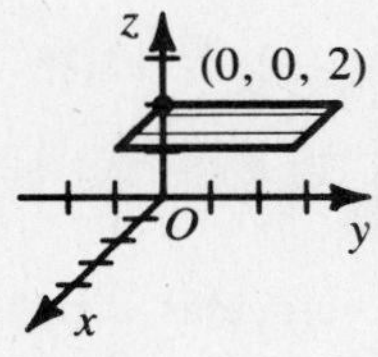

b. $\begin{cases} x = 0 \\ z = 2 \end{cases}$

$\begin{cases} y = 0 \\ z = 2 \end{cases}$

The graph has no trace in the xy-plane.

13. a. 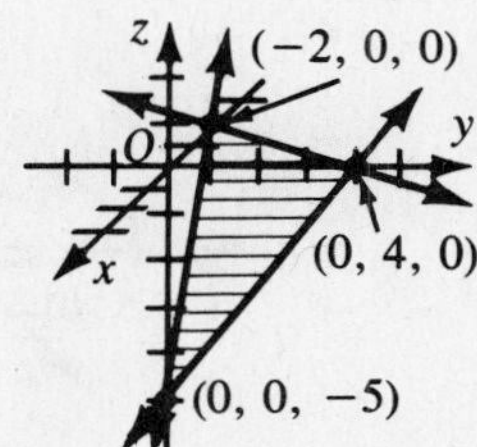

b. $\begin{cases} z = 0 \\ -10x + 5y = 20 \end{cases}$

$\begin{cases} y = 0 \\ -10x - 4z = 0 \end{cases}$

$\begin{cases} x = 0 \\ 5y - 4z = 20 \end{cases}$

14. a. 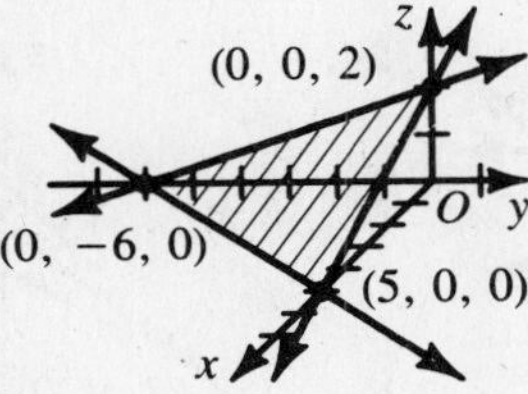

b. $\begin{cases} z = 0 \\ 6x - 5y = 30 \end{cases}$

$\begin{cases} y = 0 \\ 6x + 15z = 30 \end{cases}$

$\begin{cases} x = 0 \\ -5y + 15z = 30 \end{cases}$

15. a. 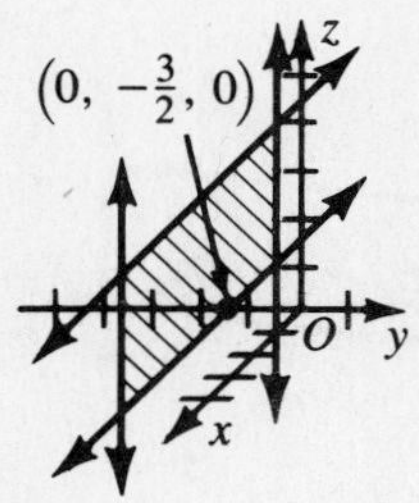

b. $\begin{cases} z = 0 \\ y = -\dfrac{3}{2} \end{cases}$

$\begin{cases} x = 0 \\ y = -\dfrac{3}{2} \end{cases}$

The graph has no trace in the xz-plane.

16. a. 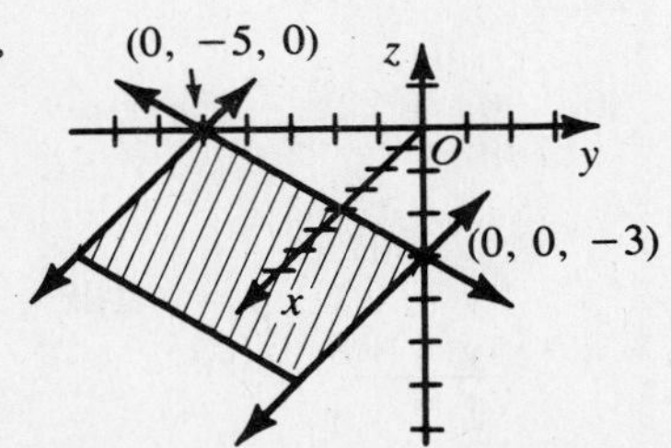

b. $\begin{cases} x = 0 \\ 3y + 5z = -15 \end{cases}$

$\begin{cases} z = 0 \\ y = -5 \end{cases}$

$\begin{cases} y = 0 \\ z = -3 \end{cases}$

17. a.

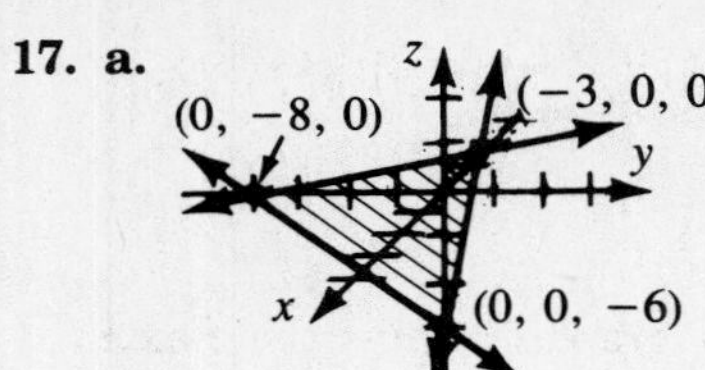

b. $\begin{cases} z = 0 \\ 8x + 3y = -24 \end{cases}$

$\begin{cases} y = 0 \\ 2x + z = -6 \end{cases}$

$\begin{cases} x = 0 \\ 3y + 4z = -24 \end{cases}$

18. a.

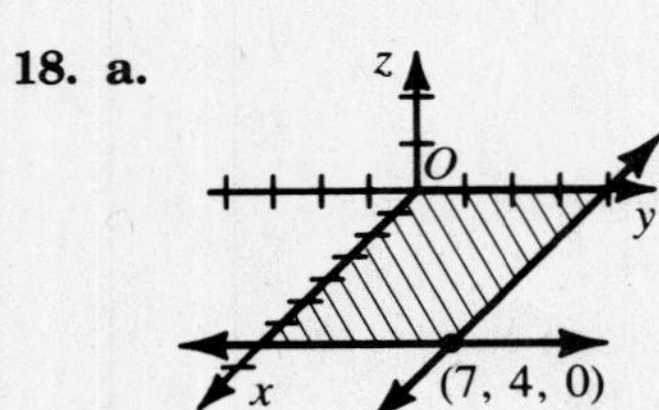

b.

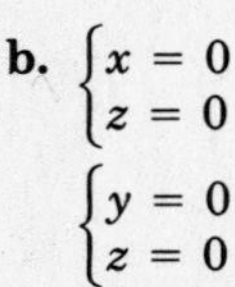

$\begin{cases} x = 0 \\ z = 0 \end{cases}$

$\begin{cases} y = 0 \\ z = 0 \end{cases}$

The graph coincides with the *xy*-plane.

B 19.

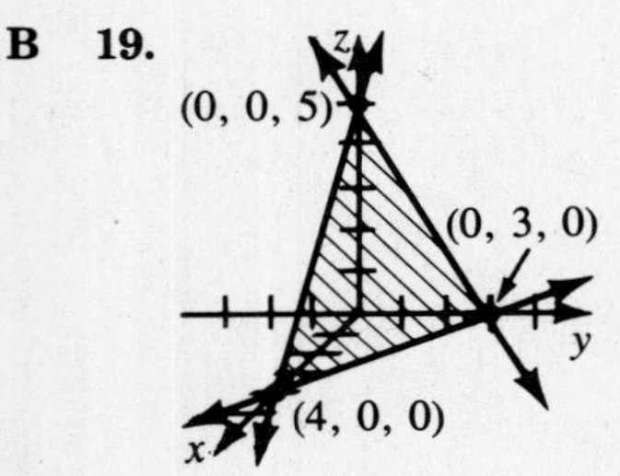

20.

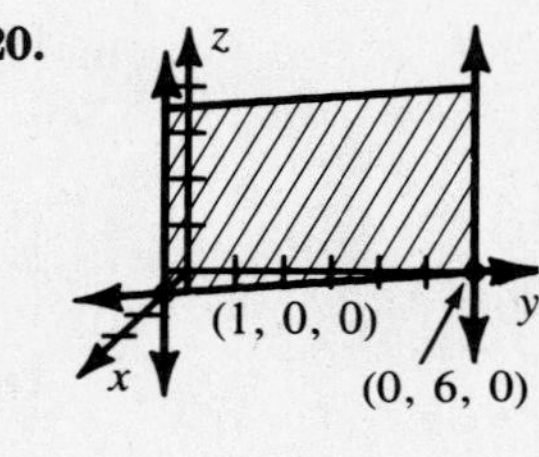

21.

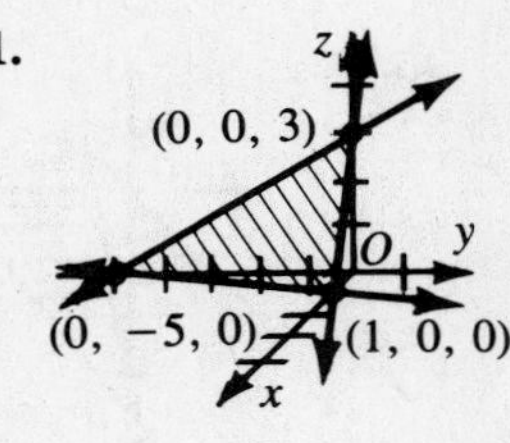

22.

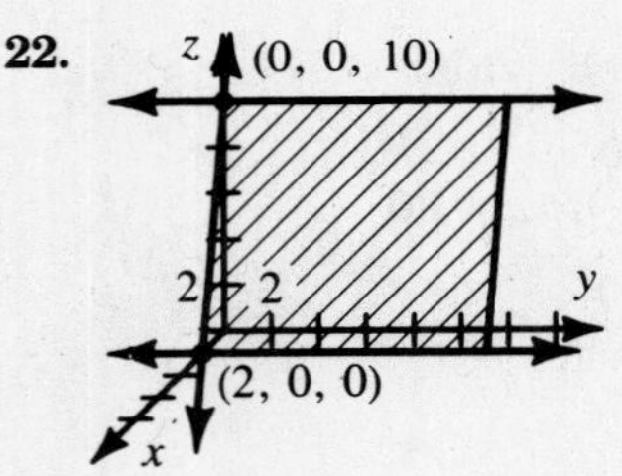

23.

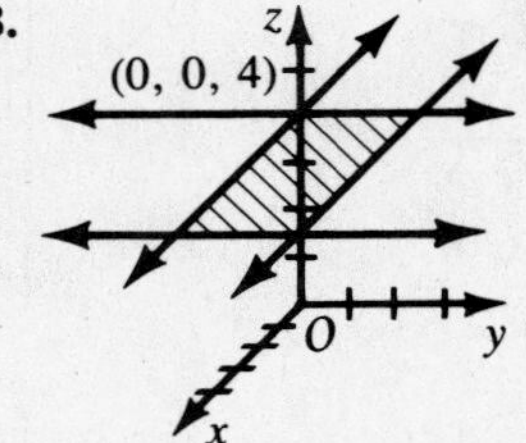

24.

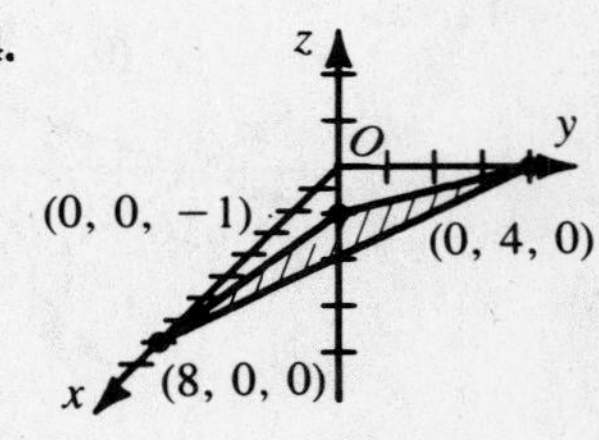

25.

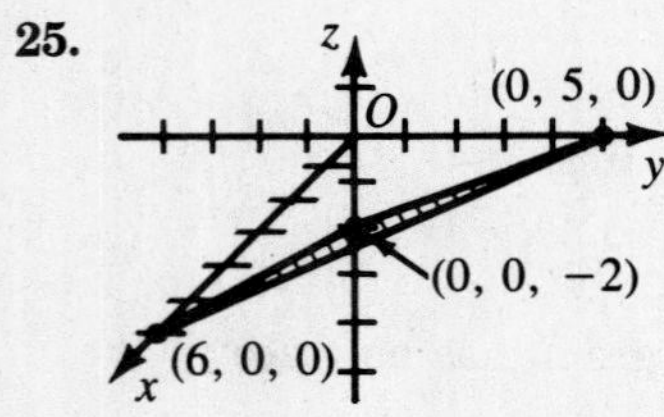

26.

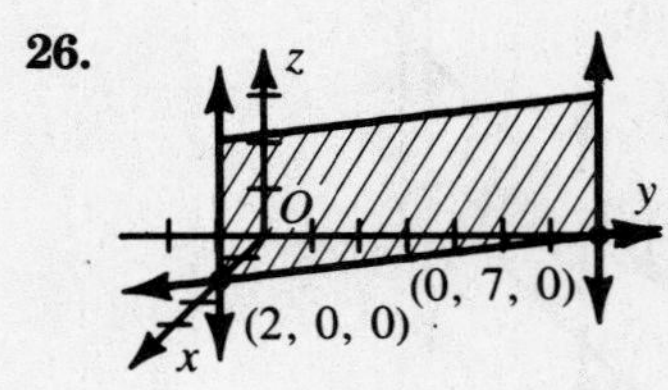

C 27. $4x + y = 4$
$z = 0$

28. $x = -6$
$y = 0$

29. $4x - 6y + 3z = 12$

30. $-6x + 20y - 15z = 60$

Pages 174–175 • WRITTEN EXERCISES

A 1. $\begin{array}{ll} (1) & 2x - y - z = 2 \\ (2) & -x + 3y + z = -1 \\ (3) & 2x - 4y - 3z = -5 \end{array}$ $\quad \begin{array}{lr} (1) + (2): & x + 2y = 1 \\ 3(2) + (3): & \underline{-x + 5y = -8} \\ & 7y = -7 \end{array}$ $\quad y = -1, x = 3, z = 5$
$\{(3, -1, 5)\}$

2. (*1*) $2x + y + 6z = -8$ (*1*) − (*3*): $5x + 10z = -10$ (*4*)
(*2*) $x - 3y - 2z = 4$ (*2*) + 3(*3*): $-8x - 14z = 10$ (*5*)
(*3*) $-3x + y - 4z = 2$

8(*4*): $40x + 80z = -80$ $z = -3, x = 4, y = 2$
5(*5*): $-40x - 70z = 50$ $\{(4, 2, -3)\}$
$10z = -30$

3. (*1*) $x - y + z = 3$ (*1*) − (*3*): $2y - 2z = -6$ (*4*)
(*2*) $3x + 6y - 2z = 2$ 3(*1*) − (*2*): $-9y + 5z = 7$ (*5*)
(*3*) $x - 3y + 3z = 9$

5(*4*): $10y - 10z = -30$ $y = 2, z = 5, x = 0$
2(*5*): $-18y + 10z = 14$ $\{(0, 2, 5)\}$
$-8y = -16$

4. (*1*) $2x + 3y + z = 11$ (*1*) + (*3*): $7x + 7y = 14$ $x = -1, y = 3, z = 4$
(*2*) $4x - y + 2z = 1$ (*2*) + 2(*3*): $14x + 7y = 7$ $\{(-1, 3, 4)\}$
(*3*) $5x + 4y - z = 3$ $-7x = 7$

5. (*1*) $x - 2y = 3$ 2(2) + (*1*): $3x - 2z = 13$ $z = -2, x = 3, y = 0$
(*2*) $x + y - z = 5$ (*3*): $3x + 4z = 1$ $\{(3, 0, -2)\}$
(*3*) $3x + 4z = 1$ $-6z = 12$

6. (*1*) $2y + z = 1$ 4(*2*) − (*3*): $-12y + 7z = 25$
(*2*) $x - 3y + z = 5$ 6(*1*): $12y + 6z = 6$
(*3*) $4x - 3z = -5$ $13z = 31$

$z = \frac{31}{13}, y = -\frac{9}{13}, x = \frac{7}{13}$ $\left\{\left(\frac{7}{13}, -\frac{9}{13}, \frac{31}{13}\right)\right\}$

7. (*1*) $x - 3y + z = 4$ 2(*1*) + (*2*): $-x - 6y = 15$ (*4*) $x = -3, z = 1$
(*2*) $-3x - 2z = 7$ 4(*4*) + (*3*): $-29y = 58; y = -2$ $\{(-3, -2, 1)\}$
(*3*) $4x - 5y = -2$

8. (*1*) $3x - y - 3z = 4$ 5(*1*) + (*3*): $16x - 16z = 16; x - z = 1$ (*4*)
(*2*) $-x + 2z = 3$ (*4*) + (*2*): $z = 4$
(*3*) $x + 5y - z = -4$

$x = 5, y = -1$ $\{(5, -1, 4)\}$

9. (*1*) $2x - 5y + 4z = 8$ (*1*) − 2(*3*): $y = -6$ $x = 7, z = -9$
(*2*) $2x + y + z = -1$ 2(*2*) − (*3*): $3x + 5y = -9$ $\{(7, -6, -9)\}$
(*3*) $x - 3y + 2z = 7$

10. (*1*) $5x - 3y + z = -6$ (*1*) + (*2*): $6x = -6; x = -1$ $y = 3, z = 8$
(*2*) $x + 3y - z = 0$ (*2*) + (*3*): $5x + 4y = 7$ $\{(-1, 3, 8)\}$
(*3*) $4x + y + z = 7$

11. (*1*) $3x + y = 5$ 3(*2*) − (*1*): $2y + 3z = -14$ (*4*) $y = -7, x = 4$
(*2*) $x + y + z = -3$ 2(*3*) + (*4*): $11z = 0; z = 0$ $\{(4, -7, 0)\}$
(*3*) $-y + 4z = 7$

12. (*1*) $2x - z = -2$ 2(*1*) + (*3*): $8x - 5y = 10$ (*4*) $y = 6, z = 12$
(*2*) $3x - y = 9$ (*4*) − 5(*2*): $-7x = -35; x = 5$ $\{(5, 6, 12)\}$
(*3*) $4x - 5y + 2z = 14$

13. (*1*) $2x + 3y - z = 4$ 2(*2*) − (*1*): $5y + 7z = 2$ (*4*)
(*2*) $x + 4y + 3z = 3$ (*4*) − 7(*3*): $-9y = -47; y = \frac{47}{9}$
(*3*) $2y + z = 7$

$x = -\frac{68}{9}, z = -\frac{31}{9}$ $\left\{\left(-\frac{68}{9}, \frac{47}{9}, -\frac{31}{9}\right)\right\}$

14. (*1*) $2x + 7y + z = 3$ (*2*) + (*3*): $2y - z = 4$ (*4*) $z = 2, x = -10$
(*2*) $-x + y - 3z = 7$ 2(*2*) + (*1*): $9y - 5z = 17$ (*5*) $\{(-10, 3, 2)\}$
(*3*) $x + y + 2z = -3$ -5(*4*) + (*5*): $-y = -3; y = 3$

15. (*1*) $x - y + 2z = 7$ (*1*) + (*3*): $-4y + 4z = 16$ (*4*) $y = -\frac{3}{2}, x = \frac{1}{2}$
(*2*) $-2x + 4y + 4z = 3$ 2(*1*) + (*2*): $2y + 8z = 17$ (*5*) $\left\{\left(\frac{1}{2}, -\frac{3}{2}, \frac{5}{2}\right)\right\}$
(*3*) $-x - 3y + 2z = 9$ (*4*) + 2(*5*): $20z = 50; z = \frac{5}{2}$

16. (*1*) $x - 3y + z = 1$ 3(*1*) + (*2*): $5x - 7y = 1$ (*4*) $x = \frac{2}{3}, z = \frac{4}{3}$
(*2*) $2x + 2y - 3z = -2$ (*4*) − (*3*): $-9y = -3; y = \frac{1}{3}$ $\left\{\left(\frac{2}{3}, \frac{1}{3}, \frac{4}{3}\right)\right\}$
(*3*) $5x + 2y = 4$

B **17.** (*1*) $2x - \frac{1}{3}y + z = 3$ 3(*1*): $6x - y + 3z = 9$ (*4*)
(*2*) $-x + \frac{1}{6}y - \frac{2}{3}z = 0$ 6(*2*): $-6x + y - 4z = 0$ (*5*)
(*3*) $\frac{1}{4}x - \frac{1}{2}y - \frac{1}{3}z = -1$ 12(*3*): $3x - 6y - 4z = -12$ (*6*)
(*4*) + (*5*): $-z = 9; z = -9$ $y = 12, x = 8$
2(*6*) + (*5*): $-11y - 12z = -24$ $\{(8, 12, -9)\}$

18. (*1*) $\frac{2}{3}x - y - z = 8$ 3(*1*): $2x - 3y - 3z = 24$ (*4*)
(*2*) $\frac{1}{6}x + 2y + \frac{3}{2}z = 6$ 6(*2*): $x + 12y + 9z = 36$ (*5*)
(*3*) $-\frac{1}{2}x - \frac{1}{3}y + \frac{1}{2}z = -2$ 6(*3*): $-3x - 2y + 3z = -12$ (*6*)
(*4*) + (*6*): $-x - 5y = 12$ (*7*) 7(*7*): $-7x - 35y = 84$
3(*4*) + (*5*): $7x + 3y = 108$ (*8*) (*8*): $7x + 3y = 108$
$-32y = 192$

$y = -6, x = 18, z = 10$
$\{(18, -6, 10)\}$

19. consistent; $\{(0, 0, z): z \in \mathscr{R}\}$ **20.** inconsistent

21. (*1*) $x + y - z = 3$ (*1*) + (*2*): $2x + 2y = 6$ consistent;
(*2*) $x + y + z = 3$ $\{(x, y, 0): x + y = 3\}$

22. (*1*) $x + y = 4$ (*1*) − (*2*): $x - z = 0$ (*4*) $z = 2, y = 2$
(*2*) $y + z = 4$ (*4*) + (*3*): $2x = 4; x = 2$ consistent;
(*3*) $x + z = 4$ $\{(2, 2, 2)\}$

C **23.** **24.** **25.** **26.**

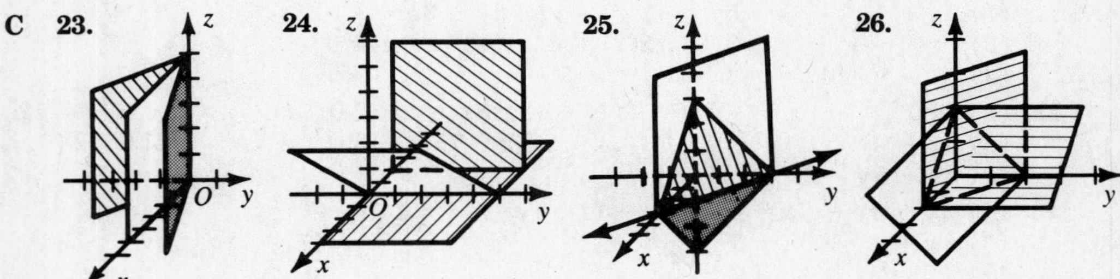

Page 175 • COMPUTER EXERCISES

1.
```
10  LET XIN = 999: LET YIN = 999: LET ZIN = 999
20  INPUT A, B, C, D
30  IF A = 0 THEN GOTO 50
40  LET XIN = D/A
50  IF B = 0 THEN GOTO 70
60  LET YIN = D/B
70  IF C = 0 THEN GOTO 90
80  LET ZIN = D/C
90  IF XIN = 0 AND YIN = 999 AND ZIN = 999
                       THEN PRINT "CONTAINS THE YZ-PLANE": GOTO 190
100  IF YIN = 0 AND XIN = 999 AND ZIN = 999
                       THEN PRINT "CONTAINS THE XZ-PLANE": GOTO 190
110  IF ZIN = 0 AND XIN = 999 AND YIN = 999
                       THEN PRINT "CONTAINS THE XY-PLANE": GOTO 190
120  IF A = 0 AND B = 0 THEN PRINT "PARALLEL TO THE
                                            XY-PLANE": GOTO 190
130  IF B = 0 AND C = 0 THEN PRINT "PARALLEL TO THE
                                            YZ-PLANE": GOTO 190
140  IF A = 0 AND C = 0 THEN PRINT "PARALLEL TO THE
                                            XZ-PLANE": GOTO 190
150  IF D = 0 THEN GOTO 190
160  IF A = 0 THEN PRINT "PARALLEL TO THE X-AXIS"
170  IF B = 0 THEN PRINT "PARALLEL TO THE Y-AXIS"
180  IF C = 0 THEN PRINT "PARALLEL TO THE Z-AXIS"
190  IF XIN = 999 THEN GOTO 210
200  PRINT "X-INTERCEPT IS ";XIN
210  IF YIN = 999 THEN GOTO 230
220  PRINT "Y-INTERCEPT IS ";YIN
230  IF ZIN = 999 THEN GOTO 250
240  PRINT "Z-INTERCEPT IS ";ZIN
250  IF XIN = 0 AND YIN = 0 THEN PRINT "CONTAINS THE Z-AXIS"
260  IF YIN = 0 AND ZIN = 0 THEN PRINT "CONTAINS THE X-AXIS"
270  IF ZIN = 0 AND XIN = 0 THEN PRINT "CONTAINS THE Y-AXIS"
280  END
```

2. contains xy-plane

3. parallel to yz-plane; $(-3, 0, 0)$

4. contains y-axis

5. parallel to z-axis; $(20, 0, 0)$, $(0, -35, 0)$

6. $(-10, 0, 0)$, $(0, 15, 0)$, $(0, 0, -6)$

7. $(2.8571, 0, 0)$, $(0, -2.5, 0)$, $(0, 0, 4)$

8.
```
10  INPUT "XY-TRACE: "; A, B, C
20  INPUT "YZ-TRACE: "; D, E, F
30  IF (C/B <> F/D) THEN PRINT "NO SOLUTION": GOTO 100
40  PRINT "THE INTERCEPTS ARE:"
50  PRINT "(";C/A;", 0, 0)"
60  PRINT "(0, ";C/B;", 0)"
70  PRINT "(0, 0, ";F/E;")"
80  PRINT "THE EQUATION OF THE XZ-TRACE IS:"
90  PRINT A * F;"X + ";C * E;"Z = ";F * C
100  END
```

9. xz-trace: $36x + 120z = 180$ (or $1.5x + 5z = 7.5$), $y = 0$; (5, 0, 0), (0, −2, 0), (0, 0, 1.5)

10. no solution

Page 176 • SELF-TEST 1

1.

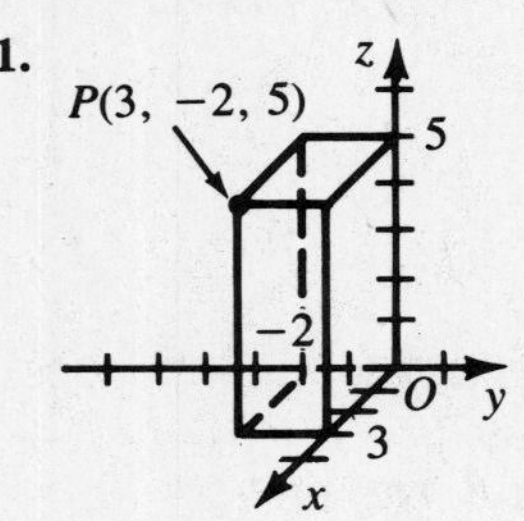

2.

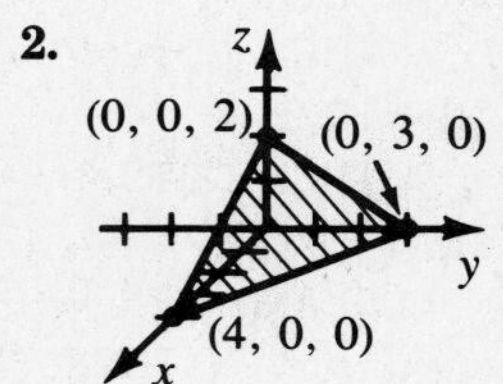

3. $4x + 13y = 24,\ z = 0$
$13y - 7z = 24,\ x = 0$
$4x - 7z = 24,\ y = 0$

4.

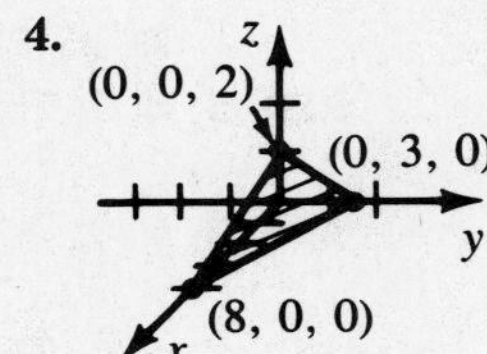

5.
$$\begin{aligned} (1)\ & 3x - y - 2z = -1 \\ (2)\ & x + y + z = 6 \\ (3)\ & 2x + 3y - z = 4 \end{aligned}$$

$(2) + (3)$: $3x + 4y = 10$ (4)
$2(2) + (1)$: $5x + y = 11$ (5)

$$\begin{array}{lrcr} (4): & 3x + 4y & = & 10 \\ -4(5): & -20x - 4y & = & -44 \\ \hline & -17x & = & -34 \end{array}$$

$x = 2,\ y = 1,\ z = 3$
$\{(2, 1, 3)\}$

Page 179 • WRITTEN EXERCISES

A

1. $5 + (-12) + 0 - 2 - 0 - 6 = -15$

2. $-16 + 0 + 9 - (-6) - 0 - (-4) = 3$

3. $10 + (-27) + (-16) - 6 - (-60) - (-12) = 33$

4. $30 + (-72) + 0 - 12 - 0 - 0 = -54$

5.
$$D = \begin{vmatrix} 1 & -3 & -1 \\ 2 & 1 & 1 \\ 1 & 2 & 2 \end{vmatrix} = (2 - 3 - 4) - (-1 + 2 - 12) = 6$$

$$D_x = \begin{vmatrix} 1 & -3 & -1 \\ 3 & 1 & 1 \\ 0 & 2 & 2 \end{vmatrix} = (2 + 0 - 6) - (0 + 2 - 18) = 12$$

$$D_y = \begin{vmatrix} 1 & 1 & -1 \\ 2 & 3 & 1 \\ 1 & 0 & 2 \end{vmatrix} = (6 + 1 + 0) - (-3 + 0 + 4) = 6$$

$$D_z = \begin{vmatrix} 1 & -3 & 1 \\ 2 & 1 & 3 \\ 1 & 2 & 0 \end{vmatrix} = (0 - 9 + 4) - (1 + 6 + 0) = -12$$

$$x = \frac{D_x}{D} = 2,\ y = \frac{D_y}{D} = 1,\ z = \frac{D_z}{D} = -2;\ \{(2, 1, -2)\}$$

6. $D = \begin{vmatrix} 1 & -1 & 2 \\ -1 & 3 & 1 \\ -2 & 1 & -1 \end{vmatrix} = (-3 + 2 - 2) - (-12 + 1 - 1) = 9$

$D_x = \begin{vmatrix} 1 & -1 & 2 \\ 5 & 3 & 1 \\ 4 & 1 & -1 \end{vmatrix} = (-3 - 4 + 10) - (24 + 1 + 5) = -27$

$D_y = \begin{vmatrix} 1 & 1 & 2 \\ -1 & 5 & 1 \\ -2 & 4 & -1 \end{vmatrix} = (-5 - 2 - 8) - (-20 + 4 + 1) = 0$

$D_z = \begin{vmatrix} 1 & -1 & 1 \\ -1 & 3 & 5 \\ -2 & 1 & 4 \end{vmatrix} = (12 + 10 - 1) - (-6 + 5 + 4) = 18$

$x = \frac{D_x}{D} = -3,\ y = \frac{D_y}{D} = 0,\ z = \frac{D_z}{D} = 2;\ \{(-3, 0, 2)\}$

7. $D = \begin{vmatrix} 2 & 0 & 1 \\ 0 & 3 & 2 \\ -3 & 1 & -1 \end{vmatrix} = (-6 + 0 + 0) - (-9 + 4 + 0) = -1$

$D_x = \begin{vmatrix} 1 & 0 & 1 \\ -2 & 3 & 2 \\ -3 & 1 & -1 \end{vmatrix} = (-3 + 0 - 2) - (-9 + 2 + 0) = 2$

$D_y = \begin{vmatrix} 2 & 1 & 1 \\ 0 & -2 & 2 \\ -3 & -3 & -1 \end{vmatrix} = (4 - 6 + 0) - (6 - 12 + 0) = 4$

$D_z = \begin{vmatrix} 2 & 0 & 1 \\ 0 & 3 & -2 \\ -3 & 1 & -3 \end{vmatrix} = (-18 + 0 + 0) - (-9 - 4 + 0) = -5$

$x = \frac{D_x}{D} = -2,\ y = \frac{D_y}{D} = -4,\ z = \frac{D_z}{D} = 5;\ \{(-2, -4, 5)\}$

8. $D = \begin{vmatrix} 2 & -4 & -1 \\ 3 & 0 & -2 \\ 1 & 1 & -1 \end{vmatrix} = (0 + 8 - 3) - (0 - 4 + 12) = -3$

$D_x = \begin{vmatrix} 5 & -4 & -1 \\ -1 & 0 & -2 \\ -3 & 1 & -1 \end{vmatrix} = (0 - 24 + 1) - (0 - 10 - 4) = -9$

$D_y = \begin{vmatrix} 2 & 5 & -1 \\ 3 & -1 & -2 \\ 1 & -3 & -1 \end{vmatrix} = (2 - 10 + 9) - (1 + 12 - 15) = 3$

$D_z = \begin{vmatrix} 2 & -4 & 5 \\ 3 & 0 & -1 \\ 1 & 1 & -3 \end{vmatrix} = (0 + 4 + 15) - (0 - 2 + 36) = -15$

$x = \frac{D_x}{D} = 3,\ y = \frac{D_y}{D} = -1,\ z = \frac{D_z}{D} = 5;\ \{(3, -1, 5)\}$

9. $D = \begin{vmatrix} 1 & -3 & 0 \\ 0 & 2 & -1 \\ -1 & 4 & 1 \end{vmatrix} = (2 - 3 + 0) - (0 - 4 + 0) = 3$

$D_x = \begin{vmatrix} 1 & -3 & 0 \\ 3 & 2 & -1 \\ -1 & 4 & 1 \end{vmatrix} = (2 - 3 + 0) - (0 - 4 - 9) = 12$

$D_y = \begin{vmatrix} 1 & 1 & 0 \\ 0 & 3 & -1 \\ -1 & -1 & 1 \end{vmatrix} = (3 + 1 + 0) - (0 + 1 + 0) = 3$

$D_z = \begin{vmatrix} 1 & -3 & 1 \\ 0 & 2 & 3 \\ -1 & 4 & -1 \end{vmatrix} = (-2 + 9 + 0) - (-2 + 12 + 0) = -3$

$x = \dfrac{D_x}{D} = 4,\ y = \dfrac{D_y}{D} = 1,\ z = \dfrac{D_z}{D} = -1;\ \{(4, 1, -1)\}$

10. $D = \begin{vmatrix} 2 & 0 & -3 \\ 1 & -2 & 0 \\ 0 & 3 & -2 \end{vmatrix} = (8 + 0 - 9) - (0 + 0 + 0) = -1$

$D_x = \begin{vmatrix} 3 & 0 & -3 \\ -4 & -2 & 0 \\ 9 & 3 & -2 \end{vmatrix} = (12 + 0 + 36) - (54 + 0 + 0) = -6$

$D_y = \begin{vmatrix} 2 & 3 & -3 \\ 1 & -4 & 0 \\ 0 & 9 & -2 \end{vmatrix} = (16 + 0 - 27) - (0 + 0 - 6) = -5$

$D_z = \begin{vmatrix} 2 & 0 & 3 \\ 1 & -2 & -4 \\ 0 & 3 & 9 \end{vmatrix} = (-36 + 0 + 9) - (0 - 24 + 0) = -3$

$x = \dfrac{D_x}{D} = 6,\ y = \dfrac{D_y}{D} = 5,\ z = \dfrac{D_z}{D} = 3;\ \{(6, 5, 3)\}$

B 11. $D = \begin{vmatrix} 1 & 1 & 0 \\ 1 & -1 & 0 \\ 1 & -2 & 0 \end{vmatrix} = (0 + 0 + 0) - (0 + 0 + 0) = 0$; infinite solution set;

$\{(0, 0, z): z \in \mathscr{R}\}$

12. $D = \begin{vmatrix} 1 & 1 & -1 \\ 2 & 4 & -1 \\ -1 & -5 & -1 \end{vmatrix} = (-4 + 1 + 10) - (4 + 5 - 2) = 0$; infinite solution set;

when Equation (3) is subtracted from Equation (1) the result is $x + 3y = 0$, which is the same result as when Equation (2) is subtracted from Equation (1).

13. $D = \begin{vmatrix} 0 & 1 & 1 \\ 0 & 1 & -1 \\ 1 & 0 & 0 \end{vmatrix} = (0 - 1 + 0) - (1 + 0 + 0) = -2;\ \{(3, 0, 4)\}$

14. $D = \begin{vmatrix} 1 & -2 & 1 \\ 2 & -4 & 2 \\ 1 & 0 & 0 \end{vmatrix} = (0 - 4 + 0) - (-4 + 0 + 0) = 0$; inconsistent; the graphs of

Equations (1) and (2) are parallel planes. Their intersection with the graph of Equation (3) yields parallel lines, so the system is inconsistent.

C 15. $\begin{vmatrix} a & b & c \\ ka & kb & kc \\ d & e & f \end{vmatrix} = (akbf + bkcd + ckae) - (dkbc + ekca + fkab) =$

$(akbf - fkab) + (bkcd - dkbc) + (ckae - ekca) = 0$

16. $\begin{vmatrix} a & b & c \\ d & e & f \\ a+d & b+e & c+f \end{vmatrix} = \begin{matrix}[ae(c+f) + bf(a+d) + cd(b+e)] - \\ [(a+d)ec + (b+e)fa + (c+f)db] =\end{matrix}$

$(aec + aef + bfa + bfd + cdb + cde) - (aec + dec + bfa + efa + cdb + fdb) =$
$(aec - aec) + (aef - efa) + (bfa - bfa) + (bfd - fdb) + (cdb - cdb) +$
$(cde - dec) = 0$

Pages 181–182 • PROBLEMS

A 1. $n + d + q = 38$ $\quad n + 2q + q = 38$
$5n + 10d + 25q = 370$ $\quad 5n + 10(2q) + 25q = 370$
$d = 2q$

$n + 3q = 38$
$5n + 45q = 370$

$$\begin{array}{r} n + 3q = 38 \\ \underline{n + 9q = 74} \\ -6q = -36 \end{array}$$

$q = 6, n = 20, d = 12$
20 nickels, 12 dimes, 6 quarters

2. $(1)\ n + d + q = 69$
$(2)\ n - d - q = 3$
$(3)\ 5n + 10d + 25q = 780$

$(1) + (2)$: $2n = 72; n = 36$

(1): $d + q = 33\ (4)$
(3): $180 + 10d + 25q = 780; 2d + 5q = 120\ (5)$

$$\begin{array}{rr} 2(4): & 2d + 2q = 66 \\ (5): & \underline{2d + 5q = 120} \\ & -3q = -54 \end{array}$$

$q = 18, d = 15$
36 nickels, 15 dimes, 18 quarters

3. $(1)\ 2l + 2w = 30$
$(2)\ 2w + 2h = 24$
$(3)\ 2l + 2h = 26$

$l + w = 15\ (4)$
$w + h = 12\ (5)$
$l + h = 13\ (6)$

$$\begin{array}{rr} (4) - (5): & l - h = 3 \\ (6): & \underline{l + h = 13} \\ & 2l \quad = 16 \end{array}$$

$l = 8, h = 5, w = 7$
8 cm by 7 cm by 5 cm

4. Let a = mass in grams of one sheet of airmail paper; let t = mass in grams of one sheet of typing paper; let c = mass in grams of one sheet of construction paper.

$(1)\ a + 3t = 25$
$(2)\ t + 2c = 25$
$(3)\ a + t + c = 20$

$$\begin{array}{rr} (1) - (3): & 2t - c = 5 \\ 2(2): & \underline{2t + 4c = 50} \\ & -5c = -45 \end{array}$$

$c = 9, t = 7, a = 4$
airmail paper, 4 g;
typing paper, 7 g;
construction paper, 9 g

5. Let m = no. of grams of protein in a glass of milk; let w = the no. of grams of protein in a serving of shredded wheat; let e = no. of grams of protein in an egg.

$(1)\ m + w + e = 17$
$(2)\ m + 3e = 29$
$(3)\ 2m + w = 18$

$$\begin{array}{rr} (3) - (1): & m - e = 1 \\ (2): & \underline{m + 3e = 29} \\ & -4e = -28 \end{array}$$

$e = 7, m = 8, w = 2$

milk, 8 g; egg, 7 g; shredded wheat, 2 g

6. Let t = no. of triangles; let s = no. of squares; let p = no. of pentagons.

$t + s + p = 40$
$3t + 4s + 5p = 153$
$2s + 5p = 72$

$$D = \begin{vmatrix} 1 & 1 & 1 \\ 3 & 4 & 5 \\ 0 & 2 & 5 \end{vmatrix} = (20 + 0 + 6) - (0 + 10 + 15) = 1$$

$$D_t = \begin{vmatrix} 40 & 1 & 1 \\ 153 & 4 & 5 \\ 72 & 2 & 5 \end{vmatrix} = (800 + 360 + 306) - (288 + 400 + 765) = 13$$

$$D_s = \begin{vmatrix} 1 & 40 & 1 \\ 3 & 153 & 5 \\ 0 & 72 & 5 \end{vmatrix} = (765 + 0 + 216) - (0 + 360 + 600) = 21$$

$$D_p = \begin{vmatrix} 1 & 1 & 40 \\ 3 & 4 & 153 \\ 0 & 2 & 72 \end{vmatrix} = (288 + 0 + 240) - (0 + 306 + 216) = 6$$

$t = \frac{D_t}{D} = 13$, $s = \frac{D_s}{D} = 21$, $p = \frac{D_p}{D} = 6$; 13 triangles, 21 squares, 6 pentagons

7. Let $x = GH$; $y = EF$; $z = EH$; then $EG = FG = 2x$

(*1*) $x + y + z = 48$
(*2*) $3x + z = 70$
(*3*) $4x + y = 80$

(*2*) − (*1*): $2x - y = 22$
(*3*): $4x + y = 80$
$6x = 102$

$x = 17$, $z = 19$, $y = 12$

$EF = 12$, $EG = 34$, $EH = 19$

8. Let $x = DX$, $y = AX$, and $z = AB$. Then $AD = DB = BC = 2x$, $AC = 2y$, and $AB = DC = z$.

(*1*) $4x + z = 18$
(*2*) $2x + 2y + z = 20$
(*3*) $x + y + z = 12$

(*1*) − (*3*): $3x - y = 6$
(*2*) − (*3*): $x + y = 8$
$4x = 14$

$x = 3.5$, $z = 4$, $y = 4.5$
$AB = 4$ m, $AD = 7$ m, $AC = 9$ m

B **9.** Let u = speed uphill in km/h; let l = speed on level ground in km/h; let d = speed downhill in km/h.

(*1*) $\frac{3}{4}u + \frac{1}{2}l = 22$ 4(*1*): $3u + 2l = 88$ (*4*)

(*2*) $\frac{1}{2}d + \frac{1}{2}l = 22$ 2(*2*): $d + l = 44$ (*5*)

(*3*) $l = \frac{u + d}{2}$ 2(*3*): $2l = u + d$; $u + d - 2l = 0$ (*6*)

(*5*) − (*6*): $-u + 3l = 44$; $-3u + 9l = 132$
(*4*): $3u + 2l = 88$
$11l = 220$

$l = 20$, $d = 24$, $u = 16$

speed uphill, 16 km/h; speed downhill, 24 km/h; speed on level ground, 20 km/h

10. (*1*) $90 = -k + v_0 + y_0$
(*2*) $80 = -4k + 2v_0 + y_0$
(*3*) $0 = -16k + 4v_0 + y_0$

(*1*) − (*2*): $3k - v_0 = 10$ (*4*)
(*2*) − (*3*): $12k - 2v_0 = 80$ (*5*)

−4(*4*): $-12k + 4v_0 = -40$
(*5*): $12k - 2v_0 = 80$
$2v_0 = 40$

$v_0 = 20$; 20 m/s

C **11.** Let x = profit from each two-pack; let y = profit from each four-pack; let z = profit from each five-pack.

(*1*) $7.5x + 3.75y = 11.70$
(*2*) $7.5x + 3z = 13.50$
(*3*) $5x + 2.5y + 2z = 12.60$

1.5(*3*) − (*2*): $3.75y = 5.4$
$y = 1.44$

$x = 0.84$, $z = 2.4$
2-pack, \$.84;
4-pack, \$1.44;
5-pack, \$2.40

12. Let x = no. of \$40 tickets; let y = no. of \$60 tickets; let z = no. of \$100 tickets.

(*1*) $40x + 60y + 100z = 3880$
(*2*) $x + y + 2z = 78$
(*3*) $y + 2z = 62$

(*2*) − (*3*): $x = 16$
(*1*): $60y + 100z = 3240$
50(*3*): $50y + 100z = 3100$

$y = 14$, $z = 24$; 16 \$40-tickets, 14 \$60-tickets, 24 \$100-tickets

13. Let x = lower speed in km/h; let y = higher speed in km/h; let z = stopping time in hours.

$\frac{48}{x} + \frac{72}{y} + z = 3$ Let $\frac{1}{x} = a$; $\frac{1}{y} = b$: $48a + 72b + z = 3$ (*1*)

$\frac{120}{y} + 3z = 3$ $120b + 3z = 3$ (*2*)

$\frac{48}{y} + \frac{72}{x} = 3$ $72a + 48b = 3$ (*3*)

$y = \frac{8}{3}x$ $b = \frac{3}{8}a$ (*4*)

Substitute (*4*) into (*3*): $72a + 48 \cdot \frac{3}{8}a = 3$; $a = \frac{1}{30}$, $x = 30$

Substitute $a = \frac{1}{30}$ into (*1*): $1.6 + 72b + z = 3$; $72b + z = 1.4$;
$216b + 3z = 4.2$ (*5*)

(*5*): $216b + 3z = 4.2$
(*2*): $\underline{120b + 3z = 3}$
$96b = 1.2$

$b = 0.0125$, $y = 80$ 30 km/h; 80 km/h

Page 183 • ON THE CALCULATOR

1. $D = (15 + 33.6 + 17.934) - (-28 + 76.86 - 4.2) = 21.874$
2. $D = (-180 - 288 + 36) - (36 - 180 - 288) = 0$
3. $$D = \begin{vmatrix} 5 & -8 & -10 \\ 1.4 & 5 & 4 \\ 8 & -3.2 & -20 \end{vmatrix} = (-500 - 256 + 44.8) - (-400 - 64 + 224) = -471.2$$

$$D_x = \begin{vmatrix} 87 & -8 & -10 \\ -3 & 5 & 4 \\ 86 & -3.2 & -20 \end{vmatrix} = (-8700 - 2752 - 96) - (-4300 - 1113.6 - 480) = -5654.4$$

$$D_y = \begin{vmatrix} 5 & 87 & -10 \\ 1.4 & -3 & 4 \\ 8 & 86 & -20 \end{vmatrix} = (300 + 2784 - 1204) - (240 + 1720 - 2436) = 2356$$

$$D_z = \begin{vmatrix} 5 & -8 & 87 \\ 1.4 & 5 & -3 \\ 8 & -3.2 & 86 \end{vmatrix} = (2150 + 192 - 389.76) - (3480 + 48 - 963.2) = -612.56$$

$x = \frac{D_x}{D} = 12$, $y = \frac{D_y}{D} = -5$, $z = \frac{D_z}{D} = 1.3$; $\{(12, -5, 1.3)\}$

4. $D = \begin{vmatrix} 0.4 & 3 & -4 \\ 5 & 0.125 & 7 \\ 0.5 & -1 & -0.75 \end{vmatrix} = (-0.0375 + 10.5 + 20) - (-0.25 - 2.8 - 11.25) = 44.7625$

$D_x = \begin{vmatrix} 17 & 3 & -4 \\ 6 & 0.125 & 7 \\ -2 & -1 & -0.75 \end{vmatrix} = (-1.59375 - 42 + 24) - (1 - 119 - 13.5) = 111.90625$

$D_y = \begin{vmatrix} 0.4 & 17 & -4 \\ 5 & 6 & 7 \\ 0.5 & -2 & -0.75 \end{vmatrix} = (-1.8 + 59.5 + 40) - (-12 - 5.6 - 63.75) = 179.05$

$D_z = \begin{vmatrix} 0.4 & 3 & 17 \\ 5 & 0.125 & 6 \\ 0.5 & -1 & -2 \end{vmatrix} = (-0.1 + 9 - 85) - (1.0625 - 2.4 - 30) = -44.7625;$

$x = \frac{D_x}{D} = 2.5,\ y = \frac{D_y}{D} = 4,\ z = \frac{D_z}{D} = -1;\ \{(2.5, 4, -1)\}$

Pages 187–188 • WRITTEN EXERCISES

A **1.** $-4\begin{vmatrix} 1 & 5 \\ -2 & -1 \end{vmatrix} - 0\begin{vmatrix} -3 & 2 \\ -2 & -1 \end{vmatrix} + 6\begin{vmatrix} -3 & 2 \\ 1 & 5 \end{vmatrix} = -4(-1 + 10) + 0 + 6(-15 - 2) = -138$

2. $0\begin{vmatrix} 2 & 34 \\ 3 & 22 \end{vmatrix} - 0\begin{vmatrix} 5 & 34 \\ -1 & 22 \end{vmatrix} + 4\begin{vmatrix} 5 & 2 \\ -1 & 3 \end{vmatrix} = 0 + 0 + 4(15 + 2) = 68$

3. $-1\begin{vmatrix} 1 & 6 \\ -2 & 3 \end{vmatrix} - 5\begin{vmatrix} 4 & 6 \\ 1 & 3 \end{vmatrix} + 2\begin{vmatrix} 4 & 1 \\ 1 & -2 \end{vmatrix} = -(3 + 12) - 5(12 - 6) + 2(-8 - 1) = -63$

4. $0\begin{vmatrix} 3 & 2 & 4 \\ -1 & 2 & 9 \\ 3 & 1 & 8 \end{vmatrix} - 0\begin{vmatrix} -1 & 2 & 4 \\ 5 & 2 & 9 \\ 2 & 1 & 8 \end{vmatrix} + 0\begin{vmatrix} -1 & 3 & 4 \\ 5 & -1 & 9 \\ -2 & 3 & 8 \end{vmatrix} + 1\begin{vmatrix} -1 & 3 & 2 \\ 5 & -1 & 2 \\ -2 & 3 & 1 \end{vmatrix}$

$= -2\begin{vmatrix} 5 & -1 \\ -2 & 3 \end{vmatrix} + 2\begin{vmatrix} -1 & 3 \\ -2 & 3 \end{vmatrix} - 1\begin{vmatrix} -1 & 3 \\ 5 & -1 \end{vmatrix} = -2(15 - 2) + 2(-3 + 6) - (1 - 15) = -6$

5. Add row 4 to row 1; expand by minors of the first column.

$\begin{vmatrix} 1 & 3 & -2 & -1 \\ 0 & 2 & 4 & 1 \\ 0 & -5 & 0 & 3 \\ -1 & -3 & -1 & 1 \end{vmatrix} = \begin{vmatrix} 0 & 0 & -3 & 0 \\ 0 & 2 & 4 & 1 \\ 0 & -5 & 0 & 3 \\ -1 & -3 & -1 & 1 \end{vmatrix} = 1\begin{vmatrix} 0 & -3 & 0 \\ 2 & 4 & 1 \\ -5 & 0 & 3 \end{vmatrix} = 0\begin{vmatrix} 4 & 1 \\ 0 & 3 \end{vmatrix} +$

$3\begin{vmatrix} 2 & 1 \\ -5 & 3 \end{vmatrix} + 0\begin{vmatrix} 2 & 4 \\ -5 & 0 \end{vmatrix} = 3(6 + 5) = 33$

6. Add row 3 to row 1; expand by minors of the third column.

$\begin{vmatrix} 3 & 4 & -2 & 1 \\ 5 & 1 & 0 & -1 \\ 0 & -3 & 2 & 3 \\ 4 & -1 & 0 & -5 \end{vmatrix} = \begin{vmatrix} 3 & 1 & 0 & 4 \\ 5 & 1 & 0 & -1 \\ 0 & -3 & 2 & 3 \\ 4 & -1 & 0 & -5 \end{vmatrix} = 2\begin{vmatrix} 3 & 1 & 4 \\ 5 & 1 & -1 \\ 4 & -1 & -5 \end{vmatrix}$

Add row 3 to row 1; add row 2 to row 3; expand by minors of the second column.

$2\begin{vmatrix} 7 & 0 & -1 \\ 5 & 1 & -1 \\ 4 & -1 & -5 \end{vmatrix} = 2\begin{vmatrix} 7 & 0 & -1 \\ 5 & 1 & -1 \\ 9 & 0 & -6 \end{vmatrix} = 2 \cdot 1\begin{vmatrix} 7 & -1 \\ 9 & -6 \end{vmatrix} = 2(-42 + 9) = -66$

B 7. $D = \begin{vmatrix} 1 & -2 & 0 & -1 \\ 2 & -1 & 1 & 0 \\ -1 & 0 & 1 & 1 \\ 0 & 3 & -1 & 1 \end{vmatrix}$ Add row 1 to row 3; add row 4 to row 1; expand by minors of the fourth column. $\begin{vmatrix} 1 & -2 & 0 & -1 \\ 2 & -1 & 1 & 0 \\ 0 & -2 & 1 & 0 \\ 0 & 3 & -1 & 1 \end{vmatrix} = \begin{vmatrix} 1 & 1 & -1 & 0 \\ 2 & -1 & 1 & 0 \\ 0 & -2 & 1 & 0 \\ 0 & 3 & -1 & 1 \end{vmatrix} =$

$\begin{vmatrix} 1 & 1 & -1 \\ 2 & -1 & 1 \\ 0 & -2 & 1 \end{vmatrix} = 1\begin{vmatrix} -1 & 1 \\ -2 & 1 \end{vmatrix} - 2\begin{vmatrix} 1 & -1 \\ -2 & 1 \end{vmatrix} = (-1 + 2) - 2(1 - 2) = 3;$

$D_x = \begin{vmatrix} 0 & -2 & 0 & -1 \\ 1 & -1 & 1 & 0 \\ 0 & 0 & 1 & 1 \\ 2 & 3 & -1 & 1 \end{vmatrix}$ Multiply column 4 by -1, add to column 3; expand by minors of the third row. $\begin{vmatrix} 0 & -2 & 1 & -1 \\ 1 & -1 & 1 & 0 \\ 0 & 0 & 0 & 1 \\ 2 & 3 & -2 & 1 \end{vmatrix} = -1\begin{vmatrix} 0 & -2 & 1 \\ 1 & -1 & 1 \\ 2 & 3 & -2 \end{vmatrix} =$

$\begin{vmatrix} 0 & 1 & -2 \\ 1 & 1 & -1 \\ 2 & -2 & 3 \end{vmatrix} = -1\begin{vmatrix} 1 & -2 \\ -2 & 3 \end{vmatrix} + 2\begin{vmatrix} 1 & -2 \\ 1 & -1 \end{vmatrix} = -(3 - 4) + 2(-1 + 2) = 3;$

$D_y = \begin{vmatrix} 1 & 0 & 0 & -1 \\ 2 & 1 & 1 & 0 \\ -1 & 0 & 1 & 1 \\ 0 & 2 & -1 & 1 \end{vmatrix}$ Add column 1 to column 4; expand by minors of the first row. $\begin{vmatrix} 1 & 0 & 0 & 0 \\ 2 & 1 & 1 & 2 \\ -1 & 0 & 1 & 0 \\ 0 & 2 & -1 & 1 \end{vmatrix} = \begin{vmatrix} 1 & 1 & 2 \\ 0 & 1 & 0 \\ 2 & -1 & 1 \end{vmatrix} = \begin{vmatrix} 1 & 2 \\ 2 & 1 \end{vmatrix} = -3;$

$D_z = \begin{vmatrix} 1 & -2 & 0 & -1 \\ 2 & -1 & 1 & 0 \\ -1 & 0 & 0 & 1 \\ 0 & 3 & 2 & 1 \end{vmatrix}$ Add column 4 to column 1; expand by minors of the third row. $\begin{vmatrix} 0 & -2 & 0 & -1 \\ 2 & -1 & 1 & 0 \\ 0 & 0 & 0 & 1 \\ 1 & 3 & 2 & 1 \end{vmatrix} = -1\begin{vmatrix} 0 & -2 & 0 \\ 2 & -1 & 1 \\ 1 & 3 & 2 \end{vmatrix} = -2\begin{vmatrix} 2 & 1 \\ 1 & 2 \end{vmatrix} = -6;$

$D_w = \begin{vmatrix} 1 & -2 & 0 & 0 \\ 2 & -1 & 1 & 1 \\ -1 & 0 & 1 & 0 \\ 0 & 3 & -1 & 2 \end{vmatrix}$ Add column 3 to column 1; expand by minors of the third row. $\begin{vmatrix} 1 & -2 & 0 & 0 \\ 3 & -1 & 1 & 1 \\ 0 & 0 & 1 & 0 \\ -1 & 3 & -1 & 2 \end{vmatrix} = 1\begin{vmatrix} 1 & -2 & 0 \\ 3 & -1 & 1 \\ -1 & 3 & 2 \end{vmatrix} = 1\begin{vmatrix} -1 & 1 \\ 3 & 2 \end{vmatrix} + 2\begin{vmatrix} 3 & 1 \\ -1 & 2 \end{vmatrix} =$

$(-2 - 3) + 2(6 + 1) = 9;$

$x = \frac{D_x}{D} = 1,\ y = \frac{D_y}{D} = -1,\ z = \frac{D_z}{D} = -2,\ w = \frac{D_w}{D} = 3$

8. $D = \begin{vmatrix} -1 & 1 & 0 & -1 \\ 0 & 2 & -1 & 1 \\ 1 & 0 & 1 & 1 \\ -3 & 1 & -2 & -1 \end{vmatrix}$ Add column 2 to column 4. $\begin{vmatrix} -1 & 1 & 0 & 0 \\ 0 & 2 & -1 & 3 \\ 1 & 0 & 1 & 1 \\ -3 & 1 & -2 & 0 \end{vmatrix}$ Add column 2 to column 1; expand by minors of the first row. $\begin{vmatrix} 0 & 1 & 0 & 0 \\ 2 & 2 & -1 & 3 \\ 1 & 0 & 1 & 1 \\ -2 & 1 & -2 & 0 \end{vmatrix}$

Multiply column 2 by -1, add to column 1; expand by minors of the first column. $-1\begin{vmatrix} 2 & -13 \\ 1 & 11 \\ -2 & -20 \end{vmatrix} = -\begin{vmatrix} 3 & -1 & 3 \\ 0 & 1 & 1 \\ 0 & -2 & 0 \end{vmatrix} = -3\begin{vmatrix} 1 & 1 \\ -2 & 0 \end{vmatrix} = -6;$

$D_x = \begin{vmatrix} 5 & 1 & 0 & -1 \\ 0 & 2 & -1 & 1 \\ 0 & 0 & 1 & 1 \\ 3 & 1 & -2 & -1 \end{vmatrix}$ Multiply column 3 by -1, add to column 4; expand by minors of the third row. $\begin{vmatrix} 5 & 1 & 0 & -1 \\ 0 & 2 & -1 & 2 \\ 0 & 0 & 1 & 0 \\ 3 & 1 & -2 & 1 \end{vmatrix} = \begin{vmatrix} 5 & 1 & -1 \\ 0 & 2 & 2 \\ 3 & 1 & 1 \end{vmatrix} = 2\begin{vmatrix} 5 & -1 \\ 3 & 1 \end{vmatrix} -$ $2\begin{vmatrix} 5 & 1 \\ 3 & 1 \end{vmatrix} = 2(5 + 3) - 2(5 - 3) = 12;$

$D_y = \begin{vmatrix} -1 & 5 & 0 & -1 \\ 0 & 0 & -1 & 1 \\ 1 & 0 & 1 & 1 \\ -3 & 3 & -2 & -1 \end{vmatrix}$ Add column 3 to column 4; expand by minors of the second row. $\begin{vmatrix} -1 & 5 & 0 & -1 \\ 0 & 0 & -1 & 0 \\ 1 & 0 & 1 & 2 \\ -3 & 3 & -2 & -3 \end{vmatrix} = \begin{vmatrix} -1 & 5 & -1 \\ 1 & 0 & 2 \\ -3 & 3 & -3 \end{vmatrix} = -1\begin{vmatrix} 5 & -1 \\ 3 & -3 \end{vmatrix} - 2\begin{vmatrix} -1 & 5 \\ -3 & 3 \end{vmatrix} =$ $-(-15 + 3) - 2(-3 + 15) = -12;$

$D_z = \begin{vmatrix} -1 & 1 & 5 & -1 \\ 0 & 2 & 0 & 1 \\ 1 & 0 & 0 & 1 \\ -3 & 1 & 3 & -1 \end{vmatrix}$ Multiply column 1 by -1, add column 1 to column 4; expand by minors of the third row. $\begin{vmatrix} -1 & 1 & 5 & 0 \\ 0 & 2 & 0 & 1 \\ 1 & 0 & 0 & 0 \\ -3 & 1 & 3 & 2 \end{vmatrix} = \begin{vmatrix} 1 & 5 & 0 \\ 2 & 0 & 1 \\ 1 & 3 & 2 \end{vmatrix} = 1\begin{vmatrix} 0 & 1 \\ 3 & 2 \end{vmatrix} -$ $5\begin{vmatrix} 2 & 1 \\ 1 & 2 \end{vmatrix} = -3 - 5(4 - 1) = -18;$

$D_w = \begin{vmatrix} -1 & 1 & 0 & 5 \\ 0 & 2 & -1 & 0 \\ 1 & 0 & 1 & 0 \\ -3 & 1 & -2 & 3 \end{vmatrix}$ Multiply column 1 by -1, add column 1 to column 3;

expand by minors of the third row. $\begin{vmatrix} -1 & 1 & 1 & 5 \\ 0 & 2 & -1 & 0 \\ 1 & 0 & 0 & 0 \\ -3 & 1 & 1 & 3 \end{vmatrix} = \begin{vmatrix} 1 & 1 & 5 \\ 2 & -1 & 0 \\ 1 & 1 & 3 \end{vmatrix} =$

$-2\begin{vmatrix} 1 & 5 \\ 1 & 3 \end{vmatrix} - 1\begin{vmatrix} 1 & 5 \\ 1 & 3 \end{vmatrix} = -2(3 - 5) - (3 - 5) = 6;$

$x = \frac{D_x}{D} = -2,\ y = \frac{D_y}{D} = 2,\ z = \frac{D_z}{D} = 3,\ w = \frac{D_w}{D} = -1$

9. a. $\begin{vmatrix} a_1 & 0 & 0 & 0 \\ 0 & b_2 & 0 & 0 \\ 0 & 0 & c_3 & 0 \\ 0 & 0 & 0 & d_4 \end{vmatrix} = a_1A_1 - 0 \cdot A_2 + 0 \cdot A_3 - 0 \cdot A_4 = a_1\begin{vmatrix} b_2 & 0 & 0 \\ 0 & c_3 & 0 \\ 0 & 0 & d_4 \end{vmatrix} =$

$a_1\left[b_2\begin{vmatrix} c_3 & 0 \\ 0 & d_4 \end{vmatrix} - 0\begin{vmatrix} 0 & 0 \\ 0 & d_4 \end{vmatrix} + 0\begin{vmatrix} 0 & c_3 \\ 0 & 0 \end{vmatrix}\right] = a_1b_2c_3d_4$

b. $\begin{vmatrix} a_1 & b_1 & 0 & 0 \\ a_2 & b_2 & 0 & 0 \\ 0 & 0 & c_3 & d_3 \\ 0 & 0 & c_4 & d_4 \end{vmatrix} = a_1A_1 - a_2A_2 + 0 \cdot A_3 - 0 \cdot A_4 = a_1\begin{vmatrix} b_2 & 0 & 0 \\ 0 & c_3 & d_3 \\ 0 & c_4 & d_4 \end{vmatrix} -$

$a_2\begin{vmatrix} b_1 & 0 & 0 \\ 0 & c_3 & d_3 \\ 0 & c_4 & d_4 \end{vmatrix} = a_1\left[b_2\begin{vmatrix} c_3 & d_3 \\ c_4 & d_4 \end{vmatrix} - 0\begin{vmatrix} 0 & 0 \\ c_4 & d_4 \end{vmatrix} + 0\begin{vmatrix} 0 & 0 \\ c_3 & d_3 \end{vmatrix}\right] -$

$a_2\left[b_1\begin{vmatrix} c_3 & d_3 \\ c_4 & d_4 \end{vmatrix} - 0\begin{vmatrix} 0 & 0 \\ c_4 & d_4 \end{vmatrix} + 0\begin{vmatrix} 0 & 0 \\ c_3 & d_3 \end{vmatrix}\right] = (a_1b_2 - a_2b_1)\begin{vmatrix} c_3 & d_3 \\ c_4 & d_4 \end{vmatrix} =$

$\begin{vmatrix} a_1 & a_2 \\ b_1 & b_2 \end{vmatrix} \cdot \begin{vmatrix} c_3 & d_3 \\ c_4 & d_4 \end{vmatrix}$

C 10. 1. $a_1, b_1, c_1, a_3, b_3, c_3 \in \mathcal{R}$ Hypothesis

2. $\begin{vmatrix} a_1 & b_1 & c_1 \\ a_1 & b_1 & c_1 \\ a_3 & b_3 & c_3 \end{vmatrix} =$

$\begin{vmatrix} a_1 & b_1 & c_1 \\ a_1 + (-1)(a_1) & b_1 + (-1)(b_1) & c_1 + (-1)(c_1) \\ a_3 & b_3 & c_3 \end{vmatrix}$ Property 5 with $k = -1$

3. $= \begin{vmatrix} a_1 & b_1 & c_1 \\ a_1 - a_1 & b_1 - b_1 & c_1 - c_1 \\ a_3 & b_3 & c_3 \end{vmatrix}$ Mult. prop. of -1

4. $= \begin{vmatrix} a_1 & b_1 & c_1 \\ 0 & 0 & 0 \\ a_3 & b_3 & c_3 \end{vmatrix}$ Ax. of add. inv.

5. $= 0$ Property 1

6. $\begin{vmatrix} a_1 & b_1 & c_1 \\ a_1 & b_1 & c_1 \\ a_3 & b_3 & c_3 \end{vmatrix} = 0$ Trans. prop. of =

The proof for two columns with corresponding elements equal is identical.

11. 1. $a_1, b_1, c_1, d_1, a_3, b_3, c_3, d_3, a_4, b_4, c_4, d_4 \in \mathscr{R}$ Hypothesis

2. $\begin{vmatrix} a_1 & b_1 & c_1 & d_1 \\ a_1 & b_1 & c_1 & d_1 \\ a_3 & b_3 & c_3 & d_3 \\ a_4 & b_4 & c_4 & d_4 \end{vmatrix} = -\begin{vmatrix} a_1 & b_1 & c_1 & d_1 \\ a_1 & b_1 & c_1 & d_1 \\ a_3 & b_3 & c_3 & d_3 \\ a_4 & b_4 & c_4 & d_4 \end{vmatrix}$ Property 2, interchanging row 1 with row 2

3. $2\begin{vmatrix} a_1 & b_1 & c_1 & d_1 \\ a_1 & b_1 & c_1 & d_1 \\ a_3 & b_3 & c_3 & d_3 \\ a_4 & b_4 & c_4 & d_4 \end{vmatrix} = 0$ Add. prop. of =, adding $\begin{vmatrix} a_1 & b_1 & c_1 & d_1 \\ a_1 & b_1 & c_1 & d_1 \\ a_3 & b_3 & c_3 & d_3 \\ a_4 & b_4 & c_4 & d_4 \end{vmatrix}$ to each side

4. $\begin{vmatrix} a_1 & b_1 & c_1 & d_1 \\ a_1 & b_1 & c_1 & d_1 \\ a_3 & b_3 & c_3 & d_3 \\ a_4 & b_4 & c_4 & d_4 \end{vmatrix} = 0$ Mult. prop. of =, multiplying both sides by $\frac{1}{2}$

The proof for two columns with corresponding elements equal is identical.

12. 1. $a_1, a_2, b_1, b_2, c_1, c_2, r, t$ all real numbers Hypothesis

2. $\begin{vmatrix} a_1 & b_1 & c_1 \\ a_2 & b_2 & c_2 \\ ra_1 + ta_2 & rb_1 + tb_2 & rc_1 + tc_2 \end{vmatrix}$ Property 5 with $k = -t$

$= \begin{vmatrix} a_1 & b_1 & c_1 \\ a_2 & b_2 & c_2 \\ ra_1 + ta_2 + (-t)a_2 & rb_1 + tb_2 + (-t)b_2 & rc_1 + tc_2 + (-t)c_2 \end{vmatrix}$

3. $= \begin{vmatrix} a_1 & b_1 & c_1 \\ a_2 & b_2 & c_2 \\ ra_1 + ta_2 - ta_2 & rb_1 + tb_1 - tb_1 & rc_1 + tc_2 - tc_2 \end{vmatrix}$ Prop. of opp. in prod.

4. $= \begin{vmatrix} a_1 & b_1 & c_1 \\ a_2 & b_2 & c_2 \\ ra_1 + 0 & rb_1 + 0 & rc_1 + 0 \end{vmatrix}$ Ax. of add. inv.

5. $= \begin{vmatrix} a_1 & b_1 & c_1 \\ a_2 & b_2 & c_2 \\ ra_1 & rb_1 & rc_1 \end{vmatrix}$ Iden. ax. for add.

6. $= \begin{vmatrix} a_1 & b_1 & c_1 \\ a_2 & b_2 & c_2 \\ ra_1 + (-r)a_1 & rb_1 + (-r)b_1 & rc_1 + (-r)c_1 \end{vmatrix}$ Property 5 with $k = -r$

7. $= \begin{vmatrix} a_1 & b_1 & c_1 \\ a_2 & b_2 & c_2 \\ ra_1 - ra_1 & rb_1 - rb_2 & rc_1 - rc_2 \end{vmatrix}$ Prop. of opp. in prod.

8. $= \begin{vmatrix} a_1 & b_1 & c_1 \\ a_2 & b_2 & c_2 \\ 0 & 0 & 0 \end{vmatrix}$ Ax. of add. inv.

9. $= 0$ Property 1

10. $\begin{vmatrix} a_1 & b_1 & c_1 \\ a_2 & b_2 & c_2 \\ ra_1 + ta_2 & rb_1 + tb_2 & rc_1 + tc_2 \end{vmatrix} = 0$ Trans. prop. of =

Pages 188–189 • COMPUTER EXERCISES

1.
```
10 READ N
20 FOR C = 1 TO N
30 FOR I = 1 TO 3
40 READ A(I), B(I), C(I), D(I)
50 NEXT I
60 LET D = (A(1) * (B(2) * C(3) - B(3) * C(2))) - (B(1) *
   (A(2) * C(3) - A(3) * C(2))) + (C(1) * (A(2) * B(3) - A(3) *
   B(2)))
70 LET DX = (D(1) * (B(2) * C(3) - B(3) * C(2))) - (B(1) *
   (D(2) * C(3) - D(3) * C(2))) + (C(1) * (D(2) * B(3) - D(3) *
   B(2)))
80 LET DY = (A(1) * (D(2) * C(3) - C(2) * D(3))) - (D(1) *
   (A(2) * C(3) - A(3) * C(2))) + (C(1) * (A(2) * D(3) - A(3) *
   D(2)))
90 LET DZ = (A(1) * (B(2) * D(3) - D(2) * B(3))) - (B(1) *
   (A(2) * D(3) - A(3) * D(2))) + (D(1) * (A(2) * B(3) - A(3) *
   B(2)))
100 IF D = 0 THEN PRINT "NO UNIQUE SOLUTION" : GOTO 150
110 LET X = DX/D
120 LET Y = DY/D
130 LET Z = DZ/D
140 PRINT "(";X;", ";Y;", ";Z;")"
150 NEXT C
160 DATA 2
170 DATA 7,8,-3,3,-5,-9,7,32,4,-2,-11,-25
180 DATA 8,-7,10,69,-12,9,-1,-10,6,-1,-5,-23
190 END
```

2. {(6, −3, 5)} **3.** {(2.5, 3, 7)} **4.** no unique solution **5.** no unique solution

6.
```
10 READ N
20 DIM A(4,4)
30 FOR K = 1 TO N
40 FOR I = 1 TO 4
50 FOR J = 1 TO 4
60 READ A(I,J)
70 NEXT J
80 NEXT I
90 LET C1 = A(4,1) * ((A(1,2) * (A(2,3) * A(3,4) - A(3,3) *
   A(2,4))) - (A(1,3) * (A(2,2) * A(3,4) - A(3,2) * A(2,4))) +
   (A(1,4) * (A(2,2) * A(3,3) - A(3,2) * A(2,3))))
100 LET C2 = A(4,2) * ((A(1,1) * (A(2,3) * A(3,4) - A(3,3) *
    A(2,4))) - (A(1,3) * (A(2,1) * A(3,4) - A(3,1) * A(2,4))) +
    (A(1,4) * (A(2,1) * A(3,3) - A(3,1) * A(2,3))))
110 LET C3 = A(4,3) * ((A(1,1) * (A(2,2) * A(3,4) - A(3,2) *
    A(2,4))) - (A(1,2) * (A(2,1) * A(3,4) - A(3,1) * A(2,4))) +
    (A(1,4) * (A(2,1) * A(3,2) - A(3,1) * A(2,2))))
```

[*Program continued on next page*]

```
120  LET C4 = A(4,4) * ((A(1,1) * (A(2,2) * A(3,3) - A(3,2) *
     A(2,3))) - (A(1,2) * (A(2,1) * A(3,3) - A(3,1) * A(2,3))) +
     (A(1,3) * (A(2,1) * A(3,2) - A(3,1) * A(2,2))))
130  LET D = -C1 + C2 - C3 + C4
140  PRINT D
150  NEXT K
160  DATA 2
170  DATA 3,-2,5,4,-7,0,6,-11,-9,1,-7,8,4,-1,9,13
180  DATA 5,-6,-9,4,-3,8,7,-3,-1,10,5,-2,2,1,0,10
190  END
```

7. 3718 **8.** 0 **9.** -6; 33; -66

Page 189 • SELF-TEST 2

1.

$$D = \begin{vmatrix} 1 & 1 & 1 \\ 2 & 3 & 0 \\ 0 & 4 & -1 \end{vmatrix} = (-3 + 0 + 8) - (0 + 0 - 2) = 7$$

$$D_x = \begin{vmatrix} 4 & 1 & 1 \\ 3 & 3 & 0 \\ -6 & 4 & -1 \end{vmatrix} = (-12 + 0 + 12) - (-18 + 0 - 3) = 21$$

$$D_y = \begin{vmatrix} 1 & 4 & 1 \\ 2 & 3 & 0 \\ 0 & -6 & -1 \end{vmatrix} = (-3 + 0 - 12) - (0 + 0 - 8) = -7$$

$$D_z = \begin{vmatrix} 1 & 1 & 4 \\ 2 & 3 & 3 \\ 0 & 4 & -6 \end{vmatrix} = (-18 + 0 + 32) - (0 + 12 - 12) = 14$$

$$x = \frac{D_x}{D} = 3,\ y = \frac{D_y}{D} = -1,\ z = \frac{D_z}{D} = 2;\ \{(3, -1, 2)\}$$

2. Let x = number of orange trees; y = number of grapefruit trees; z = number of lemon trees.

$$\begin{aligned} 5x + 3y + 2z &= 59 \\ 7x + 12y + z &= 130 \\ 4x + y &= 31 \end{aligned} \qquad D = \begin{vmatrix} 5 & 3 & 2 \\ 7 & 12 & 1 \\ 4 & 1 & 0 \end{vmatrix} = (0 + 12 + 14) - (96 + 5 + 0) = -75$$

$$D_x = \begin{vmatrix} 59 & 3 & 2 \\ 130 & 12 & 1 \\ 31 & 1 & 0 \end{vmatrix} = (0 + 93 + 260) - (744 + 59 + 0) = -450$$

$$D_y = \begin{vmatrix} 5 & 59 & 2 \\ 7 & 130 & 1 \\ 4 & 31 & 0 \end{vmatrix} = (0 + 236 + 434) - (1040 + 155 + 0) = -525$$

$$D_z = \begin{vmatrix} 5 & 3 & 59 \\ 7 & 12 & 130 \\ 4 & 1 & 31 \end{vmatrix} = (1860 + 1560 + 413) - (2832 + 650 + 651) = -300$$

$x = \frac{D_x}{D} = 6$, $y = \frac{D_y}{D} = 7$, $z = \frac{D_z}{D} = 4$; orange tree, \$6; grapefruit tree, \$7; lemon tree, \$4

3. Multiply col. 2 by 2 and add the result to col. 4; expand by minors of the second row.

$$\begin{vmatrix} -1 & -3 & 4 & -1 \\ 0 & 1 & 0 & -2 \\ 3 & 1 & 2 & 6 \\ 2 & 1 & 0 & 5 \end{vmatrix} = \begin{vmatrix} -1 & -3 & 4 & -7 \\ 0 & 1 & 0 & 0 \\ 3 & 1 & 2 & 8 \\ 2 & 1 & 0 & 7 \end{vmatrix} = \begin{vmatrix} -1 & 4 & -7 \\ 3 & 2 & 8 \\ 2 & 0 & 7 \end{vmatrix} =$$

$(-14 + 64 + 0) - (-28 + 0 + 84) = -6$

Pages 190–191 • CHAPTER REVIEW

1. c 2. d 3. a 4. a

5. $(1)\quad x + z = 9$
$(2)\quad 3x + 2y - z = -15$
$(3)\quad 5x + 3y = -7$

$(1) + (2)$: $4x + 2y = -6\ (4)$
(3): $5x + 3y = -7$

$-3(4)$: $-12x - 6y = 18$
$2(3)$: $10x + 6y = -14$
$-2x = 4$

$x = -2,\ y = 1,\ z = 11$
$\{(-2, 1, 11)\}$; c

6. b 7. $D_z = \begin{vmatrix} 3 & 2 & 6 \\ 1 & 1 & -6 \\ 0 & 2 & 7 \end{vmatrix} = (21 + 0 + 12) - (0 - 36 + 14) = 55$; d

8. $(1)\ x + y + z = 37$
$(2)\ z = x + y - 3$
$(3)\ y = 2x + 2$

$x + y + z = 37$
$x + y - z = 3$
$-2x + y = 2$

$(1) + (2)$: $2x + 2y = 40$
(3): $-2x + y = 2$
$3y = 42$

$y = 14,\ x = 6,\ z = 17$
6, 14, 17; a

9. c 10. $\begin{vmatrix} 1 & 0 & 2 \\ -1 & 0 & 0 \\ -3 & 3 & -2 \end{vmatrix} = -0\begin{vmatrix} -1 & 0 \\ -3 & -2 \end{vmatrix} + 0\begin{vmatrix} 1 & 2 \\ -3 & -2 \end{vmatrix} - 3\begin{vmatrix} 1 & 2 \\ -1 & 0 \end{vmatrix}$

$= 0 + 0 - 3(0 + 2) = -6$; b

Pages 191–192 • CHAPTER TEST

1. 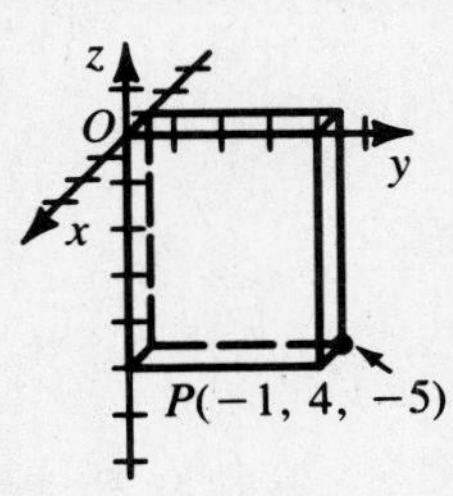

2. 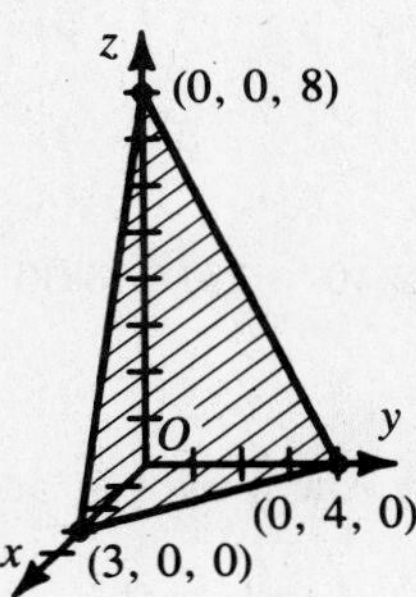

3. 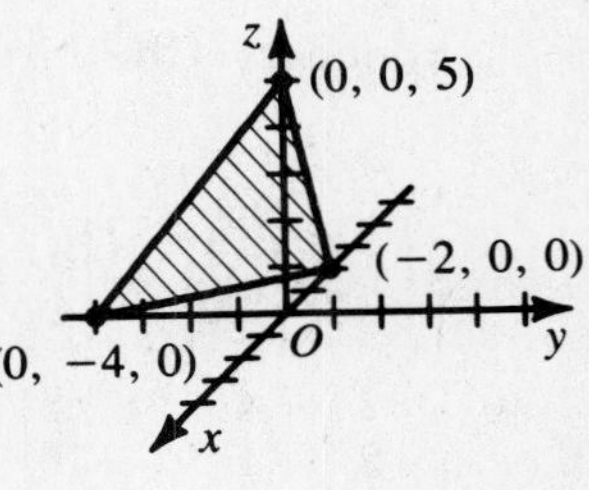

4. $(1)\quad 2x + 3y - z = 1$
$(2)\quad x - 4y + 4z = 3$
$(3)\quad -3x + y - 2z = -1$

$4(1)$: $8x + 12y - 4z = 4$
(2): $x - 4y + 4z = 3$
$9x + 8y = 7\ (4)$

(2): $x - 4y + 4z = 3$
$2(3)$: $-6x + 2y - 4z = -2$
$-5x - 2y = 1\ (5)$

(4): $9x + 8y = 7$
$4(5)$: $-20x - 8y = 4$
$-11x = 11$

$x = -1,\ y = 2,\ z = 3$
$\{(-1, 2, 3)\}$

5. $D = \begin{vmatrix} 2 & -3 & -1 \\ 1 & 2 & 3 \\ 3 & -5 & 2 \end{vmatrix} = 2\begin{vmatrix} 2 & 3 \\ -5 & 2 \end{vmatrix} - \begin{vmatrix} -3 & -1 \\ -5 & 2 \end{vmatrix} + 3\begin{vmatrix} -3 & -1 \\ 2 & 3 \end{vmatrix}$

$= 2(4 + 15) - (-6 - 5) + 3(-9 + 2) = 28;$

$D_x = \begin{vmatrix} -1 & -3 & -1 \\ -4 & 2 & 3 \\ 3 & -5 & 2 \end{vmatrix} = -\begin{vmatrix} 2 & 3 \\ -5 & 2 \end{vmatrix} + 4\begin{vmatrix} -3 & -1 \\ -5 & 2 \end{vmatrix} + 3\begin{vmatrix} -3 & -1 \\ 2 & 3 \end{vmatrix}$

$= -(4 + 15) + 4(-6 - 5) + 3(-9 + 2) = -84;$

$D_y = \begin{vmatrix} 2 & -1 & -1 \\ 1 & -4 & 3 \\ 3 & 3 & 2 \end{vmatrix} = 2\begin{vmatrix} -4 & 3 \\ 3 & 2 \end{vmatrix} - \begin{vmatrix} -1 & -1 \\ 3 & 2 \end{vmatrix} + 3\begin{vmatrix} -1 & -1 \\ -4 & 3 \end{vmatrix}$

$= 2(-8 - 9) - (-2 + 3) + 3(-3 - 4) = -56;$

$D_z = \begin{vmatrix} 2 & -3 & -1 \\ 1 & 2 & -4 \\ 3 & -5 & 3 \end{vmatrix} = 2\begin{vmatrix} 2 & -4 \\ -5 & 3 \end{vmatrix} - \begin{vmatrix} -3 & -1 \\ -5 & 3 \end{vmatrix} + 3\begin{vmatrix} -3 & -1 \\ 2 & -4 \end{vmatrix}$

$= 2(6 - 20) - (-9 - 5) + 3(12 + 2) = 28;$

$x = \frac{D_x}{D} = -3,\ y = \frac{D_y}{D} = -2,\ z = \frac{D_z}{D} = 1;\ \{(-3, -2, 1)\}$

6. $D = \begin{vmatrix} 3 & -2 & 1 \\ -1 & 3 & -1 \\ 6 & -4 & 2 \end{vmatrix} = 0$; Row 3 is twice row 1 so equations (1) and (3) represent parallel planes. Therefore the system is inconsistent.

7. $x + y + z = 26$ *(1)* $x + y + z = 26$ *(1)* − *(2)*: $2z = 24,\ z = 12$
$2x + 2y = 2z + 4$ *(2)* $x + y - z = 2$ *(1)* + *(3)*: $5x = 25,\ x = 5$
$4x = y + z - 1$ *(3)* $4x - y - z = -1$ $y = 9$

5 cm, 9 cm, 12 cm

8. $\begin{vmatrix} -1 & 2 & 3 & 1 \\ 0 & 3 & 4 & 5 \\ 1 & 0 & 0 & -2 \\ 4 & 1 & -3 & 2 \end{vmatrix}$ Multiply column 1 by 2, add column 1 to column 4; expand by minors of the third row. $\begin{vmatrix} -1 & 2 & 3 & -1 \\ 0 & 3 & 4 & 5 \\ 1 & 0 & 0 & 0 \\ 4 & 1 & -3 & 10 \end{vmatrix} = \begin{vmatrix} 2 & 3 & -1 \\ 3 & 4 & 5 \\ 1 & -3 & 10 \end{vmatrix} = 2\begin{vmatrix} 4 & 5 \\ -3 & 10 \end{vmatrix} -$

$3\begin{vmatrix} 3 & -1 \\ -3 & 10 \end{vmatrix} + \begin{vmatrix} 3 & -1 \\ 4 & 5 \end{vmatrix} = 2(40 + 15) - 3(30 - 3) + (15 + 4) = 48$

9. $\begin{vmatrix} 2 & -4 & 6 & 8 \\ -2 & 1 & 3 & 0 \\ 3 & 2 & -1 & 1 \\ 1 & -2 & 3 & 4 \end{vmatrix}$ Multiply row 4 by −2, add row 4 to row 1; by Property 1, the determinant is equal to 0. $\begin{vmatrix} 0 & 0 & 0 & 0 \\ -2 & 1 & 3 & 0 \\ 3 & 2 & -1 & 1 \\ 1 & -2 & 3 & 4 \end{vmatrix} = 0$

Page 193 • APPLICATION

1. $Q = CV$, $2 \times 10^{-6} = C(20)$, $C = 10^{-7}$. When C is doubled: $Q = (2 \times 10^{-7})25 = 5 \times 10^{-6}$ coulombs

2. **a.** The capacitance is directly proportional to the area of the plates. Therefore, the capacitance is doubled.
   b. The capacitance is directly proportional to the reciprocal of the distance between the plates. Therefore, the capacitance is doubled.

Page 194 • PREPARING FOR COLLEGE ENTRANCE EXAMS

1. $5x - y = 4$ $\quad 20x + 4y = 16$
   $-20x + 4y = -3$ $\quad \underline{-20x + 4y = -3}$
   $0 = 13$; E

2. $7x - 3y = 1$
   $-4x + 2y = -2$ $\quad D_y = \begin{vmatrix} 7 & 1 \\ -4 & -2 \end{vmatrix} = -14 + 4 = -10$; C

3. $y = mx + b_1$ $\quad y = mb_2 + b_1$; $\{(b_2, mb_2 + b_1)\}$; A
   $x = b_2$

4. $r = \dfrac{3ka}{2}$, $15 = \dfrac{3k(8)}{2}$, $15 = 12k$, $1.25 = k$; $r = \dfrac{3(1.25)(-0.5)}{2} = -\dfrac{1.875}{2} = -\dfrac{1875}{2000} = -\dfrac{15}{16}$; E

5. A 6. E 7. $-\dfrac{2}{3} = \dfrac{x+3}{3-6}$, $-\dfrac{2}{3} = \dfrac{x+3}{-3}$, $2 = x + 3$, $-1 = x$; B

8. Let x = original length of wire. $\dfrac{12}{36} = \dfrac{18}{x}$, $\dfrac{1}{3} = \dfrac{18}{x}$, $x = 54$; B

Pages 196–197 • PROGRAMMING IN PASCAL

1. **a.** See the function *det* in the program in Exercise 1b. **b.** Program follows.

```
PROGRAM cramer3 (INPUT, OUTPUT);

{* This program solves a system of three linear equations in three  *}
{* variables of the form          A1 * X + B1 * Y + C1 * Z = K1     *}
{*                                A2 * X + B2 * Y + C2 * Z = K2     *}
{*                                A3 * X + B3 * Y + C3 * Z = K3     *}
{* where A1,B1,C1,K1,A2,B2,C2,K2,A3,B3,C3, and K3 are integers,    *}
{* by Cramer's method.                                              *}

VAR
  a1,b1,c1,k1,a2,b2,c2,k2,a3,b3,c3,k3,d,dx,dy,dz : integer;
  x,y,z : real;
  answer : char;

{ ***************************************************************** }
PROCEDURE get_values;

BEGIN
  writeln('Enter coefficients and constant of 1st equation,');
  write('place a space between each of the values:');
  readln(a1,b1,c1,k1);
  writeln;
  writeln('Enter coefficients and constant of 2nd equation,');
```

[*Program continued on next page*]

```
    write('place a space between each of the values:');
    readln(a2,b2,c2,k2);
    writeln;
    writeln('Enter coefficients and constant of 3rd equation,');
    write('place a space between each of the values:');
    readln(a3,b3,c3,k3);
    writeln;
  END;

{ *************************************************************** }
FUNCTION det(s1,t1,u1,s2,t2,u2,s3,t3,u3 : integer) : integer;

BEGIN
  det := s1 * t2 * u3 + t1 * u2 * s3 + u1 * s2 * t3 - u1 * t2 * s3
                                     - s1 * u2 * t3 - t1 * s2 * u3;
END;

{ *************************************************************** }
PROCEDURE solve_system;

BEGIN
  d := det(a1,b1,c1,a2,b2,c2,a3,b3,c3);
  IF d <> 0
     THEN
       BEGIN
         dx := det(k1,b1,c1,k2,b2,c2,k3,b3,c3);
         dy := det(a1,k1,c1,a2,k2,c2,a3,k3,c3);
         dz := det(a1,b1,k1,a2,b2,k2,a3,b3,k3);
         x := dx/d;
         y := dy/d;
         z := dz/d;
       END;
END;

{ *************************************************************** }
PROCEDURE display_answer;

BEGIN
  IF d = 0
     THEN writeln('No solution or infinitely many solutions.')
     ELSE writeln('Soln. is: (',x:3:3,',',y:3:3,',',z:3:3,')');
END;

{ *************************************************************** }
BEGIN {* main *}
  REPEAT
    get_values;
    solve_system;
```

```
    display_answer;
    writeln;
    write('Do you want to solve another system < Y / N >: ');
    readln(answer);
    writeln;
  UNTIL answer in ['N','n'];
END.
```

2. See the procedure *solve_system* in Exercise 3.

3.

```
PROGRAM cramer4 (INPUT, OUTPUT);

{* This program solves a system of four linear equations in four    *}
{* variables of the form  A1 * X + B1 * Y + C1 * Z + D1 * W = K1  *}
{*                        A2 * X + B2 * Y + C2 * Z + D2 * W = K2  *}
{*                        A3 * X + B3 * Y + C3 * Z + D3 * W = K3  *}
{*                        A4 * X + B4 * Y + C4 * Z + D4 * W = K4  *}
{* where A1,B1,C1,D1,K1,A2,B2,C2,D2,K2,A3,B3,C3,D3,K3,A4,B4,C4, *}
{*                 D4 and K4 are integers, by Cramer's method. *}

VAR
  a1,b1,c1,d1,k1,a2,b2,c2,d2,k2 : integer;
  a3,b3,c3,d3,k3,a4,b4,c4,d4,k4,d,dx,dy,dz,dw : integer;
  x,y,z,w : real;
  answer : char;

{ **************************************************************}
PROCEDURE get_values;

BEGIN
  writeln('Enter coefficients and constant of 1st equation,');
  write('place a space between each of the values:');
  readln(a1,b1,c1,d1,k1);
  writeln;
  writeln('Enter coefficients and constant of 2nd equation,');
  write('place a space between each of the values:');
  readln(a2,b2,c2,d2,k2);
  writeln;
  writeln('Enter coefficients and constant of 3rd equation,');
  write('place a space between each of the values:');
  readln(a3,b3,c3,d3,k3);
  writeln;
  writeln('Enter coefficients and constant of 4th equation,');
  write('place a space between each of the values:');
  readln(a4,b4,c4,d4,k4);
  writeln;
END;
```

[*Program continued on next page*]

```
{ ***************************************************************** }
FUNCTION det(s1,t1,u1,s2,t2,u2,s3,t3,u3 : integer) : integer;

BEGIN
  det := s1 * t2 * u3 + t1 * u2 * s3 + u1 * s2 * t3 - u1 * t2 * s3 -
                                        s1 * u2 * t3 - t1 * s2 * u3;
END;

{ ***************************************************************** }
PROCEDURE solve_system;

BEGIN
  d := 0;
  d := d + a1 * det(b2,c2,d2,b3,c3,d3,b4,c4,d4);
  d := d - a2 * det(b1,c1,d1,b3,c3,d3,b4,c4,d4);
  d := d + a3 * det(b1,c1,d1,b2,c2,d2,b4,c4,d4);
  d := d - a4 * det(b1,c1,d1,b2,c2,d2,b3,c3,d3);
  IF d <> 0
     THEN
       BEGIN
         dx := 0;
         dx := dx + k1 * det(b2,c2,d2,b3,c3,d3,b4,c4,d4);
         dx := dx - k2 * det(b1,c1,d1,b3,c3,d3,b4,c4,d4);
         dx := dx + k3 * det(b1,c1,d1,b2,c2,d2,b4,c4,d4);
         dx := dx - k4 * det(b1,c1,d1,b2,c2,d2,b3,c3,d3);
         dy := 0;
         dy := dy +  a1 * det(k2,c2,d2,k3,c3,d3,k4,c4,d4);
         dy := dy -  a2 * det(k1,c1,d1,k3,c3,d3,k4,c4,d4);
         dy := dy +  a3 * det(k1,c1,d1,k2,c2,d2,k4,c4,d4);
         dy := dy -  a4 * det(k1,c1,d1,k2,c2,d2,k3,c3,d3);
         dz := 0;
         dz := dz +  a1 * det(b2,k2,d2,b3,k3,d3,b4,k4,d4);
         dz := dz -  a2 * det(b1,k1,d1,b3,k3,d3,b4,k4,d4);
         dz := dz +  a3 * det(b1,k1,d1,b2,k2,d2,b4,k4,d4);
         dz := dz -  a4 * det(b1,k1,d1,b2,k2,d2,b3,k3,d3);
         dw := 0;
         dw := dw +  a1 * det(b2,c2,k2,b3,c3,k3,b4,c4,k4);
         dw := dw -  a2 * det(b1,c1,k1,b3,c3,k3,b4,c4,k4);
         dw := dw +  a3 * det(b1,c1,k1,b2,c2,k2,b4,c4,k4);
         dw := dw -  a4 * det(b1,c1,k1,b2,c2,k2,b3,c3,k3);
         x := dx/d;
         y := dy/d;
         z := dz/d;
         w := dw/d;
       END;
END;
```

```
{ ***************************************************************}
PROCEDURE display_answer;

BEGIN
  IF d = 0
     THEN writeln('No solution or infinitely many solutions.')
     ELSE writeln('(',x:3:3,', ',y:3:3,', ',z:3:3,', ',w:3:3,')');
END;

{ ***************************************************************}
BEGIN {* main *}
  REPEAT
    get_values;
    solve_system;
    display_answer;
    writeln;
    write('Do you want to solve another system <Y / N>: ');
    readln(answer);
    writeln;
  UNTIL answer in ['N','n'];
END.
```

4. a. Both determinants equal −49.
   b. When the determinant is evaluated by the method shown on page 177, every product is equal to zero, except the product of the numbers along the main diagonal.

5.

```
PROGRAM upper_triangular (INPUT, OUTPUT);

TYPE
  determinant = ARRAY[1..3,1..3] OF real;

VAR
  situation : 1..3;
  value : determinant;
  factor : real;
  finished : boolean;

{ ***************************************************************}
PROCEDURE get_values;

VAR
  i,j : 1..3;

BEGIN
  FOR i := 1 TO 3 DO
      FOR j := 1 TO 3 DO
          BEGIN
```

[*Program continued on next page*]

```
                write('Enter element in row ',i:1,'column ',j:1,': ');
                readln(value[i,j]);
              END;
END;

{ ************************************************************** }
PROCEDURE mult_and_add(factor : real; row_or_col : char; position_1,
                                               position_2 : integer);

VAR
  i : 1..3;

BEGIN
  IF row_or_col = 'c'
     THEN FOR i := 1 TO 3 DO
              value[i,position_2] := value[i,position_1] * factor +
                                                 value[i,position_2]
     ELSE FOR i := 1 TO 3 DO
              value[position_2,i] := value[position_1,i] * factor +
                                                value[position_2,i];
END;

{ ************************************************************** }
PROCEDURE switch(row_or_col : char; place_1, place_2 : integer);

VAR
  i : 1..3;
  temp : ARRAY[1..3] OF real;

BEGIN
  IF row_or_col = 'c'
      THEN FOR i := 1 TO 3 DO
               BEGIN
                 temp[i] := value[i,place_2];
                 value[i,place_2] := value[i,place_1];
                 value[i,place_1] := temp[i];
               END
      ELSE FOR i := 1 TO 3 DO
               BEGIN
                 temp[i] := value[place_2,i];
                 value[place_2,i] := value[place_1,i];
                 value[place_1,i] := temp[i];
               END;
END;
```

```
{ ***************************************************************** }
PROCEDURE negate(row_or_col : char; place : integer);

VAR
  i : 1..3;

BEGIN
  IF row_or_col = 'c'
     THEN FOR i := 1 TO 3 DO
              value[i,place] := -value[i,place]
     ELSE FOR i := 1 TO 3 DO
              value[place,i] := -value[place,i];
END;

{ ***************************************************************** }
PROCEDURE display_det;

VAR
  i,j : 1..3;

BEGIN
  writeln;
  FOR i := 1 TO 3 DO
      BEGIN
        FOR j := 1 TO 3 DO
            write(value[i,j]:1:3,' ');
            writeln;
      END;
END;

{ ***************************************************************** }
BEGIN {* main *}
  get_values;
  IF value[1,1] <> 0
     THEN BEGIN
            IF value[2,1] <> 0
               THEN BEGIN
                      factor := -value[2,1]/value[1,1];
                      mult_and_add(factor,'r',1,2);
                    END;
            IF value[3,1] <> 0
               THEN BEGIN
                      factor := -value[3,1]/value[1,1];
                      mult_and_add(factor,'r',1,3);
                    END;
            situation := 1;
          END
```

[*Program continued on next page*]

```
ELSE IF value[3,3] <> 0
        THEN BEGIN
               IF value[3, 1] <> 0
                  THEN BEGIN
                         factor := -value[3,1]/value[3,3];
                         mult_and_add(factor,'c',3,1);
                       END;
               IF value[3,2] <> 0
                  THEN BEGIN
                         factor := -value[3,2]/value[3,3];
                         mult_and_add(factor,'c',3,2);
                       END;
               situation := 2;
             END
        ELSE BEGIN
               switch('r',1,2);
               switch('c',2,3);
               situation := 3;
             END;
CASE situation OF
     1 : BEGIN
           IF value[3,3] <> 0
              THEN BEGIN
                     IF value[3,2] <> 0
                        THEN BEGIN
                               factor := -value[3,2]/value[3,3];
                               mult_and_add(factor,'c',3,2);
                             END;
                   END
              ELSE BEGIN
                     switch('c',2,3);
                     negate('r',1);
                   END;
         END;
     2 : BEGIN
           IF value[1,1] <> 0
              THEN BEGIN
                     IF value[2,1] <> 0
                        THEN BEGIN
                               factor := -value[2,1]/value[1,1];
                               mult_and_add(factor,'r',1,2);
                             END;
                   END
              ELSE BEGIN
                     switch('r',1,2);
                     negate('r',1);
                   END;
         END;
```

```
        3 : BEGIN
              IF value[3,3] <> 0
                 THEN BEGIN
                        IF value[3,1] <> 0
                           THEN BEGIN
                                  factor := -value[3,1]/-value[3,3];
                                  mult_and_add(factor,'c',3,1);
                                END;
                      END
                 ELSE BEGIN
                        switch('c',1,3);
                        IF value[1,1] <> 0
                           THEN BEGIN
                                  IF value[2,1] <> 0
                                     THEN BEGIN
                                            factor := -value[2,1]/
                                                       value[1,1];
                                            mult_and_add(factor,
                                                         'r',1,2);
                                          END;
                                END
                           ELSE BEGIN
                                  switch('r',1,2);
                                  negate('r',1);
                                END;
                      END;
            END;
  END; {* case *}

  display_det;
END.
```

CHAPTER 6 • Polynomials and Rational Expressions

Page 203 • WRITTEN EXERCISES

A 1. $(-7u^2)(5u^8) = (-7 \cdot 5)(u^{2+8}) = -35u^{10}$

2. $(-8v^3)(-6v^{-9}) = (-8)(-6)(v^{3-9}) = 48v^{-6} = \frac{48}{v^6}$

3. $(2x^4y^{-7})(3x^{-1}y^5) = (2 \cdot 3)(x^{4-1})(y^{-7+5}) = 6x^3y^{-2} = \frac{6x^3}{y^2}$

4. $(9z^6w^8)(-5z^{-6}w^{-11}) = 9(-5)(z^{6-6})(w^{8-11}) = -45z^0w^{-3} = -\frac{45}{w^3}$

5. $(5^{-3}a^{-2}b^{-1})(-25a^6b^{-4}) = -(5^{-3} \cdot 5^2)(a^{-2+6})(b^{-1-4}) = -5^{-1}a^4b^{-5} = -\frac{a^4}{5b^5}$

6. $(-3p^4q^{-3})^2(4p^{-5}q^7) = (9p^8q^{-6})(4p^{-5}q^7) = (9 \cdot 4)(p^{8-5})(q^{-6+7}) = 36p^3q$

7. $\left(\frac{3}{4}d^3e^{-5}\right)^{-2} = \left[\left(\frac{4}{3}\right)^{-1}\right]^{-2}(d^3)^{-2}(e^{-5})^{-2} = \left(\frac{4}{3}\right)^2 d^{-6}e^{10} = \frac{16e^{10}}{9d^6}$

8. $\left(\frac{1}{2}a^2b^3\right)^4(a^{-5}b^{-10}) = \left(\frac{1}{2}\right)^4(a^2)^4a^{-5}(b^3)^4b^{-10} = \frac{1}{16}a^{8-5}b^{12-10} = \frac{a^3b^2}{16}$

9. $(6x^8y^{-9})(x^{-2}y^{-3})^3 = 6x^8(x^{-2})^3y^{-9}(y^{-3})^3 = 6x^{8-6}y^{-9-9} = 6x^2y^{-18} = \frac{6x^2}{y^{18}}$

10. $(-v^7w^{-8})^3(-v^{-9}w^6)^4 = (-v^7)^3(-v^{-9})^4(w^{-8})^3(w^6)^4 = -v^{21}v^{-36}w^{-24}w^{24} =$
$-v^{-15} = -\frac{1}{v^{15}}$

11. $(u^6)^{-4}(u^3)^8 = u^{-24}u^{24} = 1$ 12. $(ab)^{-7}(a^3b^{-2})^4 = a^{-7}a^{12}b^{-7}b^{-8} = a^5b^{-15} = \frac{a^5}{b^{15}}$

13. $\frac{c^4d^{-5}}{c^3d} = c^4c^{-3}d^{-5}d^{-1} = c^1d^{-6} = \frac{c}{d^6}$ 14. $\left(\frac{f^5g^{-3}}{g^{-5}}\right)^2 = (f^5g^{-3+5})^2 = (f^5g^2)^2 = f^{10}g^4$

15. $\frac{(2r^3s^{-3})^5}{4r^7s^{-10}} = \frac{32r^{15}s^{-15}}{4r^7s^{-10}} = 8r^{15-7}s^{-15+10} = \frac{8r^8}{s^5}$

16. $\frac{25x^{-5}y^3}{(5x^2y)^{-3}} = (25x^{-5}y^3)(5x^2y)^3 = 25 \cdot 125x^{-5}x^6y^3y^3 = 3125xy^6$

17. $\left(\frac{3u^{-4}v^3}{2u^{-5}v}\right)^{-2} = \left(\frac{2u^{-5}v}{3u^{-4}v^3}\right)^2 = \left(\frac{2}{3}u^{-5+4}v^{1-3}\right)^2 = \left(\frac{2u^{-1}v^{-2}}{3}\right)^2 = \frac{4u^{-2}v^{-4}}{9} = \frac{4}{9u^2v^4}$

18. $\left(-\frac{a^{-2}b^{-1}}{3a^{-3}b^3}\right)^{-4} = \left(\frac{3a^{-3}b^3}{a^{-2}b^{-1}}\right)^4 = (3a^{-3+2}b^{3+1})^4 = 81a^{-4}b^{16} = \frac{81b^{16}}{a^4}$

B 19. $\frac{(x+y)^{-2}}{(x+y)^{-1}} = (x+y)^{-2+1} = (x+y)^{-1} = \frac{1}{x+y}$

20. $\frac{a^{-1} - b^{-1}}{a^{-1} + b^{-1}} = \left(\frac{a^{-1} - b^{-1}}{a^{-1} + b^{-1}}\right)\frac{ab}{ab} = \frac{a^{-1+1}b - ab^{-1+1}}{a^{-1+1}b + ab^{-1+1}} = \frac{b-a}{b+a}$

21. $(z - w)(w^{-1} - z^{-1}) = (z - w)w^{-1} - (z - w)z^{-1} = zw^{-1} - 1 - 1 + wz^{-1} =$
$\frac{z}{w} + \frac{w}{z} - 2$

22. $(d^{-1} + e^{-1})(d + e)^{-1} = \left(\frac{d^{-1} + e^{-1}}{d + e}\right)\frac{de}{de} = \frac{(e + d)}{(d + e)de} = \frac{1}{de}$

23. $\frac{a^{m+1}}{a^{m-1}} = a^{(m+1)-(m-1)} = a^2$

24. $(b^{2-p}b^{2+p})^2 = (b^4)^2 = b^8$

25. $2(4^k + 4^k) = 2(2 \cdot 4^k) = 4(4^k) = 4^{k+1}$

26. $5(5^{e+1})^{e-1} = 5 \cdot 5^{(e+1)(e-1)} = 5^1 \cdot 5^{e^2-1} = 5^{e^2}$

27. Let $m < n$; $b \neq 0$.

1. $m < n;\ b \neq 0$ Hypothesis
2. $n - m > 0$ Add. prop. of order
3. $b^n = b^{n-m}b^m$ Law 1
4. $\frac{b^m}{b^n} = \frac{b^m}{b^{n-m}b^m}$ Subs. prin.
5. $= \frac{1}{b^{n-m}} \cdot \frac{b^m}{b^m}$ Corollary, page 201
6. $= \frac{1}{b^{n-m}} \cdot 1$ $\frac{b^m}{b^m} = 1$
7. $= \frac{1}{b^{n-m}}$ Iden. ax. for mult.
8. $\frac{b^m}{b^n} = \frac{1}{b^{n-m}}$ Trans. prop. of =

28.
1. $\left(\frac{a}{b}\right)^m = \left(\frac{a}{b}\right)^m \cdot 1$ Iden. ax. for mult.
2. $= \left(\frac{a}{b}\right)^m \cdot \frac{b^m}{b^m}$ $\frac{b^m}{b^m} = 1$
3. $= \left(\frac{a}{b}\right)^m \cdot \left(b^m \cdot \frac{1}{b^m}\right)$ Def. of div.
4. $= \left[\left(\frac{a}{b}\right)^m \cdot b^m\right] \cdot \frac{1}{b^m}$ Assoc. ax. for mult.
5. $= \left(\frac{a}{b} \cdot b\right)^m \cdot \frac{1}{b^m}$ Law 3
6. $= a^m \cdot \frac{1}{b^m}$ Subs. prin.
7. $= \frac{a^m}{b^m}$ Def. of div.
8. $\left(\frac{a}{b}\right)^m = \frac{a^m}{b^m}$ Trans. prop. of =

29.
1. $\frac{1}{b^{-n}} = \frac{1}{\frac{1}{b^n}}$ Def. of b^{-n}
2. $= 1\left(\frac{b^n}{1}\right)$ Ex. 34, page 28
3. $= b^n$ Iden. ax. for mult.
4. $\frac{1}{b^{-n}} = b^n$ Trans. prop. of =

30.
1. $(ab)^{-n} = \frac{1}{(ab)^n}$ Definition, page 202
2. $= \frac{1}{a^n b^n}$ Law 3
3. $= \frac{1}{a^n} \cdot \frac{1}{b^n}$ Prop. of recip. of a prod.
4. $= a^{-n}b^{-n}$ Definition, page 202
5. $(ab)^{-n} = a^{-n}b^{-n}$ Trans. prop. of =

C **31.** If $m = 0$, then $b^0b^n = b^{0+n} = b^n$ by Law 1, and $b^0 = 1$. If $n = 0$, then $b^mb^0 = b^{m+0} = b^m$ by Law 1, and $b^0 = 1$.

32. Let $m = -n$. Then by Law 1, $b^{-n}b^n = b^{-n+n} = b^0$. From the definition on page 202 and Ex. 31, $b^0 = 1$. Thus, $b^{-n} = \frac{1}{b^n}$.

33.

1. r, s, t, and $u \in \mathcal{R}$; $t \neq 0$; $u \neq 0$	Hypothesis
2. $\frac{rs}{tu} = rs \cdot \frac{1}{tu}$	Def. of div.
3. $= rs\left(\frac{1}{t} \cdot \frac{1}{u}\right)$	Prop. of recip. of a prod.
4. $= \left(r \cdot \frac{1}{t}\right)\left(s \cdot \frac{1}{u}\right)$	Assoc. and comm. axs. for mult.
5. $= \frac{r}{t} \cdot \frac{s}{u}$	Def. of div.
6. $\frac{rs}{tu} = \frac{r}{t} \cdot \frac{s}{u}$	Trans. prop. of =

Page 206 • WRITTEN EXERCISES

A **1.** $(x + 3y)(x - 2y) = x^2 - 2xy + 3xy - 6y^2 = x^2 + xy - 6y^2$
2. $(6a + 5)(a + 3) = 6a^2 + 18a + 5a + 15 = 6a^2 + 23a + 15$
3. $(t + 4)^2 = t^2 + 2(4t) + 4^2 = t^2 + 8t + 16$
4. $(3n - 2)^2 = (3n)^2 - 2 \cdot 2 \cdot 3n + 2^2 = 9n^2 - 12n + 4$
5. $(5 + s)(5 - s) = 25 - s^2$ **6.** $(2b + 1)(2b - 1) = 4b^2 - 1$
7. $(x^3 - 2y)(x^3 + 2y) = (x^3)^2 - (2y)^2 = x^6 - 4y^2$
8. $(3c + 5d)^2 = (3c)^2 + 2 \cdot 3c \cdot 5d + (5d)^2 = 9c^2 + 30cd + 25d^2$
9. $(2u^2 - 5v)^2 = (2u^2)^2 - 2 \cdot 2u^2 \cdot 5v + (5v)^2 = 4u^4 - 20u^2v + 25v^2$
10. $9x^7(x^3 - 4x) = 9x^7 \cdot x^3 - (9x^7)(4x^1) = 9x^{10} - 36x^8$
11. $(a^2 - b)(a^2 - 2b) = a^4 - a^2(2b) - a^2b + 2b^2 = a^4 - 3a^2b + 2b^2$
12. $(3t - 7)^2 = (3t)^2 - 2(3t)7 + 7^2 = 9t^2 - 42t + 49$
13. $(n^5 - 8n^3)6n^2 = 6n^5n^2 - 6 \cdot 8n^3n^2 = 6n^7 - 48n^5$
14. $(2m - 3)(m - 6) = 2m^2 - 6 \cdot 2m - 3m + 18 = 2m^2 - 15m + 18$
15. $(d^5 + e^3)^2 = (d^5)^2 + 2d^5e^3 + (e^3)^2 = d^{10} + 2d^5e^3 + e^6$
16. $(4 - 7h^3)^2 = 4^2 - 2 \cdot 4 \cdot 7h^3 + (7h^3)^2 = 16 - 56h^3 + 49h^6$
17. $(r^n - 5)(r^n + 5) = (r^n)^2 - 5^2 = r^{2n} - 25$
18. $(k^n - 3p^m)^2 = (k^n)^2 - 2k^n(3p^m) + (3p^m)^2 = k^{2n} - 6k^np^m + 9p^{2m}$
19. $(2x - 3)(2x^2 + 3x - 5) = 2x(2x^2 + 3x - 5) - 3(2x^2 + 3x - 5) =$
$(4x^3 + 6x^2 - 10x) - (6x^2 + 9x - 15) = 4x^3 + 6x^2 - 10x - 6x^2 - 9x + 15 =$
$4x^3 - 19x + 15$
20. $(a + 2b)(3a^2 - ab - 7b^2) = a(3a^2 - ab - 7b^2) + 2b(3a^2 - ab - 7b^2) =$
$(3a^3 - a^2b - 7ab^2) + (6a^2b - 2ab^2 - 14b^3) = 3a^3 + 5a^2b - 9ab^2 - 14b^3$
21. $(u - 4)(u^2 + 4u + 16) = u(u^2 + 4u + 16) - 4(u^2 + 4u + 16) =$
$(u^3 + 4u^2 + 16u) - (4u^2 + 16u + 64) = u^3 + 4u^2 + 16u - 4u^2 - 16u - 64 =$
$u^3 - 64$
22. $(5 - t)(25 + 5t + t^2) = 5(25 + 5t + t^2) - t(25 + 5t + t^2) = (125 + 25t + 5t^2) -$
$(25t + 5t^2 + t^3) = 125 + 25t + 5t^2 - 25t - 5t^2 - t^3 = 125 - t^3$

B **23.** $(z + 3)^3 = (z + 3)^2(z + 3) = (z^2 + 6z + 9)(z + 3) = (z^2 + 6z + 9)z +$
$(z^2 + 6z + 9)3 = z^3 + 6z^2 + 9z + 3z^2 + 18z + 27 = z^3 + 9z^2 + 27z + 27$
24. $(v - 4w)^3 = (v - 4w)^2(v - 4w) = (v^2 - 8vw + 16w^2)(v - 4w) =$
$(v^2 - 8vw + 16w^2)v + (v^2 - 8vw + 16w^2)(-4w) = v^3 - 8v^2w + 16vw^2 -$
$4v^2w + 32vw^2 - 64w^3 = v^3 - 12v^2w + 48vw^2 - 64w^3$

25. $[(x - y)(x + y)](x^2 + y^2) = (x^2 - y^2)(x^2 + y^2) = (x^2)^2 - (y^2)^2 = x^4 - y^4$

26. $(2c - 3d)^2(2c + 3d)^2 = [(2c - 3d)(2c + 3d)]^2 = (4c^2 - 9d^2)^2 = (4c^2)^2 - 2(4c^2)(9d^2) + (9d^2)^2 = 16c^4 - 72c^2d^2 + 81d^4$

27. $[(p + q)(p - q)](p^4 + p^2q^2 + q^4) = (p^2 - q^2)(p^4 + p^2q^2 + q^4) = p^2(p^4 + p^2q^2 + q^4) - q^2(p^4 + p^2q^2 + q^4) = (p^6 + p^4q^2 + p^2q^4) - (p^4q^2 + p^2q^4 + q^6) = p^6 + p^4q^2 + p^2q^4 - p^4q^2 - p^2q^4 - q^6 = p^6 - q^6$

28. $(h - 2k)^4 = [(h - 2k)^2]^2 = (h^2 - 4hk + 4k^2)^2$

$$\begin{array}{rrrrr}
h^2 & - 4hk & + 4k^2 & & \\
h^2 & - 4hk & + 4k^2 & & \\
\hline
h^4 & - 4h^3k & + 4h^2k^2 & & \\
 & - 4h^3k & + 16h^2k^2 & - 16hk^3 & \\
 & & 4h^2k^2 & - 16hk^3 & + 16k^4 \\
\hline
h^4 & - 8h^3k & + 24h^2k^2 & - 32hk^3 & + 16k^4
\end{array}$$

29. $(a - b)(a^2 + ab + b^2) = a^3 - b^3$ **30.** $(a + b)^3 = a^3 + 3a^2b + 3ab^2 + b^3$

C **31.**

$$\begin{array}{rrrr}
x^{2n} + & x^ny^n + & y^{2n} & \\
 & x^n + & y^n & \\
\hline
x^{3n} + & x^{2n}y^n + & x^ny^{2n} & \\
 & x^{2n}y^n + & x^ny^{2n} & + y^{3n} \\
\hline
x^{3n} + & 2x^{2n}y^n + & 2x^ny^{2n} & + y^{3n}
\end{array}$$

32. $(r^{4n} - s^{3m})^2 = (r^{4n})^2 - 2(r^{4n})(s^{3m}) + (s^{3m})^2 = r^{8n} - 2r^{4n}s^{3m} + s^{6m}$

33. $(a^{p+1} - a^2b^q)(a^{p-1} + b^q) = a^{p+1} \cdot a^{p-1} + a^{p+1}b^q - a^2b^q \cdot a^{p-1} - a^2b^q \cdot b^q = a^{2p} + a^{p+1}b^q - a^{p+1}b^q - a^2b^{2q} = a^{2p} - a^2b^{2q}$

34.

$$\begin{array}{rrrrr}
v^2 & - vw & + w^2 & & \\
v^2 & + vw & + w^2 & & \\
\hline
v^4 & - v^3w & + v^2w^2 & & \\
 & v^3w & - v^2w^2 & + vw^3 & \\
 & & v^2w^2 & - vw^3 & + w^4 \\
\hline
v^4 & & + v^2w^2 & & + w^4
\end{array}$$

35.

$$\begin{array}{rrrrrr}
a^4 + & a^3b + & a^2b^2 + & ab^3 + & b^4 & \\
 & & & a & - b & \\
\hline
a^5 + & a^4b + & a^3b^2 + & a^2b^3 + & ab^4 & \\
 & - a^4b & - a^3b^2 & - a^2b^3 & - ab^4 & - b^5 \\
\hline
a^5 & & & & & - b^5
\end{array}$$

36.

$$\begin{array}{rrrrrr}
z^9 & + z^8 & + \cdots & + z & + 1 & \\
 & & & z & - 1 & \\
\hline
z^{10} & + z^9 & + z^8 & + \cdots & + z & \\
 & - z^9 & - z^8 & - \cdots & - z & - 1 \\
\hline
z^{10} & & & & & - 1
\end{array}$$

37. $(x^n - x^{-n})^2 + 4 = (x^{2n} - 2x^nx^{-n} + x^{-2n}) + 4 = (x^{2n} - 2 + x^{-2n}) + 4 = x^{2n} + 2 + x^{-2n}$; $(x^n + x^{-n})^2 = x^{2n} + 2x^nx^{-n} + x^{-2n} = x^{2n} + 2 + x^{-2n}$

38. Let a and b be any real numbers. Then the average of their squares is $\frac{a^2 + b^2}{2}$, and their product is ab. Since $(a - b)^2 \geq 0$, $a^2 - 2ab + b^2 \geq 0$ and $a^2 + b^2 \geq 2ab$. Thus, $\frac{a^2 + b^2}{2} \geq ab$.

Pages 210–211 • WRITTEN EXERCISES

A **1.** $(5x + 3)^2$ **2.** $(5a - 3b)(a - 2b)$ **3.** $(7 - 6k)(7 + 6k)$ **4.** irreducible **5.** $(1 - 9r)^2$ **6.** $(h + 5)(h^2 - 5h + 25)$ **7.** $(3n - 4m)(9n^2 + 12mn + 16m^2)$

8. $(2c - 7d)^2$ **9.** $(3u - 5v)(u + 2v)$ **10.** $(3t + 2)(2t - 5)$
11. $(3w + 4z)(9w^2 - 12wz + 16z^2)$ **12.** $(p^2 - 11q)(p^2 + 11q)$ **13.** irreducible
14. $(2g^2 - 5)(4g^4 + 10g^2 + 25)$ **15.** $(3ab - 8)(ab + 3)$
16. $(2p^5 - 3q^4)(2p^5 + 3q^4)$

B **17.** $64xy^2 - 9x^3 = x(64y^2 - 9x^2) = x(8y - 3x)(8y + 3x)$
18. $80r^4 - 45s^2 = 5(16r^4 - 9s^2) = 5(4r^2 - 3s)(4r^2 + 3s)$
19. $81w^4 - 16 = (9w^2 - 4)(9w^2 + 4) = (3w - 2)(3w + 2)(9w^2 + 4)$
20. $3c^3 - 30c^2d + 75cd^2 = 3c(c^2 - 10cd + 25d^2) = 3c(c - 5d)^2$
21. $(z^3 - 7)^2$ **22.** $(t^2 - 9)(t^2 - 1) = (t - 3)(t + 3)(t - 1)(t + 1)$
23. $50x^3y^2 + 40x^2y^3 + 8xy^4 = 2xy^2(25x^2 + 20xy + 4y^2) = 2xy^2(5x + 2y)^2$
24. $(1 - 16a^{12}) = (1 - 4a^6)(1 + 4a^6) = (1 - 2a^3)(1 + 2a^3)(1 + 4a^6)$
25. $8b^4c + 27bc^4 = bc(8b^3 + 27c^3) = bc(2b + 3c)(4b^2 - 6bc + 9c^2)$
26. $(d^2 - 2d + 1) - 100e^2 = (d - 1)^2 - (10e)^2 = (d - 1 - 10e)(d - 1 + 10e)$
27. $(m - 1)^2 - (n + 1)^2 = [(m - 1) - (n + 1)][(m - 1) + (n + 1)] =$
$(m - n - 2)(m + n)$
28. $(r + 1)^3 - s^3 = [(r + 1) - s][(r + 1)^2 + (r + 1)s + s^2]$

C **29.** $p^{3n} - q^{3n} = (p^n)^3 - (q^n)^3 = (p^n - q^n)(p^{2n} + p^nq^n + q^{2n})$
30. $25k^{2m} - 10k^m + 1 = (5k^m - 1)^2$
31. $16v^{6n} - 49u^{4n} = (4v^{3n})^2 - (7u^{2n})^2 = (4v^{3n} - 7u^{2n})(4v^{3n} + 7u^{2n})$
32. $7x^{4y} - 11x^{2y} - 6 = (7x^{2y} + 3)(x^{2y} - 2)$
33. $c^{4n} - 13c^{2n} + 36 = (c^{2n} - 9)(c^{2n} - 4) = (c^n - 3)(c^n + 3)(c^n - 2)(c^n + 2)$
34. $d^{4k+1} - 5d^{2k+1} + 4d = d(d^{4k} - 5d^{2k} + 4) = d(d^{2k} - 4)(d^{2k} - 1) =$
$d(d^k - 2)(d^k + 2)(d^k - 1)(d^k + 1)$
35. $x^4 + 4y^4 = (x^4 + 4x^2y^2 + 4y^4) - 4x^2y^2 = (x^2 + 2y^2)^2 - (2xy)^2 =$
$(x^2 + 2y^2 - 2xy)(x^2 + 2y^2 + 2xy)$
36. $x^6 - y^6 = (x^3)^2 - (y^3)^2 = (x^3 - y^3)(x^3 + y^3) =$
$(x - y)(x^2 + xy + y^2)(x + y)(x^2 - xy + y^2)$;
$x^6 - y^6 = (x^2)^3 - (y^2)^3 = (x^2 - y^2)(x^4 + x^2y^2 + y^4) =$
$(x - y)(x + y)(x^4 + x^2y^2 + y^4)$
$\therefore x^4 + x^2y^2 + y^4 = (x^2 + xy + y^2)(x^2 - xy + y^2)$
37. Suppose m and n are not relatively prime. Then there are nonzero integers d, m_1, and n_1, such that $m = m_1d$ and $n = n_1d$ $(d \neq 1)$. Thus, by the substitution principle, $am - bn = am_1d - bn_1d = (am_1 - bn_1)d = 1$ so $am_1 - bn_1 = \frac{1}{d}$. But $am_1 - bn_1$ is an integer. $\therefore$ m and n are relatively prime.

Page 211 • COMPUTER EXERCISES

1.
```
10  INPUT A,N,B,M
20  IF ABS(A^(1/3) - INT(A^(1/3) + 0.5)) > 0.001 THEN GOTO 95
25  IF ABS(B^(1/3) - INT(B^(1/3) + 0.5)) > 0.001 THEN GOTO 95
30  IF ABS(N/3 - INT(N/3)) > 0.0001 THEN GOTO 95
35  IF ABS(M/3 - INT(M/3)) > 0.0001 THEN GOTO 95
40  LET A = INT(A^(1/3) + 0.5)
50  LET B = INT(B^(1/3) + 0.5)
60  LET N = N/3
70  LET M = M/3
80  PRINT "(";A;"X^";N;" + ";B;"Y^";M;")(";A^2;"X^";N * 2;" - ";
90  PRINT A * B;"X^";N;"Y^";M;" + ";B^2;"Y^";M * 2;")" : END
95  PRINT "IRREDUCIBLE"
100 END
```

2. $(7x^4 + 9y)(49x^8 - 63x^4y + 81y^2)$
3. $(11x^2 + 14y^5)(121x^4 - 154x^2y^5 + 196y^{10})$
4. irreducible
5. $(x^7 + 12y^{13})(x^{14} - 12x^7y^{13} + 144y^{26})$

Page 211 • SELF-TEST 1

1. $\left(\frac{2}{3}x^2y^3z\right)^2(x^{-3}y^2z^{-4}) = \frac{4}{9}x^{4-3}y^{6+2}z^{2-4} = \frac{4xy^8}{9z^2}$
2. $\frac{12a^{-3}b^{-2}c^3}{36a^{-2}b^5c^{-1}} = \frac{1}{3}a^{-3+2}b^{-2-5}c^{3+1} = \frac{c^4}{3ab^7}$
3. $(3x - 2)^4 = [(3x - 2)^2]^2 = (9x^2 - 12x + 4)^2$

$$\begin{array}{r} 9x^2 - 12x + 4 \\ 9x^2 - 12x + 4 \\ \hline 81x^4 - 108x^3 + 36x^2 \qquad\qquad \\ - 108x^3 + 144x^2 - 48x \qquad \\ 36x^2 - 48x + 16 \\ \hline 81x^4 - 216x^3 + 216x^2 - 96x + 16 \end{array}$$

4. $8x^2 - 2x - 3 = (2x + 1)(4x - 3)$
5. $8g^3 - h^3 = (2g - h)(4g^2 + 2gh + h^2)$
6. $x^2 - 20x + 64 = (x - 4)(x - 16)$
7. $z^2 - 4z + 1$ is irreducible.

Page 212 • READING ALGEBRA

1. Answers may vary. Some examples are *origin, trace,* and *minors.*

2–25. Answers may vary. Page references are given indicating where each word is defined or used in a mathematical context.

2. p. 297 **3.** p. 279 **4.** p. 505 **5.** p. 161 **6.** p. 255 **7.** p. 672 **8.** p. 252 **9.** p. 4 **10.** pp. 251, 267 **11.** p. 553 **12.** pp. 42, 579 **13.** p. 252 **14.** p. 706 **15.** p. 42 **16.** p. 459 **17.** p. 207 **18.** p. 88 **19.** p. 183 **20.** p. 5 **21.** p. 88 **22.** p. 42 **23.** p. 221 **24.** p. 42 **25.** p. 586

Page 216 • WRITTEN EXERCISES

A

1. $(x - 7)(x + 4) = 0$; $x - 7 = 0$ or $x + 4 = 0$; $x = 7$ or $x = -4$; $\{7, -4\}$
2. $(a + 12)(a - 12) = 0$; $a + 12 = 0$ or $a - 12 = 0$; $a = -12$ or $a = 12$; $\{-12, 12\}$
3. $2(5 + x)(5 - x) = 0$; $5 + x = 0$ or $5 - x = 0$; $x = -5$ or $x = 5$; $\{-5, 5\}$
4. $(x - 7)^2 = 0$; $x - 7 = 0$; $x = 7$; $\{7\}$
5. $(2y + 3)^2 = 0$; $2y + 3 = 0$; $y = -\frac{3}{2}$; $\left\{-\frac{3}{2}\right\}$
6. $(2x + 5)(x - 2) = 0$; $2x + 5 = 0$ or $x - 2 = 0$; $x = -\frac{5}{2}$ or $x = 2$; $\left\{-\frac{5}{2}, 2\right\}$
7. $5(8 + b)(8 - b) = 0$; $8 + b = 0$ or $8 - b = 0$; $b = -8$ or $b = 8$; $\{-8, 8\}$
8. $r(2r - 5) = 0$; $r = 0$ or $2r - 5 = 0$; $r = 0$ or $r = \frac{5}{2}$; $\left\{0, \frac{5}{2}\right\}$
9. $20x^2 - 45x = 0$; $5x(4x - 9) = 0$; $5x = 0$ or $4x - 9 = 0$; $x = 0$ or $x = \frac{9}{4}$; $\left\{0, \frac{9}{4}\right\}$
10. $4m^2 + 4m - 15 = 0$; $(2m + 5)(2m - 3) = 0$; $2m + 5 = 0$ or $2m - 3 = 0$; $m = -\frac{5}{2}$ or $m = \frac{3}{2}$; $\left\{-\frac{5}{2}, \frac{3}{2}\right\}$
11. $x^2 - 13x - 48 = 0$; $(x - 16)(x + 3) = 0$; $x - 16 = 0$ or $x + 3 = 0$; $x = 16$ or $x = -3$; $\{16, -3\}$

12. $(3n - 4)(n - 2) = 0$; $3n - 4 = 0$ or $n - 2 = 0$; $n = \frac{4}{3}$ or $n = 2$; $\left\{\frac{4}{3}, 2\right\}$
13. $6x^2 - x - 12 = 0$; $(3x + 4)(2x - 3) = 0$; $3x + 4 = 0$ or $2x - 3 = 0$; $x = -\frac{4}{3}$ or $x = \frac{3}{2}$; $\left\{-\frac{4}{3}, \frac{3}{2}\right\}$
14. $8d^2 - 6d - 5 = 0$; $(4d - 5)(2d + 1) = 0$; $4d - 5 = 0$ or $2d + 1 = 0$; $d = \frac{5}{4}$ or $d = -\frac{1}{2}$; $\left\{\frac{5}{4}, -\frac{1}{2}\right\}$
15. $4w^2 - 11w + 6 = 0$; $(4w - 3)(w - 2) = 0$; $4w - 3 = 0$ or $w - 2 = 0$; $w = \frac{3}{4}$ or $w = 2$; $\left\{\frac{3}{4}, 2\right\}$
16. $(x - 4)(3x + 5) - 4x = 0$; $3x^2 - 11x - 20 = 0$; $(3x + 4)(x - 5) = 0$; $3x + 4 = 0$ or $x - 5 = 0$; $x = -\frac{4}{3}$ or $x = 5$; $\left\{-\frac{4}{3}, 5\right\}$

B 17. $(5x - 6)^2 - 5x = 0$; $25x^2 - 65x + 36 = 0$; $(5x - 4)(5x - 9) = 0$; $5x - 4 = 0$ or $5x - 9 = 0$; $x = \frac{4}{5}$ or $x = \frac{9}{5}$; $\left\{\frac{4}{5}, \frac{9}{5}\right\}$
18. $(x^2 - 15x + 56) - (2x^2 + 6x) = 8 + x$; $x^2 + 22x - 48 = 0$; $(x + 24)(x - 2) = 0$; $x = -24$ or $x = 2$; $\{-24, 2\}$
19. $(2x^2 + 5x + 3) - (3x^2 + 9x + 6) = -3x - 15$; $x^2 + x - 12 = 0$; $(x + 4)(x - 3) = 0$; $x + 4 = 0$ or $x - 3 = 0$; $x = -4$ or $x = 3$; $\{-4, 3\}$
20. $z(z^2 - 4z - 21) = 0$; $z(z - 7)(z + 3) = 0$; $z = 0$ or $z - 7 = 0$ or $z + 3 = 0$; $z = 0$ or $z = 7$ or $z = -3$; $\{0, 7, -3\}$
21. $a(3a^2 - 17a + 10) = 0$; $a(3a - 2)(a - 5) = 0$; $a = 0$ or $3a - 2 = 0$ or $a - 5 = 0$; $a = 0$ or $a = \frac{2}{3}$ or $a = 5$; $\left\{0, \frac{2}{3}, 5\right\}$
22. $(x^2 - 16)(x^2 - 1) = 0$; $(x + 4)(x - 4)(x + 1)(x - 1) = 0$; $x + 4 = 0$ or $x - 4 = 0$ or $x + 1 = 0$ or $x - 1 = 0$; $x = -4$ or $x = 4$ or $x = -1$ or $x = 1$; $\{-4, -1, 1, 4\}$
23. $(x^2 - 25)(x^2 + 4) = 0$; $(x + 5)(x - 5)(x^2 + 4) = 0$; $x + 5 = 0$ or $x - 5 = 0$ or $x^2 + 4 = 0$; $x = -5$ or $x = 5$; $\{-5, 5\}$

C 24. $(x^2 - 25)(x^2 - 9) = 0$; $(x + 5)(x - 5)(x + 3)(x - 3) = 0$; $x + 5 = 0$ or $x - 5 = 0$ or $x + 3 = 0$ or $x - 3 = 0$; $x = -5$ or $x = 5$ or $x = -3$ or $x = 3$; $\{-5, -3, 3, 5\}$
25. $x^4(3x - 1) - (3x - 1) = 0$; $(x^4 - 1)(3x - 1) = 0$; $(x + 1)(x - 1)(x^2 + 1)(3x - 1) = 0$; $x + 1 = 0$ or $x - 1 = 0$ or $x^2 + 1 = 0$ or $3x - 1 = 0$; $x = -1$ or $x = 1$ or $x = \frac{1}{3}$; $\left\{-1, \frac{1}{3}, 1\right\}$
26. If $abc = 0$, then by the assoc. prop. $(ab)c = 0$. But $(ab)c = 0$ if and only if $ab = 0$ or $c = 0$ and $ab = 0$ if and only if $a = 0$ or $b = 0$. $\therefore$ if $abc = 0$, $a = 0$ or $b = 0$ or $c = 0$.

Pages 216–217 • PROBLEMS

A 1. Let l = length of rectangle; $l(25 - l) = 144$; $l^2 - 25l + 144 = 0$; $(l - 9)(l - 16) = 0$; $l = 9$ or $l = 16$; 9 cm by 16 cm
2. Let w = width of rectangle; $w(3w + 4) = 480$; $3w^2 + 4w - 480 = 0$; $(3w + 40)(w - 12) = 0$; $w = -\frac{40}{3}$ or $w = 12$; 12 cm by 40 cm
3. $60 + 32x + 4x^2 = 96$; $x^2 + 8x + 15 = 24$; $x^2 + 8x - 9 = 0$; $(x + 9)(x - 1) = 0$; $x = -9$ or $x = 1$; 1 m
4. $352 - (352 - 76x + 4x^2) = 192$; $4x^2 - 76x + 192 = 0$; $x^2 - 19x + 48 = 0$; $(x - 16)(x - 3) = 0$; $x = 16$ or $x = 3$; 3 m

5. Let x = length of shorter leg; $x^2 + (2x - 1)^2 = (x + 1)^2$; $x^2 + (4x^2 - 4x + 1) = x^2 + 2x + 1$; $4x^2 - 6x = 0$; $2x(2x - 3) = 0$; $x = 0$ or $x = \frac{3}{2}$; $x \neq 0$, so the sides are 1.5 cm, 2 cm, and 2.5 cm.

6. Let w = width of rectangle; $w^2 + (w + 0.8)^2 = 4^2$; $2w^2 + 1.6w + 0.64 = 16$; $2w^2 + 1.6w - 15.36 = 0$; $200w^2 + 160w - 1536 = 0$; $25w^2 + 20w - 192 = 0$; $(5w + 16)(5w - 12) = 0$; $w = -3.2$ or $w = 2.4$; 2.4 cm by 3.2 cm

7. Let x = width of side border; $(20 - 2x)(28 - 4x) = 270$; $560 - 136x + 8x^2 = 270$; $4x^2 - 68x + 145 = 0$; $(2x - 5)(2x - 29) = 0$; $x = 2.5$ or $x = 14.5$; since side borders 14.5 cm wide would exceed the width of the page, the borders at the top and sides are each 2.5 cm wide and the width of the bottom border is 7.5 cm.

8. Let x = width of walk; $(16 + 2x)(12 + x) - 16 \cdot 12 = 88$; $192 + 40x + 2x^2 - 192 = 88$; $x^2 + 20x - 44 = 0$; $(x + 22)(x - 2) = 0$; $x = -22$ or $x = 2$; 2 m

9. Let t = no. of hours Henry walks; $(2t)^2 + [4(t - 0.5)]^2 = 5^2$; $4t^2 + 16(t^2 - t + 0.25) = 25$; $20t^2 - 16t - 21 = 0$; $(2t - 3)(10t + 7) = 0$; $t = 1.5$ or $t = -0.7$; $10 + 1.5 = 11.5$; 11:30 A.M.

B **10.** Let x = no. of 5¢ increases; $(0.50 + 0.05x)(360 - 20x) = 196$; $180 + 8x - x^2 = 196$; $x^2 - 8x + 16 = 0$; $(x - 4)^2 = 0$; $x = 4$; fare = 50¢ + 4(5¢) = 70¢

11. Let w = width of pen; l = length of pen; $3w + 2l = 84$; $l = 42 - 1.5w$; $w(42 - 1.5w) = 240$; $-1.5w^2 + 42w - 240 = 0$; $w^2 - 28w + 160 = 0$; $(w - 8)(w - 20) = 0$; $w = 8$ or $w = 20$; 8 m by 30 m or 20 m by 12 m

12. Let x = side of smaller square; y = side of larger square; $3x + 3y + (y - x) = 80$; $x = 40 - 2y$; $(40 - 2y)^2 + y^2 = 325$; $1600 - 160y + 5y^2 = 325$; $y^2 - 32y + 255 = 0$; $(y - 15)(y - 17) = 0$; $y = 15$ or $y = 17$; 17 m and 6 m or 15 m and 10 m

13. Let x = radius of top of pan; $\frac{1}{3}\pi \cdot 3(10^2 + 10x + x^2) = 364\pi$; $\pi(x^2 + 10x + 100) = 364\pi$; $x^2 + 10x - 264 = 0$; $(x - 12)(x + 22) = 0$; $x = 12$ or $x = -22$; 12 cm

C **14.** Let r = inner radius of ball; $\frac{4}{3}\pi(r + 3)^3 - \frac{4}{3}\pi r^3 = 684\pi$; $(r + 3)^3 - r^3 = 513$; $9r^2 + 27r + 27 = 513$; $r^2 + 3r - 54 = 0$; $(r + 9)(r - 6) = 0$; $r = -9$ or $r = 6$; radius of ball = 6 cm + 3 cm = 9 cm

Pages 219–220 • WRITTEN EXERCISES

A **1.** $(x + 3)(x - 4) \geq 0$

$x + 3 \geq 0$ and $x - 4 \geq 0$ or $x + 3 \leq 0$ and $x - 4 \leq 0$

$x \geq -3$ and $x \geq 4$ $\qquad$ $x \leq -3$ and $x \leq 4$

$x \geq 4$ $\qquad$ $x \leq -3$

$\{x: x \geq 4 \text{ or } x \leq -3\}$

−3 −1 1 3

2. $(y - 5)(y - 2) \leq 0$

$y - 5 \geq 0$ and $y - 2 \leq 0$ or $y - 5 \leq 0$ and $y - 2 \geq 0$

$y \geq 5$ and $y \leq 2$ $\qquad$ $y \leq 5$ and $y \geq 2$

no such y $\qquad$ $2 \leq y \leq 5$

$\{y: 2 \leq y \leq 5\}$

0 2 4 6

3. $(a - 4)(a + 1) < 0$
$a - 4 > 0$ and $a + 1 < 0$ or $a - 4 < 0$ and $a + 1 > 0$
$a > 4$ and $a < -1$ $\quad$ $a < 4$ and $a > -1$
no such a $\quad$ $-1 < a < 4$
$\{a: -1 < a < 4\}$
$-1 \quad 1 \quad 3 \quad 5$

4. $(x - 7)(x - 2) < 0$
$x - 7 > 0$ and $x - 2 < 0$ or $x - 7 < 0$ and $x - 2 > 0$
$x > 7$ and $x < 2$ $\quad$ $x < 7$ and $x > 2$
no such x $\quad$ $2 < x < 7$
$\{x: 2 < x < 7\}$
$1 \quad 3 \quad 5 \quad 7$

5. $x(x - 7) \geq 0$
$x \geq 0$ and $x - 7 \geq 0$ or $x \leq 0$ and $x - 7 \leq 0$
$x \geq 0$ and $x \geq 7$ $\quad$ $x \leq 0$ and $x \leq 7$
$x \geq 7$ $\quad$ $x \leq 0$
$\{x: x \geq 7 \text{ or } x \leq 0\}$
$0 \quad 1 \quad 3 \quad 5 \quad 7$

6. $5(b^2 - 36) > 0$; $5(b + 6)(b - 6) > 0$; $(b + 6)(b - 6) > 0$
$b + 6 > 0$ and $b - 6 > 0$ or $b + 6 < 0$ and $b - 6 < 0$
$b > -6$ and $b > 6$ $\quad$ $b < -6$ and $b < 6$
$b > 6$ $\quad$ $b < -6$
$\{b: b > 6 \text{ or } b < -6\}$
$-6 \quad 0 \quad 6$

7. $3c^2 - 75 \leq 0$; $3(c + 5)(c - 5) \leq 0$; $(c + 5)(c - 5) \leq 0$
$c + 5 \geq 0$ and $c - 5 \leq 0$ or $c + 5 \leq 0$ and $c - 5 \geq 0$
$c \geq -5$ and $c \leq 5$ or $c \leq -5$ and $c \geq 5$
$-5 \leq c \leq 5$ $\quad$ no such c
$\{c: -5 \leq c \leq 5\}$
$-5 \quad 1 \quad 5$

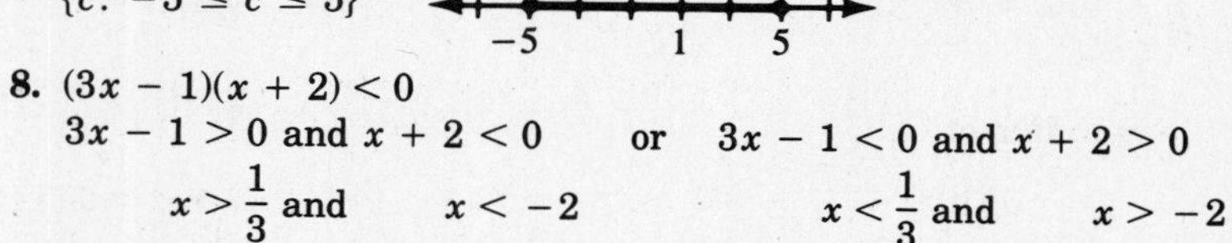

8. $(3x - 1)(x + 2) < 0$
$3x - 1 > 0$ and $x + 2 < 0$ or $3x - 1 < 0$ and $x + 2 > 0$
$x > \frac{1}{3}$ and $x < -2$ $\quad$ $x < \frac{1}{3}$ and $x > -2$
no such x $\quad$ $-2 < x < \frac{1}{3}$
$\{x: -2 < x < \frac{1}{3}\}$
$-3 \quad -2 \quad -1 \quad \frac{1}{3} \quad 1$

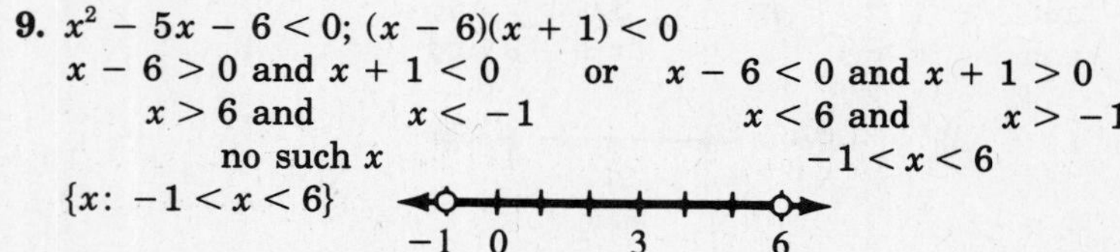

9. $x^2 - 5x - 6 < 0$; $(x - 6)(x + 1) < 0$
$x - 6 > 0$ and $x + 1 < 0$ or $x - 6 < 0$ and $x + 1 > 0$
$x > 6$ and $x < -1$ $\quad$ $x < 6$ and $x > -1$
no such x $\quad$ $-1 < x < 6$
$\{x: -1 < x < 6\}$
$-1 \quad 0 \quad 3 \quad 6$

10. $x^2 - 12x + 35 \geq 0$; $(x - 7)(x - 5) \geq 0$
$x - 7 \geq 0$ and $x - 5 \geq 0$ or $x - 7 \leq 0$ and $x - 5 \leq 0$
$x \geq 7$ and $x \geq 5$ $\quad$ $x \leq 7$ and $x \leq 5$
$x \geq 7$ $\quad$ $x \leq 5$
$\{x: x \geq 7 \text{ or } x \leq 5\}$
$3 \quad 5 \quad 7 \quad 9$

11. $(k - 3)^2 > 0$ for all k except $k = 3$; $\{k: k \neq 3\}$
$0 \quad 3 \quad 6$

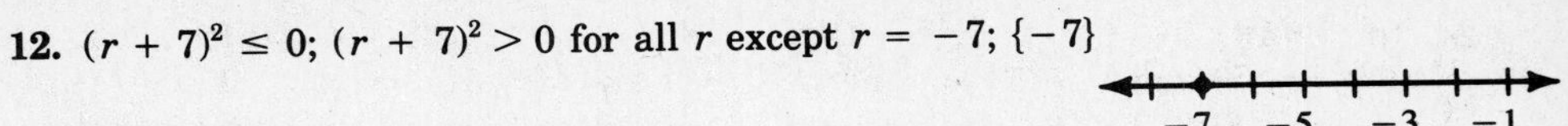

12. $(r+7)^2 \le 0$; $(r+7)^2 > 0$ for all r except $r = -7$; $\{-7\}$

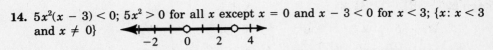

B **13.** $y^2(y-3) \ge 0$; $y^2 \ge 0$ for all y and $y - 3 \ge 0$ for $y \ge 3$; $\{y: y \ge 3\}$

14. $5x^2(x-3) < 0$; $5x^2 > 0$ for all x except $x = 0$ and $x - 3 < 0$ for $x < 3$; $\{x: x < 3$ and $x \neq 0\}$

15. $4x^3 - x^2 < 0$; $x^2(4x-1) < 0$; $x^2 > 0$ for all x except $x = 0$ and $4x - 1 < 0$ for $x < \frac{1}{4}$; $\left\{x: x < \frac{1}{4} \text{ and } x \neq 0\right\}$

16. $t(t-6)^2 \ge 0$; $(t-6)^2 \ge 0$ for all t and $t \ge 0$; $\{t: t \ge 0\}$

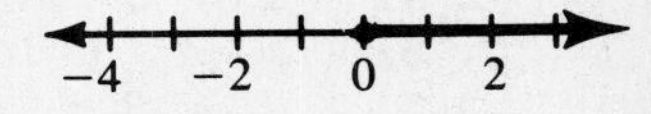

17. $(z-3)(z^2+4) > 0$; $z^2 + 4 > 0$ for all z and $z - 3 > 0$ for $z > 3$; $\{z: z > 3\}$

18. $2y(3y+1)(3y-1) \le 0$; the product is negative when either exactly one factor is negative or when exactly three factors are negative; drawing a sign graph yields the solution set $\left\{y: y \le -\frac{1}{3} \text{ or } 0 \le y \le \frac{1}{3}\right\}$.

C **19.** $x(x+3)^2 \le 0$; $(x+3)^2 \ge 0$ for all x and $x \le 0$; $\{x: x \le 0\}$

20. $x(4x^2 - 20x + 25) > 0$; $x(2x-5)^2 > 0$; $(2x-5)^2 > 0$ for all x except $x = \frac{5}{2}$ and $x > 0$; $\left\{x: x > 0 \text{ and } x \neq \frac{5}{2}\right\}$

21. $x^4 - 16 > 0$; $(x^2+4)(x^2-4) > 0$; $(x^2+4)(x+2)(x-2) > 0$; $x^2 + 4 > 0$ for all x so the product is positive when $x + 2$ and $x - 2$ are both positive or both negative:

$x + 2 > 0$ and $x - 2 > 0$ or $x + 2 < 0$ and $x - 2 < 0$

$x > -2$ and $x > 2$ $\quad$ $x < -2$ and $x < 2$

$x > 2$ $\quad$ $x < -2$

$\{x: x > 2 \text{ or } x < -2\}$

22. $x(4x^2 - 25) > 0$; $x(2x+5)(2x-5) > 0$; the product is positive when exactly one factor is positive or exactly three factors are positive. Drawing a sign graph yields the solution set $\left\{x: x > \frac{5}{2} \text{ or } -\frac{5}{2} < x < 0\right\}$.

23. $a^4 - 10a^2 + 9 \ge 0$; $(a^2-9)(a^2-1) \ge 0$; $(a+3)(a-3)(a+1)(a-1) \ge 0$; the product is positive when all factors are positive, when all factors are negative, or when 2 factors are positive; it is zero if any factor is 0. Drawing a sign graph yields the solution set $\{a: a \le -3 \text{ or } -1 \le a \le 1 \text{ or } a \ge 3\}$.

24. $(x^{2n}+4)(x^{2n}-1)<0$; $(x^{2n}+4)(x^n+1)(x^n-1)<0$; $x^{2n}+4>0$ for all x; $x^n+1>0$ and $x^n-1<0$ when $x^n>-1$ and $x^n<1$, or when $-1<x<1$; $x^n+1<0$ and $x^n-1>0$ when $x^n<-1$ and $x^n>1$, but no such x exists; $\{x: -1<x<1\}$

−1 0 1

Page 220 • SELF-TEST 2

1. $3x(x-5)=0$; $3x=0$ or $x-5=0$; $\{0, 5\}$
2. $(3x+1)(x-4)=0$; $3x+1=0$ or $x-4=0$; $\left\{-\frac{1}{3}, 4\right\}$
3. Let w = width of rectangle; $w(2w+4)=30$; $2w^2+4w-30=0$; $2(w+5)(w-3)=0$; $w=-5$ or $w=3$; if $w=3$, $2w+4=10$. ∴ the perimeter of the rectangle is 26 cm.
4. $5x^2-2x-3<0$; $(5x+3)(x-1)<0$

$5x+3>0$ and $x-1<0$ or $5x+3<0$ and $x-1>0$

$x>-\frac{3}{5}$ and $x<1$ $\quad$ $x<-\frac{3}{5}$ and $x>1$

$-\frac{3}{5}<x<1$ $\quad$ no such x

$\left\{x: -\frac{3}{5}<x<1\right\}$

$-\frac{3}{5}$ −1 0 1

5. $3x(x+2)>0$

$3x>0$ and $x+2>0$ or $3x<0$ and $x+2<0$

$x>0$ and $x>-2$ $\quad$ $x<0$ and $x<-2$

$x>0$ $\quad$ $x<-2$

$\{x: x>0 \text{ or } x<-2\}$

−2 −1 0 1

Pages 222–223 • WRITTEN EXERCISES

A

1. $8x(4x-28)^{-1}=\dfrac{8x}{4(x-7)}=\dfrac{2x}{x-7}$
2. $(6y^2)^{-1}(2y^2-10y)=\dfrac{2y(y-5)}{6y^2}=\dfrac{y-5}{3y}$
3. $(4a)^{-2}(6a^3-14a^2)=\dfrac{2a^2(3a-7)}{16a^2}=\dfrac{3a-7}{8}$
4. $(27b^2-18b)(45b)^{-1}=\dfrac{9b(3b-2)}{45b}=\dfrac{3b-2}{5}$
5. $(z^5-8z^4)(3z-24)^{-1}=\dfrac{z^4(z-8)}{3(z-8)}=\dfrac{z^4}{3}$
6. $(6r^2-15r)^{-1}(5-2r)=\dfrac{5-2r}{3r(2r-5)}=-\dfrac{1}{3r}$
7. $(9k^3+12k^2)(-6k-8)^{-1}=\dfrac{3k^2(3k+4)}{-2(3k+4)}=-\dfrac{3k^2}{2}$
8. $(h-3)^3(3-h)^{-1}=\dfrac{(h-3)^3}{-(h-3)}=-(h-3)^2$
9. $(p-7)^2(7-p)^{-3}=\dfrac{(p-7)^2}{(7-p)^3}=-\dfrac{1}{p-7}$, or $\dfrac{1}{7-p}$

10. $(x^4y^2 - 4x^3y)(x^2y^2 - 4xy)^{-1} = \dfrac{x^3y(xy - 4)}{xy(xy - 4)} = x^2$

11. $(16h - 8k)(12h^2k - 6hk^2)^{-1} = \dfrac{8(2h - k)}{6hk(2h - k)} = \dfrac{4}{3hk}$

12. $(9c^2 - d^2)(3c - d)^{-2} = \dfrac{(3c + d)(3c - d)}{(3c - d)^2} = \dfrac{3c + d}{3c - d}$

13. $\dfrac{n^3 - 8}{3n^2 - 12} = \dfrac{(n - 2)(n^2 + 2n + 4)}{3(n + 2)(n - 2)} = \dfrac{n^2 + 2n + 4}{3(n + 2)}$

14. $\dfrac{d^4 + d^2}{d^2(d + 1)^2} = \dfrac{d^2(d^2 + 1)}{d^2(d + 1)^2} = \dfrac{d^2 + 1}{(d + 1)^2}$

15. $\dfrac{c^2 - 10c + 25}{c^3 - 25c} = \dfrac{(c - 5)^2}{c(c - 5)(c + 5)} = \dfrac{c - 5}{c(c + 5)}$

16. $\dfrac{a^2 - 6ab + 8b^2}{a^2 - 3ab + 2b^2} = \dfrac{(a - 4b)(a - 2b)}{(a - 2b)(a - b)} = \dfrac{a - 4b}{a - b}$

17. $\dfrac{64u^3 + 1}{4u^2 + 5u + 1} = \dfrac{(4u + 1)(16u^2 - 4u + 1)}{(4u + 1)(u + 1)} = \dfrac{16u^2 - 4u + 1}{u + 1}$

18. $\dfrac{5m^3 - 8m^2 + 3m}{25m^3 - 9m} = \dfrac{m(5m - 3)(m - 1)}{m(5m - 3)(5m + 3)} = \dfrac{m - 1}{5m + 3}$

19. $\dfrac{r^4 - r^2 - 12}{r^4 + 3r^2} = \dfrac{(r^2 + 3)(r^2 - 4)}{r^2(r^2 + 3)} = \dfrac{r^2 - 4}{r^2}$

20. $\dfrac{4x^2 - 12xy - 7y^2}{4x^2 - 49y^2} = \dfrac{(2x + y)(2x - 7y)}{(2x + 7y)(2x - 7y)} = \dfrac{2x + y}{2x + 7y}$

B 21. $\dfrac{a^3 + b^3}{(a + b)^3} = \dfrac{(a + b)(a^2 - ab + b^2)}{(a + b)(a + b)^2} = \dfrac{a^2 - ab + b^2}{(a + b)^2}$

22. $\dfrac{p^4 - q^4}{(p + q)^2(q - p)} = \dfrac{(p^2 + q^2)(p + q)(p - q)}{(p + q)^2(-1)(p - q)} = -\dfrac{p^2 + q^2}{p + q}$

23. $\dfrac{(3x^2 - 48)(x^3 - 3x^2 - 28x)}{3(x + 4)^2(x^3 - 11x^2 + 28x)} = \dfrac{3(x + 4)(x - 4)x(x - 7)(x + 4)}{3(x + 4)^2x(x - 7)(x - 4)} = 1$

24. $\dfrac{(c^3 - d^3)(c^3 - cd^2)}{c^3 + c^2d + cd^2} = \dfrac{(c - d)(c^2 + cd + d^2)c(c + d)(c - d)}{c(c^2 + cd + d^2)} = (c + d)(c - d)^2$

25. $\dfrac{k^4 - 26k^2 + 25}{k^2 - 6k + 5} = \dfrac{(k^2 - 25)(k^2 - 1)}{(k - 5)(k - 1)} = \dfrac{(k + 5)(k - 5)(k + 1)(k - 1)}{(k - 5)(k - 1)} =$
$(k + 5)(k + 1)$

26. $\dfrac{(54 - 2b^3)(b^2 + 9)}{2b^4 - 162} = \dfrac{2(3 - b)(9 + 3b + b^2)(b^2 + 9)}{2(b + 3)(b - 3)(b^2 + 9)} = -\dfrac{9 + 3b + b^2}{b + 3}$

27. $\dfrac{r^6 - 8}{(4 - r^2)(4 + 2r + r^2)} = \dfrac{(r^2 - 2)(r^4 + 2r^2 + 4)}{(2 + r)(2 - r)(4 + 2r + r^2)}$; irreducible

28. $\dfrac{(c^2 + 4)(c - 2)^2}{c^4 - 16} = \dfrac{(c^2 + 4)(c - 2)^2}{(c^2 + 4)(c + 2)(c - 2)} = \dfrac{c - 2}{c + 2}$

C 29. $\dfrac{x^4 + 64y^4}{x^2 + 4xy + 8y^2} = \dfrac{x^4 + 16x^2y^2 + 64y^4 - 16x^2y^2}{x^2 + 4xy + 8y^2} = \dfrac{(x^2 + 8y^2)^2 - (4xy)^2}{x^2 + 4xy + 8y^2} =$
$\dfrac{(x^2 + 8y^2 + 4xy)(x^2 + 8y^2 - 4xy)}{x^2 + 4xy + 8y^2} = x^2 + 8y^2 - 4xy$

30.

Statement	Reason
1. $\frac{a+c}{b+c} = \frac{a}{b}$; $c \neq 0$; $c \in \mathscr{R}$	Hypothesis
2. $b(a + c) = a(b + c)$	Prod. of means = prod. of extremes
3. $ba + bc = ab + ac$	Dist. ax.
4. $ab + bc = ab + ac$	Comm. ax. for mult.
5. $bc = ac$	Canc. prop. of add.
6. $b = a$	Canc. prop. of mult.
7. $a = b$	Symm. prop. of =

Pages 225–226 • WRITTEN EXERCISES

A **1.** $2x^3 + 5x - 3$ **2.** $\frac{cd}{2} - \frac{4}{3} - \frac{3}{cd}$ **3.** $-b^2 - ab + a^2$ **4.** $9u + 4v + \frac{v^2}{5u}$

5.
$$\begin{array}{r} x - 2 \\ x - 4\,\overline{)\,x^2 - 6x + 3} \\ \underline{x^2 - 4x} \\ -2x + 3 \\ \underline{-2x + 8} \\ -5 \end{array}$$
$x - 2 + \frac{-5}{x - 4}$

6.
$$\begin{array}{r} -5r + 8 \\ 2r - 3\,\overline{)\,-10r^2 + 31r - 24} \\ \underline{-10r^2 + 15r} \\ 16r - 24 \\ \underline{16r - 24} \\ 0 \end{array}$$
$-5r + 8$

7.
$$\begin{array}{r} 2y - 9 \\ 3y + 5\,\overline{)\,6y^2 - 17y - 45} \\ \underline{6y^2 + 10y} \\ -27y - 45 \\ \underline{-27y - 45} \\ 0 \end{array}$$
$2y - 9$

8.
$$\begin{array}{r} 6z - 18 \\ 2z + 6\,\overline{)\,12z^2 + 0z - 108} \\ \underline{12z^2 + 36z} \\ -36z - 108 \\ \underline{-36z - 108} \\ 0 \end{array}$$
$6z - 18$

9.
$$\begin{array}{r} r^2 - 2r + 5 \\ r + 3\,\overline{)\,r^3 + r^2 - r + 15} \\ \underline{r^3 + 3r^2} \\ -2r^2 - r \\ \underline{-2r^2 - 6r} \\ 5r + 15 \\ \underline{5r + 15} \\ 0 \end{array}$$
$r^2 - 2r + 5$

10.
$$\begin{array}{r} 3u^2 + 5u - 2 \\ u - 4\,\overline{)\,3u^3 - 7u^2 - 22u + 8} \\ \underline{3u^3 - 12u^2} \\ 5u^2 - 22u \\ \underline{5u^2 - 20u} \\ -2u + 8 \\ \underline{-2u + 8} \\ 0 \end{array}$$
$3u^2 + 5u - 2$

11.
$$\begin{array}{r} -2v^2 - 4v + 3 \\ 3v - 1\,\overline{)\,-6v^3 - 10v^2 + 13v - 5} \\ \underline{-6v^3 + 2v^2} \\ -12v^2 + 13v \\ \underline{-12v^2 + 4v} \\ 9v - 5 \\ \underline{9v - 3} \\ -2 \end{array}$$
$-2v^2 - 4v + 3 + \frac{-2}{3v - 1}$

12.
$$\begin{array}{r} 2w^2 + 4w - 1 \\ 5w - 6\,\overline{)\,10w^3 + 8w^2 - 29w + 6} \\ \underline{10w^3 - 12w^2} \\ 20w^2 - 29w \\ \underline{20w^2 - 24w} \\ -5w + 6 \\ \underline{-5w + 6} \\ 0 \end{array}$$
$2w^2 + 4w - 1$

13.
$$\begin{array}{r} a^2 - 3a + 2 \\ 2a + 7\overline{)2a^3 + a^2 - 17a + 14} \\ \underline{2a^3 + 7a^2} \\ -6a^2 - 17a \\ \underline{-6a^2 - 21a} \\ 4a + 14 \\ \underline{4a + 14} \\ 0 \end{array}$$

$a^2 - 3a + 2$

14.
$$\begin{array}{r} 16t^2 + 12t + 9 \\ 4t - 3\overline{)64t^3 + 0t^2 + 0t - 25} \\ \underline{64t^3 - 48t^2} \\ 48t^2 + 0t \\ \underline{48t^2 - 36t} \\ 36t - 25 \\ \underline{36t - 27} \\ 2 \end{array}$$

$16t^2 + 12t + 9 + \dfrac{2}{4t - 3}$

15.
$$\begin{array}{r} 9n^2 - 6n + 4 \\ 3n + 2\overline{)27n^3 + 0n^2 + 0n + 8} \\ \underline{27n^3 + 18n^2} \\ -18n^2 + 0n \\ \underline{-18n^2 - 12n} \\ 12n + 8 \\ \underline{12n + 8} \\ 0 \end{array}$$

$9n^2 - 6n + 4$

16.
$$\begin{array}{r} 3x^2 + 2 \\ 6x - 5\overline{)18x^3 - 15x^2 + 12x - 10} \\ \underline{18x^3 - 15x^2} \\ 12x - 10 \\ \underline{12x - 10} \\ 0 \end{array}$$

$3x^2 + 2$

17.
$$\begin{array}{r} 3y^2 + 2y - 4 \\ 4y - 1\overline{)12y^3 + 5y^2 - 18y + 4} \\ \underline{12y^3 - 3y^2} \\ 8y^2 - 18y \\ \underline{8y^2 - 2y} \\ -16y + 4 \\ \underline{-16y + 4} \\ 0 \end{array}$$

$3y^2 + 2y - 4$

18.
$$\begin{array}{r} 2z^2 - 5z + 3 \\ 2z + 5\overline{)4z^3 + 0z^2 - 19z + 15} \\ \underline{4z^3 + 10z^2} \\ -10z^2 - 19z \\ \underline{-10z^2 - 25z} \\ 6z + 15 \\ \underline{6z + 15} \\ 0 \end{array}$$

$2z^2 - 5z + 3$

B 19.
$$\begin{array}{r} 16b^4 - 8b^3 + 4b^2 - 2b + 1 \\ 2b + 1\overline{)32b^5 + 0b^4 + 0b^3 + 0b^2 + 0b + 9} \\ \underline{32b^5 + 16b^4} \\ -16b^4 + 0b^3 \\ \underline{-16b^4 - 8b^3} \\ 8b^3 + 0b^2 \\ \underline{8b^3 + 4b^2} \\ -4b^2 + 0b \\ \underline{-4b^2 - 2b} \\ 2b + 9 \\ \underline{2b + 1} \\ 8 \end{array}$$

$16b^4 - 8b^3 + 4b^2 - 2b + 1 + \dfrac{8}{2b + 1}$

$$\begin{array}{r}
m^3 - 2m^2 - 5m - 3 \\
\textbf{20.}\ 5m - 3\,\overline{)\,5m^4 - 13m^3 - 19m^2 + 0m + 9} \\
\underline{5m^4 - 3m^3} \qquad\qquad\qquad\qquad \\
-10m^3 - 19m^2 \qquad\qquad\quad \\
\underline{-10m^3 + 6m^2} \qquad\qquad\quad \\
-25m^2 + 0m \qquad \\
\underline{-25m^2 + 15m} \qquad \\
-15m + 9 \\
\underline{-15m + 9} \\
0
\end{array}$$

$m^3 - 2m^2 - 5m - 3$

$$\begin{array}{r}
2w^2 + 4w - 5 \\
\textbf{21.}\ w^2 - 3w + 2\,\overline{)\,2w^4 - 2w^3 - 13w^2 + 23w - 10} \\
\underline{2w^4 - 6w^3 + 4w^2} \qquad\qquad\qquad \\
4w^3 - 17w^2 + 23w \qquad \\
\underline{4w^3 - 12w^2 + 8w} \qquad \\
-5w^2 + 15w - 10 \\
\underline{-5w^2 + 15w - 10} \\
0
\end{array}$$

$2w^2 + 4w - 5$

$$\begin{array}{r}
4x^2 + 3x + 2 \\
\textbf{22.}\ x^2 - 2x + 1\,\overline{)\,4x^4 - 5x^3 + 0x^2 - x + 2} \\
\underline{4x^4 - 8x^3 + 4x^2} \qquad\qquad \\
3x^3 - 4x^2 - x \qquad \\
\underline{3x^3 - 6x^2 + 3x} \qquad \\
2x^2 - 4x + 2 \\
\underline{2x^2 - 4x + 2} \\
0
\end{array}$$

$4x^2 + 3x + 2$

$$\begin{array}{r}
2v^2 + v - 5 \\
\textbf{23.}\ v^2 + 0v - 6\,\overline{)\,2v^4 + v^3 - 17v^2 - 5v + 0} \\
\underline{2v^4 + 0v^3 - 12v^2} \qquad\qquad\qquad \\
v^3 - 5v^2 - 5v \qquad\quad \\
\underline{v^3 + 0v^2 - 6v} \qquad\quad \\
-5v^2 + v + 0 \\
\underline{-5v^2 + 0v + 30} \\
v - 30
\end{array}$$

$2v^2 + v - 5 + \dfrac{v - 30}{v^2 - 6}$

$$\begin{array}{r}
2a + 3 \\
\textbf{24.}\ 4a^2 - 6a + 9\,\overline{)\,8a^3 + 0a^2 + 0a + 27} \\
\underline{8a^3 - 12a^2 + 18a} \qquad \\
12a^2 - 18a + 27 \\
\underline{12a^2 - 18a + 27} \\
0
\end{array}$$

$2a + 3$

C 25. $\frac{a^n - b^n}{a - b} = a^{n-1} + ba^{n-2} + b^2a^{n-3} + \cdots + b^{n-3}a^2 + b^{n-2}a + b^{n-1}$

26. $\frac{a^n + b^n}{a + b} = a^{n-1} - ba^{n-2} + b^2a^{n-3} - \cdots + b^{n-3}a^2 - b^{n-2}a + b^{n-1}$

Page 228 • WRITTEN EXERCISES

A 1. $\frac{5c - 15}{c} \cdot \frac{c^2 + 3c}{c^2 - 9} = \frac{5(c - 3)c(c + 3)}{c(c + 3)(c - 3)} = 5$

2. $\frac{7b^2 - 42b}{49b^3} \cdot \frac{b}{(b - 6)^2} = \frac{7b^2(b - 6)}{49b^3(b - 6)^2} = \frac{1}{7b(b - 6)}$

3. $\frac{16x^2 - 1}{(4x + 1)^2} \cdot \frac{3x^2}{12x^2 - 3x} = \frac{(4x + 1)(4x - 1)3x^2}{(4x + 1)^2 3x(4x - 1)} = \frac{x}{4x + 1}$

4. $\frac{5}{p^2 - 5p} \cdot \frac{p^2 + 3p - 10}{5p - 10} = \frac{5(p + 5)(p - 2)}{p(p - 5)5(p - 2)} = \frac{p + 5}{p(p - 5)}$

5. $\frac{7v - 63}{12} \cdot \frac{6}{v^2 - 81} = \frac{7(v - 9)6}{12(v + 9)(v - 9)} = \frac{7}{2(v + 9)}$

6. $\frac{y^2 + 5y}{y^2 + 6y + 5} \cdot \frac{3y + 3}{y^3} = \frac{y(y + 5)3(y + 1)}{(y + 5)(y + 1)y^3} = \frac{3}{y^2}$

7. $\frac{3a - 3b}{a^2 + 2ab + b^2} \cdot \frac{a^2 - ab - 2b^2}{a^2 - 3ab + 2b^2} = \frac{3(a - b)(a - 2b)(a + b)}{(a + b)^2(a - 2b)(a - b)} = \frac{3}{a + b}$

8. $\frac{t^3 - t^2 - 2t}{t^2 - 3t + 2} \cdot \frac{t^2 + 2t}{t^2 + 3t + 2} = \frac{t(t - 2)(t + 1)t(t + 2)}{(t - 2)(t - 1)(t + 2)(t + 1)} = \frac{t^2}{t - 1}$

9. $\frac{u^3 - 3u^2 - 4u}{u^2 - 8u + 16} \cdot \frac{u^2 - 16}{3u^2 + 3u} = \frac{u(u - 4)(u + 1)(u + 4)(u - 4)}{(u - 4)^2 3u(u + 1)} = \frac{u + 4}{3}$

10. $\frac{(r^2 - s^2)^2}{r^2 - rs} \cdot \frac{r^2 + s^2}{r^4 - s^4} = \frac{(r + s)^2(r - s)^2(r^2 + s^2)}{r(r - s)(r^2 + s^2)(r + s)(r - s)} = \frac{r + s}{r}$

B 11. $\frac{x^4 - 81}{x^2 + 9} \cdot \frac{x^3 - 6x^2 + 9x}{(x - 3)^3} = \frac{(x^2 + 9)(x + 3)(x - 3)x(x - 3)^2}{(x^2 + 9)(x - 3)^3} = x(x + 3)$

12. $\frac{6m^4 - 6n^4}{m^3 + n^3} \cdot \frac{m^2 - mn + n^2}{3m^2 + 3n^2} = \frac{6(m^2 + n^2)(m + n)(m - n)(m^2 - mn + n^2)}{(m + n)(m^2 - mn + n^2)3(m^2 + n^2)} =$
$2(m - n)$

13. $\frac{(a - b)^2}{5a^2 - 3ab - 2b^2} \cdot \frac{4b^2 - 25a^2}{a^3 - b^3} = \frac{(a - b)^2(2b + 5a)(2b - 5a)}{(5a + 2b)(a - b)(a - b)(a^2 + ab + b^2)} =$
$\frac{2b - 5a}{a^2 + ab + b^2}$

14. $\frac{c^3(c - d)^3}{c^3 - d^3} \cdot \frac{c^2 + cd + d^2}{c^3 - 2c^2d + cd^2} = \frac{c^3(c - d)^3(c^2 + cd + d^2)}{(c - d)(c^2 + cd + d^2)c(c - d)^2} = c^2$

15. $x\left(\frac{x^2 - y^2}{x}\right) \div \left(\frac{x - y}{xy}\right) = \frac{x(x + y)(x - y)}{x} \cdot \frac{xy}{x - y} = xy(x + y)$

16. $\left(\frac{p^2 - q^2}{q}\right)\left(\frac{p - q}{q}\right)^{-2} = \frac{(p + q)(p - q)q^2}{q(p - q)^2} = \frac{q(p + q)}{p - q}$

17. $\frac{6s + r}{3} \div \frac{36s^2 - r^2}{rs} = \frac{6s + r}{3} \cdot \frac{rs}{(6s + r)(6s - r)} = \frac{rs}{3(6s - r)}$

18. $\frac{x^2 - 3(2x - 3)}{3x} \div \frac{x^2 - 9}{3x} = \frac{x^2 - 6x + 9}{3x} \cdot \frac{3x}{(x + 3)(x - 3)} =$
$\frac{(x - 3)^2}{3x} \cdot \frac{3x}{(x + 3)(x - 3)} = \frac{x - 3}{x + 3}$

19. $\frac{k(k+2)+1}{k+2}\cdot\frac{k^2+k-2}{k^2+k}=\frac{(k+1)^2(k+2)(k-1)}{(k+2)k(k+1)}=\frac{(k+1)(k-1)}{k}$

20. $\frac{x^3-y^3}{y^3}\cdot\left(\frac{x-y}{y}\right)^{-1}\div\frac{x^2-y^2}{y^2}=\frac{(x-y)(x^2+xy+y^2)}{y^3}\cdot\frac{y}{x-y}\cdot\frac{y^2}{(x+y)(x-y)}=$
$\frac{x^2+xy+y^2}{(x-y)(x+y)}$

C 21. $\frac{a^3-b^3}{a^4+a^2b^2+b^4}\cdot\frac{a^3+b^3}{a^2-b^2}=\frac{a^6-b^6}{a^6-b^6}=1$

22. $\frac{(c-2d)(c+d)+3d^2}{c+d}\div\left(\frac{d^3+c^3}{c^3d^3}\right)=\frac{c^2-cd+d^2}{c+d}\cdot\frac{c^3d^3}{(d+c)(d^2-cd+c^2)}=$
$\frac{c^3d^3}{(c+d)^2}$

Pages 230–231 • WRITTEN EXERCISES

A 1. $\frac{5x-20}{x-4}=\frac{5(x-4)}{x-4}=5$ 2. $\frac{3\cdot3-5\cdot2b}{24b^2}=\frac{9-10b}{24b^2}$ 3. $\frac{3d-9c+6cd}{c^2d^2}$

4. $\frac{2\cdot4+3r+7}{2(r+5)}=\frac{3(r+5)}{2(r+5)}=\frac{3}{2}$

5. $\frac{2y+y(y-3)-(5y-9)}{(y-3)^2}=\frac{y^2-6y+9}{(y-3)^2}=\frac{(y-3)^2}{(y-3)^2}=1$

6. $\frac{(t-5)6t-2(3t^2)-7(4)}{24t^3}=\frac{-30t-28}{24t^3}=\frac{2(-15t-14)}{2\cdot12t^3}=\frac{-15t-14}{12t^3}$

7. $\frac{a(a-6)-(30-7a)}{(a+6)(a-6)}=\frac{a^2+a-30}{(a+6)(a-6)}=\frac{(a+6)(a-5)}{(a+6)(a-6)}=\frac{a-5}{a-6}$

8. $\frac{x(x-3y)-6xy+3y(x+3y)}{(x+3y)(x-3y)}=\frac{x^2-6xy+9y^2}{(x+3y)(x-3y)}=\frac{(x-3y)^2}{(x+3y)(x-3y)}=\frac{x-3y}{x+3y}$

9. $\frac{3(2-x)-3(x+2)-3x^2}{(2+x)(2-x)}=\frac{6-3x-3x-6-3x^2}{(2+x)(2-x)}=\frac{-3x(2+x)}{(2+x)(2-x)}=$
$-\frac{3x}{2-x}=\frac{3x}{x-2}$

10. $\frac{b(b+3)-(b+1)(b+2)}{(b+2)(b+3)}=\frac{-2}{(b+2)(b+3)}$

11. $\frac{(x+y)(x+5y)-(x-y)(x-5y)}{(x-5y)^2(x+5y)}=\frac{12xy}{(x-5y)^2(x+5y)}$

12. $\frac{2c(c-d)-(c+d)(c+d)}{(c-d)^2(c+d)}=\frac{c^2-4cd-d^2}{(c-d)^2(c+d)}$

13. $\frac{(x-3)(x-2)-(x+3)(x+2)}{(x-3)(x+2)(x-2)}=\frac{-10x}{(x-3)(x+2)(x-2)}$

14. $\frac{(x-8)(x+3)-(x+6)(x-4)}{(x-4)(x-2)(x+3)}=\frac{-7x}{(x-4)(x-2)(x+3)}$

15. $\frac{36p^2+(3p+q)^2-(3p-q)^2}{(3p+q)(3p-q)}=\frac{36p^2+12pq}{(3p+q)(3p-q)}=\frac{12p(3p+q)}{(3p+q)(3p-q)}=\frac{12p}{3p-q}$

16. $\frac{16c^2+d^2-(4c+d)(4c-d)}{(4c-d)^3}=\frac{2d^2}{(4c-d)^3}$

17. $\frac{a^2b^{-1}-b}{ab^{-1}+1}\cdot\frac{b}{b}=\frac{a^2-b^2}{a+b}=\frac{(a-b)(a+b)}{a+b}=a-b$

18. $\left(\frac{r}{s} - \frac{s}{r}\right) \div \left(\frac{1}{r} - \frac{1}{s}\right) = \left(\frac{r^2 - s^2}{rs}\right) \div \left(\frac{s - r}{rs}\right) = \frac{(r + s)(r - s)}{rs} \cdot \frac{rs}{s - r} =$
$-(r + s) = -r - s$

B 19. $\frac{x(x + 2) - (4x + 3)}{x + 2} \div \frac{x(x + 2) - (6x + 3)}{x + 2} = \frac{x^2 - 2x - 3}{x + 2} \div \frac{x^2 - 4x - 3}{x + 2} =$
$\frac{(x - 3)(x + 1)}{x + 2} \cdot \frac{x + 2}{x^2 - 4x - 3} = \frac{(x - 3)(x + 1)}{x^2 - 4x - 3}$

20. $\frac{2x(x - 1) - x(x - 2)}{(x - 2)(x - 1)} \div \frac{3x(x - 2) - 2x(x - 3)}{(x - 3)(x - 2)} =$
$\frac{x^2}{(x - 2)(x - 1)} \div \frac{x^2}{(x - 3)(x - 2)} = \frac{x^2}{(x - 2)(x - 1)} \cdot \frac{(x - 3)(x - 2)}{x^2} = \frac{x - 3}{x - 1}$

21. $\frac{(a + b)(a^{-1} - b^{-1})}{(a - b)(a^{-1} + b^{-1})} \cdot \frac{ab}{ab} = \frac{(a + b)(b - a)}{(a - b)(b + a)} = -1$

22. $\frac{y(y - 3) + 2y - 2}{y - 3} \cdot \frac{y(y + 3) - (2y + 2)}{y + 3} \div \frac{y^2(y^2 - 9) + 4y^2 + 4}{(y + 3)(y - 3)} =$
$\frac{y^2 - y - 2}{y - 3} \cdot \frac{y^2 + y - 2}{y + 3} \cdot \frac{(y + 3)(y - 3)}{y^4 - 5y^2 + 4} =$
$\frac{(y - 2)(y + 1)(y + 2)(y - 1)(y + 3)(y - 3)}{(y - 3)(y + 3)(y + 2)(y - 2)(y + 1)(y - 1)} = 1$

23. $\frac{(u - v)(u + v) + 2v^2}{u + v} \cdot \frac{v^2 - u^2}{u^2v^2} \div \frac{u^3(u + v) - (u^3v + v^4)}{u + v} =$
$\frac{u^2 + v^2}{u + v} \cdot \frac{v^2 - u^2}{u^2v^2} \cdot \frac{u + v}{u^4 - v^4} = \frac{(u^2 + v^2)(v^2 - u^2)(u + v)}{(u + v)u^2v^2(u^2 + v^2)(u^2 - v^2)} = -\frac{1}{u^2v^2}$

C 24. $\frac{Ax + B}{(x - 1)(x - 2)} = \frac{p}{x - 1} + \frac{q}{x - 2} = \frac{p(x - 2) + q(x - 1)}{(x - 1)(x - 2)}$

Setting the numerators equal, we have $Ax + B = (p + q)x + (-2p - q)$.

Setting the coefficients of x equal: $A = p + q$ (1)
Setting the constant terms equal: $B = -2p - q$ (2)

(1) + (2): $A + B = -p$, or $p = -A - B$
(1): $A = (-A - B) + q$, or $q = 2A + B$

$\therefore \frac{Ax + B}{(x - 1)(x - 2)} = \frac{-A - B}{x - 1} + \frac{2A + B}{x - 2}$

Pages 232–233 • PROBLEMS

A 1. Let t = no. of minutes it would take the pumps to fill the tank; $1 = \frac{1}{30}t + \frac{1}{45}t$; $90 = 3t + 2t$; $t = 18$; 18 min

2. Let t = no. of minutes it would take the pumps to fill the tank with the drain pipe open; $1 = \frac{1}{30}t + \frac{1}{45}t - \frac{1}{90}t$; $90 = 3t + 2t - t$; $t = 22.5$; 22.5 min

3. Let s = no. of seconds it would take the machines to make 50 parts; $50 = \frac{1}{18}s + \frac{1}{12}s$; $36 \cdot 50 = 2s + 3s$; $s = 360$; 360 s, or 6 min

4. Let x = no. of seconds it would take the machines to fill 70 molds; $70 = 3\left(\frac{1}{10}x\right) + \frac{1}{6}x$; $30 \cdot 70 = 9x + 5x$; $x = 150$; 150 s, or 2.5 min

5. Let x = no. of hours Mary traveled; $15x = \left(x - \frac{1}{4}\right)30$; $60x = (4x - 1)30$; $2x = 4x - 1$; $x = \frac{1}{2}$; each traveled $15 \cdot \frac{1}{2}$, or 7.5, km.

6. Let x = no. of hours it took Isabel to run the first leg; $9x + 10\left(\frac{5}{4} - x\right) = 12$; $9x - 10x + 12.5 = 12$; $x = 0.5$; (9 km/h)(0.5 h) = 4.5 km

7. Let x = no. of miles Frank will have to drive on the highway; $20\left(\frac{150}{18} + \frac{x}{24}\right) = 150 + x$; $20\left(\frac{600 + 3x}{72}\right) = 150 + x$; $12{,}000 + 60x = 10{,}800 + 72x$; $x = 100$; 100 mi

8. $20\left(\frac{150}{18} + \frac{x}{24} + \frac{1}{2}\right) = 150 + x$; $20\left(\frac{636 + 3x}{72}\right) = 150 + x$; $12{,}720 + 60x = 10{,}800 + 72x$; $x = 160$; 160 mi

9. Let x = no. of grams of pure hydrogen peroxide that must be added; $240(0.45) + x = (240 + x)0.70$; $108 + x = 168 + 0.7x$; $x = 200$; 200 g

10. Let x = no. of grams of 80% nitric acid solution that must be added; $0.80x + (0.12)175 = 0.60(x + 175)$; $0.80x + 21 = 0.60x + 105$; $x = 420$; 420 g

B 11. Let x = no. of minutes it takes to process one truckload of cans after the first two machines start working; $2x \cdot \frac{1}{25} + (x - 6)\frac{1}{20} = 1$; $8x + 5(x - 6) = 100$; $13x = 130$; $x = 10$; 10 min

12. Let t = no. of minutes the first machine worked; $\frac{t}{18} + \frac{2(20 - t)}{60} = 1$; $5t + 3(20 - t) = 90$; $t = 15$; 15 min

C 13. $\frac{1}{2} \cdot 2x + \frac{4}{5}(x^2 + 4) = 6.5$; $8x^2 + 10x - 33 = 0$; $(2x - 3)(4x + 11) = 0$; $x = \frac{3}{2}$; 1.5 m

14. Let x = distance Lydia traveled at 60 km/h; total distance = average rate · total time; $30 + x = 54\left(\frac{3}{4} + \frac{x}{60}\right)$; $30 + x = 40.5 + 0.9x$; $x = 105$; 105 km

Pages 235–236 • WRITTEN EXERCISES

A 1. LCD = $12x$; $9 + 2 = 12x$; $x = \frac{11}{12}$; $\left\{\frac{11}{12}\right\}$

2. LCD = $10b^2$; $9b - 4b^2 = 5$; $4b^2 - 9b + 5 = 0$; $(4b - 5)(b - 1) = 0$; $b = \frac{5}{4}$ or $b = 1$; $\left\{\frac{5}{4}, 1\right\}$

3. LCD = $6y$; $15 = 8 - 12y$; $12y = -7$; $y = -\frac{7}{12}$; $\left\{-\frac{7}{12}\right\}$

4. LCD = $3(5c - 3)$; $20c - 12 = 12c + 18$; $8c = 30$; $c = \frac{15}{4}$; $\left\{\frac{15}{4}\right\}$

5. LCD = $(z + 2)(z - 2)$; $3(z + 2) - 5z(z - 2) = 0$; $3z + 6 - 5z^2 + 10z = 0$; $5z^2 - 13z - 6 = 0$; $(5z + 2)(z - 3) = 0$; $z = -\frac{2}{5}$ or $z = 3$; $\left\{-\frac{2}{5}, 3\right\}$

6. LCD $= (x - 3)(x + 3)$; $5(x + 3) - 2x(x - 3) - 9 = 0$; $5x + 15 - 2x^2 + 6x - 9 = 0$; $2x^2 - 11x - 6 = 0$; $(2x + 1)(x - 6) = 0$; $x = -\frac{1}{2}$ or $x = 6$; $\left\{-\frac{1}{2}, 6\right\}$

7. LCD $= (x - 5)(x + 5)$; $17 - x(x + 5) = -1(x - 5)$; $17 - x^2 - 5x = -x + 5$; $x^2 + 4x - 12 = 0$; $(x + 6)(x - 2) = 0$; $x = -6$ or $x = 2$; $\{-6, 2\}$

8. LCD $= 3(p - 2)$; $p^2 - 2 - p(p - 2) = 2 \cdot 3$; $2p - 2 = 6$; $p = 4$; $\{4\}$

9. LCD $= (y - 4)(2y + 1)$; $(y - 2)(y - 4) + (y - 2)(2y + 1) = 2(2y + 1)(y - 4)$; $y^2 - 6y + 8 + 2y^2 - 3y - 2 = 4y^2 - 14y - 8$; $y^2 - 5y - 14 = 0$; $(y - 7)(y + 2) = 0$; $y = 7$ or $y = -2$; $\{7, -2\}$

10. LCD $= (2n - 1)(2n + 1)$; $n(2n + 1) - 1 = 4(2n - 1)$; $2n^2 + n - 1 = 8n - 4$; $2n^2 - 7n + 3 = 0$; $(2n - 1)(n - 3) = 0$; $n = \frac{1}{2}$ or $n = 3$; $\frac{1}{2}$ is not an allowable root; $\{3\}$

11. LCD $= (a - 1)(a + 1)$; $4 + (a - 2)(a + 1) = (a - 3)(a - 1)$; $4 + a^2 - a - 2 = a^2 - 4a + 3$; $3a = 1$; $a = \frac{1}{3}$; $\left\{\frac{1}{3}\right\}$

12. LCD $= (k - 1)(k - 2)$; $k^2 + 9 - 6(k - 2) = 2k(k - 1)$; $k^2 - 6k + 21 = 2k^2 - 2k$; $k^2 + 4k - 21 = 0$; $(k + 7)(k - 3) = 0$; $k = -7$ or $k = 3$; $\{-7, 3\}$

B

13. LCD $= (n + 3)(n - 1)$; $(n + 4)(n - 1) - (n - 3)(n + 3) = 3 - n^2$; $n^2 + 3n - 4 - n^2 + 9 = 3 - n^2$; $n^2 + 3n + 2 = 0$; $(n + 2)(n + 1) = 0$; $n = -2$ or $n = -1$; $\{-2, -1\}$

14. LCD $= (2v - 1)(v + 3)$; $(2v + 4)(v + 3) - 2(2v^2 + 5v - 3) = 17 - v$; $2v^2 + 10v + 12 - 4v^2 - 10v + 6 = 17 - v$; $2v^2 - v - 1 = 0$; $(2v + 1)(v - 1) = 0$; $v = -\frac{1}{2}$ or $v = 1$; $\left\{-\frac{1}{2}, 1\right\}$

15. LCD $= (y - 3)(2y - 1)$; $(2y - 5)(2y - 1) = 3(2y^2 - 7y + 3) + 3$; $4y^2 - 12y + 5 = 6y^2 - 21y + 12$; $2y^2 - 9y + 7 = 0$; $(2y - 7)(y - 1) = 0$; $y = \frac{7}{2}$ or $y = 1$; $\left\{\frac{7}{2}, 1\right\}$

16. LCD $= (x - 2)^3$; $5(x^2 - 4x + 4) - (4x + 1)(x - 2) + 3x + 2 = 0$; $5x^2 - 20x + 20 - 4x^2 + 7x + 2 + 3x + 2 = 0$; $x^2 - 10x + 24 = 0$; $(x - 6)(x - 4) = 0$; $x = 6$ or $x = 4$; $\{6, 4\}$

17. LCD $= r(r - 2)(r + 2)$; $(r + 3)(r + 2) - (r + 4)(r - 2) = r(r - 10)$; $r^2 + 5r + 6 - r^2 - 2r + 8 = r^2 - 10r$; $r^2 - 13r - 14 = 0$; $(r - 14)(r + 1) = 0$; $r = 14$ or $r = -1$; $\{14, -1\}$

18. LCD $= (p + 1)(p - 1)(p + 3)$; $(p - 2)(p + 3) - 3(p - 1) = (2p - 1)(p + 1)$; $p^2 + p - 6 - 3p + 3 = 2p^2 + p - 1$; $p^2 + 3p + 2 = 0$; $(p + 2)(p + 1) = 0$; $p = -2$ or $p = -1$; -1 is not an allowable root; $\{-2\}$

19. LCD $= (k + 1)(k^2 - k + 1)$; $2k^2 - 1 + (k + 2)(k + 1) = 7(k^2 - k + 1)$; $2k^2 - 1 + k^2 + 3k + 2 = 7k^2 - 7k + 7$; $4k^2 - 10k + 6 = 0$; $2(2k - 3)(k - 1) = 0$; $k = \frac{3}{2}$ or $k = 1$; $\left\{\frac{3}{2}, 1\right\}$

20. LCD $= (a - 3)(a^2 + 3a + 9)$; $4(a^2 + 3a + 9) - (a + 2)(a - 3) = 2a^2 + 6$; $4a^2 + 12a + 36 - a^2 + a + 6 = 2a^2 + 6$; $a^2 + 13a + 36 = 0$; $(a + 9)(a + 4) = 0$; $a = -9$ or $a = -4$; $\{-9, -4\}$

C

21. LCD $= (c^2 - 4)(c^2 + 1)$; $(2c^2 - 5)(c^2 + 1) - (c^2 - 4)(c^2 + 1) = (c^2 + 7)(c^2 - 4)$; $2c^4 - 3c^2 - 5 - c^4 + 3c^2 + 4 = c^4 + 3c^2 - 28$; $3c^2 - 27 = 0$; $3(c + 3)(c - 3) = 0$; $c = 3$ or $c = -3$; $\{3, -3\}$

22. LCD $= (d + 2)(d + 1)(d - 3)$; $(3d - 2)(d + 1)(d - 3) - 3d(d + 2)(d - 3) = (d + 2)(d + 1)$; $-5d^2 + 13d + 6 = d^2 + 3d + 2$; $6d^2 - 10d - 4 = 0$; $2(3d + 1)(d - 2) = 0$; $d = -\frac{1}{3}$ or $d = 2$; $\left\{-\frac{1}{3}, 2\right\}$

23. LCD $= x^2$; $(x - 1)^2 - x(x - 1) - 6x^2 = 0$; $x^2 - 2x + 1 - x^2 + x - 6x^2 = 0$; $6x^2 + x - 1 = 0$; $(3x - 1)(2x + 1) = 0$; $x = \frac{1}{3}$ or $x = -\frac{1}{2}$; $\left\{\frac{1}{3}, -\frac{1}{2}\right\}$

24. LCD $= y^2$; $(y^2 - 36)^2 = 25y^2$; $y^4 - 72y^2 + 1296 = 25y^2$; $y^4 - 97y^2 + 1296 = 0$; $(y^2 - 81)(y^2 - 16) = 0$; $(y + 9)(y - 9)(y + 4)(y - 4) = 0$; $y = -9$, $y = 9$, $y = -4$, or $y = 4$; $\{-9, -4, 4, 9\}$

Pages 236–237 • PROBLEMS

A **1.** $(p - 1)\left(\frac{240}{p} + 20\right) = 240$; $240 - \frac{240}{p} + 20p - 20 = 240$; $20p^2 - 20p - 240 = 0$; $20(p - 4)(p + 3) = 0$; $p = 4$ or $p = -3$; 4 kPa

2. $\frac{1}{60} + \frac{1}{f + 3} = \frac{1}{f}$; LCD $= 60f(f + 3)$; $f(f + 3) + 60f = 60(f + 3)$; $f^2 + 3f - 180 = 0$; $(f + 15)(f - 12) = 0$; $f = -15$ or $f = 12$; 12 cm

3. $\frac{1}{C_1} + \frac{1}{5C_1} + \frac{1}{5C_1 - 5} = \frac{1}{2}$; LCD $= 10C_1(C_1 - 1)$; $10(C_1 - 1) + 2(C_1 - 1) + 2C_1 = 5C_1(C_1 - 1)$; $14C_1 - 12 = 5C_1^2 - 5C_1$; $5C_1^2 - 19C_1 + 12 = 0$; $(5C_1 - 4)(C_1 - 3) = 0$; $C_1 = \frac{4}{5}$ or $C_1 = 3$; $C_1 \neq \frac{4}{5}$ because $C_3 \neq -1$; $C_1 = 3\ \mu F$, $C_2 = 15\ \mu F$, $C_3 = 10\ \mu F$

4. Let t = no. of minutes it would take the second copier alone; $18 \cdot \frac{1}{30} + 18 \cdot \frac{1}{t} = 1$; LCD $= 30t$; $18t + 540 = 30t$; $t = 45$; 45 min

5. Let t = no. of hours it would take the newer machine alone; $3 \cdot \frac{2}{15} + 3 \cdot \frac{1}{t} = 1$; LCD $= 15t$; $6t + 45 = 15t$; $t = 5$; 5 h

6. Let t = no. of hours it would take each new machine alone; $3 \cdot \frac{1}{t + 6} + 2 \cdot \frac{1}{t} = 1$; LCD $= t(t + 6)$; $3t + 2(t + 6) = t(t + 6)$; $5t + 12 = t^2 + 6t$; $t^2 + t - 12 = 0$; $(t + 4)(t - 3) = 0$; $t = -4$ or $t = 3$; 3 h

7. Let r = Brett's speed for the first 7 km in km/h; $\frac{D}{r} = t$; $\frac{7}{r} + \frac{9}{r + 3} = \frac{4}{3}$; LCD $= 3r(r + 3)$; $21(r + 3) + 27r = 4r(r + 3)$; $48r + 63 = 4r^2 + 12r$; $4r^2 - 36r - 63 = 0$; $(2r + 3)(2r - 21) = 0$; $r = -\frac{3}{2}$ or $r = \frac{21}{2}$; 10.5 km/h

8. Let r = Lucy's speed on the way to the meeting in mi/h; 12 min $= \frac{1}{5}$ h; $\frac{D}{r} = t$; $\frac{72}{r - 5} - \frac{72}{r} = \frac{1}{5}$; LCD $= 5r(r - 5)$; $72 \cdot 5r - 72 \cdot 5(r - 5) = r(r - 5)$; $r^2 - 5r - 1800 = 0$; $(r - 45)(r + 40) = 0$; $r = 45$ or $r = -40$; 45 mi/h

9. Let t = no. of hours John worked before he received his raise; total earned ÷ hours worked = hourly wage; $\frac{63}{t} + \frac{3}{4} = \frac{63}{t - 2}$; LCD $= 4t(t - 2)$; $63 \cdot 4(t - 2) + 3t(t - 2) = 63 \cdot 4t$; $3t^2 - 6t - 504 = 0$; $t^2 - 2t - 168 = 0$; $(t - 14)(t + 12) = 0$; $t = 14$ or $t = -12$; 14 h

B **10.** Let c = rate of current in km/h; $\frac{6}{15 + c} + \frac{6}{15 - c} = \frac{5}{6}$; LCD = $6(15 + c)(15 - c)$; $36(15 - c) + 36(15 + c) = 5(15 + c)(15 - c)$; $540 - 36c + 540 + 36c = 1125 - 5c^2$; $5c^2 = 45$; $c^2 = 9$; $c = 3$ or $c = -3$; 3 km/h

11. Let r = the canoeist's rate in still water in km/h; $\frac{12}{r + 2} + \frac{12}{r - 2} = \frac{25}{r}$; LCD = $r(r + 2)(r - 2)$; $12r(r - 2) + 12r(r + 2) = 25(r^2 - 4)$; $r^2 - 100 = 0$; $(r + 10)(r - 10) = 0$; $r = -10$ or $r = 10$; 10 km/h

12. Let t = no. of days it would take the apprentice to paint the house alone; rate of apprentice = $\frac{1}{t}$; rate of day laborer = $\frac{1}{t + 3}$; rate of master painter = $\frac{1}{t - 3}$; $\frac{7}{t - 3} = \frac{6}{t} + \frac{6}{t + 3}$; LCD = $t(t + 3)(t - 3)$; $7t(t + 3) = 6(t^2 - 9) + 6t(t - 3)$; $5t^2 - 39t - 54 = 0$; $(5t + 6)(t - 9) = 0$; $t = -\frac{6}{5}$ or $t = 9$; 9 days

C **13.** Let t = average no. of hours it takes for the population to increase by 1 person; $\frac{5}{2}t - 2t - \frac{2}{3}t + \frac{2}{9}t = 1$; LCD = 18; $(45 - 36 - 12 + 4)t = 18$; $t = 18$; 18 h

Page 238 • SELF-TEST 3

1. $\frac{3y^2 + y^{-2}}{y^3 - y} \cdot \frac{y^2}{y^2} = \frac{3y^4 + 1}{y^5 - y^3}$

2.
$$\begin{array}{r|l} & 2x^2 + 2x - 5 \\ 3x + 5 & 6x^3 + 16x^2 - 5x - 12 \\ & \underline{6x^3 + 10x^2} \\ & \quad 6x^2 - 5x \\ & \quad \underline{6x^2 + 10x} \\ & \qquad -15x - 12 \\ & \qquad \underline{-15x - 25} \\ & \qquad\qquad 13 \end{array}$$
; $2x^2 + 2x - 5 + \frac{13}{3x + 5}$

3. $\frac{(2y + 3)(y - 3)}{(y - 3)(y + 1)} \cdot \frac{(2y + 3)(y + 3)}{y(y + 3)(y - 3)} = \frac{(2y + 3)^2}{y(y - 3)(y + 1)}$

4. LCD = $(4x - 5)(x + 1)$; $\frac{(4x + 9)(x + 1)}{(4x - 5)(x + 1)} + \frac{(x - 3)(4x - 5)}{(4x - 5)(x + 1)} - \frac{(34 - 2x)}{(4x - 5)(x + 1)} = \frac{4x^2 + 13x + 9 + 4x^2 - 17x + 15 - 34 + 2x}{(4x - 5)(x + 1)} = \frac{8x^2 - 2x - 10}{(4x - 5)(x + 1)} = \frac{2(4x - 5)(x + 1)}{(4x - 5)(x + 1)} = 2$

5. LCD = $(x + 4)(4 - x)$; $2x - 32 + (x + 14)(4 - x) - 3(16 - x^2) = 0$; $2x^2 - 8x - 24 = 0$; $x^2 - 4x - 12 = 0$; $(x - 6)(x + 2) = 0$; $x = 6$ or $x = -2$; $\{-2, 6\}$

6. Let x = number of students that had originally planned to go; number of students × cost per student = total cost = \$600; $(x - 2)\left(\frac{600}{x} + 15\right) = 600$; $600 - \frac{1200}{x} + 15x - 30 = 600$; LCD = x; $x^2 - 2x - 80 = 0$; $(x - 10)(x + 8) = 0$; $x = 10$ or $x = -8$; 10 students

Pages 239–240 • CHAPTER REVIEW

1. $(x^{-9}y^5)^{-1}(2^{-1}x^{-4}y)^2 = x^9y^{-5} \cdot 2^{-2}x^{-8}y^2 = 2^{-2}xy^{-3} = \frac{x}{4y^3}$; b

2. $$\begin{array}{r} y^2 - 4y + 2 \\ 2y - 3 \\ \hline 2y^3 - 8y^2 + 4y \\ -3y^2 + 12y - 6 \\ \hline 2y^3 - 11y^2 + 16y - 6 \end{array}$$; c

3. $24a^3b - 32a^2b^2 - 6ab^3 = 2ab(12a^2 - 16ab - 3b^2) = 2ab(6a + b)(2a - 3b)$; d

4. $x^3 - 3x^2 - 4x = 0$; $x(x^2 - 3x - 4) = 0$; $x(x - 4)(x + 1) = 0$; $x = 0$, $x = 4$, or $x = -1$; c

5. $x^2 - 5x - 6 \le 0$; $(x - 6)(x + 1) \le 0$; $x - 6 \ge 0$ and $x + 1 \le 0$ or $x - 6 \le 0$ and $x + 1 \ge 0$; $x \ge 6$ and $x \le -1$ or $x \le 6$ and $x \ge -1$; no such x or $-1 \le x \le 6$; a

6. $\frac{(x^2 - 4)(x^2 - 16)}{(x + 2)(x + 4)(x - 2)} = \frac{(x + 2)(x - 2)(x + 4)(x - 4)}{(x + 2)(x + 4)(x - 2)} = x - 4$; b

7. $$\begin{array}{r} 4y^3 + 8y^2 + 16y + 32 \\ y - 2\overline{)4y^4 + 0y^3 + 0y^2 + 0y - 68} \\ \underline{4y^4 - 8y^3} \\ 8y^3 + 0y^2 \\ \underline{8y^3 - 16y^2} \\ 16y^2 + 0y \\ \underline{16y^2 - 32y} \\ 32y - 68 \\ \underline{32y - 64} \\ -4 \end{array}$$; $4y^3 + 8y^2 + 16y + 32 + \frac{-4}{y - 2}$; d

8. $\frac{(y + 5)(y - 5)}{(y + 5)(y + 6)} \cdot \frac{(y + 6)(y - 3)}{(y - 5)(y - 3)} = 1$; c

9. $\frac{2x - 3}{(x - 4)(x - 1)} - \frac{x + 1}{(x - 4)(x + 1)} = \frac{(2x - 3)(x + 1) - (x + 1)(x - 1)}{(x - 4)(x - 1)(x + 1)} =$
$\frac{x^2 - x - 2}{(x - 4)(x + 1)(x - 1)} = \frac{(x - 2)(x + 1)}{(x - 4)(x + 1)(x - 1)} = \frac{x - 2}{(x - 4)(x - 1)}$; a, d

10. Let x = no. of grams of salt that must be added; $0.22(220) + x = 0.34(220 + x)$; $48.4 + x = 74.8 + 0.34x$; $0.66x = 26.4$; $x = 40$; 40 g; d

11. $\frac{x + 1}{(x + 2)(x - 2)} + \frac{x - 1}{(x + 2)(x - 1)} = \frac{1}{x + 2}$; LCD $= (x + 2)(x - 2)(x - 1)$;
$(x + 1)(x - 1) + (x - 1)(x - 2) = (x - 2)(x - 1)$; $x^2 - 1 = 0$;
$(x + 1)(x - 1) = 0$; $x = -1$ or $x = 1$; 1 is not an allowable root; $\{-1\}$; b

Pages 240–241 • CHAPTER TEST

1. $(2x^{-3}y^{-1})(2^{-4}x^4y^{-8}) = 2^{-3}xy^{-9} = \frac{x}{8y^9}$

2. $\frac{m^{-2} - n^{-2}}{m^{-1} + n^{-1}} \cdot \frac{m^2n^2}{m^2n^2} = \frac{n^2 - m^2}{mn^2 + m^2n} = \frac{(n + m)(n - m)}{mn(n + m)} = \frac{n - m}{mn}$

3. $(2x - 1)(3x^2 - 5x + 8) = 6x^3 - 13x^2 + 21x - 8$

4. $[(3x - 2)(3x + 2)]^2 = (9x^2 - 4)^2 = 81x^4 - 72x^2 + 16$

5. $2g^3h - 6g^2h^2 + 4gh^3 = 2gh(g^2 - 3gh + 2h^2) = 2gh(g - 2h)(g - h)$

6. $x^6 - y^6 = (x^3 + y^3)(x^3 - y^3) = (x + y)(x^2 - xy + y^2)(x - y)(x^2 + xy + y^2)$

7. $2w^2 + 7w - 15 = 0$; $(2w - 3)(w + 5) = 0$; $w = \frac{3}{2}$ or $w = -5$; $\left\{\frac{3}{2}, -5\right\}$

8. $3x^2 - 7x - 6 - 10x = 0$; $3x^2 - 17x - 6 = 0$; $(3x + 1)(x - 6) = 0$; $x = -\frac{1}{3}$ or $x = 6$; $\left\{-\frac{1}{3}, 6\right\}$

9. Let x = length of side of square in centimeters; $2x(x + 6) = x^2 + 13$; $x^2 + 12x - 13 = 0$; $(x + 13)(x - 1) = 0$; $x = -13$ or $x = 1$; 1 cm

10. $(4x + 1)(x - 2) \geq 0$

$4x + 1 \geq 0$ and $x - 2 \geq 0$ or $4x + 1 \leq 0$ and $x - 2 \leq 0$

$x \geq -\frac{1}{4}$ and $x \geq 2$ $\qquad$ $x \leq -\frac{1}{4}$ and $x \leq 2$

$x \geq 2$ $\qquad$ $x \leq -\frac{1}{4}$

$\left\{x: x \geq 2 \text{ or } x \leq -\frac{1}{4}\right\}$

$-\frac{1}{4}$ 0 1 2

11. $\frac{2(m^2 - 3m - 10)}{4(m^2 - 25)} = \frac{2(m - 5)(m + 2)}{4(m - 5)(m + 5)} = \frac{m + 2}{2(m + 5)}$

12.
$$
\begin{array}{r}
4y^3 - 3y^2 + 2y - 1 \\
2y + 1 \overline{)\, 8y^4 - 2y^3 + y^2 + 0y - 4} \\
\underline{8y^4 + 4y^3} \qquad\qquad\qquad \\
-6y^3 + y^2 \qquad\qquad \\
\underline{-6y^3 - 3y^2} \qquad\qquad \\
4y^2 + 0y \qquad \\
\underline{4y^2 + 2y} \qquad \\
-2y - 4 \\
\underline{-2y - 1} \\
-3
\end{array}
$$
; $4y^3 - 3y^2 + 2y - 1 + \frac{-3}{2y + 1}$

13. $\frac{xy^2 - x}{y^2} \div \frac{xy - x}{y} = \frac{x(y + 1)(y - 1)}{y^2} \cdot \frac{y}{x(y - 1)} = \frac{y + 1}{y}$

14. $\frac{x^2 - 5x}{x - 1} + \frac{4x}{x - 1} + \frac{3(x - 1)}{x - 1} = \frac{x^2 + 2x - 3}{x - 1} = \frac{(x + 3)(x - 1)}{x - 1} = x + 3$

15. Let x = no. of hours the printers work together; $\frac{x}{4} + \frac{x}{6} = 1$; $3x + 2x = 12$; $x = \frac{12}{5}$; 2.4 h

16. $\frac{x^2 - 6x - 11}{(x - 3)(x + 1)} + \frac{x + 2}{x - 3} = \frac{3x - 1}{x + 1}$; LCD $= (x - 3)(x + 1)$; $x^2 - 6x - 11 + (x + 2)(x + 1) = (3x - 1)(x - 3)$; $x^2 - 7x + 12 = 0$; $(x - 4)(x - 3) = 0$; $x = 4$ or $x = 3$; 3 is not an allowable root; $\{4\}$

Pages 241–242 • MIXED REVIEW

1. $14 - 8 \div 4 \cdot 3 + (1 + 2)^2 = 14 - 2 \cdot 3 + 3^2 = 14 - 6 + 9 = 8 + 9 = 17$

2. $-[11 + (-z)] - z + (-2) = -11 + z - z + (-2) = -13$

3. $-4[-(2x + y) + (-3)(x - y)] = -4[-2x - y - 3x + 3y] = -4[-5x + 2y] = 20x - 8y$

4. $5s^2 - (3st - 2t)^2 - (-3s^2 + 5st) = 5s^2 - (9s^2t^2 - 12st^2 + 4t^2) + 3s^2 - 5st = 8s^2 - 9s^2t^2 + 12st^2 - 4t^2 - 5st$

5. $x - 4 = 2x - 6$; $x = 2$; $\{2\}$ **6.** $-2x - 3 = x + 6$; $3x = -9$; $\{-3\}$

7. $2m - 6 - 4m + 1 < 1$; $-2m < 6$; $m > -3$; $\{m: m > -3\}$

8. $-12 < 3d - 6 < 6$; $-6 < 3d < 12$; $-2 < d < 4$; $\{d: -2 < d < 4\}$

9. $|2x - 3| = 5$; $2x - 3 = 5$ or $2x - 3 = -5$; $x = 4$ or $x = -1$; $\{-1, 4\}$

10. $y - 3 \geq 6$ or $y - 3 \leq -6$; $y \geq 9$ or $y \leq -3$; $\{y: y \leq -3 \text{ or } y \geq 9\}$

11. $b(4a - 5) = 3a^2$; $b = \dfrac{3a^2}{4a - 5}$ **12.** $b - n^2 = 2cn$; $b = n^2 + 2cn$

13. (*1*): $4\left(\dfrac{x + 1}{2}\right) = 4\left(\dfrac{y - 3}{4}\right)$; $2x + 2 = y - 3$; $2x = y - 5$

(*2*): $x = 5 - 2y$

Substitute (*2*) into (*1*): $2(5 - 2y) = y - 5$; $5y = 15$; $y = 3$; $x = -1$; $\{(-1, 3)\}$

14. $x = -2$ **15. a.** $[f \circ g](-2) = f(8) = 5$ **b.** $[g \circ f](-2) = g(5) = 29$

16. a. $D = \begin{vmatrix} 4 & -1 \\ 2 & 3 \end{vmatrix} = 12 + 2 = 14$; $D_x = \begin{vmatrix} 3 & -1 \\ 5 & 3 \end{vmatrix} = 9 + 5 = 14$

$D_y = \begin{vmatrix} 4 & 3 \\ 2 & 5 \end{vmatrix} = 20 - 6 = 14$; $x = \dfrac{D_x}{D} = 1$; $y = \dfrac{D_y}{D} = 1$; $\{(1, 1)\}$

b. $\begin{matrix} 2x + 3y = 3 \\ 12x + 18y = 18 \end{matrix}$; $D = \begin{vmatrix} 2 & 3 \\ 12 & 18 \end{vmatrix} = 36 - 36 = 0$; the equations represent the same line; $\left\{(x, y): y = \dfrac{-2}{3}x + 1\right\}$

17. $m = \dfrac{5 - (-2)}{-4 - 10} = -\dfrac{1}{2}$; $y = -\dfrac{1}{2}x + b$; at $(10, -2)$ we have $-2 = -\dfrac{1}{2}(10) + b$, or $b = 3$; $y = -\dfrac{1}{2}x + 3$

18. $\dfrac{270}{3} = \dfrac{585}{x}$; $270x = 3(585)$; $x = 6.5$; 6.5 h

19. Let d = no. of dimes and q = no. of quarters.

$d + q = 39$; $d = 39 - q$ (*1*)

$10d + 25q = 405$ (*2*)

Substitute (*1*) into (*2*): $10(39 - q) + 25q = 405$; $15q = 15$; $q = 1$; $d = 38$;

1 quarter, 38 dimes

20.

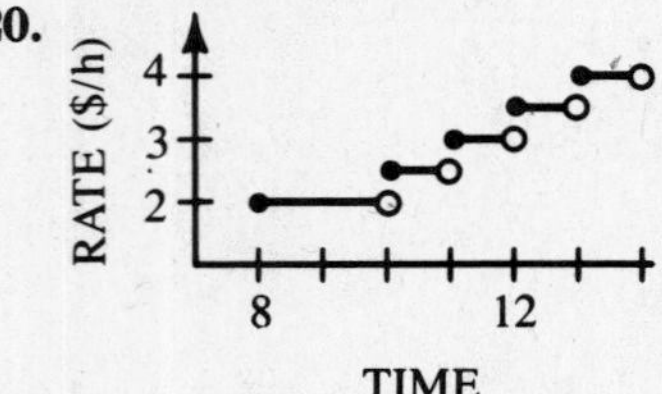

Page 242 • CONTEST PROBLEMS

Possible amount: \$564; Alvarez originally took \$396, Baird originally took \$156, and Curtis originally took \$12.

Pages 244–245 • PROGRAMMING IN PASCAL

1. Answers will vary depending on the integers used.
2. **a.**
```
PROCEDURE check_it_out

BEGIN
  factor := 0;
  write('The pairs of positive, integral factors of');
  writeln(num:1,'are:');
  REPEAT
    factor := factor + 1;
    IF (num MOD factor = 0)
       THEN writeln(factor:1,' ':5,(num DIV factor):1);
  UNTIL (factor = num) OR (num = 2);
END;
```
   b. See the procedure in Exercise 2c.

   c.
```
PROCEDURE check_it_out;

VAR
  count : integer;

BEGIN
  count := 0;
  factor := 0;
  write('The pairs of positive, integral factors of');
  writeln(num:1,' are:');
  REPEAT
    factor := factor + 1;
    IF (num MOD factor = 0) AND (factor <= sqrt(num))
       THEN
         BEGIN
           count := count + 1;
           writeln(factor:1,' ':5,(num DIV factor):1);
         END;
  UNTIL (factor >= sqrt(num)) OR (num = 2);
  write('The number of pairs of different factors');
  writeln(is:',count:1);
END;  {* check_it_out *}
```
3. **a.**
```
PROGRAM prime_1000 (INPUT,OUTPUT);

VAR
  num,factor : integer;
  prime : boolean;
```
[Program continued on next page]

```
{********************************************************}
PROCEDURE check_it_out;

BEGIN
  prime := TRUE;
  factor := 1;
  REPEAT
    factor := factor + 1;
    IF (num MOD factor = 0) AND (num <> 2)        {* 2 is an
                                                   exception *}
       THEN prime := FALSE;
  UNTIL (NOT prime) OR (factor >= sqrt(num));
END;  {* check_it_out *}

{********************************************************}
BEGIN  {* main *}
  FOR num := 1 TO 1000 DO
      BEGIN
        check_it_out;
        IF prime
           THEN writeln(num:1);
      END;
END.
```

b. See the main body of the program in Exercise 3c.

c.

```
PROGRAM prime_1000 (INPUT,OUTPUT);

VAR
  num,factor,count : integer;
  lower_bound,upper_bound : integer;
  prime : boolean;

{********************************************************}
PROCEDURE check_it_out;

BEGIN
  prime := TRUE;
  factor := 1;
  REPEAT
    factor := factor + 1;
    IF (num MOD factor = 0) AND (num <> 2)        {* 2 is an
                                                   exception *}
       THEN prime := FALSE;
  UNTIL (NOT prime) OR (factor >= sqrt(num));
END;  {* check_it_out *}
```

```
{***********************************************************}
PROCEDURE get_bounds;

BEGIN
  write('What is the lower boundary:');
  readln(lower_bound);
  write('What is the upper boundary:');
  readln(upper_bound);
END;  {* get_bounds *}

{***********************************************************}
BEGIN  {* main *}
  get_bounds;
  count := 0;
  FOR num := lower_bound TO upper_bound DO
      BEGIN
        check_it_out;
        IF (prime) AND (num > 1)
           THEN
             BEGIN
               writeln(num:1);
               count := count + 1;
             END;
      END;
  write('There are ',count:1,' prime numbers between ');
  writeln(lower_bound:1,' and ',upper_bound:1);
END.
```

4. a. According to the Fundamental Rule of Arithmetic, every integer greater than one can be expressed as a product of primes. If no such product exists, that number must be prime. As a result, the prime numbers less than the given number are the only values that could possibly be divisors of the given number.

b.

```
PROGRAM primes (INPUT,OUTPUT);

VAR
  num,factor,count : integer;
  lower_bound,upper_bound : integer;
  no_of_primes : integer;
  prime : boolean;
  primes : ARRAY[1..200] OF integer;
```

[*Program continued on next page*]

```
{**********************************************************}
PROCEDURE check_it_out;

VAR
  i : integer;

BEGIN
  prime := TRUE;
  FOR i := 1 TO no_of_primes DO
      BEGIN
        IF num MOD primes[i] = 0
           THEN prime := FALSE;
      END;
  IF (prime) AND (num > 1)
     THEN
       BEGIN
         no_of_primes := no_of_primes + 1;
         primes[no_of_primes] := num;
       END;
END;   {* check_it_out *}

{**********************************************************}
PROCEDURE get_bounds;

BEGIN
  write('What is the lower boundary:');
  readln(lower_bound);
  write('What is the upper boundary:');
  readln(upper_bound);
END;  {* get_bounds *}

{**********************************************************}
BEGIN  {* main *}
  get_bounds;
  count := 0;
  FOR num := lower_bound TO upper_bound DO
      BEGIN
        check_it_out;
        IF (prime) AND (num > 1)
           THEN
             BEGIN
               writeln(num:1);
               count := count + 1;
             END;
      END;
  write('There are ',count:1,' prime numbers between ');
  writeln(lower_bound:1,' and ',upper_bound:1);
END.
```

5.
```
PROGRAM sieve (INPUT,OUTPUT);

VAR
  num : integer;
  primes : ARRAY[2..1000] OF integer;

{**********************************************************}
PROCEDURE replace_multiples;

VAR
  index : integer;

BEGIN
  index := 2 * num;
  WHILE index <= 2000 DO
    BEGIN
      primes[index] := 0;
      index := index + num;
    END;
END;

{**********************************************************}
BEGIN  {* main *}
  FOR num := 2 TO 2000 DO
      primes[num] := num;
  FOR num := 2 TO 2000 DO
      replace_multiples;
  FOR num := 2 TO 2000 DO
      BEGIN
        IF primes[num] <> 0
           THEN writeln(primes[num]);
      END;
END.
```

Pages 249–250 • WRITTEN EXERCISES

A 1.
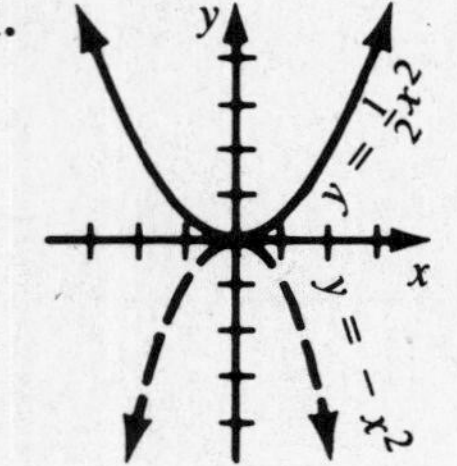

2.
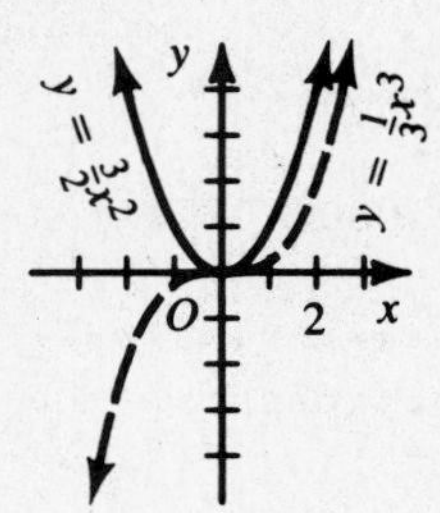

3.
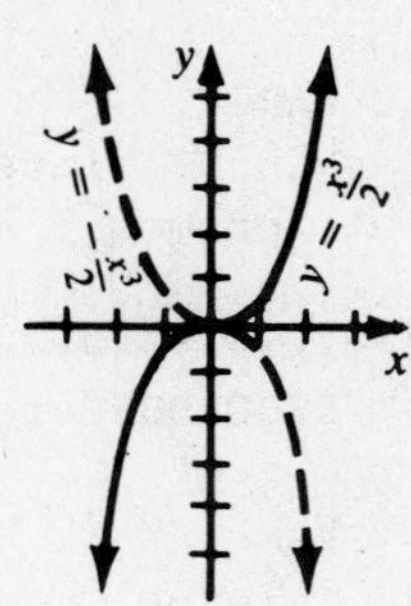

4.
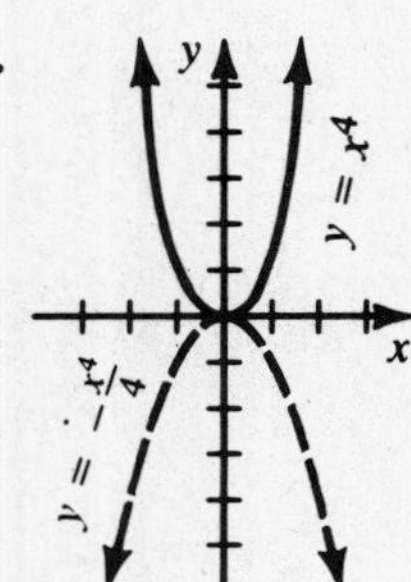

5.
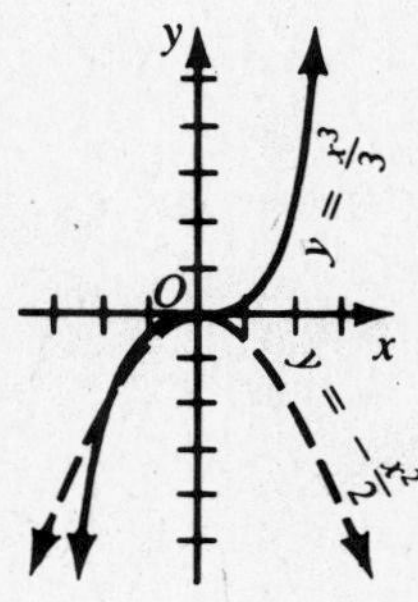

6.
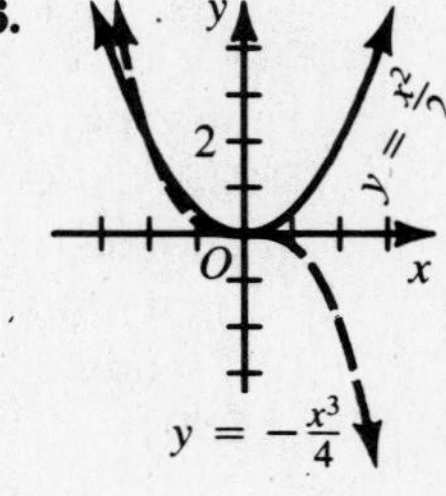

7. $6 = 9k;\ k = \dfrac{2}{3}$ 8. $48 = 8k;\ k = 6$ 9. $54 = -27k;\ k = -2$

10. $-\dfrac{5}{6} = \dfrac{k}{9};\ k = -\dfrac{15}{2}$ 11. $\dfrac{3}{8} = \dfrac{k}{64};\ k = 24$ 12. $\dfrac{7}{24} = \dfrac{k}{16};\ k = \dfrac{14}{3}$

13. $\dfrac{12}{8^2} = \dfrac{y}{28^2};\ y = \dfrac{12 \times 784}{64} = 147$ 14. $\dfrac{36}{9^2} = \dfrac{4}{x^2};\ x^2 = \dfrac{81 \times 4}{36} = 9;\ x = 3 \text{ or } x = -3$

15. $\dfrac{\frac{2}{3}}{(\frac{1}{6})^2} = \dfrac{24}{x^2};\ \dfrac{2}{3}x^2 = \dfrac{1}{36}(24);\ x^2 = 1;\ x = 1 \text{ or } x = -1$

16. $\dfrac{-8}{6^2} = \dfrac{y}{2^2};\ y = \dfrac{-8 \times 4}{36} = -\dfrac{8}{9}$

B 17. $f(2x) = 5(2x)^3 = 8(5x^3) = 8(f(x))$; multiplied by 8

18. $f(3x) = 4(3x)^2 = 9(4x^2) = 9(f(x))$; multiplied by 9

19. $f\left(\dfrac{x}{2}\right) = 6\left(\dfrac{x}{2}\right)^2 = \dfrac{1}{4}(6x^2) = \dfrac{1}{4}(f(x))$; divided by 4

20. $f\left(\dfrac{x}{2}\right) = 10\left(\dfrac{x}{2}\right)^3 = \dfrac{1}{8}(10x^3) = \dfrac{1}{8}(f(x))$; divided by 8

21. $f(4x) = 3(4x)^2 = 16(3x^2) = 16(f(x))$; multiplied by 16

22. $f(kx) = (kx)^3 = k^3(x^3) = k^3(f(x))$; multiplied by k^3

Pages 250–251 • PROBLEMS

A 1. $\dfrac{l_1}{(t_1)^2} = \dfrac{l_2}{(t_2)^2};\ \dfrac{1.2}{(2.2)^2} = \dfrac{l_2}{(1.1)^2};\ l_2 = \dfrac{1.2 \times 1.21}{4.84} = 0.3$; 0.3 m

2. $\dfrac{d_1}{(w_1)^4} = \dfrac{d_2}{(w_2)^4};\ \dfrac{0.162}{3^4} = \dfrac{d_2}{2^4};\ d_2 = \dfrac{0.162 \times 16}{81} = 0.032$; 0.032 cm

3. $\frac{p_1}{(c_1)^2} = \frac{p_2}{(c_2)^2}$; $\frac{100}{500^2} = \frac{p_2}{4^2}$; $p_2 = \frac{100 \times 16}{250{,}000} = 0.0064$; 0.0064 kW, or 6.4 W

4. $\frac{m_1}{(d_1)^3} = \frac{m_2}{(d_3)^3}$; $\frac{0.057}{(0.2)^3} = \frac{m_2}{(0.4)^3}$; $m_2 = \frac{0.057 \times 0.064}{0.008} = 0.456$; 0.456 g

5. $\frac{p_1}{(s_1)^3} = \frac{p_2}{(s_2)^3}$; $\frac{160}{40^3} = \frac{p_2}{30^3}$; $p_2 = \frac{160 \times 27{,}000}{64{,}000} = 67.5$; 67.5 W

B 6. $\frac{r_1}{(t_1)^4} = \frac{r_2}{(t_2)^4}$; $\frac{324}{300^4} = \frac{r_2}{200^4}$; $r_2 = \frac{324 \times 1{,}600{,}000{,}000}{8{,}100{,}000{,}000} = 64$; 64 (J/s)/m^2

7. $\frac{e_1}{(p_1)^2} = \frac{e_2}{(p_2)^2}$ **a.** $\frac{0.00009}{6^2} = \frac{e_2}{8^2}$; $e_2 = \frac{0.00009 \times 64}{36} = 0.00016$; 0.00016 J

b. $a = \frac{0.00009}{6^2} = 0.0000025$

8. $\frac{f_1}{(s_1)^2} = \frac{f_2}{(s_2)^2}$; $\frac{2.7}{(3.6)^2} = \frac{7.5}{(s_2)^2}$; $(s_2)^2 = \frac{12.96 \times 7.5}{2.7} = 36$; $s_2 = \pm 6$; 6 m/s

9. $\frac{(p_1)^2}{(d_1)^3} = \frac{(p_2)^2}{(d_2)^3}$; $\frac{8^2}{20{,}000^3} = \frac{(p_2)^2}{30{,}000^3}$; $(p_2)^2 = \frac{64 \times 27{,}000{,}000{,}000{,}000}{8{,}000{,}000{,}000{,}000} = 216$;
$p_2 \approx \pm 14.70$; 14.70 h

Page 253 • WRITTEN EXERCISES

A 1. $\{-12, 12\}$ 2. $x^2 = 25$; $\{-5, 5\}$ 3. $x^4 = 1$; $\{-1, 1\}$ 4. $\{-3, 3\}$

5. $x^3 = -8$; $\{-2\}$ 6. $27 = x^3$; $\{3\}$ 7. $x^2 = -9$; $\emptyset$ 8. $x^3 = \frac{1}{64}$; $\left\{\frac{1}{4}\right\}$

9. $\frac{49}{16} = x^2$; $\left\{\frac{7}{4}, -\frac{7}{4}\right\}$ 10. $x^3 = -\frac{125}{27}$; $\left\{-\frac{5}{3}\right\}$ 11. $x^5 = \frac{1}{32}$; $\left\{\frac{1}{2}\right\}$

12. $x^2 = -\frac{36}{121}$; $\emptyset$ 13. $-\frac{1}{2} + \frac{7}{2} = 3$ 14. $-2 - 5 = -7$ 15. $(-2)(-1) = 2$

16. $-\frac{3}{4} \div \frac{5}{2} = -\frac{3}{10}$ 17. $x - 4 = 0$, $x = 4$; $\{4\}$ 18. $|x| = 5$, $x = \pm 5$; $\{-5, 5\}$

19. $-x = 6$, $x = -6$; $\{-6\}$ 20. $|x| = -x$; $\{x: x \le 0\}$ 21. $|x| = x$; $\{x: x \ge 0\}$

22. $-x = -|x|$, $x = |x|$; $\{x: x \ge 0\}$

B 23. $x - 1 = 7$, $x = 8$; $\{8\}$

24. $|x - 3| = 4$; $x - 3 = 4$ or $x - 3 = -4$; $x = 7$ or $x = -1$; $\{-1, 7\}$

25. $|y + 1| = y + 1$; $y + 1 \ge 0$; $y \ge -1$; $\{y: y \ge -1\}$

26. $\sqrt{((y + 1)^2)^2} = (y + 1)^2$; $(y + 1)^2 = (y + 1)^2$; $y \in \mathscr{R}$

27. $\sqrt{((x - 2)^3)^2} = (x - 2)^3$; $|x - 2|^3 = (x - 2)^3$; $|x - 2| = x - 2$; $x - 2 \ge 0$; $x \ge 2$;
$\{x: x \ge 2\}$

28. $x - 8 = 8 - x$, $2x = 16$, $x = 8$; $\{8\}$

C 29.

1. $x \ge 0, y \ge 0$ $\sqrt{x^2 + y^2} > x + y$	Hypothesis
2. $\sqrt{x^2 + y^2} > 0$	Def. of principal root
3. $x + y > 0$	Closure ax. for $\mathscr{R}_+$
4. $(\sqrt{x^2 + y^2})^2 > (x + y)^2$	Ex. 34, page 60
5. $x^2 + y^2 > x^2 + 2xy + y^2$	Subs. prin.
6. $0 > 2xy$	Add. prop. of order
7. $x < 0$ or $y < 0$	Prop. of neg. in prod.
8. Contradiction	Hypothesis
9. $\therefore \sqrt{x^2 + y^2} \le x + y$	By indirect proof

30. If $\sqrt{x^2 + y^2} = x + y$, then $(\sqrt{x^2 + y^2})^2 = (x + y)^2$, or $x^2 + y^2 = x^2 + 2xy + y^2$. Then by the add. prop. of order, $2xy = 0$, so $xy = 0$, which is true only if $x = 0$ or $y = 0$.

Pages 255–256 • WRITTEN EXERCISES

A **1.** $f(1) = 0$; $f(-1) = 0$; $f(2) = 0$; $f(-2) = -12$; roots: $1, -1, 2$

2. $f(1) = -7$; $f(-1) = -9$; $f(2) = 0$; $f(-2) = -16$; $f(4) = 56$; $f(-4) = -72$; $f(8) = 504$; $f(-8) = -520$; root: 2

3. $f(1) = 0$; $f(-1) = 6$; $f(2) = 0$; $f(-2) = 0$; $f(4) = 36$; $f(-4) = -60$; roots: $1, 2, -2$

4. $f(1) = 6$; $f(-1) = 0$; $f\left(\frac{1}{2}\right) = 0$; $f\left(-\frac{1}{2}\right) = 0$; $f\left(\frac{1}{4}\right) = -\frac{15}{16}$; $f\left(-\frac{1}{4}\right) = -\frac{9}{16}$; roots: $-1, \frac{1}{2}, -\frac{1}{2}$

5. $f(1) = 12$; $f(-1) = 0$; $f\left(\frac{1}{2}\right) = \frac{3}{2}$; $f\left(-\frac{1}{2}\right) = 0$; $f\left(\frac{1}{3}\right) = 0$; $f\left(-\frac{1}{3}\right) = -\frac{4}{9}$; $f\left(\frac{1}{6}\right) = -\frac{7}{9}$; $f\left(-\frac{1}{6}\right) = -\frac{31}{36}$; roots: $-1, -\frac{1}{2}, \frac{1}{3}$

6. $f(1) = 8$; $f(-1) = 0$; $f(3) = 0$; $f(-3) = -72$; $f(9) = 360$; $f(-9) = -1152$; roots: $-1, 3$

7. $f(1) = 18$; $f(-1) = -4$; $f(2) = 80$; $f(-2) = 0$; $f(4) = 396$; $f(-4) = -52$; $f\left(\frac{1}{3}\right) = 0$; $f\left(-\frac{1}{3}\right) = -\frac{50}{9}$; $f\left(\frac{2}{3}\right) = \frac{64}{9}$; $f\left(-\frac{2}{3}\right) = -\frac{48}{9}$; $f\left(\frac{4}{3}\right) = \frac{100}{3}$; $f\left(-\frac{4}{3}\right) = \frac{124}{3}$; roots: $-2, \frac{1}{3}$

8. $f(1) = 0$; $f(-1) = -24$; $f(3) = 0$; $f(-3) = -168$; $f\left(\frac{1}{2}\right) = 0$; $f\left(-\frac{1}{2}\right) = -\frac{21}{2}$; $f\left(\frac{3}{2}\right) = -\frac{3}{2}$; $f\left(-\frac{3}{2}\right) = -45$; roots: $1, 3, \frac{1}{2}$

9. $f(1) = f(-1) = 0$; $f(5) = f(-5) = 0$; $f(25) = f(-25) = 374{,}400$; roots: $1, -1, 5, -5$

10. $f(1) = f(-1) = 0$; $f\left(\frac{1}{2}\right) = f\left(-\frac{1}{2}\right) = 0$; $f\left(\frac{1}{4}\right) = f\left(-\frac{1}{4}\right) = \frac{45}{64}$; roots: $1, -1, \frac{1}{2}, -\frac{1}{2}$

11. $p \in \{\pm 1, \pm 2, \pm 4\}$, $q \in \{\pm 1, \pm 2\}$; possible rational roots are $\pm 1, \pm 2, \pm 4, \pm\frac{1}{2}$; $f(1) = 0$; $f(-1) = -10$; $f(2) = -10$; $f(-2) = -54$; $f(4) = 0$; $f(-4) = -280$; $f\left(\frac{1}{2}\right) = \frac{7}{2}$; $f\left(-\frac{1}{2}\right) = 0$; roots: $1, 4, -\frac{1}{2}$

12. $p \in \{\pm 1, \pm 2\}$, $q \in \{\pm 1, \pm 2, \pm 3, \pm 6\}$; possible rational roots are $\pm 1, \pm 2, \pm\frac{1}{2}$, $\pm\frac{1}{3}, \pm\frac{2}{3}, \pm\frac{1}{6}$; $f(1) = 6$; $f(-1) = 12$; $f(2) = 60$; $f(-2) = 0$; $f\left(\frac{1}{2}\right) = 0$; $f\left(-\frac{1}{2}\right) = \frac{15}{2}$; $f\left(\frac{1}{3}\right) = 0$; $f\left(-\frac{1}{3}\right) = \frac{50}{9}$; $f\left(\frac{2}{3}\right) = \frac{8}{9}$; $f\left(-\frac{2}{3}\right) = \frac{28}{3}$; $f\left(\frac{1}{6}\right) = \frac{13}{18}$; $f\left(-\frac{1}{6}\right) = \frac{11}{3}$; roots: $-2, \frac{1}{2}, \frac{1}{3}$

13. $p \in \{\pm 1, \pm 2, \pm 3, \pm 6\}$, $q \in \{\pm 1, \pm 2, \pm 4\}$; possible rational roots are $\pm 1, \pm 2$, $\pm 3, \pm 6, \pm\frac{1}{2}, \pm\frac{3}{2}, \pm\frac{1}{4}, \pm\frac{3}{4}$; $f(1) = -3$; $f(-1) = 15$; $f(2) = 12$; $f(-2) = 0$;

$f(3) = 75$; $f(-3) = -63$; $f(6) = 792$; $f(-6) = -780$; $f\left(\frac{1}{2}\right) = 0$; $f\left(-\frac{1}{2}\right) = 12$; $f\left(\frac{3}{2}\right) = 0$; $f\left(-\frac{3}{2}\right) = 12$; $f\left(\frac{1}{4}\right) = \frac{45}{16}$; $f\left(-\frac{1}{4}\right) = \frac{147}{16}$; $f\left(\frac{3}{4}\right) = -\frac{33}{16}$; $f\left(-\frac{3}{4}\right) = \frac{225}{16}$; roots: $-2, \frac{1}{2}, \frac{3}{2}$

14. $p \in \{\pm1, \pm5\}$, $q \in \{\pm1, \pm2, \pm4\}$; possible rational roots are $\pm1, \pm5, \pm\frac{1}{2}, \pm\frac{5}{2}, \pm\frac{1}{4}, \pm\frac{5}{4}$; $f(1) = -6$, $f(-1) = 0$; $f(5) = 270$; $f(-5) = -660$; $f\left(\frac{1}{2}\right) = 0$; $f\left(-\frac{1}{2}\right) = 6$; $f\left(\frac{5}{2}\right) = 0$; $f\left(-\frac{5}{2}\right) = -90$; $f\left(\frac{1}{4}\right) = \frac{45}{16}$; $f\left(-\frac{1}{4}\right) = \frac{99}{16}$; $f\left(\frac{5}{4}\right) = -\frac{135}{16}$; $f\left(-\frac{5}{4}\right) = -\frac{105}{16}$; roots: $-1, \frac{1}{2}, \frac{5}{2}$

15. $p \in \{\pm1, \pm2, \pm4, \pm8\}$, $q \in \{\pm1, \pm2\}$; possible rational roots are $\pm1, \pm2, \pm4, \pm8, \pm\frac{1}{2}$; $f(1) = 15$; $f(-1) = -15$; $f(2) = 96$; $f(-2) = 0$; $f(4) = 840$; $f(-4) = 360$; $f(8) = 10{,}200$; $f(-8) = 6936$; $f\left(\frac{1}{2}\right) = 0$; $f\left(-\frac{1}{2}\right) = -\frac{51}{4}$; roots: $-2, \frac{1}{2}$

16. $p \in \{\pm1, \pm3\}$, $q \in \{\pm1, \pm2, \pm4\}$; possible rational roots are $\pm1, \pm3, \pm\frac{1}{2}, \pm\frac{3}{2}, \pm\frac{1}{4}, \pm\frac{3}{4}$; $f(1) = -6$; $f(-1) = 10$; $f(3) = 210$; $f(-3) = 450$; $f\left(\frac{1}{2}\right) = -5$; $f\left(-\frac{1}{2}\right) = 0$; $f\left(\frac{3}{2}\right) = 0$; $f\left(-\frac{3}{2}\right) = 39$; $f\left(\frac{1}{4}\right) = -\frac{63}{16}$; $f\left(-\frac{1}{4}\right) = -\frac{23}{16}$; $f\left(\frac{3}{4}\right) = -\frac{375}{64}$; $f\left(-\frac{3}{4}\right) = \frac{225}{64}$; roots: $-\frac{1}{2}, \frac{3}{2}$

B 17. $x^2 - 5 = 0$; possible rational roots are $\pm1, \pm5$; $f(1) = -4$; $f(-1) = -4$; $f(5) = 20$; $f(-5) = 20$; no rational roots so $\sqrt{5}$ is irrational.

18. $x^3 - 2 = 0$; possible rational roots are $\pm1, \pm2$; $f(1) = -1$; $f(-1) = -3$; $f(2) = 6$; $f(-2) = -10$; no rational roots so $\sqrt[3]{2}$ is irrational.

19. $x^2 - 7 = 0$; possible rational roots are $\pm1, \pm7$; $f(1) = -6$; $f(-1) = -6$; $f(7) = 42$; $f(-7) = 42$; no rational roots so $\sqrt{7}$ is irrational.

20. $x^3 + 6 = 0$; possible rational roots are $\pm1, \pm2, \pm3, \pm6$; $f(1) = 7$; $f(-1) = 5$; $f(2) = 14$; $f(-2) = -2$; $f(3) = 33$; $f(-3) = -21$; $f(6) = 222$; $f(-6) = -210$; no rational roots so $-\sqrt[3]{6}$ is irrational.

21. $2x^3 - 1 = 0$; possible rational roots are $\pm1, \pm\frac{1}{2}$; $f(1) = 1$; $f(-1) = -3$; $f\left(\frac{1}{2}\right) = -\frac{3}{4}$; $f\left(-\frac{1}{2}\right) = -\frac{5}{4}$; no rational roots so $\sqrt[3]{\frac{1}{2}}$ is irrational.

22. $3x^2 - 5 = 0$; possible rational roots are $\pm1, \pm5, \pm\frac{1}{3}, \pm\frac{5}{3}$; $f(1) = -2$; $f(-1) = -2$; $f(5) = 70$; $f(-5) = 70$; $f\left(\frac{1}{3}\right) = -\frac{14}{3}$; $f\left(-\frac{1}{3}\right) = -\frac{14}{3}$; $f\left(\frac{5}{3}\right) = \frac{10}{3}$; $f\left(-\frac{5}{3}\right) = \frac{10}{3}$; no rational roots so $\sqrt{\frac{5}{3}}$ is irrational.

23. Suppose $\sqrt{5} + 9$ is rational. Then there must be integers a, b $(b \neq 0)$ such that $\sqrt{5} + 9 = \frac{a}{b}$, or $\sqrt{5} = \frac{a - 9b}{b}$. Since $a - 9b$ and b are integers, $\frac{a - 9b}{b}$ is rational. Ex. 17 shows that $\sqrt{5}$ is irrational. Hence, $\sqrt{5} + 9$ is irrational.

24. Suppose $\frac{\sqrt{7}}{2}$ is rational. Then there must be integers a, b $(b \neq 0)$ such that $\frac{\sqrt{7}}{2} = \frac{a}{b}$, or $\sqrt{7} = \frac{2a}{b}$. Since $2a$ and b are integers, $\frac{2a}{b}$ is rational. Ex. 19 shows that $\sqrt{7}$ is irrational. Hence, $\frac{\sqrt{7}}{2}$ is irrational.

25. Suppose $8\sqrt[3]{2}$ is rational. Then there must be integers a, b $(b \neq 0)$ such that $8\sqrt[3]{2} = \frac{a}{b}$, or $\sqrt[3]{2} = \frac{a}{8b}$. Since a and $8b$ are integers, $\frac{a}{8b}$ is rational. Ex. 18 shows that $\sqrt[3]{2}$ is irrational. Hence, $8\sqrt[3]{2}$ is irrational.

26. Suppose $6 - \sqrt[3]{2}$ is rational. Then there must be integers a, b $(b \neq 0)$ such that $6 - \sqrt[3]{2} = \frac{a}{b}$, or $\sqrt[3]{2} = \frac{6b - a}{b}$. Since $6b - a$ and b are integers, $\frac{6b - a}{b}$ is rational. Ex. 18 shows that $\sqrt[3]{2}$ is irrational. Hence, $6 - \sqrt[3]{2}$ is irrational.

27. Suppose $\frac{4 + \sqrt{7}}{3}$ is rational. Then there must be integers a, b $(b \neq 0)$ such that $\frac{4 + \sqrt{7}}{3} = \frac{a}{b}$, or $\sqrt{7} = \frac{3a - 4b}{b}$. Since $3a - 4b$ and b are integers, $\frac{3a - 4b}{b}$ is rational. Ex. 19 shows that $\sqrt{7}$ is irrational. Hence, $\frac{4 + \sqrt{7}}{3}$ is irrational.

28. Suppose $\frac{3\sqrt{5} - 1}{4}$ is rational. Then there must be integers a, b $(b \neq 0)$ such that $\frac{3\sqrt{5} - 1}{4} = \frac{a}{b}$, or $\sqrt{5} = \frac{4a + b}{3b}$. Since $4a + b$ and $3b$ are integers, $\frac{4a + b}{3b}$ is rational. Ex. 17 shows that $\sqrt{5}$ is irrational. Hence, $\frac{3\sqrt{5} - 1}{4}$ is irrational.

C **29.** **a.** Since n is a root of $ax^3 + bx^2 + cx + d = 0$, then n must satisfy the equation. Thus, $an^3 + bn^2 + cn + d = 0$, or $an^3 + bn^2 + cn = -d$. Factoring on the left, we have $n(an^2 + bn + c) = -d$. Since $an^2 + bn + c$ and n are both integers, n must be a factor of $-d$ and, hence, of d.

b. By similar reasoning, since $\frac{1}{n}$ is a root of the equation, $\frac{a}{n^3} = -\frac{b}{n^2} - \frac{c}{n} - d$. Multiplying both sides by n^3, we have $a = n(-b - cn - dn^2)$. Since n and $(-b - cn - dn^2)$ are both integers, n must be a factor of a.

30. Assume $at + b$ is rational. Then $at + b = \frac{c}{d}$ where c, d are integers, $d \neq 0$.

Solving for t results in $t = \frac{c - bd}{ad}$. Since $c - bd$ and ad are integers, $\frac{c - bd}{ad}$ is rational. That contradicts the hypothesis that t is irrational. Hence, $at + b$ is irrational.

Page 256 • SELF-TEST 1

1. $\frac{5}{2^3} = \frac{y}{4^3}$; $y = \frac{5 \times 64}{8} = 40$ **2. a.** 12 **b.** -3 **3.** $\left\{-\frac{7}{2}\right\}$

4. $p \in \{\pm 1, \pm 2\}$, $q \in \{\pm 1\}$; possible roots are $\{\pm 1, \pm 2\}$; $f(1) = 4$; $f(-1) = 0$; $f(2) = 12$; $f(-2) = -8$; root: -1

Page 260 • WRITTEN EXERCISES

A **1.** 3.4×10^4 **2.** 4.67×10^{-2} **3.** 5.1×10^{-4} **4.** 3.82×10^2 **5.** 2.785×10^3 **6.** 5.01×10^{-2} **7.** 2.023×10^1 **8.** 4.006×10^{-3} **9.** 35,000 **10.** 0.0041 **11.** 0.0965 **12.** 287,000 **13.** 0.303 **14.** 0.0004007 **15.** 6,781,000 **16.** 0.00007146 **17.** $\frac{(4 \times 10^4)(6 \times 10^{-2})}{3 \times 10^{-3}} = 8 \times 10^5$, or 800,000

18. $\frac{(7 \times 10^{-2})(9 \times 10^3)}{3 \times 10^6} = 21 \times 10^{-5} \approx 2 \times 10^{-4}$, or 0.0002

B **19.** $\frac{(400{,}000)(7000)}{(0.007)(0.0002)} = \frac{(4 \times 10^5)(7 \times 10^3)}{(7 \times 10^{-3})(2 \times 10^{-4})} = 2 \times 10^{15}$, or 2,000,000,000,000,000

20. $\frac{(40{,}000)(0.8)}{(0.5)(2)} = \frac{(4 \times 10^4)(8 \times 10^{-1})}{(5 \times 10^{-1})(2)} = 3.2 \times 10^4 \approx 3 \times 10^4$, or 30,000

21. $\frac{(70{,}000)(0.6)}{(0.09)(20)} = \frac{(7 \times 10^4)(6 \times 10^{-1})}{(9 \times 10^{-2})(2 \times 10^1)} \approx 2.3 \times 10^4 \approx 2 \times 10^4$, or 20,000

22. $\frac{(40{,}000)(0.2)}{(0.08)(50)} = \frac{(4 \times 10^4)(2 \times 10^{-1})}{(8 \times 10^{-2})(5 \times 10^1)} = 0.2 \times 10^4 = 2 \times 10^3$, or 2000

23. $\frac{(20{,}000)(0.9)}{(2)(0.3)} = \frac{(2 \times 10^4)(9 \times 10^{-1})}{(2)(3 \times 10^{-1})} = 3 \times 10^4$, or 30,000

24. $\frac{(700)(2)(2000)}{(0.03)(0.8)} = \frac{(7 \times 10^2)(2)(2 \times 10^3)}{(3 \times 10^{-2})(8 \times 10^{-1})} \approx 1.17 \times 10^8 \approx 1 \times 10^8$, or 100,000,000

C **25.** $a \approx 4$, $b \approx 3$; maximum possible error is 0.5; $a + b \approx 7$, accuracy: $\frac{0.5}{7} \approx 0.071$ or 7.1%; $a - b \approx 1$, accuracy: $\frac{0.5}{1} = 0.5$ or 50%; $ab \approx 10$, accuracy: $\frac{0.5}{10} = 0.05$ or 5%; $a \div b \approx 1$, accuracy: $\frac{0.5}{1} = 0.5$ or 50%; ab is most accurate.

26. $a = 4.135$, $b = 2.692$; maximum possible error is 0.0005; $a + b = 6.827$, accuracy: $\frac{0.0005}{6.827} \approx 0.00007$ or 0.007%; $a - b = 1.443$, accuracy: $\frac{0.0005}{1.443} \approx 0.00035$ or 0.035%; $ab \approx 11.13$, accuracy: $\frac{0.0005}{11.13} \approx 0.00004$ or 0.004%; $a \div b \approx 1.536$, accuracy: $\frac{0.0005}{1.536} \approx 0.00033$ or 0.033%; ab is most accurate.

Page 261 • ON THE CALCULATOR

Answers may vary depending on the calculators used.

1. 1.3995×10^{14} **2.** 4.5213×10^8 **3.** 1.5665×10^{-2} **4.** 3.919×10^{-10} **5.** 3.2088×10^{-8} **6.** 1.2131×10^8

Pages 264–265 • WRITTEN EXERCISES

A **1.** 0.4375 **2.** 0.325 **3.** $0.1\overline{6}$ **4.** $-0.\overline{54}$ **5.** $0.74\overline{6}$ **6.** $-0.\overline{714285}$ **7.** $0.\overline{216}$ **8.** $0.\overline{56097}$

Answers to Exercises 9–12 may vary.

9. 1.35; 1.313113111 . . . **10.** $\frac{7}{10} = 0.7$, $\frac{5}{8} = 0.625$: $0.\overline{6}$ or $\frac{2}{3}$; 0.6252252225 . . .

11. $\frac{7}{9} = \frac{70}{90}$, $0.8 = \frac{80}{90}$: $\frac{71}{90}$; $\frac{7}{9} = 0.\overline{7}$, 0.8: 0.7891011 . . . **12.** $0.1\overline{32}$; 0.131331333 . . .

13. $0.475 = \frac{475}{1000} = \frac{19}{40}$ **14.** $0.0062 = \frac{62}{10{,}000} = \frac{31}{5000}$ **15.** $5.072 = \frac{5072}{1000} = \frac{634}{125}$

16. $-3.0084 = -\frac{30{,}084}{10{,}000} = -\frac{7521}{2500}$

17.
$$\begin{aligned} 100N &= 45.\overline{45} \\ N &= 0.\overline{45} \\ \hline 99N &= 45 \\ N &= \frac{45}{99} = \frac{5}{11} \end{aligned}$$

18.
$$\begin{aligned} 100N &= 87.\overline{87} \\ N &= 0.\overline{87} \\ \hline 99N &= 87 \\ N &= \frac{87}{99} = \frac{29}{33} \end{aligned}$$

19.
$$\begin{aligned} 1000N &= 135.\overline{135} \\ N &= 0.\overline{135} \\ \hline 999N &= 135 \\ N &= \frac{135}{999} = \frac{5}{37} \end{aligned}$$

20.
$$\begin{aligned} 100N &= 281.\overline{81} \\ N &= 2.\overline{81} \\ \hline 99N &= 279 \\ N &= \frac{279}{99} = \frac{31}{11} \end{aligned}$$

B **21.**
$$\begin{aligned} 100N &= 238.8\overline{8} \\ N &= 2.3\overline{8} \\ \hline 99N &= 236.5 \\ 990N &= 2365 \\ N &= \frac{2365}{990} = \frac{43}{18} \end{aligned}$$

22.
$$\begin{aligned} 1000N &= -3360.\overline{360} \\ N &= -3.\overline{360} \\ \hline 999N &= -3357 \\ N &= -\frac{3357}{999} = -\frac{373}{111} \end{aligned}$$

23.
$$\begin{aligned} 1000N &= 416.66\overline{6} \\ N &= 0.41\overline{6} \\ \hline 999N &= 416.25 \\ 99900N &= 41625 \\ N &= \frac{41625}{99900} = \frac{5}{12} \end{aligned}$$

24.
$$\begin{aligned} 100N &= -459.0\overline{90} \\ N &= -4.5\overline{90} \\ \hline 99N &= -454.5 \\ 990N &= -4545 \\ N &= -\frac{4545}{990} = -\frac{101}{22} \end{aligned}$$

25.
$$\begin{aligned} 1000N &= 2037.\overline{037} \\ N &= 2.\overline{037} \\ \hline 999N &= 2035 \\ N &= \frac{2035}{999} = \frac{55}{27} \end{aligned}$$

26.
$$\begin{aligned} 1000N &= 1256.7\overline{567} \\ N &= 1.2\overline{567} \\ \hline 999N &= 1255.5 \\ 9990N &= 12555 \\ N &= \frac{12555}{9990} = \frac{93}{74} \end{aligned}$$

27.
$$\begin{aligned} 10N &= -3.9\overline{9} \\ N &= -0.3\overline{9} \\ \hline 9N &= -3.6 \\ 90N &= -36 \\ N &= -\frac{36}{90} = -\frac{2}{5} \end{aligned}$$

28.
$$\begin{aligned} 10N &= 1.249\overline{9} \\ N &= 0.124\overline{9} \\ \hline 9N &= 1.125 \\ 9000N &= 1125 \\ N &= \frac{1125}{9000} = \frac{1}{8} \end{aligned}$$

C **29. a.** Since a, b, c, and d are positive integers, if $ad < bc$, then $\frac{1}{bd}(ad) < \frac{1}{bd}(bc)$. Thus, $\frac{a}{b} < \frac{c}{d}$. Conversely, if $\frac{a}{b} < \frac{c}{d}$, then $bd\left(\frac{a}{b}\right) < bd\left(\frac{c}{d}\right)$. Thus, $ad < bc$.

b. By Exercise 29(a), if $\frac{a}{b} < \frac{c}{d}$, then $ad < bc$. By the addition property of order, $ab + ad < ab + bc$, or $a(b + d) < b(a + c)$. Again by Exercise 29(a), $\frac{a}{b} < \frac{a + c}{b + d}$.

30. By Exercise 29(a), if $\frac{a}{b} < \frac{c}{d}$, then $ad < bc$. By the addition property of order, $ad + cd < bc + cd$, or $d(a + c) < c(b + d)$. Again by Exercise 29(a), $\frac{a + c}{b + d} < \frac{c}{d}$.

31. Let $0.\overline{abc} = N$. Then $1000N = abc.\overline{abc}$, and $999N = abc$. Therefore, $N = \frac{abc}{999}$, which is in the form $\frac{p}{q}$, where q can have at most 3 digits.

32. If $\frac{a}{b}$ is a fraction in lowest terms and can be represented by a terminating decimal, then $\frac{a}{b} = \frac{c}{10^n} = \frac{c}{(2 \cdot 5)^n} = \frac{c}{2^n \cdot 5^n}$. Since a and b are relatively prime, b must be a factor of $2^n \cdot 5^n$. Thus the only possible prime factors of b are 2 and 5.

33. a. If $r < s$, then $s - r > 0$, so $r + \frac{\sqrt{2}}{2}(s - r) > r$. Since $r < s$ and $1 - \frac{\sqrt{2}}{2} > 0$, $r\left(1 - \frac{\sqrt{2}}{2}\right) < s\left(1 - \frac{\sqrt{2}}{2}\right)$, $r - \frac{\sqrt{2}}{2}r < s - \frac{\sqrt{2}}{2}s$, $r - \frac{\sqrt{2}}{2}r + \frac{\sqrt{2}}{2}s < s$, $r + \frac{\sqrt{2}}{2}(s - r) < s$. $\therefore r < r + \frac{\sqrt{2}}{2}(s - r) < s$.

b. $\sqrt{2}$ is irrational. Thus by Exercise 30, page 256, $r + \frac{\sqrt{2}}{2}(s - r)$ is irrational.

34. Assume the sum of a rational number $\frac{p}{q}$ and an irrational number t is a rational number $\frac{r}{s}$, where r and s are integers, $s \neq 0$. $\frac{p}{q} + t = \frac{r}{s}$, $t = \frac{r}{s} - \frac{p}{q}$, $t = \frac{rq - ps}{qs}$. Since $rq - ps$ and qs are integers, $\frac{rq - ps}{qs}$ is rational. That contradicts the hypothesis. Hence the sum of a rational number and an irrational number is irrational.

Page 265 • SELF-TEST 2

1. 5.614×10^3 **2.** 8.37×10^{-3} **3.** 3.492×10^1

4. $\frac{(4000)(50)}{(4)(0.007)} = \frac{(4 \times 10^3)(5 \times 10^1)}{(4)(7 \times 10^{-3})} \approx 0.7 \times 10^7 = 7 \times 10^6$, or 7,000,000 **5.** $0.\overline{4}$

6.
$$\begin{aligned} 100N &= 31.\overline{31} \\ N &= 0.\overline{31} \\ \hline 99N &= 31 \\ N &= \frac{31}{99} \end{aligned}$$

7. Answers may vary. $\frac{5}{11} = 0.\overline{45}$: 0.451; 0.454454445 . . .

Page 269 • WRITTEN EXERCISES

A **1.** $\sqrt{192} = \sqrt{64} \cdot \sqrt{3} = 8\sqrt{3}$ **2.** $\sqrt{320} = \sqrt{64} \cdot \sqrt{5} = 8\sqrt{5}$

3. $\sqrt{\frac{72}{49}} = \frac{\sqrt{36} \cdot \sqrt{2}}{\sqrt{49}} = \frac{6\sqrt{2}}{7}$ **4.** $\sqrt{\frac{121}{3}} = \frac{\sqrt{121}}{\sqrt{3}} \cdot \frac{\sqrt{3}}{\sqrt{3}} = \frac{11\sqrt{3}}{3}$

5. $\sqrt[3]{125^2} = (\sqrt[3]{125})^2 = 5^2 = 25$ **6.** $\sqrt[6]{(-64)^2} = \sqrt[6]{4096} = \sqrt[6]{4^6} = 4$

7. $-\sqrt[3]{27^{-2}} = -\left(\sqrt[3]{\frac{1}{27}}\right)^2 = -\left(\frac{1}{3}\right)^2 = -\frac{1}{9}$ **8.** $\sqrt[4]{(-81)^{-2}} = \left(\sqrt[4]{\frac{1}{81}}\right)^2 = \left(\frac{1}{3}\right)^2 = \frac{1}{9}$

9. $\sqrt[6]{\frac{27}{64}} = \sqrt[2\cdot 3]{\left(\frac{3}{4}\right)^3} = \sqrt{\frac{3}{4}} = \frac{\sqrt{3}}{2}$ **10.** $\sqrt[4]{\frac{36}{25}} = \sqrt[2\cdot 2]{\left(\frac{6}{5}\right)^2} = \sqrt{\frac{6}{5}} = \frac{\sqrt{6}}{\sqrt{5}} \cdot \frac{\sqrt{5}}{\sqrt{5}} = \frac{\sqrt{30}}{5}$

11. $\sqrt[3]{\frac{27}{125}} = \sqrt[3]{\left(\frac{3}{5}\right)^3} = \frac{3}{5}$ **12.** $\sqrt[6]{8} \cdot \sqrt[8]{81} = \sqrt[2\cdot 3]{2^3} \cdot \sqrt[2\cdot 4]{3^4} = \sqrt{2} \cdot \sqrt{3} = \sqrt{6}$

13. $7\sqrt{48} = 7 \cdot \sqrt{16} \cdot \sqrt{3} = 7 \cdot 4 \cdot \sqrt{3} = 28\sqrt{3} \approx 48.50$

14. $\frac{2\sqrt{150}}{5} = \frac{2 \cdot \sqrt{25} \cdot \sqrt{6}}{5} = \frac{2 \cdot 5 \cdot \sqrt{6}}{5} = 2\sqrt{6} \approx 4.90$

15. $\frac{2}{3}\sqrt{24} = \frac{2}{3} \cdot \sqrt{4} \cdot \sqrt{6} = \frac{2}{3} \cdot 2 \cdot \sqrt{6} = \frac{4}{3}\sqrt{6} \approx 3.27$

16. $\sqrt{0.98} = \sqrt{\frac{98}{100}} = \frac{\sqrt{49} \cdot \sqrt{2}}{\sqrt{100}} = \frac{7\sqrt{2}}{10} \approx 0.99$

17. $\sqrt{\frac{27}{125}} \cdot \sqrt{80} = \frac{\sqrt{27}}{\sqrt{125}} \cdot \sqrt{80} = \frac{3\sqrt{3}}{5\sqrt{5}} \cdot 4\sqrt{5} = \frac{12\sqrt{3}}{5} \approx 4.16$

18. $\sqrt[3]{\frac{640}{25}} = \frac{\sqrt[3]{64} \cdot \sqrt[3]{10}}{\sqrt[3]{25}} \cdot \frac{\sqrt[3]{5}}{\sqrt[3]{5}} = \frac{4\sqrt[3]{50}}{\sqrt[3]{125}} = \frac{4\sqrt[3]{50}}{5} \approx 2.95$

19. $\sqrt{\frac{3}{28}} \cdot \sqrt{\frac{175}{3}} = \sqrt{\frac{3}{28} \cdot \frac{175}{3}} = \sqrt{\frac{175}{28}} = \sqrt{\frac{25}{4}} = \frac{5}{2} = 2.50$

20. $\sqrt[3]{-72} \cdot \sqrt[3]{375} = \sqrt[3]{-8} \cdot \sqrt[3]{9} \cdot \sqrt[3]{125} \cdot \sqrt[3]{3} = -2\sqrt[3]{9} \cdot 5\sqrt[3]{3} = -10\sqrt[3]{27} = -30$

21. $\sqrt{36x^{16}} = 6x^8$ **22.** $\sqrt{75x^{12}y^5} = \sqrt{25x^{12}y^4} \cdot \sqrt{3y} = 5x^6y^2\sqrt{3y}$

23. $\sqrt[4]{48x^3y^{20}} = \sqrt[4]{16y^{20}} \cdot \sqrt[4]{3x^3} = 2|y|^5\sqrt[4]{3x^3}$ **24.** $\sqrt[5]{-32a^{10}b^{15}} = -2a^2b^3$

25. $\sqrt{\frac{a^6}{b^5}} = \frac{\sqrt{a^6}}{\sqrt{b^5}} \cdot \frac{\sqrt{b}}{\sqrt{b}} = \frac{|a|^3\sqrt{b}}{b^3}$ **26.** $\sqrt[3]{\frac{24xy^6}{z^2}} = \frac{\sqrt[3]{8y^6} \cdot \sqrt[3]{3x}}{\sqrt[3]{z^2}} \cdot \frac{\sqrt[3]{z}}{\sqrt[3]{z}} = \frac{2y^2\sqrt[3]{3xz}}{z}$

27. $\frac{\sqrt[5]{96c^9}}{\sqrt[5]{3c^2}} = \sqrt[5]{\frac{96c^9}{3c^2}} = \sqrt[5]{32c^7} = \sqrt[5]{32c^5} \cdot \sqrt[5]{c^2} = 2c\sqrt[5]{c^2}$

28. $\frac{x+3}{\sqrt{x^2+9}} \cdot \frac{\sqrt{x^2+9}}{\sqrt{x^2+9}} = \frac{(x+3)\sqrt{x^2+9}}{x^2+9}$

B **29.** $\sqrt{c^{-2} - (5d)^{-2}} = \sqrt{\frac{1}{c^2} - \frac{1}{25d^2}} = \sqrt{\frac{25d^2 - c^2}{25c^2d^2}} = \frac{\sqrt{25d^2 - c^2}}{5|cd|}$

30. $\sqrt{9x^4 + 9x^6y^{-2}} = \sqrt{9x^4 + \frac{9x^6}{y^2}} = \sqrt{\frac{9x^4y^2 + 9x^6}{y^2}} = \frac{\sqrt{9x^4} \cdot \sqrt{y^2 + x^2}}{\sqrt{y^2}} = \frac{3x^2\sqrt{x^2 + y^2}}{|y|}$

31. $\sqrt[3]{x^3(x-y)^{-3} - x^2(x-y)^{-2}} = \sqrt[3]{\frac{x^3}{(x-y)^3} - \frac{x^2}{(x-y)^2}} = \sqrt[3]{\frac{x^3 - x^2(x-y)}{(x-y)^3}} = \frac{\sqrt[3]{x^2y}}{x-y}$

32. $\sqrt{16x^{-1} + 16y^{-1}} \cdot \sqrt{(x+y)^{-1}} = \sqrt{\frac{16}{x} + \frac{16}{y}} \cdot \sqrt{\frac{1}{x+y}} = \sqrt{\frac{16x+16y}{xy}} \cdot \frac{1}{\sqrt{x+y}} =$

$\frac{\sqrt{16} \cdot \sqrt{x+y}}{\sqrt{xy}} \cdot \frac{1}{\sqrt{x+y}} = \frac{4}{\sqrt{xy}} = \frac{4\sqrt{xy}}{xy}$

33. $\sqrt[8]{a^4(a+b)^{-4}} \cdot \sqrt[6]{a^3(a+b)^{-3}} = \sqrt[2\cdot4]{\frac{a^4}{(a+b)^4}} \cdot \sqrt[2\cdot3]{\frac{a^3}{(a+b)^3}} = \sqrt{\frac{a}{a+b}} \cdot \sqrt{\frac{a}{a+b}} =$

$\frac{a}{a+b}$ **34.** $\sqrt[5]{(3c)^{-5} + 3c^{-4}} = \sqrt[5]{\frac{1}{3^5c^5} + \frac{3}{c^4}} = \sqrt[5]{\frac{1+3(3)^5c}{3^5c^5}} = \frac{\sqrt[5]{1+729c}}{3c}$

C **35.** It has been proved that $\sqrt[n]{a} \cdot \sqrt[n]{b}$ is one of the nth roots of ab. Since exactly one of the numbers a, b is negative, exactly one of $\sqrt[n]{a}$ and $\sqrt[n]{b}$ is negative, and $\sqrt[n]{a} \cdot \sqrt[n]{b}$ is nonpositive. Also $\sqrt[n]{ab}$ is nonpositive since exactly one of a, b is negative. $\therefore$ $\sqrt[n]{ab}$, the single real root of ab (since n is odd), is equal to $\sqrt[n]{a} \cdot \sqrt[n]{b}$.

36. $\left(\frac{\sqrt[n]{a}}{\sqrt[n]{b}}\right)^n = \frac{(\sqrt[n]{a})^n}{(\sqrt[n]{b})^n} = \frac{a}{b}$. Thus, $\frac{\sqrt[n]{a}}{\sqrt[n]{b}}$ is an nth root of $\frac{a}{b}$. Since $a \geq 0$ and $b > 0$, then if n is even, the principal nth root of $\frac{a}{b}$ is nonnegative. $\frac{\sqrt[n]{a}}{\sqrt[n]{b}}$ is nonnegative and represents the principal nth root of $\frac{a}{b}$, $\sqrt[n]{\frac{a}{b}}$. Since $\frac{a}{b} \geq 0$, then if n is odd, $\frac{\sqrt[n]{a}}{\sqrt[n]{b}} \geq 0$ and $\sqrt[n]{\frac{a}{b}} \geq 0$. So $\frac{\sqrt[n]{a}}{\sqrt[n]{b}} = \sqrt[n]{\frac{a}{b}}$.

37. If $\frac{a}{b} < 0$, then $a < 0$ or $b < 0$, but not both. Hence $\sqrt[n]{\frac{a}{b}} < 0$ and $\sqrt[n]{a} < 0$ or $\sqrt[n]{b} < 0$. This means $\frac{\sqrt[n]{a}}{\sqrt[n]{b}} < 0$, since n is odd. $\left(\frac{\sqrt[n]{a}}{\sqrt[n]{b}}\right)^n = \frac{(\sqrt[n]{a})^n}{(\sqrt[n]{b})^n} = \frac{a}{b}$, so $\sqrt[n]{\frac{a}{b}} = \frac{\sqrt[n]{a}}{\sqrt[n]{b}}$ if $\frac{a}{b} < 0$ and n is odd.

38. Using the first theorem on page 267, $\sqrt[n]{b^m} = \underbrace{\sqrt[n]{b} \cdot \sqrt[n]{b} \cdot \ldots \cdot \sqrt[n]{b}}_{m \text{ factors}} = (\sqrt[n]{b})^m$.

Page 271 • WRITTEN EXERCISES

A **1.** $3\sqrt{5} + \sqrt{125} = 3\sqrt{5} + 5\sqrt{5} = 8\sqrt{5}$

2. $4\sqrt{98} - 3\sqrt{72} = 4 \cdot \sqrt{49} \cdot \sqrt{2} - 3 \cdot \sqrt{36} \cdot \sqrt{2} = 4 \cdot 7\sqrt{2} - 3 \cdot 6\sqrt{2} = 10\sqrt{2}$

3. $\frac{1}{3}\sqrt{27} - 2\sqrt{75} + 5\sqrt{12} = \frac{1}{3} \cdot 3\sqrt{3} - 2 \cdot 5\sqrt{3} + 5 \cdot 2\sqrt{3} = \sqrt{3}$

4. $3\sqrt{\frac{9}{5}} - \frac{2}{5}\sqrt{80} + \sqrt{\frac{81}{5}} = 3 \cdot \frac{3}{\sqrt{5}} \cdot \frac{\sqrt{5}}{\sqrt{5}} - \frac{2}{5} \cdot 4\sqrt{5} + \frac{9}{\sqrt{5}} \cdot \frac{\sqrt{5}}{\sqrt{5}} =$

$\frac{9\sqrt{5}}{5} - \frac{8\sqrt{5}}{5} + \frac{9\sqrt{5}}{5} = \frac{10\sqrt{5}}{5} = 2\sqrt{5}$

5. $5\sqrt[3]{24} - 2\sqrt[3]{54} + \sqrt[3]{3000} = 5 \cdot \sqrt[3]{8} \cdot \sqrt[3]{3} - 2 \cdot \sqrt[3]{27} \cdot \sqrt[3]{2} + \sqrt[3]{1000} \cdot \sqrt[3]{3} =$
$5 \cdot 2\sqrt[3]{3} - 2 \cdot 3\sqrt[3]{2} + 10\sqrt[3]{3} = 20\sqrt[3]{3} - 6\sqrt[3]{2}$

6. $\dfrac{2\sqrt[3]{5}}{\sqrt[3]{1000}} + \dfrac{\sqrt[3]{135}}{\sqrt[3]{1000}} = \dfrac{2\sqrt[3]{5}}{10} + \dfrac{\sqrt[3]{27} \cdot \sqrt[3]{5}}{10} = \dfrac{2\sqrt[3]{5} + 3\sqrt[3]{5}}{10} = \dfrac{\sqrt[3]{5}}{2} = 0.5\sqrt[3]{5}$

7. $4\sqrt{9b^3} + b\sqrt{49b} - \sqrt{b^3} = 4 \cdot \sqrt{9b^2} \cdot \sqrt{b} + b \cdot \sqrt{49} \cdot \sqrt{b} - \sqrt{b^2} \cdot \sqrt{b} =$
$12b\sqrt{b} + 7b\sqrt{b} - b\sqrt{b} = 18b\sqrt{b}$

8. $\sqrt{18y^6} + y\sqrt{128y^4} - 5y^2\sqrt{162y^2} = \sqrt{9y^6} \cdot \sqrt{2} + y \cdot \sqrt{64y^4} \cdot \sqrt{2} -$
$5y^2 \cdot \sqrt{81y^2} \cdot \sqrt{2} = 3|y|^3\sqrt{2} + 8y^3\sqrt{2} - 45|y|^3\sqrt{2} = -42|y|^3\sqrt{2} + 8y^3\sqrt{2}$

9. $3x\sqrt{80x^3} - \dfrac{1}{3}\sqrt{180x^5} + \sqrt{320x} = 3x \cdot \sqrt{16x^2} \cdot \sqrt{5x} - \dfrac{1}{3} \cdot \sqrt{36x^4} \cdot \sqrt{5x} +$
$\sqrt{64} \cdot \sqrt{5x} = 3x \cdot 4x\sqrt{5x} - \dfrac{1}{3} \cdot 6x^2\sqrt{5x} + 8\sqrt{5x} = 10x^2\sqrt{5x} + 8\sqrt{5x} =$
$\sqrt{5x}(10x^2 + 8)$

10. $3\sqrt[3]{16r^2} + \sqrt[3]{250r^8} - 5\sqrt[3]{2r^5} = 3 \cdot \sqrt[3]{8} \cdot \sqrt[3]{2r^2} + \sqrt[3]{125r^6} \cdot \sqrt[3]{2r^2} -$
$5 \cdot \sqrt[3]{r^3} \cdot \sqrt[3]{2r^2} = 3 \cdot 2\sqrt[3]{2r^2} + 5r^2\sqrt[3]{2r^2} - 5r\sqrt[3]{2r^2} = \sqrt[3]{2r^2}(6 + 5r^2 - 5r)$

11. $2\sqrt{15}(4\sqrt{3} - 3\sqrt{12}) = 8\sqrt{45} - 6\sqrt{180} = 8 \cdot \sqrt{9} \cdot \sqrt{5} - 6 \cdot \sqrt{36} \cdot \sqrt{5} =$
$24\sqrt{5} - 36\sqrt{5} = -12\sqrt{5}$

12. $\sqrt{10}\left(\dfrac{5}{\sqrt{2}} - \dfrac{\sqrt{2}}{4}\right) = \sqrt{10}\left(\dfrac{5}{\sqrt{2}}\right) - \sqrt{10}\left(\dfrac{\sqrt{2}}{4}\right) = 5\sqrt{5} - \dfrac{\sqrt{20}}{4} = 5\sqrt{5} - \dfrac{2\sqrt{5}}{4} =$
$\dfrac{9\sqrt{5}}{2}$

13. $\sqrt{\dfrac{5}{3}}\left(4\sqrt{3} - \dfrac{11}{\sqrt{3}}\right) = \dfrac{\sqrt{5}}{\sqrt{3}} \cdot 4\sqrt{3} - \dfrac{\sqrt{5}}{\sqrt{3}} \cdot \dfrac{11}{\sqrt{3}} = 4\sqrt{5} - \dfrac{11\sqrt{5}}{3} = \dfrac{\sqrt{5}}{3}$

14. $(\sqrt{5} - \sqrt{13})(\sqrt{5} + \sqrt{13}) = (\sqrt{5})^2 - (\sqrt{13})^2 = 5 - 13 = -8$

15. $(\sqrt{2} - \sqrt{7})^2 = (\sqrt{2})^2 - 2(\sqrt{2} \cdot \sqrt{7}) + (\sqrt{7})^2 = 2 - 2\sqrt{14} + 7 = 9 - 2\sqrt{14}$

16. $(2\sqrt{3} - \sqrt{6})^2 = (2\sqrt{3})^2 - 2(2\sqrt{3})(\sqrt{6}) + (\sqrt{6})^2 =$
$4 \cdot 3 - 4\sqrt{18} + 6 = 12 - 4 \cdot 3\sqrt{2} + 6 = 18 - 12\sqrt{2}$

17. $(3\sqrt{5} - 4)(3\sqrt{5} + 4) = (3\sqrt{5})^2 - 4^2 = 9 \cdot 5 - 16 = 45 - 16 = 29$

18. $(5\sqrt{2} - 3\sqrt{7})(5\sqrt{2} + 3\sqrt{7}) = (5\sqrt{2})^2 - (3\sqrt{7})^2 =$
$25 \cdot 2 - 9 \cdot 7 = 50 - 63 = -13$

19. $(\sqrt[3]{9} + 2\sqrt[3]{3})(\sqrt[3]{3} - 3) = \sqrt[3]{27} - 3\sqrt[3]{9} + 2\sqrt[3]{9} - 6\sqrt[3]{3} =$
$3 - 3\sqrt[3]{9} + 2\sqrt[3]{9} - 6\sqrt[3]{3} = 3 - \sqrt[3]{9} - 6\sqrt[3]{3}$

20. $(\sqrt[3]{3} - 3\sqrt[3]{2})(\sqrt[3]{4} + 2\sqrt[3]{9}) = \sqrt[3]{12} + 2\sqrt[3]{27} - 3\sqrt[3]{8} - 6\sqrt[3]{18} =$
$\sqrt[3]{12} + 2 \cdot 3 - 3 \cdot 2 - 6\sqrt[3]{18} = \sqrt[3]{12} - 6\sqrt[3]{18}$

21. $\dfrac{4}{(\sqrt{10} - 3)} \cdot \dfrac{(\sqrt{10} + 3)}{(\sqrt{10} + 3)} = \dfrac{4\sqrt{10} + 12}{10 - 9} = 4\sqrt{10} + 12$

22. $\dfrac{6}{(\sqrt{3} + 2)} \cdot \dfrac{(\sqrt{3} - 2)}{(\sqrt{3} - 2)} = \dfrac{6\sqrt{3} - 12}{3 - 4} = -6\sqrt{3} + 12$

23. $\frac{(\sqrt{3}-5)}{(\sqrt{3}+2)} \cdot \frac{(\sqrt{3}-2)}{(\sqrt{3}-2)} = \frac{3-2\sqrt{3}-5\sqrt{3}+10}{3-4} = 7\sqrt{3}-13$

24. $\frac{(4+\sqrt{5})}{(6-2\sqrt{5})} \cdot \frac{(6+2\sqrt{5})}{(6+2\sqrt{5})} = \frac{24+8\sqrt{5}+6\sqrt{5}+10}{36-20} = \frac{34+14\sqrt{5}}{16} = \frac{17+7\sqrt{5}}{8}$

B 25. $(\sqrt[3]{3}-1)(\sqrt[3]{9}+\sqrt[3]{3}+1) = \sqrt[3]{27}+\sqrt[3]{9}+\sqrt[3]{3}-\sqrt[3]{9}-\sqrt[3]{3}-1 = 3-1 = 2$

26. $(\sqrt[3]{2}+\sqrt[3]{5})(\sqrt[3]{4}-\sqrt[3]{10}+\sqrt[3]{25}) = \sqrt[3]{8}-\sqrt[3]{20}+\sqrt[3]{50}+\sqrt[3]{20}-\sqrt[3]{50}+\sqrt[3]{125} = 2+5 = 7$

27. $(\sqrt[3]{a}+\sqrt[3]{b})(\sqrt[3]{a^2}-\sqrt[3]{ab}+\sqrt[3]{b^2}) = \sqrt[3]{a^3}-\sqrt[3]{a^2b}+\sqrt[3]{ab^2}+\sqrt[3]{a^2b}-\sqrt[3]{ab^2}+\sqrt[3]{b^3} = a+b$

28. $(\sqrt[3]{a}-\sqrt[3]{b})(\sqrt[3]{a^2}+\sqrt[3]{ab}+\sqrt[3]{b^2}) = \sqrt[3]{a^3}+\sqrt[3]{a^2b}+\sqrt[3]{ab^2}-\sqrt[3]{a^2b}-\sqrt[3]{ab^2}-\sqrt[3]{b^3} = a-b$

29. $\frac{\sqrt[3]{4}}{(\sqrt[3]{2}-1)} \cdot \frac{(\sqrt[3]{4}+\sqrt[3]{2}+1)}{(\sqrt[3]{4}+\sqrt[3]{2}+1)} = \frac{\sqrt[3]{16}+\sqrt[3]{8}+\sqrt[3]{4}}{2-1} = 2\sqrt[3]{2}+2+\sqrt[3]{4}$

30. $\frac{\sqrt[3]{5}}{(\sqrt[3]{5}-\sqrt[3]{4})} \cdot \frac{(\sqrt[3]{25}+\sqrt[3]{20}+\sqrt[3]{16})}{(\sqrt[3]{25}+\sqrt[3]{20}+\sqrt[3]{16})} = \frac{\sqrt[3]{125}+\sqrt[3]{100}+\sqrt[3]{80}}{5-4} = 5+\sqrt[3]{100}+2\sqrt[3]{10}$

31. $\frac{\sqrt[3]{4}}{(\sqrt[3]{3}-\sqrt[3]{2})} \cdot \frac{(\sqrt[3]{9}+\sqrt[3]{6}+\sqrt[3]{4})}{(\sqrt[3]{9}+\sqrt[3]{6}+\sqrt[3]{4})} = \frac{\sqrt[3]{36}+\sqrt[3]{24}+\sqrt[3]{16}}{3-2} = \sqrt[3]{36}+2\sqrt[3]{3}+2\sqrt[3]{2}$

32. $\frac{\sqrt[3]{3}}{(\sqrt[3]{6}+\sqrt[3]{3})} \cdot \frac{(\sqrt[3]{36}-\sqrt[3]{18}+\sqrt[3]{9})}{(\sqrt[3]{36}-\sqrt[3]{18}+\sqrt[3]{9})} = \frac{\sqrt[3]{108}-\sqrt[3]{54}+\sqrt[3]{27}}{6+3} = \frac{3\sqrt[3]{4}-3\sqrt[3]{2}+3}{9} = \frac{\sqrt[3]{4}-\sqrt[3]{2}+1}{3}$

33. $(x\sqrt{3}+y)(x\sqrt{3}-y)$ 34. $(5a-\sqrt{7})(5a+\sqrt{7})$ 35. $(x+\sqrt{5})^2$
36. $(x\sqrt{3}-y\sqrt{2})^2$ 37. $(a-\sqrt[3]{10})(a^2+a\sqrt[3]{10}+\sqrt[3]{100})$
38. $(c+\sqrt[3]{6})(c^2-c\sqrt[3]{6}+\sqrt[3]{36})$ 39. $(x+5\sqrt{2})^2$ 40. $(a+\sqrt{6})^2$

C 41.

	Statement	Reason
1.	$a, b, c, d \in \mathcal{R}$	Hypothesis
2.	$(a+b\sqrt{2})(c+d\sqrt{2}) = ac+(bc+ad)\sqrt{2}+2bd$	Dist. ax.
3.	$(a+b\sqrt{2})(c+d\sqrt{2}) = ac+2bd+(bc+ad)\sqrt{2}$	Comm. ax. for add.
4.	$ac+2bd \in \mathcal{R}$, $bc+ad \in \mathcal{R}$	Closure
5.	The set of numbers of the form $a+b\sqrt{2}$, $a, b \in \mathcal{R}$, is closed under multiplication.	Def. of closure

42.

	Statement	Reason
1.	$a, b, c, d \in \mathcal{R}$, $c \pm d\sqrt{2} \neq 0$	Hypothesis
2.	$\frac{a+b\sqrt{2}}{c+d\sqrt{2}} = \frac{(a+b\sqrt{2})(c-d\sqrt{2})}{(c+d\sqrt{2})(c-d\sqrt{2})}$	Iden. ax. for mult.

[*Proof continued on next page*]

3. $\dfrac{a + b\sqrt{2}}{c + d\sqrt{2}} = \dfrac{ac + (bc - ad)\sqrt{2} - 2bd}{c^2 - 2d^2}$ Dist. ax.

4. $\dfrac{a + b\sqrt{2}}{c + d\sqrt{2}} = \dfrac{ac - 2bd}{c^2 - 2d^2} + \left(\dfrac{bc - ad}{c^2 - 2d^2}\right)\sqrt{2}$ Theorem, p. 228

5. $\dfrac{ac - 2bd}{c^2 - 2d^2} \in \mathscr{R}, \dfrac{bc - ad}{c^2 - 2d^2} \in \mathscr{R}$ Closure

6. The set of numbers of the form $a + b\sqrt{2}$, $a, b \in \mathscr{R}$, is closed under division except by 0. Def. of closure

43. $x^4 + 1 = (x^4 + 2x^2 + 1) - 2x^2 = (x^2 + 1)^2 - (x\sqrt{2})^2 = [(x^2 + 1) - x\sqrt{2}][(x^2 + 1) + x\sqrt{2}] = (x^2 - x\sqrt{2} + 1)(x^2 + x\sqrt{2} + 1)$

Pages 272–273 • PROBLEMS

A 1. $P = I^2R$; $1960 = I^2(10)$, $I^2 = 196$, $I = \sqrt{196}$, $I = \pm 14$; 14 A

2. $d = v_0\sqrt{\dfrac{h}{4.9}}$; $d = 35\sqrt{\dfrac{10}{4.9}} = 35\sqrt{\dfrac{100}{49}} = 35 \cdot \dfrac{10}{7} = 50$; 50 m

3. $f = \dfrac{1}{2L}\sqrt{\dfrac{10^5F}{m}}$; $f = \dfrac{1}{2(50)}\sqrt{\dfrac{10^5(484)}{1.25}} = \dfrac{1}{100}\sqrt{\dfrac{10^7(484)}{125}} = \dfrac{1}{100}\dfrac{\sqrt{10^6 \cdot 484} \cdot \sqrt{10}}{\sqrt{25} \cdot \sqrt{5}} = \dfrac{1}{100} \cdot \dfrac{10^3 \cdot 22}{5} \cdot \sqrt{2} = 44\sqrt{2} \approx 62.2$; 62.2 Hz

4. $f = \dfrac{1}{2\pi\sqrt{LC}}$; $f = \dfrac{1}{2\pi\sqrt{1.2 \times 10^{-2} \times 50 \times 10^{-6}}} = \dfrac{1}{2\pi\sqrt{60 \times 10^{-8}}} = \dfrac{1}{2\pi\sqrt{4 \times 10^{-8}} \cdot \sqrt{15}} = \dfrac{1}{2\pi(2 \times 10^{-4})\sqrt{15}} = \dfrac{10^4}{4\pi\sqrt{15}} \cdot \dfrac{\sqrt{15}}{\sqrt{15}} = \dfrac{10^4\sqrt{15}}{60\pi} \approx 210$; 210 Hz

5. $V = \dfrac{4}{3}\pi r^3$; $92 \approx \dfrac{4}{3} \cdot \dfrac{22}{7} \cdot r^3$, $r^3 \approx 92 \cdot \dfrac{21}{88}$, $r \approx \sqrt[3]{21.95} \approx 2.8$; 2.8 cm

B 6. $\dfrac{s_1}{\sqrt{k_1}} = \dfrac{s_2}{\sqrt{k_2}}$; $\dfrac{320}{\sqrt{256}} = \dfrac{s_2}{\sqrt{300}}$, $s_2 = \dfrac{320\sqrt{300}}{\sqrt{256}} = \dfrac{3200\sqrt{3}}{16} \approx 346$; 346 m/s

7. $k_1(\sqrt[3]{v_1})^2 = k_2(\sqrt[3]{v_2})^2$; $300\left(\sqrt[3]{\dfrac{8}{125}}\right)^2 = k_2(\sqrt[3]{1})^2$, $k_2 = 300\left(\dfrac{2}{5}\right)^2 = 48$; 48°

8. $s^3 = \dfrac{4}{3}\pi r^3$; $s^3 = \dfrac{4}{3} \cdot \dfrac{22}{7}(25)^3 = \dfrac{88(25^3)}{21}$, $s = \dfrac{\sqrt[3]{25^3} \cdot \sqrt[3]{8} \cdot \sqrt[3]{11}}{\sqrt[3]{21}} \cdot \dfrac{\sqrt[3]{21^2}}{\sqrt[3]{21^2}} = \dfrac{25 \cdot 2 \cdot \sqrt[3]{11} \cdot \sqrt[3]{21^2}}{21} = \dfrac{50\sqrt[3]{4851}}{21} \approx 40$

9. $\dfrac{{d_m}^3}{{t_m}^2} = \dfrac{{d_s}^3}{{t_s}^2}$; $d_m = \dfrac{1}{6}d_s$; $\dfrac{\left(\dfrac{1}{6}d_s\right)^3}{{t_m}^2} = \dfrac{{d_s}^3}{{t_s}^2}$, $\dfrac{{t_m}^2}{{t_s}^2} = \dfrac{\dfrac{1}{216}{d_s}^3}{{d_s}^3}$, $\dfrac{t_m}{t_s} = \sqrt{\dfrac{1}{216}} = \sqrt{\dfrac{1}{36} \cdot \dfrac{1}{6}} = \dfrac{1}{6}\sqrt{\dfrac{1}{6}} \cdot \dfrac{\sqrt{6}}{\sqrt{6}} = \dfrac{\sqrt{6}}{36} \approx 0.07$; ratio is 0.07.

C **10.** $m = \dfrac{m_0}{\sqrt{1 - \dfrac{v^2}{c^2}}}$; $v = \dfrac{2}{3}c$; $\dfrac{m}{m_0} = \dfrac{1}{\sqrt{1 - \dfrac{(\frac{2}{3}c)^2}{c^2}}} = \dfrac{1}{\sqrt{1 - \dfrac{4}{9}}} = \dfrac{1}{\sqrt{\dfrac{5}{9}}} = \dfrac{3}{\sqrt{5}} = \dfrac{3\sqrt{5}}{5}$;

increased by a factor of $\dfrac{3\sqrt{5}}{5}$

Page 275 • WRITTEN EXERCISES

A **1.** $\sqrt{x + 3} = 7$, $x + 3 = 49$, $x = 46$ *Check:* $\sqrt{46 + 3} \stackrel{?}{=} 7$, $7 = 7$; $\{46\}$

2. $8 - \sqrt{4 - y} = 0$, $\sqrt{4 - y} = 8$, $4 - y = 64$, $y = -60$
Check: $8 - \sqrt{4 - (-60)} \stackrel{?}{=} 0$, $8 - \sqrt{64} \stackrel{?}{=} 0$, $0 = 0$; $\{-60\}$

3. $\sqrt[3]{5n} = 3$, $5n = 3^3$, $n = \dfrac{27}{5}$ *Check:* $\sqrt[3]{5\left(\dfrac{27}{5}\right)} \stackrel{?}{=} 3$, $\sqrt[3]{27} \stackrel{?}{=} 3$, $3 = 3$; $\left\{\dfrac{27}{5}\right\}$

4. $\sqrt[3]{2r - 1} = 5$, $2r - 1 = 125$, $r = 63$ *Check:* $\sqrt[3]{2(63) - 1} \stackrel{?}{=} 5$, $\sqrt[3]{125} \stackrel{?}{=} 5$, $5 = 5$; $\{63\}$

5. $\sqrt{4c - 1} + 2 = 9$, $\sqrt{4c - 1} = 7$, $4c - 1 = 49$, $c = \dfrac{25}{2}$
Check: $\sqrt{4\left(\dfrac{25}{2}\right) - 1} + 2 \stackrel{?}{=} 9$, $\sqrt{49} + 2 \stackrel{?}{=} 9$, $9 = 9$; $\left\{\dfrac{25}{2}\right\}$

6. $-2 + \sqrt[4]{\dfrac{a}{3}} = 1$, $\sqrt[4]{\dfrac{a}{3}} = 3$, $\dfrac{a}{3} = 81$, $a = 243$ *Check:* $-2 + \sqrt[4]{\dfrac{243}{3}} \stackrel{?}{=} 1$,
$-2 + \sqrt[4]{81} \stackrel{?}{=} 1$, $1 = 1$; $\{243\}$

7. $3 + \sqrt[3]{\dfrac{x}{2}} = 7$, $\sqrt[3]{\dfrac{x}{2}} = 4$, $\dfrac{x}{2} = 64$, $x = 128$ *Check:* $3 + \sqrt[3]{\dfrac{128}{2}} \stackrel{?}{=} 7$, $3 + \sqrt[3]{64} \stackrel{?}{=} 7$,
$7 = 7$; $\{128\}$

8. $4 + \sqrt{3n + 10} = 9$, $\sqrt{3n + 10} = 5$, $3n + 10 = 25$, $n = 5$
Check: $4 + \sqrt{3(5) + 10} \stackrel{?}{=} 9$, $4 + \sqrt{25} \stackrel{?}{=} 9$, $9 = 9$; $\{5\}$

9. $\sqrt{k^2 + 9} - k = 1$, $\sqrt{k^2 + 9} = k + 1$, $k^2 + 9 = k^2 + 2k + 1$, $k = 4$
Check: $\sqrt{4^2 + 9} - 4 \stackrel{?}{=} 1$, $\sqrt{25} - 4 \stackrel{?}{=} 1$, $1 = 1$; $\{4\}$

10. $p - \sqrt{p^2 - 16} = 2$, $\sqrt{p^2 - 16} = p - 2$, $p^2 - 16 = p^2 - 4p + 4$, $p = 5$
Check: $5 - \sqrt{5^2 - 16} \stackrel{?}{=} 2$, $5 - \sqrt{9} \stackrel{?}{=} 2$, $2 = 2$; $\{5\}$

11. $\dfrac{5}{2}\sqrt[3]{3m} = 15$, $\sqrt[3]{3m} = 15 \cdot \dfrac{2}{5}$, $\sqrt[3]{3m} = 6$, $3m = 216$, $m = 72$
Check: $\dfrac{5}{2}\sqrt[3]{3(72)} \stackrel{?}{=} 15$, $\dfrac{5}{2}\sqrt[3]{216} \stackrel{?}{=} 15$, $15 = 15$; $\{72\}$

12. $\dfrac{2}{3}\sqrt[4]{x} = 20$, $\sqrt[4]{x} = 20 \cdot \dfrac{3}{2}$, $\sqrt[4]{x} = 30$, $x = 810{,}000$ *Check:* $\dfrac{2}{3}\sqrt[4]{810{,}000} \stackrel{?}{=} 20$,
$\dfrac{2}{3}(30) \stackrel{?}{=} 20$, $20 = 20$; $\{810{,}000\}$

13. $2\sqrt{u + 5} = u + 2$, $4(u + 5) = u^2 + 4u + 4$, $u^2 = 16$, $u = \pm 4$
Check: $2\sqrt{4 + 5} \stackrel{?}{=} 4 + 2$, $2\sqrt{9} \stackrel{?}{=} 6$, $6 = 6$; $2\sqrt{-4 + 5} \stackrel{?}{=} -4 + 2$, $2 \neq -2$; $\{4\}$

14. $6 - \sqrt{5t} = -4$, $\sqrt{5t} = 10$, $5t = 100$, $t = 20$ *Check:* $6 - \sqrt{5(20)} \stackrel{?}{=} -4$,
$6 - \sqrt{100} \stackrel{?}{=} -4$, $-4 = -4$; $\{20\}$

15. $7 + \sqrt{8 - r} = 9 - r$, $\sqrt{8 - r} = 2 - r$, $8 - r = 4 - 4r + r^2$, $r^2 - 3r - 4 = 0$, $(r - 4)(r + 1) = 0$, $r = 4$ or $r = -1$ *Check:* $7 + \sqrt{8 - 4} \stackrel{?}{=} 9 - 4$, $7 + \sqrt{4} \stackrel{?}{=} 5$, $9 \neq 5$; $7 + \sqrt{8 + 1} \stackrel{?}{=} 9 - (-1)$, $7 + \sqrt{9} \stackrel{?}{=} 10$, $10 = 10$; $\{-1\}$

16. $\sqrt[3]{n^2 - 9} - 6 = 0$, $\sqrt[3]{n^2 - 9} = 6$, $n^2 - 9 = 216$, $n^2 = 225$, $n = \pm 15$
Check: $\sqrt[3]{(15)^2 - 9} - 6 \stackrel{?}{=} 0$, $\sqrt[3]{216} - 6 \stackrel{?}{=} 0$, $0 = 0$;
$\sqrt[3]{(-15)^2 - 9} - 6 \stackrel{?}{=} 0$, $\sqrt[3]{216} - 6 \stackrel{?}{=} 0$, $0 = 0$; $\{-15, 15\}$

17. $3 - 2\sqrt[3]{3n + 5} = 0$, $\sqrt[3]{3n + 5} = \frac{3}{2}$, $3n + 5 = \frac{27}{8}$, $n = -\frac{13}{24}$
Check: $3 - 2\sqrt[3]{3\left(-\frac{13}{24}\right) + 5} \stackrel{?}{=} 0$, $3 - 2\sqrt[3]{\frac{27}{8}} \stackrel{?}{=} 0$, $0 = 0$; $\left\{-\frac{13}{24}\right\}$

18. $-2x = \sqrt{6x + 4}$, $4x^2 = 6x + 4$, $2(2x + 1)(x - 2) = 0$, $x = -\frac{1}{2}$ or $x = 2$
Check: $-2\left(-\frac{1}{2}\right) \stackrel{?}{=} \sqrt{6\left(-\frac{1}{2}\right) + 4}$, $1 \stackrel{?}{=} \sqrt{-3 + 4}$, $1 = 1$; $-2(2) \stackrel{?}{=} \sqrt{6(2) + 4}$, $-4 \stackrel{?}{=} \sqrt{16}$, $-4 \neq 4$; $\left\{-\frac{1}{2}\right\}$

19. $\sqrt{k + 9} = \sqrt{k} + 1$, $k + 9 = k + 2\sqrt{k} + 1$, $4 = \sqrt{k}$, $k = 16$ *Check:* $\sqrt{16 + 9} \stackrel{?}{=} \sqrt{16} + 1$, $\sqrt{25} \stackrel{?}{=} 4 + 1$, $5 = 5$; $\{16\}$

20. $\sqrt{2z} - \sqrt{2z - 7} = 1$, $\sqrt{2z - 7} = \sqrt{2z} - 1$, $2z - 7 = 2z - 2\sqrt{2z} + 1$, $\sqrt{2z} = 4$, $2z = 16$, $z = 8$ *Check:* $\sqrt{2(8)} - \sqrt{2(8) - 7} \stackrel{?}{=} 1$, $\sqrt{16} - \sqrt{9} = 1$, $1 = 1$; $\{8\}$

21. $\sqrt{r - 10} = 5 + \sqrt{r}$, $r - 10 = 25 + 10\sqrt{r} + r$, $\sqrt{r} = -\frac{7}{2}$, $r = \frac{49}{4}$
Check: $\sqrt{\frac{49}{4} - 10} \stackrel{?}{=} 5 + \sqrt{\frac{49}{4}}$, $\sqrt{\frac{9}{4}} \stackrel{?}{=} 5 + \frac{7}{2}$, $\frac{3}{2} \neq \frac{17}{2}$; $\emptyset$

22. $\sqrt{c - 6} + 3 = \sqrt{c}$, $c - 6 + 6\sqrt{c - 6} + 9 = c$, $\sqrt{c - 6} = -\frac{1}{2}$, $c - 6 = \frac{1}{4}$, $c = \frac{25}{4}$ *Check:* $\sqrt{\frac{25}{4} - 6} + 3 \stackrel{?}{=} \sqrt{\frac{25}{4}}$, $\sqrt{\frac{1}{4}} + 3 \stackrel{?}{=} \frac{5}{2}$, $\frac{7}{2} \neq \frac{5}{2}$; $\emptyset$

23. $\sqrt{x - 1} = \sqrt{x + 4} - 1$, $x - 1 = x + 4 - 2\sqrt{x + 4} + 1$, $\sqrt{x + 4} = 3$, $x + 4 = 9$, $x = 5$ *Check:* $\sqrt{5 - 1} \stackrel{?}{=} \sqrt{5 + 4} - 1$, $\sqrt{4} \stackrel{?}{=} \sqrt{9} - 1$, $2 = 2$; $\{5\}$

24. $\sqrt{y + 6} - \sqrt{4 - y} = 2$, $\sqrt{y + 6} = 2 + \sqrt{4 - y}$, $y + 6 = 4 + 4\sqrt{4 - y} + 4 - y$, $y - 1 = 2\sqrt{4 - y}$, $y^2 - 2y + 1 = 16 - 4y$, $y^2 + 2y - 15 = 0$, $(y + 5)(y - 3) = 0$, $y = -5$ or $y = 3$ *Check:* $\sqrt{-5 + 6} - \sqrt{4 - (-5)} \stackrel{?}{=} 2$, $\sqrt{1} - \sqrt{9} \stackrel{?}{=} 2$, $-2 \neq 2$; $\sqrt{3 + 6} - \sqrt{4 - 3} \stackrel{?}{=} 2$, $\sqrt{9} - \sqrt{1} \stackrel{?}{=} 2$, $2 = 2$; $\{3\}$

25. $\sqrt{2x + 1} = \sqrt{x - 3} + 2$, $2x + 1 = x - 3 + 4\sqrt{x - 3} + 4$, $x = 4\sqrt{x - 3}$, $x^2 = 16x - 48$, $x^2 - 16x + 48 = 0$, $(x - 4)(x - 12) = 0$, $x = 4$ or $x = 12$
Check: $\sqrt{2(4) + 1} \stackrel{?}{=} \sqrt{4 - 3} + 2$, $\sqrt{9} \stackrel{?}{=} 1 + 2$, $3 = 3$; $\sqrt{2(12) + 1} \stackrel{?}{=} \sqrt{12 - 3} + 2$, $\sqrt{25} \stackrel{?}{=} \sqrt{9} + 2$, $5 = 5$; $\{4, 12\}$

26. $\sqrt{2x + 3} - \sqrt{x + 1} = 1$, $\sqrt{2x + 3} = 1 + \sqrt{x + 1}$, $2x + 3 = 1 + 2\sqrt{x + 1} + x + 1$, $x + 1 = 2\sqrt{x + 1}$, $x^2 + 2x + 1 = 4x + 4$, $x^2 - 2x - 3 = 0$, $(x - 3)(x + 1) = 0$, $x = 3$ or $x = -1$ *Check:* $\sqrt{2(3) + 3} - \sqrt{3 + 1} \stackrel{?}{=} 1$,

$\sqrt{9} - \sqrt{4} \stackrel{?}{=} 1, 1 = 1; \sqrt{2(-1) + 3} - \sqrt{-1 + 1} \stackrel{?}{=} 1, \sqrt{1} - \sqrt{0} \stackrel{?}{=} 1, 1 = 1;$
$\{3, -1\}$

27. $\sqrt{2x + 8} - \sqrt{x + 2} = 2, \sqrt{2x + 8} = 2 + \sqrt{x + 2}, 2x + 8 =$
$4 + 4\sqrt{x + 2} + x + 2, x + 2 = 4\sqrt{x + 2}, x^2 + 4x + 4 = 16x + 32,$
$x^2 - 12x - 28 = 0, (x - 14)(x + 2) = 0, x = 14$ or $x = -2$
Check: $\sqrt{2(14) + 8} - \sqrt{14 + 2} \stackrel{?}{=} 2, \sqrt{36} - \sqrt{16} \stackrel{?}{=} 2, 2 = 2;$
$\sqrt{2(-2) + 8} - \sqrt{-2 + 2} \stackrel{?}{=} 2, \sqrt{4} - \sqrt{0} \stackrel{?}{=} 2, 2 = 2; \{14, -2\}$

28. $\sqrt{x + 3} + \sqrt{3x + 10} = 3, \sqrt{3x + 10} = 3 - \sqrt{x + 3}, 3x + 10 =$
$9 - 6\sqrt{x + 3} + x + 3, x - 1 = -3\sqrt{x + 3}, x^2 - 2x + 1 = 9x + 27,$
$x^2 - 11x - 26 = 0, (x - 13)(x + 2) = 0, x = 13$ or $x = -2$
Check: $\sqrt{13 + 3} + \sqrt{3(13) + 10} \stackrel{?}{=} 3, \sqrt{16} + \sqrt{49} \stackrel{?}{=} 3, 4 + 7 \neq 3;$
$\sqrt{-2 + 3} + \sqrt{3(-2) + 10} \stackrel{?}{=} 3, \sqrt{1} + \sqrt{4} \stackrel{?}{=} 3, 3 = 3; \{-2\}$

B 29. $x\sqrt{x} = \frac{1}{8}, x^2 \cdot x = \frac{1}{64}, x^3 = \frac{1}{64}, x = \frac{1}{4}$ *Check:* $\frac{1}{4}\sqrt{\frac{1}{4}} \stackrel{?}{=} \frac{1}{8}, \frac{1}{8} = \frac{1}{8}; \left\{\frac{1}{4}\right\}$

30. $x\sqrt{x - 2} = x, x^2(x - 2) = x^2, x^3 - 2x^2 = x^2, x^3 - 3x^2 = 0, x^2(x - 3) = 0, x = 0$ or
$x = 3$ *Check:* $0\sqrt{0 - 2} \stackrel{?}{=} 0, \sqrt{-2}$ is not real; $3\sqrt{3 - 2} \stackrel{?}{=} 3, 3 = 3; \{3\}$

31. $\frac{2x - 3}{\sqrt{x} + 2} = \sqrt{x}, 2x - 3 = x + 2\sqrt{x}, x - 3 = 2\sqrt{x}, x^2 - 6x + 9 = 4x,$
$x^2 - 10x + 9 = 0, (x - 9)(x - 1) = 0, x = 9$ or $x = 1$
Check: $\frac{2(9) - 3}{\sqrt{9} + 2} \stackrel{?}{=} \sqrt{9}, \frac{15}{3 + 2} \stackrel{?}{=} 3, 3 = 3; \frac{2(1) - 3}{\sqrt{1} + 2} \stackrel{?}{=} \sqrt{1}, \frac{-1}{3} \neq 1; \{9\}$

32. $\frac{4x + 2}{\sqrt{x} + 1} = 3\sqrt{x}, 4x + 2 = 3x + 3\sqrt{x}, x + 2 = 3\sqrt{x}, x^2 + 4x + 4 = 9x,$
$x^2 - 5x + 4 = 0, (x - 4)(x - 1) = 0, x = 4$ or $x = 1$
Check: $\frac{4(4) + 2}{\sqrt{4} + 1} \stackrel{?}{=} 3\sqrt{4}, \frac{18}{3} \stackrel{?}{=} 3(2), 6 = 6; \frac{4(1) + 2}{\sqrt{1} + 1} \stackrel{?}{=} 3(1), \frac{6}{2} = 3; \{4, 1\}$

33. $\frac{5x + 5}{\sqrt{2x} + 5} = \sqrt{2x} + 1, 5x + 5 = 2x + 6\sqrt{2x} + 5, x = 2\sqrt{2x}, x^2 = 8x,$
$x^2 - 8x = 0, x(x - 8) = 0, x = 0$ or $x = 8$ *Check:* $\frac{5(0) + 5}{\sqrt{2(0)} + 5} \stackrel{?}{=} \sqrt{2(0)} + 1,$
$\frac{5}{5} = 1; \frac{5(8) + 5}{\sqrt{2(8)} + 5} \stackrel{?}{=} \sqrt{2(8)} + 1, \frac{45}{\sqrt{16} + 5} \stackrel{?}{=} \sqrt{16} + 1, 5 = 5; \{0, 8\}$

34. $\frac{3 - 2x}{\sqrt{2x} - 3} = \sqrt{2x} - 2, 3 - 2x = 2x - 5\sqrt{2x} + 6, 4x + 3 = 5\sqrt{2x},$
$16x^2 + 24x + 9 = 50x, 16x^2 - 26x + 9 = 0, (8x - 9)(2x - 1) = 0, x = \frac{9}{8}$ or
$x = \frac{1}{2}$ *Check:* $\frac{3 - 2\left(\frac{9}{8}\right)}{\sqrt{2\left(\frac{9}{8}\right)} - 3} \stackrel{?}{=} \sqrt{2\left(\frac{9}{8}\right)} - 2, \frac{\frac{3}{4}}{\frac{3}{2} - 3} \stackrel{?}{=} \frac{3}{2} - 2, -\frac{1}{2} = -\frac{1}{2};$

$$\frac{3 - 2\left(\frac{1}{2}\right)}{\sqrt{2\left(\frac{1}{2}\right) - 3}} \stackrel{?}{=} \sqrt{2\left(\frac{1}{2}\right) - 2}, \frac{2}{-2} \stackrel{?}{=} 1 - 2, -1 = -1; \left\{\frac{9}{8}, \frac{1}{2}\right\}$$

C 35. a. (1) If $r = 0$, $r^3 = 0$ by the mult. prop. of 0. Hence $s^3 = 0$ and $s = \sqrt[3]{0} = 0$. (2) If $r > 0$, $r^3 > 0$ and $s^3 > 0$. Hence $s > 0$ since the nth root of a positive number is positive when n is odd. (3) If $r < 0$, $r^3 < 0$ since the product of an odd number of negative numbers is negative. Hence $s^3 < 0$ and $s < 0$ since the nth root of a negative number is negative when n is odd.
b. $r^3 = s^3$; $r^3 - s^3 = 0$; $(r - s)(r^2 + rs + s^2) = 0$; $r = s$ or $r^2 + rs + s^2 = 0$. But $r^2 + rs + s^2 = 0$ only if $r = s = 0$ since we know from (a) that if r and s are not both zero, then $r^2 > 0$, $rs > 0$, and $s^2 > 0$ because r and s are both positive or both negative. So if $r^3 = s^3$, $r = s$.

36. a. (1) If $r = 0$, $r^4 = 0$ by the mult. prop. of 0. Thus $s^4 = 0$ and $s = \sqrt[4]{0} = 0$. (2) If $r \neq 0$, $r^2 \neq 0$ and $r^4 = (r^2)^2 \neq 0$ using the fact that a product is 0 if and only if one factor is 0. Hence $s^4 \neq 0$ and $s \neq 0$ since $0^4 = 0$.
b. $r^4 = s^4$; $r^4 - s^4 = 0$; $(r^2 - s^2)(r^2 + s^2) = 0$; $(r - s)(r + s)(r^2 + s^2) = 0$; $r = s$ or $r = -s$ or $r^2 + s^2 = 0$. But $r^2 + s^2 = 0$ only if $r = s = 0$. So if $r^4 = s^4$, $r = s$ or $r = -s$.

Page 275 • SELF-TEST 3

1. $\sqrt{28} - 2\sqrt{63} - 3\sqrt{175} = \sqrt{4}\cdot\sqrt{7} - 2\cdot\sqrt{9}\cdot\sqrt{7} - 3\cdot\sqrt{25}\cdot\sqrt{7} = 2\sqrt{7} - 2\cdot 3\sqrt{7} - 3\cdot 5\sqrt{7} = -19\sqrt{7}$
2. $\sqrt[3]{32xy^4} - \sqrt[3]{108x^4y} = \sqrt[3]{8y^3}\cdot\sqrt[3]{4xy} - \sqrt[3]{27x^3}\cdot\sqrt[3]{4xy} = 2y\sqrt[3]{4xy} - 3x\sqrt[3]{4xy} = \sqrt[3]{4xy}(2y - 3x)$
3. $(2\sqrt{5} - 3)(\sqrt{5} + 4) = 2\cdot 5 + 8\sqrt{5} - 3\sqrt{5} - 12 = 5\sqrt{5} - 2$
4. $\frac{3}{(6 - \sqrt{7})}\cdot\frac{(6 + \sqrt{7})}{(6 + \sqrt{7})} = \frac{18 + 3\sqrt{7}}{36 - 7} = \frac{18 + 3\sqrt{7}}{29}$
5. $\sqrt{\frac{2x}{5}} = 4$, $\frac{2x}{5} = 16$, $x = 40$ *Check:* $\sqrt{\frac{2(40)}{5}} \stackrel{?}{=} 4$, $\sqrt{16} \stackrel{?}{=} 4$, $4 = 4$; $\{40\}$
6. $\sqrt{6 - x} + 1 = \sqrt{3 - x}$, $6 - x + 2\sqrt{6 - x} + 1 = 3 - x$, $\sqrt{6 - x} = -2$, $6 - x = 4$, $x = 2$ *Check:* $\sqrt{6 - 2} + 1 \stackrel{?}{=} \sqrt{3 - 2}$, $\sqrt{4} + 1 \stackrel{?}{=} \sqrt{1}$, $3 \neq 1$; $\emptyset$

Pages 278–279 • WRITTEN EXERCISES

A 1. $i^{10} = i^8 \cdot i^2 = 1(-1) = -1$ 2. $i^7 = i^4 \cdot i^3 = 1(-i) = -i$ 3. $i^{12} = (i^4)^3 = 1$
4. $i^{21} = i^{20} \cdot i = 1 \cdot i = i$ 5. $\sqrt{-50} = i\sqrt{50} = 5i\sqrt{2}$ 6. $\sqrt{-45} = i\sqrt{45} = 3i\sqrt{5}$
7. $3\sqrt{-98} = 3i\sqrt{98} = 3i \cdot 7\sqrt{2} = 21i\sqrt{2}$
8. $-2\sqrt{-75} = -2i\sqrt{75} = -2i \cdot 5\sqrt{3} = -10i\sqrt{3}$ 9. $\sqrt{-\frac{3}{16}} = \frac{i\sqrt{3}}{\sqrt{16}} = \frac{\sqrt{3}}{4}i$
10. $\sqrt{\frac{-4}{7}} = \frac{2i}{\sqrt{7}}\cdot\frac{\sqrt{7}}{\sqrt{7}} = \frac{2\sqrt{7}}{7}i$ 11. $-\sqrt{-\frac{5}{8}} = -\frac{i\sqrt{5}}{2\sqrt{2}}\cdot\frac{\sqrt{2}}{\sqrt{2}} = -\frac{\sqrt{10}}{4}i$

12. $\sqrt{-2} \cdot \sqrt{-18} = i\sqrt{2} \cdot 3i\sqrt{2} = 6i^2 = -6$

13. $3\sqrt{-10} \cdot \sqrt{-15} = 3i\sqrt{10} \cdot i\sqrt{15} = 3i^2\sqrt{150} = -3 \cdot 5\sqrt{6} = -15\sqrt{6}$

14. $5\sqrt{-12} \cdot 2\sqrt{-6} = 5 \cdot 2i\sqrt{3} \cdot 2i\sqrt{6} = 20i^2\sqrt{18} = -20 \cdot 3\sqrt{2} = -60\sqrt{2}$

15. $\dfrac{\sqrt{-20}}{\sqrt{-5}} = \dfrac{2i\sqrt{5}}{i\sqrt{5}} = 2$ **16.** $\dfrac{8\sqrt{-63}}{\sqrt{-7}} = \dfrac{8 \cdot 3i\sqrt{7}}{i\sqrt{7}} = 24$

17. $\dfrac{6}{\sqrt{-10}} = \dfrac{6}{i\sqrt{10}} = -\dfrac{6i}{\sqrt{10}} \cdot \dfrac{\sqrt{10}}{\sqrt{10}} = -\dfrac{6i\sqrt{10}}{10} = -\dfrac{3\sqrt{10}}{5}i$

18. $\dfrac{5}{\sqrt{-6}} = \dfrac{5}{i\sqrt{6}} = -\dfrac{5i}{\sqrt{6}} \cdot \dfrac{\sqrt{6}}{\sqrt{6}} = -\dfrac{5\sqrt{6}}{6}i$ **19.** $\dfrac{3\sqrt{7}}{\sqrt{-28}} = \dfrac{3\sqrt{7}}{2i\sqrt{7}} = -\dfrac{3}{2}i$

20. $\dfrac{\sqrt{5}}{-4i} = \dfrac{\sqrt{5}}{4}i$ **21.** $\dfrac{39i}{-3i} = -13$ **22.** $\dfrac{4}{i^3} = \dfrac{4}{-i} = 4i$

23. $\dfrac{-8}{i^6} = \dfrac{-8}{-1} = 8$ **24.** $\dfrac{9i}{i^{14}} = \dfrac{9i}{-1} = -9i$

B 25. $3\sqrt{-48} + 5\sqrt{-3} = 3 \cdot 4i\sqrt{3} + 5 \cdot i\sqrt{3} = 12i\sqrt{3} + 5i\sqrt{3} = 17i\sqrt{3}$

26. $2\sqrt{-7} - 3\sqrt{-175} = 2 \cdot i\sqrt{7} - 3 \cdot 5i\sqrt{7} = 2i\sqrt{7} - 15i\sqrt{7} = -13i\sqrt{7}$

27. $\sqrt{-192} + \sqrt{-300} = 8i\sqrt{3} + 10i\sqrt{3} = 18i\sqrt{3}$

28. $i^7 + i^8 + i^9 = -i + 1 + i = 1$

29. $i^{11} + i^{12} - i^{13} + i^{14} = -i + 1 - i - 1 = -2i$

30. $\dfrac{1}{i^5} + \dfrac{1}{i^6} + \dfrac{1}{i^7} = \dfrac{1}{i} + \dfrac{1}{-1} + \dfrac{1}{-i} = -i - 1 + i = -1$

31. $\sqrt{-\dfrac{4}{5}} - \sqrt{-\dfrac{3}{10}} = \dfrac{2i}{\sqrt{5}} - \dfrac{i\sqrt{3}}{\sqrt{10}} = \dfrac{2i\sqrt{5}}{5} - \dfrac{i\sqrt{30}}{10} = i\left(\dfrac{2}{5}\sqrt{5} - \dfrac{1}{10}\sqrt{30}\right)$

32. $3\sqrt{-\dfrac{1}{6}} + \sqrt{-\dfrac{2}{3}} = \dfrac{3i}{\sqrt{6}} + \dfrac{i\sqrt{2}}{\sqrt{3}} = \dfrac{3i\sqrt{6}}{6} + \dfrac{i\sqrt{6}}{3} = \dfrac{3i\sqrt{6}}{6} + \dfrac{2i\sqrt{6}}{6} = \dfrac{5\sqrt{6}}{6}i$

33. $3\sqrt{-\dfrac{1}{48}} - 2\sqrt{-\dfrac{5}{12}} = \dfrac{3i}{4\sqrt{3}} - \dfrac{2i\sqrt{5}}{2\sqrt{3}} = \dfrac{3i\sqrt{3}}{12} - \dfrac{4i\sqrt{15}}{12} = i\left(\dfrac{1}{4}\sqrt{3} - \dfrac{1}{3}\sqrt{15}\right)$

C 34. Let $r = -a$ and $s = -b$. Then $r, s > 0$ and $-r = a$, $-s = b$, $(-1)r = -r$, and $(-1)s = -s$. Then $\sqrt{a} = \sqrt{(-1)r} = i\sqrt{r}$; $\sqrt{b} = \sqrt{(-1)s} = i\sqrt{s}$. Thus $\sqrt{a} \cdot \sqrt{b} = i\sqrt{r} \cdot i\sqrt{s} = i^2\sqrt{rs} = -\sqrt{rs}$. Also $ab = (-r)(-s) = rs$. $\therefore -\sqrt{ab} = -\sqrt{rs} = \sqrt{a} \cdot \sqrt{b}$

35. Let $r = -a$ and $s = -b$. Then $r, s > 0$ and $-r = a$, $-s = b$, $(-1)r = -r$, and $(-1)s = -s$. Then $\sqrt{a} = \sqrt{(-1)r} = i\sqrt{r}$; $\sqrt{b} = \sqrt{(-1)s} = i\sqrt{s}$. Thus $\dfrac{\sqrt{a}}{\sqrt{b}} = \dfrac{i\sqrt{r}}{i\sqrt{s}} = \sqrt{\dfrac{r}{s}}$. Also $\dfrac{a}{b} = \dfrac{-r}{-s} = \dfrac{r}{s}$. $\therefore \sqrt{\dfrac{a}{b}} = \sqrt{\dfrac{r}{s}} = \dfrac{\sqrt{a}}{\sqrt{b}}$

36. yes; yes

$\times$	1	-1	i	$-i$
1	1	-1	i	$-i$
-1	-1	1	$-i$	i
i	i	$-i$	-1	1
$-i$	$-i$	i	1	-1

37. $ax^2 + b = 0$, $x^2 = -\frac{b}{a}$, $x = \sqrt{-\frac{b}{a}}$, $x = \pm\frac{i\sqrt{b}}{\sqrt{a}}\cdot\frac{\sqrt{a}}{\sqrt{a}} = \pm\frac{i\sqrt{ab}}{a}$; $\frac{i\sqrt{ab}}{a}$ and $-\frac{i\sqrt{ab}}{a}$

Page 282 • WRITTEN EXERCISES

A **1.** $(3 - 6i) + (-5 + 8i) = -2 + 2i$ **2.** $(3 - 6i) + (4 + 7i) = 7 + i$
3. $(3 - 6i) - (4 + 7i) = -1 - 13i$ **4.** $(4 + 7i) - (-6i) = 4 + 13i$
5. $(4 - 7i) - (-5 + 8i) = 9 - 15i$ **6.** $-6i - (3 + 6i) = -3 - 12i$
7. $\left(\frac{1}{3} + \frac{3}{2}i\right) + \left(-\frac{5}{6} + \frac{1}{2}i\right) = -\frac{1}{2} + 2i$ **8.** $\left(-\frac{5}{6} + \frac{1}{2}i\right) - \left(\frac{2}{3} + \frac{3}{4}i\right) = -\frac{3}{2} - \frac{1}{4}i$
9. $-\frac{3}{8} + \left(\frac{2}{3} - \frac{3}{4}i\right) = \frac{7}{24} - \frac{3}{4}i$ **10.** $\left(\frac{1}{3} - \frac{3}{2}i\right) - \left(\frac{2}{3} - \frac{3}{4}i\right) = -\frac{1}{3} - \frac{3}{4}i$
11. $\left(\frac{1}{3} + \frac{3}{2}i\right) + \left(\frac{2}{3} + \frac{3}{4}i\right) = 1 + \frac{9}{4}i$ **12.** $\left(\frac{1}{3} + \frac{3}{2}i\right) - \left(-\frac{5}{6} + \frac{1}{2}i\right) = \frac{7}{6} + i$
13. $(3 - 6i) + (3 + 6i) = 6$ **14.** $\left(-\frac{5}{6} - \frac{1}{2}i\right) + \left(-\frac{5}{6} - \frac{1}{2}i\right) = -\frac{5}{3} - i$
15. $\left(\frac{1}{3} + \frac{3}{2}i\right) - \left(\frac{1}{3} - \frac{3}{2}i\right) = 3i$ **16.** $\left(\frac{2}{3} - \frac{3}{4}i\right) - \left(\frac{2}{3} + \frac{3}{4}i\right) = -\frac{3}{2}i$
17. $\left(-\frac{5}{6} + \frac{1}{2}i\right) - \left(-\frac{3}{8}\right) = -\frac{11}{24} + \frac{1}{2}i$ **18.** $\left(\frac{2}{3} + \frac{3}{4}i\right) - (6i) = \frac{2}{3} - \frac{21}{4}i$

B **19.** $x + y = 2$ and $y = -3$; $x + (-3) = 2$, $x = 5$; $\therefore x = 5, y = -3$
20. $x - y = 5$ and $x + y = 9$; solving the system results in $2x = 14$, $x = 7$; $7 - y = 5$, $y = 2$; $\therefore x = 7, y = 2$
21. $2x - y = 7$ and $x + 3y = 0$; $x = -3y$; $2(-3y) - y = 7$, $y = -1$; $x = -3(-1) = 3$; $\therefore x = 3, y = -1$
22. $3x + 5y = -2$ and $-x + 7y = 10$; $\begin{array}{r} 3x + 5y = -2 \\ -3x + 21y = 30 \\ \hline 26y = 28 \end{array}$; $y = \frac{28}{26} = \frac{14}{13}$;
$-x + 7\left(\frac{14}{13}\right) = 10$, $-x = \frac{32}{13}$, $x = -\frac{32}{13}$; $\therefore x = -\frac{32}{13}$, $y = \frac{14}{13}$
23. $z = a + bi$ so $\bar{z} = a - bi$. But $z = \bar{z}$ so $a + bi = a - bi$. Thus $a = a$, $bi = -bi$ or $2bi = 0$ so $b = 0$. $\therefore z = a + 0i = a$, a real number.
24. $z = a + bi$ so $\bar{z} = a - bi$. But $z = \bar{z}$ so $a + bi = -(a - bi)$, $a + bi = -a + bi$. Thus $a = -a$ or $2a = 0$ so $a = 0$. $\therefore z = 0 + bi = bi$, a pure imaginary number.
25. $z = a + bi$ so $z + (0 + 0i) = (a + bi) + (0 + 0i) = (a + 0) + (b + 0)i = a + bi = z$. Also, $(0 + 0i) + z = (0 + 0i) + (a + bi) = (0 + a) + (0 + b)i = a + bi = z$.
26. $z = a + bi$ so $-z = -a - bi$; $z + (-z) = (a + bi) + (-a - bi) = (a - a) + (b - b)i = 0 + 0i$
27. For $a \in \mathcal{R}$, $b \in \mathcal{R}$, $(a + bi) + (a - bi) = (a + a) + (b - b)i = 2a + 0i = 2a$. Also $(a + bi) - (a - bi) = (a - a) + (b + b)i = 0 + 2bi = 2bi$.
28. Prove the conjugate of $(a + bi) + (c + di)$ is the same as $(a - bi) + (c - di)$. $(a + bi) + (c + di) = (a + c) + (b + d)i$; thus the conjugate of the sum is $(a + c) - (b + d)i$. On the other hand, the sum of the conjugates is $(a - bi) + (c - di) = (a + c) + (-b - d)i = (a + c) - (b + d)i$.

Page 284 • WRITTEN EXERCISES

A **1.** $(5 - 9i)(2 + i) = (10 - (-9)) + (5 - 18)i = 19 - 13i$
2. $(-7 + 3i)(8 - 2i) = (-56 - (-6)) + (14 + 24)i = -50 + 38i$

3. $(5-2i)^2=(25-4)+(-10-10)i=21-20i$

4. $\left(\frac{\sqrt{2}}{2}+\frac{\sqrt{2}}{2}i\right)^2=\left(\frac{2}{4}-\frac{2}{4}\right)+\left(\frac{2}{4}+\frac{2}{4}\right)i=0+1i=i$

5. $\left(-\frac{1}{2}+\frac{\sqrt{3}}{2}i\right)\left(-\frac{1}{2}-\frac{\sqrt{3}}{2}i\right)=\left(-\frac{1}{2}\right)^2+\left(\frac{\sqrt{3}}{2}\right)^2=\frac{1}{4}+\frac{3}{4}=1$

6. $(-1+i\sqrt{7})^2=(1-7)+(-\sqrt{7}-\sqrt{7})i=-6-2i\sqrt{7}$

7. $\left(-\frac{1}{2}+\frac{\sqrt{3}}{2}i\right)^2=\left(\frac{1}{4}-\frac{3}{4}\right)+\left(-\frac{\sqrt{3}}{4}-\frac{\sqrt{3}}{4}\right)i=-\frac{1}{2}-\frac{\sqrt{3}}{2}i$

8. $\frac{6+2i}{5-i}\cdot\frac{5+i}{5+i}=\frac{(30-2)+(6+10)i}{25+1}=\frac{28+16i}{26}=\frac{14}{13}+\frac{8i}{13}$

9. $\frac{10-5i}{1+2i}\cdot\frac{1-2i}{1-2i}=\frac{(10-10)+(-20-5)i}{1+4}=\frac{-25i}{5}=-5i$

10. $\frac{7-5i}{4+3i}\cdot\frac{4-3i}{4-3i}=\frac{(28-15)+(-21-20)i}{16+9}=\frac{13}{25}-\frac{41}{25}i$

11. $\frac{8-3i}{-1-6i}\cdot\frac{-1+6i}{-1+6i}=\frac{(-8-(-18))+(48+3)i}{1+36}=\frac{10}{37}+\frac{51}{37}i$

12. $\frac{7}{5-i\sqrt{3}}\cdot\frac{5+i\sqrt{3}}{5+i\sqrt{3}}=\frac{7(5+i\sqrt{3})}{25+3}=\frac{5+i\sqrt{3}}{4}=\frac{5}{4}+\frac{\sqrt{3}}{4}i$

B 13. $\frac{5+2i}{4-i}-\frac{3-i}{2+3i}=\frac{5+2i}{4-i}\cdot\frac{4+i}{4+i}-\frac{3-i}{2+3i}\cdot\frac{2-3i}{2-3i}=$

$\frac{(20-2)+(5+8)i}{16+1}-\frac{(6-3)+(-9-2)i}{4+9}=\frac{18+13i}{17}-\frac{3-11i}{13}=$

$\frac{234+169i}{221}-\frac{51-187i}{221}=\frac{183}{221}+\frac{356}{221}i$

14. $\left(-\frac{1}{2}+\frac{\sqrt{3}}{2}i\right)^2=\left(\frac{1}{4}-\frac{3}{4}\right)+\left(-\frac{\sqrt{3}}{4}-\frac{\sqrt{3}}{4}\right)i=-\frac{1}{2}-\frac{\sqrt{3}}{2}i,$

$\left(-\frac{1}{2}-\frac{\sqrt{3}}{2}i\right)\left(-\frac{1}{2}+\frac{\sqrt{3}}{2}i\right)=\left(-\frac{1}{2}\right)^2+\left(\frac{\sqrt{3}}{2}\right)^2=\frac{1}{4}+\frac{3}{4}=1$

15. $\left(\frac{\sqrt{2}}{2}+i\frac{\sqrt{2}}{2}\right)^4=\left[\left(\frac{\sqrt{2}}{2}+i\frac{\sqrt{2}}{2}\right)^2\right]^2=\left[\left(\frac{2}{4}-\frac{2}{4}\right)+\left(\frac{2}{4}+\frac{2}{4}\right)i\right]^2=i^2=-1$

16. $x^2+9=(x+3i)(x-3i)$ 17. $4y^2+49=(2y+7i)(2y-7i)$

18. $v^2+5=(v+i\sqrt{5})(v-i\sqrt{5})$

19. $80a^2+15b^2=(4a\sqrt{5}+bi\sqrt{15})(4a\sqrt{5}-bi\sqrt{15})$, or

$80a^2+15b^2=5(16a^2+3b^2)=5(4a+bi\sqrt{3})(4a-bi\sqrt{3})$

20. $50x^2+12=(5x\sqrt{2}+2i\sqrt{3})(5x\sqrt{2}-2i\sqrt{3})$, or

$50x^2+12=2(25x^2+6)=2(5x+i\sqrt{6})(5x-i\sqrt{6})$

21. $2z^2+27=(z\sqrt{2}+3i\sqrt{3})(z\sqrt{2}-3i\sqrt{3})$

C 22. $\frac{1}{a+bi}=\frac{1}{(a+bi)}\cdot\frac{(a-bi)}{(a-bi)}=\frac{a-bi}{a^2+b^2}=\frac{a}{a^2+b^2}-\frac{b}{a^2+b^2}i$

23. $\frac{a+bi}{c+di}=\frac{a+bi}{c+di}\cdot\frac{c-di}{c-di}=\frac{(ac+bd)+(bc-ad)i}{c^2+d^2}=\frac{ac+bd}{c^2+d^2}+\frac{bc-ad}{c^2+d^2}i$

24. Prove the conjugate of $(a + bi)(c + di) = (a - bi)(c - di)$. $(a + bi)(c + di) = (ac - bd) + (ad + bc)i$; thus the conjugate of the product is $(ac - bd) - (ad + bc)i$. The product of the conjugates is $(a - bi)(c - di) = (ac - bd) + (-ad - bc)i = (ac - bd) - (ad + bc)i$.

25. $(x^2)^2 + 1^2 = (x^2 + i)(x^2 - i) = \left[x^2 - \left(\frac{\sqrt{2}}{2} + \frac{\sqrt{2}}{2}i\right)^2\right]\left[x^2 + \left(\frac{\sqrt{2}}{2} + \frac{\sqrt{2}}{2}i\right)^2\right] =$

$\left(x - \frac{\sqrt{2}}{2} - \frac{\sqrt{2}}{2}i\right)\left(x + \frac{\sqrt{2}}{2} + \frac{\sqrt{2}}{2}i\right)\left[x + \left(\frac{\sqrt{2}}{2} + \frac{\sqrt{2}}{2}i\right)i\right]\left[x - \left(\frac{\sqrt{2}}{2} + \frac{\sqrt{2}}{2}i\right)i\right] =$

$\left(x - \frac{\sqrt{2}}{2} - \frac{\sqrt{2}}{2}i\right)\left(x + \frac{\sqrt{2}}{2} + \frac{\sqrt{2}}{2}i\right)\left(x - \frac{\sqrt{2}}{2} + \frac{\sqrt{2}}{2}i\right)\left(x + \frac{\sqrt{2}}{2} - \frac{\sqrt{2}}{2}i\right)$

26. If $z_1 + z_2 = (a + c) + (b + d)i$ is real, $b + d = 0$, or $b = -d$. If $z_1z_2 = (ac - bd) + (ad + bc)i$ is real, $ad + bc = 0$. Since $b = -d$, $ad - dc = 0$, or $d(a - c) = 0$. Since $d \neq 0$, $a - c = 0$, or $a = c$. $\therefore z_2 = c + di = a - bi$, the complex conjugate of z_1.

Page 285 • COMPUTER EXERCISES

1.

```
10  INPUT M
20  LET R = M - (INT(M/4) * 4)
30  IF R = 1 THEN PRINT "I"
40  IF R = 2 THEN PRINT "-1"
50  IF R = 3 THEN PRINT "-I"
60  IF R = 0 THEN PRINT "1"
70  END
```

2. i **3.** 1 **4.** -1 **5.** $-i$

6.

```
10  INPUT A,B,N
20  LET M1 = A: LET M2 = B
30  FOR I = 1 TO N - 1
40  LET M3 = M1 * A - M2 * B
50  LET M2 = M1 * B + M2 * A
60  LET M1 = M3
70  NEXT I
80  IF M1 = 0 THEN PRINT M2;"I":END
90  PRINT M1;
100  IF M2 < 0 THEN PRINT M2;"I":END
110  PRINT " + ";M2;"I"
120  END
```

7. $32i$ **8.** $-16 + 16i$ **9.** $-278 + 29i$ **10.** $-8432 - 5376i$

11.

```
10  INPUT A,B,C,D
20  LET R = (A * C + B * D)/(C^2 + D^2)
30  LET I = (B * C - A * D)/(C^2 + D^2)
40  IF I < 0 THEN PRINT R;" - ";ABS(I);"I":END
50  PRINT R;" + ";I;"I"
60  END
```

12. $-1 - 2i$ **13.** $4 + i$ **14.** $-0.4864 + 0.4752i$ **15.** $-1.423077 + 0.1153846i$

Page 285 • SELF-TEST 4

1. $2\sqrt{-25} = 2 \cdot 5i = 10i$ **2.** $3\sqrt{-\frac{64}{3}} = 3 \cdot \frac{i\sqrt{64}}{\sqrt{3}} = \frac{24i}{\sqrt{3}} \cdot \frac{\sqrt{3}}{\sqrt{3}} = \frac{24i\sqrt{3}}{3} = 8i\sqrt{3}$

3. $\sqrt{-54} = i \cdot \sqrt{9} \cdot \sqrt{6} = 3i\sqrt{6}$ **4.** $\sqrt{-17} = i\sqrt{17}$
5. $(2 + 3i) + (-7 - 2i) = -5 + i$ **6.** $5i - (6 + 4i) = -6 + i$
7. $(4 - 3i)(5 + 2i) = (20 - (-6)) + (8 - 15)i = 26 - 7i$
8. $\frac{3 + 2i}{6 - 5i} \cdot \frac{6 + 5i}{6 + 5i} = \frac{(18 - 10) + (15 + 12)i}{36 + 25} = \frac{8}{61} + \frac{27}{61}i$
9. $(2 - 6i)^2 = (4 - 36) + (-12 - 12)i = -32 - 24i$
10. $(4 - 2i)^{-1} = \frac{1}{4 - 2i} \cdot \frac{4 + 2i}{4 + 2i} = \frac{4 + 2i}{16 + 4} = \frac{4}{20} + \frac{2i}{20} = \frac{1}{5} + \frac{1}{10}i$

Pages 287–288 • CHAPTER REVIEW

1. $28 = k(4)$; $k = 7$; b **2.** $\frac{\sqrt{27}}{-3} = \frac{\sqrt{9}\sqrt{3}}{-3} = \frac{3\sqrt{3}}{-3} = -\sqrt{3}$; a
3. $p \in \{\pm 1, \pm 2, \pm 4\}$, $q \in \{\pm 1\}$; possible roots are $\pm 1, \pm 2, \pm 4$; $f(1) = -7$; $f(-1) = -3$; $f(2) = 0$; $f(-2) = -8$; $f(4) = 52$; $f(-4) = -60$; c
4. c **5.** $\frac{(4 \times 10^2)(2)}{2 \times 10^3} = 4 \times 10^{-1}$; d **6.** b **7.**
$$\begin{aligned} 1000N &= 923.\overline{923} \\ N &= 0.\overline{923} \\ \hline 999N &= 923 \\ N &= \frac{923}{999}; \text{ a} \end{aligned}$$
8. $\sqrt[6]{\frac{27}{b^3}} = \sqrt[2\cdot 3]{\frac{3^3}{b^3}} = \frac{\sqrt{3}}{\sqrt{b}} = \frac{\sqrt{3b}}{b}$; d **9.** $\frac{(\sqrt{a} + 2)}{(\sqrt{a} - 5)} \cdot \frac{(\sqrt{a} + 5)}{(\sqrt{a} + 5)} = \frac{a + 7\sqrt{a} + 10}{a - 25}$; b
10. $6 - \sqrt{5y} = 1$, $\sqrt{5y} = 5$, $5y = 25$, $y = 5$ *Check:* $6 - \sqrt{5(5)} \stackrel{?}{=} 1$, $6 - \sqrt{25} \stackrel{?}{=} 1$, $1 = 1$; $\{5\}$; a
11. $\frac{\sqrt{-20}}{i^3\sqrt{5}} = \frac{2i\sqrt{5}}{-i\sqrt{5}} = -2$; b
12. $x + y = 3$ and $3x = 6$; $x = 2$; $2 + y = 3$, $y = 1$; d
13. $\frac{5 + 2i}{6 + 3i} - \frac{1 + i}{15} = \frac{5 + 2i}{6 + 3i} \cdot \frac{6 - 3i}{6 - 3i} - \frac{1 + i}{15} = \frac{(30 + 6) + (12 - 15)i}{45} - \frac{1 + i}{15} =$
$\frac{36 - 3i}{45} - \frac{1 + i}{15} = \frac{12 - i}{15} - \frac{1 + i}{15} = \frac{11 - 2i}{15} = \frac{11}{15} - \frac{2}{15}i$; a
14. $(2 + i\sqrt{5})^2 = (4 - 5) + (2\sqrt{5} + 2\sqrt{5})i = -1 + 4i\sqrt{5}$; d

Page 288 • CHAPTER TEST

1. $\frac{48}{4^2} = \frac{y}{3^2}$, $y = \frac{48 \times 9}{16} = 27$ **2.** $\frac{d_1}{(t_1)^2} = \frac{d_2}{(t_2)^2}$; $\frac{54}{3^2} = \frac{d_2}{4^2}$; $d_2 = \frac{54 \times 16}{9} = 96$; 96 ft
3. $x^2 = 1.21$; $\{1.1, -1.1\}$
4. $p \in \{\pm 1, \pm 2\}$, $q \in \{\pm 1\}$; possible roots are $\{\pm 1, \pm 2\}$; $f(1) = 3$; $f(-1) = -1$; $f(2) = 20$; $f(-2) = 0$; root: -2
5. 3.907×10^5 **6.** $\frac{(2 \times 10^2)(8 \times 10^{-2})}{4} = 4 \times 10^0 = 4$ **7.** $0.\overline{428571}$
8.
$$\begin{aligned} 100N &= 162.62\overline{62} \\ N &= 1.62\overline{62} \\ \hline 99N &= 161 \\ N &= \frac{161}{99} \end{aligned}$$
9. Answers may vary. $\frac{3}{5} = 0.6$: $0.\overline{6}$ or $\frac{2}{3}$; 0.61661666 . . .

10. $\dfrac{2\sqrt{147}}{\sqrt[4]{9}} = \dfrac{2 \cdot \sqrt{49} \cdot \sqrt{3}}{\sqrt[2\cdot 2]{9}} = \dfrac{14\sqrt{3}}{\sqrt{3}} = 14$

11. $\dfrac{\sqrt[3]{81a^4}}{\sqrt[3]{30a^3}} = \sqrt[3]{\dfrac{27a}{10}} = \dfrac{\sqrt[3]{27} \cdot \sqrt[3]{a}}{\sqrt[3]{10}} \cdot \dfrac{\sqrt[3]{100}}{\sqrt[3]{100}} = \dfrac{3\sqrt[3]{100a}}{10}, a \neq 0$

12. $\sqrt[4]{\dfrac{(64a)^2}{c^{10}}} = \sqrt[2\cdot 2]{\dfrac{(64a)^2}{(c^5)^2}} = \sqrt{\dfrac{64|a|}{|c|^5}} = \dfrac{\sqrt{64} \cdot \sqrt{|a|}}{\sqrt{|c|^5}} \cdot \dfrac{\sqrt{|c|}}{\sqrt{|c|}} = \dfrac{8\sqrt{|ac|}}{|c|^3}$

13. $\sqrt{25a^2b + 25b^2} \cdot \sqrt[4]{16a^6} = \sqrt{25a^2b + 25b^2} \cdot \sqrt[2\cdot 2]{16a^6} = \sqrt{25a^2b + 25b^2} \cdot \sqrt{4|a^3|} =$
$\sqrt{100|a^5b| + 100|a^3b^2|} = \sqrt{100a^2} \cdot \sqrt{|a^3b| + |ab^2|} = 10|a|\sqrt{|a^3b| + |ab^2|}$

14. $b\sqrt[3]{27} + \sqrt{b^2a} + 2b\sqrt{144a} = 3b + |b|\sqrt{a} + 24b\sqrt{a}$

15. $\dfrac{(\sqrt{5} + 3)}{(\sqrt{5} - 2)} \cdot \dfrac{(\sqrt{5} + 2)}{(\sqrt{5} + 2)} = \dfrac{5 + 2\sqrt{5} + 3\sqrt{5} + 6}{5 - 4} = 11 + 5\sqrt{5}$

16. $x - 4 - \sqrt{x - 4} = 0, x - 4 = \sqrt{x - 4}, x^2 - 8x + 16 = x - 4,$
$x^2 - 9x + 20 = 0, (x - 5)(x - 4) = 0, x = 5$ or $x = 4$
Check: $5 - 4 - \sqrt{5 - 4} \stackrel{?}{=} 0, 0 = 0; 4 - 4 - \sqrt{4 - 4} \stackrel{?}{=} 0, 0 = 0; \{4, 5\}$

17. $\dfrac{2i}{\sqrt{-16}} + i\sqrt{-27} = \dfrac{2i}{4i} + i \cdot 3i\sqrt{3} = \dfrac{1}{2} - 3i^2\sqrt{3} = \dfrac{1}{2} - 3\sqrt{3}$

18. $\bar{a} = 2 + 3i, \bar{a} + b = (2 + 3i) + \left(\dfrac{2}{3}i\right) = 2 + \dfrac{11}{3}i$

19. $\dfrac{6 - 4i}{1 + 2i} \cdot \dfrac{1 - 2i}{1 - 2i} = \dfrac{(6 - 8) + (-12 - 4)i}{1 + 4} = -\dfrac{2}{5} - \dfrac{16}{5}i$

Page 289 • PREPARING FOR COLLEGE ENTRANCE EXAMS

1. $(a^{5p} + 4a^{-3p})(a^{5p} - 4a^{-3p}) = a^{10p} - 4a^{2p} + 4a^{2p} - 16a^{-6p} = a^{10p} - 16a^{-6p}$; C

2. E **3.** $x^2 = 10x + 416, x^2 - 10x - 416 = 0, (x - 26)(x + 16) = 0, x = 26$ or $x = -16$; D **4.** C

5. (2, 4) satisfies the function. Since $(-2, -4)$ satisfies the function, the function is odd. Since $(-2, 4)$ does not satisfy the function, the function is not even. D

6. $y = ax^2z^3; y = a(2x)^2(2z)^3, y = a(4x^2)(8z^3), y = 32ax^2z^3$; A

7. $f(1) = 2 + 1 - 2 - 1 = 0; f(-1) = -2 + 1 + 2 - 1 = 0; f\left(-\dfrac{1}{2}\right) =$
$-\dfrac{1}{4} + \dfrac{1}{4} + 1 - 1 = 0$; E

8. $\sqrt{2x + 7} = 1 + \sqrt{x + 3}, 2x + 7 = 1 + 2\sqrt{x + 3} + x + 3, x + 3 = 2\sqrt{x + 3},$
$x^2 + 6x + 9 = 4(x + 3), x^2 + 2x - 3 = 0, (x + 3)(x - 1) = 0, x = -3$ or
$x = 1$ *Check:* $\sqrt{2(-3) + 7} \stackrel{?}{=} 1 + \sqrt{-3 + 3}, \sqrt{1} \stackrel{?}{=} 1 + 0, 1 = 1; \sqrt{2(1) + 7} \stackrel{?}{=}$
$1 + \sqrt{1 + 3}, \sqrt{9} \stackrel{?}{=} 1 + \sqrt{4}, 3 = 3; \{-3, 1\}$; A

Page 290 • CONTEST PROBLEMS

1. Use letters to represent the digits in the problem. Note that digits named by the same letter are not necessarily the same number.

```
        a b 8 c d
eee ) f f f f g h i j
      k k k
        l l g h
        m m m
          n n i j
          p p p p
                0
```

8 times a 3-digit divisor (*eee*) results in a 3-digit number (*mmm*); thus the divisor is between 99 and 125. The divisor times *d* results in a 4-digit number (*pppp*); thus $d = 9$. Since $d = 9$ and *pppp* is a 4-digit number, the divisor must be between 111 and 125. Since *gh* and *ij* are both brought down before any subtraction occurs, the divisor does not divide into either *llg* or *nni*; thus *b* and *c* are both 0. The divisor multiplied by *a* results in a 3-digit number (*kkk*), and when *kkk* is subtracted from a 4-digit number (*ffff*), the remainder is a 2-digit number (*ll*). Thus *a* must equal 8 because any number less than 8 multiplied by the divisor would result, when subtracted, in a 3-digit remainder. Therefore, the quotient is 80809 and the divisor is between 111 and 125. Using trial and error results in

```
        80809
124 ) 10020316
       992
        1003
         992
          1116
          1116
             0
```

Page 291 • PROGRAMMING IN PASCAL

1. a. See the function *power* in the program in Exercise 1b.

b.

```
PROGRAM integral_powers (INPUT,OUTPUT);

VAR
  b,e : integer;

{*********************************************************}
FUNCTION power (x : real; n : integer) : real;

BEGIN
  IF n >= 0
     THEN IF n = 0
             THEN power := 1
             ELSE power := power(x,n - 1) * x
     ELSE BEGIN
            n := -n;
            power := 1/power(x,n);
          END;
END;
```

[*Program continued on next page*]

```
{**********************************************************}
BEGIN  {* main *}
  writeln('Enter a base and an exponent: ');
  readln(b,e);
  IF (b = 0) AND (e < 0)
     THEN writeln('0 raised to a negative power is undefined.')
     ELSE writeln(power(b,e));
END.
```

c. Answers will vary depending on numbers chosen.

d. Differences in answers are a result of internal rounding errors.

2. a.

```
PROGRAM roots (INPUT,OUTPUT);

TYPE
  factors = ARRAY[1..20] OF integer;

VAR
  coefficient,exponent : ARRAY[1..20] OF integer;
  leading_term,constant : factors;
  error : boolean;
  divisor,last,count : integer;

{**********************************************************}
PROCEDURE get_terms;

VAR
   i : integer;

BEGIN
  error := FALSE;
  write('How many terms are in the polynomial: ');
  readln(last);
  FOR i := 1 TO last DO
      BEGIN
        write('Enter coefficient: ');
        readln(coefficient[i]);
        write('Enter degree: ');
        readln(exponent[i]);
      END;
  IF coefficient[1] = 0
     THEN error := TRUE;
  IF coefficient[last] = 0
     THEN error := TRUE;
  IF exponent[last] <> 0
     THEN error := TRUE;
END;

{**********************************************************}
PROCEDURE get_factors (num : integer; VAR list : factors);
```

```
BEGIN
  num := abs(num);
  divisor := 0;
  count := 0;
  REPEAT
    divisor := divisor + 1;
    IF (num MOD divisor) = 0
       THEN BEGIN
              count := count + 1;
              list[count + 1] := divisor;
            END;
  UNTIL divisor = num;
  list[1] := count;
END;

{**********************************************************}
PROCEDURE write_results;

VAR
  i,j : integer;

BEGIN
  writeln;
  FOR i := 2 TO leading_term[1] + 1 DO
      BEGIN
        FOR j := 2 TO constant[1] + 1 DO
            BEGIN
              write(constant[j],'/',leading_term[i],'     ');
              writeln(-constant[j],'/',leading_term[i]);
            END;
      END;
END;

{**********************************************************}
BEGIN  {* main *}
  get_terms;
  IF NOT (error)
     THEN BEGIN
            get_factors(coefficient[1],leading_term);
            get_factors(coefficient[last],constant);
            write_results;
          END
     ELSE writeln('ILLEGAL POLYNOMIAL');
END.
```

b. In the program, modify the procedure *write_results* as shown. Then insert the function *power* and the procedure *test* so they are positioned before *write_results*.

```
{***********************************************************}
FUNCTION power (base : real; degree : integer) : real;

BEGIN
  IF degree >= 0
     THEN IF degree = 0
             THEN power := 1
             ELSE power := power(base,degree - 1) * base
     ELSE BEGIN
            degree := - degree;
            power := 1/power(base,degree);
          END;
END;

{***********************************************************}
PROCEDURE test (numerator,denominator : integer);

VAR
  x,p_of_x : real;
  i : integer;

BEGIN
  p_of_x := 0;
  x := numerator/denominator;
  FOR i := 1 TO last DO
      p_of_x := p_of_x + coefficient[i]*power(x,exponent[i];
  write(numerator,'/',denominator);
  IF abs(p_of_x) < 0.00001
     THEN writeln(' is a root.')
     ELSE writeln(' is not a root.');
END;

{***********************************************************}
PROCEDURE write_results;

VAR
  i,j : integer;

BEGIN
  writeln;
  FOR i := 2 TO leading_term[1] + 1 DO
      BEGIN
        FOR j := 2 TO constant[1] + 1 DO
            BEGIN
              test(constant[j],leading_term[i]);
              test(-constant[j],leading_term[i]);
            END;
      END;
END;
```

CHAPTER 8 • Sequences and Series

Pages 295–296 • WRITTEN EXERCISES

A **1.** 2, −6, 18, −54 **2.** −3, 4, 11, 18; $d = 7$ **3.** $\frac{5}{2}, 2, \frac{3}{2}, 1; d = -\frac{1}{2}$

4. 2, 4, 16, 256 **5.** $-4, -4 + k, -4 + 2k, -4 + 3k; d = k$ **6.** 8, −5, 8, −5

7. 2, 5, 10, 17 **8.** −1, 3, 7, 11; $d = 4$ **9.** 5, 2, −1, −4; $d = -3$

10. $\frac{1}{3}, -\frac{1}{3}, -1, -\frac{5}{3}; d = -\frac{2}{3}$ **11.** 5, 10, 20, 40 **12.** −1, 2, −3, 4

13. $a_1 = 2, a_{n+1} = a_n - 6$ **14.** $a_1 = 7, a_{n+1} = 2a_n$ **15.** $a_1 = k, a_{n+1} = a_nk^2$

16. $a_1 = 6, a_{n+1} = \frac{1}{2}a_n$ **17.** $a_1 = -\frac{1}{2}, a_{n+1} = a_n + 3$ **18.** $a_1 = -9, a_{n+1} = -a_n$

19. $a_n = n^2$ **20.** $a_n = 3n + 1$ **21.** $a_n = 11\frac{1}{2} - \frac{7}{2}n$ **22.** $a_n = c + nb^2$

23. $a_n = 2 \cdot 5^{n-1}$ **24.** $a_n = ar^{n-1}$

B **25.** $a_n = \frac{n(n-1)}{2} + 1; a_{n+1} = \frac{(n+1)(n+1-1)}{2} + 1 = \frac{n(n+1)+2}{2} = \frac{n^2+n+2}{2};$
$a_{n+1} - a_n = \frac{n^2+n+2}{2} - \left[\frac{n(n-1)}{2} + \frac{2}{2}\right] = \frac{n^2+n+2-n^2+n-2}{2} =$
$\frac{2n}{2} = n; a_1 = 1, a_{n+1} = a_n + n$

26. $a_n = n^2; a_{n+1} = (n+1)^2; a_{n+1} - a_n = n^2 + 2n + 1 - n^2 = 2n + 1; a_1 = 1,$
$a_{n+1} = a_n + 2n + 1$

27. $a_n = \frac{n}{n+1}; a_{n+1} = \frac{n+1}{n+2}; a_{n+1} - a_n = \frac{n+1}{n+2} - \frac{n}{n+1} = \frac{(n+1)^2 - n(n+2)}{(n+2)(n+1)} =$
$\frac{n^2+2n+1-n^2-2n}{(n+1)(n+2)} = \frac{1}{(n+1)(n+2)}; a_1 = \frac{1}{2}, a_{n+1} = a_n + \frac{1}{(n+1)(n+2)}$

28. $a_1 = 1, a_2 = 1, a_n = a_{n-1} + a_{n-2}$

29. Since $d = a_2 - a_1$ and $d = a_3 - a_2$, $a_3 - a_2 = a_2 - a_1$. Thus, $a_1 + a_3 = 2a_2$.
$\therefore \frac{a_1 + a_3}{2} = a_2$.

C **30.** Since $d = a_3 - a_2$ and $d = a_2 - a_1$, $a_3 = 2a_2 - a_1$. Also, $a_4 - a_3 = a_2 - a_1$.
Substitute for a_3: $a_1 - 2a_2 + a_1 = a_2 - a_4$. Then $2a_1 + a_4 = 3a_2$. $\therefore \frac{2a_1 + a_4}{3} =$
a_2.

31. $a_2 - a_1 = d = a_3 - a_2$; $5x - 3 - (2x - 1) = 4x + 3 - (5x - 3)$, $3x - 2 =$
$-x + 6$; $x = 2$

Pages 298–299 • WRITTEN EXERCISES

A **1.** $a_{16} = 17 + 15(-3) = -28$ **2.** $a_{15} = 3 + 14(50) = 703$

3. $a_{12} = -8 + 11(5) = 47$ **4.** $a_{21} = 1 + 20(-4) = -79$

5. $a_{19} = -4 + 18(4.5) = 77$ **6.** $a_{17} = 0.1 + 16(0.05) = 0.9$

7. $a_{23} = \frac{1}{6} + 22\left(-\frac{1}{2}\right) = -\frac{65}{6}$ **8.** $a_{18} = 1 + 17(101) = 1718$

9. $a_{51} = 4 + 50(8) = 404$ **10.** $95 = 11 + 14d; d = 6$

11. $472 = a_1 + 29(19); a_1 = -3$ **12.** $-28 = 172 + 50d; d = -4$

13. $-82 = -2 + 16d; d = -5$ **14.** $a_{32} = \frac{3}{2} + 31\left(-\frac{5}{2}\right) = -76$

15. $63 = a_1 + 28(4); a_1 = -49$ **16.** $-27 = a_1 + 44(-0.75); a_1 = 6$

17. $-114 = 12 + (n - 1)(-7); n = 19$ **18.** $5 = 19 + (n - 1)\left(-\frac{2}{3}\right); n = 22$

19. $-5 = -44 + (n - 1)\left(\frac{3}{2}\right); n = 27$ **20.** $117 = 1.8 + (n - 1)(3.2); n = 37$

21. $a_5 = a_1 + 4d; 91 = 7 + 4d; d = 21;$ 28, 49, 70

22. $a_7 = a_1 + 6d; 37 = -5 + 6d; d = 7;$ 2, 9, 16, 23, 30

23. $a_5 = a_1 + 4d; 21 = 3 + 4d; d = 4.5;$ 7.5, 12, 16.5

24. $a_9 = a_1 + 8d; 13 = -1 + 8d; d = 1.75;$ 0.75, 2.5, 4.25, 6, 7.75, 9.5, 11.25

25. $a_{11} = a_1 + 10d; -3 = -18 + 10d; d = 1.5;$ −16.5, −15, −13.5, −12, −10.5, −9, −7.5, −6, −4.5

26. $a_7 = a_1 + 6d; 2 = \frac{1}{5} + 6d; d = \frac{3}{10}; \frac{1}{2}, \frac{4}{5}, \frac{11}{10}, \frac{7}{5}, \frac{17}{10}$

27. $a_1 + 11d = 8$
$a_1 + 16d = 23$
$-5d = -15; d = 3; a_1 + 11(3) = 8; a_1 = -25$

28. $a_1 + 6d = 3$
$a_1 + 15d = -15$
$-9d = 18; d = -2; a_1 + 6(-2) = 3; a_1 = 15$

29. $a_1 + 9d = 25$
$a_1 + 15d = 67$
$-6d = -42; d = 7; a_1 + 9(7) = 25; a_1 = -38$

30. $a_1 + 4d = 9$
$a_1 + 12d = 21$
$-8d = -12; d = 1.5; a_1 + 4(1.5) = 9; a_1 = 3$

31. $a_1 + 6d = -1$
$a_1 + 10d = 9$
$-4d = -10; d = 2.5; a_1 + 6(2.5) = -1; a_1 = -16$

32. $a_1 + 9d = -2$
$a_1 + 15d = -5$
$-6d = 3; d = -0.5; a_1 + 9(-0.5) = -2; a_1 = 2.5$

B **33.** $a_1 + d = -3$
$a_1 + 6d = -18$
$-5d = 15; d = -3; a_1 + (-3) = -3; a_1 = 0; a_3 = 0 + 2(-3) = -6$

34. $a_1 + 5d = 18$
$a_1 + 9d = 6$
$-4d = 12; d = -3; a_1 + 5(-3) = 18; a_1 = 33; a_9 = 33 + 8(-3) = 9$

35. $a_1 + 12d = 4$
$a_1 + 20d = 8$
$-8d = -4; d = 0.5; a_1 + 12(0.5) = 4; a_1 = -2; a_5 = -2 + 4(0.5) = 0$

36. $a_1 + 3d = 2$
$a_1 + 24d = 65$
$-21d = -63; d = 3; a_1 + 3(3) = 2; a_1 = -7; a_{30} = -7 + 29(3) = 80$

37. $\frac{1}{2}(x + 4 + 4x + 5) = 12; 5x + 9 = 24; x = 3$

38. $\frac{1}{2}(x - 5 + 2x + 1) = x + 2; 3x - 4 = 2x + 4; x = 8$

39. $a_5 = a_1 + 4d; 2x = x - 2 + 4(x - 7); 2x = 5x - 30; x = 10$

40. $d = a_2 - a_1 = (3x - 2) - (x + 2) = 2x - 4; a_6 = a_1 + 5d;$
$5x = x + 2 + 5(2x - 4); 5x = 11x - 18; x = 3$

41. $a_1 = -1; d = 6; a_n = a_1 + (n - 1)d; 245 = -1 + (n - 1)6; n = 42;$ 42nd

42. $a_1 = -3; d = -7; a_n = a_1 + (n - 1)d; -143 = -3 + (n - 1)(-7); n = 21;$ 21st

C **43.** 1. $c - a = b - c$ Def. of arith. seq.
2. $2c = a + b$ Add. prop. of =
3. $c = \frac{1}{2}(a + b)$ Mult. prop. of =

44. 1. $a_n - a_1 = (n - 1)d$ Def. of arith. seq.
2. $a_{2n-1} - a_n = (2n - 1 - n)d$ Def. of arith. seq.
3. $a_{2n-1} - a_n = (n - 1)d$ Simplifying
4. $a_n - a_1 = a_{2n-1} - a_n$ Trans. prop. of =
5. $2a_n = a_1 + a_{2n-1}$ Add. prop. of =
6. $a_n = \frac{1}{2}(a_1 + a_{2n-1})$ Mult. prop. of =

Pages 299–300 • PROBLEMS

A **1.** $a_1 = 17$; $d = 8$; $a_{12} = 17 + 11(8) = 105$; 105 apples
2. $a_1 = 24$; $d = 16.5$; $a_{15} = 24 + 14(16.5) = 255$; 255 cm
3. $a_1 = 12.4$; $a_8 = 4$; $4 = 12.4 + 7d$; $d = -1.2$; $a_2 = 12.4 - 1.2 = 11.2$; 11.2 cm
4. $a_1 = 12{,}000$; $d = 1150$; $28{,}100 = 12{,}000 + (n - 1)1150$; $n = 15$; 15 years
5. $a_1 = 6\frac{33}{60}$; $d = -\frac{1.6}{60} = -\frac{2}{75}$; $a_{26} = 6\frac{33}{60} + 25\left(-\frac{2}{75}\right) = 6 + \left(-\frac{7}{60}\right) = 5\frac{53}{60}$; 5:53 A.M. $6\frac{9}{60} = 6\frac{33}{60} + (n - 1)\left(-\frac{2}{75}\right)$; $n = 16$; March 16

B **6.** $a_1 = 49$; $d = -9.8$; $a_8 = 49 + 7(-9.8) = -19.6$; 19.6 m/s downward; $-49 = 49 + (n - 1)(-9.8)$; $n = 11$; 11 s
7. $14{,}000 + (n - 1)550 = 15{,}000 + (n - 5)700$; $1000 = (n - 1)550 - (n - 5)700$; $1000 = -150n + 2950$; $n = 13$; 1992
8. $a_1 = 229.22$; $d = -1.57$; $103.62 = 229.22 + (n - 1)(-1.57)$; $n = 81$; 81 m high; $0 = 229.22 + (n - 1)(-1.57)$; $n = 147$; 146 m high

Page 303 • WRITTEN EXERCISES

A **1.** $S_7 = \frac{7}{2}(3 + 21) = 84$ **2.** $S_{15} = \frac{15}{2}(-7 + 29) = 165$
3. $S_{32} = \frac{32}{2}(-86 - 14) = -1600$ **4.** $S_{14} = \frac{14}{2}[2 \cdot 33 + 13(-4)] = 98$
5. $S_{52} = \frac{52}{2}[2(-12) + 51(3)] = 3354$ **6.** $S_{27} = \frac{27}{2}(9 + 45) = 729$
7. $S_{11} = \frac{11}{2}[2(14) + (11 - 1)9] = 649$
8. $59 = 68 + 4d$; $d = -2.25$; $S_{17} = \frac{17}{2}[2(68) + 16(-2.25)] = 850$
9. $S_{11} = \frac{11}{2}(9 - 37) = -154$ **10.** $S_{21} = \frac{21}{2}[2(3) + 20(2)] = 483$
11. $72 = 4 + (n - 1)4$; $n = 18$; $S_{18} = \frac{18}{2}(4 + 72) = 684$
12. $S_{25} = \frac{25}{2}(31 - 19) = 150$ **13.** $S_{10} = \frac{10}{2}\left[2\left(\frac{1}{2}\right) + 9\left(\frac{1}{3}\right)\right] = 20$
14. $48 = 15 + (n - 1)3$; $n = 12$; $S_{12} = \frac{12}{2}(15 + 48) = 378$

15. $63 = 41 + (n - 1)2$; $n = 12$; $S_{12} = \frac{12}{2}(41 + 63) = 624$

16. $S_{15} = \frac{15}{2}[2(-13) + 14(6)] = 435$ **17.** $S_{14} = \frac{14}{2}[2(9) + 13(-5)] = -329$

18. $S_{17} = \frac{17}{2}(5 + 37) = 357$ **19.** $S_{12} = \frac{12}{2}(5 + 82) = 522$

20. $S_{10} = \frac{10}{2}(1 - 35) = -170$ **21.** $\sum_{i=1}^{4} 2i + 7$ **22.** $\sum_{i=1}^{4} 7i - 6$

23. $\sum_{i=1}^{5} 4i - 9$ **24.** $\sum_{i=1}^{5} 20 - 9i$ **25.** $\sum_{i=1}^{19} 2i + 2$ **26.** $\sum_{i=1}^{10} 7i - 16$

B **27.** $-170 = \frac{17}{2}(4 + 16d)$; $d = -\frac{3}{2}$ **28.** $-288 = \frac{16}{2}(-21 + a_{16})$; $a_{16} = -15$

29. $360 = \frac{n}{2}(42 - 26)$; $n = 45$ **30.** $324 = \frac{12}{2}(a_1 + 59)$; $a_1 = -5$

31. $198 = \frac{11}{2}[2a_1 + 10(6)]$; $a_1 = -12$

32. $-8 = \frac{n}{2}[-30 + (n - 1)4]$; $-16 = n(4n - 34)$; $4n^2 - 34n + 16 = 0$; $2(2n - 1)(n - 8) = 0$; $n = 8$

33. $210 = 3 + (n - 1)3$; $n = 70$; $S_{70} = \frac{70}{2}(3 + 210) = 7455$

34. $10 = 100 + (n - 1)(-5)$; $n = 19$; $S_{19} = \frac{19}{2}(100 + 10) = 1045$

C **35.** $S_n = \frac{n}{2}[2(1) + (n - 1)1] = 4950$; $n(1 + n) = 9900$; $n^2 + n - 9900 = 0$; $(n + 100)(n - 99) = 0$; $n = 99$

36. $S_n = \frac{n}{2}[2(3) + (n - 1)3] = 315$; $n(3n + 3) = 630$; $3(n^2 + n - 210) = 0$; $3(n + 15)(n - 14) = 0$; $n = 14$

37. $a_1 = 1$; $d = 2$; $S_n = \frac{n}{2}[2 \cdot 1 + (n - 1)2] = \frac{n}{2}(2 + 2n - 2) = n^2$

38. $a_1 = m$, $a_n = m^2$, $d = 1$, so $m^2 = m + (n - 1)1$ or $n = m^2 - m + 1$. Thus $S_n = \frac{m^2 - m + 1}{2}(m + m^2) = \frac{m^4 + m}{2} = \frac{m(m^3 + 1)}{2}$.

Pages 303–305 • PROBLEMS

A **1.** $a_1 = 120$; $d = 5$; $n = 24$; $S_{24} = \frac{24}{2}[2(120) + 23(5)] = 4260$; \$4260

2. $a_1 = 35$; $d = 4$; $n = 14$; $S_{14} = \frac{14}{2}[2(35) + 13(4)] = 854$; 854 bu

3. Jack: $a_1 = 15{,}000$; $d = 1000$; $n_1 = 10$; $n_2 = 20$; $S_{10} = \frac{10}{2}[30{,}000 + 9(1000)] = 195{,}000$; $S_{20} = \frac{20}{2}[30{,}000 + 19(1000)] = 490{,}000$; Kate: $a_1 = 7500$, $d = 250$; $n_1 = 20$; $n_2 = 40$; $S_{20} = \frac{20}{2}[15{,}000 + 19(250)] = 197{,}500$; $S_{40} = \frac{40}{2}[15{,}000 + 39(250)] = 495{,}000$; Kate with \$197,500; Kate with \$495,000

4. $a_1 = 23$; $d = 1$; $n = 16$; $S_{16} = \frac{16}{2}[2(23) + 15(1)] = 488$; 488 degree-days

B **5.** $a_1 = 4.9$; $d = 9.8$; $S_n = 313.6$; $313.6 = \frac{n}{2}[2(4.9) + (n - 1)9.8]$; $n = 8$; 8 s

6. $S_{16} = 40{,}000$; $a_1 = 1600$; $n = 16$; $40{,}000 = \frac{16}{2}[2(1600) + 15d]$; $d = 120$; \$120

7. Point A is the endpoint of 25 different segments, point B is the endpoint of 24 different segments, point C the endpoint of 23 different segments, on to point Y which is the endpoint of 1 segment. Thus, the number of different segments is the sum $25 + 24 + 23 + \cdots + 1$, which is $S_{25} = \frac{25}{2}(25 + 1) = 325$; 325 segments

8. Sum of horiz. dividers $= 2 + 3 + 4 + \cdots + 25 = \frac{24}{2}(2 + 25) = 324$; sum of vert. dividers $= 1 + 2 + 3 + \cdots + 23 = \frac{23}{2}(1 + 23) = 276$; there is 1 additional 23 cm divider; thus total length $= 324 + 276 + 23 = 623$; 623 cm

C **9.** Let $a_1 = 1250$ L (the flow for 30 min); $d = 1250$. Let $n = 9$ (the number of 30 min intervals until all pipes are open). Then $a_9 = 1250 + 8(1250) = 11{,}250$ L that has flowed through the first pipe in $4\frac{1}{2}$ h (nine 30 min intervals). $S_9 = \frac{9}{2}(1250 + 11{,}250) = 56{,}250$ L, the amount that has flowed through all 9 open pipes. Let t = number of 30 min intervals after all 9 pipes are open until capacity of 180,000 L is reached. $56{,}250 + 11{,}250t = 180{,}000$; $t = 11$; eleven 30 min intervals equals $5\frac{1}{2}$ h; thus it takes $4\frac{1}{2}$ h until all pipes are open and $5\frac{1}{2}$ h until capacity of 180,000 is reached. $4\frac{1}{2} + 5\frac{1}{2} = 10$; 10 h to fill tank

10. Let x = her first contribution; then $x + (x + 1) + (x + 2) + \cdots + 2x = 165$. $a_1 = x$; $d = 1$; $a_n = 2x$; $a_n = a_1 + (n - 1)d$ or $2x = x + (n - 1)1$ so $n = x + 1$; $S = \frac{n}{2}[a_1 + a_n] = \frac{x + 1}{2}(x + 2x) = 165$; $3x(x + 1) = 330$; $3(x^2 + x - 110) = 0$; $3(x + 11)(x - 10) = 0$; $x = 10$; May 10

11. For T_n: $a_1 = 1$, $d = 1$, $n = n$. $T_n = \frac{n}{2}[2(1) + (n - 1)1] = \frac{n^2 + n}{2}$. For T_{n-1}: $n = n - 1$. $T_{n-1} = \frac{n - 1}{2}[2(1) + (n - 2)1] = \frac{n^2 - n}{2}$.

$T_n + T_{n-1} = \frac{n^2 + n}{2} + \frac{n^2 - n}{2} = n^2$.

12. $T_{2n+1} = \frac{2n + 1}{2}[2(1) + (2n)1] = (2n + 1)(n + 1) = 2n^2 + 3n + 1$; $2T_n = 2\left(\frac{n^2 + n}{2}\right) = n^2 + n$; $T_{2n+1} - 2T_n = (2n^2 + 3n + 1) - (n^2 + n) = n^2 + 2n + 1 = (n + 1)^2$. The number of dots in each small triangle represents T_5. The number of dots in each large triangle represents T_{11}, which can also be written as $T_{2(5)+1}$. The difference between the number of dots in the large triangle and the number of dots in the 2 small triangles is the number of dots in the 6×6 square. Thus $T_{2(5)+1} - 2T_5 = (5 + 1)^2$.

Page 305 • SELF-TEST 1

1. a. $a_1 = 4,\ a_{n+1} = a_n + 4$ **b.** $a_n = 4n$ **2. a.** $a_1 = 3,\ a_{n+1} = 3a_n$ **b.** $a_n = 3^n$

3. $-22 = a_1 + 40(-3),\ a_1 = 98$

4. $$\begin{array}{r} a_1 + 14d = 9 \\ \underline{a_1 + 56d = 30} \\ -42d = -21 \end{array};\ d = \frac{1}{2};\ a_1 + 14\left(\frac{1}{2}\right) = 9;\ a_1 = 2$$

5. $a_3 = 10;\ a_1 = -2;\ 10 = -2 + 2d;\ d = 6;\ 4$

6. $a_7 = 10;\ a_1 = -2;\ 10 = -2 + 6d;\ d = 2;\ 0, 2, 4, 6, 8$

7. $S_{12} = \frac{12}{2}[2(2) + 11(-5)] = -306$ **8.** $S_{10} = \frac{10}{2}(4 + 58) = 310$

9. $a_1 = 15;\ d = 2;\ a_n = 45 = 15 + (n - 1)2;\ n = 16;$ March 17

10. $a_1 = 200;\ d = 10;\ n = 36;\ S_{36} = \frac{36}{2}[2(200) + 35(10)] = 13{,}500;\ \$13{,}500$

Pages 309–310 • WRITTEN EXERCISES

A **1.** 5, 25, 125, 625 **2.** 3, −9, 27, −81 **3.** 40, 20, 10, 5 **4.** $\frac{9}{25}, -\frac{3}{5}, 1, -\frac{5}{3}$

5. $\frac{c^2}{d}, 2c, 4d, \frac{8d^2}{c}$ **6.** $p, q, \frac{q^2}{p}, \frac{q^3}{p^2}$ **7.** $a_4 = 20\left(\frac{1}{4}\right)^3 = \frac{5}{16}$

8. $a_5 = \left(-\frac{1}{9}\right)(-3)^4 = -9$ **9.** $a_5 = 18\left(-\frac{1}{3}\right)^4 = \frac{2}{9}$ **10.** $a_6 = \frac{1}{3}\left(\frac{3}{2}\right)^5 = \frac{81}{32}$

11. $a_6 = \frac{b}{c^2}\left(\frac{c}{b}\right)^5 = \frac{c^3}{b^4}$ **12.** $a_7 = \frac{1}{4n^3}(2n)^6 = 16n^3$ **13.** $a_8 = \frac{8}{3}\left(\frac{1}{2}\right)^7 = \frac{1}{48}$

14. $a_7 = \frac{1}{20}(2^6) = \frac{16}{5}$ **15.** $a_6 = 250\left(\frac{1}{5}\right)^5 = \frac{2}{25}$ **16.** $a_7 = \frac{27}{2}\left(-\frac{1}{3}\right)^6 = \frac{3^3}{2 \cdot 3^6} = \frac{1}{54}$

17. $a_8 = \frac{1000}{3}\left(\frac{1}{10}\right)^7 = \frac{10^3}{3 \cdot 10^7} = \frac{1}{30{,}000}$ **18.** $a_6 = (0.0008)10^5 = 80$

B **19.** $\frac{3}{2} = \frac{3}{4}r,\ r = 2;\ \frac{3}{4} = a_1 \cdot 2^4,\ a_1 = \frac{3}{4 \cdot 16} = \frac{3}{64}$

20. $\frac{1}{2} = \left(-\frac{5}{2}\right)r,\ r = -\frac{1}{5};\ -\frac{5}{2} = a_1\left(-\frac{1}{5}\right)^3,\ a_1 = \frac{5 \cdot 125}{2} = \frac{625}{2};$

$a_2 = \frac{625}{2}\left(-\frac{1}{5}\right) = -\frac{125}{2}$

21. arith.; $d = 12;\ a_7 = -16 + 6(12) = 56$

22. geom.; $r = \frac{1}{3};\ a_6 = \left(-\frac{9}{2}\right)\left(\frac{1}{3}\right)^5 = -\frac{3^2}{2 \cdot 3^5} = -\frac{1}{54}$

23. geom.; $r = \frac{1}{10};\ a_6 = \frac{2000}{9}\left(\frac{1}{10}\right)^5 = \frac{2 \cdot 10^3}{9 \cdot 10^5} = \frac{2}{900} = \frac{1}{450}$

24. arith.; $d = \frac{3}{4};\ a_7 = \frac{3}{2} + 6\left(\frac{3}{4}\right) = 6$

25. arith.; $d = 0.48;\ a_8 = -0.64 + 7(0.48) = 2.72$

26. geom.; $r = \frac{1}{ab^2};\ a_7 = a^7b^6\left(\frac{1}{ab^2}\right)^6 = \frac{a^7b^6}{a^6b^{12}} = \frac{a}{b^6}$

C **27.** Since $a_n = a_{n-1} \cdot r$, then $a_{n-1} = \frac{a_n}{r}$. By the same formula, $a_{n+1} = a_n \cdot r$. Thus

$a_{n-1}a_{n+1} = \frac{a_n}{r}(a_n \cdot r) = (a_n)^2$.

28. If $a_1, a_2, a_3, \ldots$ is arithmetic, then it can be written as $a_1, a_1 + d, a_1 + 2d, \ldots$ If the series is geometric, then $r = \frac{a_1 + d}{a_1} = \frac{a_1 + 2d}{a_1 + d}$. $(a_1 + d)^2 = a_1(a_1 + 2d)$; $(a_1)^2 + 2a_1d + d^2 = (a_1)^2 + 2a_1d$; $d^2 = 0$; and $d = 0$. Thus, $a_n = a_1 + 0(n - 1) = a_1$. $\therefore$ the terms must all be equal.

29. $\frac{a_{n-1}}{a_n} = \frac{1}{\frac{a_n}{a_{n-1}}} = \frac{1}{r}$ **30.** $\frac{2a_2}{2a_1} = \frac{a_2}{a_1} = r$ **31.** $\frac{a_3}{a_1} = \frac{a_1r^2}{a_1} = r^2$ **32.** $\frac{3a_2}{a_1} = \frac{3 \cdot a_1r}{a_1} = 3r$

Pages 310–311 • PROBLEMS

A **1.** $a_5 = 3 \cdot 3^4 = 243$; 243 neutrons
2. $a_{64} = 1 \cdot 2^{63} = (2^{10})^6 \cdot 2^3 \approx (1000)^6 \cdot 8 = 8 \times 10^{18}$
3. $a_5 = 5000(1.2)^4 = 10{,}368$; 10,368 insects
4. $a_5 = 10{,}000(0.8)^4 = 4096$; \$4096
5. $a_5 = 6 \cdot 2^4 = 96$; 96 great-great-grandparents
6. $a_4 = 128(0.75)^3 = \$54$; $a_5 = 128(0.75)^4 = \$40.50$

B **7.** $a_8 = 16\left(\frac{1}{2}\right)^7 = 0.125$; 0.125 mg

8. $a_1 = 96(3) = 288$; $r = \frac{1}{2}$; $a_8 = 288\left(\frac{1}{2}\right)^7 = 2.25$; 2.25 cm

9. **a.** $A = 8000(1 + 0.2)^1 = \$9600$ **b.** $A = 8000\left(1 + \frac{0.2}{2}\right)^2 = \$9680$

c. $A = 8000\left(1 + \frac{0.2}{4}\right)^4 \approx \$9724.05$

10. $A = 5000\left(1 + \frac{0.08}{2}\right)^{2(1.5)} = 5000(1.04)^3 = \$5624.32$

11. $A = 2000(1 + 0.1)^5 = \$3221.02$; $\$3221.02 - \$2000 = \$1221.02$

Pages 314–315 • WRITTEN EXERCISES

A **1.** $108 = 4 \cdot r^3$; $27 = r^3$; $r = 3$ **2.** $250 = (-2)r^3$; $-125 = r^3$; $r = -5$

3. $-9 = 72 \cdot r^3$; $-\frac{1}{8} = r^3$; $r = -\frac{1}{2}$ **4.** $8 = 27 \cdot r^3$; $\frac{8}{27} = r^3$; $r = \frac{2}{3}$

5. $a_1r^2 = 7 \longrightarrow a_1 = \frac{7}{r^2}$; $a_1r^4 = 112$; $\left(\frac{7}{r^2}\right)r^4 = 112$; $r^2 = 16$; $r = \pm 4$; 4 or -4

6. $a_1r = -162 \longrightarrow a_1 = -\frac{162}{r}$; $a_1r^5 = -2$; $\left(-\frac{162}{r}\right)r^5 = -2$; $r^4 = \frac{1}{81}$; $r = \pm\frac{1}{3}$; $\frac{1}{3}$ or $-\frac{1}{3}$

7. $a_1r^2 = 15 \longrightarrow a_1 = \frac{15}{r^2}$; $a_1r^5 = 120$; $\left(\frac{15}{r^2}\right)r^5 = 120$; $r^3 = 8$; $r = 2$; $a_1(2)^2 = 15$; $a_1 = \frac{15}{4}$; $a_2 = \frac{15}{2}$

8. $a_1r^2 = -2 \longrightarrow a_1 = -\frac{2}{r^2}$; $a_1r^5 = 54$; $\left(-\frac{2}{r^2}\right)r^5 = 54$; $r^3 = -27$; $r = -3$; $a_1(-3)^2 = -2$; $a_1 = -\frac{2}{9}$; $a_2 = \frac{2}{3}$

9. $a_1r^3 = -10 \longrightarrow a_1 = -\frac{10}{r^3}$; $a_1r^6 = -80$; $\left(-\frac{10}{r^3}\right)r^6 = -80$; $r^3 = 8$; $r = 2$; $a_1(2)^3 = -10$; $a_1 = -\frac{5}{4}$; $a_2 = -\frac{5}{2}$

10. $a_1r^2 = 6 \longrightarrow a_1 = \frac{6}{r^2}$; $a_1r^5 = \frac{3}{4}$; $\left(\frac{6}{r^2}\right)r^5 = \frac{3}{4}$; $r^3 = \frac{1}{8}$; $r = \frac{1}{2}$; $a_1\left(\frac{1}{2}\right)^2 = 6$; $a_1 = 24$; $a_2 = 12$

11. $a_1r^4 = 20 \longrightarrow a_1 = \frac{20}{r^4}$; $a_1r^7 = \frac{4}{25}$; $\left(\frac{20}{r^4}\right)r^7 = \frac{4}{25}$; $r^3 = \frac{1}{125}$; $r = \frac{1}{5}$; $a_1\left(\frac{1}{5}\right)^4 = 20$; $a_1 = 12{,}500$; $a_2 = 2500$

12. $a_1r^2 = -18 \longrightarrow a_1 = -\frac{18}{r^2}$; $a_1r^5 = \frac{16}{3}$; $\left(-\frac{18}{r^2}\right)r^5 = \frac{16}{3}$; $r^3 = -\frac{8}{27}$; $r = -\frac{2}{3}$; $a_1\left(-\frac{2}{3}\right)^2 = -18$; $a_1 = -\frac{81}{2}$; $a_2 = 27$

13. $a_1 = 4$; $a_4 = 500$; $500 = 4r^3$; $r^3 = 125$; $r = 5$; 4, 20, 100, 500

14. $a_1 = -22$; $a_4 = -176$; $-176 = -22(r^3)$; $r^3 = 8$; $r = 2$; $-22, -44, -88, -176$

15. $a_1 = -5$; $a_4 = 135$; $135 = -5(r^3)$; $r^3 = -27$; $r = -3$; $-5, 15, -45, 135$

16. $a_1 = \frac{3}{2}$; $a_4 = -96$; $-96 = \frac{3}{2}(r^3)$; $r^3 = -64$; $r = -4$; $\frac{3}{2}, -6, 24, -96$

17. $a_1 = \frac{7}{4}$; $a_5 = 28$; $28 = \frac{7}{4}r^4$; $r^4 = 16$; $r = \pm 2$; $\frac{7}{4}, \frac{7}{2}, 7, 14, 28$ or $\frac{7}{4}, -\frac{7}{2}, 7, -14, 28$

18. $a_1 = -\frac{25}{3}$; $a_5 = -\frac{27}{25}$; $-\frac{27}{25} = -\frac{25}{3}(r^4)$; $r^4 = \frac{81}{625}$; $r = \pm\frac{3}{5}$; $-\frac{25}{3}, -5, -3, -\frac{9}{5}, -\frac{27}{25}$ or $-\frac{25}{3}, 5, -3, \frac{9}{5}, -\frac{27}{25}$

19. $a_1 = 4$; $a_5 = \frac{81}{4}$; $\frac{81}{4} = 4r^4$; $r^4 = \frac{81}{16}$; $r = \pm\frac{3}{2}$; $4, 6, 9, \frac{27}{2}, \frac{81}{4}$ or $4, -6, 9, -\frac{27}{2}, \frac{81}{4}$

20. $a_1 = -125$, $a_5 = -\frac{16}{5}$, $-\frac{16}{5} = -125(r^4)$, $r^4 = \frac{16}{625}$, $r = \pm\frac{2}{5}$; $-125, -50, -20, -8, -\frac{16}{5}$ or $-125, 50, -20, 8, -\frac{16}{5}$

B 21. $r = \frac{5}{2}$; $\frac{625}{8} = 2\left(\frac{5}{2}\right)^{n-1}$; $\frac{5^4}{2^4} = \left(\frac{5}{2}\right)^{n-1}$; $n - 1 = 4$; $n = 5$

22. $r = \dfrac{\frac{27}{2}}{\frac{81}{4}} = \frac{2}{3}$; $4 = \frac{81}{4}\left(\frac{2}{3}\right)^{n-1}$; $\frac{2^4}{3^4} = \left(\frac{2}{3}\right)^{n-1}$; $n - 1 = 4$; $n = 5$

23. $a_1r^4 = \frac{1}{2} \longrightarrow a_1 = \frac{1}{2r^4}$; $a_1r^7 = -\frac{1}{128}$; $\left(\frac{1}{2r^4}\right)r^7 = -\frac{1}{128}$; $r^3 = -\frac{1}{64}$; $r = -\frac{1}{4}$; $a_1\left(-\frac{1}{4}\right)^4 = \frac{1}{2}$; $a_1 = 128$; $a_2 = -32$

24. $a_1r^5 = 250 \longrightarrow a_1 = \frac{250}{r^5}$; $a_1r^7 = 6250$; $\left(\frac{250}{r^5}\right)r^7 = 6250$; $r^2 = 25$; $r = \pm 5$; $a_1(5)^5 = 250$; $a_1 = 0.08$; $a_3 = (0.08)(5)^2$; $a_3 = 2$; or $a_1(-5)^5 = 250$; $a_1 = -0.08$; $a_3 = (-0.08)(-5)^2$; $a_3 = -2$

25. $a_1r^3 = \frac{x^3}{y^2} \longrightarrow a_1 = \frac{x^3}{r^3y^2}$; $a_1r^6 = \frac{y}{x^3}$; $\left(\frac{x^3}{r^3y^2}\right)r^6 = \frac{y}{x^3}$; $r^3 = \frac{y^3}{x^6}$; $r = \frac{y}{x^2}$; $a_1\left(\frac{y}{x^2}\right)^3 = \frac{x^3}{y^2}$; $a_1\left(\frac{y^3}{x^6}\right) = \frac{x^3}{y^2}$; $a_1 = \frac{x^9}{y^5}$

26. $a_1r^4 = \frac{k^2}{n^4} \longrightarrow a_1 = \frac{k^2}{r^4n^4}$; $a_1r^7 = \frac{k^5}{n^{10}}$; $\left(\frac{k^2}{r^4n^4}\right)r^7 = \frac{k^5}{n^{10}}$; $r^3 = \frac{k^3}{n^6}$; $r = \frac{k}{n^2}$; $a_1\left(\frac{k}{n^2}\right)^4 = \frac{k^2}{n^4}$; $a_1\left(\frac{k^4}{n^8}\right) = \frac{k^2}{n^4}$; $a_1 = \frac{n^4}{k^2}$

27. $a_1 = 4800$; $a_3 = 7500$; $7500 = 4800r^2$; $r^2 = \frac{25}{16}$; $r = \frac{5}{4}$

28. Find a_1 when $a_3 = 26 - 20 = 6$ and $a_4 = 22 - 20 = 2$. $r = \frac{a_4}{a_3} = \frac{1}{3}$; $6 = a_1\left(\frac{1}{3}\right)^2$; $a_1 = 54$; 74°C

29. $\frac{x-1}{2x} = \frac{2x}{5x+3}$; $5x^2 - 2x - 3 = 4x^2$; $x^2 - 2x - 3 = 0$; $(x-3)(x+1) = 0$; $x = 3$ or $x = -1$

30. $\frac{x-2}{x+2} = \frac{x+2}{3x-2}$; $3x^2 - 8x + 4 = x^2 + 4x + 4$; $2x^2 - 12x = 0$; $2x(x-6) = 0$; $x = 6$ or $x = 0$

C **31.** The common ratio of the sequence $a_1^2, a_1^2r^2, a_1^2r^4, a_1^2r^6, \ldots$ is r^2, so the sequence is geometric.

32. The sequence is $\frac{1}{a_1}, \frac{1}{a_1r}, \frac{1}{a_1r^2}, \frac{1}{a_1r^3}, \ldots$. The common ratio is $\frac{1}{r}$, so the sequence is geometric.

33. Given: $t^a, t^b, t^c, t^d, \ldots$ is geometric. Therefore, $\frac{t^b}{t^a} = \frac{t^c}{t^b} = \frac{t^d}{t^c} = \cdots$, or $t^{b-a} = t^{c-b} = t^{d-c} = \cdots$. Thus, $b - a = c - b = d - c = \cdots$ and the sequence $a, b, c, d, \ldots$ is arithmetic.

34. Let r_1 = the common ratio of $a_1, a_2, a_3, \ldots$ and r_2 = the common ratio of $b_1, b_2, b_3 \cdots$. So $r_1 = \frac{a_2}{a_1} = \frac{a_3}{a_2} = \cdots$ and $r_2 = \frac{b_2}{b_1} = \frac{b_3}{b_2} = \cdots$. Then $r_1r_2 = \frac{a_2b_2}{a_1b_1} = \frac{a_3b_3}{a_2b_2} = \cdots$. Thus r_1r_2 is the common ratio of $a_1b_1, a_2b_2, a_3b_3, \ldots$, and $a_1b_1, a_2b_2, a_3b_3 \ldots$ is a geometric sequence.

35. Since the square of any real number is nonnegative, $(a - b)^2 \geq 0$. $a^2 - 2ab + b^2 \geq 0$ means that $a^2 + b^2 \geq 2ab$ or $\frac{a^2 + b^2}{2} \geq ab$. Now, $\frac{a^2 + b^2}{2}$ is the arithmetic mean of a^2 and b^2 and ab is the geometric mean of a^2 and b^2 $(a, b \geq 0)$. Therefore, the arithmetic mean of a^2 and b^2 is greater than or equal to the geometric mean of a^2 and b^2.

36. For $n > 0$, $a_n = a_1r^{n-1}$ and $a_{2n-1} = a_1r^{2n-2}$. Hence, $\frac{a_{2n-1}}{a_n} = \frac{a_1r^{2n-2}}{a_1r^{n-1}} = r^{n-1}$. Also $\frac{a_n}{a_1} = r^{n-1}$. Thus $\frac{a_{2n-1}}{a_n} = \frac{a_n}{a_1}$. Therefore a_n is the geometric mean of a_1 and a_{2n-1}.

Pages 317–318 • WRITTEN EXERCISES

A 1. $S_7 = \dfrac{\frac{3}{8} - \frac{3}{8}\cdot 2^7}{1 - 2} = \dfrac{\frac{3}{8} - 48}{-1} = 47\frac{5}{8}$ 2. $S_6 = \dfrac{1 - 1\cdot(-3)^6}{1 - (-3)} = \dfrac{1 - 729}{4} = -182$

3. $S_6 = \dfrac{324 - 324\left(\frac{1}{3}\right)^6}{1 - \frac{1}{3}} = \dfrac{324 - \frac{4}{9}}{\frac{2}{3}} = 485\frac{1}{3}$

4. $S_5 = \dfrac{128 - 128\left(\frac{3}{4}\right)^5}{1 - \frac{3}{4}} = \dfrac{128 - \frac{243}{8}}{\frac{1}{4}} = 390\frac{1}{2}$

5. $a_1 = -64$; $a_n = -1$; $r = -\frac{1}{2}$; $S_n = \dfrac{-64 - \left(-\frac{1}{2}\right)(-1)}{1 - \left(-\frac{1}{2}\right)} = \dfrac{-64 - \frac{1}{2}}{\frac{3}{2}} = -\dfrac{129}{2}\cdot\dfrac{2}{3} =$

$-\dfrac{129}{3} = -43$

6. $a_1 = 100$; $a_n = \frac{4}{125}$, $r = \frac{1}{5}$; $S_n = \dfrac{100 - \frac{1}{5}\left(\frac{4}{125}\right)}{1 - \frac{1}{5}} = \dfrac{100 - \frac{4}{625}}{\frac{4}{5}} = \dfrac{62{,}496}{625}\cdot\dfrac{5}{4} =$

$\dfrac{15{,}624}{125} = 124\dfrac{124}{125}$

7. $S_n = 1 + 4 + 16 + \cdots + 256 = \dfrac{1 - 4(256)}{1 - 4} = 341$

8. $S_5 = \dfrac{3 - 3(3)^5}{1 - 3} = \dfrac{3 - 729}{-2} = 363$

9. $S_n = \dfrac{5 - 2(640)}{1 - 2} = 1275$ 10. $S_5 = \dfrac{405 - 405\left(-\frac{1}{3}\right)^5}{1 - \left(-\frac{1}{3}\right)} = 305$

11. $S_6 = \dfrac{2 - 2(3)^6}{1 - 3} = 728$ 12. $S_5 = \dfrac{\frac{1}{2} - \frac{1}{2}(6)^5}{1 - 6} = 777\frac{1}{2}$

13. $S_7 = \dfrac{-4 - (-4)\left(\frac{1}{2}\right)^7}{1 - \frac{1}{2}} = -\dfrac{127}{16} = -7\dfrac{15}{16}$ 14. $S_8 = \dfrac{\frac{1}{6} - \frac{1}{6}(-2)^8}{1 - (-2)} = -\dfrac{85}{6} = -14\dfrac{1}{6}$

15. $S_5 = \dfrac{\frac{2}{9} - \frac{2}{9}(3)^5}{1 - 3} = \dfrac{242}{9} = 26\dfrac{8}{9}$ 16. $S_5 = \dfrac{100 - 100\left(\frac{1}{5}\right)^5}{1 - \frac{1}{5}} = \dfrac{3124}{25} = 124\dfrac{24}{25}$

17. $S_6 = \dfrac{27 - 27\left(\frac{2}{3}\right)^6}{1 - \frac{2}{3}} = 73\frac{8}{9}$ **18.** $S_5 = \dfrac{8 - 8\left(\frac{3}{4}\right)^5}{1 - \frac{3}{4}} = \dfrac{781}{32} = 24\dfrac{13}{32}$

B **19.** $393 = \dfrac{a_1 - a_1 \cdot 2^6}{1 - 2}$; $-393 = -63a_1$; $a_1 = \dfrac{131}{21}$

20. $410 = \dfrac{a_1 - a_1\left(-\frac{1}{4}\right)^5}{\frac{5}{4}}$; $\dfrac{1025}{2} = \dfrac{1025}{1024}a_1$; $a_1 = 512$

21. $\dfrac{305}{3} = \dfrac{a_1 - \left(-\frac{1}{3}\right)\frac{5}{3}}{\frac{4}{3}}$; $\dfrac{305}{3} \cdot \dfrac{4}{3} = a_1 + \dfrac{5}{9}$; $a_1 = \dfrac{1215}{9} = 135$

22. $484 = \dfrac{4 - 4(3)^n}{-2}$; $-968 = 4 - 4(3)^n$; $243 = 3^n$; $n = 5$

23. $812 = \dfrac{32 - 32\left(\frac{5}{2}\right)^n}{-\frac{3}{2}}$; $-1218 = 32 - 32\left(\dfrac{5}{2}\right)^n$; $-1250 = -32\left(\dfrac{5}{2}\right)^n$; $\dfrac{625}{16} = \left(\dfrac{5}{2}\right)^n$;

$n = 4$

24. $\dfrac{189}{2} = \dfrac{a_1 - \frac{1}{2} \cdot \frac{3}{2}}{\frac{1}{2}}$; $\dfrac{189}{4} = a_1 - \dfrac{3}{4}$; $a_1 = 48$; $a_n = a_1 r^{n-1}$; $\dfrac{3}{2} = 48\left(\dfrac{1}{2}\right)^{n-1}$;

$\dfrac{1}{32} = \left(\dfrac{1}{2}\right)^{n-1}$; $n - 1 = 5$; $n = 6$

25. $148.75 = \dfrac{a_1 - a_1(0.25)^4}{0.75}$; $111.5625 = a_1 - 0.00390625a_1$; $111.5625 =$
$-0.99609375a_1$; $a_1 = 112$; $a_4 = a_1 r^3 = 112(0.25)^3 = 1.75$

26. $126 = \dfrac{a_1 - a_1(-0.5)^6}{1.5}$; $189 = a_1 - 0.015625a_1$; $189 = 0.984375a_1$; $a_1 = 192$;
$a_6 = a_1 r^5 = 192(-0.5)^5 = -6$

27. $\dfrac{95}{2} = \dfrac{10 - 10r^3}{1 - r} = \dfrac{10(1 - r^3)}{1 - r} = 10(1 + r + r^2)$; $\dfrac{19}{4} = 1 + r + r^2$;
$r^2 + r - \dfrac{15}{4} = 0$; $4r^2 + 4r - 15 = 0$; $(2r - 3)(2r + 5) = 0$; $r = \dfrac{3}{2}$ or $r = -\dfrac{5}{2}$

28. $\dfrac{38}{3} = \dfrac{6 - 6r^3}{1 - r} = \dfrac{6(1 - r^3)}{1 - r} = 6(1 + r + r^2)$; $\dfrac{19}{9} = 1 + r + r^2$; $r^2 + r - \dfrac{10}{9} = 0$;
$9r^2 + 9r - 10 = 0$; $(3r + 5)(3r - 2) = 0$; $r = -\dfrac{5}{3}$ or $r = \dfrac{2}{3}$

C **29.** $2 + 2r + 2r^2 = 2r^2 - r + 3$; $3r = 1$; $r = \dfrac{1}{3}$

30. $a_1 + a_1 r = 90$; $a_1 r^5 + a_1 r^6 = -\dfrac{10}{27}$; $r^5(a_1 + a_1 r) = -\dfrac{10}{27}$; $90r^5 = -\dfrac{10}{27}$;

$r^5 = -\frac{1}{243}$; $r = -\frac{1}{3}$; $a_1 - \frac{1}{3}a_1 = 90$; $a_1 = 135$; $S_7 = \dfrac{135 - 135\left(-\frac{1}{3}\right)^7}{\frac{4}{3}} =$

$$\frac{135 + \frac{135}{2187}}{\frac{4}{3}} = \frac{135 + \frac{5}{81}}{\frac{4}{3}} = \frac{10{,}940}{108} = \frac{2735}{27} = 101\frac{8}{27}$$

31. $\sum_{k=1}^{n}(2^{-k}) = \frac{1}{2} + \frac{1}{4} + \cdots + \frac{1}{2^n}$; $a_1 = \frac{1}{2}$; $r = \frac{1}{2}$; $a_n = \frac{1}{2^n}$; so $S_n = \dfrac{\frac{1}{2} - \frac{1}{2}\left(\frac{1}{2}\right)^n}{1 - \frac{1}{2}} =$

$$\frac{\frac{1}{2} - \frac{1}{2}\left(\frac{1}{2^n}\right)}{\frac{1}{2}} = 1 - \frac{1}{2^n}$$

32. Prove that $\dfrac{a_1 - a_1r^n}{1 - r} = a_1 + a_1r + a_1r^2 + \cdots + a_1r^{n-1}$. $S_n = \dfrac{a_1 - a_1r^n}{1 - r} =$ $\dfrac{a_1(1 - r^n)}{1 - r} = a_1\left(\dfrac{r^n - 1}{r - 1}\right)$. By long division, $\dfrac{r^n - 1}{r - 1}$ is

$$\begin{array}{r|l} & r^{n-1} + r^{n-2} + \cdots + r + 1 \\ r - 1 & r^n \qquad\qquad\qquad\qquad - 1 \\ & \underline{r^n - r^{n-1}} \\ & \quad r^{n-1} \\ & \quad \underline{r^{n-1} - r^{n-2}} \\ & \qquad r^{n-2} \\ & \qquad \vdots \\ & \qquad\qquad\qquad \overline{r - 1} \\ & \qquad\qquad\qquad \underline{r - 1} \\ & \qquad\qquad\qquad\quad 0 \end{array}$$

So $S_n = a_1(r^{n-1} + r^{n-2} + \cdots + r + 1) = a_1 + a_1r + \cdots + a_1r^{n-2} + a_1r^{n-1}$.

33. Given: $a_1 + a_1r + a_1r^2 + \cdots + a_1r^{n-1} = S_n$. Multiply both sides of the equation by $\frac{1}{r^{n-1}}$. $\frac{1}{r^{n-1}}(a_1 + a_1r + a_1r^2 + \cdots + a_1r^{n-1}) = \frac{1}{r^{n-1}} \cdot S_n$; $\frac{a_1}{r^{n-1}} + \frac{a_1r}{r^{n-1}} + \frac{a_1r^2}{r^{n-1}} + \cdots + \frac{a_1r^{n-1}}{r^{n-1}} = \frac{1}{r^{n-1}} \cdot S_n$; $\frac{a_1}{r^{n-1}} + \frac{a_1}{r^{n-2}} + \frac{a_1}{r^{n-3}} + \cdots + a_1 = \frac{1}{r^{n-1}} \cdot S_n$ or $a_1 + \cdots + \frac{a_1}{r^{n-2}} + \frac{a_1}{r^{n-1}} = \frac{1}{r^{n-1}} \cdot S_n$

Pages 318–319 • PROBLEMS

A **1.** $a_1 = 2$; $r = 2$; $n = 6$; $S_6 = \dfrac{2 - 2 \cdot 2^6}{1 - 2} = 126$; 126 ancestors

2. $a_1 = 100{,}000$; $r = 0.9$; $n = 4$; $S_4 = \dfrac{100{,}000 - 100{,}000(0.9)^4}{1 - 0.9} = 343{,}900$; \$343,900

3. $S_5 = 21.01$; $r = 0.8$; $n = 5$; $21.01 = \dfrac{a_1 - a_1(0.8)^5}{1 - 0.8}$; $4.202 = a_1(1 - 0.32768)$; $a_1 = 6.25$; 6.25 cm

4. $a_1 = 200(1.1)$; $r = 1.1$; $n = 4$; $S_4 = \dfrac{200(1.1) - 200(1.1)(1.1)^4}{1 - 1.1} \approx 1021$; about 1021 fish

B **5. a.** Let x = original amt. of air, then $\frac{3}{4}x = a_1$ = amt. remaining after 1 min; $\frac{3}{4}x - \frac{1}{4}\cdot\frac{3}{4}x = \frac{9}{16}x = a_2$ = amt. remaining after 2 min; $\frac{9}{16}x - \frac{1}{4}\cdot\frac{9}{16}x = \frac{27}{64}x = a_3, \ldots$ The sequence $a_1, a_2, a_3, \ldots$ is $\frac{3}{4}x, \frac{9}{16}x, \frac{27}{64}x, \ldots$, where $r = \frac{3}{4}$. $a_5 = a_1r^4 = \frac{3}{4}x \cdot \left(\frac{3}{4}\right)^4 = \frac{243}{1024}x$; $\frac{243}{1024}$

b. $a_1 = \frac{1}{4}x$; $a_2 = \frac{1}{4}x \cdot \frac{3}{4} = \frac{3}{16}x$; $a_3 = \frac{3}{16}x \cdot \frac{3}{4} = \frac{9}{64}x, \ldots$, so $a_1 = \frac{1}{4}x$ and $r = \frac{3}{4}$.

$$S_5 = \frac{\frac{1}{4}x - \frac{1}{4}x\left(\frac{3}{4}\right)^5}{1 - \frac{3}{4}} = \frac{781}{1024}x;\ \frac{781}{1024}$$

6. $a_1 = 1280$; $r = 1.25$; $S_n = 7380$; $7380 = \dfrac{1280 - 1280(1.25)^n}{1 - 1.25}$; $-1845 = 1280 - 1280(1.25)^n$, $-3125 = -1280(1.25)^n$; $\frac{625}{256} = \left(\frac{5}{4}\right)^n$; $n = 4$; 4 yr

7. $a_1 = \frac{1}{3}$; $r = \frac{2}{3}$; $S_n = \dfrac{\frac{1}{3} - \frac{1}{3}\left(\frac{2}{3}\right)^n}{1 - \frac{2}{3}} \geq \frac{4}{5}$; $1 - \left(\frac{2}{3}\right)^n \geq \frac{4}{5}$; $\left(\frac{2}{3}\right)^n \leq \frac{1}{5}$; $\left(\frac{2}{3}\right)^4 = \frac{16}{81}$, $\frac{16}{81} < \frac{16}{80}$, $\frac{16}{80} = \frac{1}{5}$; thus $\left(\frac{2}{3}\right)^4 \leq \frac{1}{5}$ and $n = 4$; 4 s

C **8.** $P = \frac{m}{1+r} + \frac{m}{(1+r)^2} + \cdots + \frac{m}{(1+r)^k} = S_k$, where $a_1 = \frac{m}{1+r}$, the common ratio is $\frac{1}{1+r}$, and $n = k$. Thus $P = \dfrac{\frac{m}{1+r} - \frac{m}{1+r}\cdot\frac{1}{(1+r)^k}}{1 - \frac{1}{1+r}} =$

$$\frac{m}{1+r}\left[\frac{1 - \frac{1}{(1+r)^k}}{1 - \frac{1}{1+r}}\right] = m\left[\frac{1 - \frac{1}{(1+r)^k}}{1 + r - 1}\right] = m\left(\frac{1 - (1+r)^{-k}}{r}\right);\ \therefore m = \frac{rP}{1 - (1+r)^{-k}}.$$

Page 320 • COMPUTER EXERCISES

1.
```
10  INPUT A,R,N
20  LET S = A
30  FOR I = 2 TO N
```

[*Program continued on next page*]

```
40 LET A = A * R
50 LET S = S + A
60 NEXT I
70 PRINT S
80 END
```

2. .333333333 **3.** 3071.25 **4.** 33.3007813 **5.** -520.8336

Page 320 • SELF-TEST 2

1. $a_{10} = 6(2)^9 = 3072$

2. $a_1r^3 = -9 \rightarrow a_1 = \frac{-9}{r^3}$; $a_1r^6 = 243$; $\left(\frac{-9}{r^3}\right)r^6 = 243$; $r^3 = -27$; $r = -3$;

$a_1(-3)^3 = -9$; $a_1 = \frac{1}{3}$

3. $\frac{100}{x} = \frac{x}{10{,}000}$; $x^2 = 1{,}000{,}000$; $x = \pm 1000$; 1000 or -1000

4. $a_1 = 15$; $a_6 = 480$; $480 = 15r^5$; $r^5 = 32$; $r = 2$; 30, 60, 120, 240

5. $S_5 = \frac{20 - 20(-4)^5}{1 - (-4)} = \frac{20 - 20(-1024)}{5} = 4100$

6. $a_1 = 1$; $r = -\frac{1}{3}$; $a_n = -\frac{1}{243}$; $S_n = \frac{1 - \left(-\frac{1}{3}\right)\left(-\frac{1}{243}\right)}{1 - \left(-\frac{1}{3}\right)} = \frac{\frac{728}{729}}{\frac{4}{3}} = \frac{182}{243}$

7. $a_1 = 20{,}000$; $r = 1.1$; $a_5 = 20{,}000(1.1)^4 = 29{,}282$; 29,282 people

8. $a_1 = 1$; $r = 2$; $S_{14} = \frac{1 - 1(2)^{14}}{1 - 2} = \frac{1 - 16{,}384}{-1} = 16{,}383$; \$16,383

Pages 323–324 • WRITTEN EXERCISES

A

1. $a_1 = \frac{3 + 1}{1} = 4$; $a_2 = \frac{6 + 1}{2} = 3\frac{1}{2}$; $a_3 = \frac{9 + 1}{3} = 3\frac{1}{3}$; $a_4 = \frac{12 + 1}{4} = 3\frac{1}{4}$; $L = 3$

2. $a_1 = \frac{1 + 1}{2} = 1$; $a_2 = \frac{4 + 1}{2(4)} = \frac{5}{8}$; $a_3 = \frac{9 + 1}{2(9)} = \frac{5}{9}$; $a_4 = \frac{16 + 1}{2(16)} = \frac{17}{32}$; $L = \frac{1}{2}$

3. $a_1 = \frac{1}{20}$; $a_2 = \frac{4}{20} = \frac{1}{5}$; $a_3 = \frac{9}{20}$; $a_4 = \frac{16}{20} = \frac{4}{5}$; not convergent

4. $a_1 = 2 - \frac{1}{3} = 1\frac{2}{3}$; $a_2 = 2 - \frac{1}{9} = 1\frac{8}{9}$; $a_3 = 2 - \frac{1}{27} = 1\frac{26}{27}$; $a_4 = 2 - \frac{1}{81} = 1\frac{80}{81}$;

$L = 2$

5. $a_1 = (-1)\frac{2}{2 + 1} = -\frac{2}{3}$; $a_2 = (-1)^2\frac{4}{4 + 1} = \frac{4}{5}$; $a_3 = (-1)^3\frac{8}{8 + 1} = -\frac{8}{9}$;

$a_4 = (-1)^4\frac{16}{16 + 1} = \frac{16}{17}$; not convergent

6. $a_1 = 5(1 - 1) = 0$; $a_2 = 5\left(1 - \frac{1}{2}\right) = \frac{5}{2} = 2\frac{1}{2}$; $a_3 = 5\left(1 - \frac{1}{3}\right) = \frac{10}{3} = 3\frac{1}{3}$;

$a_4 = 5\left(1 - \frac{1}{4}\right) = \frac{15}{4} = 3\frac{3}{4}$; $L = 5$

7.

n	a_n	$\|L - a_n\|$
1	3	1
2	$\frac{7}{2}$	$\frac{1}{2}$
3	$\frac{11}{3}$	$\frac{1}{3}$
4	$\frac{15}{4}$	$\frac{1}{4}$

$|L - a_n| = \frac{1}{n}$

8.

n	a_n	$\|L - a_n\|$
1	$-\frac{1}{2}$	$\frac{1}{2}$
2	$-\frac{3}{4}$	$\frac{1}{4}$
3	$-\frac{7}{8}$	$\frac{1}{8}$
4	$-\frac{15}{16}$	$\frac{1}{16}$

$|L - a_n| = \left(\frac{1}{2}\right)^n$

9.

n	a_n	$\|L - a_n\|$
1	2	$\frac{1}{2}$
2	$\frac{7}{4}$	$\frac{1}{4}$
3	$\frac{5}{3}$	$\frac{1}{6}$
4	$\frac{13}{8}$	$\frac{1}{8}$

$|L - a_n| = \frac{1}{2n}$

10.

n	a_n	$\|L - a_n\|$
1	4	1
2	$5\frac{1}{4}$	$\frac{1}{4}$
3	$4\frac{8}{9}$	$\frac{1}{9}$
4	$5\frac{1}{16}$	$\frac{1}{16}$

$|L - a_n| = \frac{1}{n^2}$

B **11.**

n	a_n	$\|L - a_n\|$
1	$\frac{2}{5}$	$\frac{1}{2} - \frac{2}{5} = \frac{1}{10}$
2	$\frac{4}{9}$	$\frac{1}{2} - \frac{4}{9} = \frac{1}{18}$
3	$\frac{6}{13}$	$\frac{1}{2} - \frac{6}{13} = \frac{1}{26}$
4	$\frac{8}{17}$	$\frac{1}{2} - \frac{8}{17} = \frac{1}{34}$

$|L - a_n| = \frac{1}{8n + 2}$

12.

n	a_n	$\|L - a_n\|$
1	7	$7 - 3 = 4$
2	$\frac{25}{7}$	$\frac{25}{7} - 3 = \frac{4}{7}$
3	$\frac{55}{17}$	$\frac{55}{17} - 3 = \frac{4}{17}$
4	$\frac{97}{31}$	$\frac{97}{31} - 3 = \frac{4}{31}$

$|L - a_n| = \frac{4}{2n^2 - 1}$

13. $\frac{1}{n} < \frac{1}{10}$ when $n > 10$; if $n = 11$, $|L - a_n| = \frac{1}{11}$

14. $\frac{1}{2^n} < \frac{1}{10}$ when $n > 3$; if $n = 4$, $|L - a_n| = \frac{1}{16}$

15. $\frac{1}{2n} < \frac{1}{10}$ when $n > 5$; if $n = 6$, $|L - a_n| = \frac{1}{12}$

16. $\frac{1}{n^2} < \frac{1}{10}$ when $n > 3$; if $n = 4$, $|L - a_n| = \frac{1}{16}$

17. $\frac{1}{2(4n + 1)} < \frac{1}{10}$; $2(4n + 1) > 10$; $4n + 1 > 5$; $n > 1$; if $n = 2$, $|L - a_n| = \frac{1}{18}$

18. $\frac{4}{2n^2 - 1} < \frac{1}{10}$; $2n^2 - 1 > 40$; $2n^2 > 41$; $n^2 > 20\frac{1}{2}$; $n > 4$; if $n = 5$, $|L - a_n| = \frac{4}{49}$

19. $2 - \frac{1}{n} < a_n$ and $a_n < 2$. Thus, $2 - a_n < \frac{1}{n}$ and $2 - a_n > 0$. That is, $0 < 2 - a_n < \frac{1}{n}$, or $|2 - a_n| < \frac{1}{n}$. Since $\frac{1}{n}$ is a positive number that is as small as you wish by choosing n great enough, $\lim_{n\to\infty} a_n = 2$.

C **20.** Since $|0 - (L - a_n)|$ is less than some positive number for n as great as you wish, then the limit of $|a_n - L|$ is 0. Also $|k| \cdot 0 = 0$, so $\lim_{n\to\infty} |k| \cdot |a_n - L| = 0$. Furthermore, $|k| \cdot |a_n - L| = |ka_n - kL| = |b_n - kL|$. $\therefore$ $|b_n - kL|$ has limit 0 and the sequence $b_1, b_2, b_3, \ldots$ has limit kL.

Pages 327–328 • WRITTEN EXERCISES

A **1.** $r = \frac{1}{3}$; $S = \frac{36}{1 - \frac{1}{3}} = 54$ **2.** $r = -\frac{1}{5}$; $S = \frac{250}{1 + \frac{1}{5}} = \frac{625}{3} = 208\frac{1}{3}$

3. $r = \frac{5}{6}$; $S = \frac{6}{1 - \frac{5}{6}} = 36$ **4.** $r = -1$; divergent **5.** $r = -\frac{3}{4}$; $S = \frac{\frac{4}{3}}{1 + \frac{3}{4}} = \frac{16}{21}$

6. $r = \frac{1}{6}$; $S = \frac{\frac{8}{3}}{1 - \frac{1}{6}} = \frac{16}{5} = 3\frac{1}{5}$ **7.** $r = -2$; divergent

8. $r = \frac{3}{7}$; $S = \frac{14}{1 - \frac{3}{7}} = \frac{49}{2} = 24\frac{1}{2}$ **9.** $r = 0.1$; $S = \frac{0.8}{1 - 0.1} = 0.\overline{8}$ or $\frac{8}{9}$

10. $r = 0.01$; $S = \frac{0.6}{1 - 0.01} = 0.\overline{60}$ or $\frac{20}{33}$ **11.** $a_1 = 4$; $r = \frac{1}{3}$; $S = \frac{4}{1 - \frac{1}{3}} = 6$

12. $a_1 = 3$; $r = -\frac{2}{5}$; $S = \frac{3}{1 + \frac{2}{5}} = \frac{15}{7} = 2\frac{1}{7}$

13. $a_1 = \frac{5}{6}$; $r = -\frac{3}{4}$; $S = \frac{\frac{5}{6}}{1 + \frac{3}{4}} = \frac{10}{21}$ **14.** $a_1 = \frac{1}{10}$; $r = \frac{6}{5}$; divergent

15. $a_1 = (1 - r)S = \frac{2}{3} \cdot 15 = 10$ **16.** $a_1 = (1 - r)S = \frac{11}{6} \cdot 9 = \frac{33}{2} = 16\frac{1}{2}$

17. $a_1 = (1 - r)S = \frac{5}{4} \cdot 24 = 30$ **18.** $r = 1 - \frac{a_1}{S} = 1 - \frac{12}{16} = \frac{1}{4}$

19. $r = 1 - \frac{a_1}{S} = 1 - \frac{35}{30} = -\frac{1}{6}$ **20.** $r = 1 - \frac{a_1}{S} = 1 - \frac{a_1}{\frac{4}{3}a_1} = 1 - \frac{3}{4} = \frac{1}{4}$

21. $0.\overline{5} = 0.5 + 0.5(0.1) + 0.5(0.1)^2 + \cdots$; $S = \frac{0.5}{1 - 0.1} = \frac{5}{9}$

22. $0.\overline{36} = 0.36 + 0.36(0.01) + 0.36(0.01)^2 + \cdots$; $S = \frac{0.36}{1 - 0.01} = \frac{36}{99} = \frac{4}{11}$

23. $0.\overline{21} = 0.21 + 0.21(0.01) + 0.21(0.01)^2 + \cdots; S = \frac{0.21}{1 - 0.01} = \frac{21}{99} = \frac{7}{33}$

24. $0.\overline{162} = 0.162 + 0.162(0.001) + 0.162(0.001)^2 + \cdots; S = \frac{0.162}{1 - 0.001} = \frac{162}{999} = \frac{6}{37}$

25. $0.\overline{117} = 0.117 + 0.117(0.001) + 0.117(0.001)^2 + \cdots; S = \frac{0.117}{1 - 0.001} = \frac{117}{999} = \frac{13}{111}$

26. $0.\overline{108} = 0.108 + 0.108(0.001) + 0.108(0.001)^2 + \cdots; S = \frac{0.108}{1 - 0.001} = \frac{108}{999} = \frac{4}{37}$

27. $0.0\overline{3} = 0.03 + 0.03(0.1) + 0.03(0.1)^2 + \cdots; S = \frac{0.03}{1 - 0.1} = \frac{3}{90} = \frac{1}{30}$

28. $0.00\overline{72} = 0.0072 + 0.0072(0.01) + 0.0072(0.01)^2 + \cdots;$

$S = \frac{0.0072}{1 - 0.01} = \frac{72}{9900} = \frac{2}{275}$

B **29.** $\frac{2}{x} = \frac{3}{1 - x}; 2(1 - x) = 3x; 2 - 2x = 3x; x = \frac{2}{5}$

30. $9x = \frac{-10}{1 - x}; 9x(1 - x) = -10; 9x^2 - 9x - 10 = 0; (3x + 2)(3x - 5) = 0;$

$x = -\frac{2}{3}$ or $x = \frac{5}{3}$; but $|x| < 1$, so $x = -\frac{2}{3}$

31. $a_1 = x; r = x; S = 5; 5 = \frac{x}{1 - x}; 5 - 5x = x; x = \frac{5}{6}$

32. $a_1 = x; r = x^2; S = \frac{4}{15}; \frac{4}{15} = \frac{x}{1 - x^2}; 4(1 - x^2) = 15x; 4x^2 + 15x - 4 = 0;$

$(4x - 1)(x + 4) = 0; x = \frac{1}{4}$ or $x = -4$; but $|x^2| < 1$, so $x = \frac{1}{4}$

C **33.** $a_1 = 5; r = 3x^2; S = \frac{5}{2x}; \frac{5}{2x} = \frac{5}{1 - 3x^2}; 5(1 - 3x^2) = 5(2x); 15x^2 + 10x - 5 = 0;$

$5(3x - 1)(x + 1) = 0; x = \frac{1}{3}$ or $x = -1$; but $|3x^2| < 1$, so $x = \frac{1}{3}$

34. Sum of first series $= \frac{a_1}{1 - r}$; sum of second series $= \frac{2a_1}{1 - s}$. Since the sums are equal, $\frac{a_1}{1 - r} = \frac{2a_1}{1 - s}$, or $a_1(1 - s) = 2a_1(1 - r)$. If $a_1 \neq 0$, $1 - s = 2 - 2r$, or $s = 2r - 1$. Example: $a_1 = 1$, $r = \frac{1}{4}$, and $s = 2\left(\frac{1}{4}\right) - 1 = -\frac{1}{2}$. The series are $1 + \frac{1}{4} + \frac{1}{16} + \cdots$ and $2 - 1 + \frac{1}{2} + \cdots; S = \frac{4}{3}$.

Pages 328–330 • PROBLEMS

A **1.** $a_1 = 16; r = 0.9; S = \frac{16}{1 - 0.9} = 160$; 160 cm

2. $a_1 = 3(4 \cdot 30); a_2 = 3(4 \cdot 15); a_3 = 3(4 \cdot 75); \cdots; a_1 = 360; r = \frac{1}{2};$

$S = \frac{360}{1 - \frac{1}{2}} = 720$; 720 cm

3. $a_1 = 2.4; r = 0.98; S = \frac{2.4}{1 - 0.98} = 120$; 120 mm

4. Downward: $a_1 = 120$; $r = \frac{3}{8}$; $S = \frac{120}{1 - \frac{3}{8}} = 192$; upward: $a_1 = 45$; $r = \frac{3}{8}$;

$S = \frac{45}{1 - \frac{3}{8}} = 72$; $192 + 72 = 264$; 264 cm

B 5. Let r = radius of original circle. Then $\frac{1}{3}r$ = radius of each of first 6 circles, $\frac{1}{9}r$ = radius of each of second 6 circles, and so on. The sum of their areas is:

$6\pi\left(\frac{1}{3}r\right)^2 + 6\pi\left(\frac{1}{9}r\right)^2 + 6\pi\left(\frac{1}{27}r\right)^2 + \cdots = 6\pi r^2\left(\frac{1}{9} + \frac{1}{81} + \frac{1}{729} + \cdots\right)$. Let $a_1 = \frac{1}{9}$

and $r = \frac{1}{9}$, then $S = \frac{\frac{1}{9}}{1 - \frac{1}{9}} = \frac{1}{8}$, so the sum of their areas $= 6\pi r^2 \cdot \frac{1}{8} = \frac{3}{4}\pi r^2$;

$\frac{3}{4}$ of the original circle.

6. $a_1 = 4 \cdot 7$, $a_2 = 4 \cdot 5$, $r = \frac{5}{7}$, $S = \frac{4 \cdot 7}{1 - \frac{5}{7}} = 98$; 98

C 7. In the first operation $\frac{1}{4}$ of the area is removed, and $\frac{3}{4}$ of the area remains.

In the second operation, $\frac{1}{4}\left(\frac{3}{4}\right) = \frac{3}{16}$ of the remaining area is removed, and $1 - \left(\frac{1}{4} + \frac{3}{16}\right) = \frac{9}{16}$ remains. The amount removed is $\frac{1}{4} + \frac{3}{16} + \cdots$, where

$a_1 = \frac{1}{4}$ and $r = \frac{3}{4}$, so $S = \frac{\frac{1}{4}}{1 - \frac{3}{4}} = 1$. $\therefore$ the entire area is removed; none remains.

8. First the bee travels 60 km while the car travels $\frac{1}{2} \cdot 60 = 30$ km. Then the bee reverses direction until it meets the car. Let d = distance car travels after bee gets to P; $2d$ = distance bee travels in the reverse direction. Thus $2d + d = 30$, $d = 10$. Car travels 10 km (for a total of 40 km) while bee travels $2(10) = 20$ km back to car. Next the bee travels 20 km to P while the car travels 10 km (total: 50 km). Then the bee travels $\frac{2}{3} \cdot 10 = \frac{20}{3}$ km and so on. The total distance the bee travels can be expressed as the sum of two infinite series, one that expresses the distance traveled in the original direction $(60 + 20 + \cdots)$, and the other the distance traveled in the reverse direction $\left(20 + \frac{20}{3} + \cdots\right)$. $S_1 = \frac{60}{1 - \frac{1}{3}} = 90$;

$S_2 = \frac{20}{1 - \frac{1}{3}} = 30$; $S_1 + S_2 = 120$ km. An easier way to solve the problem: The car travels 60 km at 40 km/h in 1.5 h. In 1.5 h, the bee travels 80(1.5) = 120 km.

9. Area of original triangle $= s^2\frac{\sqrt{3}}{4} = \frac{\sqrt{3}}{4}$. Area after first trisection $=$

$3\left(\frac{1}{3}\right)^2\frac{\sqrt{3}}{4} + \frac{\sqrt{3}}{4}$. Area after second trisection $= 12\left(\frac{1}{9}\right)^2\frac{\sqrt{3}}{4} + 3\left(\frac{1}{3}\right)^2\frac{\sqrt{3}}{4} + \frac{\sqrt{3}}{4}$.

Area enclosed by curve $= \frac{\sqrt{3}}{4} + 3\left(\frac{1}{3}\right)^2\frac{\sqrt{3}}{4} + 12\left(\frac{1}{9}\right)^2\frac{\sqrt{3}}{4} + 48\left(\frac{1}{27}\right)^2\frac{\sqrt{3}}{4} + \cdots =$

$$\frac{\sqrt{3}}{4}\left[1 + \frac{1}{3} + \frac{4}{27} + \frac{16}{243} + \cdots\right] = \frac{\sqrt{3}}{4}\left[1 + \left(\frac{1}{3} + \frac{1}{3}\left(\frac{4}{9}\right) + \frac{1}{3}\left(\frac{4}{9}\right)^2 + \cdots\right)\right] =$$

$$\frac{\sqrt{3}}{4}\left[1 + \frac{\frac{1}{3}}{1 - \frac{4}{9}}\right] = \frac{\sqrt{3}}{4}\left[1 + \frac{3}{5}\right] = \frac{\sqrt{3}}{4}\left(\frac{8}{5}\right) = \frac{2\sqrt{3}}{5} \text{ square units}$$

10. Before the first trisection, there are 3 sides each of length 1; $P = 3$. After the first trisection, there are 12 sides each of length $\frac{1}{3}$; $P = 12 \cdot \frac{1}{3} = \frac{4}{3} \cdot 3$.

After the second trisection, there are 48 sides each of length $\frac{1}{9}$;

$P = 48 \cdot \frac{1}{9} = \frac{16}{9} \cdot 3 = \left(\frac{4}{3}\right)^2 \cdot 3$. The perimeters after each trisection form a geometric sequence $3, 3\left(\frac{4}{3}\right), 3\left(\frac{4}{3}\right)^2, \ldots$. Since $r = \frac{4}{3} > 1$, a_n increases without bound as n grows larger.

Page 330 • COMPUTER EXERCISES

1.

```
10  INPUT A,R
20  IF ABS(R) >= 1 THEN PRINT "ABS. VALUE OF R >= 1" : END
30  LET S = A/(1 - R)
40  PRINT S
50  END
```

2. $0.\overline{3}$ **3.** $|r| \geq 1$ **4.** $33.\overline{3}$

Page 331 • SELF-TEST 3

1. $a_1 = 3$; $a_2 = 2$; $a_3 = \frac{5}{3}$; $a_4 = \frac{3}{2}$; $L = 1$; $|L - a_n| = \frac{2}{n}$

2. $a_1 = 2\frac{2}{3}$; $a_2 = 2\frac{8}{9}$; $a_3 = 2\frac{26}{27}$; $a_4 = 2\frac{80}{81}$; $L = 3$; $|L - a_n| = \frac{1}{3^n}$

3. $a_1 = 9$; $r = -\frac{1}{3}$; $S = \frac{9}{1 + \frac{1}{3}} = \frac{27}{4}$ **4.** $a_1 = \frac{2}{5}$; $r = \frac{1}{2}$; $S = \frac{\frac{2}{5}}{1 - \frac{1}{2}} = \frac{4}{5}$

5. $0.\overline{213} = 0.213 + 0.213(0.001) + 0.213(0.001)^2 + \cdots$; $a_1 = 0.213$; $r = 0.001$;

$S = \frac{0.213}{1 - 0.001} = \frac{213}{999} = \frac{71}{333}$

Pages 332–333 • CHAPTER REVIEW

1. $a_1 = 8$; $a_2 = 4$; $a_3 = 0$; since $a_3 - a_2 = a_2 - a_1 = -4$, the sequence is arithmetic with $d = -4$; $a_{n+1} = a_n - 4$; b
2. $a_{24} = a_1 + 23(-1) = -18$; $a_1 = 5$; d
3. $$\begin{array}{r} a_1 + 3d = -2 \\ \underline{a_1 + 8d = 13} \\ -5d = -15 \end{array}$$ $d = 3$; $a_1 + 3(3) = -2$; $a_1 = -11$; $a_7 = -11 + 6(3) = 7$; a
4. $a_4 = 12 + 3(d) = -6$; $d = -6$; $S_{10} = \frac{10}{2}[2(12) + 9(-6)] = 5[-30] = -150$; c
5. $\sum_{i=1}^{6}(13 + 2i) = 15 + 17 + 19 + 21 + 23 + 25$; $S_6 = \frac{6}{2}(15 + 25) = \frac{6}{2}(40) = 120$; d
6. $a_1 = 100$; $a_2 = 50$; $r = \frac{50}{100} = \frac{1}{2}$; $a_8 = 100\left(\frac{1}{2}\right)^7 = \frac{25}{32}$; b
7. $81r^3 = -3$; $r^3 = -\frac{1}{27}$; $r = -\frac{1}{3}$; $a_2 = -27$; $a_3 = 9$; c
8. $a_1r = \frac{3}{4} \longrightarrow a_1 = \frac{3}{4r}$; $a_1r^6 = -\frac{3}{128}$; $\left(\frac{3}{4r}\right)r^6 = -\frac{3}{128}$; $r^5 = -\frac{1}{32}$; $r = -\frac{1}{2}$; $a_1\left(-\frac{1}{2}\right) = \frac{3}{4}$; $a_1 = -\frac{3}{2}$; $a_4 = a_1r^3 = -\frac{3}{2}\left(-\frac{1}{2}\right)^3 = \frac{3}{16}$; a
9. $S_5 = \dfrac{-2 - (-2)\left(-\frac{1}{3}\right)^5}{1 - \left(-\frac{1}{3}\right)} = \dfrac{-2 - (-2)\left(-\frac{1}{243}\right)}{\frac{4}{3}} = -\frac{488}{243} \cdot \frac{3}{4} = -\frac{122}{81}$; c
10. $\sum_{n=1}^{4} 100\left(\frac{1}{5}\right)^{n-1} = 100 + 20 + 4 + \frac{4}{5} = 124\frac{4}{5}$; b
11. $\lim_{n\to\infty}\left(\frac{2n+1}{n}\right) = \lim_{n\to\infty}\left(2 + \frac{1}{n}\right)$. For large values of n, $\frac{1}{n}$ is close to 0. Thus, $\lim_{n\to\infty}\left(\frac{2n+1}{n}\right) = 2$; c
12. $a_1 = 180$; $r = -\frac{1}{3}$; $S = \dfrac{180}{\frac{4}{3}} = 135$; a
13. $a_1 = 2$; $r = \frac{1}{4}$; $S = \dfrac{2}{1 - \frac{1}{4}} = \frac{8}{3}$; d

Pages 333–334 • CHAPTER TEST

1. $a_1 = -1$; $a_2 = -1 + \frac{1}{4} = -\frac{3}{4}$; $a_3 = -\frac{3}{4} + \frac{1}{4} = -\frac{1}{2}$; $a_4 = -\frac{1}{2} + \frac{1}{4} = -\frac{1}{4}$
2. $a_n = 7 - 2n$; $a_1 = 5$, $a_{n+1} = a_n - 2$
3. $$\begin{array}{r} a_1 + 4d = 5 \\ \underline{a_1 + 7d = -13} \\ -3d = 18 \end{array}$$ $d = -6$; $a_1 + 4(-6) = 5$; $a_1 = 29$
4. $a_1 = 10$; $d = 15$; $100 = 10 + (n - 1)15$; $105 = 15n$; $n = 7$; $S_7 = \frac{7}{2}(10 + 100) = 385$
5. $27 = \frac{6}{2}[2(2) + 5d]$; $9 = 4 + 5d$; $d = 1$
6. $a_{10} = -120 \cdot \left(-\frac{1}{2}\right)^9 = \frac{120}{512} = \frac{15}{64}$
7. $a_1 = 25$; $r = 0.8$; $a_5 = 25(0.8)^4 = \$10.24$
8. $\frac{3}{x} = \frac{x}{48}$; $x^2 = 144$; $x = 12$

9. $S_6 = \dfrac{81 - 81\left(\frac{1}{3}\right)^6}{1 - \frac{1}{3}} = \dfrac{728}{6} = 121\frac{1}{3}$

10. $|L - a_n|$: $\dfrac{1}{2} - \left(\dfrac{n^2 - n}{2n^2}\right) = \dfrac{n^2 - (n^2 - n)}{2n^2} = \dfrac{n}{2n^2} = \dfrac{1}{2n}$

11. $0.\overline{24} = 0.24 + 0.24(0.01) + 0.24(0.01)^2 + \cdots$; $S = \dfrac{0.24}{1 - 0.01} = \dfrac{24}{99} = \dfrac{8}{33}$

Pages 334–336 • CUMULATIVE REVIEW

1.

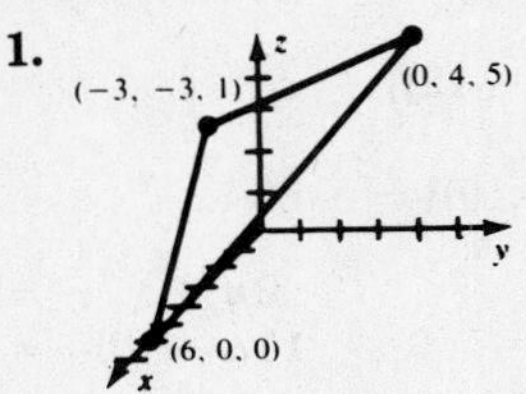

2.

3. Add twice column 1 to column 2; expand by minors of the first row.

$$D = \begin{vmatrix} 1 & -2 & 3 \\ 2 & 5 & -1 \\ 3 & 3 & 2 \end{vmatrix} = \begin{vmatrix} 1 & 0 & 3 \\ 2 & 9 & -1 \\ 3 & 9 & 2 \end{vmatrix} = 1\begin{vmatrix} 9 & -1 \\ 9 & 2 \end{vmatrix} - 0 + 3\begin{vmatrix} 2 & 9 \\ 3 & 9 \end{vmatrix} =$$

$(18 + 9) + 3(18 - 27) = 0$; solution is $\emptyset$.

4. Add column 3 to column 2; expand by minors of the third row.

$$D = \begin{vmatrix} -1 & 2 & 3 \\ 2 & 7 & 5 \\ 3 & 4 & -4 \end{vmatrix} = \begin{vmatrix} -1 & 5 & 3 \\ 2 & 12 & 5 \\ 3 & 0 & -4 \end{vmatrix} = 3\begin{vmatrix} 5 & 3 \\ 12 & 5 \end{vmatrix} + (-4)\begin{vmatrix} -1 & 5 \\ 2 & 12 \end{vmatrix} + 0 =$$

$3(25 - 36) + (-4)(-12 - 10) = -33 + 88 = 55.$

Add column 2 to column 3; subtract column 1 from column 2; expand by minors of the third row.

$$D_x = \begin{vmatrix} -12 & 2 & 3 \\ -2 & 7 & 5 \\ 4 & 4 & -4 \end{vmatrix} = \begin{vmatrix} -12 & 2 & 5 \\ -2 & 7 & 12 \\ 4 & 4 & 0 \end{vmatrix} = \begin{vmatrix} -12 & 14 & 5 \\ -2 & 9 & 12 \\ 4 & 0 & 0 \end{vmatrix} = 4\begin{vmatrix} 14 & 5 \\ 9 & 12 \end{vmatrix} = 492.$$

Add column 2 to column 3; expand by minors of the third row.

$$D_y = \begin{vmatrix} -1 & -12 & 3 \\ 2 & -2 & 5 \\ 3 & 4 & -4 \end{vmatrix} = \begin{vmatrix} -1 & -12 & -9 \\ 2 & -2 & 3 \\ 3 & 4 & 0 \end{vmatrix} = 3\begin{vmatrix} -12 & -9 \\ -2 & 3 \end{vmatrix} - 4\begin{vmatrix} -1 & -9 \\ 2 & 3 \end{vmatrix} =$$

$3(-36 - 18) - 4(-3 + 18) = -162 - 60 = -222.$

Subtract column 3 from column 2; expand by minors of the third row.

$$D_z = \begin{vmatrix} -1 & 2 & -12 \\ 2 & 7 & -2 \\ 3 & 4 & 4 \end{vmatrix} = \begin{vmatrix} -1 & 14 & -12 \\ 2 & 9 & -2 \\ 3 & 0 & 4 \end{vmatrix} = 3\begin{vmatrix} 14 & -12 \\ 9 & -2 \end{vmatrix} + 4\begin{vmatrix} -1 & 14 \\ 2 & 9 \end{vmatrix} =$$

$3(-28 + 108) + 4(-9 - 28) = 240 - 148 = 92$; $\left\{\left(\dfrac{492}{55}, -\dfrac{222}{55}, \dfrac{92}{55}\right)\right\}$

5. Expand by minors of the fourth row.

$$\begin{vmatrix} 3 & 2 & 1 & 1 \\ 2 & 1 & 2 & 5 \\ 4 & -3 & -1 & 2 \\ 0 & 3 & 0 & 1 \end{vmatrix} = 3\begin{vmatrix} -3 & 1 & 1 \\ 2 & 2 & 5 \\ 4 & -1 & 2 \end{vmatrix} + 1\begin{vmatrix} -3 & 2 & 1 \\ 2 & 1 & 2 \\ 4 & -3 & -1 \end{vmatrix}$$

In first matrix, subtract column 2 from column 3; in second matrix, add row 3 to row 1.

$$3\begin{vmatrix} -3 & 1 & 0 \\ 2 & 2 & 3 \\ 4 & -1 & 3 \end{vmatrix} + 1\begin{vmatrix} 1 & -1 & 0 \\ 2 & 1 & 2 \\ 4 & -3 & -1 \end{vmatrix} = 3\left[-3\begin{vmatrix} 2 & 3 \\ -1 & 3 \end{vmatrix} - 1\begin{vmatrix} 2 & 3 \\ 4 & 3 \end{vmatrix}\right] +$$

$$1\left[1\begin{vmatrix} 1 & 2 \\ -3 & -1 \end{vmatrix} + 1\begin{vmatrix} 2 & 2 \\ 4 & -1 \end{vmatrix}\right] = 3[-3(6+3) - 1(6-12)] +$$

$$1[1(-1+6) + 1(-2-8)] = 3[-27+6] + [5 + (-10)] = -68$$

6. $n + d + q = 46$ $\quad n + (n+5) + q = 46$ $\quad 2n + q = 41$

$5n + 10d + 25q = 550$ $\quad 5n + 10(n+5) + 25q = 550$ $\quad 15n + 25q = 500$

$d = n + 5$

$$\begin{aligned} -50n - 25q &= -1025 \\ 15n + 25q &= 500 \\ \hline -35n &= -525 \end{aligned}$$

$n = 15$; $d = 20$; $q = 11$; 15 nickels, 20 dimes, 11 quarters

7. $(-r)(rs^2)(r^2s)^3 = (-r)(rs^2)(r^6s^3) = -r^8s^5$

8. $(5m - 3)(3m - 5) = 15m^2 - 34m + 15$

9. $\dfrac{(-2x^2)^3(7x)^2}{(4x)^2(7x)(x^2)} = \dfrac{(-8x^6)(49x^2)}{(16x^2)(7x)(x^2)} = -\dfrac{7}{2}x^3$

10. $\dfrac{(c+6)(c-1)}{(c+4)(c+3)} \cdot \dfrac{(c-5)(c+4)}{(c-5)(c-1)} \cdot \dfrac{(c+3)(c-3)}{(c+6)(c-6)} = \dfrac{c-3}{c-6}$

11. $\dfrac{12w^2 + 4w - 18}{2w + 3}$

12. $\dfrac{11r^3s^3}{11r^2s^2} - \dfrac{33rs}{11r^2s^2} + \dfrac{66r^2s^2}{11r^2s^2} = \dfrac{r^2s^2 - 3 + 6rs}{rs} = \dfrac{r^2s^2 + 6rs - 3}{rs}$

13. $\dfrac{9(2-a)}{4-a^2} + \dfrac{7(2+a)}{4-a^2} - \dfrac{3a-1}{4-a^2} = \dfrac{18 - 9a + 14 + 7a - 3a + 1}{4-a^2} = \dfrac{33 - 5a}{(2+a)(2-a)}$

14. $\dfrac{\frac{1}{x} + \frac{1}{y}}{\frac{x}{y} - \frac{y}{x}} \cdot \dfrac{xy}{xy} = \dfrac{y + x}{x^2 - y^2} = \dfrac{x + y}{(x+y)(x-y)} = \dfrac{1}{x - y}$

15. $2x^2 + 5x + 3 = 0$; $(2x + 3)(x + 1) = 0$; $\left\{-\dfrac{3}{2}, -1\right\}$

16. $y^2 - 3y + 2 = 0$; $(y - 2)(y - 1) = 0$; $\{1, 2\}$

17. $\dfrac{d-5}{d^2-25} = \dfrac{-17}{d^2-25} + \dfrac{d(d+5)}{d^2-25}$; $d \neq \pm 5$; $d - 5 = -17 + d^2 + 5d$;
$d^2 + 4d - 12 = 0$; $(d + 6)(d - 2) = 0$; $\{-6, 2\}$

18. $4y^2 - 3y - 1 > 0$; $(4y + 1)(y - 1) > 0$; $4y + 1 > 0$ and $y - 1 > 0$ or $4y + 1 < 0$ and $y - 1 < 0$; $y > -\frac{1}{4}$ and $y > 1$ or $y < -\frac{1}{4}$ and $y < 1$; $y > 1$ or $y < -\frac{1}{4}$; $\{y\colon y < -\frac{1}{4} \text{ or } y > 1\}$

19. $l = 2w - 5$, $(2w - 5)w = 187$; $2w^2 - 5w - 187 = 0$; $(2w + 17)(w - 11) = 0$; $w = 11$; $l = 17$; 11 cm by 17 cm

20. $y = kx^3$; $28 = k(2^3)$; $k = \frac{28}{8} = \frac{7}{2}$

21. $p \in \{\pm 1, \pm 2, \pm 3, \pm 6\}$; $q \in \{\pm 1\}$; possible rational roots are $\pm 1, \pm 2, \pm 3, \pm 6$; $f(1) = -10$; $f(-1) = -4$; $f(2) = -10$; $f(-2) = -10$; $f(3) = 0$; $f(-3) = -30$; $f(6) = 150$; $f(-6) = -234$; root: 3

22.
$$\begin{aligned} 100N &= 37.\overline{37} \\ N &= 0.\overline{37} \\ \hline 99N &= 37 \\ N &= \frac{37}{99} \end{aligned}$$

23. $-\frac{1}{3} + \frac{9}{2} = \frac{-2 + 27}{6} = \frac{25}{6}$

24. $\sqrt{cd^2} - d\sqrt{c} + \sqrt{16cd^2} = |d|\sqrt{c} - d\sqrt{c} + 4|d|\sqrt{c} = 5|d|\sqrt{c} - d\sqrt{c}$

25. $\frac{2}{1 - \sqrt{3}} \cdot \frac{1 + \sqrt{3}}{1 + \sqrt{3}} = \frac{2(1 + \sqrt{3})}{1 - 3} = -1 - \sqrt{3}$

26. $(\sqrt{10} - 4)(2\sqrt{10} + 7) = 2 \cdot 10 + 7\sqrt{10} - 8\sqrt{10} - 4 \cdot 7 = -8 - \sqrt{10}$

27. $81x^4 - 16 = 0$; $x^4 = \frac{16}{81}$; $x = \pm\frac{2}{3}$; $\left\{\frac{2}{3}, -\frac{2}{3}\right\}$

28. $2n^3 + 16 = 0$; $n^3 = -\frac{16}{2}$; $n^3 = -8$; $n = -2$; $\{-2\}$

29. $b - 3 = \sqrt{6b + 9}$; $b^2 - 6b + 9 = 6b + 9$; $b^2 - 12b = 0$; $b(b - 12) = 0$; $b = 0$ or $b = 12$ *Check:* $0 - 3 - \sqrt{6(0) + 9} \stackrel{?}{=} 0$, $-3 - 3 \neq 0$; $12 - 3 - \sqrt{6(12) + 9} \stackrel{?}{=} 0$, $9 - 9 = 0$; $\{12\}$

30. $\sqrt{v + 5} = 5 - \sqrt{8 - v}$; $v + 5 = 25 - 10\sqrt{8 - v} + (8 - v)$; $2v - 28 = -10\sqrt{8 - v}$; $v - 14 = -5\sqrt{8 - v}$; $v^2 - 28v + 196 = 25(8 - v)$; $v^2 - 3v - 4 = 0$; $(v - 4)(v + 1) = 0$; $v = 4$ or $v = -1$
Check: $\sqrt{4 + 5} + \sqrt{8 - 4} \stackrel{?}{=} 5$, $3 + 2 = 5$; $\sqrt{-1 + 5} + \sqrt{8 + 1} \stackrel{?}{=} 5$, $2 + 3 = 5$; $\{4, -1\}$

31. $8 - 10i$ **32.** $3 - 14i$

33. $(7 - i)(-6 + 2i) = (-42 + 2) + (14 + 6)i = -40 + 20i$

34. $\frac{2 - 3i}{7 + 4i} \cdot \frac{7 - 4i}{7 - 4i} = \frac{(14 - 12) + (-8 - 21)i}{49 + 16} = \frac{2 - 29i}{65} = \frac{2}{65} - \frac{29}{65}i$

35. $-1, 2, -4, 8$ **36.** $3, -1, -5, -9$

37. $5 = 12 + (n - 1)\left(-\frac{1}{3}\right)$; $15 = 36 + (1 - n)$; $n = 22$

38.
$$\begin{aligned} a_1 + 9d &= 7 \\ a_1 + 13d &= 15 \\ \hline -4d &= -8 \end{aligned}$$
$d = 2$; $a_1 + 9(2) = 7$; $a_1 = -11$

39. $345 = \frac{n}{2}(2 + 44)$; $345 = n(23)$; $n = 15$

40. $10 - (x + 3) = (2x + 5) - 10$; $7 - x = 2x - 5$; $3x = 12$; $x = 4$

41. $a_4 = 32\left(\frac{1}{2}\right)^3 = \frac{32}{8} = 4$ **42.** $-\frac{2}{9} = 54(r)^5$; $-\frac{1}{243} = r^5$; $r = -\frac{1}{3}$

43. $S_5 = \dfrac{36 - 36\left(-\frac{1}{3}\right)^5}{1 - \left(-\frac{1}{3}\right)} = \dfrac{36 + \frac{36}{243}}{\frac{4}{3}} = \dfrac{\frac{8784}{243}}{\frac{4}{3}} = \dfrac{2196}{81} = 27\frac{1}{9}$

44. $S_6 = \dfrac{a_1 - a_1\left(\frac{3}{2}\right)^6}{1 - \frac{3}{2}} = 498\frac{3}{4}$; $\dfrac{a_1\left(1 - \frac{729}{64}\right)}{-\frac{1}{2}} = \dfrac{1995}{4}$; $\dfrac{a_1\left(-\frac{665}{64}\right)}{-\frac{1}{2}} = \dfrac{1995}{4}$; $a_1 = 24$

45. From a height of 128 ft, the first rebound is 64 ft. Thus, $a_1 = 64$;
$a_9 = 64\left(\frac{1}{2}\right)^8 = \frac{1}{4}$; $\frac{1}{4}$ ft

Page 337 • APPLICATION

1. a. $\frac{1}{125} = 2 \cdot \frac{1}{250}$; exposure is doubled. **b.** $22 \approx \sqrt{2} \cdot 16$; exposure is halved.
c. The two changes compensate for each other. Thus exposure is not changed.

2. $\dfrac{\frac{1}{500}}{x} = \dfrac{x}{\frac{1}{30}}$; $x^2 = \dfrac{1}{15{,}000}$; $x \approx \pm\dfrac{1}{122.5}$; $\dfrac{1}{122.5} \approx \dfrac{1}{125}$

Pages 338–339 • PROGRAMMING IN PASCAL

1. See answers to Exercises 21–26 on page 168 of this solution key.

2. a.
```
PROGRAM arith_sq (INPUT,OUTPUT);

VAR
  first_term,difference : real;
  num : integer;

{************************************************************}
PROCEDURE display_result;

BEGIN
  write('The value of the term number ',num:1,' is: ');
  writeln(first_term + (num - 1) * difference:1:4);
END;

{************************************************************}
PROCEDURE get_first_term_and_difference;

BEGIN
  write('Enter the value of the first term: ');
  readln(first_term);
```

```
    write('Enter the common difference: ');
    readln(difference);
  END;

  {**********************************************************}
  BEGIN  {* main *}
    write('Which term is to be found: ');
    readln(num);
    get_first_term_and_difference;
    display_result;
  END.
```

b.

```
  PROGRAM geo_sq (INPUT,OUTPUT);

  VAR
    first_term,ratio : real;
    num : integer;

  {**********************************************************}
  FUNCTION power (base : real; degree : integer) : real;

  BEGIN
    IF degree >= 0
       THEN IF degree = 0
               THEN power := 1
               ELSE power := power(base,degree - 1) * base
       ELSE BEGIN
              degree := -degree;
              power := power(1/base,degree);
            END;
  END;

  {**********************************************************}
  PROCEDURE display_result;

  BEGIN
    write('The value of term number ',num:1,' is: ');
    writeln(first_term * power(ratio,num - 1):1:4);
  END;

  {**********************************************************}
  PROCEDURE get_first_term_and_ratio;

  BEGIN
    write('Enter the value of the first term: ');
    readln(first_term);
    write('Enter the common ratio: ');
    readln(ratio);
  END;
```

[Program continued on next page]

```
  {*************************************************************}
  BEGIN {* main *}
    write('Which term is to be found: ');
    readln(num);
    get_first_term_and_ratio;
    display_result;
  END.
```

3. PROGRAM sequence (INPUT,OUTPUT);

```
VAR
  first_term,difference,ratio : real;
  num : integer;
  ans : char;

{*************************************************************}
FUNCTION power (base : real; degree : integer) : real;
BEGIN
  IF degree >= 0
     THEN IF degree = 0
             THEN power := 1
             ELSE power := power(base,degree - 1) * base
     ELSE BEGIN
            degree := -degree;
            power := power(1/base,degree);
          END;
END;

{*************************************************************}
PROCEDURE display_result;

BEGIN
  write('The value of term number ',num:1,' is: ');
  IF ans IN ['a','A']
     THEN writeln(first_term + (num - 1) * difference:1:4)
     ELSE writeln(first_term * power(ratio, num - 1):1:4);
END;

{*************************************************************}
PROCEDURE get_values;

BEGIN
  write('Enter the value of the first term: ');
  readln(first_term);
  IF ans IN ['a','A']
     THEN BEGIN
            write('Enter the common difference: ');
            readln(difference);
          END
```

```
      ELSE BEGIN
             write('Enter the common ratio: ');
             readln(ratio);
           END;
  END;

  {*************************************************************}
  BEGIN {* main *}
    write('Which term is to be found: ');
    readln(num);
    REPEAT
      write('What kind of sequence is used, <A>rithmetic or ');
      write('<G>eometric: ');
      readln(ans);
    UNTIL ans IN ['a','A','g','G'];
    get_values;
    display_result;
  END.
```

4.
```
PROGRAM sum (INPUT,OUTPUT);

VAR
  sum, temp_sum : real;
  done : boolean;
  num, term : integer;

BEGIN
  sum := 0;
  done := FALSE;
  term := 1;
  write('How many terms are to be summed: ');
  readln(num);
  WHILE (NOT done) AND (term <= num) DO
    BEGIN
      temp_sum := sum;
      sum := sum + (1/(term * term * term));
      IF abs(temp_sum - sum) < 0.00005
         THEN done := TRUE;
      term := term + 1;
    END;
   writeln('The sum is: ',sum:2:9);
END.
```

5.
```
PROGRAM terms (INPUT, OUTPUT);

VAR
  i, place_on_line : integer;
```

[*Program continued on next page*]

```
{***********************************************************}
FUNCTION term (n : integer) : real;

BEGIN
  term := (n * n + 1)/(2 * n * n);
END;

{***********************************************************}
BEGIN {* main *}
  FOR i := 1 TO 50 DO
      BEGIN
        FOR place_on_line := 1 TO 5 DO
            write(term(i):2:4,'  ');
        writeln;
      END;
END.
```

For Exercise 3 on page 323, the function *term* should be:

```
FUNCTION term (n : integer) : real;

BEGIN
  term := (n * n)/20;
END;
```

For Exercise 6 on page 323, the function *term* should be:

```
FUNCTION term (n : integer) : real;

BEGIN
  term := 5 * (1 - 1/n);
END;
```

CHAPTER 9 • Polynomial Functions

Pages 345–346 • WRITTEN EXERCISES

A **1.** $x^2 - 10x + 25 = 2 + 25$; $(x - 5)^2 = 27$; $x - 5 = \pm 3\sqrt{3}$; $x = 5 \pm 3\sqrt{3}$; $\{5 + 3\sqrt{3}, 5 - 3\sqrt{3}\}$

2. $x^2 + 12x + 36 = -4 + 36$; $(x + 6)^2 = 32$; $x + 6 = \pm 4\sqrt{2}$; $x = -6 \pm 4\sqrt{2}$; $\{-6 + 4\sqrt{2}, -6 - 4\sqrt{2}\}$

3. $y^2 - 5y + \frac{25}{4} = -\frac{28}{4} + \frac{25}{4}$; $\left(y - \frac{5}{2}\right)^2 = -\frac{3}{4}$; $y - \frac{5}{2} = \pm\frac{i\sqrt{3}}{2}$; $y = \frac{5 \pm i\sqrt{3}}{2}$; $\left\{\frac{5 + i\sqrt{3}}{2}, \frac{5 - i\sqrt{3}}{2}\right\}$

4. $t^2 - \frac{4}{3}t + \frac{4}{9} = -\frac{9}{9} + \frac{4}{9}$; $\left(t - \frac{2}{3}\right)^2 = -\frac{5}{9}$; $t - \frac{2}{3} = \pm\frac{i\sqrt{5}}{3}$; $t = \frac{2 \pm i\sqrt{5}}{3}$; $\left\{\frac{2 + i\sqrt{5}}{3}, \frac{2 - i\sqrt{5}}{3}\right\}$

5. $r^2 - \frac{2}{5}r + \frac{1}{25} = \frac{5}{25} + \frac{1}{25}$; $\left(r - \frac{1}{5}\right)^2 = \frac{6}{25}$; $r - \frac{1}{5} = \pm\frac{\sqrt{6}}{5}$; $r = \frac{1 \pm \sqrt{6}}{5}$; $\left\{\frac{1 + \sqrt{6}}{5}, \frac{1 - \sqrt{6}}{5}\right\}$

6. $n^2 - \frac{5}{2}n + \frac{25}{16} = -\frac{20}{16} + \frac{25}{16}$; $\left(n - \frac{5}{4}\right)^2 = \frac{5}{16}$; $n - \frac{5}{4} = \frac{\pm\sqrt{5}}{4}$; $n = \frac{5 \pm \sqrt{5}}{4}$; $\left\{\frac{5 + \sqrt{5}}{4}, \frac{5 - \sqrt{5}}{4}\right\}$

7. $2u^2 - 8u + 9 = 0$; $u^2 - 4u = -\frac{9}{2}$; $u^2 - 4u + 4 = 4 - \frac{9}{2}$; $(u - 2)^2 = -\frac{1}{2}$; $u - 2 = \pm\frac{i\sqrt{2}}{2}$; $u = 2 \pm \frac{i\sqrt{2}}{2}$; $\left\{\frac{4 + i\sqrt{2}}{2}, \frac{4 - i\sqrt{2}}{2}\right\}$

8. $8v^2 + 4v = 3$; $v^2 + \frac{v}{2} = \frac{3}{8}$; $v^2 + \frac{v}{2} + \frac{1}{16} = \frac{6}{16} + \frac{1}{16}$; $\left(v + \frac{1}{4}\right)^2 = \frac{7}{16}$; $v + \frac{1}{4} = \frac{\pm\sqrt{7}}{4}$; $v = \frac{-1 \pm \sqrt{7}}{4}$; $\left\{\frac{-1 + \sqrt{7}}{4}, \frac{-1 - \sqrt{7}}{4}\right\}$

9. $3y^2 + 16y = -8$; $y^2 + \frac{16}{3}y = -\frac{8}{3}$; $y^2 + \frac{16}{3}y + \frac{64}{9} = -\frac{24}{9} + \frac{64}{9}$; $\left(y + \frac{8}{3}\right)^2 = \frac{40}{9}$; $y + \frac{8}{3} = \pm\frac{2\sqrt{10}}{3}$; $y = \frac{-8 \pm 2\sqrt{10}}{3}$; $\left\{\frac{-8 + 2\sqrt{10}}{3}, \frac{-8 - 2\sqrt{10}}{3}\right\}$

10. $2x^2 - 6x = -5$; $x^2 - 3x = -\frac{5}{2}$; $x^2 - 3x + \frac{9}{4} = -\frac{10}{4} + \frac{9}{4}$; $\left(x - \frac{3}{2}\right)^2 = -\frac{1}{4}$; $x - \frac{3}{2} = \pm\frac{i}{2}$; $x = \frac{3 \pm i}{2}$; $\left\{\frac{3 + i}{2}, \frac{3 - i}{2}\right\}$

11. $k^2 - 6k + 4 = 0$; $a = 1$, $b = -6$, $c = 4$; $k = \frac{6 \pm \sqrt{36 - 4(4)}}{2(1)} = \frac{6 \pm \sqrt{20}}{2} = \frac{6 \pm 2\sqrt{5}}{2} = 3 \pm \sqrt{5}$; $\{3 + \sqrt{5}, 3 - \sqrt{5}\}$

12. $3y^2 - 4y - 1 = 0$; $a = 3$, $b = -4$, $c = -1$; $y = \dfrac{4 \pm \sqrt{16 - 4(3)(-1)}}{6} = \dfrac{4 \pm \sqrt{16 + 12}}{6} = \dfrac{4 \pm \sqrt{28}}{6} = \dfrac{4 \pm 2\sqrt{7}}{6} = \dfrac{2 \pm \sqrt{7}}{3}$; $\left\{\dfrac{2 + \sqrt{7}}{3}, \dfrac{2 - \sqrt{7}}{3}\right\}$

13. $2x^2 - 8x + 11 = 0$; $a = 2$, $b = -8$, $c = 11$; $x = \dfrac{8 \pm \sqrt{64 - 4(2)(11)}}{4} = \dfrac{8 \pm \sqrt{64 - 88}}{4} = \dfrac{8 \pm \sqrt{-24}}{4} = \dfrac{8 \pm 2i\sqrt{6}}{4}$; $\left\{\dfrac{4 + i\sqrt{6}}{2}, \dfrac{4 - i\sqrt{6}}{2}\right\}$

14. $8v^2 + 12v - 3 = 0$; $a = 8$, $b = 12$, $c = -3$; $v = \dfrac{-12 \pm \sqrt{144 - 4(8)(-3)}}{16} = \dfrac{-12 \pm \sqrt{240}}{16} = \dfrac{-12 \pm 4\sqrt{15}}{16} = \dfrac{-3 \pm \sqrt{15}}{4}$; $\left\{\dfrac{-3 + \sqrt{15}}{4}, \dfrac{-3 - \sqrt{15}}{4}\right\}$

15. $7z^2 - 6z = 0$; $a = 7$, $b = -6$, $c = 0$; $z = \dfrac{6 \pm \sqrt{36 - 28(0)}}{14} = \dfrac{6 \pm 6}{14} = 0$ or $\dfrac{6}{7}$; $\{0, \frac{6}{7}\}$

16. $9x^2 - 8x + 4 = 0$; $a = 9$, $b = -8$, $c = 4$; $x = \dfrac{8 \pm \sqrt{64 - 4(9)(4)}}{18} = \dfrac{8 \pm \sqrt{-80}}{18} = \dfrac{8 \pm 4i\sqrt{5}}{18} = \dfrac{4 \pm 2i\sqrt{5}}{9}$; $\left\{\dfrac{4 + 2i\sqrt{5}}{9}, \dfrac{4 - 2i\sqrt{5}}{9}\right\}$

17. $2k^2 - 10k + 5 = 0$; $a = 2$, $b = -10$, $c = 5$; $k = \dfrac{10 \pm \sqrt{100 - 4(2)(5)}}{4} = \dfrac{10 \pm \sqrt{60}}{4} = \dfrac{10 \pm 2\sqrt{15}}{4} = \dfrac{5 \pm \sqrt{15}}{2}$; $\left\{\dfrac{5 + \sqrt{15}}{2}, \dfrac{5 - \sqrt{15}}{2}\right\}$

18. $6v^2 + 16v - 3 = 0$; $a = 6$, $b = 16$, $c = -3$; $v = \dfrac{-16 \pm \sqrt{256 - 4(6)(-3)}}{12} = \dfrac{-16 \pm \sqrt{328}}{12} = \dfrac{-16 \pm 2\sqrt{82}}{12} = \dfrac{-8 \pm \sqrt{82}}{6}$; $\left\{\dfrac{-8 + \sqrt{82}}{6}, \dfrac{-8 - \sqrt{82}}{6}\right\}$

19. $n^2 - 2n + 45 = 0$; $a = 1$, $b = -2$, $c = 45$; $n = \dfrac{2 \pm \sqrt{4 - 4(45)}}{2} = \dfrac{2 \pm \sqrt{-176}}{2} = \dfrac{2 \pm 4i\sqrt{11}}{2} = 1 \pm 2i\sqrt{11}$; $\{1 + 2i\sqrt{11}, 1 - 2i\sqrt{11}\}$

20. $5(t^2 - 4t) = -7(3)$; $5t^2 - 20t + 21 = 0$; $a = 5$, $b = -20$, $c = 21$; $t = \dfrac{20 \pm \sqrt{400 - 4(5)(21)}}{10} = \dfrac{20 \pm \sqrt{-20}}{10} = \dfrac{20 \pm 2i\sqrt{5}}{10} = \dfrac{10 \pm i\sqrt{5}}{5}$; $\left\{2 + \dfrac{i\sqrt{5}}{5}, 2 - \dfrac{i\sqrt{5}}{5}\right\}$

21. $2(x + 3)^2 + 7 = 0$; $(x + 3)^2 = -\dfrac{7}{2}$; $x + 3 = \pm\dfrac{i\sqrt{14}}{2}$; $x = -3 \pm \dfrac{i\sqrt{14}}{2}$; $\left\{-3 + \dfrac{i\sqrt{14}}{2}, -3 - \dfrac{i\sqrt{14}}{2}\right\}$

22. $\dfrac{2}{5}x^2 - 3x = 0$; $2x^2 - 15x = 0$; $x(2x - 15) = 0$; $x = 0$ or $x = \dfrac{15}{2}$; $\left\{0, \dfrac{15}{2}\right\}$

23. $7x^2 - 12x + 4 = 0$; $x = \dfrac{12 \pm \sqrt{144 - 4(7)(4)}}{14} = \dfrac{12 \pm \sqrt{144 - 112}}{14} = \dfrac{12 \pm \sqrt{32}}{14} = \dfrac{12 \pm 4\sqrt{2}}{14} = \dfrac{6 \pm 2\sqrt{2}}{7}$; $\left\{\dfrac{6 + 2\sqrt{2}}{7}, \dfrac{6 - 2\sqrt{2}}{7}\right\}$

24. $\dfrac{(3x - 1)^2}{2} = 10$; $(3x - 1)^2 = 20$; $3x - 1 = \pm 2\sqrt{5}$; $3x = 1 \pm 2\sqrt{5}$; $x = \dfrac{1 \pm 2\sqrt{5}}{3}$; $\left\{\dfrac{1 + 2\sqrt{5}}{3}, \dfrac{1 - 2\sqrt{5}}{3}\right\}$

25. $3x^2 - 2x + 1 = 0$; $x = \dfrac{2 \pm \sqrt{4 - 4(3)}}{6} = \dfrac{2 \pm \sqrt{-8}}{6} = \dfrac{2 \pm 2i\sqrt{2}}{6} = \dfrac{1 \pm i\sqrt{2}}{3}$; $\left\{\dfrac{1 + i\sqrt{2}}{3}, \dfrac{1 - i\sqrt{2}}{3}\right\}$

26. $-2x^2 + 2x - 5 = 0$; $2x^2 - 2x + 5 = 0$; $x = \dfrac{2 \pm \sqrt{4 - 4(2)(5)}}{4} = \dfrac{2 \pm \sqrt{-36}}{4} = \dfrac{2 \pm 6i}{4} = \dfrac{1 \pm 3i}{2}$; $\left\{\dfrac{1 + 3i}{2}, \dfrac{1 - 3i}{2}\right\}$

27. $4x^2 - 12x + 7 = 0$; $x = \dfrac{12 \pm \sqrt{144 - 4(4)(7)}}{8} = \dfrac{12 \pm \sqrt{144 - 112}}{8} = \dfrac{12 \pm \sqrt{32}}{8} = \dfrac{12 \pm 4\sqrt{2}}{8} = \dfrac{3 \pm \sqrt{2}}{2}$; $\left\{\dfrac{3 + \sqrt{2}}{2}, \dfrac{3 - \sqrt{2}}{2}\right\}$

28. $2x^2 + 15x + 29 = 0$; $x = \dfrac{-15 \pm \sqrt{225 - 4(2)(29)}}{4} = \dfrac{-15 \pm \sqrt{-7}}{4} = \dfrac{-15 \pm i\sqrt{7}}{4}$; $\left\{\dfrac{-15 + i\sqrt{7}}{4}, \dfrac{-15 - i\sqrt{7}}{4}\right\}$

B **29.** $x^2\sqrt{3} + 6x + 7\sqrt{3} = 0$; $x = \dfrac{-6 \pm \sqrt{36 - 4(\sqrt{3})(7\sqrt{3})}}{2\sqrt{3}} = \dfrac{-6 \pm \sqrt{36 - 84}}{2\sqrt{3}} = \dfrac{-6 \pm \sqrt{-48}}{2\sqrt{3}} = \dfrac{-6 \pm 4i\sqrt{3}}{2\sqrt{3}} = -\sqrt{3} \pm 2i$; $\{-\sqrt{3} + 2i, -\sqrt{3} - 2i\}$

30. $8x^2 - 12\sqrt{5}x + 9 = 0$; $x = \dfrac{12\sqrt{5} \pm \sqrt{720 - 4(8)(9)}}{16} = \dfrac{12\sqrt{5} \pm \sqrt{432}}{16} = \dfrac{12\sqrt{5} \pm 12\sqrt{3}}{16} = \dfrac{3\sqrt{5} \pm 3\sqrt{3}}{4}$; $\left\{\dfrac{3\sqrt{5} + 3\sqrt{3}}{4}, \dfrac{3\sqrt{5} - 3\sqrt{3}}{4}\right\}$

31. $29 + 2x(x + 3) = 10(x + 3)$; $2x^2 + 6x + 29 = 10x + 30$; $2x^2 - 4x - 1 = 0$; $x^2 - 2x = \dfrac{1}{2}$; $x^2 - 2x + 1 = \dfrac{1}{2} + 1$; $(x - 1)^2 = \dfrac{3}{2}$; $x - 1 = \dfrac{\pm\sqrt{6}}{2}$; $x = 1 \pm \dfrac{\sqrt{6}}{2}$; $\left\{1 + \dfrac{\sqrt{6}}{2}, 1 - \dfrac{\sqrt{6}}{2}\right\}$

32. $(x + 4)(x + 8) = (x - 2)(3x - 4)$; $x^2 + 12x + 32 = 3x^2 - 10x + 8$; $2x^2 - 22x - 24 = 0$; $2(x - 12)(x + 1) = 0$; $x = 12$ or $x = -1$; $\{-1, 12\}$

33. $5x + 4(x - 3) + 3x(x - 3) = 0$; $3x^2 - 12 = 0$; $3(x + 2)(x - 2) = 0$; $x = 2$ or $x = -2$; $\{-2, 2\}$ **34.** $7(x + 1) - 3(x^2 - 1) = 2(x - 1)$; $-3x^2 + 7x + 10 = 2x - 2$; $3x^2 - 5x - 12 = 0$; $(3x + 4)(x - 3) = 0$; $x = -\frac{4}{3}$ or $x = 3$; $\{-\frac{4}{3}, 3\}$

35. $x^4 - 5x^2 - 36 = 0$; $(x^2 - 9)(x^2 + 4) = (x + 3)(x - 3)(x^2 + 4) = 0$; $x + 3 = 0$, $x - 3 = 0$, or $x^2 + 4 = 0$; $x = \pm 3$ or $x = \pm 2i$; $\{-3, 3, -2i, 2i\}$

36. $x^4 - 6x^2 + 5 = 0$; $(x^2 - 1)(x^2 - 5) = 0$; $(x - 1)(x + 1)(x - \sqrt{5})(x + \sqrt{5}) = 0$; $x = \pm 1$ or $x = \pm\sqrt{5}$; $\{-\sqrt{5}, -1, \sqrt{5}, 1\}$

37. $x^6 - 7x^3 - 8 = 0$; $(x^3 - 8)(x^3 + 1) = 0$; $(x - 2)(x^2 + 2x + 4)(x + 1)(x^2 - x + 1) = 0$; $x - 2 = 0$, $x^2 + 2x + 4 = 0$, $x + 1 = 0$, $x^2 - x + 1 = 0$; $x = 2$, $x = \dfrac{-2 \pm \sqrt{4 - 16}}{2} = -1 \pm i\sqrt{3}$, $x = -1$, $x = \dfrac{1 \pm \sqrt{1 - 4}}{2} = \dfrac{1 \pm i\sqrt{3}}{2}$; $\left\{-1, 2, -1 + i\sqrt{3}, -1 - i\sqrt{3}, \dfrac{1 + i\sqrt{3}}{2}, \dfrac{1 - i\sqrt{3}}{2}\right\}$

38. $x^6 + 65x^3 + 64 = 0$; $(x^3 + 64)(x^3 + 1) = 0$; $(x + 4)(x^2 - 4x + 16)(x + 1)(x^2 - x + 1) = 0$; $x + 4 = 0$, $x^2 - 4x + 16 = 0$, $x + 1 = 0$, $x^2 - x + 1 = 0$; $x = -4$, $x = \dfrac{4 \pm \sqrt{16 - 64}}{2} = 2 \pm 2i\sqrt{3}$, $x = -1$, $x = \dfrac{1 \pm \sqrt{1 - 4}}{2} = \dfrac{1 \pm i\sqrt{3}}{2}$; $\left\{-4, -1, 2 + 2i\sqrt{3}, 2 - 2i\sqrt{3}, \dfrac{1 + i\sqrt{3}}{2}, \dfrac{1 - i\sqrt{3}}{2}\right\}$

39. $\left(\dfrac{1}{x}\right)^2 - 2\left(\dfrac{1}{x}\right) - 1 = 0$; $1 - 2x - x^2 = 0$; $x^2 + 2x = 1$; $x^2 + 2x + 1 = 2$; $(x + 1)^2 = 2$; $x + 1 = \pm\sqrt{2}$; $x = -1 \pm \sqrt{2}$; $\{-1 + \sqrt{2}, -1 - \sqrt{2}\}$

40. $\left(\dfrac{1}{x + 1}\right)^2 - \left(\dfrac{1}{x + 1}\right) = 5$; $1 - (x + 1) = 5(x + 1)^2$; $-x = 5x^2 + 10x + 5$; $5x^2 + 11x + 5 = 0$; $x = \dfrac{-11 \pm \sqrt{121 - 100}}{10} = \dfrac{-11 \pm \sqrt{21}}{10}$; $\left\{\dfrac{-11 + \sqrt{21}}{10}, \dfrac{-11 - \sqrt{21}}{10}\right\}$

41. $x^2 + ix + 6 = 0$; $x = \dfrac{-i \pm \sqrt{-1 - 24}}{2} = \dfrac{-i \pm 5i}{2}$; $x = -3i$ or $x = 2i$; $\{2i, -3i\}$

42. $x^2 - 2ix + 3 = 0$; $x = \dfrac{2i \pm \sqrt{-4 - 12}}{2} = \dfrac{2i \pm 4i}{2}$; $x = 3i$ or $x = -i$; $\{-i, 3i\}$

43. $x = \dfrac{2 + 3i \pm \sqrt{(2 + 3i)^2 - 4(3 + 3i)}}{2} = \dfrac{2 + 3i \pm \sqrt{-17}}{2} = \dfrac{2 + 3i \pm i\sqrt{17}}{2} = 1 + \dfrac{3 \pm \sqrt{17}}{2}i$; $\left\{1 + \dfrac{3 + \sqrt{17}}{2}i, 1 + \dfrac{3 - \sqrt{17}}{2}i\right\}$

44. $x = \dfrac{2i - 1 \pm \sqrt{(1 - 2i)^2 - 4i(3i - 1)}}{2i} = \dfrac{2i - 1 \pm \sqrt{9}}{2i} = \dfrac{2i - 1 \pm 3}{2i}$; $x = \dfrac{2i + 2}{2i}$ or $x = \dfrac{2i - 4}{2i}$; $x = 1 + \dfrac{1}{i}$ or $x = 1 - \dfrac{2}{i}$; $x = 1 - i$ or $x = 1 + 2i$; $\{1 - i, 1 + 2i\}$

C 45. a. $x^4 + 4 = (x^4 + 4x^2 + 4) - 4x^2 = (x^2 + 2)^2 - 4x^2 = (x^2 + 2x + 2)(x^2 - 2x + 2)$

b. $x^4 + 1 = (x^4 + 2x^2 + 1) - 2x^2 = (x^2 + 1)^2 - (x\sqrt{2})^2 = (x^2 + 1 + x\sqrt{2})(x^2 + 1 - x\sqrt{2}) = (x^2 + x\sqrt{2} + 1)(x^2 - x\sqrt{2} + 1)$

46. $\sqrt{2x - x^2} = (x-1)^2$; $\sqrt{1-(1-2x+x^2)} = (x-1)^2$; $\sqrt{1-(x-1)^2} = (x-1)^2$. Let $u = (x-1)^2$, then $\sqrt{1-u} = u$, or $1 - u = u^2$. So, $u^2 + u - 1 = 0$, and $u = \frac{-1 \pm \sqrt{5}}{2}$. Substituting for u, $(x-1)^2 = \frac{-1 \pm \sqrt{5}}{2}$; $x - 1 = \pm\sqrt{\frac{-1 \pm \sqrt{5}}{2}}$;

$x - 1 = \frac{\pm\sqrt{2}\sqrt{-1 \pm \sqrt{5}}}{2}$; $x - 1 = \frac{\pm\sqrt{-2 \pm 2\sqrt{5}}}{2}$; $x = 1 \pm \frac{\sqrt{-2 \pm 2\sqrt{5}}}{2}$;

$\left\{1 \pm \frac{\sqrt{-2 \pm 2\sqrt{5}}}{2}\right\}$

Pages 346–347 • PROBLEMS

A **1.** $7 = -4.9t^2 + 14t$; $4.9t^2 - 14t + 7 = 0$; $49t^2 - 140t + 70 = 0$;
$7(7t^2 - 20t + 10) = 0$; $t = \frac{20 \pm \sqrt{120}}{14} = \frac{20 \pm 2\sqrt{30}}{14}$
a. $\frac{10 + \sqrt{30}}{7}$ s, $\frac{10 - \sqrt{30}}{7}$ s **b.** 2.2 s, 0.6 s

2. $R_1 + R_2 = 10$; $\frac{R_1R_2}{R_1 + R_2} = 2$. Substitute $10 - R_2$ for R_1 and 10 for $R_1 + R_2$ in the second equation to get $\frac{R_2(10 - R_2)}{10} = 2$, or $R_2^{\,2} - 10R_2 + 20 = 0$;
$R_2 = \frac{10 \pm \sqrt{100 - 80}}{2} = \frac{10 \pm 2\sqrt{5}}{2} = 5 \pm \sqrt{5}$; $R_1 = 5 \mp \sqrt{5}$ **a.** $(5 + \sqrt{5})\ \Omega$, $(5 - \sqrt{5})\ \Omega$ **b.** 7.2 Ω, 2.8 Ω

3. $40wh = 11{,}000$; $40 + 2(w + h) = 108$; $w = \frac{275}{h}$; $40 + 2\left(\frac{275}{h} + h\right) = 108$;
$\frac{275}{h} + h = 34$; $h^2 - 34h + 275 = 0$; $h = \frac{34 \pm \sqrt{1156 - 1100}}{2} = 17 \pm \sqrt{14}$;
$w = 17 \mp \sqrt{14}$ **a.** $(17 + \sqrt{14})$ in., $(17 - \sqrt{14})$ in. **b.** 20.7 in., 13.3 in.

4. $r_2 = r_1 - 3$; $h_2 = 2h_1$; $V_1 = V_2$; $\frac{1}{3}\pi r_1^{\,2}h_1 = \frac{1}{3}\pi r_2^{\,2}h_2$; $r_1^{\,2}h_1 = r_2^{\,2}h_2$; $r_1^{\,2}h_1 = (r_1 - 3)^2(2h_1)$. $h_1 \neq 0$, so $r_1^{\,2} = 2(r_1 - 3)^2$; $r_1^{\,2} = 2r_1^{\,2} - 12r_1 + 18$;
$r_1^{\,2} - 12r_1 + 36 = -18 + 36$; $(r_1 - 6)^2 = 18$; $r_1 = 6 \pm 3\sqrt{2}$; $r_1 = 6 + 3\sqrt{2}$ and $r_2 = 3 + 3\sqrt{2}$ or $r_1 = 6 - 3\sqrt{2}$ and $r_2 = 3 - 3\sqrt{2} < 0$
a. $(6 + 3\sqrt{2})$ cm and $(3 + 3\sqrt{2})$ cm **b.** 10.2 cm and 7.2 cm

5. Let x = edge of closed box, then its surface area $= 6x^2$. Let $x + 1$ = edge of open box, then its surface area $= 5(x + 1)^2$; $6x^2 = 5(x + 1)^2$; $x^2 - 10x = 5$; $x^2 - 10x + 25 = 5 + 25$; $(x - 5)^2 = 30$; $x = 5 \pm \sqrt{30}$; $x = 5 + \sqrt{30}$ and $x + 1 = 6 + \sqrt{30}$; $x = 5 - \sqrt{30} < 0$ **a.** $(6 + \sqrt{30})$ cm and $(5 + \sqrt{30})$ cm
b. 11.5 cm and 10.5 cm

6. Let w = width of frame. $(16 + 2w)(20 + 2w) = 464$; $2(8 + w)2(10 + w) = 464$; $(8 + w)(10 + w) = 116$; $w^2 + 18w = 36$; $w^2 + 18w + 81 = 117$; $(w + 9)^2 = 117$; $w = -9 \pm 3\sqrt{13}$ **a.** $(-9 + 3\sqrt{13})$ cm **b.** 1.8 cm

B **7.** Let r = radius of cone and let $r + 2$ = radius of cylinder. $V_{cyl} = \pi(r + 2)^2(6)$; $V_{cone} = \frac{1}{3}\pi r^2(6)$; $V_{entire\ part} = 102\pi = 6\pi(r + 2)^2 + 2\pi r^2$;

$102\pi = 8\pi r^2 + 24\pi r + 24\pi$; $8\pi r^2 + 24\pi r - 78\pi = 0$; $2\pi(4r^2 + 12r - 39) = 0$;
$r = \dfrac{-12 \pm \sqrt{768}}{8} = \dfrac{-12 \pm 16\sqrt{3}}{8} = \dfrac{-3 \pm 4\sqrt{3}}{2}$ **a.** $\dfrac{1 + 4\sqrt{3}}{2}$ cm **b.** 4.0 cm

8. Let r = outside radius. $V_{bottom} = \pi r^2(1) = \pi r^2$. $V_{shell} = \pi r^2(r - 1) - \pi(r - 1)^2(r - 1) = \pi(r - 1)[r^2 - (r^2 - 2r + 1)] = \pi(r - 1)(2r - 1)$. $V_{total} = \pi r^2 + \pi(r - 1)(2r - 1) = 55\pi$; $3\pi r^2 - 3\pi r + \pi = 55\pi$; $3\pi r^2 - 3\pi r - 54\pi = 3\pi(r^2 - r - 18) = 0$; $r = \dfrac{1 \pm \sqrt{73}}{2}$ **a.** $\dfrac{1 + \sqrt{73}}{2}$ cm **b.** 4.8 cm

9. Let $2l$ = length of pen; w = width. $2(2l + w) = 30$; $l^2 + w^2 = 100$; $w = 15 - 2l$; $l^2 + (15 - 2l)^2 = 100$; $5l^2 - 60l + l^2 - 12l; -25$; $l^2 - 12l + 36 = 11$; $(l - 6)^2 = 11$; $l = 6 \pm \sqrt{11}$; $w = 3 \mp 2\sqrt{11}$ **a.** $(3 + 2\sqrt{11})$ m by $(12 - 2\sqrt{11})$ m **b.** 9.6 m by 5.4 m

10. Let b, h = length of legs. $\frac{1}{2}bh = 15$ or $h = \frac{30}{b}$; $b^2 + h^2 = 100$;
$b^2 + \left(\frac{30}{b}\right)^2 = 100$; $b^2 + \frac{900}{b^2} = 100$; $b^4 - 100b^2 = -900$;
$b^4 - 100b^2 + 2500 = 1600$; $(b^2 - 50)^2 = 1600$; $b^2 = 50 \pm 40$; $b^2 = 90$ or $b^2 = 10$;
$b = \sqrt{90} = \pm 3\sqrt{10}$ or $b = \sqrt{10}$ **a.** $3\sqrt{10}$ cm, $\sqrt{10}$ cm **b.** 9.5 cm, 3.2 cm

Pages 349–350 • WRITTEN EXERCISES

A

1. $D = 9 - 4(3) = -3$; 2 imaginary roots
2. $D = 100 - 4(3)(-2) = 124$; 2 real irrational roots
3. $D = 121 - 4(5)(6) = 1$; 2 real rational roots
4. $D = 225 - 4(7)(0) = 225$; 2 real rational roots
5. $D = 25 - 4(2)(8) = 89$; 2 real irrational roots
6. $D = 16 - 4(-4)(-5) = -64$; 2 imaginary roots
7. $D = 144 - 4(8)\left(\frac{9}{2}\right) = 0$; 1 double root
8. $D = 121 - 4(5)(2) = 81$; 2 real rational roots
9. $D = 0 - 4(-6)(72) = 1728$; 2 real irrational roots
10. $D = 1 - 4\left(\frac{3}{2}\right)(-20) = 121$; 2 real rational roots

B

11. $D = 100 - 4(2)(5k)$; $100 - 40k = 0$; $k = 2\frac{1}{2}$
12. $D = k^2 - 4(5)(8)$; $k^2 - 160 = 0$; $k = \pm 4\sqrt{10}$
13. $D = 4k^2 - 4(k + 6)$; $4(k^2 - k - 6) = 0$; $(k - 3)(k + 2) = 0$; $k = 3$ or $k = -2$
14. $D = k^2 - 4(3 - 2k)$; $k^2 + 8k - 12 = 0$;
$k = \dfrac{-8 \pm \sqrt{64 + 48}}{2} = \dfrac{-8 \pm \sqrt{112}}{2} = \dfrac{-8 \pm 4\sqrt{7}}{2}$; $k = -4 \pm 2\sqrt{7}$
15. $D = 36 - 4(3)(k)$; $36 - 12k > 0$, $k < 3$; $\{k: k < 3\}$
16. $D = 16 - 4(5k)(2)$; $16 - 40k > 0$, $k < 0.4$; $\{k: k < 0.4\}$
17. $D = 64 - 4(2k - 1)$; $68 - 8k < 0$, $k > \frac{17}{2}$; $\left\{k: k > \frac{17}{2}\right\}$
18. $D = [4(k + 1)]^2 - 4(4k^2) = 16(k^2 + 2k + 1) - 16k^2$; $32k + 16 < 0$; $k < -\frac{1}{2}$; $\left\{k: k < -\frac{1}{2}\right\}$

C

19. $c = 0$; $D = b^2 - 4ac = b^2 - 4(a)(0) = b^2$; D is the square of a rational number and the coefficients of the equation are rational. Therefore, the equation has two rational roots.

20. $a = -c$; $D = b^2 - 4ac = b^2 - 4(-c)(c) = b^2 + 4c^2 > 0$. Therefore, the equation has two real roots.

Page 352 • WRITTEN EXERCISES

A

1. $-2 + 7 = -\frac{b}{a}$, $-2(7) = \frac{c}{a}$; $\frac{b}{a} = -5$, $\frac{c}{a} = -14$; let $a = 1$, then $b = -5$ and $c = -14$; $x^2 - 5x - 14 = 0$

2. $-3 - 5 = -\frac{b}{a}$, $-3(-5) = \frac{c}{a}$; $\frac{b}{a} = 8$, $\frac{c}{a} = 15$; let $a = 1$, then $b = 8$ and $c = 15$; $x^2 + 8x + 15 = 0$

3. $-\frac{3}{2} - \frac{3}{2} = -\frac{b}{a}$, $\left(-\frac{3}{2}\right)\left(-\frac{3}{2}\right) = \frac{c}{a}$; $-\frac{b}{a} = -3$, $\frac{c}{a} = \frac{9}{4}$; let $a = 4$, then $b = 12$ and $c = 9$; $4x^2 + 12x + 9 = 0$

4. $(1 + \sqrt{7}) + (1 - \sqrt{7}) = -\frac{b}{a}$, $(1 + \sqrt{7})(1 - \sqrt{7}) = \frac{c}{a}$; $-\frac{b}{a} = 2$, $\frac{c}{a} = -6$; let $a = 1$, then $b = -2$ and $c = -6$; $x^2 - 2x - 6 = 0$

5. $6\sqrt{2} - 6\sqrt{2} = -\frac{b}{a}$, $6\sqrt{2}(-6\sqrt{2}) = \frac{c}{a}$; $\frac{b}{a} = 0$, $\frac{c}{a} = -72$; let $a = 1$, then $b = 0$ and $c = -72$; $x^2 - 72 = 0$

6. $(1 + 3i) + (1 - 3i) = -\frac{b}{a}$, $(1 + 3i)(1 - 3i) = \frac{c}{a}$; $\frac{b}{a} = -2$, $\frac{c}{a} = 10$; let $a = 1$, then $b = -2$ and $c = 10$; $x^2 - 2x + 10 = 0$

7. $(3 + i\sqrt{5}) + (3 - i\sqrt{5}) = -\frac{b}{a}$, $(3 + i\sqrt{5})(3 - i\sqrt{5}) = \frac{c}{a}$; $\frac{b}{a} = -6$, $\frac{c}{a} = 14$; let $a = 1$, then $b = -6$ and $c = 14$; $x^2 - 6x + 14 = 0$

8. $(5 + \sqrt{3}) + (5 - \sqrt{3}) = -\frac{b}{a}$, $(5 + \sqrt{3})(5 - \sqrt{3}) = \frac{c}{a}$; $\frac{b}{a} = -10$, $\frac{c}{a} = 22$; let $a = 1$, then $b = -10$ and $c = 22$; $x^2 - 10x + 22 = 0$

9. $\left(\frac{3}{2} - \frac{\sqrt{5}}{2}i\right) + \left(\frac{3}{2} + \frac{\sqrt{5}}{2}i\right) = -\frac{b}{a}$, $\left(\frac{3}{2} - \frac{\sqrt{5}}{2}i\right)\left(\frac{3}{2} + \frac{\sqrt{5}}{2}i\right) = \frac{c}{a}$; $\frac{b}{a} = -3$, $\frac{c}{a} = \frac{7}{2}$; let $a = 2$, then $b = -6$ and $c = 7$; $2x^2 - 6x + 7 = 0$

10. $\left(\frac{4}{3} - \frac{2}{3}i\right) + \left(\frac{4}{3} + \frac{2}{3}i\right) = -\frac{b}{a}$, $\left(\frac{4}{3} - \frac{2}{3}i\right)\left(\frac{4}{3} + \frac{2}{3}i\right) = \frac{c}{a}$; $\frac{b}{a} = -\frac{8}{3}$, $\frac{c}{a} = \frac{20}{9}$; let $a = 9$, then $b = -24$ and $c = 20$; $9x^2 - 24x + 20 = 0$

11. $(7 - 2\sqrt{3}) + (7 + 2\sqrt{3}) = -\frac{b}{a}$, $(7 - 2\sqrt{3})(7 + 2\sqrt{3}) = \frac{c}{a}$; $\frac{b}{a} = -14$, $\frac{c}{a} = 37$; let $a = 1$, then $b = -14$ and $c = 37$; $x^2 - 14x + 37 = 0$

12. $(\sqrt{2} + 3i) + (\sqrt{2} - 3i) = -\frac{b}{a}$, $(\sqrt{2} + 3i)(\sqrt{2} - 3i) = \frac{c}{a}$; $\frac{b}{a} = -2\sqrt{2}$, $\frac{c}{a} = 11$; let $a = 1$, then $b = -2\sqrt{2}$ and $c = 11$; $x^2 - 2x\sqrt{2} + 11 = 0$

B

13. $r + \frac{1}{2} = -\frac{1}{6}$, so $r = -\frac{2}{3}$; $\frac{1}{2}r = \frac{k}{6}$, $\frac{1}{2}\left(-\frac{2}{3}\right)(6) = k$, $k = -2$

14. $r + 2 + \sqrt{11} = 4$, so $r = 2 - \sqrt{11}$; $(2 - \sqrt{11})(2 + \sqrt{11}) = \frac{k}{1}$, $k = -7$

15. $-\frac{3}{2}r = -\frac{21}{2}$, so $r = 7$; $7 - \frac{3}{2} = -\frac{k}{2}$, $k = -11$

16. $\left(\frac{3}{2} + 2i\right)r = \frac{25}{4}$, $r = \frac{25}{4} \cdot \frac{1}{3 + 4i} \cdot \frac{3 - 4i}{3 - 4i} = \frac{3}{2} - 2i$; $\left(\frac{3}{2} + 2i\right) + \left(\frac{3}{2} - 2i\right) = -\frac{k}{4}$, $3 = -\frac{k}{4}$, $k = -12$

17. $r_2 = 2r_1$; $r_1 + 2r_1 = \frac{3}{k}$, $r_1 = \frac{1}{k}$; $r_1(2r_1) = \frac{10}{k}$, $2\left(\frac{1}{k^2}\right) = \frac{10}{k}$, $k = \frac{1}{5}$

18. $r_2 = 3r_1$; $r_1 + 3r_1 = \frac{9}{16}$, $r_1 = \frac{9}{64}$; $r_1(3r_1) = \frac{k}{16}$, $\frac{9}{64} \cdot \frac{27}{64} = \frac{k}{16}$, $k = \frac{243}{256}$

C **19.** According to the quadratic formula, the roots of such an equation are $\frac{-b + \sqrt{b^2 - 4ac}}{2a}$ and $\frac{-b - \sqrt{b^2 - 4ac}}{2a}$. Let $D = \sqrt{b^2 - 4ac}$. Adding the roots gives $\frac{-b + D - b - D}{2a} = \frac{-2b}{2a} = -\frac{b}{a}$. Multiplying the two roots gives $\frac{(-b + D)(-b - D)}{4a^2} = \frac{b^2 - D^2}{4a^2}$; substituting for D then gives $\frac{b^2 - (b^2 - 4ac)}{4a^2} = \frac{b^2 - b^2 + 4ac}{4a^2} = \frac{4ac}{4a^2} = \frac{c}{a}$.

20.

1. $r_1 = \frac{-b + \sqrt{b^2 - 4ac}}{2a}$, $r_2 = \frac{-b - \sqrt{b^2 - 4ac}}{2a}$	Def. of r_1, r_2
2. $\frac{1}{r_1} + \frac{1}{r_2} = \frac{r_1 + r_2}{r_1 r_2}$	Add. of fractions
3. $r_1 + r_2 = -\frac{b}{a}$, $r_1 r_2 = \frac{c}{a}$	Results of Ex. 19
4. $\frac{r_1 + r_2}{r_1 r_2} = \frac{-\frac{b}{a}}{\frac{c}{a}} = -\frac{b}{c}$	Substitution and simplifying

21. $(r_1 + r_2)^2 = r_1^{\,2} + 2r_1r_2 + r_2^{\,2}$, so $r_1^{\,2} + r_2^{\,2} = (r_1 + r_2)^2 - 2r_1r_2$; $r_1^{\,2} + r_2^{\,2} = \left(-\frac{b}{a}\right)^2 - 2\left(\frac{c}{a}\right) = \frac{b^2}{a^2} - \frac{2c}{a} = \frac{b^2 - 2ac}{a^2}$

22. Let z_1 and z_2 be the two imaginary roots. Then $z_1 + z_2 = -\frac{b}{a}$ and $z_1z_2 = \frac{c}{a}$, by Exercise 19. Since $a, b, c \in \mathcal{R}$, $a \neq 0$, $-\frac{b}{a}$ is real, and $\frac{c}{a}$ is real by closure. Therefore, $z_1 + z_2$ and z_1z_2 are both real. By Exercise 26, p. 284, it follows that z_1 and z_2 are complex conjugates.

23. Let $r_1 = p + q\sqrt{2}$ and $r_2 = r + s\sqrt{2}$; $r_1 + r_2 = -\frac{b}{a}$, so $p + q\sqrt{2} + r + s\sqrt{2} = -\frac{b}{a}$. Setting the rational and irrational parts equal, we have $p + r = -\frac{b}{a}$ and (1) $q + s = 0$; $r_1r_2 = \frac{c}{a}$, so $(p + q\sqrt{2})(r + s\sqrt{2}) = \frac{c}{a}$, or $(pr + 2qs) + \sqrt{2}(qr + ps) = \frac{c}{a}$. Setting the rational and irrational parts equal, we have $pr + 2qs = \frac{c}{a}$ and (2) $qr + ps = 0$. By Eq. (1), $q = -s$. Substitution into Eq. (2) yields: $-sr + ps = 0$, $s(p - r) = 0$, so $p = r$ or $s = 0$. But $s \neq 0$ so $p = r$ and $q = -s$.

Page 353 • COMPUTER EXERCISES

1.
```
10  INPUT A, B, C
20  LET D = B^2 - 4 * A * C
25  IF ABS(D - INT(D)) < 0.00001 THEN LET D = INT(D)
30  IF D = 0 THEN PRINT "ONE ROOT: X = ";-B/(2 * A) : END
40  IF D < 0 THEN GOTO 80
50  LET R1 = (-B + D^.5)/(2 * A)
60  LET R2 = (-B - D^.5)/(2 * A)
70  PRINT "X = ";R1;" OR X = ";R2 : END
80  LET D = -D
90  PRINT "COMPLEX CONJUGATE ROOTS"
100  PRINT "X = ";-B/(2 * A);" + ";D^.5/(2 * A);"I OR ";
110  PRINT "X = ";-B/(2 * A);" - ";D^.5/(2 * A);"I"
120  END
```

2. $\{3, 4\}$ **3.** $\{-2.5\}$ **4.** $\{3 + 2i, 3 - 2i\}$ **5.** $\{-0.3333, -2\}$

6.
```
10  PRINT "1 = TWO REAL ROOTS OR 2 = COMPLEX CONJUGATE ROOTS"
20  INPUT "WHAT KIND OF ROOTS ";S
30  IF S = 1 THEN GOTO 90
40  IF S <> 2 THEN GOTO 10
50  INPUT M, N
60  LET MID = -2 * M
70  LET LAST = M^2 + N^2
80  GOTO 120
90  INPUT R1, R2
100  LET MID = -R2 - R1
110  LET LAST = R1 * R2
120  PRINT "X^2";
130  IF MID < 0 THEN PRINT MID;"X";  : GOTO 160
140  IF MID = 0 THEN GOTO 160
150  PRINT " + ";MID;"X";
160  IF LAST < 0 THEN PRINT LAST; : GOTO 190
170  IF LAST = 0 THEN GOTO 190
180  PRINT " + ";LAST;
190  PRINT " = 0"
200  END
```

7. $x^2 + 2x - 15 = 0$ **8.** $x^2 - 2x + 5 = 0$ **9.** $x^2 + 5x + 6.25 = 0$
10. $x^2 + 6x = 0$ **11.** $x^2 + 6x + 25 = 0$ **12.** $x^2 + 64 = 0$

13.
```
10  INPUT P, Q, R, S
20  PRINT "1 = TWO RATIONAL NUMBERS"
30  PRINT "2 = COMPLEX CONJUGATES"
40  INPUT "WHAT KIND OF NUMBERS ";Z
50  IF Z = 2 THEN GOTO 110
60  IF Z <> 1 THEN GOTO 20
70  LET A = Q * S
80  LET B = -P * S - Q * R
90  LET C = P * R
100  GOTO 180
110  LET A1 = Q
```

[*Program continued on next page*]

```
120 LET A2 = Q^2 * S^2
130 LET B = -2 * P
140 LET C = (S * P)^2 + (Q * R)^2
150 IF A1/A2 = INT(A1/A2) THEN LET B = B/(A1/A2) : LET A = A2
160 IF A2/A1 = INT(A2/A1) AND C/(A2/A1) = INT(C/(A2/A1)) THEN
    LET C = C/(A2/A1) : LET A = A1
170 IF A2/A1 = INT(A2/A1) THEN LET B = B * (A2/A1) : LET A = A2
180 PRINT A;"X^2";
190 IF B < 0 THEN PRINT B;"X"; : GOTO 220
200 IF B = 0 THEN GOTO 220
210 PRINT " + ";B;"X";
220 IF C < 0 THEN PRINT C; : GOTO 250
230 IF C = 0 THEN GOTO 250
240 PRINT " + ";C;
250 PRINT " = 0"
260 END
```

14. $12x^2 - 11x + 2 = 0$ **15.** $4x^2 - 12x + 13 = 0$ **16.** $4x^2 + 17x - 15 = 0$
17. $100x^2 + 100x + 61 = 0$

Pages 353–354 • SELF-TEST 1

1. $2x^2 - 6x + 9 = 0$; $2x^2 - 6x = -9$; $x^2 - 3x = -\frac{9}{2}$; $x^2 - 3x + \frac{9}{4} = -\frac{9}{2} + \frac{9}{4}$; $\left(x - \frac{3}{2}\right)^2 = -\frac{9}{4}$; $x - \frac{3}{2} = \pm\frac{3i}{2}$; $x = \frac{3 \pm 3i}{2}$; $\left\{\frac{3 + 3i}{2}, \frac{3 - 3i}{2}\right\}$

2. $5x^2 + 10x - 2 = 0$; $x = \frac{-10 \pm \sqrt{100 - 4(5)(-2)}}{2(5)} = \frac{-10 \pm \sqrt{140}}{10} = \frac{-10 \pm 2\sqrt{35}}{10} = \frac{-5 \pm \sqrt{35}}{5}$; $\left\{\frac{-5 + \sqrt{35}}{5}, \frac{-5 - \sqrt{35}}{5}\right\}$

3. $x^2 - 6x + 29 = 0$; $x = \frac{6 \pm \sqrt{36 - 4(29)}}{2} = \frac{6 \pm \sqrt{-80}}{2} = \frac{6 \pm 4i\sqrt{5}}{2} = 3 \pm 2i\sqrt{5}$; $\{3 + 2i\sqrt{5}, 3 - 2i\sqrt{5}\}$

4. $D = 8^2 - 4(4)(5) = 64 - 80 < 0$; 2 complex conjugate roots

5. $D = 30^2 - 4(9)(25) = 900 - 900 = 0$; 1 double rational root

6. a. $r_1 + r_2 = \frac{5}{2}$ **b.** $r_1 r_2 = \frac{3}{2}$

7. $(2 + 5i) + (2 - 5i) = -\frac{b}{a}$, $(2 + 5i)(2 - 5i) = \frac{c}{a}$; $-\frac{b}{a} = 4$, $\frac{c}{a} = 29$; let $a = 1$, then $b = -4$, and $c = 29$; $x^2 - 4x + 29 = 0$

Pages 356–357 • WRITTEN EXERCISES

A 1.

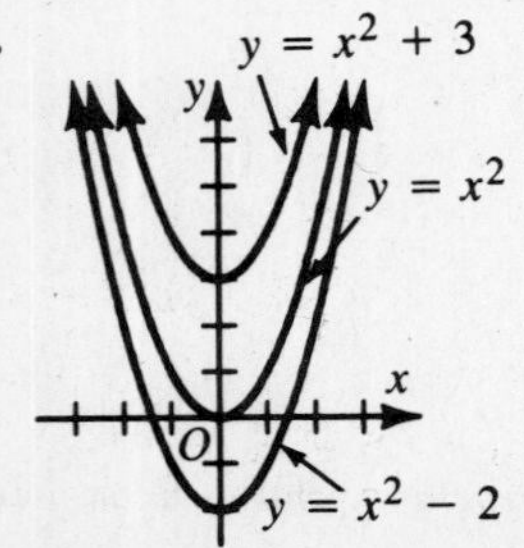

2.

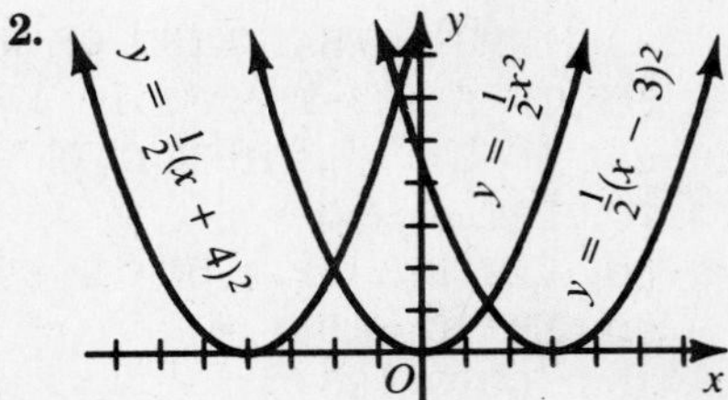

3. 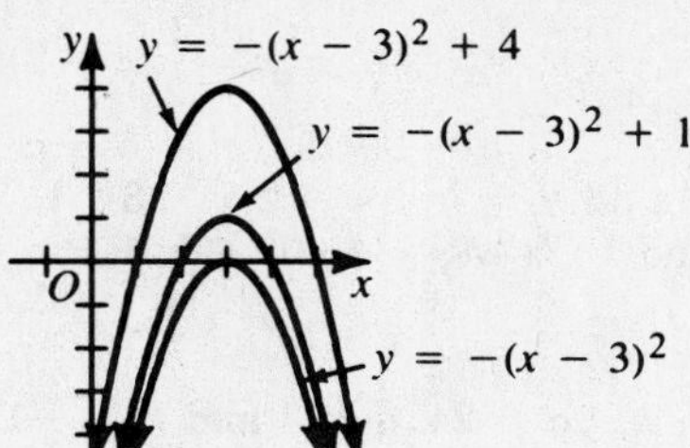

4. 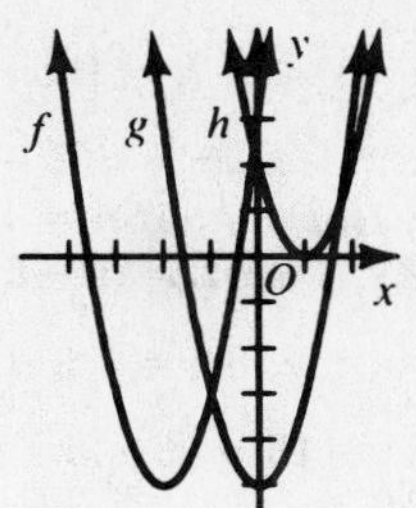

5. 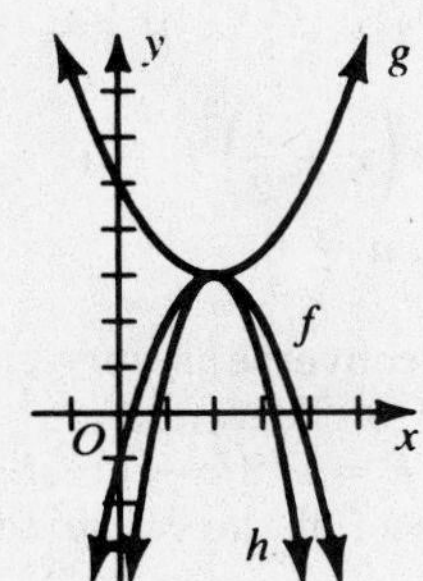

6. 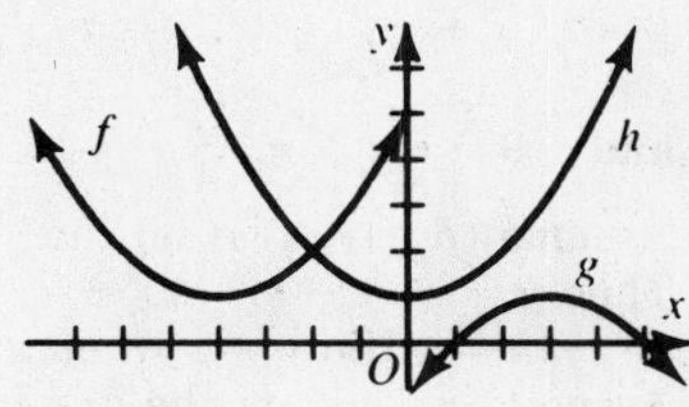

7. 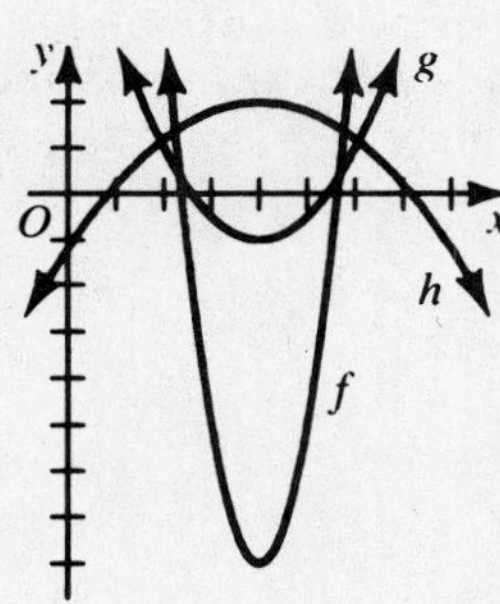

8. 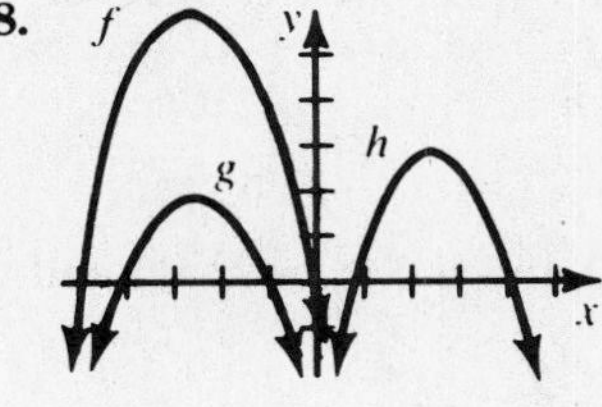

9. $y = a(x - 1)^2 + 3;\ 21 = a(4 - 1)^2 + 3,\ a = 2;\ y = 2(x - 1)^2 + 3$

10. $y = a(x + 2)^2 - 5;\ 0 = a(-3 + 2)^2 - 5,\ a = 5;\ y = 5(x + 2)^2 - 5$

11. $y = a(x + 1)^2 + 4;\ -14 = a(5 + 1)^2 + 4,\ a = -\frac{1}{2};\ y = -\frac{1}{2}(x + 1)^2 + 4$

12. $y = a\left(x - \frac{1}{2}\right)^2 - \frac{3}{2};\ 12 = a\left(-1 - \frac{1}{2}\right)^2 - \frac{3}{2},\ a = 6;\ y = 6\left(x - \frac{1}{2}\right)^2 - \frac{3}{2}$

13. $(-3, -4)$ is vertex, so $y = 3(x + 3)^2 - 4$

14. $y = -1\left(x - \frac{1}{2}\right)^2 + k;\ -8 = -\left(3 - \frac{1}{2}\right)^2 + k,\ k = -\frac{7}{4};\ y = -\left(x - \frac{1}{2}\right)^2 - \frac{7}{4}$

15. Since $(-2, 1)$ and $(2, 1)$ are on the graph, the axis of sym. is $x = 0$; $y = ax^2 + k$; $-5 = a(0) + k,\ k = -5;\ 1 = 4a - 5,\ a = \frac{3}{2};\ y = \frac{3}{2}x^2 - 5$

16. Since $(-3, 2)$ and $(1, 2)$ are on the graph, the axis of sym. is $x = -1$; $y = a(x + 1)^2 + k$; $2 = a(-3 + 1)^2 + k$ and $8 = a(-2 + 1)^2 + k$, $2 = 4a + k$ and $8 = a + k$, $3a = -6$, $a = -2$ and $k = 10$; $y = -2(x + 1)^2 + 10$

B 17. $13 = (2 - h)^2 + 4,\ 13 = 4 - 4h + h^2 + 4,\ h^2 - 4h - 5 = 0,\ (h - 5)(h + 1) = 0,$ $h = 5$ or $h = -1$; $y = (x - 5)^2 + 4$ or $y = (x + 1)^2 + 4$

18. $26 = \frac{1}{2}(1 - h)^2 - 6,\ 52 = h^2 - 2h + 1 - 12,\ h^2 - 2h - 63 = 0,$

$(h - 9)(h + 7) = 0$, $h = 9$ or $h = -7$; $y = \frac{1}{2}(x - 9)^2 - 6$ or $y = \frac{1}{2}(x + 7)^2 - 6$

19. $y = a(x - 3)^2 + k$; $5 = a(2 - 3)^2 + k$ and $-25 = a(-1 - 3)^2 + k$, $5 = a + k$ and $-25 = 16a + k$, $15a = -30$, $a = -2$ and $k = 7$; $y = -2(x - 3)^2 + 7$

20. $y = a(x + 2)^2 + k$; $1 = a(-3 + 2)^2 + k$ and $13 = a(-5 + 2)^2 + k$, $1 = a + k$ and $13 = 9a + k$, $8a = 12$, $a = \frac{3}{2}$ and $k = -\frac{1}{2}$; $y = \frac{3}{2}(x + 2)^2 - \frac{1}{2}$

21. $y = ax^2 + k$; $-13 = a + k$ and $11 = 9a + k$, $8a = 24$, $a = 3$ and $k = -16$; $y = 3x^2 - 16$

22. $y = a\left(x + \frac{1}{2}\right)^2 + k$; $6 = a\left(\frac{3}{2} + \frac{1}{2}\right)^2 + k$ and $-9 = a\left(-1 + \frac{1}{2}\right)^2 + k$, $6 = 4a + k$ and $-9 = \frac{1}{4}a + k$, $\frac{15}{4}a = 15$, $a = 4$ and $k = -10$; $y = 4\left(x + \frac{1}{2}\right)^2 - 10$

C 23. Assume $(h + r, s)$ is on the graph of $y = a(x - h)^2 + k$, $a \neq 0$. Then $y = a(h - r - h)^2 + k = ar^2 + k$. But $s = a(h + r - h)^2 + k = ar^2 + k = y$. So $y = s$ when $x = h - r$. The proof of the converse is very similar.

24. Assume (r, s) is on the graph of $y = a(x - h)^2 + k$, $a \neq 0$. Therefore: $s = a(r - h)^2 + k = a(-1)^2(r - h)^2 + k = a(h - r)^2 + k = a(2h - r - h)^2 + k$. This shows that $(2h - r, s)$ is also on the graph. The proof of the converse is very similar.

25. Since $(r_1, 0)$ and $(r_2, 0)$ lie on the graph, $0 = a(r_1 - h)^2 + k = a(r_2 - h)^2 + k$. Simplifying, $a(r_1 - h)^2 = a(r_2 - h)^2$. Since $a \neq 0$, $(r_1 - h)^2 = (r_2 - h)^2$ and $r_1 - h = \pm(r_2 - h)$. If $r_1 - h = r_2 - h$, that would imply $r_1 = r_2$. But $r_1 \neq r_2$.

26. $k < 0$ and $a > 0$, or $k > 0$ and $a < 0$.

27.

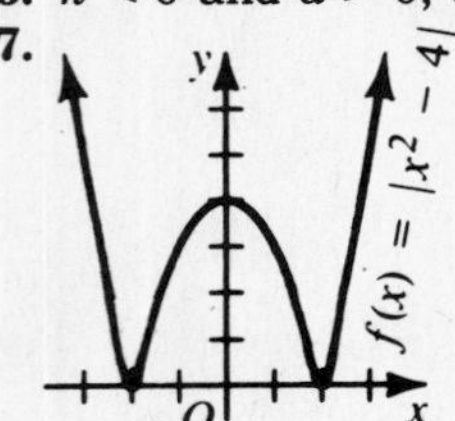

Pages 360–361 • WRITTEN EXERCISES

A 1. $y = x^2 - 10x$
$= (x^2 - 10x + 25) - 25$
$= (x - 5)^2 - 25$
$x = 5$; $(5, -25)$

2. $y = x^2 + 3x$
$= \left(x^2 + 3x + \frac{9}{4}\right) - \frac{9}{4}$
$= \left(x + \frac{3}{2}\right)^2 - \frac{9}{4}$
$x = -\frac{3}{2}$; $\left(-\frac{3}{2}, -\frac{9}{4}\right)$

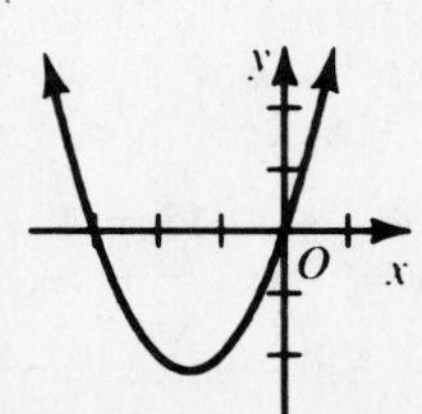

3. $y = 2x^2 - 12x$

$\frac{y}{2} = (x^2 - 6x + 9) - 9$

$\frac{y}{2} = (x - 3)^2 - 9$

$y = 2(x - 3)^2 - 18$

$x = 3$; $(3, -18)$

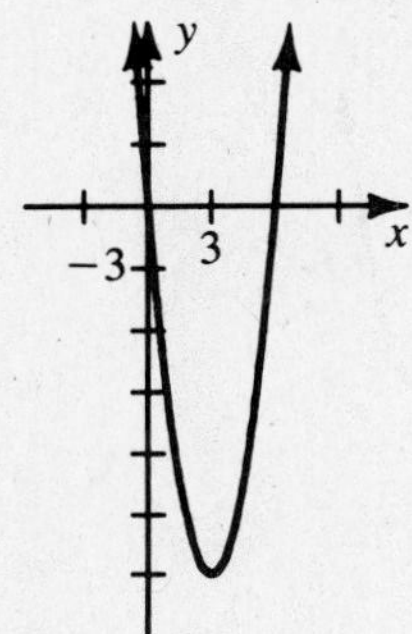

4. $y = -x^2 + 5x$

$-y = \left(x^2 - 5x + \frac{25}{4}\right) - \frac{25}{4}$

$-y = \left(x - \frac{5}{2}\right)^2 - \frac{25}{4}$

$y = -\left(x - \frac{5}{2}\right)^2 + \frac{25}{4}$

$x = \frac{5}{2}$; $\left(\frac{5}{2}, \frac{25}{4}\right)$

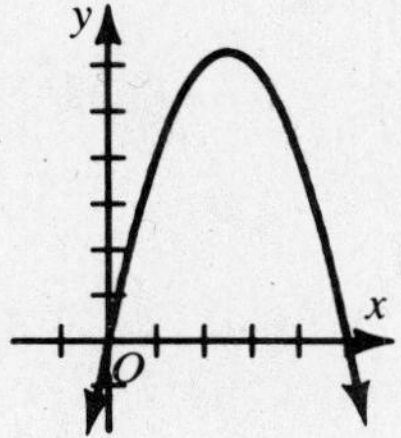

5. $y = -2x^2 - 6$

$x = 0$; $(0, -6)$

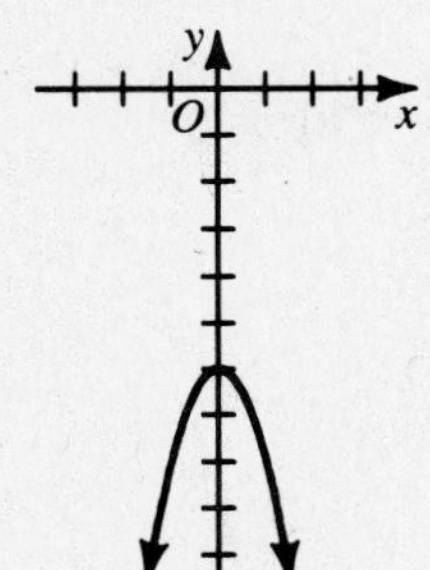

6. $y = \frac{1}{2}x^2 - 5$

$x = 0$; $(0, -5)$

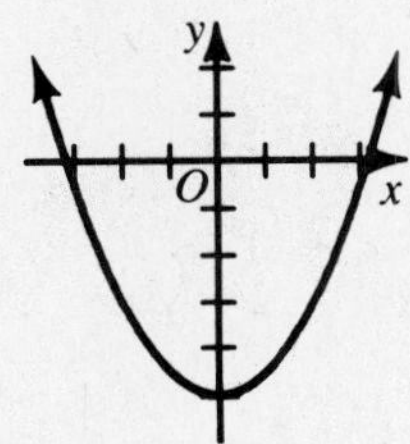

7. $y = x^2 - 6x + 7$

$= (x^2 - 6x + 9) - 2$

$= (x - 3)^2 - 2$

$x = 3$; $(3, -2)$

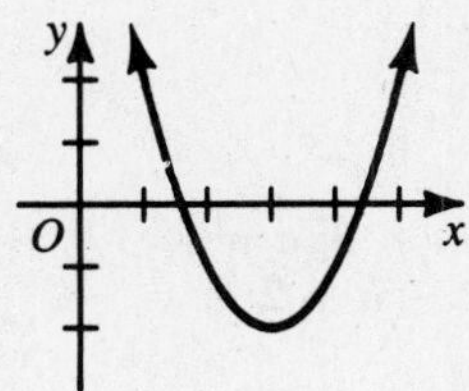

8. $y = x^2 + 4x + 1$

$= (x^2 + 4x + 4) - 3$

$= (x + 2)^2 - 3$

$x = -2$; $(-2, -3)$

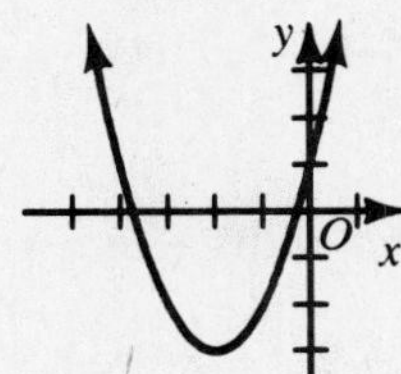

9. $y = 2x^2 - 8x - 3$

$$\frac{y+3}{2} = (x^2 - 4x + 4) - 4$$

$$\frac{y+3}{2} = (x-2)^2 - 4$$

$$y = 2(x-2)^2 - 2(4) - 3$$
$$= 2(x-2)^2 - 11$$

$x = 2$; $(2, -11)$

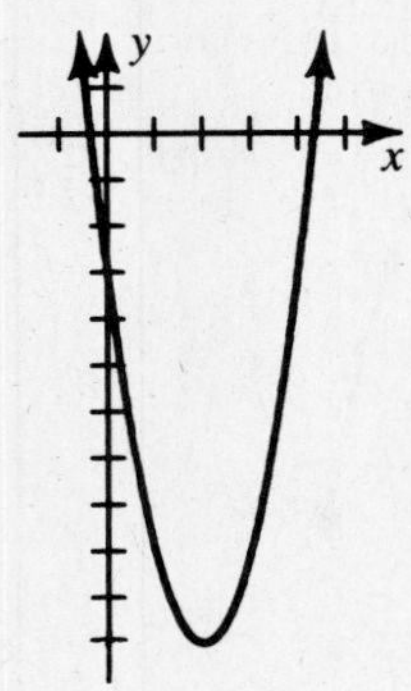

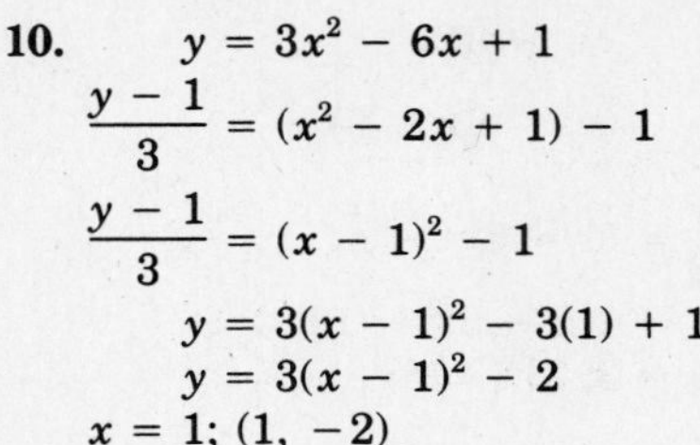

10. $y = 3x^2 - 6x + 1$

$$\frac{y-1}{3} = (x^2 - 2x + 1) - 1$$

$$\frac{y-1}{3} = (x-1)^2 - 1$$

$$y = 3(x-1)^2 - 3(1) + 1$$
$$y = 3(x-1)^2 - 2$$

$x = 1$; $(1, -2)$

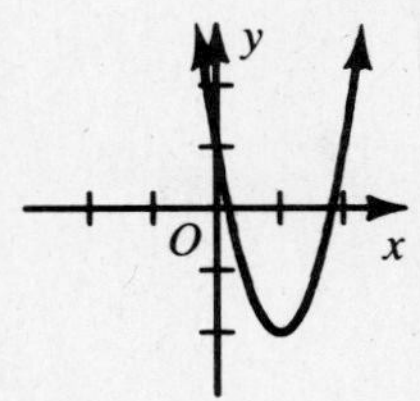

11. $y = -2x^2 + 12x - 9$

$$\frac{y+9}{-2} = (x^2 - 6x + 9) - 9$$

$$\frac{y+9}{-2} = (x-3)^2 - 9$$

$$y = -2(x-3)^2 - 9(-2) - 9$$
$$y = -2(x-3)^2 + 9$$

$x = 3$; $(3, 9)$

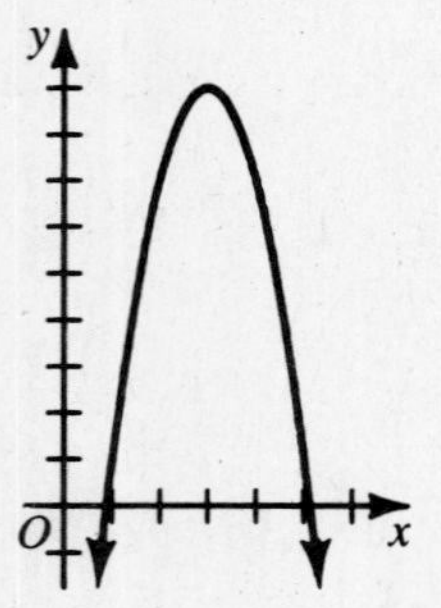

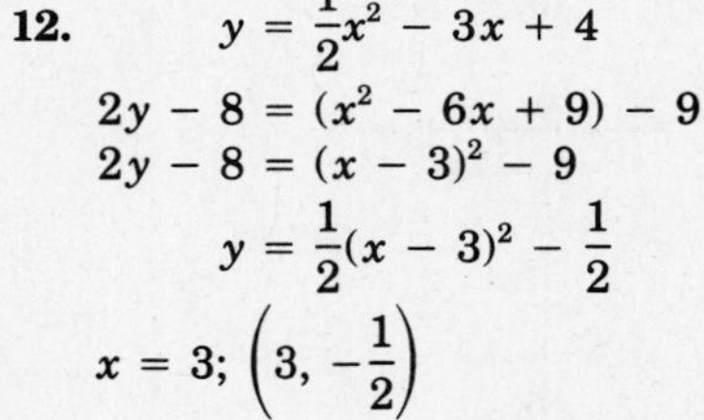

12. $y = \frac{1}{2}x^2 - 3x + 4$

$$2y - 8 = (x^2 - 6x + 9) - 9$$
$$2y - 8 = (x-3)^2 - 9$$
$$y = \frac{1}{2}(x-3)^2 - \frac{1}{2}$$

$x = 3$; $\left(3, -\frac{1}{2}\right)$

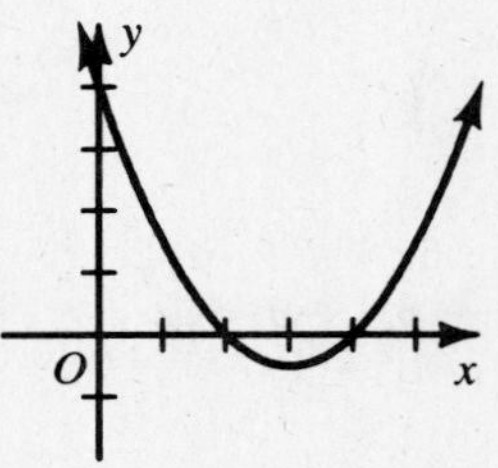

B **13.** (1): $2 = a(0) + b(0) + c$; $c = 2$
(2): $9 = a - b + 2$
(3): $6 = 4a + 2b + 2$
$(3) + 2(2)$: $24 = 6a + 6$; $a = 3$
(2): $9 = 3 - b + 2$; $b = -4$
$y = 3x^2 - 4x + 2$

14. (1): $2 = 9a - 3b + c$
(2): $6 = a - b + c$
(3): $2 = a + b + c$
$(3) - (2)$: $-4 = 2b$; $b = -2$
(1): $2 = 9a + c + 6$
(2): $6 = a + c + 2$
$(1) - (2)$: $-4 = 8a + 4$; $a = -1$
(2): $6 = -1 + 2 + c$; $c = 5$
$y = -x^2 - 2x + 5$

15. (1): $9 = 4a - 2b + c$ $(3) - (1)$: $-6 = 4b;\ b = -\frac{3}{2}$

(2): $-3 = a + b + c$ (1): $9 = 4a + 3 + c;\ 6 = 4a + c$

(3): $3 = 4a + 2b + c$ (2): $-3 = a - \frac{3}{2} + c;\ -\frac{3}{2} = a + c$

$(1) - (2)$: $\frac{15}{2} = 3a;\ a = \frac{5}{2}$ (1): $6 = 10 + c;\ c = -4$ $y = \frac{5}{2}x^2 - \frac{3}{2}x - 4$

16. (1): $2 = a + b + c$ $(2) - (1)$: $-1 = 3a - 3b$; $-1 = 3a - 3b$

(2): $1 = 4a - 2b + c$ $(3) - (2)$: $5 = 5a + 5b$; $3 = 3a + 3b$

(3): $6 = 9a + 3b + c$ $2 = 6a$ $a = \frac{1}{3}$

$(2) - (1)$: $-1 = 1 - 3b;\ b = \frac{2}{3}$ (1): $2 = \frac{1}{3} + \frac{2}{3} + c;\ c = 1$ $y = \frac{1}{3}x^2 + \frac{2}{3}x + 1$

17. $y = r(x^2 - s);\ y = rx^2 - rs;\ x = 0;\ (0, -rs)$

18. $y = rx^2 + 2rx;\ \frac{y}{r} = (x^2 + 2x + 1) - 1;\ y = r(x + 1)^2 - r;\ x = -1;\ (-1, -r)$

19. $y = rx^2 - 2rsx - 2s^2;\ \frac{y + 2s^2}{r} = (x^2 - 2sx + s^2) - s^2;\ y + 2s^2 = r(x - s)^2 - rs^2;$

$y = r(x - s)^2 - (rs^2 + 2s^2);\ x = s;\ (s, -rs^2 - 2s^2)$

20. $y = \frac{s^2}{2} + rsx - rx^2;\ y - \frac{s^2}{2} = -r(x^2 - sx);\ y - \frac{s^2}{2} = -r\left(x^2 - sx + \frac{s^2}{4}\right) + \frac{rs^2}{4};$

$y = -r\left(x - \frac{s}{2}\right)^2 + \frac{rs^2}{4} + \frac{s^2}{2};\ x = \frac{s}{2};\ \left(\frac{s}{2}, \frac{s^2}{2} + \frac{rs^2}{4}\right)$

C 21.
1. $y - k = a_1(x - h)^2,\ y - k = a_2(x - h)^2$, (m, n) satisfies both equations, $m \neq h$ — Hypothesis
2. $n - k = a_1(m - h)^2$ and $n - k = a_2(m - h)^2$ — Subs. prin.
3. $a_1(m - h)^2 = a_2(m - h)^2$ — Trans. prop. of =
4. $a_1 = a_2$ — Mult. prop. of =

22. By completing the square for $y = ax^2 + bx + c$, the equation becomes $y + \frac{b^2 - 4ac}{4a} = a\left(x + \frac{b}{2a}\right)^2$; see Eq. (3), p. 358. Substituting for x and since $\left(-\frac{b}{2a} + r + \frac{b}{2a}\right)^2 = \left(-\frac{b}{2a} - r + \frac{b}{2a}\right)^2$, the points $\left(-\frac{b}{2a} + r, s\right)$ and $\left(-\frac{b}{2a} - r, s\right)$ are both on the graph.

Pages 361–362 • PROBLEMS

A 1. Let x = one number, $24 - x$ = other number; $f(x) = x(24 - x) = 24x - x^2$; $a = -1, b = 24;\ x = -\frac{b}{2a} = -\frac{24}{-2} = 12$; 12 and 12

2. Let x = length; $18 - x$ = width; $f(x) = x(18 - x) = 18x - x^2$; $a = -1, b = 18$; $x = -\frac{b}{2a} = 9$; 9 m by 9 m

3. Let x = width; $72 - 2x$ = length; $f(x) = x(72 - 2x) = 72x - 2x^2$; $a = -2$, $b = 72;\ x = -\frac{b}{2a} = 18$; 18 m

4. $W(I) = 120I - 12I^2;\ a = -12, b = 120;\ I = -\frac{b}{2a} = 5;\ W(5) = 300$; 300 W

B **5.** $h(t) = 30 + 29.4t - 4.9t^2$; $a = -4.9$, $b = 29.4$; $t = -\frac{b}{2a} = 3$; $h(3) = 74.1$; 74.1 m

6.

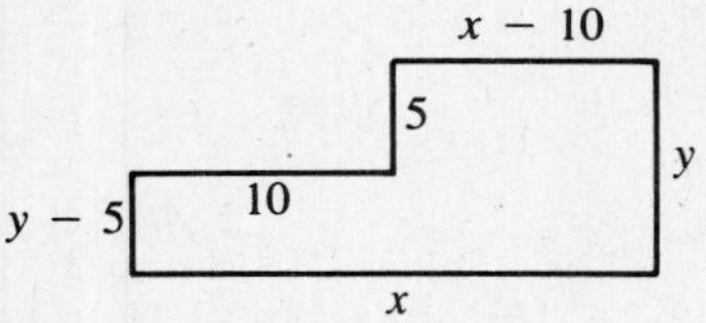
x − 10
5
y
10
y − 5
x

$x + (x - 10) + y + (y - 5) = 60$; $2x + 2y = 75$; $y = \frac{75}{2} - x$; area $= xy - 50 = x\left(\frac{75}{2} - x\right) - 50$; $f(x) = -x^2 + \frac{75}{2}x - 50$; $a = -1$, $b = \frac{75}{2}$; $x = -\frac{b}{2a} = \frac{75}{4}$; $x = \frac{75}{4}$ and $y = \frac{75}{4}$; 18.75 m, 18.75 m

7.

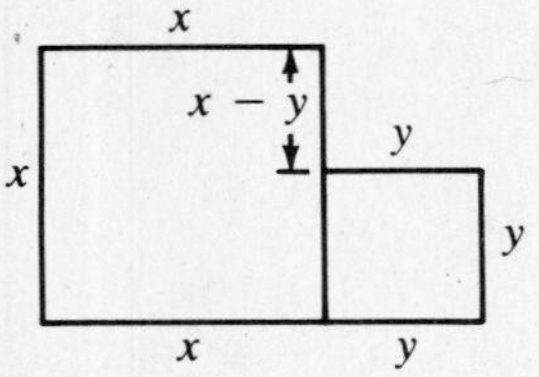
x
x − y
y
x
y
x
y

$3x + (x - y) + 3y = 30$; $4x + 2y = 30$; $y = 15 - 2x$; area $= x^2 + y^2$; $f(x) = x^2 + (15 - 2x)^2 = 5x^2 - 60x + 225$; $x = -\frac{b}{2a} = \frac{60}{10} = 6$; 6 m

8. $v = 29.4 - 9.8t$; $t = \frac{v - 29.4}{-9.8} = 3 - \frac{v}{9.8}$; $h = 30 + 29.4t - 4.9t^2$; $h = 30 + 29.4\left(3 - \frac{v}{9.8}\right) - 4.9\left(3 - \frac{v}{9.8}\right)^2 = 30 + 88.2 - 3v - 4.9\left(9 - \frac{6v}{9.8} + \frac{v^2}{96.04}\right) = 118.2 - 3v - 44.1 + 3v - \frac{4.9}{96.04}v^2 = -\frac{4.9}{96.04}v^2 + 74.1$; $a = -\frac{4.9}{96.04}$, $b = 0$; $v = -\frac{b}{2a} = 0$; hence h is a maximum when $v = 0$.

C **9.** Let x = length of one piece, $60 - x$ = length of other piece. Area $= f(x) = \left(\frac{x}{4}\right)^2 + \left(\frac{60}{4} - \frac{x}{4}\right)^2 = \frac{x^2}{16} + \left(\frac{3600}{16} - \frac{120x}{16} + \frac{x^2}{16}\right) = \frac{x^2}{8} - 7.5x + 225$; $a = \frac{1}{8}$, $b = -7.5$; $x = -\frac{b}{2a} = 30$; $60 - x = 30$; cut the wire at its midpoint.

10. Let x = number of weeks from now; $12 + x$ = fare per person; $400 - 10x$ = no. of passengers. Income $= f(x) = (12 + x)(400 - 10x)$; roots of $(12 + x)(400 - 10x) = 0$ are -12 and 40; $\frac{-12 + 40}{2} = 14$; 14 weeks; fare: \$26

11. **a.** $\triangle CDE \sim \triangle CAB$, $\frac{CD}{CA} = \frac{DE}{AB}$; $\frac{8 - y}{8} = \frac{x}{12}$; $96 - 12y = 8x$; $12y = 96 - 8x$; $y = \frac{24 - 2x}{3}$ **b.** area $= xy = x\left(\frac{24 - 2x}{3}\right) = \frac{24x - 2x^2}{3}$

c. $f(x) = \frac{24x - 2x^2}{3} = -\frac{2}{3}x^2 + 8x$; $x = -\frac{b}{2a} = -8 \div \left(-\frac{4}{3}\right) = 6$

Page 364 • WRITTEN EXERCISES

A **1.** **a.** $x^2 - 16 > 0$
b. $(x + 4)(x - 4) = 0$; $x = -4$ or $x = 4$; $\{-4, 4\}$
d. $\{x: x < -4\} \cup \{x: x > 4\}$
c.

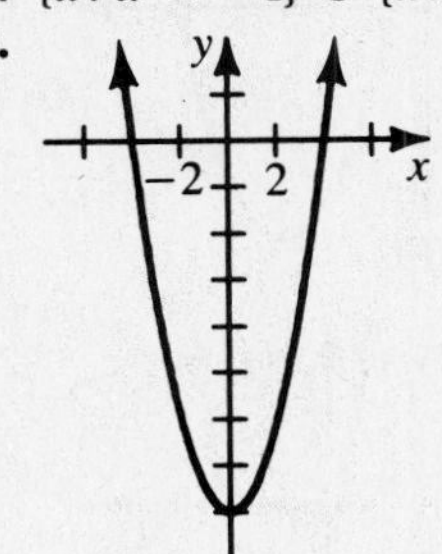

2. **a.** $9x^2 - 4 \le 0$
b. $(3x + 2)(3x - 2) = 0$; $x = -\frac{2}{3}$ or $x = \frac{2}{3}$; $\left\{-\frac{2}{3}, \frac{2}{3}\right\}$
d. $\left\{x: -\frac{2}{3} \le x \le \frac{2}{3}\right\}$
c.

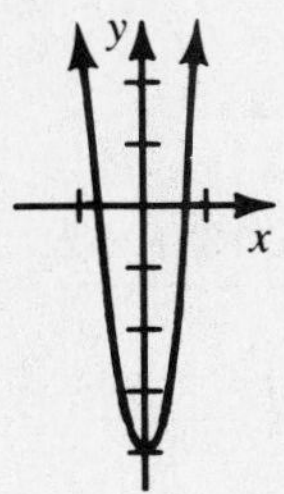

3. **a.** $5x - x^2 < 0$
b. $x(5 - x) = 0$; $x = 0$ or $x = 5$; $\{0, 5\}$
d. $\{x: x < 0\} \cup \{x: x > 5\}$
c.

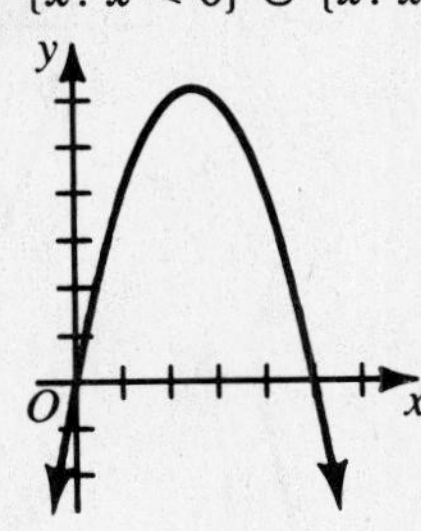

4. **a.** $3x^2 - 8x < 0$
b. $x(3x - 8) = 0$; $x = 0$ or $x = \frac{8}{3}$; $\left\{0, \frac{8}{3}\right\}$
d. $\left\{x: 0 < x < \frac{8}{3}\right\}$
c.

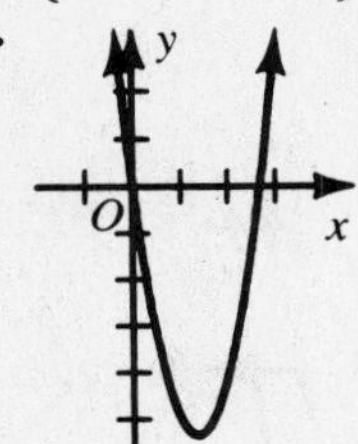

5. **a.** $x^2 - x - 12 \ge 0$
b. $(x - 4)(x + 3) = 0$; $x = 4$ or $x = -3$; $\{-3, 4\}$
d. $\{x: x \le -3\} \cup \{x: x \ge 4\}$
c.

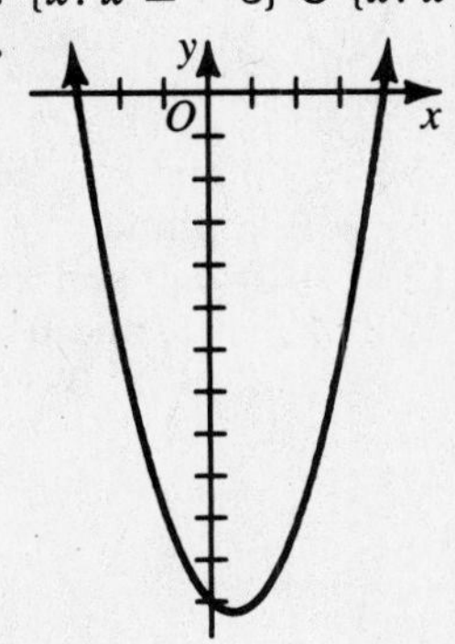

6. **a.** $2x^2 - 9x + 7 \le 0$
b. $(2x - 7)(x - 1) = 0$; $x = \frac{7}{2}$ or $x = 1$; $\left\{1, \frac{7}{2}\right\}$
d. $\left\{x: 1 \le x \le \frac{7}{2}\right\}$
c.

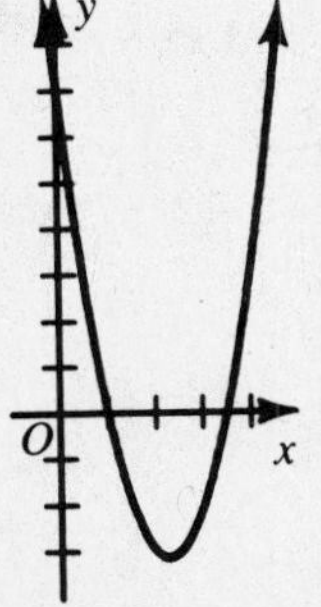

B

7. **a.** $x^2 - 6x + 4 \le 0$

b. $x = \dfrac{6 \pm \sqrt{36 - 4(4)}}{2} =$

$3 \pm \sqrt{5}$; $\{3 - \sqrt{5}, 3 + \sqrt{5}\}$

c.

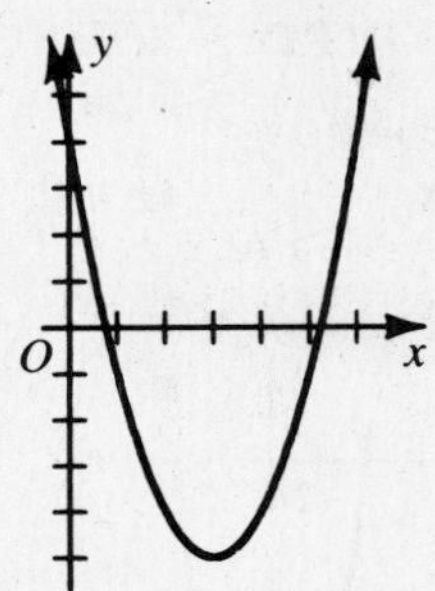

d. $\{x\colon 3 - \sqrt{5} \le x \le 3 + \sqrt{5}\}$

8. **a.** $2x^2 - 4x - 1 \ge 0$

b. $x = \dfrac{4 \pm \sqrt{16 + 4(2)}}{4} =$

$\dfrac{2 \pm \sqrt{6}}{2}$; $\left\{\dfrac{2 + \sqrt{6}}{2}, \dfrac{2 - \sqrt{6}}{2}\right\}$

c.

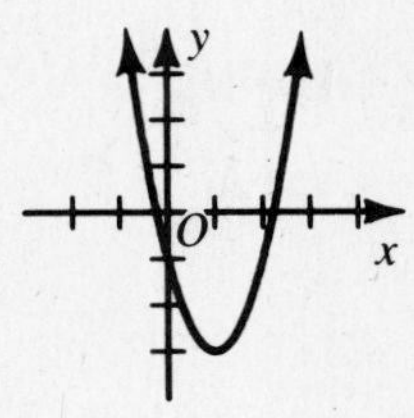

d. $\left\{x\colon x \le \dfrac{2 - \sqrt{6}}{2}\right\} \cup \left\{x\colon x \ge \dfrac{2 + \sqrt{6}}{2}\right\}$

9. **a.** $x^2 + 4x - 1 > 0$

b. $x = \dfrac{-4 \pm \sqrt{16 + 4}}{2} = -2 \pm \sqrt{5}$; $\{-2 + \sqrt{5}, -2 - \sqrt{5}\}$

c.

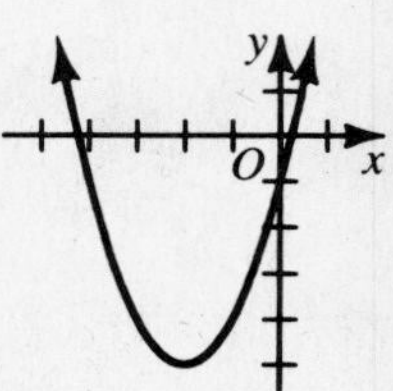

d. $\{x\colon x < -2 - \sqrt{5}\} \cup \{x\colon x > -2 + \sqrt{5}\}$

C

10.

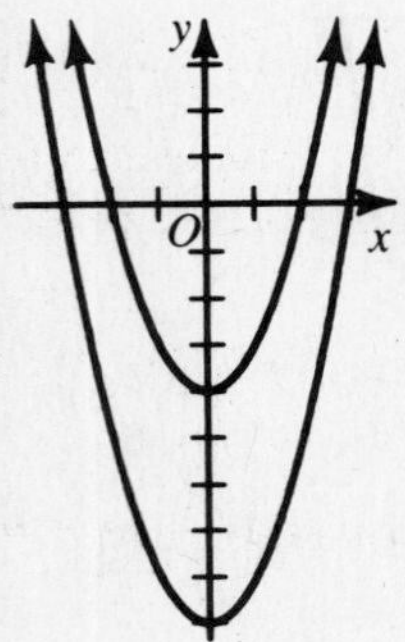

$-5 < x^2 - 9$ and $x^2 - 9 < 0$;
$x^2 - 4 > 0$ and $x^2 - 9 < 0$;
$(x - 2)(x + 2) > 0$, $x < -2$ or $x > 2$;
and $(x - 3)(x + 3) < 0$, $-3 < x < 3$;
$\{x\colon -3 < x < -2\} \cup \{x\colon 2 < x < 3\}$

11.

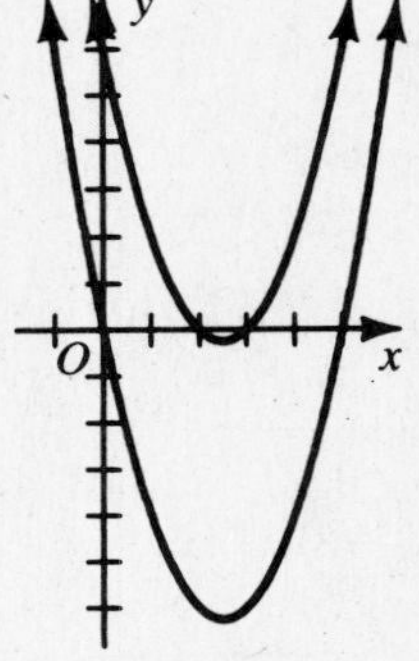

$x^2 - 5x \ge -6$ and $x^2 - 5x \le 0$;
$x^2 - 5x + 6 \ge 0$ and $x^2 - 5x \le 0$;
$(x - 2)(x - 3) \ge 0$ and $x(x - 5) \le 0$; $x \le 2$ or $x \ge 3$, and $0 \le x \le 5$;
$\{x\colon 0 \le x \le 2\} \cup \{x\colon 3 \le x \le 5\}$

Page 364 • SELF-TEST 2

1. $x = 2$; $(2, 3)$

2. maximum

3. 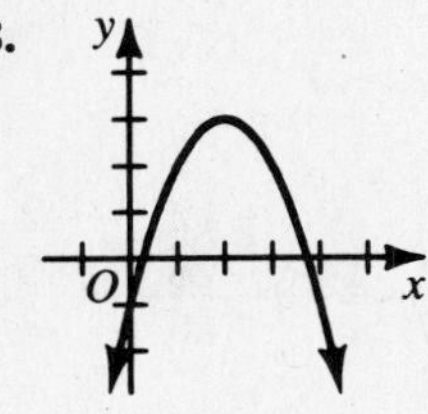

4. $\frac{y-1}{3} = (x^2 + 2x + 1) - 1$
$\frac{y-1}{3} = (x+1)^2 - 1$
$y - 1 = 3(x+1)^2 - 3$
$y = 3(x+1)^2 - 2$

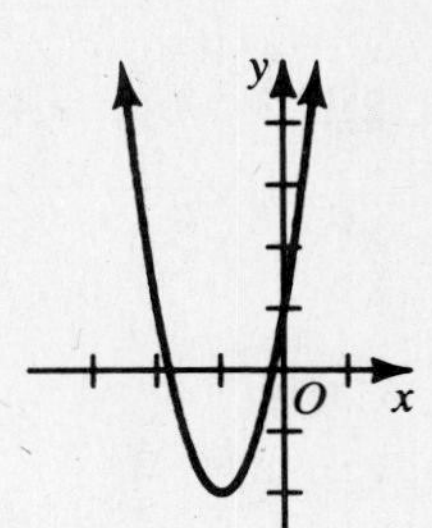

5. Let x = length of one leg; $12 - x$ = length of other leg. $f(x) = \frac{1}{2}x(12 - x) = -\frac{1}{2}x^2 + 6x$; $a = -\frac{1}{2}$, $b = 6$; $x = -\frac{b}{2a} = 6$; $f(x) = \frac{1}{2} \cdot 6(12 - 6) = 18$; 18 cm^2

6. $x^2 + 2x - 2 > 0$; $x = \frac{-2 \pm \sqrt{4 - 4(-2)}}{2} = \frac{-2 \pm \sqrt{12}}{2} = -1 \pm \sqrt{3}$;
$\{x: x < -1 - \sqrt{3}\} \cup \{x: x > -1 + \sqrt{3}\}$

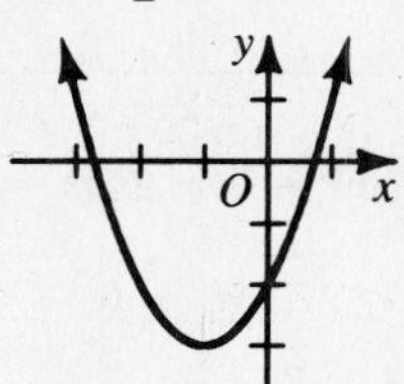

Pages 367–368 • WRITTEN EXERCISES

A

1. $$\begin{array}{r|rrrr} 2 & 1 & -2 & -9 & 18 \\ & & 2 & 0 & -18 \\ \hline & 1 & 0 & -9 & 0 \end{array}$$ 0; zero

2. $$\begin{array}{r|rrrr} -1 & 1 & -2 & -9 & 18 \\ & & -1 & 3 & 6 \\ \hline & 1 & -3 & -6 & 24 \end{array}$$

3. $$\begin{array}{r|rrrr} -3 & 1 & -2 & -9 & 18 \\ & & -3 & 15 & -18 \\ \hline & 1 & -5 & 6 & 0 \end{array}$$ 0; zero

4. $$\begin{array}{r|rrrr} 3 & 1 & -2 & -9 & 18 \\ & & 3 & 3 & -18 \\ \hline & 1 & 1 & -6 & 0 \end{array}$$ 0; zero

5. $$\begin{array}{r|rrrr} -2 & 1 & -2 & -9 & 18 \\ & & -2 & 8 & 2 \\ \hline & 1 & -4 & -1 & 20 \end{array}$$

6. $$\begin{array}{r|rrrrr} -1 & 1 & 1 & -3 & -5 & -6 \\ & & -1 & 0 & 3 & 2 \\ \hline & 1 & 0 & -3 & -2 & -4 \end{array}$$

7. $$\begin{array}{r|rrrrr} -2 & 1 & 1 & -3 & -5 & -6 \\ & & -2 & 2 & 2 & 6 \\ \hline & 1 & -1 & -1 & -3 & 0 \end{array}$$ 0; zero

8. $$\begin{array}{r|rrrrr} 2 & 1 & 1 & -3 & -5 & -6 \\ & & 2 & 6 & 6 & 2 \\ \hline & 1 & 3 & 3 & 1 & -4 \end{array}$$

9. $$\begin{array}{r|rrrrr} 3 & 1 & 1 & -3 & -5 & -6 \\ & & 3 & 12 & 27 & 66 \\ \hline & 1 & 4 & 9 & 22 & 60 \end{array}$$

10. $$\begin{array}{r|rrrrr} -3 & 1 & 1 & -3 & -5 & -6 \\ & & -3 & 6 & -9 & 42 \\ \hline & 1 & -2 & 3 & -14 & 36 \end{array}$$

11. $$\begin{array}{r|rrrrrr} -1 & 3 & 0 & -4 & -10 & -7 & -10 \\ & & -3 & 3 & 1 & 9 & -2 \\ \hline & 3 & -3 & -1 & -9 & 2 & -12 \end{array}$$

12. $$\begin{array}{r|rrrrrr} i & 3 & 0 & -4 & -10 & -7 & -10 \\ & & 3i & -3 & 0 - 7i & 7 - 10i & 10 \\ \hline & 3 & 3i & -7 & -10 - 7i & -10i & 0 \end{array}$$ 0; zero

13. $$\begin{array}{r|rrrrrr} -i & 3 & 0 & -4 & -10 & -7 & -10 \\ & & -3i & -3 & 0 + 7i & 7 + 10i & 10 \\ \hline & 3 & -3i & -7 & -10 + 7i & 10i & 0 \end{array}$$ 0; zero

14. $$\begin{array}{r|rrrrrr} 2 & 3 & 0 & -4 & -10 & -7 & -10 \\ & & 6 & 12 & 16 & 12 & 10 \\ \hline & 3 & 6 & 8 & 6 & 5 & 0 \end{array}$$ 0; zero

15. $$\begin{array}{r|rrrrrr} 2i & 3 & 0 & -4 & -10 & -7 & -10 \\ & & 6i & -12 & 0-32i & 64-20i & 40+114i \\ \hline & 3 & 6i & -16 & -10-32i & 57-20i & 30+114i \end{array}$$

16. $$\begin{array}{r|rrrrr} 3 & 2 & -3 & 0 & -12 & -32 \\ & & 6 & 9 & 27 & 45 \\ \hline & 2 & 3 & 9 & 15 & 13 \end{array}$$

17. $$\begin{array}{r|rrrrr} -2 & 2 & -3 & 0 & -12 & -32 \\ & & -4 & 14 & -28 & 80 \\ \hline & 2 & -7 & 14 & -40 & 48 \end{array}$$

18. $$\begin{array}{r|rrrrr} 2i & 2 & -3 & 0 & -12 & -32 \\ & & 0+4i & -8-6i & 12-16i & 32 \\ \hline & 2 & -3+4i & -8-6i & -16i & 0 \end{array}$$ 0; zero

19. $$\begin{array}{r|rrrrr} -2i & 2 & -3 & 0 & -12 & -32 \\ & & 0-4i & -8+6i & 12+16i & 32 \\ \hline & 2 & -3-4i & -8+6i & 16i & 0 \end{array}$$ 0; zero

20. $$\begin{array}{r|rrrrr} i & 2 & -3 & 0 & -12 & -32 \\ & & 0+2i & -2-3i & 3-2i & 2-9i \\ \hline & 2 & -3+2i & -2-3i & -9-2i & -30-9i \end{array}$$

B

21. $G(1) = 8 + m - 5 = 2$, $m = -1$

22. $$\begin{array}{r|rrrr} 3 & 2 & -7 & 5 & m \\ & & 6 & -3 & +6 \\ \hline & 2 & -1 & 2 & m+6 \end{array}$$ $m + 6 = -1$; $m = -7$

23. $$\begin{array}{r|rrrr} -2 & 3 & -1 & m & 10 \\ & & -6 & +14 & -28-2m \\ \hline & 3 & -7 & m+14 & -18-2m \end{array}$$ $-18 - 2m = 0$; $m = -9$

24. $$\begin{array}{r|rrrr} 3 & 2 & m & -8 & 6 \\ & & +6 & 18+3m & 30+9m \\ \hline & 2 & m+6 & 10+3m & 36+9m \end{array}$$ $36 + 9m = 0$; $m = -4$

25. possible rational roots: ± 1, ± 2, ± 5, ± 10

$$\begin{array}{r|rrrr} -1 & 1 & 2 & -13 & 10 \\ & & -1 & -1 & 14 \\ \hline & 1 & 1 & -14 & 24 \end{array}$$

$$\begin{array}{r|rrrr} 1 & 1 & 2 & -13 & 10 \\ & & 1 & 3 & -10 \\ \hline & 1 & 3 & -10 & 0 \end{array}$$

$$\begin{array}{r|rrrr} -2 & 1 & 2 & -13 & 10 \\ & & -2 & 0 & 26 \\ \hline & 1 & 0 & -13 & 36 \end{array}$$

$$\begin{array}{r|rrrr} 2 & 1 & 2 & -13 & 10 \\ & & 2 & 8 & -10 \\ \hline & 1 & 4 & -5 & 0 \end{array}$$

$$\begin{array}{r|rrrr} -5 & 1 & 2 & -13 & 10 \\ & & -5 & 15 & -10 \\ \hline & 1 & -3 & 2 & 0 \end{array}$$

Since we have found 3 roots, there is no need to test the other possible roots; $\{-5, 1, 2\}$

26. possible rational roots: ± 1, ± 2, ± 7, ± 14

$$\begin{array}{r|rrrr} -1 & 1 & -8 & 5 & 14 \\ & & -1 & 9 & -14 \\ \hline & 1 & -9 & 14 & 0 \end{array}$$

$$\begin{array}{r|rrrr} 1 & 1 & -8 & 5 & 14 \\ & & 1 & -7 & -2 \\ \hline & 1 & -7 & -2 & 12 \end{array}$$

$$\begin{array}{r|rrrr} -2 & 1 & -8 & 5 & 14 \\ & & -2 & 20 & -50 \\ \hline & 1 & -10 & 25 & -36 \end{array}$$

$$\begin{array}{r|rrrr} 2 & 1 & -8 & 5 & 14 \\ & & 2 & -12 & -14 \\ \hline & 1 & -6 & -7 & 0 \end{array}$$

$$\begin{array}{r|rrrr} -7 & 1 & -8 & 5 & 14 \\ & & -7 & 105 & -770 \\ \hline & 1 & -15 & 110 & -756 \end{array}$$

$$\begin{array}{r|rrrr} 7 & 1 & -8 & 5 & 14 \\ & & 7 & -7 & -14 \\ \hline & 1 & -1 & -2 & 0 \end{array}$$

Since we have found 3 roots, there is no need to test the other possible roots; $\{-1, 2, 7\}$

27. possible rational roots: $\pm\frac{1}{2}, \pm 1, \pm\frac{3}{2}, \pm 2, \pm 3, \pm 6$

$$\begin{array}{r|rrrr} -0.5 & 2 & -1 & -7 & 6 \\ & & -1 & 1 & 3 \\ \hline & 2 & -2 & -6 & 9 \end{array} \qquad \begin{array}{r|rrrr} 0.5 & 2 & -1 & -7 & 6 \\ & & 1 & 0 & -3.5 \\ \hline & 2 & 0 & -7 & 2.5 \end{array}$$

$$\begin{array}{r|rrrr} -1 & 2 & -1 & -7 & 6 \\ & & -2 & 3 & 4 \\ \hline & 2 & -3 & -4 & 10 \end{array} \qquad \begin{array}{r|rrrr} 1 & 2 & -1 & -7 & 6 \\ & & 2 & 1 & -6 \\ \hline & 2 & 1 & -6 & 0 \end{array}$$

$$\begin{array}{r|rrrr} -1.5 & 2 & -1 & -7 & 6 \\ & & -3 & 6 & 1.5 \\ \hline & 2 & -4 & -1 & 7.5 \end{array} \qquad \begin{array}{r|rrrr} 1.5 & 2 & -1 & -7 & 6 \\ & & 3 & 3 & -6 \\ \hline & 2 & 2 & -4 & 0 \end{array}$$

$$\begin{array}{r|rrrr} -2 & 2 & -1 & -7 & 6 \\ & & -4 & 10 & -6 \\ \hline & 2 & -5 & 3 & 0 \end{array}$$

Since we have found 3 roots, there is no need to test the other possible roots; $\left\{-2, 1, \frac{3}{2}\right\}$

28. possible rational roots: $\pm 1, \pm\frac{1}{2}, \pm\frac{1}{5}, \pm\frac{1}{10}$

$$\begin{array}{r|rrrr} -1 & 10 & 3 & -6 & 1 \\ & & -10 & 7 & -1 \\ \hline & 10 & -7 & 1 & 0 \end{array} \qquad \begin{array}{r|rrrr} 1 & 10 & 3 & -6 & 1 \\ & & 10 & 13 & 7 \\ \hline & 10 & 13 & 7 & 8 \end{array}$$

$$\begin{array}{r|rrrr} -0.5 & 10 & 3 & -6 & 1 \\ & & -5 & 1 & 2.5 \\ \hline & 10 & -2 & -5 & 3.5 \end{array} \qquad \begin{array}{r|rrrr} 0.5 & 10 & 3 & -6 & 1 \\ & & 5 & 4 & -1 \\ \hline & 10 & 8 & -2 & 0 \end{array}$$

$$\begin{array}{r|rrrr} -0.2 & 10 & 3 & -6 & 1 \\ & & -2 & -0.2 & 1.24 \\ \hline & 10 & 1 & -6.2 & 2.24 \end{array} \qquad \begin{array}{r|rrrr} 0.2 & 10 & 3 & -6 & 1 \\ & & 2 & 1 & -1 \\ \hline & 10 & 5 & -5 & 0 \end{array}$$

Since we have found 3 roots, there is no need to test the other possible roots; $\left\{-1, \frac{1}{5}, \frac{1}{2}\right\}$

C

29.
$$\begin{array}{r|lllll} i & 5 & -3 & a & b & -7 \\ & & 0 + 5i & -5 - 3i & 3 + (a - 5)i & (5 - a) + (3 + b)i \\ \hline & 5 & -3 + 5i & (a - 5) - 3i & (3 + b) + (a - 5)i & (-2 - a) + (3 + b)i \end{array}$$

$-2 - a = 0$ and $3 + b = 0$; $a = -2$ and $b = -3$

30.
$$\begin{array}{r|llll} -2i & 2 & 6 & a & b \\ & & 0 - 4i & -8 - 12i & -24 + (16 - 2a)i \\ \hline & 2 & 6 - 4i & (a - 8) - 12i & (b - 24) + (16 - 2a)i \end{array}$$

[*cont. on next line*]

$$\begin{array}{ll} 4 & -36 \\ (32 - 4a) + (48 - 2b)i & (96 - 4b) + (8a - 72)i \\ \hline (36 - 4a) + (48 - 2b)i & (60 - 4b) + (8a - 72)i \end{array}$$

$60 - 4b = 0$ and $8a - 72 = 0$; $a = 9$ and $b = 15$

31.
$$\begin{array}{r|llll} 0 & a & b & c & d \\ & & 0 & 0 & 0 \\ \hline & a & b & c & d \end{array}$$
; $P(0) = d$

32.
$$\begin{array}{r|llll} 2 & a & b & c & d \\ & & 2a & 4a + 2b & 8a + 4b + 2c \\ \hline & a & 2a + b & 4a + 2b + c & 8a + 4b + 2c + d \end{array}$$
; $8a + 4b + 2c + d = 0$

33. **a.** $P(i) = -ai - b + ci + d = 0$; $(c - a)i - b + d = 0$; $\therefore a = c$ and $b = d$

b. $P(x) = 0$; $P(x) = ax^3 + bx^2 + cx + d = ax^3 + bx^2 + ax + b = ax(x^2 + 1) + b(x^2 + 1) = (ax + b)(x^2 + 1)$; $ax + b = 0$ or $x^2 + 1 = 0$; $x = -\frac{b}{a}$ or $x^2 = -1$; $x = \pm i$

c. If i is a zero of $P(x)$, then $-i$ is a zero of $P(x)$.

Page 368 • COMPUTER EXERCISES

1.
```
10  INPUT "ENTER THE NUMBER OF TERMS IN THE POLYNOMIAL"; D
20  FOR I = D TO 1 STEP -1
30  INPUT "ENTER THE COEFFICIENTS"; P(I)
40  NEXT I
50  INPUT "ENTER THE VALUE OF X"; X
60  LET S = P(D + 1)
70  FOR I = D TO 1 STEP -1
80  LET S = P(I) + (S * X)
90  NEXT I
100  PRINT S
110  END
```
2. -41 **3.** -2 **4.** 411.25

5.
```
10  INPUT "ENTER THE NUMBER OF TERMS IN THE POLYNOMIAL"; D
20  FOR I = D TO 1 STEP -1
30  INPUT "ENTER THE COEFFICIENTS"; P(I)
40  NEXT I
50  INPUT "ENTER A AND B FOR THE COMPLEX NUMBER X"; A, B
60  LET M1 = A : LET M2 = B
70  FOR C = D TO 2 STEP -1
80  IF C = 2 THEN GOTO 140
90  FOR I = 1 TO C - 2
100  LET M3 = M1 * A - M2 * B
110  LET M2 = M1 * B + M2 * A
120  LET M1 = M3
130  NEXT I
140  LET R = P(C) * M1 + R
150  LET IMAG = P(C) * M2 + IMAG
160  LET M1 = A : LET M2 = B
170  NEXT C
180  LET R = R + P(1)
190  IF R = 0 THEN PRINT IMAG;"I" : END
200  PRINT R;
210  IF IMAG < 0 THEN PRINT IMAG;"I" : END
220  IF IMAG = 0 THEN END
230  PRINT " + ";IMAG;"I"
240  END
```
6. a. -13 **b.** $2 - 3i$ **7. a.** $-76 - 168i$ **b.** $-76 + 168i$

Pages 371–372 • WRITTEN EXERCISES

A 1. a.
$$\begin{array}{r|rrrr} 2 & 3 & -8 & 5 & 6 \\ & & 6 & -4 & 2 \\ \hline & 3 & -2 & 1 & 8 \end{array}$$
$3x^3 - 8x^2 + 5x + 6 = (x - 2)(3x^2 - 2x + 1) + 8$

b.
$$\begin{array}{r|rrrr} -1 & 3 & -8 & 5 & 6 \\ & & -3 & 11 & -16 \\ \hline & 3 & -11 & 16 & -10 \end{array}$$
$3x^3 - 8x^2 + 5x + 6 = (x + 1)(3x^2 - 11x + 16) - 10$

2. a. $\underline{3|}\ \begin{array}{rrrr} 2 & -7 & -1 & 10 \\ & 6 & -3 & -12 \\ \hline 2 & -1 & -4 & -2 \end{array}$; $2x^3 - 7x^2 - x + 10 = (x - 3)(2x^2 - x - 4) - 2$

b. $\underline{4|}\ \begin{array}{rrrr} 2 & -7 & -1 & 10 \\ & 8 & 4 & 12 \\ \hline 2 & 1 & 3 & 22 \end{array}$; $2x^3 - 7x^2 - x + 10 = (x - 4)(2x^2 + x + 3) + 22$

3. a. $\underline{-3|}\ \begin{array}{rrrr} 4 & 8 & -7 & 5 \\ & -12 & 12 & -15 \\ \hline 4 & -4 & 5 & -10 \end{array}$; $4x^3 + 8x^2 - 7x + 5 = (x + 3)(4x^2 - 4x + 5) - 10$

b. $\underline{\tfrac{1}{2}|}\ \begin{array}{rrrr} 4 & 8 & -7 & 5 \\ & 2 & 5 & -1 \\ \hline 4 & 10 & -2 & 4 \end{array}$; $4x^3 + 8x^2 - 7x + 5 = \left(x - \frac{1}{2}\right)(4x^2 + 10x - 2) + 4$

4. a. $\underline{\tfrac{3}{2}|}\ \begin{array}{rrrr} 8 & 0 & -12 & -9 \\ & 12 & 18 & 9 \\ \hline 8 & 12 & 6 & 0 \end{array}$; $8x^3 - 12x - 9 = \left(x - \frac{3}{2}\right)(8x^2 + 12x + 6)$

b. $\underline{i|}\ \begin{array}{rrrr} 8 & 0 & -12 & -9 \\ & 0 + 8i & -8 & 0 - 20i \\ \hline 8 & 8i & -20 & -9 - 20i \end{array}$; $8x^3 - 12x - 9 =$
$(x - i)(8x^2 + 8ix - 20) - 9 - 20i$

5. a. $\underline{-4|}\ \begin{array}{rrrr} 1 & 0 & -13 & 18 \\ & -4 & 16 & -12 \\ \hline 1 & -4 & 3 & 6 \end{array}$; $\dfrac{x^3 - 13x + 18}{x + 4} = x^2 - 4x + 3 + \dfrac{6}{x + 4}$

b. $\underline{2|}\ \begin{array}{rrrr} 1 & 0 & -13 & 18 \\ & 2 & 4 & -18 \\ \hline 1 & 2 & -9 & 0 \end{array}$; $\dfrac{x^3 - 13x + 18}{x - 2} = x^2 + 2x - 9$

6. a. $\underline{3|}\ \begin{array}{rrrr} -2 & 3 & -2 & 3 \\ & -6 & -9 & -33 \\ \hline -2 & -3 & -11 & -30 \end{array}$; $\dfrac{-2x^3 + 3x^2 - 2x + 3}{x - 3} = -2x^2 - 3x - 11 + \dfrac{-30}{x - 3}$

b. $\underline{-i|}\ \begin{array}{rrrr} -2 & 3 & -2 & 3 \\ & 0 + 2i & 2 - 3i & -3 \\ \hline -2 & 3 + 2i & -3i & 0 \end{array}$; $\dfrac{-2x^3 + 3x^2 - 2x + 3}{x + i} =$
$-2x^2 + (3 + 2i)x - 3i$

7. a. $\underline{5|}\ \begin{array}{rrrrr} 1 & -4 & 0 & -3 & -9 \\ & 5 & 5 & 25 & 110 \\ \hline 1 & 1 & 5 & 22 & 101 \end{array}$; $\dfrac{x^4 - 4x^3 - 3x - 9}{x - 5} = x^3 + x^2 + 5x + 22 + \dfrac{101}{x - 5}$

b. $\underline{2i|}\ \begin{array}{rrrrr} 1 & -4 & 0 & -3 & -9 \\ & 0 + 2i & -4 - 8i & 16 - 8i & 16 + 26i \\ \hline 1 & -4 + 2i & -4 - 8i & 13 - 8i & 7 + 26i \end{array}$

$$\frac{x^4 - 4x^3 - 3x - 9}{x - 2i} = x^3 + (-4 + 2i)x^2 - (4 + 8i)x + 13 - 8i + \frac{7 + 26i}{x - 2i}$$

8. a. $\underline{-\tfrac{3}{2}|}\ \begin{array}{rrrr} 8 & 10 & -7 & -5 \\ & -12 & 3 & 6 \\ \hline 8 & -2 & -4 & 1 \end{array}$; $\dfrac{8x^3 + 10x^2 - 7x - 5}{x + \frac{3}{2}} = 8x^2 - 2x - 4 + \dfrac{1}{x + \frac{3}{2}}$

b. $\begin{array}{r|rrrr} \frac{1}{2} & 8 & 10 & -7 & -5 \\ & & 4 & 7 & 0 \\ \hline & 8 & 14 & 0 & -5; \end{array}$ $\dfrac{8x^3 + 10x^2 - 7x - 5}{x - \frac{1}{2}} = 8x^2 + 14x + \dfrac{-5}{x - \frac{1}{2}}$

9. $\begin{array}{r|rrrrr} -2 & 1 & 1 & 3 & 7 & -6 \\ & & -2 & 2 & -10 & 6 \\ \hline & 1 & -1 & 5 & -3 & 0; \end{array}$ yes

10. $\begin{array}{r|rrrrr} 3 & 1 & -5 & 10 & -6 & -18 \\ & & 3 & -6 & 12 & 18 \\ \hline & 1 & -2 & 4 & 6 & 0; \end{array}$ yes

11. $\begin{array}{r|rrrrr} -i & 1 & 0 & 7 & 0 & -4 \\ & & 0 - i & -1 & 0 - 6i & -6 \\ \hline & 1 & -i & 6 & -6i & -10; \end{array}$ no; $R = -10$

12. $\begin{array}{r|rrrr} 2i & 2 & -5 & 8 & 20 \\ & & 0 + 4i & -8 - 10i & 20 \\ \hline & 2 & -5 + 4i & -10i & 40; \end{array}$ no; $R = 40$

13. $\begin{array}{r|rrrr} \frac{1}{2} & 2 & -7 & 5 & -3 \\ & & 1 & -3 & 1 \\ \hline & 2 & -6 & 2 & -2; \end{array}$ no; $R = -2$

14. $\begin{array}{r|rrrrr} \frac{1}{3} & 3 & -7 & 0 & 1 & -2 \\ & & 1 & -2 & -\frac{2}{3} & \frac{1}{9} \\ \hline & 3 & -6 & -2 & \frac{1}{3} & -\frac{17}{9}; \end{array}$ no; $R = -\frac{17}{9}$

15. $\begin{array}{r|rrrr} & 1 & -3 & -10 & 24 \\ \hline 2 & 1 & -1 & -12 & 0 \end{array}$ $x^2 - x - 12 = (x - 4)(x + 3) = 0;\ -3, 4$

16. $\begin{array}{r|rrrr} & 1 & 3 & -2 & -6 \\ \hline -3 & 1 & 0 & -2 & 0 \end{array}$ $x^2 - 2 = (x + \sqrt{2})(x - \sqrt{2}) = 0;\ \sqrt{2}, -\sqrt{2}$

17. $\begin{array}{r|rrrr} & 2 & 3 & -2 & -3 \\ \hline 1 & 2 & 5 & 3 & 0 \end{array}$ $2x^2 + 5x + 3 = (2x + 3)(x + 1) = 0;\ -1, -\frac{3}{2}$

18. $\begin{array}{r|rrrr} & 2 & -1 & -25 & -12 \\ \hline -\frac{1}{2} & 2 & -2 & -24 & 0 \end{array}$ $2x^2 - 2x - 24 = 2(x - 4)(x + 3) = 0;\ -3, 4$

19. $\begin{array}{r|rrrr} & 1 & 5 & 9 & 45 \\ \hline -5 & 1 & 0 & 9 & 0 \end{array}$ $x^2 + 9 = (x + 3i)(x - 3i) = 0;\ 3i, -3i$

20. $\begin{array}{r|rrrr} & 1 & -7 & 17 & -15 \\ \hline 3 & 1 & -4 & 5 & 0 \end{array}$ $x^2 - 4x + 5 = 0;\ x = \dfrac{4 \pm \sqrt{-4}}{2} = 2 \pm i;\ 2 + i,\ 2 - i$

B 21. $\begin{array}{r|rrrrr} & 1 & 1 & -11 & -9 & 18 \\ \hline -2 & 1 & -1 & -9 & 9 & 0 \\ 1 & 1 & 0 & -9 & 0 & \end{array}$ $x^2 - 9 = (x + 3)(x - 3) = 0;\ 1, 3, -3$

22. $$\begin{array}{r|rrrrr} & 1 & 2 & -2 & 2 & -3 \\ \hline 1 & 1 & 3 & 1 & 3 & 0 \\ -3 & 1 & 0 & 1 & 0 & \end{array} \quad x^2 + 1 = (x + i)(x - i) = 0;\ -3,\ i,\ -i$$

23. $x^3 + x^2 - 4x - 4 = 0;\ \dfrac{p}{q} \in \{\pm 1, \pm 2, \pm 4\};$

$$\begin{array}{r|rrrr} & 1 & 1 & -4 & -4 \\ \hline -1 & 1 & 0 & -4 & 0 \end{array} \quad x^2 - 4 = (x + 2)(x - 2) = 0;\ \{-1, -2, 2\}$$

24. $x^3 + 4x^2 + x - 6 = 0;\ \dfrac{p}{q} \in \{\pm 1, \pm 2, \pm 3, \pm 6\};$

$$\begin{array}{r|rrrr} & 1 & 4 & 1 & -6 \\ \hline 1 & 1 & 5 & 6 & 0 \end{array} \quad x^2 + 5x + 6 = (x + 3)(x + 2) = 0;\ \{1, -2, -3\}$$

25. $x^3 - 5x^2 + 10x - 8 = 0;\ \dfrac{p}{q} \in \{\pm 1, \pm 2, \pm 4, \pm 8\};$

$$\begin{array}{r|rrrr} & 1 & -5 & 10 & -8 \\ \hline 2 & 1 & -3 & 4 & 0 \end{array} \quad x^2 - 3x + 4 = 0,\ x = \frac{3 \pm \sqrt{-7}}{2};\ \left\{2, \frac{3 + i\sqrt{7}}{2}, \frac{3 - i\sqrt{7}}{2}\right\}$$

26. $3x^3 + 5x^2 - 26x + 8 = 0;\ \dfrac{p}{q} \in \left\{\pm 1, \pm 2, \pm 4, \pm 8, \pm\frac{1}{3}, \pm\frac{2}{3}, \pm\frac{4}{3}, \pm\frac{8}{3}\right\};$

$$\begin{array}{r|rrrr} & 3 & 5 & -26 & 8 \\ \hline 2 & 3 & 11 & -4 & 0 \end{array} \quad 3x^2 + 11x - 4 = (3x - 1)(x + 4) = 0;\ \left\{2, \frac{1}{3}, -4\right\}$$

27. $x^4 - x^3 - 5x^2 - x - 6 = 0;\ \dfrac{p}{q} \in \{\pm 1, \pm 2, \pm 3, \pm 6\};$

$$\begin{array}{r|rrrrr} & 1 & -1 & -5 & -1 & -6 \\ \hline -2 & 1 & -3 & 1 & -3 & 0 \\ 3 & 1 & 0 & 1 & 0 & \end{array} \quad x^2 + 1 = (x + i)(x - i) = 0;\ \{-2, 3, i, -i\}$$

28. $2x^4 + 3x^3 + 6x^2 + 12x - 8 = 0;\ \dfrac{p}{q} \in \left\{\pm 1, \pm 2, \pm 4, \pm 8, \pm\frac{1}{2}\right\};$

$$\begin{array}{r|rrrrr} & 2 & 3 & 6 & 12 & -8 \\ \hline -2 & 2 & -1 & 8 & -4 & 0 \\ \frac{1}{2} & 2 & 0 & 8 & 0 & \end{array} \quad 2x^2 + 8 = 2(x + 2i)(x - 2i) = 0;\ \left\{-2, \frac{1}{2}, 2i, -2i\right\}$$

29. $$\begin{array}{r|rrrr} 3 & 1 & -4 & m & -9 \\ & & 3 & -3 & 3m - 9 \\ \hline & 1 & -1 & m - 3 & 3m - 18 \end{array} \qquad \begin{array}{l} 3m - 18 = 0 \\ m = 6 \end{array}$$

30. $$\begin{array}{r|rrrr} -2 & 1 & 3 & -7 & m \\ & & -2 & -2 & 18 \\ \hline & 1 & 1 & -9 & 18 + m \end{array} \qquad \begin{array}{l} 18 + m = 0 \\ m = -18 \end{array}$$

31. $$\begin{array}{r|rrrr} -3 & 1 & 2 & m & -3 \\ & & -3 & 3 & -3m - 9 \\ \hline & 1 & -1 & m + 3 & -3m - 12 \end{array} \qquad \begin{array}{l} -3m - 12 = -3 \\ m = -3 \end{array}$$

C **32.** $P(x) = x^{85} + mx - 4$; $P(-1) = (-1)^{85} - m - 4 = -m - 5$; by the Factor Theorem, $P(-1) = 0$ so $-m - 5 = 0$ or $m = -5$

33. Let $P(x) = 2x^{18} + mx^2 - 1$, then $P(i) = 2i^{18} + mi^2 - 1 = -2 - m - 1 = -3 - m$; since $(x - i)$ is a factor of $P(x)$, $P(i) = 0$. Thus $-3 - m = 0$ or

$m = 3$. Substituting for m and evaluating $P(-i)$ gives $P(-i) = 2(-i)^{18} - 3(-i)^2 - 1 = -2 + 3 - 1 = 0$. Therefore by the Factor Theorem, $x + i$ is a factor of $P(x)$.

34. $P(x) = ax^3 + bx^2 + cx + d$; $P(ki) = a(ki)^3 + b(ki)^2 + c(ki) + d = ak^3i^3 + bk^2i^2 + cki + d = (-bk^2 + d) + (-ak^3 + ck)i$. $P(-ki) = a(-ki)^3 + b(-ki)^2 + c(-ki) + d = -ak^3i^3 + bk^2i^2 - cki + d = (-bk^2 + d) + (ak^3 - ck)i = (-bk^2 + d) - (-ak^3 + ck)i$. Since $a, b, c, d \in \mathcal{R}$, then $P(ki)$ is the complex conjugate of $P(-ki)$, $k \in \mathcal{R}$.

Page 373 • READING ALGEBRA

1. set **2.** summation **3.** approximately equal **4.** set of real numbers **5.** $|c|$ **6.** $\mathcal{C}$ **7.** $\lim\limits_{n \to \infty}$

Page 375 • WRITTEN EXERCISES

A

1. $(x - 3)(x + 1)(x - 2) = 0$; $x^3 - 4x^2 + x + 6 = 0$

2. $(x - \sqrt{6})(x + \sqrt{6})(x + 2) = 0$; $x^3 + 2x^2 - 6x - 12 = 0$

3. $(x - 3i)(x + 3i)(x - 4) = 0$; $x^3 - 4x^2 + 9x - 36 = 0$

4. $(x + 5)(x - 2 - i)(x - 2 + i) = 0$; $x^3 + x^2 - 15x + 25 = 0$

5. $(x - 2)(x + 2)(x - i\sqrt{3})(x + i\sqrt{3}) = 0$; $x^4 - x^2 - 12 = 0$

6. $(x - 6)(x - 1 - i\sqrt{5})(x - 1 + i\sqrt{5}) = 0$; $x^3 - 8x^2 + 18x - 36 = 0$

7. Since $2 + 3i$ is a root, $2 - 3i$ is a root; $x^3 - 3x^2 + 9x + 13 = (x + 1)(x - 2 - 3i)(x - 2 + 3i)$

8. Since $-1 + i\sqrt{3}$ is a root, $-1 - i\sqrt{3}$ is a root; $x^3 - 2x^2 - 4x - 16 = (x - 4)(x + 1 - i\sqrt{3})(x + 1 + i\sqrt{3})$

9. Since $i\sqrt{5}$ is a root, $-i\sqrt{5}$ is a root: $(x - i\sqrt{5})(x + i\sqrt{5}) = x^2 + 5$; $(x^3 - 3x^2 + 5x - 15) \div (x^2 + 5) = x - 3$, $x = 3$; $x^3 - 3x^2 + 5x - 15 = (x + i\sqrt{5})(x - i\sqrt{5})(x - 3)$

10. Since $-3i$ is a root, $3i$ is a root: $(x - 3i)(x + 3i) = x^2 + 9$; $(x^4 + 11x^2 + 18) \div (x^2 + 9) = x^2 + 2 = (x + i\sqrt{2})(x - i\sqrt{2})$; $x = \pm i\sqrt{2}$; $x^4 + 11x^2 + 18 = (x + 3i)(x - 3i)(x + i\sqrt{2})(x - i\sqrt{2})$

11. Since $3 + i$ is a root, $3 - i$ is a root: $(x - 3 - i)(x - 3 + i) = x^2 - 6x + 10$; $(x^3 - 2x^2 - 14x + 40) \div (x^2 - 6x + 10) = x + 4$, $x = -4$; $x^3 - 2x^2 - 14x + 40 = (x - 3 - i)(x - 3 + i)(x + 4)$

12. Since $-2 - i$ is a root, $-2 + i$ is a root: $(x + 2 + i)(x + 2 - i) = x^2 + 4x + 5$; $(x^4 + 4x^3 + 2x^2 - 12x - 15) \div (x^2 + 4x + 5) = x^2 - 3 = (x + \sqrt{3})(x - \sqrt{3})$, $x = \pm\sqrt{3}$; $x^4 + 4x^3 + 2x^2 - 12x - 15 = (x + 2 + i)(x + 2 - i)(x + \sqrt{3})(x - \sqrt{3})$

13. Since $-2 + i$ is a root, $-2 - i$ is a root: $(x + 2 + i)(x + 2 - i) = x^2 + 4x + 5$; $(x^4 + 4x^3 + 11x^2 + 24x + 30) \div (x^2 + 4x + 5) = x^2 + 6 = (x + i\sqrt{6})(x - i\sqrt{6})$, $x = \pm i\sqrt{6}$; $x^4 + 4x^3 + 11x^2 + 24x + 30 = (x + 2 + i)(x + 2 - i)(x + i\sqrt{6})(x - i\sqrt{6})$

14. Since $-i\sqrt{3}$ is a root, $i\sqrt{3}$ is a root: $(x - i\sqrt{3})(x + i\sqrt{3}) = x^2 + 3$; $(x^4 - 6x^3 + 27x^2 - 18x + 72) \div (x^2 + 3) = x^2 - 6x + 24$, $x = 3 \pm i\sqrt{15}$; $x^4 - 6x^3 + 27x^2 - 18x + 72 = (x - i\sqrt{3})(x + i\sqrt{3})(x - 3 + i\sqrt{15})(x - 3 - i\sqrt{15})$

B **15.**

Statement	Reason
1. $P(x)$ is a polynomial of degree $2n - 1$, $n > 0$, with real coefficients.	Hypothesis
2. Suppose $P(x)$ has no real roots.	Hypothesis
3. $P(x)$ has $2n - 1$ complex roots.	First theorem, page 374
4. All roots are of the form $a + bi$, $b \neq 0$.	Def. of imaginary number
5. If $a + bi$ is a root, $a - bi$ is a root.	Second theorem, page 374
6. There must be an even number of roots.	All the roots occur in pairs.
7. $P(x)$ has at least one real root.	Since (6) contradicted (3), the assumption (2) must be false.

16.

Statement	Reason
1. $P(x)$ is a polynomial of degree $2n$, $n \geq 0$, with real coefficients.	Hypothesis
2. Suppose $P(x)$ has at least one real root, r_1.	Hypothesis
3. $P_1(x) = \dfrac{P(x)}{x - r_1}$ is a polynomial with real coefficients of degree $2n - 1$.	Factor Theorem, canc. prop. of mult.
4. $P_1(x)$ has at least one real root, r_2.	Ex. 15, p. 375
5. $P_2(x) = \dfrac{P_1(x)}{x - r_2}$ is a polynomial with real coefficients and degree $2n - 2$.	Factor Theorem, canc. prop. of mult.
6. Suppose $P_2(x)$ has at least one real root, r_3.	Hypothesis
7. $P_3(x) = \dfrac{P_2(x)}{x - r_3}$ is a polynomial with degree $2(n - 1) - 1$.	Factor Theorem, canc. prop. of mult.
8. $P_3(x)$ has at least one real root, r_4.	Ex. 15, p. 375
9. $P(x)$ has no roots or it has an even number of roots.	Continuing as above, for each root found for an even-degree polynomial, by Ex. 15, the resulting depressed equation of odd degree has at least one real root. This continues until an even-degree polynomial with no roots occurs, leaving a set of an even number of real roots.

17. From Ex. 24, p. 284, $\overline{c(a + bi)^k} \cdot \overline{c(a - bi)^k} = \overline{[c(a + bi)^k\, c(a - bi)^k]} = \overline{c^2[(a + bi)(a - bi)]^k} = \overline{c^2(a^2 + b^2)^k} = c^2(a^2 + b^2)^k$. Then $\overline{c(a - bi)^k} = \dfrac{c^2(a^2 + b^2)^k}{c(a + bi)^k} = \dfrac{c^2}{c}\left[\dfrac{a^2 + b^2}{a + bi}\right]^k = c\left[\dfrac{(a^2 + b^2)(a + bi)}{(a - bi)(a + bi)}\right]^k =$ $c\left[\dfrac{(a^2 + b^2)(a + bi)}{a^2 + b^2}\right]^k = c(a + bi)^k$

C **18.** $P(x)$ is a polynomial with real coefficients, say, $P(x) = a_0x^k + a_1x^{k-1} + a_2x^{k-2} + \cdots + a_{k-1}x + a_k$. Then the conjugate of $P(a - bi)$ equals the conjugate of $a_0(a - bi)^k + a_1(a - bi)^{k-1} + \cdots + a_{k-1}(a - bi) + a_k$, and by Ex. 17 above and by Ex. 28 on page 282 this is equal to $a_0(a + bi)^k + a_1(a + bi)^{k-1} + \cdots + a_{k-1}(a + bi) + a_k = P(a + bi)$.

19. $P(x)$ is a polynomial with real coefficients and $P(a + bi) = 0$. By Ex. 18 above, $P(a + bi)$ is the complex conjugate of $P(a - bi)$. But $P(a + bi)$ is a real number, since 0 is real, therefore the complex conjugate of $P(a + bi)$, $P(a - bi)$ is equal to $P(a + bi) = 0$.

Page 376 • SELF-TEST 3

1. a.
$$\begin{array}{r|rrrrr} -2 & 1 & -5 & 3 & -4 & -1 \\ & & -2 & 14 & -34 & 76 \\ \hline & 1 & -7 & 17 & -38 & 75 \end{array};\ P(-2) = 75$$

b.
$$\begin{array}{r|rrrrr} i & 1 & -5 & 3 & -4 & -1 \\ & & 0 + i & -1 - 5i & 5 + 2i & -2 + i \\ \hline & 1 & -5 + i & 2 - 5i & 1 + 2i & -3 + i \end{array};\ P(i) = -3 + i$$

2.
$$\begin{array}{r|rrrr} 3 & 2 & -5 & 3 & 4 \\ & & 6 & 3 & 18 \\ \hline & 2 & 1 & 6 & 22 \end{array} \quad \frac{2x^3 - 5x^2 + 3x + 4}{x - 3} = 2x^2 + x + 6 + \frac{22}{x - 3}$$

3. $\frac{p}{q} \in \left\{\pm 1, \pm 2, \pm 5, \pm 10, \pm\frac{1}{2}, \pm\frac{5}{2}\right\}$;
$$\begin{array}{r|rrrr} & 2 & -7 & 1 & 10 \\ \hline -1 & 2 & -9 & 10 & 0 \end{array} \quad 2x^2 - 9x + 10 = (2x - 5)(x - 2) = 0;\ \left\{-1, 2, \frac{5}{2}\right\}$$

4. Since $3 - 2i$ is a root, $3 + 2i$ is a root. $(x - 3 + 2i)(x - 3 - 2i) = x^2 - 6x + 13$; $(x^3 - 8x^2 + 25x - 26) \div (x^2 - 6x + 13) = x - 2$; $x = 2$; $\{3 - 2i, 3 + 2i, 2\}$

Page 379 • WRITTEN EXERCISES

A 1. $P(x) = x^3 - 5x^2 + x + 2$ (sign changes 1, 2)

$P(-x) = -x^3 - 5x^2 - x + 2$ (sign change 1)

pos.	neg.	imag.
2	1	0
0	1	2

2. $P(x) = -x^3 - 4x^2 + 6x - 3$ (sign changes 1, 2)

$P(-x) = x^3 - 4x^2 - 6x - 3$ (sign change 1)

pos.	neg.	imag.
2	1	0
0	1	2

3. $P(x) = -2x^4 + x^3 - 7x^2 + 4x + 1$ (sign changes 1, 2, 3)

$P(-x) = -2x^4 - x^3 - 7x^2 - 4x + 1$ (sign change 1)

pos.	neg.	imag.
3	1	0
1	1	2

4. $P(x) = x^4 + 5x^3 - x^2 - 7x + 8$ (sign changes 1, 2)

$P(-x) = x^4 - 5x^3 - x^2 + 7x + 8$ (sign changes 1, 2)

pos.	neg.	imag.
2	2	0
2	0	2
0	2	2
0	0	4

5. $P(y) = y^5 + y^4 - 2y^3 - y^2 + y - 10$ (sign changes 1, 2, 3)

$P(-y) = -y^5 + y^4 + 2y^3 - y^2 - y - 10$ (sign changes 1, 2)

pos.	neg.	imag.
3	2	0
3	0	2
1	2	2
1	0	4

6. $P(v) = \underbrace{v^6 - 2v^5}_{1} - \underbrace{4v^4 - v^3 + v}_{2} + 1$

$P(-v) = v^6 + \underbrace{2v^5 - 4v^4}_{1}\underbrace{\;+ v^3}_{2}\underbrace{\; - v}_{3}\underbrace{\; + 1}_{4}$

Sign changes: $P(v)$: 1, 2; $P(-v)$: 1, 2, 3, 4

pos.	neg.	imag.
2	4	0
2	2	2
2	0	4
0	4	2
0	2	4
0	0	6

7.

x				$P(x)$
1	1	0	2	18
−1	1	−2	4	5
−2	1	−3	8	−7

$M = 1, L = -2$

8.

x				$P(x)$
1	1	−1	−2	−5
2	1	0	−1	−5
3	1	1	2	3
−1	1	−3	2	−5

$M = 3, L = -1$

9.

y				$P(y)$
1	1	0	−5	−3
2	1	1	−3	−4
3	1	2	1	5
−1	1	−2	−3	5
−2	1	−3	1	0

$M = 3, L = -2$

10.

z				$P(z)$
1	2	1	−3	0
2	2	3	2	7
−1	2	−3	−1	4
−2	2	−5	6	−9

$M = 2, L = -2$

B 11.

u					$P(u)$
1	1	−1	−4	0	8
2	1	0	−3	−2	4
3	1	1	0	4	20
−1	1	−3	0	4	4
−2	1	−4	5	−6	20

$M = 3, L = -2$

12.

w					$P(w)$
1	2	3	0	−1	4
2	2	5	7	13	31
−1	2	−1	−2	1	4
−2	2	−3	3	−7	19

$M = 2, L = -2$

13.

x					$P(x)$
1	1	1	−2	−9	−3
2	1	2	1	−5	−4
3	1	3	6	11	39
−1	1	−1	−2	−5	11
−2	1	−2	1	−9	24

$M = 3, L = -2$

14.

v					$P(v)$
1	1	3	2	9	4
−1	1	1	−2	9	−14
−2	1	0	−1	9	−23
−3	1	−1	2	1	−8
−4	1	−2	7	−21	79

$M = 1, L = -4$

C 15. Assume that $P(x_0) = 0$ and $x_0 > M$. Since $M \geq 0$, then $x_0 - M > 0$ and $x_0 > 0$. Since $Q(x)$ has nonnegative coefficients and $x_0 > 0$, then each term of $Q(x)$ is greater than or equal to 0 so $Q(x) \geq 0$. Given that the degree of $P(x) \geq 1$, then the degree of $Q(x) > 0$ and $Q(x) \neq 0$ (i.e., $Q(x)$ is not the zero polynomial). The product $(x_0 - M)Q(x) > 0$ since each of the factors is greater than 0. It is given that $R \geq 0$, so $P(x_0) = (x_0 - M)Q(x) + R > 0$. This contradicts the assumption that $P(x_0) = 0$ when $x_0 > M$. Thus when $x_0 > M$, x_0 cannot be a root of $P(x) = 0$.

16. Assume that $P(x_0) = 0$ and $x_0 < L$. Since $L \le 0$, then $x_0 - L < 0$ and $x_0 < 0$. Since x_0 is a negative number, then $x_0^n > 0$ when n is even and $x_0^n < 0$ when n is odd. *Case I:* The constant term of $Q(x) \ge 0$. Thus, because of alternating signs, $R \le 0$, terms of $Q(x)$ with even exponents have coefficients ≥ 0, and terms with odd exponents have coefficients ≤ 0. Thus, when n is even, $a_n \ge 0$ and $x_0^n > 0$ so $a_n x_0^n \ge 0$. When n is odd, $a_n \le 0$ and $x_0^n < 0$ so $a_n x_0^n \ge 0$. Therefore, $Q(x) > 0$. The product $(x_0 - L)Q(x) < 0$ since $x_0 - L < 0$. It has been shown that $R \le 0$, so $P(x_0) = (x_0 - L)Q(x) + R < 0$. This contradicts the assumption that $P(x_0) = 0$ when $x_0 < L$. Thus when $x_0 < L$, x_0 cannot be a root of $P(x) = 0$. *Case II:* The constant term of $Q(x) \le 0$. Thus, because of alternating signs, $R \ge 0$, terms of $Q(x)$ with even exponents have coefficients ≤ 0, and terms with odd exponents have coefficients ≥ 0. Thus, when n is even, $a_n \le 0$ and $x_0^n > 0$ so $a_n x_0^n \le 0$. When n is odd, $a_n \ge 0$ and $x_0^n < 0$ so $a_n x_0^n \le 0$. Therefore, $Q(x) < 0$. The product $(x_0 - L)Q(x) > 0$ since $x_0 - L < 0$. It has been shown that $R \ge 0$, so $P(x_0) = (x_0 - L)Q(x) + R > 0$. This contradicts the assumption that $P(x_0) = 0$ when $x_0 < L$. Thus when $x_0 < L$, x_0 cannot be a root of $P(x) = 0$.

Page 382 • WRITTEN EXERCISES

A 1.

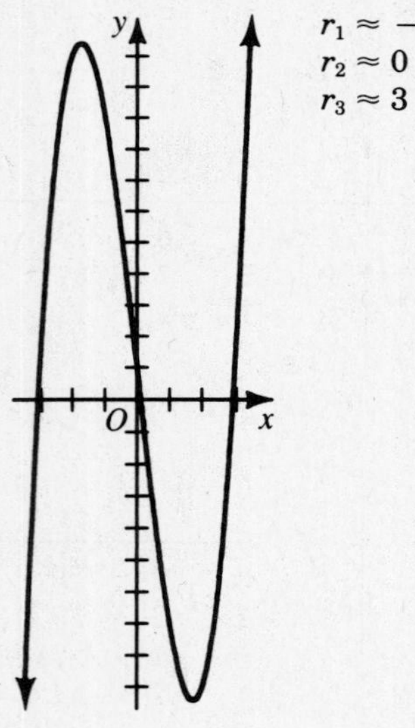

$r_1 \approx -3$
$r_2 \approx 0$
$r_3 \approx 3$

2.

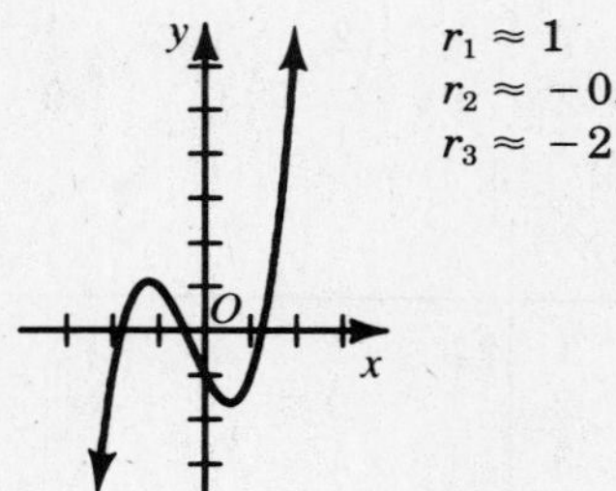

$r_1 \approx 1$
$r_2 \approx -0.5$
$r_3 \approx -2$

3.

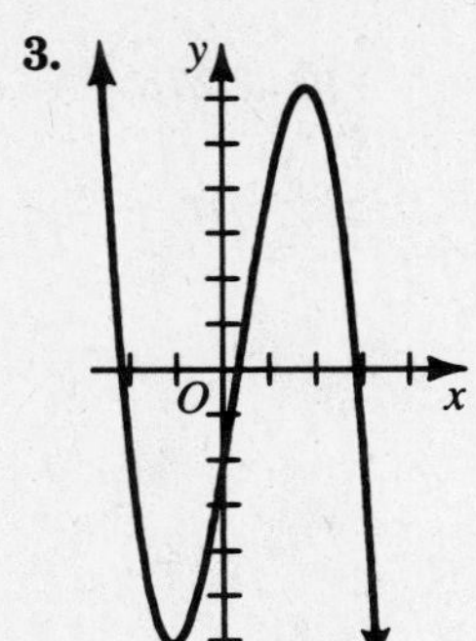

$r_1 \approx -2$
$r_2 \approx 0.5$
$r_3 \approx 3$

4.

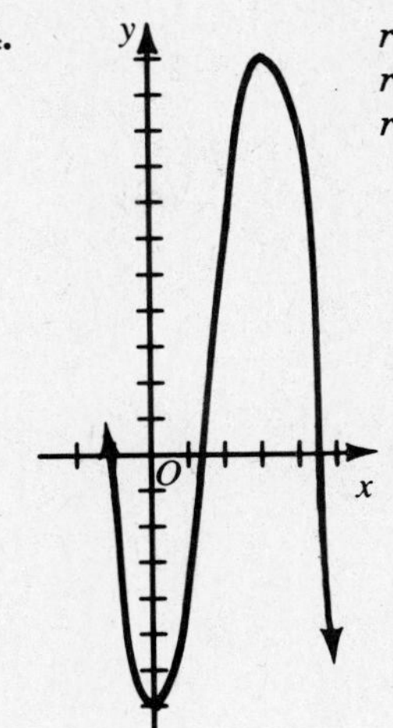

$r_1 \approx -1$
$r_2 \approx 1.5$
$r_3 \approx 4.5$

5.

x	$P(x)$
-2	-1
-1	3
0	1
1	-1
2	3

$-2 < r_1 < -1$, $0 < r_2 < 1$, $1 < r_3 < 2$; $P\left(-\frac{3}{2}\right) = 2.125$, $P\left(\frac{1}{2}\right) = -0.375$, $P\left(\frac{3}{2}\right) = -0.125$; $-2 < r_1 < -\frac{3}{2}$, $0 < r_2 < \frac{1}{2}$, $\frac{3}{2} < r_3 < 2$

6.

x	$P(x)$
-2	-2
-1	1
0	-2
1	-5
2	-2
3	13

$-2 < r_1 < -1$, $-1 < r_2 < 0$, $2 < r_3 < 3$; $P\left(-\frac{3}{2}\right) = 0.625$, $P\left(-\frac{1}{2}\right) = -0.125$, $P\left(\frac{5}{2}\right) = 3.625$; $-2 < r_1 < -\frac{3}{2}$, $-1 < r_2 < -\frac{1}{2}$, $2 < r_3 < \frac{5}{2}$

7.

x	$P(x)$
-2	-7
-1	1
0	1
1	-1
2	1

$-2 < r_1 < -1$, $0 < r_2 < 1$, $1 < r_3 < 2$; $P\left(-\frac{3}{2}\right) = -1.625$, $P\left(\frac{1}{2}\right) = -0.125$, $P\left(\frac{3}{2}\right) = -0.825$; $-\frac{3}{2} < r_1 < -1$, $0 < r_2 < \frac{1}{2}$, $\frac{3}{2} < r_3 < 2$

8.

x	$P(x)$
-2	-5
-1	5
0	1
1	-5
2	-1
3	25

$-2 < r_1 < -1$, $0 < r_2 < 1$, $2 < r_3 < 3$; $P\left(-\frac{3}{2}\right) = 2.5$, $P\left(\frac{1}{2}\right) = -2.5$, $P\left(\frac{5}{2}\right) = 8.5$; $-2 < r_1 < -\frac{3}{2}$, $0 < r_2 < \frac{1}{2}$, $2 < r_3 < \frac{5}{2}$

9.

x	$P(x)$
-2	-1
-1	5
0	3
1	-1
2	-1
3	9

$-2 < r_1 < -1$, $0 < r_2 < 1$, $2 < r_3 < 3$; $P\left(-\frac{3}{2}\right) = 3.375$, $P\left(\frac{1}{2}\right) = 0.875$, $P\left(\frac{5}{2}\right) = 2.375$; $-2 < r_1 < -\frac{3}{2}$, $\frac{1}{2} < r_2 < 1$, $2 < r_3 < \frac{5}{2}$

10.

x	$P(x)$
-2	-1
-1	8
0	7
1	2
2	-1
3	4

$-2 < r_1 < -1$, $1 < r_2 < 2$, $2 < r_3 < 3$; $P\left(-\frac{3}{2}\right) = 5.125$, $P\left(\frac{3}{2}\right) = -0.125$, $P\left(\frac{5}{2}\right) = 0.125$; $-2 < r_1 < -\frac{3}{2}$, $1 < r_2 < \frac{3}{2}$, $2 < r_3 < \frac{5}{2}$

B **11.** By Descartes' Rule of Signs, there is 1 positive root and 0 negative roots. $2 < r < 3$; $P\left(\frac{5}{2}\right) = 3.375$; $2 < r < \frac{5}{2}$

x	$P(x)$
0	-6
1	-6
2	-2
3	12

12. By Descartes' Rule of Signs, there are 0 positive roots and 1 negative root. $-1 < r < 0$; $P\left(-\frac{1}{2}\right) = -1.75$; $-\frac{1}{2} < r < 0$

x	$P(x)$
0	1
-1	-6

13. $P(x) = x^3 - x^2 - 6$; $2 < r < 2.5$; check values by tenths; $P(2.2) = -0.192$, $P(2.3) = 0.877$; $\frac{0 - (-0.192)}{h} \approx \frac{0.877 - (-0.192)}{2.3 - 2.2}$; $\frac{0.192}{h} \approx \frac{1.069}{0.1}$;
$h \approx \frac{0.0192}{1.069} \approx 0.02$; $r \approx 2.2 + 0.02 = 2.22$

14. $P(x) = x^3 - x^2 - 4x + 3$ For r_1: $P(-2) = -1$, $P(-1.9) = 0.131$;
$\frac{0 - (-1)}{h} \approx \frac{0.131 - (-1)}{-1.9 - (-2)}$; $\frac{1}{h} \approx \frac{1.131}{0.1}$; $h \approx \frac{0.1}{1.131} \approx 0.09$;
$r_1 \approx -2 + 0.09 = -1.91$ For r_2: $P(0.7) = 0.053$, $P(0.8) = -0.328$;
$\frac{0 - 0.053}{h} \approx \frac{-0.328 - 0.053}{0.8 - 0.7}$; $\frac{-0.053}{h} \approx \frac{-0.381}{0.1}$; $h \approx \frac{-0.0053}{-0.381} \approx 0.01$;
$r_2 \approx 0.7 + 0.01 = 0.71$ For r_3: $P(2.1) = -0.549$, $P(2.2) = 0.008$;
$\frac{0 - (-0.549)}{h} \approx \frac{0.008 - (-0.549)}{2.2 - 2.1}$; $\frac{0.549}{h} \approx \frac{0.557}{0.1}$; $h \approx \frac{0.0549}{0.557} \approx 0.1$;
$r_3 \approx 2.1 + 0.10 = 2.20$

15. $P(x) = x^3 - 9x + 1$ For r_1: $P(-3.1) = -0.891$, $P(-3) = 1$;
$\frac{0 - (-0.891)}{h} \approx \frac{1 - (-0.891)}{-3 - (-3.1)}$; $\frac{0.891}{h} \approx \frac{1.891}{0.1}$; $h \approx \frac{0.0891}{1.891} \approx 0.05$;
$r_1 \approx -3.1 + 0.05 = -3.05$ For r_2: $P(0) = 1$, $P(0.1) = 0.101$, $P(0.2) = -0.792$;
$\frac{0 - 0.101}{h} \approx \frac{-0.792 - 0.101}{0.2 - 0.1}$; $\frac{-0.101}{h} \approx \frac{-0.893}{0.1}$; $h \approx \frac{-0.0101}{-0.893} \approx 0.01$;

$r_2 \approx 0.1 + 0.01 = 0.11$ For r_3: $P(2.9) = -0.711, P(3) = 1;$
$\frac{0-(-0.711)}{h} \approx \frac{1-(-0.711)}{3-2.9}; \frac{0.711}{h} \approx \frac{1.711}{0.1}; h \approx \frac{0.0711}{1.711} \approx 0.04;$
$r_3 \approx 2.9 + 0.04 = 2.94$

16. $P(x) = x^3 - 3x + 1$ For r_1: $P(-1.9) = -0.159, P(-1.8) = 0.568;$
$\frac{0-(-0.159)}{h} \approx \frac{0.568-(-0.159)}{-1.8-(-1.9)}; \frac{0.159}{h} \approx \frac{0.727}{0.1}; h \approx \frac{0.0159}{0.727} \approx 0.02;$
$r_1 \approx -1.9 + 0.02 = -1.88$ For r_2: $P(0.3) = 0.127, P(0.4) = -0.136;$
$\frac{0-0.127}{h} \approx \frac{-0.136-0.127}{0.4-0.3}; \frac{-0.127}{h} \approx \frac{-0.263}{0.1}; h \approx \frac{-0.0127}{-0.263} \approx 0.05;$
$r_2 \approx 0.3 + 0.05 = 0.35$ For r_3: $P(1.5) = -0.125, P(1.6) = 0.296;$
$\frac{0-(-0.125)}{h} \approx \frac{0.296-(-0.125)}{1.6-1.5}; \frac{0.125}{h} \approx \frac{0.421}{0.1}; h \approx \frac{0.0125}{0.421} \approx 0.03;$
$r_3 \approx 1.5 + 0.03 = 1.53$

C 17.

x					$P(x)$
−2	16	−16	1	−34	66
−1.5	16	−8	−19	−3.5	3.25
−1	16	0	−31	−1	−1
−0.5	16	8	−35	−14.5	5.25
0	16	16	−31	−32	−2
0.5	16	24	−19	−41.5	−22.75
1	16	32	1	−31	−2
1.5	16	40	29	11.5	15.25
2	16	48	65	98	194

$-\frac{3}{2} < r_1 < -1, -1 < r_2 < -\frac{1}{2},$
$-\frac{1}{2} < r_3 < 0, 1 < r_4 < \frac{3}{2}$

18. Answers will vary. An example follows: $r_1 = \frac{1}{2}, r_2 = \frac{2}{5}, r_3 = \frac{3}{5};$
$P(x) = (2x - 1)(5x - 2)(5x - 3) = 0; 50x^3 - 75x^2 + 37x - 6 = 0$

Page 383 • SELF-TEST 4

1. $P(x) = x^5 - 2x^4 - 3x^3 - x^2 + x - 2$ (sign changes: 1, 2, 3)

$P(-x) = -x^5 - 2x^4 + 3x^3 - x^2 - x - 2$ (sign changes: 1, 2)

pos.	neg.	imag.
3	0	2
1	2	2
1	0	4
3	2	0

2.

y				$P(y)$
1	1	−2	0	−2
2	1	−1	0	−2
3	1	0	2	4
−1	1	−4	6	−8

$M = 3, L = -1$

3.

x	$P(x)$
−2	−8
−1	2
0	2
1	−2
2	−4
3	2

$-2 < r_1 < -1, 0 < r_2 < 1,$
$2 < r_3 < 3; P\left(-\frac{3}{2}\right) = -1.375,$
$P\left(\frac{1}{2}\right) = 0.125, P\left(\frac{5}{2}\right) = -2.375;$
$-\frac{3}{2} < r_1 < -1, \frac{1}{2} < r_2 < 1,$
$\frac{5}{2} < r_3 < 3$

Pages 384–385 • CHAPTER REVIEW

1. $x = \frac{4 \pm \sqrt{16 - 48}}{8} = \frac{4 \pm 4i\sqrt{2}}{8}$; $\left\{\frac{1}{2} + \frac{\sqrt{2}}{2}i, \frac{1}{2} - \frac{\sqrt{2}}{2}i\right\}$; c

2. $D = 11^2 - 4(3)(-4) = 169 = 13^2$; c 3. $-\frac{b}{a} = \frac{5}{3}$; d 4. $\frac{c}{a} = \frac{1}{3}$; c 5. d

6. $y = (x - 3)^2$; $(3, 0)$; b 7. $x^2 - 5x - 6 \le 0$, $(x - 6)(x + 1) \le 0$, $\{x\colon -1 \le x \le 6\}$; c

8. $\begin{array}{r|rrrr} 3 & 2 & -3 & -5 & -1 \\ & & 6 & 9 & 12 \\ \hline & 2 & 3 & 4 & 11 \end{array}$; b 9. $\begin{array}{r|rrrr} -2 & 3 & 0 & -5 & 2 \\ & & -6 & 12 & -14 \\ \hline & 3 & -6 & 7 & -12 \end{array}$;

$(x + 2)(3x^2 - 6x + 7) - 12$; c

10. $-2 + i$ and $-2 - i$ are roots; $(x + 2 - i)(x + 2 + i) = x^2 + 4x + 5$; $(x^3 + 3x^2 + x - 5) \div (x^2 + 4x + 5) = x - 1$, $x = 1$; $\{-2 + i, -2 - i, 1\}$; a

11. $P(x) = 3x^5 + 2x^4 + x^3 - 4x^2 + x - 1$; (sign changes 1, 2, 3)

$P(-x) = -3x^5 + 2x^4 - x^3 - 4x^2 - x - 1$; c (sign changes 1, 2)

12. $P(-3) = -49$, $p(-2) = -16$, $P(-1) = -1$, $P(2) = -4$, $P(3) = -1$, $P(4) = 14$; $3 < x < 4$; d

Pages 385–386 • CHAPTER TEST

1. $x^2 - 10x + 25 = -8 + 25$, $(x - 5)^2 = 17$, $x - 5 = \pm\sqrt{17}$; $\{5 + \sqrt{17}, 5 - \sqrt{17}\}$

2. $4y^2 - 4y + 5 = 0$, $y = \frac{4 \pm \sqrt{16 - 4(4)(5)}}{8} = \frac{4 \pm 8i}{8}$; $\left\{\frac{1}{2} + i, \frac{1}{2} - i\right\}$

3. $D = 4^2 - 4(4)(-15) = 256$; since $256 = 16^2$, roots are rational; 2 real, rational roots

4. $-\frac{b}{a} = -\frac{15}{3} = -5$; $\frac{c}{a} = \frac{1}{3}$

5. $-\frac{b}{a} = 2$, $\frac{c}{a} = -2$; let $a = 1$, then $b = -2$ and $c = -2$; $x^2 - 2x - 2 = 0$

6. $y = a(x - 2)^2 + 5$, $2 = a(-1 - 2)^2 + 5$, $9a + 5 = 2$, $a = -\frac{1}{3}$;

$y = -\frac{1}{3}(x - 2)^2 + 5$

7. $y + 3 = x^2 + 6x$
$y + 12 = x^2 + 6x + 9$
$y + 12 = (x + 3)^2$
$y = (x + 3)^2 - 12$
$x = -3$, $(-3, -12)$

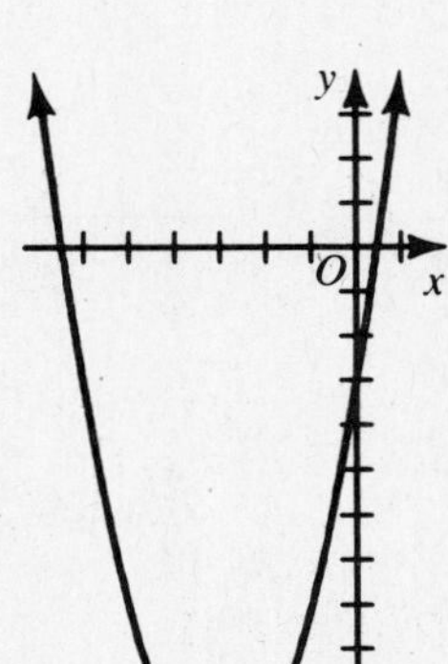

8. a.

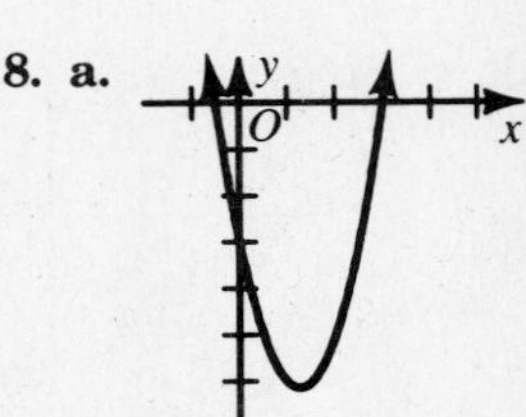

b. $\left\{x\colon -\frac{1}{2} < x < 3\right\}$

9. a. $\underline{-1|}$ 1 2 −5 3

-1	1	2	−5	3
		−1	−1	6
	1	1	−6	9; $P(-1) = 9$

b.

$2i$	1	2	−5	3
		$0 + 2i$	$-4 + 4i$	$-8 - 18i$
	1	$2 + 2i$	$-9 + 4i$	$-5 - 18i$; $P(2i) = -5 - 18i$

10.

3	3	−7	−20	3
		9	6	−42
	3	2	−14	−39

$$\frac{3x^3 - 7x^2 - 20x + 3}{x - 3} = 3x^2 + 2x - 14 + \frac{-39}{x - 3}$$

11. $i\sqrt{2}$ and $-i\sqrt{2}$ are roots; $(x - i\sqrt{2})(x + i\sqrt{2}) = x^2 + 2$; $(x^4 - 5x^3 + 8x^2 - 10x + 12) \div (x^2 + 2) = x^2 - 5x + 6 = (x - 3)(x - 2)$; $\{3, 2, i\sqrt{2}, -i\sqrt{2}\}$

12. $P(x) = 2x^4 + 3x^3 - x^2 - 2x + 1$ (sign changes: 1, 2)

$P(-x) = 2x^4 - 3x^3 - x^2 + 2x + 1$ (sign changes: 1, 2)

pos.	neg.	imag.
2	2	0
2	0	2
0	2	2
0	0	4

13.

x	$P(x)$
−1	−4
0	5
1	4
2	−1
3	−4
4	1

$-1 < r_1 < 0$, $1 < r_2 < 2$, $3 < r_3 < 4$; $P\left(-\frac{1}{2}\right) = 2.125$, $P\left(\frac{3}{2}\right) = 1.625$, $P\left(\frac{7}{2}\right) = -2.875$; $-1 < r_1 < -\frac{1}{2}$, $\frac{3}{2} < r_2 < 2$, $\frac{7}{2} < r_3 < 4$

Page 387 • APPLICATION

1. The frequency of string t is twice the frequency of string s.
2. $27.5(2^x) = 4186$, $2^x \approx 152$, $x \approx 7$; about 7 octaves

Page 388 • PREPARING FOR COLLEGE ENTRANCE EXAMS

1. E **2.** $430 = \frac{10}{2}(a_1 + 79)$; $a_1 = 7$; A **3.** $\frac{x - 1}{x + 2} = \frac{x + 2}{3x}$, $3x^2 - 3x = x^2 + 4x + 4$, $2x^2 - 7x - 4 = 0$, $(2x + 1)(x - 4) = 0$; $x = -\frac{1}{2}$ or $x = 4$; D

4. $r = 4 \geq 1$; divergent; E **5.** $6i\sqrt{2} + 3i\sqrt{2} = 9i\sqrt{2}$; B

6. $9x^2 - 6x\sqrt{2} + 2 = (3x - \sqrt{2})(3x - \sqrt{2}) = 0$; $x = \frac{\sqrt{2}}{3}$; A

7. $a^2 - ka + (k + 8) = 0$; $(-k)^2 - 4(k + 8) = 0$; $k^2 - 4k - 32 = 0$; $(k - 8)(k + 4) = 0$; $k = 8$ or $k = -4$; E

8. Corner points: (0, 0), (2, 0), (0, 4); values: 0, −4, 4; B

Page 391 • PROGRAMMING IN PASCAL

1. b. In the TYPE declaration, make the following change:

```
complex = ARRAY[1..2] OF real;
```

In the procedures and main program, make the following replacements:

z1.a replace with z1[1]
z1.b replace with z1[2]
z2.a replace with z2[1]
z2.b replace with z2[2]
z3.a replace with z3[1]
z3.b replace with z3[2]

c.

```
PROGRAM complex_arith (INPUT,OUTPUT);

TYPE
  complex = RECORD
              a : real;
              b : real;
            END;

VAR
  z1,z2,z3 : complex;
  operation : char;
  denominator : real;

{*************************************************************}
PROCEDURE obtain_nums;

BEGIN
  writeln('Enter the real and imaginary parts of the first');
  write('complex number, with a space between them:');
  readln(z1.a, z1.b);
  writeln('Enter the real and imaginary parts of the second');
  write('complex number, with a space between them:');
  readln(z2.a, z2.b);
END; {* obtain_nums *}

{*************************************************************}
PROCEDURE which_operation;

BEGIN
  writeln('Do you wish to multiply <*>, divide </>,');
  write('add <+>, or subtract <-> the first by the second:');
  readln(operation);
END; {* which_operation *}

{*************************************************************}
PROCEDURE do_operation;

VAR
  denominator : real;
```

```
BEGIN
  CASE operation OF
      '*' : BEGIN
              z3.a := (z1.a * z2.a) - (z1.b * z2.b);
              z3.b := (z1.a * z2.b) + (z1.b * z2.a);
            END;
      '/' : BEGIN
              denominator := z2.a * z2.a + z2.b * z2.b;
              z3.a := (z1.a * z2.a + z1.b * z2.b)/denominator;
              z3.b := (z1.b * z2.a - z1.a * z2.b)/denominator;
            END;
      '+' : BEGIN
              z3.a := z1.a + z2.a;
              z3.b := z1.b + z2.b;
            END;
      '-' : BEGIN
              z3.a := z1.a - z2.a;
              z3.b := z1.b - z2.b;
            END;
  END; {* case *}
END; {* do_operation *}

{**************************************************************}
PROCEDURE display_result;

BEGIN
  writeln(z3.a:1:4, ' + ', z3.b:1:4, 'i');
END; {* display_result *}

{**************************************************************}
BEGIN {* main *}
  obtain_nums;
  which_operation;
  do_operation;
  display_result;
END.
```

2.
```
PROGRAM synthetic_substitution (INPUT,OUTPUT);

TYPE
  term = RECORD
           coefficient : integer;
           degree : integer;
         END;

VAR
  polynomial : ARRAY[1..4] OF term;
  x,p_of_x : real;
  i : integer;
```

[*Program continued on next page*]

```
{****************************************************************}
PROCEDURE initialize_polynomial;

BEGIN
  polynomial[1].coefficient := 5;
  polynomial[1].degree := 3;
  polynomial[2].coefficient := -12;
  polynomial[2].degree := 2;
  polynomial[3].coefficient := -20;
  polynomial[3].degree := 1;
  polynomial[4].coefficient := 1;
  polynomial[4].degree := 0;
END;

{****************************************************************}
BEGIN {* main *}
  initialize_polynomial;
  write('Enter the value of x: ');
  readln(x);
  p_of_x := polynomial[1].coefficient;
  FOR i := 2 TO 4 DO
      p_of_x := x * p_of_x + polynomial[i].coefficient;
  writeln('P(x) = ', p_of_x:1:4);
END.
```

3.

```
PROGRAM ply_root (INPUT,OUTPUT);

TYPE
  link = ^polynomial;
  polynomial = RECORD
                 coefficient : integer;
                 degree : integer;
                 next : link;
               END;

VAR
  first_term : link;
  point_1,point_2,point_3 : real;
  total : integer;

{****************************************************************}
PROCEDURE get_polynomial (VAR term : link; num : integer);

BEGIN
  IF num > total
     THEN term := NIL
     ELSE BEGIN
            new(term);
            writeln;
            writeln('For term number ', num);
```

```
            write('Enter the coefficient: ');
            readln(term^.coefficient);
            write('Enter the degree: ');
            readln(term^.degree);
            get_polynomial(term^.next, num + 1);
          END;
END;

{************************************************************}
FUNCTION power (x : real; degree : integer) : real;

BEGIN
  IF degree >= 0
     THEN IF degree = 0
             THEN power := 1
             ELSE power := power(x, degree - 1) * x
     ELSE BEGIN
            degree := -degree;
            power := 1/power(x, degree);
          END;
END;

{************************************************************}
FUNCTION p_of_x (x : real) : real;

VAR
  sum : real;
  term : link;

BEGIN
  sum := 0;
  term := first_term;
  WHILE term <> NIL DO
        BEGIN
          sum := sum + term^.coefficient * power(x,term^.degree);
          term := term^.next;
        END;
  p_of_x := sum;
END;

{************************************************************}
PROCEDURE get_interval;

VAR
  done : boolean;

BEGIN
  done := FALSE;
```

[*Program continued on next page*]

```
  REPEAT
    writeln;
    write('Enter an endpoint of the interval: ');
    readln(point_1);
    write('Enter the other endpoint of the interval: ');
    readln(point_2);
    IF (p_of_x(point_1) > 0) AND (p_of_x(point_2) < 0)
       THEN done := TRUE;
    IF (p_of_x(point_1) < 0) AND (p_of_x(point_2) > 0)
       THEN done := TRUE;
    IF NOT done
       THEN BEGIN
              writeln;
              write('The graph doesn''t cross the ');
              writeln('x-axis in the given interval.');
              writeln('RE-ENTER.');
            END;
  UNTIL done;
END;

{************************************************************}
PROCEDURE interpolate;

BEGIN
  REPEAT
    point_3 := (point_1 - point_2)/(p_of_x(point_1) -
                                              p_of_x(point_2));
    point_3 := point_3 * (-p_of_x(point_1)) + point_1;
    IF (p_of_x(point_3) > 0)
       THEN IF (p_of_x(point_1) > 0)
               THEN point_1 := point_3
               ELSE point_2 := point_3
       ELSE IF (p_of_x(point_1) < 0)
               THEN point_1 := point_3
               ELSE point_2 := point_3;
  UNTIL abs(p_of_x(point_3) - 0) < 0.000005;
END;

{************************************************************}
BEGIN {* main *}
  write('How many terms are in the polynomial: ');
  readln(total);
  get_polynomial(first_term, 1);
  get_interval;
  interpolate;
  writeln('The answer is: ', point_3:2:6);
END.
```

Pages 396–397 • WRITTEN EXERCISES

A **1.** $d = \sqrt{[4-(-1)]^2 + (3-3)^2} = \sqrt{25} = 5$

2. $d = \sqrt{(5-5)^2 + [-9-(-2)]^2} = \sqrt{49} = 7$

3. $d = \sqrt{(4-7)^2 + (2-6)^2} = \sqrt{25} = 5$

4. $d = \sqrt{(-2-10)^2 + [1-(-4)]^2} = \sqrt{169} = 13$

5. $d = \sqrt{[-2-(-3)]^2 + (6-8)^2} = \sqrt{5}$

6. $d = \sqrt{[5-(-4)]^2 + (-2-1)^2} = \sqrt{90} = 3\sqrt{10}$

7. $d = \sqrt{[5-(-1)]^2 + \left[\frac{1}{2}-(-4)\right]^2} = \sqrt{36 + \frac{81}{4}} = \sqrt{\frac{225}{4}} = \frac{15}{2}$

8. $d = \sqrt{(-\sqrt{3} - 5\sqrt{3})^2 + [-2-(-8)]^2} = \sqrt{144} = 12$

9. $d = \sqrt{(a-b)^2 + (b-a)^2} = \sqrt{2(a-b)^2} = \sqrt{2}\,|a-b|$

10. $M = (4, 5)$. $\sqrt{(3-4)^2 + (1-5)^2} = \sqrt{17}$; $\sqrt{(5-4)^2 + (9-5)^2} = \sqrt{17}$

11. $M = (0, 4)$. $\sqrt{(-2-0)^2 + (7-4)^2} = \sqrt{13}$; $\sqrt{(2-0)^2 + (1-4)^2} = \sqrt{13}$

12. $M = \left(\frac{7}{2}, -\frac{1}{2}\right)$. $\sqrt{\left(\frac{7}{2}-3\right)^2 + \left(-\frac{1}{2}-(-3)\right)^2} = \sqrt{\frac{1}{4} + \frac{25}{4}} = \sqrt{\frac{13}{2}}$;

$\sqrt{\left(\frac{7}{2}-4\right)^2 + \left(-\frac{1}{2}-2\right)^2} = \sqrt{\frac{1}{4} + \frac{25}{4}} = \sqrt{\frac{13}{2}}$

13. a.

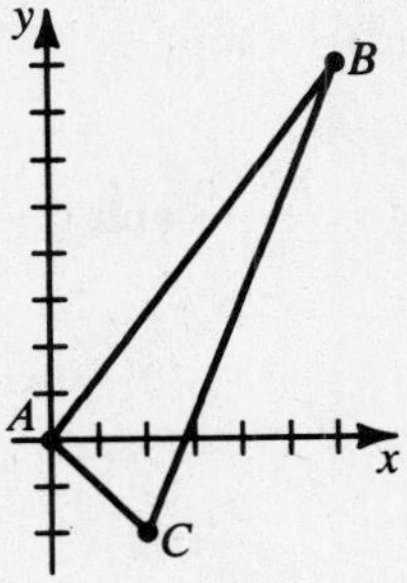

b.–c. $AB = \sqrt{(6-0)^2 + (8-0)^2} = \sqrt{100} = 10$;
$BC = \sqrt{(6-2)^2 + (8+2)^2} = \sqrt{116} = 2\sqrt{29}$;
$AC = \sqrt{(2-0)^2 + (-2-0)^2} = \sqrt{8} = 2\sqrt{2}$;
the triangle is neither isosceles nor right.

14. a.

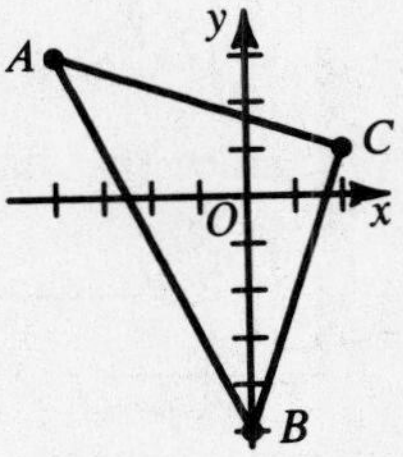

b.–c. $AB = \sqrt{(-4-0)^2 + (3+5)^2} = \sqrt{80} = 4\sqrt{5}$;
$BC = \sqrt{(2-0)^2 + (1+5)^2} = \sqrt{40} = 2\sqrt{10}$;
$AC = \sqrt{(-4-2)^2 + (3-1)^2} = \sqrt{40} = 2\sqrt{10}$.
$BC = AC$; the triangle is isosceles.
$(AC)^2 + (BC)^2 = (AB)^2$; the triangle is right.

15. a.

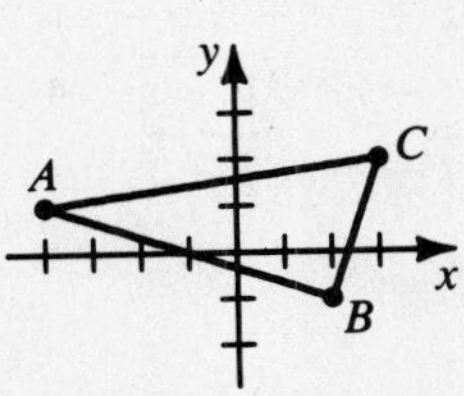

b.–c. $AB = \sqrt{(-4-2)^2 + (1+1)^2} = \sqrt{40} = 2\sqrt{10}$;
$BC = \sqrt{(3-2)^2 + (2+1)^2} = \sqrt{10}$;
$AC = \sqrt{(-4-3)^2 + (1-2)^2} = \sqrt{50} = 5\sqrt{2}$.
$(AB)^2 + (BC)^2 = (AC)^2$; the triangle is not isosceles but is right.

16. a.

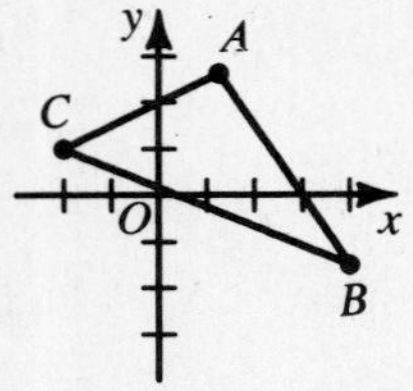

b.–c. $AB = \sqrt{(4-1)^2 + \left(-\frac{3}{2} - \frac{5}{2}\right)^2} = \sqrt{25} = 5;$

$BC = \sqrt{(-2-4)^2 + \left(1 + \frac{3}{2}\right)^2} = \sqrt{\frac{169}{4}} = \frac{13}{2};$

$AC = \sqrt{(-2-1)^2 + \left(1 - \frac{5}{2}\right)^2} = \sqrt{\frac{45}{4}} = \frac{3\sqrt{5}}{2};$

the triangle is neither isosceles nor right.

17. $F(x, y)$: $2 = \frac{7+x}{2}, 3 = \frac{4+y}{2}; x = -3, y = 2; F(-3, 2)$

18. $F(x, y)$: $-2 = \frac{3+x}{2}, 5 = \frac{-1+y}{2}; x = -7, y = 11; F(-7, 11)$

19. $F(x, y)$: $\frac{3}{2} = \frac{2+x}{2}, -\frac{1}{2} = \frac{-6+y}{2}; x = 1, y = 5; F(1, 5)$

20. $F(x, y)$: $h = \frac{h+m+x}{2}, k = \frac{k-n+y}{2}; x = h - m, y = k + n; F(h - m, k + n)$

B 21. $M\left(\frac{5-1}{2}, \frac{8+2}{2}\right) = M(2, 5); MZ = \sqrt{(-3-2)^2 + (-7-5)^2} = \sqrt{169} = 13$

22. $M\left(\frac{9+5}{2}, \frac{-3-15}{2}\right) = M(7, -9); MZ = \sqrt{(7+1)^2 + (-9-6)^2} = \sqrt{289} = 17$

23. $P_1P_2 = \sqrt{(b-a)^2 + (c-c)^2} = \sqrt{(b-a)^2} = |a - b|$

24. $P_1P_2 = \sqrt{(b-a)^2 + (2\sqrt{ab} - 0)^2} = \sqrt{(b^2 - 2ab + a^2) + (4ab)} =$
$\sqrt{a^2 + 2ab + b^2} = \sqrt{(a+b)^2} = |a + b|$

25. Let $B = (a, b)$. Then $(p, q) = \left(\frac{r+a}{2}, \frac{s+b}{2}\right)$; thus $p = \frac{r+a}{2}$ and $q = \frac{s+b}{2}$.
So $a = 2p - r$, $b = 2q - s$, and the coordinates of B are $(2p - r, 2q - s)$.

26. $AB = \sqrt{(3-x)^2 + (-4-8)^2} = 13; (3-x)^2 + (-12)^2 = 13^2; (9 - 6x + x^2) +$
$144 = 169; x^2 - 6x - 16 = 0; (x - 8)(x + 2) = 0; x = 8$ or $x = -2; \{-2, 8\}$

27. $\sqrt{(-3-x)^2 + (4-y)^2} = \sqrt{(5-x)^2 + (-2-y)^2}$.
Square both sides of the equation and simplify.
$(9 + 6x + x^2) + (16 - 8y + y^2) = (25 - 10x + x^2) + (4 + 4y + y^2);$
$25 + 6x - 8y = 29 - 10x + 4y; 16x - 12y = 4; 4x - 3y = 1$

C 28. Midpt. of $\overline{AC} = \left(\frac{a+b}{2}, \frac{c}{2}\right)$, midpt. of $\overline{BD} = \left(\frac{a+b}{2}, \frac{c}{2}\right)$; since the diagonals have the same midpoint, they must bisect each other.

29. $AB = \sqrt{(s-r)^2 + (ms - mr)^2} = \sqrt{(s-r)^2 + m^2(s-r)^2} = \sqrt{(s-r)^2(1+m^2)} =$
$(s - r)\sqrt{1+m^2}; BC = \sqrt{(t-s)^2 + (mt - ms)^2} = \sqrt{(t-s)^2 + m^2(t-s)^2} =$
$\sqrt{(t-s)^2(1+m^2)} = (t - s)\sqrt{1+m^2}; AC = \sqrt{(t-r)^2 + (mt - mr)^2} =$
$\sqrt{(t-r)^2 + m^2(t-r)^2} = \sqrt{(t-r)^2(1+m^2)} = (t - r)\sqrt{1+m^2}; AB + BC =$
$(s - r)\sqrt{1+m^2} + (t - s)\sqrt{1+m^2} = (s - r + t - s)\sqrt{1+m^2} =$
$(t - r)\sqrt{1+m^2} = AC$

Page 399 • WRITTEN EXERCISES

A 1. $3y = x - 5, y = \frac{1}{3}x - \frac{5}{3}; m = -3; y - 7 = -3(x + 2), y = -3x + 1$

2. $5y + 3x = 6, y = -\frac{3}{5}x + \frac{6}{5}; m = \frac{5}{3}; y + 1 = \frac{5}{3}(x - 3), 3y + 3 = 5x - 15,$
$5x - 3y = 18$

3. $4x + 5 = 3y, y = \frac{4}{3}x + \frac{5}{3}; m = -\frac{3}{4}; y + 11 = -\frac{3}{4}(x - 8),$
$4y + 44 = -3x + 24, 3x + 4y = -20$

4. $4y = 7x + 1, y = \frac{7}{4}x + \frac{1}{4}; m = -\frac{4}{7}; y + \frac{2}{3} = -\frac{4}{7}\left(x + \frac{1}{3}\right),$
$7y + \frac{14}{3} = -4x - \frac{4}{3}, 4x + 7y = -6$

5. slope of $\overline{AB} = \frac{6 - 4}{7 + 1} = \frac{1}{4}; m_1 = -4; y - 3 = -4(x + 3), y = -4x - 9$

6. slope of $\overline{AB} = \frac{5 + 1}{-2 - 6} = -\frac{3}{4}; m_1 = \frac{4}{3}; y + 3 = \frac{4}{3}(x + 4), 3y = 4x + 7$

7. slope of $\overline{AB} = \frac{-7 - 5}{-8 - 2} = \frac{6}{5}; m_1 = -\frac{5}{6}; y + 3 = -\frac{5}{6}\left(x - \frac{2}{3}\right), 18y = -15x - 44,$
$15x + 18y = -44$

8. slope of $\overline{AB} = \frac{\frac{7}{4} - \frac{1}{4}}{-\frac{1}{5} - \frac{2}{5}} = \frac{6}{4}\left(-\frac{5}{3}\right) = -\frac{5}{2}; m_1 = \frac{2}{5}; y - 4 = \frac{2}{5}\left(x - \frac{3}{2}\right),$
$5y = 2x + 17$

9. $m = \frac{7 + 5}{3 - 1} = 6; m_1 = -\frac{1}{6}; M = \left(\frac{1 + 3}{2}, \frac{-5 + 7}{2}\right) = (2, 1); y - 1 =$
$-\frac{1}{6}(x - 2), 6y = -x + 8, x + 6y = 8$

10. $m = \frac{-13 + 7}{10 - 6} = -\frac{3}{2}; m_1 = \frac{2}{3}; M = \left(\frac{6 + 10}{2}, \frac{-7 - 13}{2}\right) =$
$(8, -10); y + 10 = \frac{2}{3}(x - 8), 3y = 2x - 46, 2x - 3y = 46$

11. $m = \frac{4 - 1}{-3 - 2} = -\frac{3}{5}; m_1 = \frac{5}{3}; M = \left(\frac{-3 + 2}{2}, \frac{4 + 1}{2}\right) =$
$\left(-\frac{1}{2}, \frac{5}{2}\right); y - \frac{5}{2} = \frac{5}{3}\left(x + \frac{1}{2}\right), 6y = 10x + 20, 5x - 3y = -10$

12. $m = \frac{\frac{5}{6} - \frac{1}{2}}{\frac{1}{3} + 1} = \frac{2}{6} \cdot \frac{3}{4} = \frac{1}{4}; m_1 = -4; M = \left(\frac{-1 + \frac{1}{3}}{2}, \frac{\frac{1}{2} + \frac{5}{6}}{2}\right) =$
$\left(-\frac{1}{3}, \frac{2}{3}\right); y - \frac{2}{3} = -4\left(x + \frac{1}{3}\right), 12x + 3y = -2$

B **13.** $AB = \sqrt{(-5 - 1)^2 + (1 - 4)^2} = \sqrt{45} = 3\sqrt{5}; CD = \sqrt{(-7 + 1)^2 + (5 - 8)^2} =$
$\sqrt{45} = 3\sqrt{5}; BC = \sqrt{(1 + 1)^2 + (4 - 8)^2} = \sqrt{20} = 2\sqrt{5};$
$AD = \sqrt{(-5 + 7)^2 + (1 - 5)^2} = \sqrt{20} = 2\sqrt{5}$; slope of $\overline{AB} = \frac{4 - 1}{1 + 5} = \frac{1}{2};$
slope of $\overline{BC} = \frac{8 - 4}{-1 - 1} = -2; \overline{AB} \perp \overline{BC}$

14. $AB = \sqrt{(-4-2)^2 + (-2+6)^2} = \sqrt{52} = 2\sqrt{13}$;
$CD = \sqrt{(4+2)^2 + (-3-1)^2} = \sqrt{52} = 2\sqrt{13}$;
$BC = \sqrt{(2-4)^2 + (-6+3)^2} = \sqrt{13}$;
$AD = \sqrt{(-4+2)^2 + (-2-1)^2} = \sqrt{13}$;
slope of $\overline{AB} = \dfrac{-2+6}{-4-2} = -\dfrac{2}{3}$; slope of $\overline{BC} = \dfrac{-3+6}{4-2} = \dfrac{3}{2}$; $\overline{AB} \perp \overline{BC}$

15. $A(0, 0)$; $B(\sqrt{a^2+b^2}, 0)$; $C(\sqrt{a^2+b^2} + b, a)$; $D(b, a)$

a. $AB = \sqrt{(\sqrt{a^2+b^2} - 0)^2 + (0-0)^2} = \sqrt{a^2+b^2}$;

$BC = \sqrt{(\sqrt{a^2+b^2} + b - \sqrt{a^2+b^2})^2 + (a-0)^2} = \sqrt{a^2+b^2}$;

$CD = \sqrt{(b - (\sqrt{a^2+b^2} + b))^2 + (a-a)^2} = \sqrt{a^2+b^2}$;

$AD = \sqrt{(b-0)^2 + (a-0)^2} = \sqrt{a^2+b^2}$

b. $m_1 =$ slope of $\overline{BD} = \dfrac{a-0}{b-\sqrt{a^2+b^2}} = \dfrac{a}{(b-\sqrt{a^2+b^2})}$; $m_2 =$ slope of $\overline{AC} =$
$\dfrac{(a-0)(\sqrt{a^2+b^2} - b)}{(\sqrt{a^2+b^2} + b)(\sqrt{a^2+b^2} - b)} = \dfrac{a(\sqrt{a^2+b^2} - b)}{a^2+b^2-b^2} = \dfrac{-(b-\sqrt{a^2+b^2})}{a} =$
$-\dfrac{1}{m_1}$

16. $m_1 = \dfrac{2b-0}{0-2a} = -\dfrac{b}{a}$; $M = \left(\dfrac{2a+0}{2}, \dfrac{0+2b}{2}\right) = (a, b)$, $m_2 = \dfrac{a}{b}$.
$y - b = \dfrac{a}{b}(x-a)$, $by - b^2 = ax - a^2$, $ax - by = a^2 - b^2$.

C **17.** midpoint, M, of $\overline{BC} = \left(\dfrac{a+2b+a+2c}{2}, \dfrac{a+2c+a+2b}{2}\right) =$
$(a+b+c, a+b+c)$; slope of $\overline{BC} = \dfrac{a+2c-(a+2b)}{a+2b-(a+2c)} = \dfrac{2(c-b)}{2(b-c)} = -1$;
slope of $\overline{AM} = \dfrac{a+b+c-a}{a+b+c-a} = 1$; $\therefore \overline{BC} \perp \overline{AM}$.

18. slope of $\overline{AC} = \dfrac{s-0}{r-a} = \dfrac{s}{r-a}$; slope of $\overline{BC} = \dfrac{s-0}{r-(-a)} = \dfrac{s}{r+a}$;
since $\overline{AC} \perp \overline{BC}$, $\dfrac{s}{r-a} \cdot \dfrac{s}{r+a} = -1$. Thus $s^2 = -1(r^2 - s^2)$,
or $s^2 + r^2 = a^2$. $OC = \sqrt{(r-0)^2 + (s-0)^2} = \sqrt{r^2+s^2} = \sqrt{a^2} = a$

Pages 399–400 • COMPUTER EXERCISES

1.
```
10  FOR I = 1 TO 4
20  INPUT X(I), Y(I)
30  NEXT I
35  PRINT "THE SLOPES ARE:"
40  FOR I = 1 TO 3
50  LET M(I) = (Y(I + 1) - Y(I))/(X(I + 1) - X(I))
60  PRINT M(I);" ";
70  NEXT I
```

```
80  LET M(4) = (Y(4) - Y(1))/(X(4) - X(1))
85  PRINT M(4) : PRINT
90  FOR I = 1 TO 3
100 FOR J = I + 1 TO 4
110 IF M(I) = M(J) THEN C1 = C1 + 1 : GOTO 130
120 NEXT J
130 NEXT I
140 IF C1 = 0 THEN PRINT "QUADRILATERAL" : GOTO 230
150 IF C1 = 1 THEN PRINT "TRAPEZOID" : GOTO 230
160 FOR I = 1 TO 3
170 FOR J = I + 1 TO 4
180 IF M(I) = -1/M(J) THEN 220
190 NEXT J
200 NEXT I
210 PRINT "PARALLELOGRAM" : GOTO 230
220 PRINT "RECTANGLE"
230 END
```

2. parallelogram **3.** rectangle **4.** trapezoid **5.** quadrilateral

Page 400 • SELF-TEST 1

1. $P_1P_2 = \sqrt{(-3 - 7)^2 + (4 - 3)^2} = \sqrt{101}$ **2.** $\left(\frac{3 + 6}{2}, \frac{5 + 1}{2}\right) = \left(\frac{9}{2}, 3\right)$

3. $m_1 = \frac{8 - 2}{-3 - 0} = -2$; $m_2 = \frac{1}{2}$; $y - 2 = \frac{1}{2}(x - 4)$; $y = \frac{1}{2}x$

Pages 402–403 • WRITTEN EXERCISES

A

1. $x^2 + y^2 = 4^2$; $x^2 + y^2 - 16 = 0$

2. $(x - 1)^2 + (y - 5)^2 = 6^2$; $x^2 + y^2 - 2x - 10y - 10 = 0$

3. $(x + 3)^2 + (y - 7)^2 = 2^2$; $x^2 + y^2 + 6x - 14y + 54 = 0$

4. $(x + 2)^2 + (y + 4)^2 = 7^2$; $x^2 + y^2 + 4x + 8y - 29 = 0$

5. $\left(x - \frac{3}{2}\right)^2 + \left(y + \frac{5}{2}\right)^2 = 2^2$; $x^2 + y^2 - 3x + 5y + \frac{9}{2} = 0$

6. $(x + 3)^2 + \left(y - \frac{1}{2}\right)^2 = \left(\frac{3}{2}\right)^2$; $x^2 + y^2 + 6x - y + 7 = 0$

7. $(x - a)^2 + (y - b)^2 = b^2$; $x^2 + y^2 - 2ax - 2by + a^2 = 0$

8. $(x + c)^2 + (y + d)^2 = (c + d)^2$; $x^2 + y^2 + 2cx + 2dy - 2cd = 0$

9. $2x^2 + 2y^2 = 72$

$(x - 0)^2 + (y - 0)^2 = 6^2$

$C(0, 0)$; $r = 6$

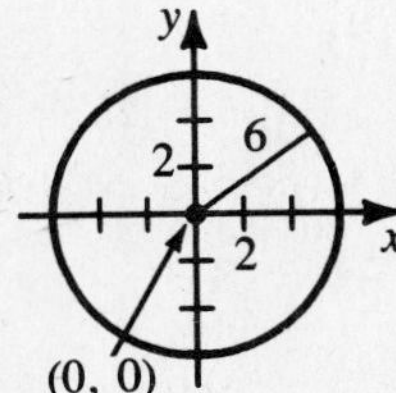

10.
$$x^2 + y^2 - 8x = 0$$
$$(x^2 - 8x + 16) + y^2 = 16$$
$$(x - 4)^2 + (y - 0)^2 = 4^2$$
$C(4, 0)$; $r = 4$

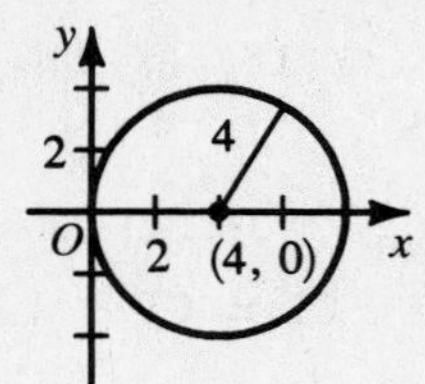

11.
$$x^2 + y^2 + 4y = 0$$
$$x^2 + (y^2 + 4y + 4) = 4$$
$$(x - 0)^2 + (y + 2)^2 = 2^2$$
$C(0, -2)$; $r = 2$

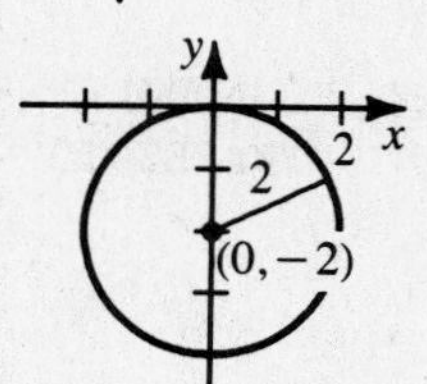

12.
$$x^2 + y^2 + 2x + 6y = 15$$
$$(x^2 + 2x + 1) + (y^2 + 6y + 9) = 15 + 1 + 9$$
$$(x + 1)^2 + (y + 3)^2 = 5^2$$
$C(-1, -3)$; $r = 5$

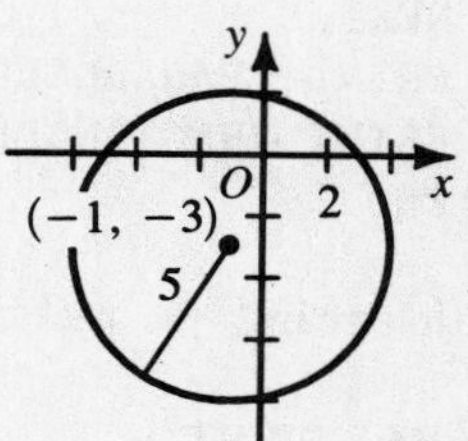

13.
$$x^2 + y^2 + 4x - 12y - 9 = 0$$
$$(x^2 + 4x + 4) + (y^2 - 12y + 36) = 9 + 4 + 36$$
$$(x + 2)^2 + (y - 6)^2 = 7^2$$
$C(-2, 6)$; $r = 7$

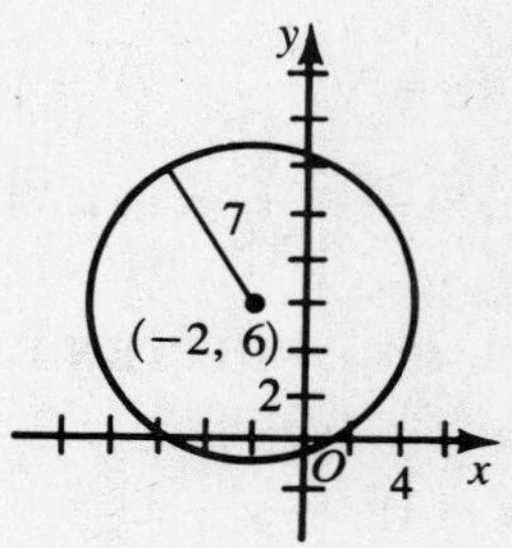

14.
$$x^2 + y^2 - 4x - 3y = 0$$
$$(x^2 - 4x + 4) + \left(y^2 - 3y + \frac{9}{4}\right) = 4 + \frac{9}{4}$$
$$(x - 2)^2 + \left(y - \frac{3}{2}\right)^2 = \left(\frac{5}{2}\right)^2$$
$C\left(2, \frac{3}{2}\right)$; $r = \frac{5}{2}$

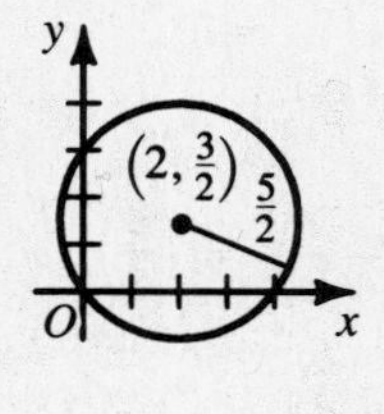

15.
$$2x^2 + 2y^2 - 10x + 2y - 5 = 0$$
$$\left(x^2 - 5x + \frac{25}{4}\right) + \left(y^2 + y + \frac{1}{4}\right) = \frac{5}{2} + \frac{25}{4} + \frac{1}{4}$$
$$\left(x - \frac{5}{2}\right)^2 + \left(y + \frac{1}{2}\right)^2 = 3^2$$
$C\left(\frac{5}{2}, -\frac{1}{2}\right)$; $r = 3$

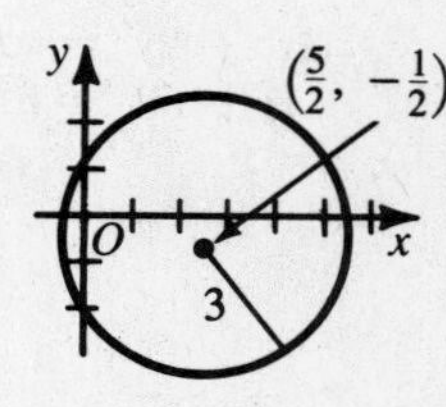

16. $4x^2 + 4y^2 - 24x - 12y + 29 = 0$

$$(x^2 - 6x + 9) + \left(y^2 - 3y + \frac{9}{4}\right) = -\frac{29}{4} + 9 + \frac{9}{4}$$

$$(x - 3)^2 + \left(y - \frac{3}{2}\right)^2 = 2^2$$

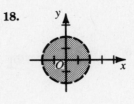

$C\left(3, \frac{3}{2}\right)$; $r = 2$

17.

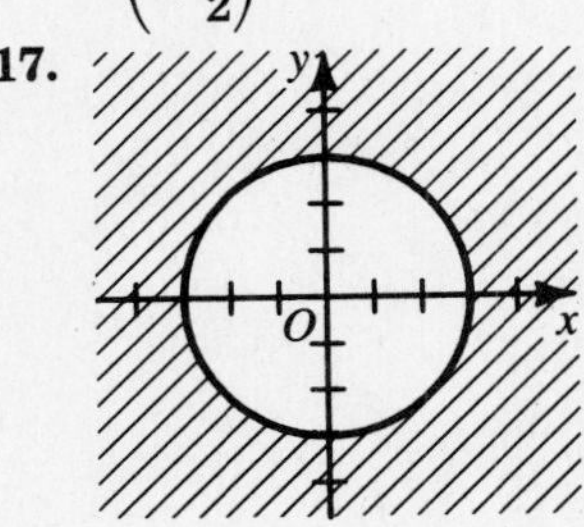

18.

B **19.** $x^2 + (y^2 + 4y + 4) \le 4$

$x^2 + (y + 2)^2 \le 4$

$C(0, -2)$; $r = 2$

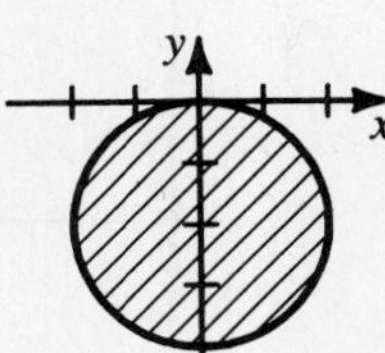

20. $(x^2 - 6x + 9) + y^2 > 9$

$(x - 3)^2 + y^2 > 9$

$C(3, 0)$; $r = 3$

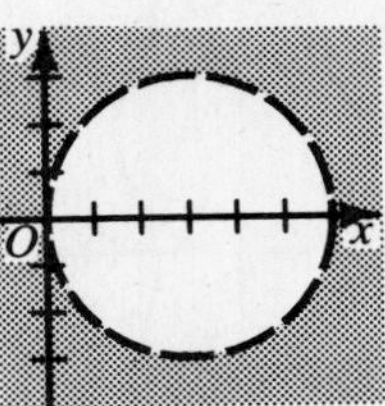

21. $C = \left(\frac{5 + 5}{2}, \frac{9 + 3}{2}\right) = (5, 6)$; $2r = 9 - 3 = 6$; $r = 3$; $(x - 5)^2 + (y - 6)^2 = 9$

22. $C = \left(\frac{-2 + 8}{2}, \frac{4 + 4}{2}\right) = (3, 4)$; $2r = 8 - (-2) = 10$; $r = 5$;
$(x - 3)^2 + (y - 4)^2 = 25$

23. $C = \left(\frac{-4 + 2}{2}, \frac{1 + 9}{2}\right) = (-1, 5)$; $2r = \sqrt{(-4 - 2)^2 + (1 - 9)^2} =$
$\sqrt{36 + 64} = 10$; $r = 5$; $(x + 1)^2 + (y - 5)^2 = 25$

24. $C = \left(\frac{-5 - 1}{2}, \frac{6 - 2}{2}\right) = (-3, 2)$;
$2r = \sqrt{(-5 + 1)^2 + (6 + 2)^2} = \sqrt{80} = 4\sqrt{5}$; $r = 2\sqrt{5}$; $(x + 3)^2 + (y - 2)^2 = 20$

25. $(x - 3)^2 + (y + 2)^2 = r^2$; $(7 - 3)^2 + (1 + 2)^2 = r^2$; $25 = r^2$;
$(x - 3)^2 + (y + 2)^2 = 25$; $x^2 + y^2 - 6x + 4y - 12 = 0$

26. $(x + 1)^2 + (y - 5)^2 = r^2$; $(-1 + 1)^2 + (8 - 5)^2 = r^2$; $9 = r^2$;
$(x + 1)^2 + (y - 5)^2 = 9$; $x^2 + y^2 + 2x - 10y + 17 = 0$

27. $(x - k)^2 + y^2 = r^2$; $(2k - k)^2 + 0^2 = r^2$; $k^2 = r^2$;
$(x - k)^2 + y^2 = k^2$; $x^2 + y^2 - 2kx = 0$

28. Let R = radius of circle. $(x - r)^2 + (y - s)^2 = R^2$; $(0 - r)^2 + (0 - s)^2 = R^2$;
$r^2 + s^2 = R^2$; $(x - r)^2 + (y - s)^2 = r^2 + s^2$; $x^2 + y^2 - 2rx - 2sy = 0$

29. $x^2 = 9 - y^2, x \geq 0$
$x^2 + y^2 = 9, x \geq 0$
$C(0, 0); r = 3$

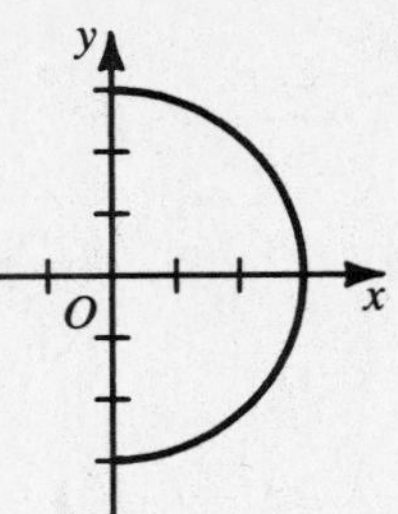

30. $y^2 = 16 - x^2, y \leq 0$
$x^2 + y^2 = 16, y \leq 0$
$C(0, 0); r = 4$

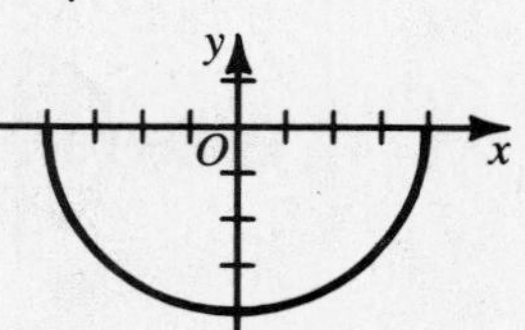

C **31.** $y - 1 = \sqrt{6x - x^2}, y \geq 1$
$y^2 - 2y + 1 = 6x - x^2, y \geq 1$
$(x^2 - 6x + 9) + (y^2 - 2y + 1) = 9$
$(x - 3)^2 + (y - 1)^2 = 9, y \geq 1$
$C(3, 1); r = 3$

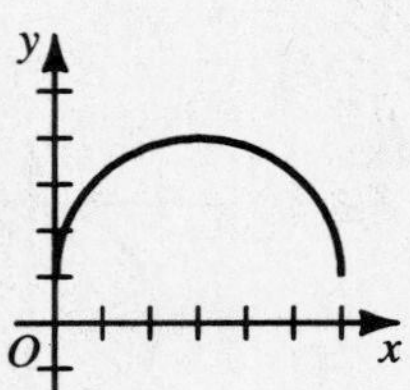

32. $x^2 = 4y - y^2, x \geq 0$
$x^2 + (y^2 - 4y + 4) = 4, x \geq 0$
$x^2 + (y - 2)^2 = 4, x \geq 0$
$C(0, 2); r = 2$

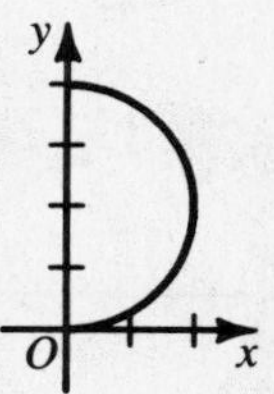

33. $C = (4, -1)$; tangent to the y-axis at $(0, -1)$; $r = 4 - 0 = 4$;
$(x - 4)^2 + (y + 1)^2 = 16$

34. Slope of radius from $(0, 0)$ to $(-4, 3)$ is $-\frac{3}{4}$; slope of tangent line is $\frac{4}{3}$;

$y - 3 = \frac{4}{3}(x + 4)$, or $4x - 3y = -25$

35. $x^2 + y^2 - 8y = 9$; $x^2 + (y - 4)^2 = 5^2$; $C = (0, 4)$

a. Midpoint of chord $= \left(\frac{3 + 5}{2}, \frac{8 + 4}{2}\right) = (4, 6)$; slope of radius $=$

$\frac{6 - 4}{4 - 0} = \frac{1}{2}$; slope of chord $= \frac{8 - 4}{3 - 5} = -2$;

$\therefore$ radius and chord are perpendicular.

b. Since the radius is $\perp$ to the chord, the slope of the radius must equal $\frac{1}{2}$;

equation of line containing the radius is $y - 4 = \frac{1}{2}x$, or $x - 2y = -8$;

$(4, 6)$ satisfies this equation, therefore the radius contains the midpoint of the chord.

36. $x^2 + y^2 + ax + by + c = 0$

$$3^2 + 0^2 + 3a + 0(b) + c = 0 \longrightarrow \quad 3a + c = -9$$
$$(-3)^2 + 0^2 + (-3)a + 0(b) + c = 0 \longrightarrow \quad -3a + c = -9$$
$$2c = -18;\ c = -9;$$

$3a + (-9) = -9;\ 3a = 0;\ a = 0;$
$0^2 + (9)^2 + 0(0) + 9b + (-9) = 0;\ 9b = -72;\ b = -8;$
$x^2 + y^2 - 8y - 9 = 0$

37. Distance from (x, y) to $(6, 0) = \sqrt{(x-6)^2 + y^2}$, distance from (x, y) to $(3, 0) = \sqrt{(x-3)^2 + y^2}$. $\sqrt{(x-6)^2 + y^2} = 2\sqrt{(x-3)^2 + y^2}$, or
$(x-6)^2 + y^2 = 4((x-3)^2 + y^2);\ x^2 - 12x + 36 + y^2 = 4x^2 - 24x + 36 + 4y^2;$
$3x^2 - 12x + 3y^2 = 0;\ x^2 - 4x + y^2 = 0;\ (x-2)^2 + y^2 = 4;\ C(2, 0),\ r = 2$

38. $x^2 + y^2 + ax + by + c = 0$ is equivalent to
$$\left(x + \frac{a}{2}\right)^2 + \left(y + \frac{b}{2}\right)^2 = -c + \frac{a^2}{4} + \frac{b^2}{4}.$$
Thus $r^2 = -c + \frac{a^2}{4} + \frac{b^2}{4} = \frac{a^2 + b^2 - 4c}{4}$, and $r = \sqrt{\frac{a^2 + b^2 - 4c}{4}} =$

$\frac{\sqrt{a^2 + b^2 - 4c}}{2}$, a nonnegative number. $\therefore a^2 + b^2 - 4c \geq 0$.

Page 404 • COMPUTER EXERCISES

1.
```
10 INPUT A, B, C
20 LET C = -C + (A/2)^2 + (B/2)^2
30 IF C < 0 THEN PRINT "EMPTY SET" : GOTO 70
40 IF C = 0 THEN PRINT "SINGLE POINT" : GOTO 70
50 PRINT "CIRCLE WITH CENTER ("; -A/2; ", "; -B/2; ")";
60 PRINT " AND RADIUS "; C^0.5
70 END
```

2. Circle; $(2, -3)$; $r = 5$ **3.** Circle; $(5, 0)$; $r = 5$ **4.** Circle; $(-1, 6)$; $r \approx 8.6$
5. Empty set

Pages 407–408 • WRITTEN EXERCISES

A **1.** $y^2 = 4px;\ p = 10;\ y^2 = 40x$ **2.** $y^2 = 4px;\ p = -2;\ y^2 = -8x$
3. $x^2 = 4py;\ p = -\frac{1}{2};\ x^2 = -2y$ **4.** $x^2 = 4py;\ p = \frac{2}{3};\ x^2 = \frac{8}{3}y$
5. $x^2 = 4py;\ p = 2;\ x^2 = 8y$ **6.** $y^2 = 4px;\ p = -5;\ y^2 = -20x$

B **7.** $y = \frac{1}{2}(x^2 - 6)$
$y = \frac{1}{2}(x - 0)^2 - 3$

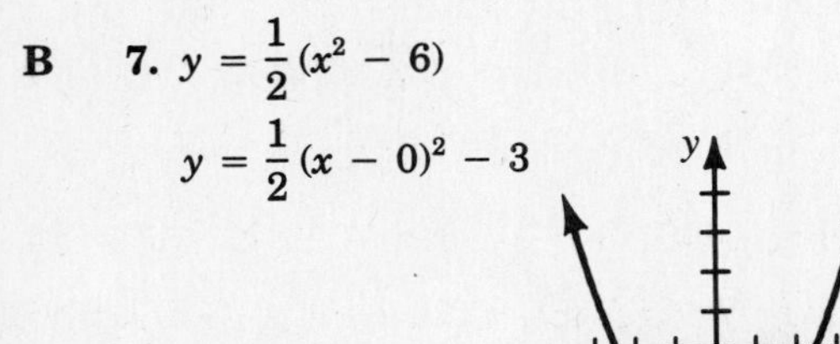

8. $x = -2(4 - y^2)$
$x = 2(y - 0)^2 - 8$

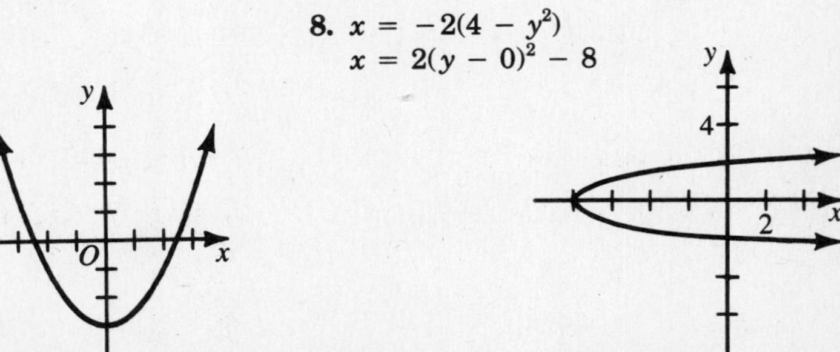

9.
$$x + 3 = y^2 - 2y$$
$$x + 3 + 1 = y^2 - 2y + 1$$
$$x = (y - 1)^2 - 4$$

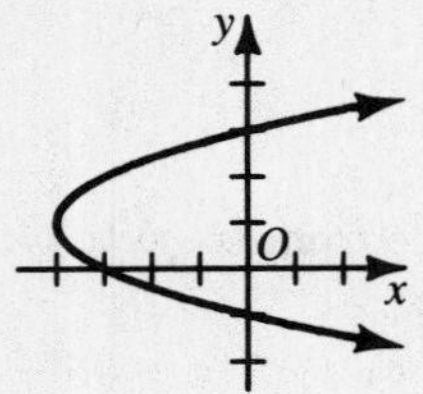

10.
$$y = 2x^2 - 12x$$
$$y = 2(x^2 - 6x + 9) - 18$$
$$y = 2(x - 3)^2 - 18$$

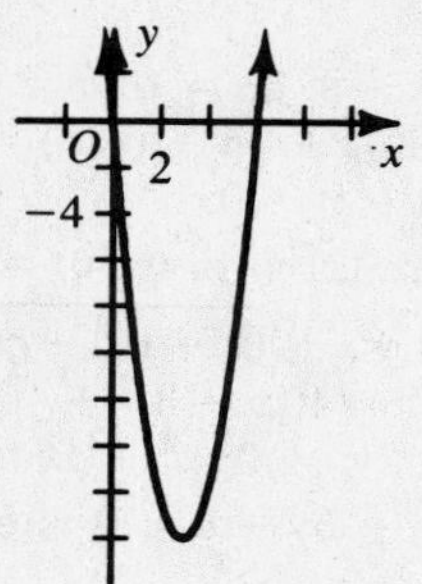

11.
$$y = \frac{1}{4}x^2 + 3x$$
$$4y = x^2 + 12x$$
$$= (x + 6)^2 - 36$$
$$y = \frac{1}{4}(x + 6)^2 - 9$$

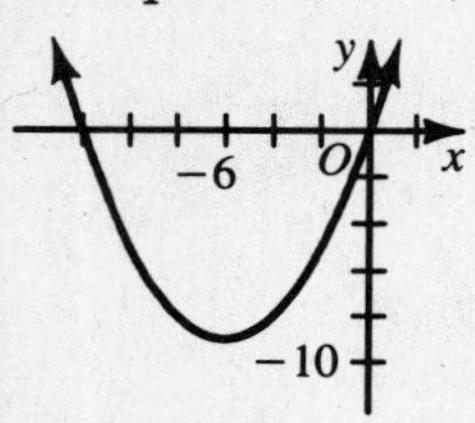

12.
$$x = \frac{1}{2}y^2 + 4y$$
$$2x = y^2 + 8y$$
$$= (y + 4)^2 - 16$$
$$x = \frac{1}{2}(y + 4)^2 - 8$$

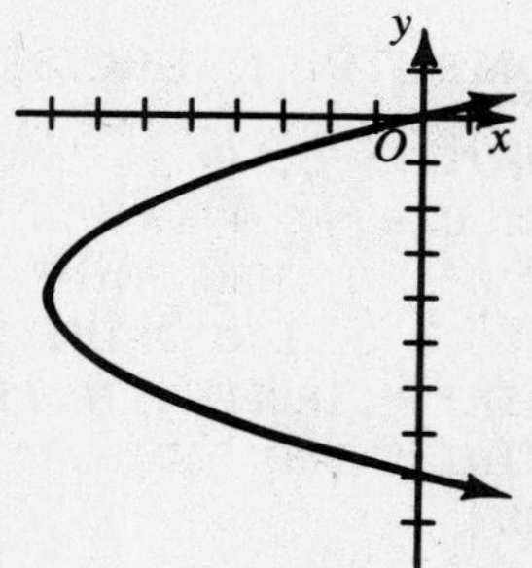

13.
$$y = 2x^2 - 8x + 5$$
$$y = 2(x^2 - 4x + 4) + 5 - 8$$
$$y = 2(x - 2)^2 - 3$$

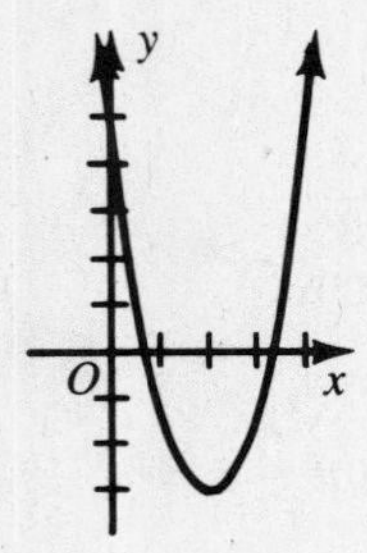

14.
$$x = 3y^2 - 6y - 5$$
$$x = 3(y^2 - 2y + 1) - 5 - 3$$
$$x = 3(y - 1)^2 - 8$$

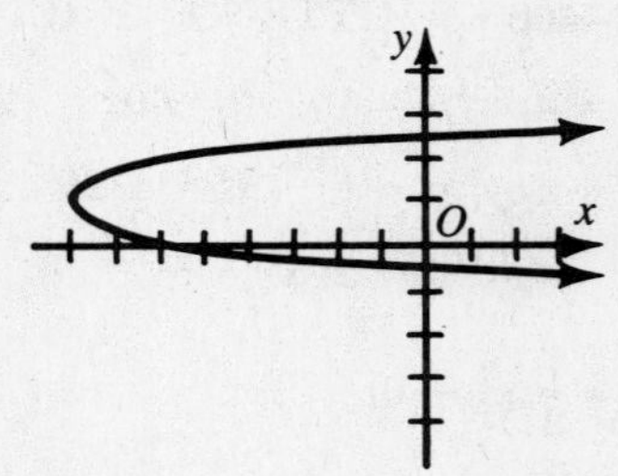

15. $3x = y^2 - 2y + 10$
$3x = (y - 1)^2 + 9$
$x = \frac{1}{3}(y - 1)^2 + 3$

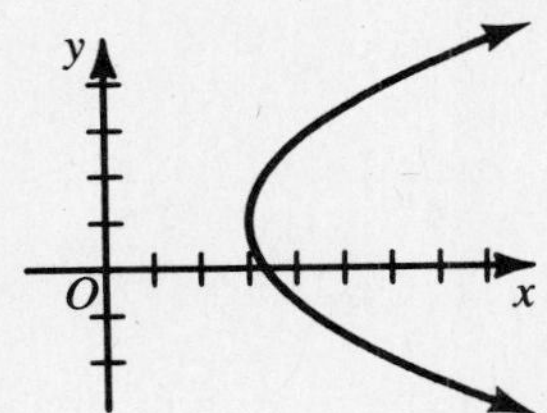

16. $x = \sqrt{y + 3}; x \geq 0$
$x^2 = y + 3$
$y = x^2 - 3, x \geq 0$

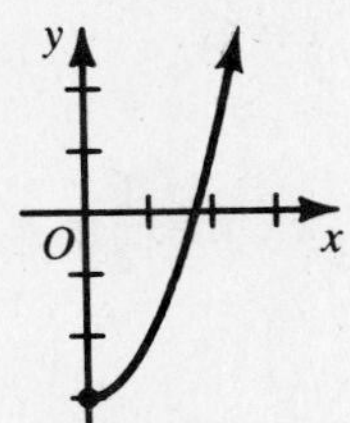

17. $x = 3 + \sqrt{y - 2}; x \geq 3$
$x - 3 = \sqrt{y - 2}$
$(x - 3)^2 = y - 2$
$y = (x - 3)^2 + 2, x \geq 3$

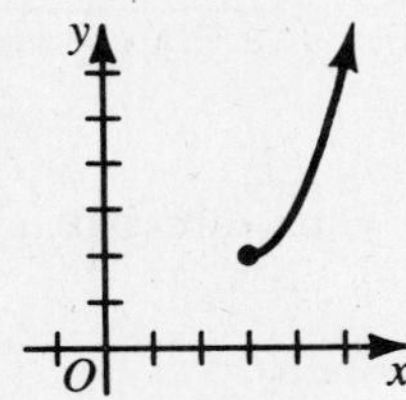

18. $y = -4 + \sqrt{x - 1}; y \geq -4$
$y + 4 = \sqrt{x - 1}$
$(y + 4)^2 = x - 1$
$x = (y + 4)^2 + 1, y \geq -4$

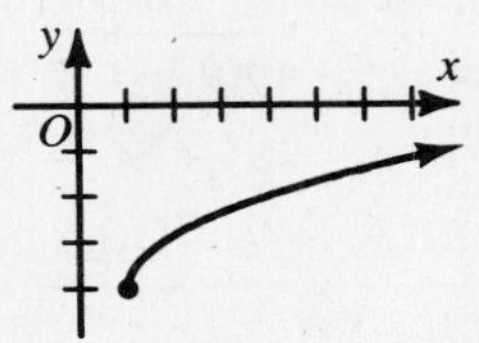

C **19.** $F(0, 1)$, $y = -5$: $(FP)^2 = (x - 0)^2 + (y - 1)^2$, $(PD)^2 = (y + 5)^2$;
$(y + 5)^2 = (x - 0)^2 + (y - 1)^2$; $y^2 + 10y + 25 = x^2 + y^2 - 2y + 1$; $12y = x^2 - 24$;
$y = \frac{1}{12}x^2 - 2$

20. $F(-2, 6)$, $y = 4$: $(FP)^2 = (x + 2)^2 + (y - 6)^2$, $(PD)^2 = (y - 4)^2$;
$(y - 4)^2 = (x + 2)^2 + (y - 6)^2$; $y^2 - 8y + 16 = x^2 + 4x + 4 + y^2 - 12y + 36$;
$4y = x^2 + 4x + 24$; $y = \frac{1}{4}x^2 + x + 6$

21. $F(3, 4)$, $x = -1$: $(FP)^2 = (x - 3)^2 + (y - 4)^2$, $(PD)^2 = (x + 1)^2$;
$(x + 1)^2 = (x - 3)^2 + (y - 4)^2$; $x^2 + 2x + 1 = x^2 - 6x + 9 + y^2 - 8y + 16$;
$8x = y^2 - 8y + 24$; $x = \frac{1}{8}y^2 - y + 3$

22. $F(-3, 3)$, $x = -1$: $(FP)^2 = (x + 3)^2 + (y - 3)^2$, $(PD)^2 = (x + 1)^2$;
$(x + 1)^2 = (x + 3)^2 + (y - 3)^2$; $x^2 + 2x + 1 = x^2 + 6x + 9 + y^2 - 6y + 9$;
$-4x = y^2 - 6y + 17$; $x = -\frac{1}{4}y^2 + \frac{3}{2}y - \frac{17}{4}$

23. $F\left(4, \frac{5}{2}\right)$, $y = \frac{3}{2}$: $(FP)^2 = (x - 4)^2 + \left(y - \frac{5}{2}\right)^2$, $(PD)^2 = \left(y - \frac{3}{2}\right)^2$; $\left(y - \frac{3}{2}\right)^2 =$
$(x - 4)^2 + \left(y - \frac{5}{2}\right)^2$; $y^2 - 3y + \frac{9}{4} = x^2 - 8x + 16 + y^2 - 5y + \frac{25}{4}$;
$2y = x^2 - 8x + 20$; $y = \frac{1}{2}x^2 - 4x + 10$

24. $F\left(\frac{7}{8}, 3\right), x = \frac{9}{8}$: $(FP)^2 = \left(x - \frac{7}{8}\right)^2 + (y - 3)^2$, $(PD)^2 = \left(x - \frac{9}{8}\right)^2$;
$\left(x - \frac{9}{8}\right)^2 = \left(x - \frac{7}{8}\right)^2 + (y - 3)^2$; $x^2 - \frac{9}{4}x + \frac{81}{64} = x^2 - \frac{7}{4}x + \frac{49}{64} + y^2 - 6y + 9$;
$-\frac{1}{2}x = y^2 - 6y + \frac{17}{2}$; $x = -2y^2 + 12y - 17$

25. $V\left(-\frac{1}{2}, \frac{1}{4}\right) = (h, k)$; $y = \frac{1}{2} = k - m = \frac{1}{4} - m$; $m = -\frac{1}{4}$; $y - k = \frac{1}{4m}(x - h)^2$;
$y - \frac{1}{4} = -\left(x + \frac{1}{2}\right)^2$; $y = -\left(x + \frac{1}{2}\right)^2 + \frac{1}{4}$

26. $V\left(-\frac{9}{4}, \frac{3}{2}\right) = (h, k)$; $F\left(-2, \frac{3}{2}\right)$; m = dist. between focus and vertex =
$-2 - \left(-\frac{9}{4}\right) = \frac{1}{4}$; $x - h = \frac{1}{4m}(y - k)^2$; $x + \frac{9}{4} = \left(y - \frac{3}{2}\right)^2$; $x = \left(y - \frac{3}{2}\right)^2 - \frac{9}{4}$

27. $x^2 = 4py$; $y = \frac{1}{4p}(x - 0)^2 + 0$; $a = \frac{1}{4p}$; $p = \frac{1}{4a}$

28. $F(0, m)$; directrix, $y = -m$: $P(x, y)$ on parabola, $D(x, -m)$ on directrix, $PF = PD$;
$PF = \sqrt{(x - 0)^2 + (y - m)^2}$, $PD = |y - (-m)|$; $|y + m|^2 = \left(\sqrt{x^2 + (y - m)^2}\right)^2$;
$y^2 + 2my + m^2 = x^2 + y^2 - 2my + m^2$; $4my = x^2$; $y = \frac{1}{4m}x^2$.

29. $F(m, k)$; directrix, $x = -m$: $P(x, y)$ on parabola, $D(-m, y)$ on directrix, $PF = PD$;
$PF = \sqrt{(x - m)^2 + (y - k)^2}$, $PD = |x - (-m)|$;
$|x + m|^2 = \left(\sqrt{(x - m)^2 + (y - k)^2}\right)^2$;
$x^2 + 2mx + m^2 = x^2 - 2mx + m^2 + (y - k)^2$; $4mx = (y - k)^2$; $x = \frac{1}{4m}(y - k)^2$.

30. vertex, (h, k); directrix, $y = k - m$: The distance from the point (h, k) to the directrix is the same as the distance from (h, k) to the focus:
$d = |k - (k - m)| = |m|$, so $F = (h, k + m)$. $F(h, k + m)$ is focus, $P(x, y)$ on parabola, $D(x, k - m)$ on directrix, $PF = PD$;
$PF = \sqrt{(x - h)^2 + (y - (k + m))^2}$, $PD = |y - (k - m)|$; $|y - (k - m)|^2 =$
$\left(\sqrt{(x - h)^2 + (y - (k + m))^2}\right)^2$; $y^2 - 2(k - m)y + (k - m)^2 =$
$(x - h)^2 + y^2 - 2(k + m)y + (k + m)^2$; $y^2 - 2ky + 2my + k^2 - 2km + m^2 =$
$(x - h)^2 + y^2 - 2ky - 2my + k^2 + 2km + m^2$; $4my - 4km = (x - h)^2$;
$y = \frac{1}{4m}(x - h)^2 + k$

31. By Ex. 30, the equation of the parabola is $y = \frac{1}{4m}(x - h)^2 + k$. Substituting in $(h + 2m, k + m)$ gives $k + m = \frac{1}{4m}(h + 2m - h)^2 + k$, $k + m = \frac{1}{4m}(2m)^2 + k$, $k + m = m + k$. Substituting in $(h - 2m, k + m)$ gives $k + m =$ $\frac{1}{4m}(h - 2m - h)^2 + k$, $k + m = \frac{1}{4m}(-2m)^2 + k$, $k + m = m + k$. $\therefore$ points are on parabola. $P_1P_2 = \sqrt{[(h + 2m) - (h - 2m)]^2 + [(k + m) - (k + m)]^2} =$ $|(h + 2m) - (h - 2m)| = |4m| = 4|m|$.

Pages 411–412 • WRITTEN EXERCISES

A **1.** $c^2 = a^2 - b^2 = 25 - 16 = 9,$
$c = \pm 3; F_1(0, 3), F_2(0, -3)$

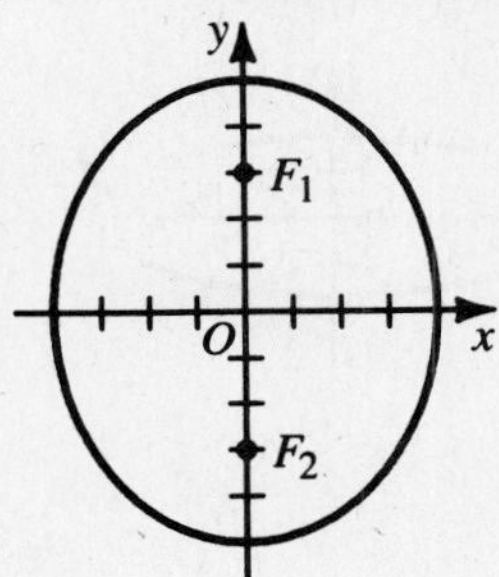

2. $c^2 = a^2 - b^2 = 25 - 9 = 16,$
$c = \pm 4; F_1(-4, 0), F_2(4, 0)$

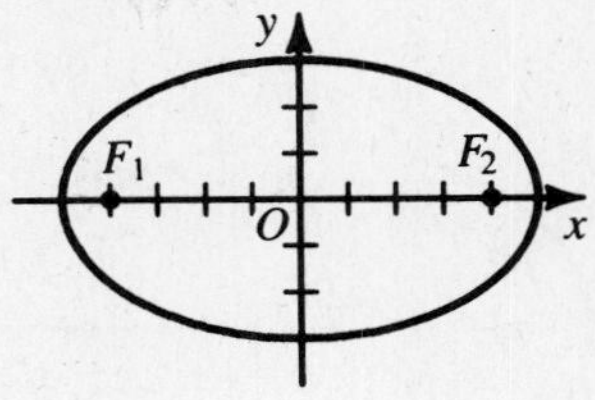

3. $c^2 = a^2 - b^2 = 9 - 1 = 8,$
$c = \pm 2\sqrt{2}; F_1(0, 2\sqrt{2}),$
$F_2(0, -2\sqrt{2})$

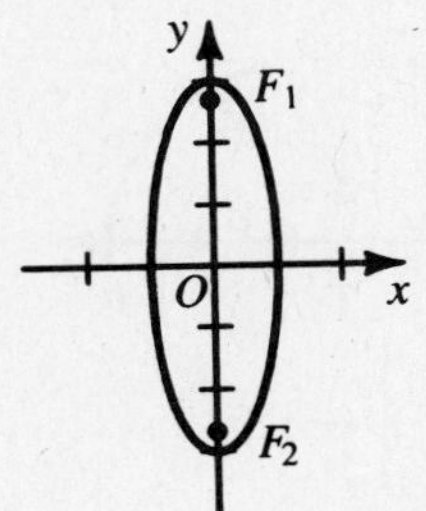

4. $c^2 = a^2 - b^2 = 8 - 1 = 7,$
$c = \pm\sqrt{7}; F_1(-\sqrt{7}, 0), F_2(\sqrt{7}, 0)$

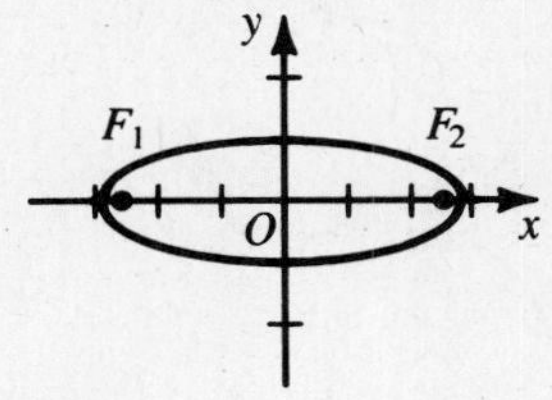

5. $c^2 = a^2 - b^2 = 169 - 144 = 25,$
$c = \pm 5; F_1(-5, 0), F_2(5, 0)$

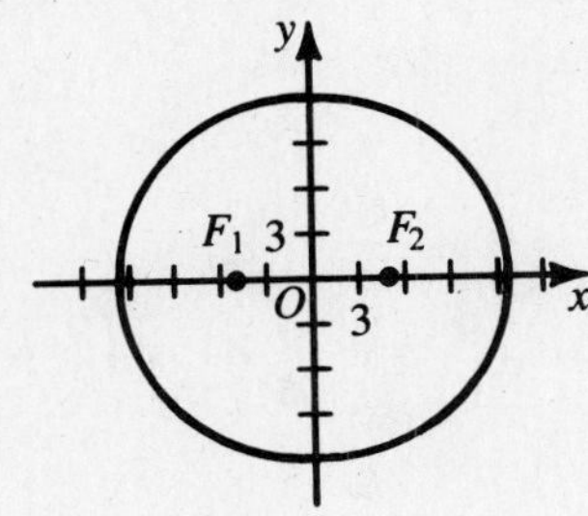

6. $c^2 = a^2 - b^2 = 225 - 81 = 144,$
$c = \pm 12; F_1(0, 12), F_2(0, -12)$

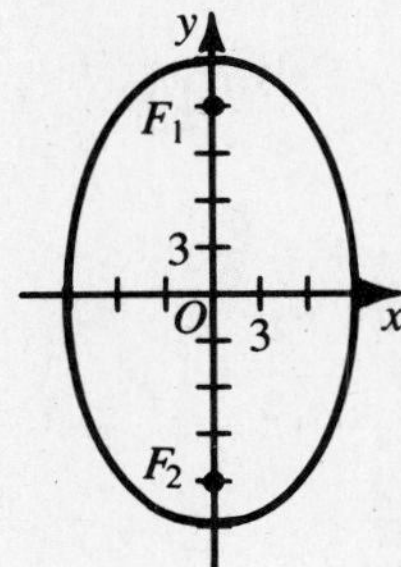

7. $4x^2 + 9y^2 = 144$, $\frac{x^2}{36} + \frac{y^2}{16} = 1$;
$c^2 = 36 - 16 = 20$,
$c = \pm 2\sqrt{5}$; $F_1(-2\sqrt{5}, 0)$,
$F_2(2\sqrt{5}, 0)$

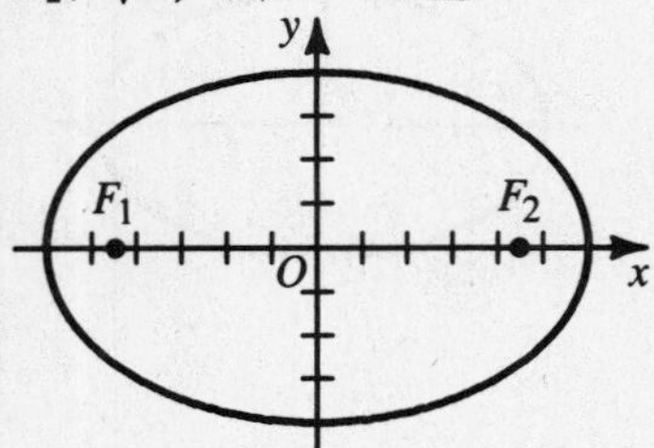

8. $x^2 + 9y^2 = 225$, $\frac{x^2}{225} + \frac{y^2}{25} = 1$;
$c^2 = 225 - 25 = 200$, $c = \pm 10\sqrt{2}$;
$F_1(-10\sqrt{2}, 0)$, $F_2(10\sqrt{2}, 0)$

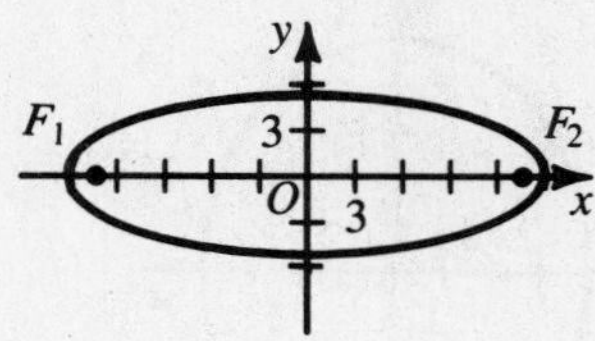

9. $9x^2 + 3y^2 = 900$, $\frac{x^2}{100} + \frac{y^2}{300} = 1$;
$c^2 = 300 - 100 = 200$,
$c = \pm 10\sqrt{2}$; $F_1(0, 10\sqrt{2})$,
$F_2(0, -10\sqrt{2})$

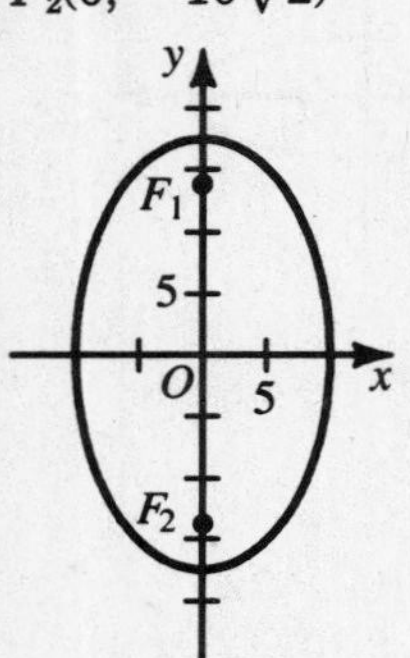

10. $16x^2 + 25y^2 = 400$, $\frac{x^2}{25} + \frac{y^2}{16} = 1$;
$c^2 = 25 - 16 = 9$,
$c = \pm 3$; $F_1(-3, 0)$, $F_2(3, 0)$

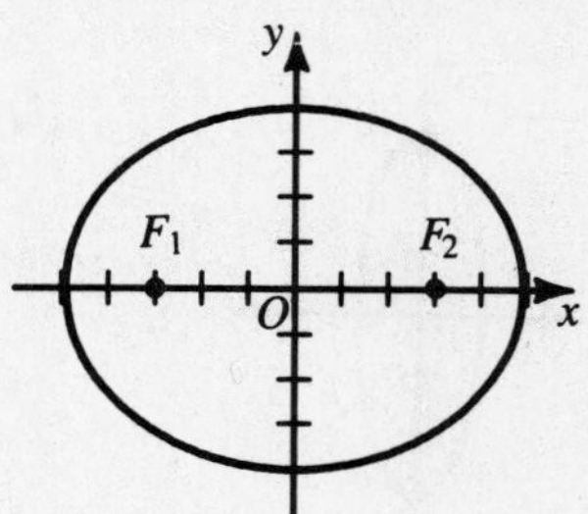

11. $9x^2 + 25y^2 = 100$, $\frac{x^2}{\frac{100}{9}} + \frac{y^2}{4} = 1$;

$c^2 = \frac{100}{9} - 4 = \frac{64}{9}$,
$c = \pm\frac{8}{3}$; $F_1\left(-\frac{8}{3}, 0\right)$, $F_2\left(\frac{8}{3}, 0\right)$

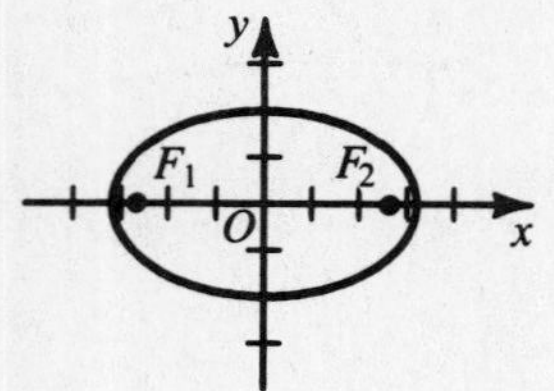

12. $16x^2 + 25y^2 = 144$, $\frac{x^2}{9} + \frac{y^2}{\frac{144}{25}} = 1$;

$c^2 = 9 - \frac{144}{25} = \frac{81}{25}$,
$c = \pm\frac{9}{5}$; $F_1\left(-\frac{9}{5}, 0\right)$, $F_2\left(\frac{9}{5}, 0\right)$

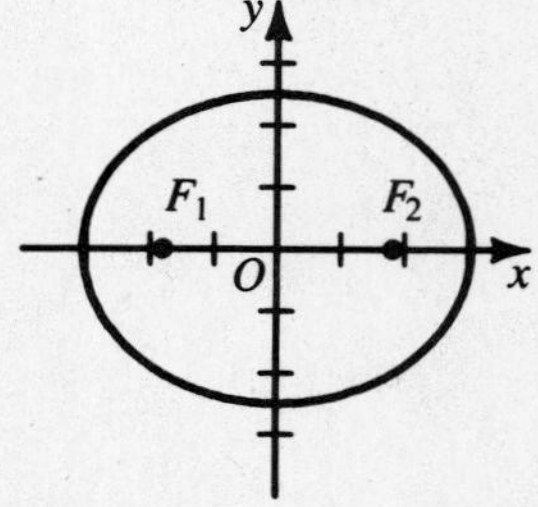

13. $\frac{(x-3)^2}{16}+\frac{(y-2)^2}{9}=1$;
$c^2 = 16 - 9 = 7$,
$c = \pm\sqrt{7}$; $F_1(3-\sqrt{7}, 2)$,
$F_2(3+\sqrt{7}, 2)$

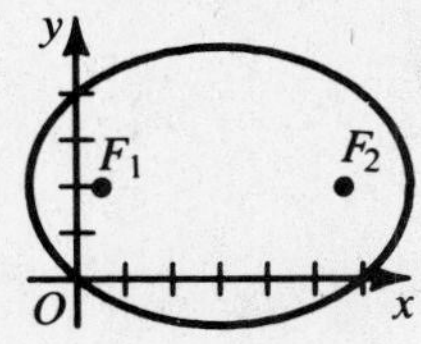

14. $\frac{(x+1)^2}{4}+\frac{(y-6)^2}{36}=1$;
$c^2 = 36 - 4 = 32$, $c =$
$\pm 4\sqrt{2}$, $F_1(-1, 6+4\sqrt{2})$,
$F_2(-1, 6-4\sqrt{2})$

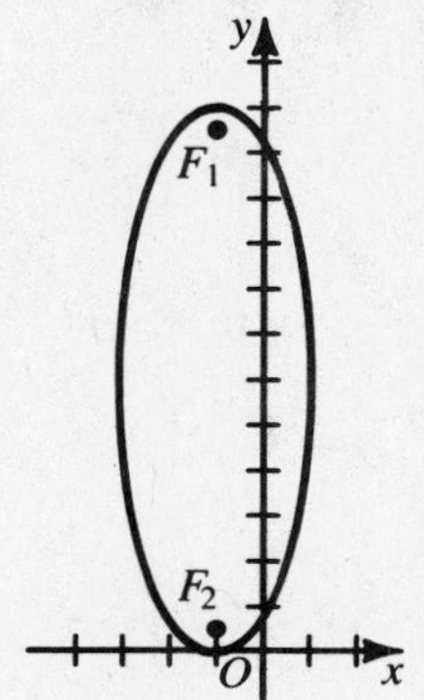

15. $\frac{(x+3)^2}{9}+(y-1)^2=1$;
$c^2 = 9 - 1 = 8$,
$c = \pm 2\sqrt{2}$;
$F_1(-3-2\sqrt{2}, 1)$,
$F_2(-3+2\sqrt{2}, 1)$

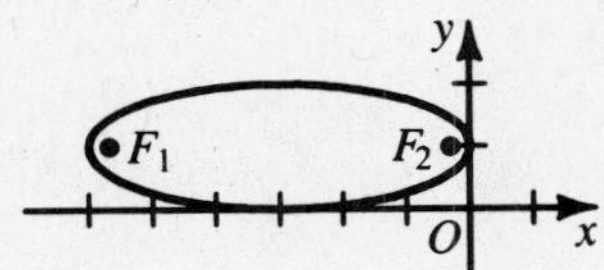

16. $\frac{(x-5)^2}{25}+\frac{(y+3)^2}{4}=1$;
$c^2 = 25 - 4 = 21$, $c = \pm\sqrt{21}$;
$F_1(5-\sqrt{21}, -3)$, $F_2(5+\sqrt{21}, -3)$

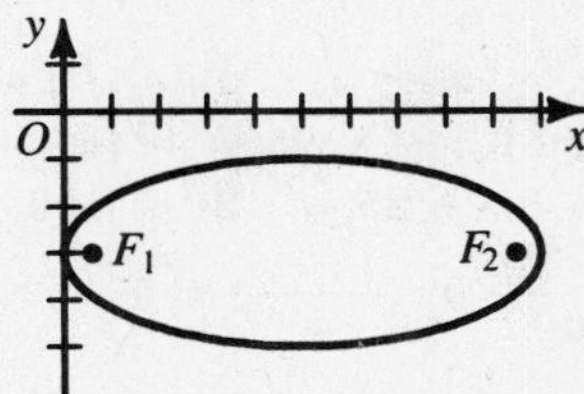

B 17.

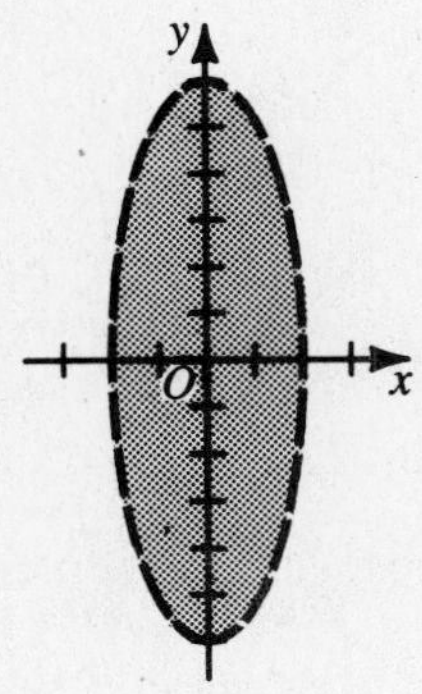

18.

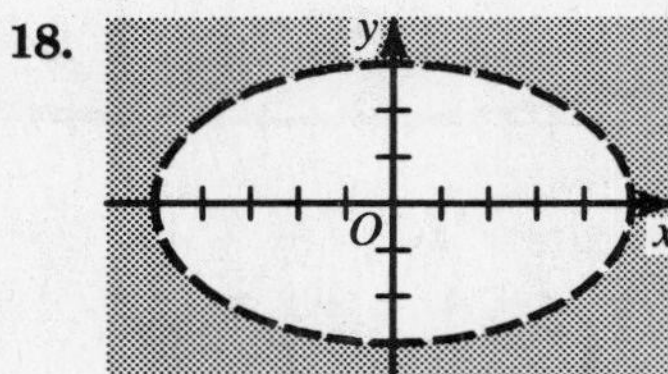

19.

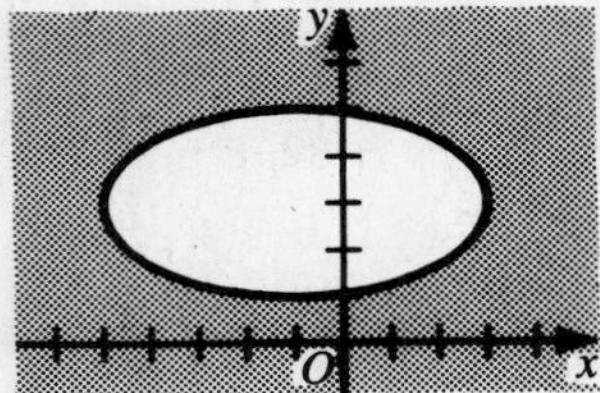

20.

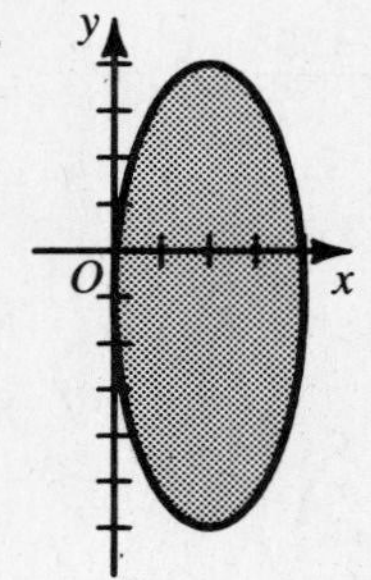

21. $\frac{y}{3} = \sqrt{1 - x^2},\ y \geq 0$

$\frac{y^2}{9} = 1 - x^2,\ y \geq 0$

$x^2 + \frac{y^2}{9} = 1,\ y \geq 0$

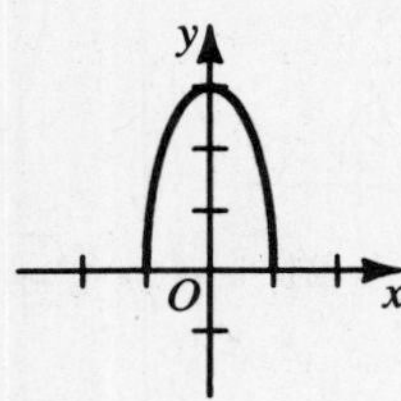

22. $\frac{x}{6} = \sqrt{1 - y^2},\ x \geq 0$

$\frac{x^2}{36} = 1 - y^2,\ x \geq 0$

$\frac{x^2}{36} + y^2 = 1,\ x \geq 0$

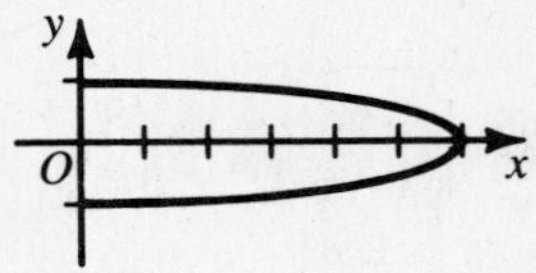

23. $x^2 + 6x + 25y^2 - 100y + 9 = 0$

$x^2 + 6x + 9 + 25(y^2 - 4y + 4) = -9 + 9 + 100$

$(x + 3)^2 + 25(y - 2)^2 = 100$

$\frac{(x + 3)^2}{100} + \frac{(y - 2)^2}{4} = 1$

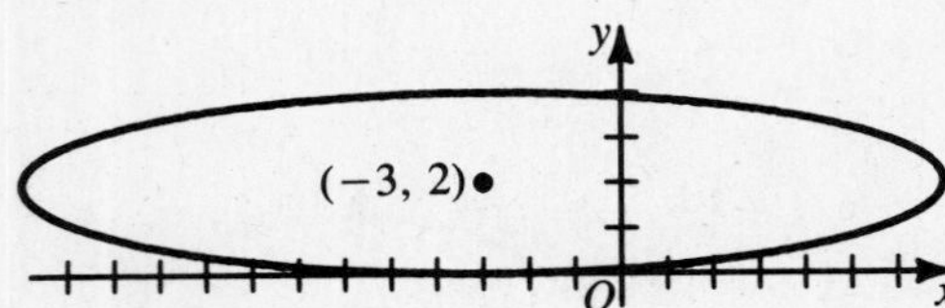

24. $9x^2 + 36x + 4y^2 - 8y + 4 = 0$

$9(x^2 + 4x + 4) + 4(y - 2y + 1) = -4 + 36 + 4$

$9(x + 2)^2 + 4(y - 1)^2 = 36$

$\frac{(x + 2)^2}{4} + \frac{(y - 1)^2}{9} = 1$

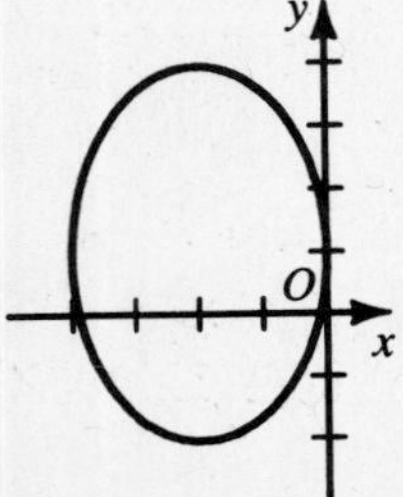

25. $c = 5$; $b = 12$; $a^2 = b^2 + c^2 = 169$; $\frac{x^2}{169} + \frac{y^2}{144} = 1$

26. $a = 13$; $b = 9$; $\frac{x^2}{81} + \frac{y^2}{169} = 1$

27. $b = 3$; $c = 4$; $a^2 = b^2 + c^2 = 25$; $\frac{x^2}{9} + \frac{y^2}{25} = 1$

28. $2a = 20$; $a = 10$; $b = 7$; $\frac{x^2}{100} + \frac{y^2}{49} = 1$

29. $2a = 14$; $a = 7$; $c = \sqrt{13}$; $b^2 = a^2 - c^2 = 36$; $\frac{x^2}{36} + \frac{y^2}{49} = 1$

30. $a = 11$; $c = 4\sqrt{6}$; $b^2 = a^2 - c^2 = 25$; $\frac{x^2}{121} + \frac{y^2}{25} = 1$

31. Center is midpoint of foci: $C = \left(\frac{0 + 8}{2}, \frac{-7 - 7}{2}\right) = (4, -7)$; c = distance from center to a focus $= 4$; $a = 5$; $b^2 = a^2 - c^2 = 9$; $\frac{(x - 4)^2}{25} + \frac{(y + 7)^2}{9} = 1$

32. Center is midpoint of foci: $C = \left(\frac{-3 - 3}{2}, \frac{5 - 1}{2}\right) = (-3, 2)$; c = distance from center to focus $= 3$; $b = 4$; $a^2 = b^2 + c^2 = 25$; $\frac{(x + 3)^2}{16} + \frac{(y - 2)^2}{25} = 1$

C **33.** $F_1(-3, 0)$, $F_2(3, 0)$, $P(2, 2\sqrt{6})$: $(F_1P)^2 = (-3 - 2)^2 + (0 - 2\sqrt{6})^2 = 25 + 24 = 49$, $F_1P = 7$; $(F_2P)^2 = (3 - 2)^2 + (0 - 2\sqrt{6})^2 = 1 + 24 = 25$, $F_2P = 5$; $F_1P + F_2P = 2a = 12$, $a = 6$; but $c = 3$, so $b^2 = 36 - 9 = 27$; $\frac{x^2}{36} + \frac{y^2}{27} = 1$

34. $P_1(x, y)$, line $x = 8$, $P_2(2, 0)$: dist. from P_1 to line $= 2 \cdot P_1P_2$;
$\sqrt{(x - 8)^2 + (y - y)^2} = 2\sqrt{(x - 2)^2 + (y - 0)^2}$; $(x - 8)^2 = 4((x - 2)^2 + y^2)$;
$x^2 - 16x + 64 = 4x^2 - 16x + 16 + 4y^2$; $3x^2 + 4y^2 = 48$; $\frac{x^2}{16} + \frac{y^2}{12} = 1$

35. $F_1(c, 0)$, $F_2(-c, 0)$; sum of focal radii $= 2a$; point on ellipse, $P(x, y)$:
$PF_1 = \sqrt{(x - c)^2 + (y - 0)^2} = \sqrt{(x - c)^2 + y^2}$; $PF_2 = \sqrt{(x - (-c))^2 + (y - 0)^2} = \sqrt{(x + c)^2 + y^2}$; $PF_1 + PF_2 = 2a$, so $\sqrt{(x - c)^2 + y^2} + \sqrt{(x + c)^2 + y^2} = 2a$;
$\left(\sqrt{(x - c)^2 + y^2}\right)^2 = \left(2a - \sqrt{(x + c)^2 + y^2}\right)^2$;
$(x - c)^2 + y^2 = 4a^2 - 4a\sqrt{(x + c)^2 + y^2} + (x + c)^2 + y^2$;
$a^2 + xc = a\sqrt{(x + c)^2 + y^2}$; $a^4 + 2a^2xc + x^2c^2 = a^2(x^2 + 2xc + c^2) + a^2y^2$;
$x^2(a^2 - c^2) + a^2y^2 = a^4 - a^2c^2$; $x^2(a^2 - c^2) + a^2y^2 = a^2(a^2 - c^2)$; $\frac{x^2}{a^2} + \frac{y^2}{a^2 - c^2} = 1$

36. $F_1(0, c)$, $F_2(0, -c)$; sum of focal radii $= 2a$; point on ellipse $P(x, y)$:
$PF_1 = \sqrt{(x - 0)^2 + (y - c)^2} = \sqrt{x^2 + (y - c)^2}$; $PF_2 = \sqrt{(x - 0)^2 + (y - (-c))^2} = \sqrt{x^2 + (y + c)^2}$; $PF_1 + PF_2 = 2a$, so $\sqrt{x^2 + (y - c)^2} + \sqrt{x^2 + (y + c)^2} = 2a$;
$\left(\sqrt{x^2 + (y - c)^2}\right)^2 = \left(2a - \sqrt{x^2 + (y + c)^2}\right)^2$;
$x^2 + y^2 - 2yc + c^2 = 4a^2 - 4a\sqrt{x^2 + (y + c)^2} + x^2 + y^2 + 2yc + c^2$;
$yc + a^2 = a\sqrt{x^2 + (y + c)^2}$; $y^2c^2 + 2a^2yc + a^4 = a^2x^2 + a^2(y^2 + 2yc + c^2)$;
$a^2x^2 + y^2(a^2 - c^2) = a^4 - a^2c^2$; $ax^2 + y^2(a^2 - c^2) = a^2(a^2 - c^2)$; $\frac{x^2}{a^2 - c^2} + \frac{y^2}{a^2} = 1$

Pages 416–417 • WRITTEN EXERCISES

A **1.** $a^2 = 9$; $b^2 = 16$; $c^2 = 9 + 16$; $c = \pm 5$; $F_1(-5, 0)$, $F_2(5, 0)$

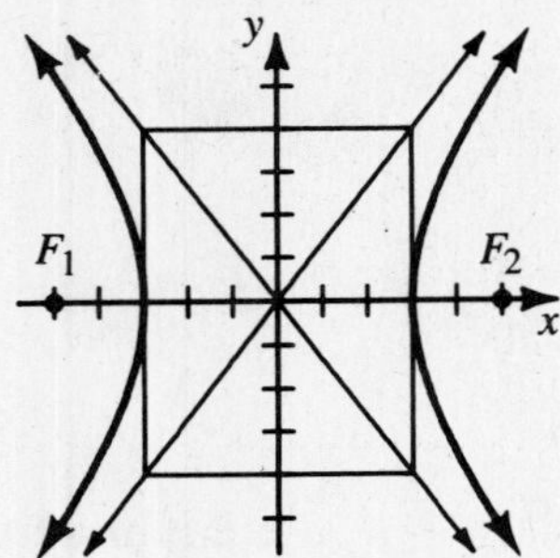

2. $a^2 = 64$; $b^2 = 36$; $c^2 = 64 + 36$; $c = \pm 10$; $F_1(-10, 0)$, $F_2(10, 0)$

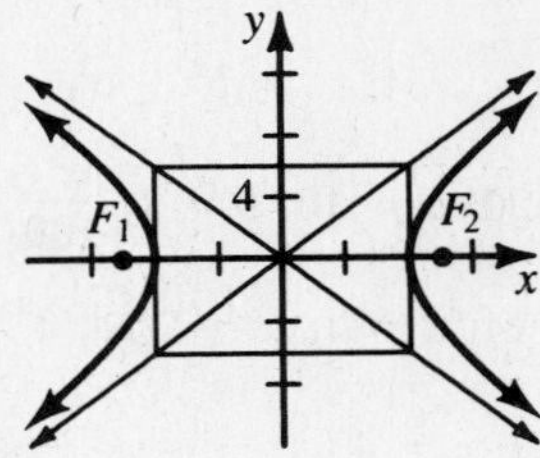

3. $a^2 = 25$; $b^2 = 100$; $c^2 = 100 + 25$; $c = \pm 5\sqrt{5}$; $F_1(0, 5\sqrt{5})$, $F_2(0, -5\sqrt{5})$

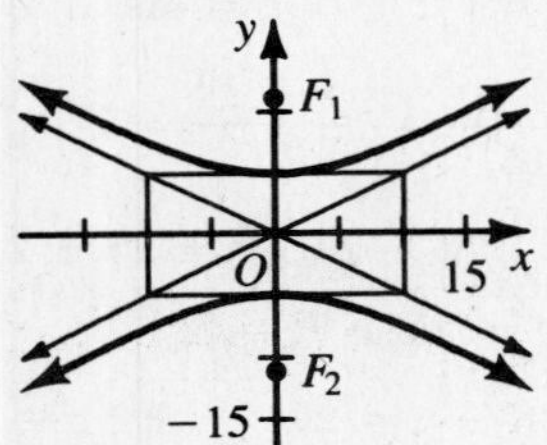

4. $a^2 = 9$; $b^2 = 1$; $c^2 = 9 + 1$; $c = \pm\sqrt{10}$; $F_1(-\sqrt{10}, 0)$, $F_2(\sqrt{10}, 0)$

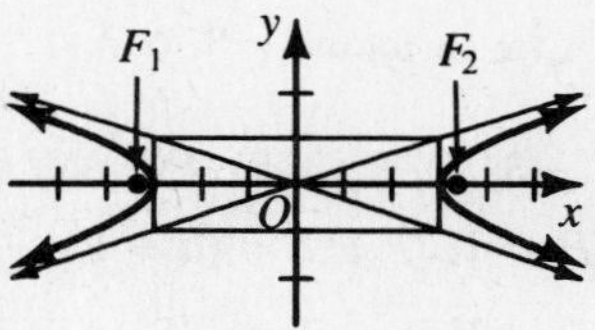

5. $a^2 = 49$; $b^2 = 49$; $c^2 = 49 + 49$; $c = \pm 7\sqrt{2}$; $F_1(0, 7\sqrt{2})$, $F_2(0, -7\sqrt{2})$

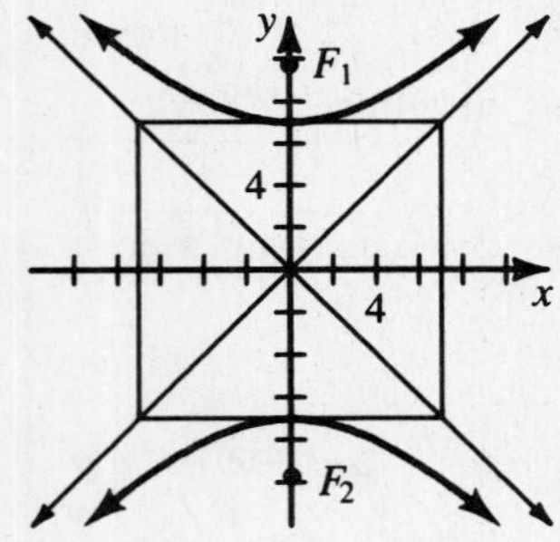

6. $a^2 = 4$; $b^2 = 36$; $c^2 = 36 + 4$; $c = \pm 2\sqrt{10}$; $F_1(-2\sqrt{10}, 0)$, $F_2(2\sqrt{10}, 0)$

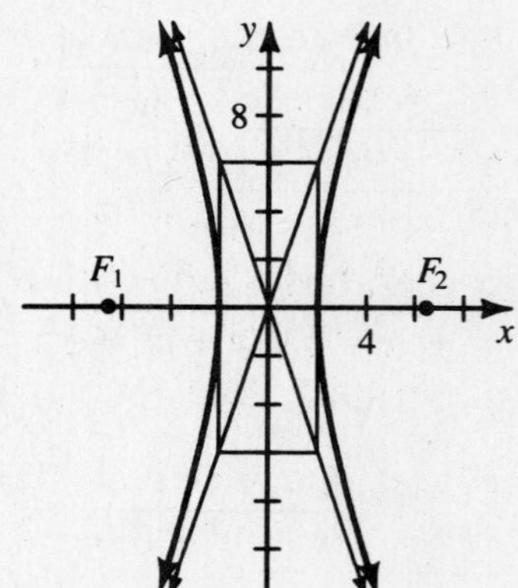

7. $2x^2 - 2y^2 = 50, \frac{x^2}{25} - \frac{y^2}{25} = 1;$

$a^2 = 25; b^2 = 25; c^2 = 50;$

$c = \pm 5\sqrt{2}; F_1(-5\sqrt{2}, 0), F_2(5\sqrt{2}, 0)$

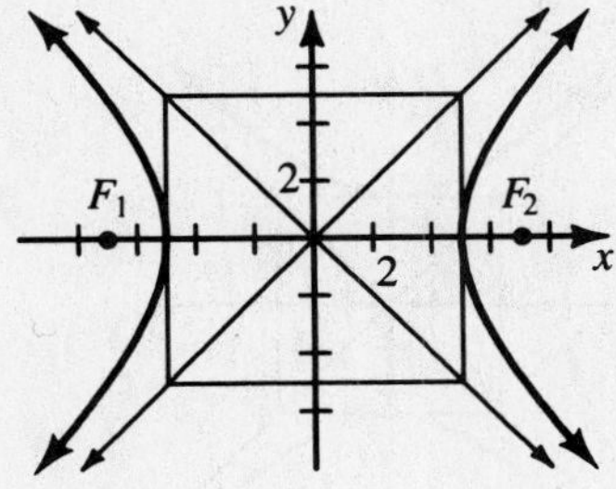

8. $4y^2 - 16x^2 = 64, \frac{y^2}{16} - \frac{x^2}{4} = 1;$

$a^2 = 16; b^2 = 4;$

$c^2 = 20; c = \pm 2\sqrt{5};$

$F_1(0, 2\sqrt{5}), F_2(0, -2\sqrt{5})$

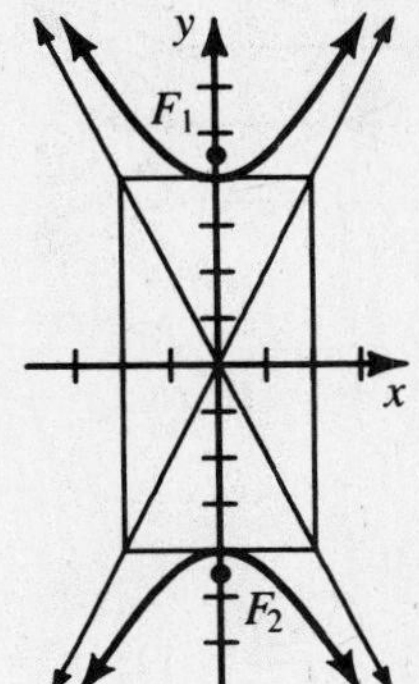

9. $144y^2 - 25x^2 = 900, \frac{y^2}{\frac{25}{4}} - \frac{x^2}{36} = 1;$

$a^2 = \frac{25}{4}; b^2 = 36; c^2 = \frac{169}{4};$

$c = \pm\frac{13}{2}; F_1\left(0, \frac{13}{2}\right), F_2\left(0, -\frac{13}{2}\right)$

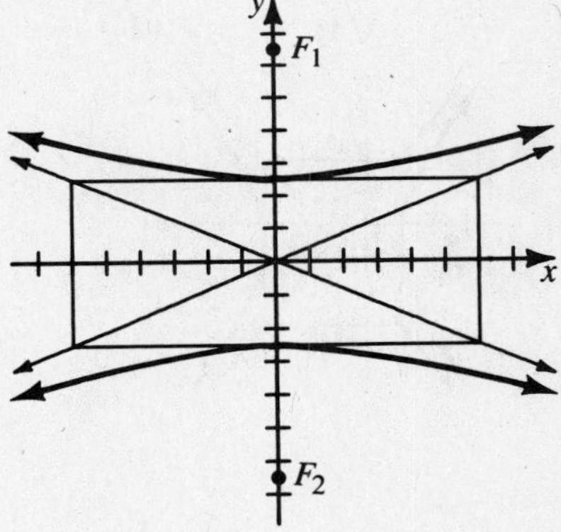

10. $9x^2 - 16y^2 = 16, \frac{x^2}{\frac{16}{9}} - y^2 = 1;$

$a^2 = \frac{16}{9}; b^2 = 1; c^2 = \frac{25}{9};$

$c = \pm\frac{5}{3}; F_1\left(-\frac{5}{3}, 0\right), F_2\left(\frac{5}{3}, 0\right)$

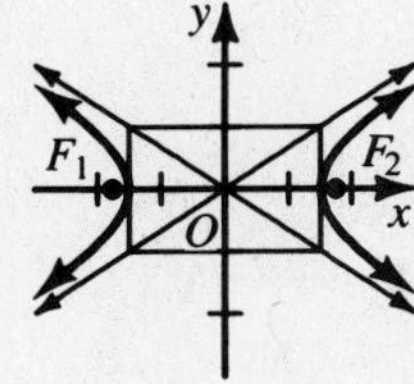

11. $6x^2 - 30y^2 = 180$, $\frac{x^2}{30} - \frac{y^2}{6} = 1$;
$a^2 = 30$; $b^2 = 6$; $c^2 = 36$;
$c = \pm 6$; $F_1(-6, 0)$, $F_2(6, 0)$

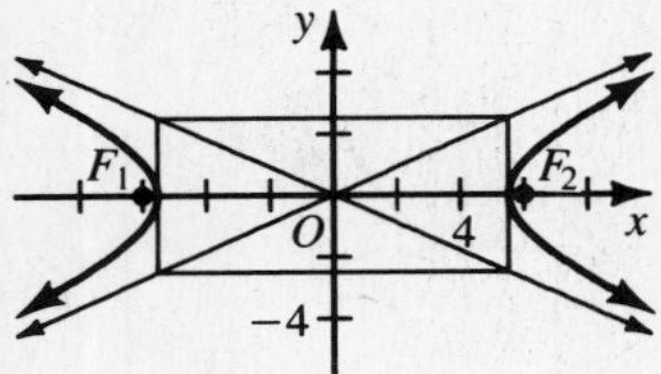

12. $16y^2 - 20x^2 = 400$, $\frac{y^2}{25} - \frac{x^2}{20} = 1$;
$a^2 = 25$; $b^2 = 20$; $c^2 = 45$;
$c = \pm 3\sqrt{5}$; $F_1(0, 3\sqrt{5})$,
$F_2(0, -3\sqrt{5})$

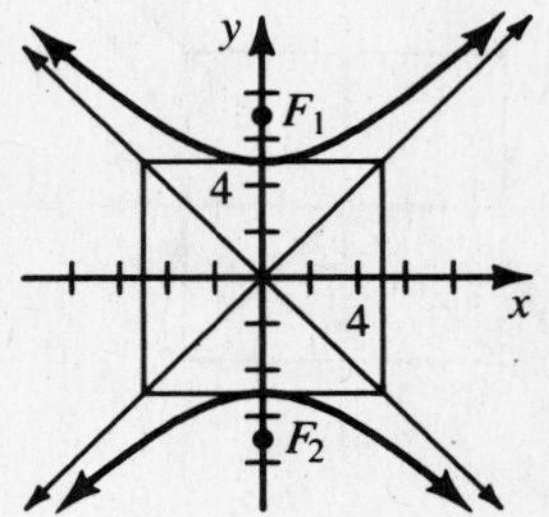

13. $\frac{(x-2)^2}{16} - \frac{(y+1)^2}{25} = 1$; $a^2 = 16$;
$b^2 = 25$; $c^2 = 41$; $c = \pm\sqrt{41}$;
$F_1(2 - \sqrt{41}, -1)$, $F_2(2 + \sqrt{41}, -1)$

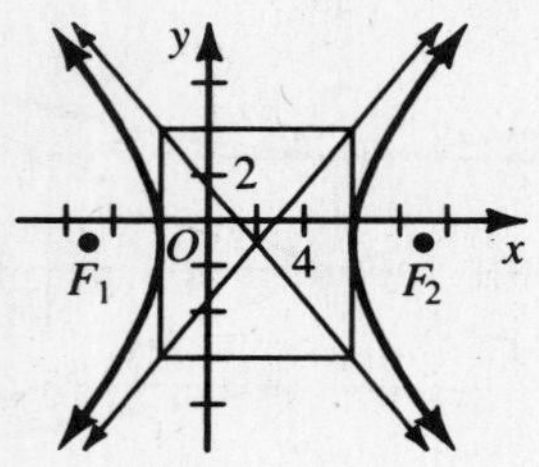

14. $\frac{(y-3)^2}{9} - (x-1)^2 = 1$; $a^2 = 9$;
$b^2 = 1$; $c^2 = 10$; $c = \pm\sqrt{10}$;
$F_1(1, 3 + \sqrt{10})$, $F_2(1, 3 - \sqrt{10})$

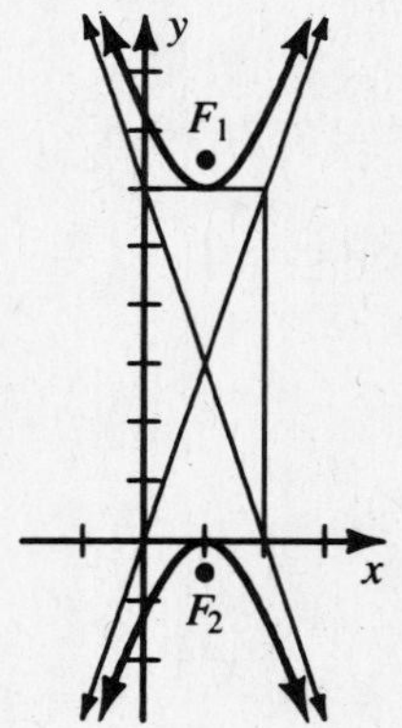

15. $\frac{(y-3)^2}{4} - \frac{(x+4)^2}{9} = 1$; $a^2 = 4$;
$b^2 = 9$; $c^2 = 13$; $c = \pm\sqrt{13}$;
$F_1(-4, 3 + \sqrt{13})$, $F_2(-4, 3 - \sqrt{13})$

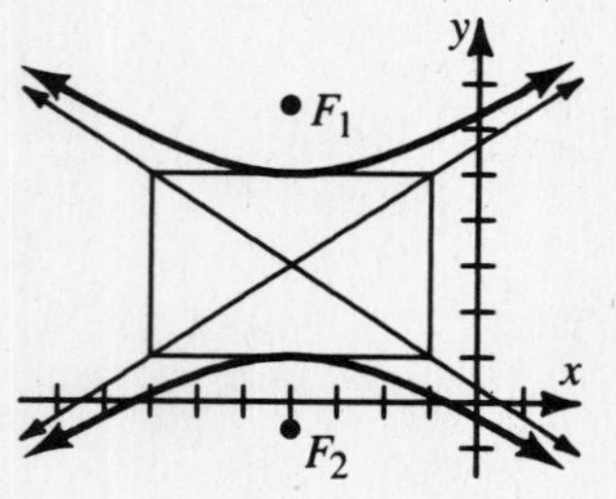

16. $x^2 - y^2 + 6y = 10$, $x^2 - (y^2 - 6y + 9) = 10 - 9$, $x^2 - (y-3)^2 = 1$; $a^2 = 1$;
$b^2 = 1$; $c^2 = 2$; $c = \pm\sqrt{2}$;
$F_1(-\sqrt{2}, 3)$, $F_2(\sqrt{2}, 3)$

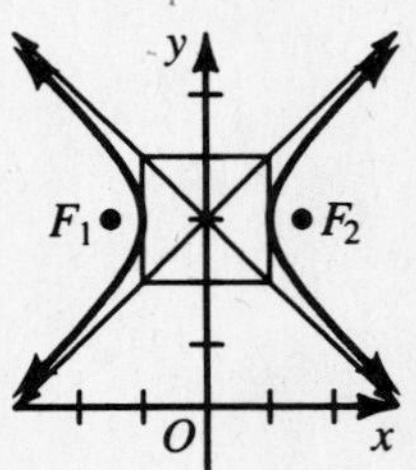

17. $y^2 - x^2 \geq 9, \frac{y^2}{9} - \frac{x^2}{9} \geq 1$

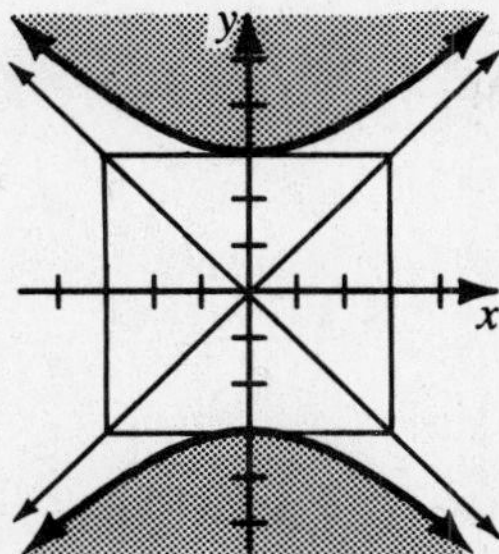

18. $y^2 - x^2 \leq 4, \frac{y^2}{4} - \frac{x^2}{4} \leq 1$

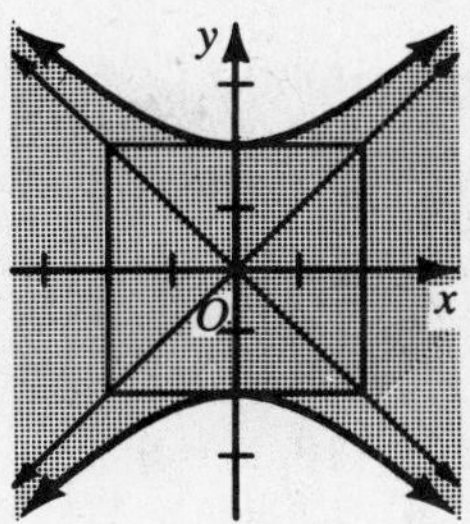

19. $x^2 < y^2 + 4, \frac{x^2}{4} - \frac{y^2}{4} < 1$

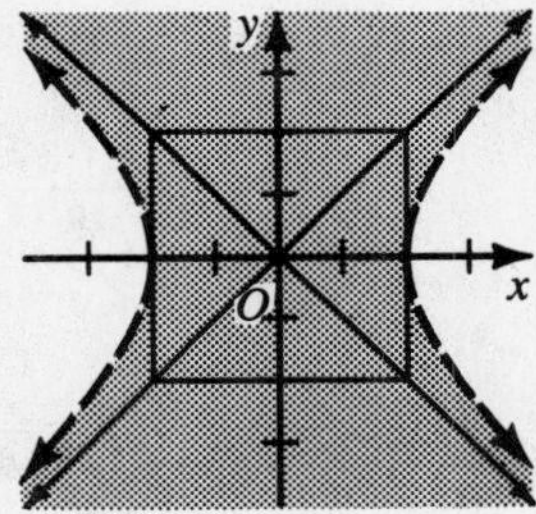

20. $y^2 < x^2 - 1, x^2 - y^2 > 1$

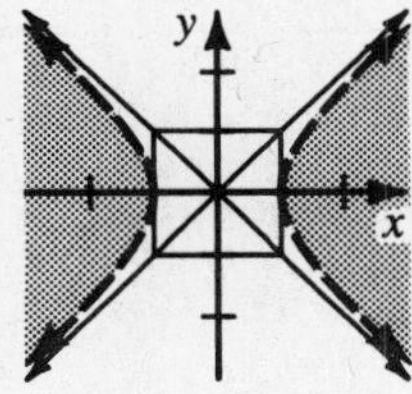

B **21.** $x = \sqrt{y^2 + 1}$;
$x \geq 1, y \in \mathcal{R}; x^2 - y^2 = 1$

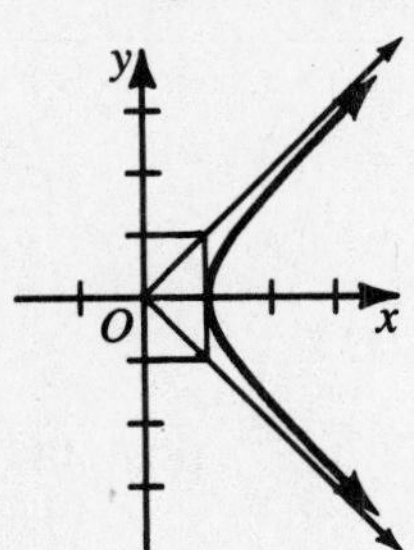

22. $y = \sqrt{x^2 - 9}$;
$x \leq -3$ or $x \geq 3, y \geq 0; \frac{x^2}{9} - \frac{y^2}{9} = 1$

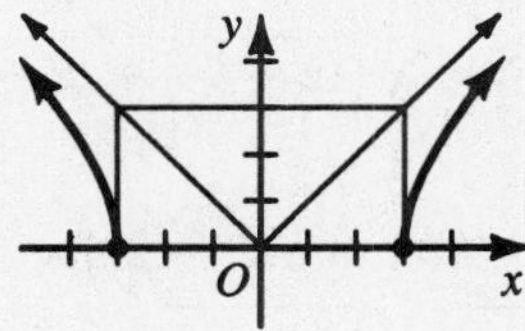

23. $y = \sqrt{x^2 - 16}; x \leq -4$
or $x \geq 4, y \geq 0; \frac{x^2}{16} - \frac{y^2}{16} = 1$

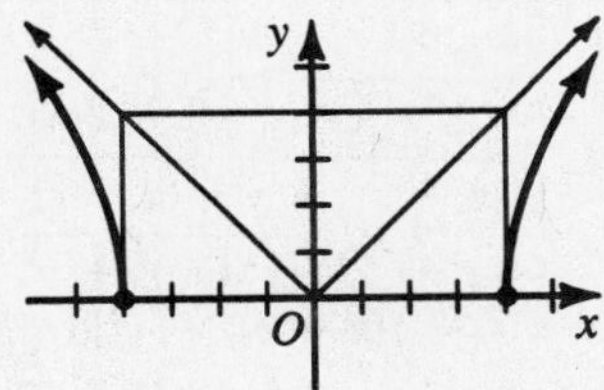

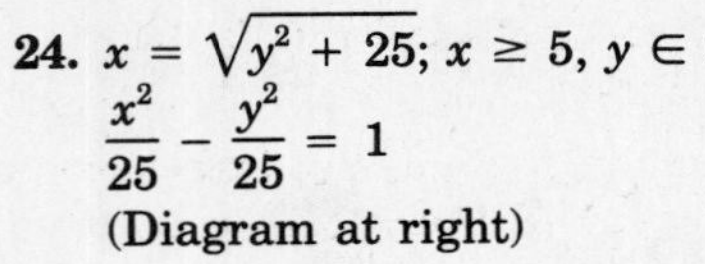

24. $x = \sqrt{y^2 + 25}$; $x \geq 5$, $y \in \mathscr{R}$;
$\frac{x^2}{25} - \frac{y^2}{25} = 1$
(Diagram at right)

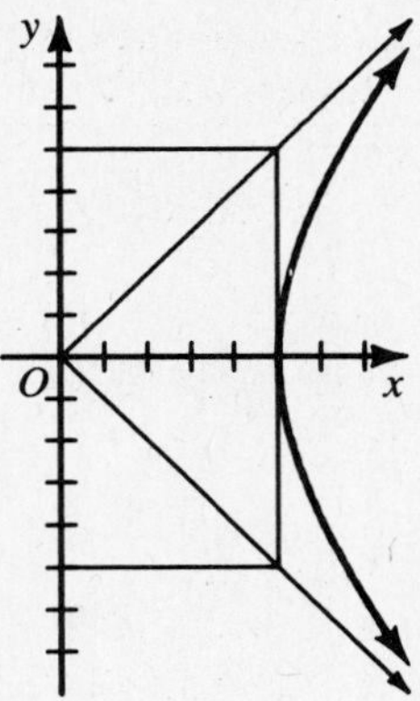

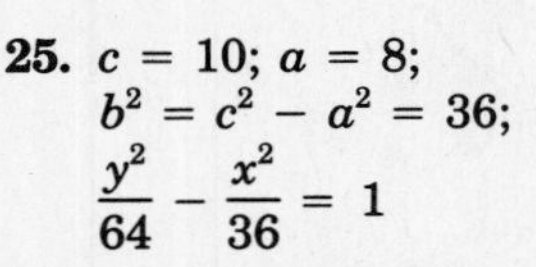

25. $c = 10$; $a = 8$;
$b^2 = c^2 - a^2 = 36$;
$\frac{y^2}{64} - \frac{x^2}{36} = 1$

26. $c = 12$; $2a = 18$, $a = 9$;
$b^2 = c^2 - a^2 = 63$;
$\frac{x^2}{81} - \frac{y^2}{63} = 1$

C **27.** $2a = 6$; $a = 3$; $F_1F_2 = 2c = 10$; $c = 5$; $b^2 = c^2 - a^2 = 16$; center = midpoint of $\overline{F_1F_2} = \left(\frac{-2 + 8}{2}, \frac{1 + 1}{2}\right) = (3, 1)$; $\frac{(x - 3)^2}{9} - \frac{(y - 1)^2}{16} = 1$

28. $F_1F_2 = 2c = 16$; $c = 8$; $V_1V_2 = 2a = 8$; $a = 4$; $b^2 = c^2 - a^2 = 48$; center = midpoint of $\overline{F_1F_2} = \left(\frac{3 + 3}{2}, \frac{11 - 5}{2}\right) = (3, 3)$; $\frac{(y - 3)^2}{16} - \frac{(x - 3)^2}{48} = 1$

29. a. $c = 15$; $a^2 + b^2 = 15^2$; $y = \frac{b}{a}x$; $\frac{b}{a} = \frac{4}{3}$, or $b = \frac{4a}{3}$

b. $a^2 + \left(\frac{4a}{3}\right)^2 = 225$; $9a^2 + 16a^2 = 2025$; $a^2 = 81$; $a = 9$; $b = 12$; $\frac{x^2}{81} - \frac{y^2}{144} = 1$

30. $c = 5$; $a^2 + b^2 = 25$; $\frac{a}{b} = 2$, or $a = 2b$; $(2b)^2 + b^2 = 25$; $b^2 = 5$; $b = \sqrt{5}$; $a = 2\sqrt{5}$; $\frac{y^2}{20} - \frac{x^2}{5} = 1$

31. $P(12, 2)$, $F_1(0, 7)$, $F_2(0, -7)$; $PF_2 - PF_1 = 2a$; $PF_2 = \sqrt{(12 - 0)^2 + (2 - (-7))^2} = 15$; $PF_1 = \sqrt{(12 - 0)^2 + (2 - 7)^2} = 13$; $2a = 2$, $a = 1$; $c = 7$; $b^2 = c^2 - a^2 = 48$; $y^2 - \frac{x^2}{48} = 1$

32. $P(x, y)$, $A(8, 0)$: $AP = 2$(dist. from P to line); $\sqrt{(x - 8)^2 + (y - 0)^2} = 2\sqrt{(x - 2)^2 + (y - y)^2}$; $(x - 8)^2 + y^2 = 4(x - 2)^2$; $3x^2 - y^2 = 48$; $\frac{x^2}{16} - \frac{y^2}{48} = 1$

33. Foci: $(c, 0)$, $(-c, 0)$; |difference of focal radii| $= 2a > 0$, $a < c$; $F_1 = (c, 0)$, $F_2(-c, 0)$, $P(x, y)$ on hyperbola: $|PF_2 - PF_1| = 2a$;
$\left|\sqrt{(x - (-c))^2 + (y - 0)^2} - \sqrt{(x - c)^2 + (y - 0)^2}\right| = 2a$;
$\left|\sqrt{(x + c)^2 + y^2} - \sqrt{(x - c)^2 + y^2}\right| = 2a$;
$(x + c)^2 + y^2 - 2\sqrt{(x + c)^2 + y^2}\sqrt{(x - c)^2 + y^2} + (x - c)^2 + y^2 = 4a^2$;
$2x^2 + 2c^2 + 2y^2 - 4a^2 = 2\sqrt{(x + c)^2 + y^2}\sqrt{(x - c)^2 + y^2}$;
$(x^2 + c^2 + y^2 - 2a^2)^2 = [(x + c)^2 + y^2][(x - c)^2 + y^2]$;
$x^4 + 2c^2x^2 + 2x^2y^2 - 4a^2x^2 + c^4 + 2c^2y^2 - 4a^2c^2 + y^4 - 4a^2y^2 + 4a^4 =$
$x^4 - 2c^2x^2 + c^4 + 2x^2y^2 + 2c^2y^2 + y^4$; $4c^2x^2 - 4a^2x^2 - 4a^2c^2 - 4a^2y^2 + 4a^4 = 0$;
$4c^2x^2 - 4a^2x^2 + 4a^2y^2 = 4a^2c^2 - 4a^4$; $x^2(c^2 - a^2) - a^2y^2 = a^2(c^2 - a^2)$;
$\frac{x^2}{a^2} - \frac{y^2}{c^2 - a^2} = 1$

34. Foci: $(0, c)$, $(0, -c)$; |difference of focal radii| $= 2a > 0$, $a < c$; $F_1 = (0, c)$, $F_2 = (0, -c)$, $P(x, y)$ on hyperbola: $|PF_2 - PF_1| = 2a$; $|\sqrt{(x - 0)^2 + (y - (-c))^2} - \sqrt{(x - 0)^2 + (y - c)^2}| = 2a$. Squaring both sides and proceeding as in Exercise 33 yields $y^2(c^2 - a^2) - a^2x^2 = a^2(c^2 - a^2)$, or $\frac{y^2}{a^2} - \frac{x^2}{c^2 - a^2} = 1.$

Page 417 • COMPUTER EXERCISES

1.
```
10   INPUT A,B,C
20   IF C = 0 THEN GOTO 100
30   IF A = B and A * C > 0 THEN GOTO 200
40   IF A/C > 0 AND B/C > 0 THEN GOTO 300
50   IF A/C < 0 AND B/C < 0 THEN GOTO 400
60   GOTO 500
100 IF A * B > 0 THEN PRINT "ONE POINT" : GOTO 1000
110 PRINT "A PAIR OF LINES" : GOTO 1000
200 PRINT "CIRCLE WITH RADIUS ";(C/B)^.5 : GOTO 1000
300 PRINT "ELLIPSE WITH FOCI ";
310 IF C/A > C/B THEN LET F = SQR(C/A - C/B) : GOTO 340
320 LET F = SQR(C/B - C/A)
330 PRINT "(0, ";F;") AND (0, ";-F;")" : GOTO 1000
340 PRINT "(";F;", 0) AND (";-F;", 0)" : GOTO 1000
400 PRINT "EMPTY SET" : GOTO 1000
500 PRINT "HYPERBOLA WITH FOCI ";
510 LET F = SQR(ABS(C/A) + ABS(C/B))
520 IF C/B > 0 THEN GOTO 540
530 PRINT "(";F;", 0) AND (";-F;", 0)" : GOTO 1000
540 PRINT "(0, ";F;") AND (0, ";-F;")"
1000 END
```

2. Hyperbola; $(5, 0)$, $(-5, 0)$ **3.** Ellipse; $(0, 3)$, $(0, -3)$ **4.** Empty set **5.** One point **6.** A pair of lines **7.** Hyperbola; $(0, 13)$, $(0, -13)$ **8.** Circle; $r = 5$ **9.** Empty set

Page 420 • WRITTEN EXERCISES

A **1.** $xy = 7 \cdot 4$; $xy = 28$ **2.** $xy = 6\left(-\frac{1}{3}\right)$; $xy = -2$ **3.** $x^2y = 1^2(5)$; $x^2y = 5$

4. $x^3y = \left(-\frac{3}{4}\right)^3(6)$; $x^3y = -\frac{81}{32}$ **5.** $y\sqrt{x} = \frac{-3}{2} \cdot \sqrt{25}$; $y\sqrt{x} = \frac{-15}{2}$

6. $y\sqrt[3]{x} = 250 \cdot \sqrt[3]{0.027}$; $y\sqrt[3]{x} = 75$ **7.** $x_1y_1 = x_2y_2$; $\frac{1}{5} \cdot 5 = x_2 \cdot 4$; $x_2 = \frac{1}{4}$

8. $x_1y_1 = x_2y_2$; $-5 \cdot 2 = 3 \cdot y_2$; $y_2 = -\frac{10}{3}$ **9.** $x_1y_1 = x_2y_2$; $\frac{4}{5} \cdot 30 = 16 \cdot y_2$; $y_2 = \frac{3}{2}$

10. $x_1^2y_1 = x_2^2y_2$; $2^2 \cdot 9 = x_2^2 \cdot 4$; $x_2^2 = 9$; $x_2 = 3$ or $x_2 = -3$

11. $x_1^2y_1 = x_2^2y_2$; $\left(-\frac{2}{3}\right)^2(27) = 4^2 \cdot y_2$; $\frac{4}{9} \cdot 27 = 16y_2$; $y_2 = \frac{3}{4}$

12. $x_1^2y_1 = x_2^2y_2$; $(3\sqrt{3})^2\left(\frac{1}{4}\right) = x_2^2 \cdot 12$; $\frac{27}{4} = 12x_2^2$; $x_2^2 = \frac{9}{16}$; $x_2 = \frac{3}{4}$ or $x_2 = -\frac{3}{4}$

13. $z = kxy$; $12 = k(9 \cdot 4)$; $k = \frac{1}{3}$; $z = \frac{1}{3} \cdot 7 \cdot 2$; $z = \frac{14}{3}$

14. $z = kxy$; $\frac{2}{3} = k\left(27 \cdot \frac{1}{2}\right)$; $k = \frac{4}{81}$; $z = \frac{4}{81} \cdot 9 \cdot 18$; $z = 8$

B **15.** $z = \frac{kx}{y}$; $6 = \frac{4k}{5}$; $k = \frac{15}{2}$; $z = \frac{15}{2} \cdot \frac{8}{3}$; $z = 20$

16. $z = \frac{kx}{y^3}$; $9 = \frac{2k}{\frac{1}{27}}$; $k = \frac{1}{6}$; $z = \frac{1}{6} \cdot \frac{3}{\frac{1}{8}}$; $z = 4$

17. $z = \frac{kxy}{w}$; $5 = \frac{k \cdot \frac{1}{2} \cdot 3}{6}$; $k = 20$; $z = \frac{20 \cdot 2 \cdot \frac{1}{3}}{8}$; $z = \frac{5}{3}$

18. $z = \frac{kxy}{w^2}$; $72 = \frac{k \cdot 80 \cdot 30}{25}$; $k = \frac{3}{4}$; $z = \frac{\frac{3}{4} \cdot 20 \cdot 60}{81} = \frac{100}{9}$

19.

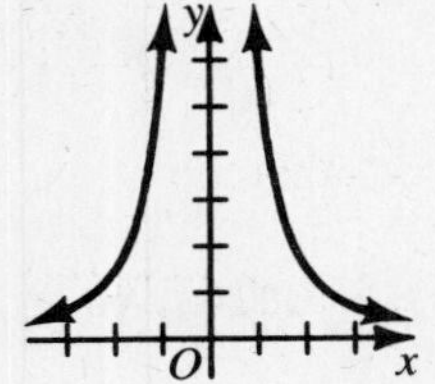

20.

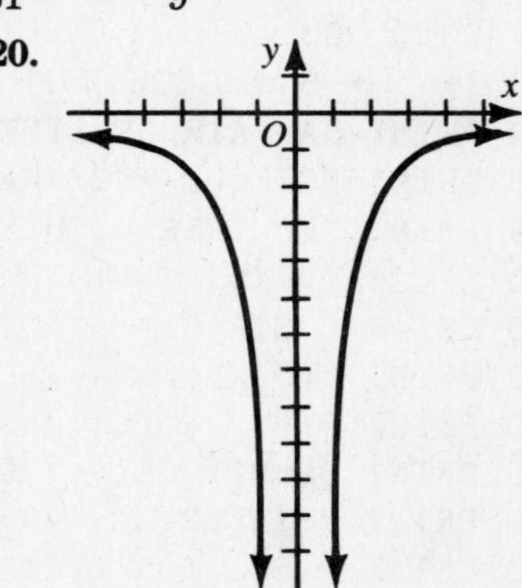

C **21.** $F_1(2, 2)$, $F_2(-2, -2)$, $P(x, y)$, $|PF_2 - PF_1| = 4$; $PF_2 - PF_1 = 4$ or $PF_2 - PF_1 = -4$. Using either equation produces the same result. Thus, we will work with $PF_2 - PF_1 = 4$. $\sqrt{(x + 2)^2 + (y + 2)^2} - \sqrt{(x - 2)^2 + (y - 2)^2} = 4$;
$\sqrt{(x + 2)^2 + (y + 2)^2} = 4 + \sqrt{(x - 2)^2 + (y - 2)^2}$;
$(x + 2)^2 + (y + 2)^2 = 16 + (x - 2)^2 + (y - 2)^2 + 8\sqrt{(x - 2)^2 + (y - 2)^2}$;
$8x + 8y - 16 = 8\sqrt{(x - 2)^2 + (y - 2)^2}$; $x + y - 2 = \sqrt{(x - 2)^2 + (y - 2)^2}$;
$x^2 + 2xy - 4x + y^2 - 4y + 4 = (x - 2)^2 + (y - 2)^2$;
$x^2 + 2xy - 4x + y^2 - 4y + 4 = x^2 - 4x + 4 + y^2 - 4y + 4$;
$2xy = 4$; $xy = 2$

22. $\frac{b}{a} = 1$; $2a = 4$; $a = 2$; $b = 2$; $\frac{x^2}{4} - \frac{y^2}{4} = 1$

Pages 421–422 • PROBLEMS

A **1.** Let n = no. of oscillations per min; m = mass of object. $n\sqrt{m} = k$;
$60\sqrt{400} = n\sqrt{625}$; $n = 48$; 48 oscillations per minute

2. Let m = slope of curve; r = radius of curve;
v = maximum speed of cars. $\frac{mr}{v^2} = k$; $\frac{(0.049)(1250)}{(24.5)^2} = \frac{m(765)}{15^2}$;
$m = \frac{(0.049)(1250)15^2}{(24.5)^2(765)} \approx 0.030$

3. Let v = volume; t = temperature; p = pressure. $\frac{vp}{t} = k$;
$\frac{(342)(200)}{300} = \frac{v(400)}{320}$; $v = \frac{(342)(200)(320)}{(300)(400)} \approx 182$; 182 m^3

4. Let v = speed; d = distance from Earth's center.
$v\sqrt{d} = k$; $(1.58 \times 10^5)\sqrt{16{,}000} = v\sqrt{400{,}000}$;
$v = \frac{(1.58 \times 10^5)\sqrt{16{,}000}}{\sqrt{400{,}000}} \approx 3.16 \times 10^4$; 3.16×10^4 m/s

5. Let W = power in watts; V = voltage in volts; R = resistance in ohms.
$\frac{WR}{V^2} = k$; $\frac{(576)(25)}{120^2} = \frac{(605)(20)}{V^2}$; $V^2 = \frac{(120^2)(605)(20)}{(576)(25)} = 12{,}100$; $V = 110$; 110 V

B 6. Let s = speed of sound in helium; d = density of carbon dioxide;
$0.09d$ = density of helium. $2904\sqrt{d} = s\sqrt{0.09d}$; $\frac{2904\sqrt{d}}{0.3\sqrt{d}} = s$; $s = 9680$;
9680 m/s

7. Let m_1, m_2 = masses of two objects; d = distance between them;
F = force. $\frac{Fd^2}{m_1m_2} = k$; $\frac{(6 \times 10^{-8})(1^2)}{(30)(30)} = \frac{(9.8)(6 \times 10^6)^2}{(1)(m_2)}$;
$m_2 = \frac{(9.8)(6 \times 10^6)^2(30)(30)}{6 \times 10^{-8}} \approx 5 \times 10^{24}$; 5×10^{24} kg

Page 422 • SELF-TEST 2

1. $x = y^2 + 4y + 1$
$x = (y^2 + 4y + 4) + 1 - 4$
$x = (y + 2)^2 - 3$

2. $9x^2 - 4y^2 = 36$
$\frac{x^2}{4} - \frac{y^2}{9} = 1$

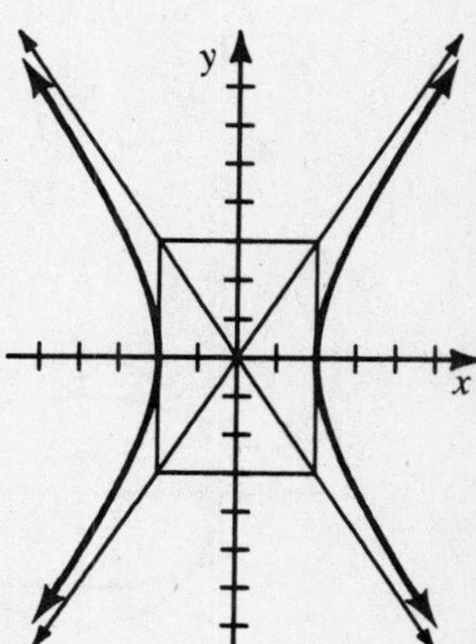

3. $(x - 1)^2 + (y + 2)^2 = 3^2$; $x^2 + y^2 - 2x + 4y - 4 = 0$

4. $2a = 26$, $a = 13$; $c = 5$; $b^2 = a^2 - c^2 = 144$; $\frac{x^2}{144} + \frac{y^2}{169} = 1$

5. $a = \frac{kb^2}{c}$; $\frac{9}{2} = \frac{k(3^2)}{4}$; $k = 2$; $a = \frac{2(6^2)}{12} = 6$

6. Let v = speed; n = number of teeth. $vn = k$; $(30)(36) = v(90)$; $v = 12$;
12 rev/min

Page 424 • WRITTEN EXERCISES

A 1.

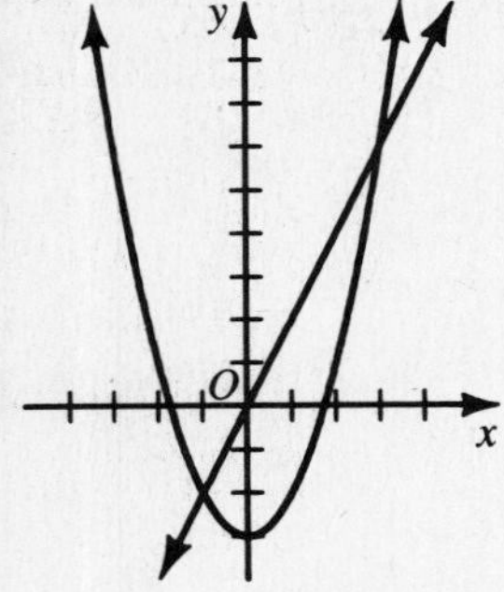

{(−1, −2), (3, 6)}

Check:

$y = x^2 - 3$ $\quad y = 2x$

$-2 = (-1)^2 - 3$ $\quad -2 = 2(-1)$

$6 = 3^2 - 3$ $\quad 6 = 2(3)$

2.

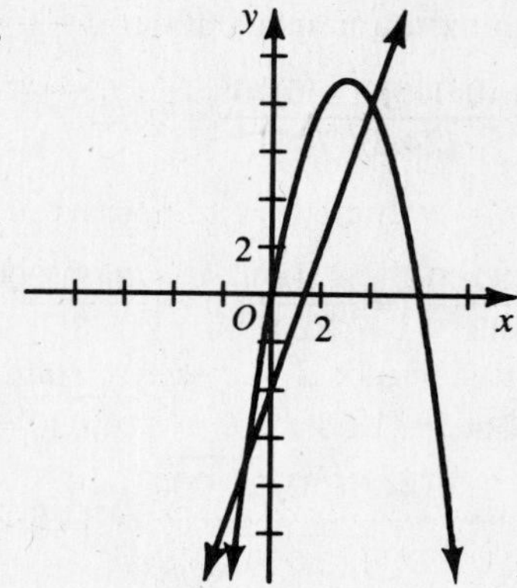

{(−1, −7), (4, 8)}

Check:

$y = 6x - x^2$ $\quad y = 3x - 4$

$-7 = 6(-1) - (-1)^2$ $\quad -7 = 3(-1) - 4$

$8 = 6(4) - 4^2$ $\quad 8 = 3(4) - 4$

3.

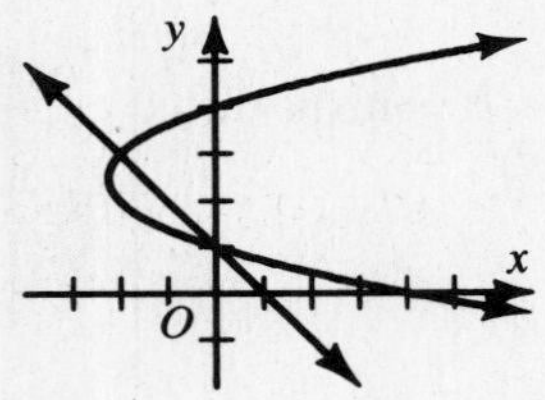

{(−2, 3), (0, 1)}

Check:

$x = y^2 - 5y + 4$ $\quad x + y = 1$

$-2 = 3^2 - 5(3) + 4$ $\quad -2 + 3 = 1$

$0 = 1^2 - 5(1) + 4$ $\quad 0 + 1 = 1$

4.

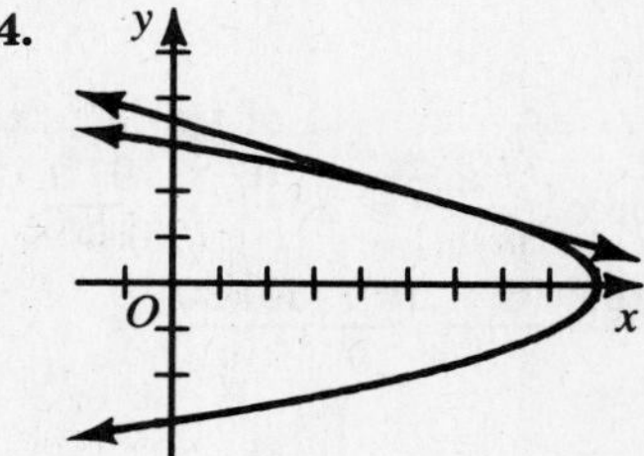

{(5, 2)}

Check:

$x = 9 - y^2$ $\quad x = -4y + 13$

$5 = 9 - 2^2$ $\quad 5 = -4(2) + 13$

5.

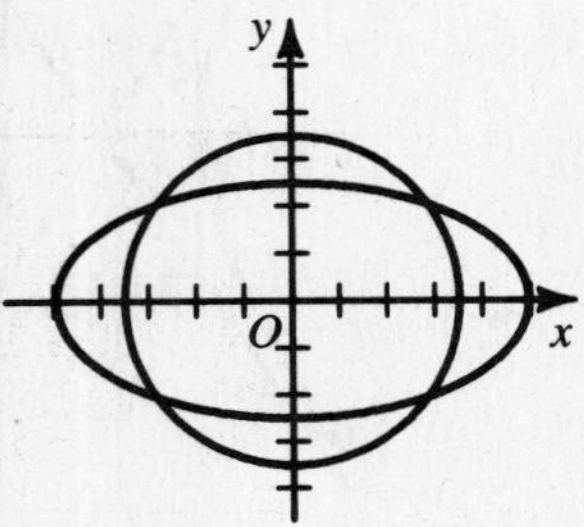

{(−3, −2), (−3, 2), (3, −2), (3, 2)}

Check:

$x^2 + 4y^2 = 25$ $\quad x^2 + y^2 = 13$

$(-3)^2 + 4(-2)^2 = 25$ $\quad (-3)^2 + (-2)^2 = 13$

$(-3)^2 + 4(2)^2 = 25$ $\quad (-3)^2 + 2^2 = 13$

$(3)^2 + 4(-2)^2 = 25$ $\quad 3^2 + (-2)^2 = 13$

$(3)^2 + 4(2)^2 = 25$ $\quad 3^2 + 2^2 = 13$

6. 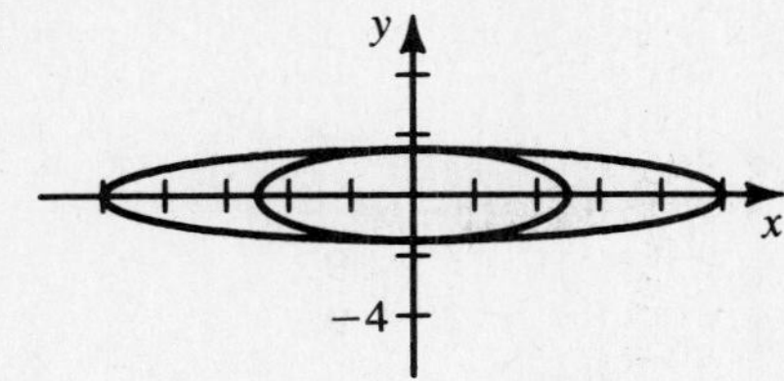

$\left\{\left(0, \frac{5}{3}\right), \left(0, -\frac{5}{3}\right)\right\}$

Check:

$$x^2 + 9y^2 = 25 \qquad x^2 + 36y^2 = 100$$
$$0^2 + 9\left(\frac{5}{3}\right)^2 = 25 \qquad 0^2 + 36\left(\frac{5}{3}\right)^2 = 100$$
$$0^2 + 9\left(-\frac{5}{3}\right)^2 = 25 \qquad 0^2 + 36\left(-\frac{5}{3}\right)^2 = 100$$

7. 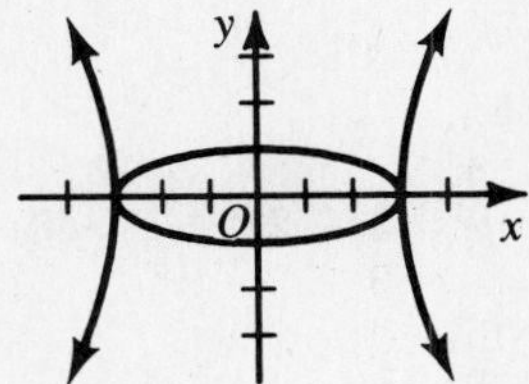

$\{(3, 0), (-3, 0)\}$

Check:

$$x^2 + 9y^2 = 9 \qquad 4x^2 - y^2 = 36$$
$$3^2 + 9(0^2) = 9 \qquad 4(3)^2 - 0^2 = 36$$
$$(-3)^2 + 9(0^2) = 9 \qquad 4(-3)^2 - 0^2 = 36$$

8. 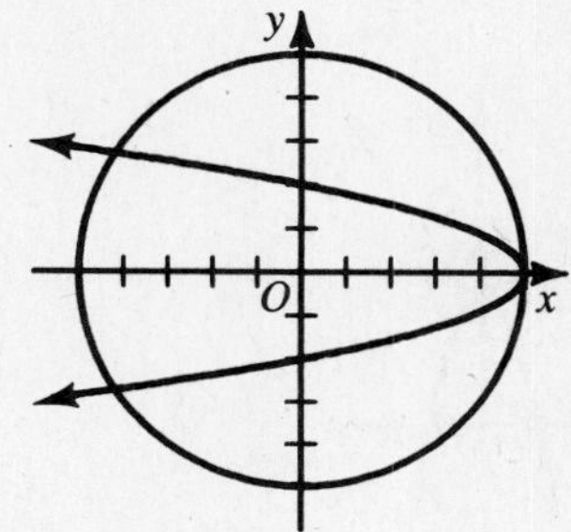

$\{(-4, 3), (-4, -3), (5, 0)\}$

Check:

$$x = 5 - y^2 \qquad x^2 + y^2 = 25$$
$$-4 = 5 - 3^2 \qquad (-4)^2 + 3^2 = 25$$
$$-4 = 5 - (-3)^2 \qquad (-4)^2 + (-3)^2 = 25$$
$$5 = 5 - 0^2 \qquad 5^2 + 0^2 = 25$$

9. 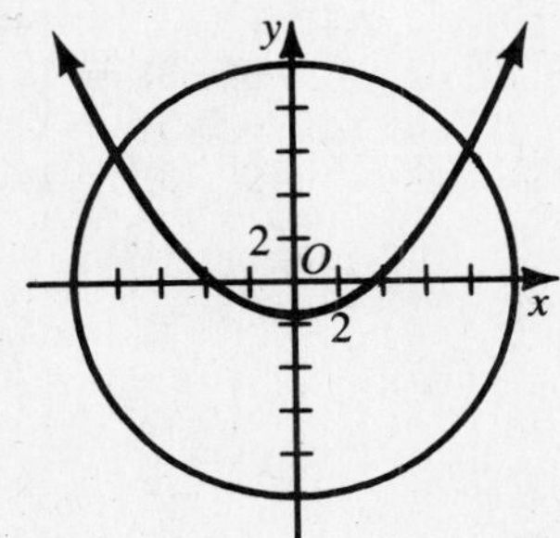

$\{(8, 6), (-8, 6)\}$

Check:

$$x^2 + y^2 = 100 \qquad x^2 - 10 = 9y$$
$$8^2 + 6^2 = 100 \qquad 8^2 - 10 = 9(6)$$
$$(-8)^2 + 6^2 = 100 \qquad (-8)^2 - 10 = 9(6)$$

B 10. 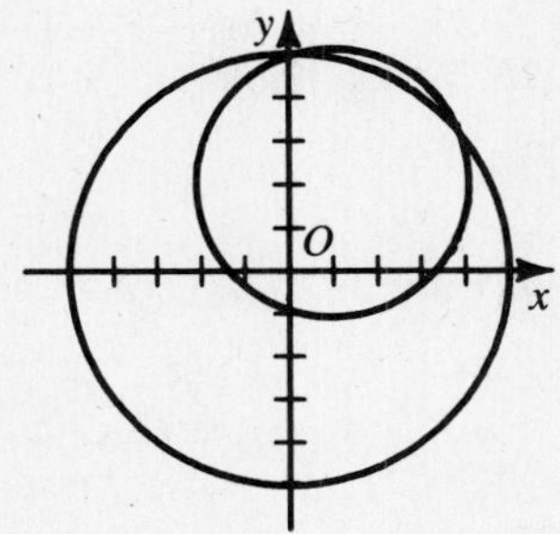

$\{(0, 5), (4, 3)\}$

Check:

$$x^2 - 2x + y^2 - 4y - 5 = 0 \qquad x^2 + y^2 = 25$$
$$(x - 1)^2 + (y - 2)^2 = 10 \qquad 0^2 + 5^2 = 25$$
$$(0 - 1)^2 + (5 - 2)^2 = 10 \qquad 4^2 + 3^2 = 25$$
$$(4 - 1)^2 + (3 - 2)^2 = 10$$

11. 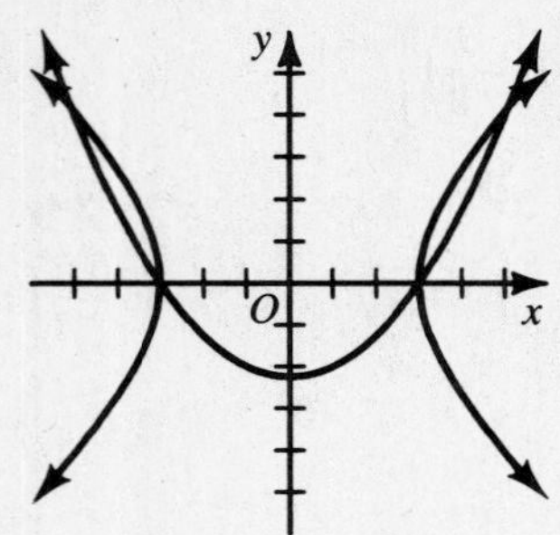

$\{(-5, 4), (5, 4), (3, 0), (-3, 0)\}$

Check:

$x^2 - y^2 = 9$	$4y = x^2 - 9$
$(-5)^2 - 4^2 = 9$	$4(4) = (-5)^2 - 9$
$5^2 - 4^2 = 9$	$4(4) = 5^2 - 9$
$3^2 - 0^2 = 9$	$4(0) = 3^2 - 9$
$(-3)^2 - 0^2 = 9$	$4(0) = (-3)^2 - 9$

12. 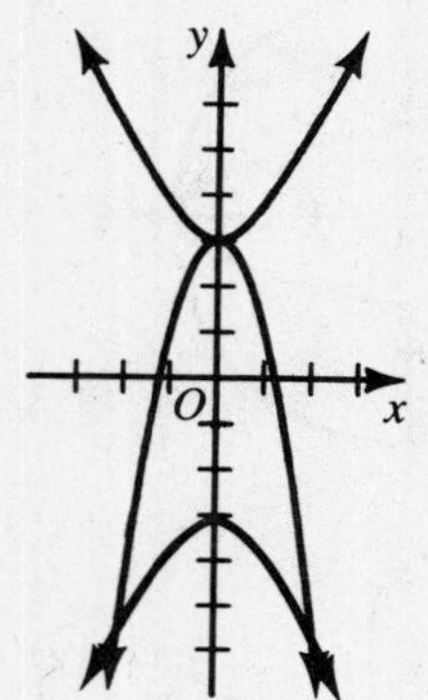

$\{(-2, -5), (2, -5), (0, 3)\}$

Check:

$y^2 - 4x^2 = 9$	$y = 3 - 2x^2$
$(-5)^2 - 4(-2)^2 = 9$	$-5 = 3 - 2(-2)^2$
$(-5)^2 - 4(2)^2 = 9$	$-5 = 3 - 2(2)^2$
$3^2 - 0^2 = 9$	$3 = 3 - 2(0)^2$

C 13. 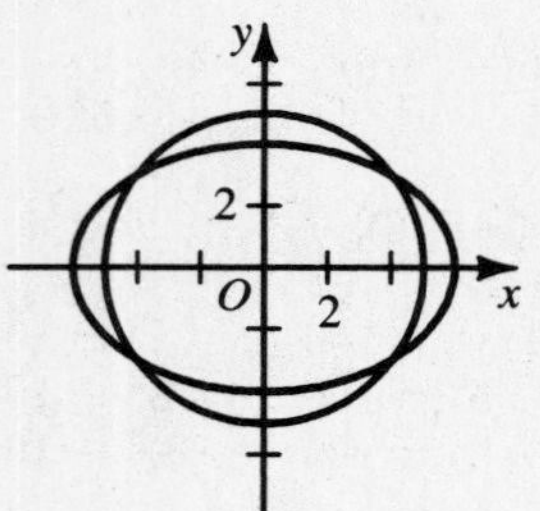

$\{(4, 3), (4, -3), (-4, 3), (-4, -3)\}$

Check:

$x^2 + y^2 = 25$	$4x^2 + 9y^2 = 144$
$4^2 + 3^2 = 25$	$4(4)^2 + 9(3)^2 = 145$
$4^2 + (-3)^2 = 25$	$4(4)^2 + 9(-3)^2 = 145$
$(-4)^2 + 3^2 = 25$	$4(-4)^2 + 9(3)^2 = 145$
$(-4)^2 + (-3)^2 = 25$	$4(-4)^2 + 9(-3)^2 = 145$

14. 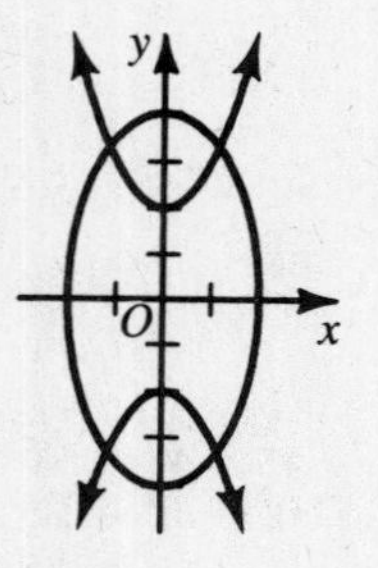

$\{(1, 3.5), (1, -3.5), (-1, -3.5), (-1, 3.5)\}$

Check:

$4x^2 + y^2 = 16$	$y^2 - 8x^2 = 4$
$4(1)^2 + (3.5)^2 = 16.25$	$(3.5)^2 - 8(1)^2 = 4.25$
$4(1)^2 + (-3.5)^2 = 16.25$	$(-3.5)^2 - 8(1)^2 = 4.25$
$4(-1)^2 + (-3.5)^2 = 16.25$	$(-3.5)^2 - 8(-1)^2 = 4.25$
$4(-1)^2 + (3.5)^2 = 16.25$	$(3.5)^2 - 8(-1)^2 = 4.25$

15. 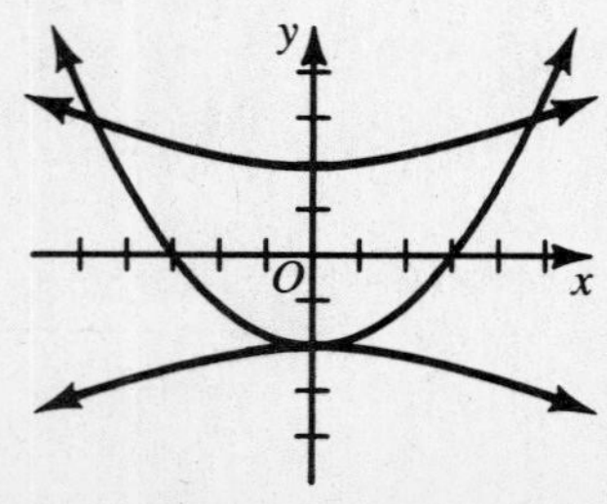

$\{(4.5, 3), (-4.5, 3), (0, -2)\}$

Check:

$4y = x^2 - 8$	$4y^2 - x^2 = 16$
$x^2 - 4y = 8$	$4(3)^2 - (4.5)^2 = 15.75$
$(4.5)^2 - 4(3) = 8.25$	$4(3) - (-4.5)^2 = 15.75$
$(-4.5)^2 - 4(3) = 8.25$	$4(-2)^2 - 0^2 = 16$
$0^2 - 4(-2) = 8$	

Page 427 • WRITTEN EXERCISES

A **1.** $\left.\begin{array}{l} y = x^2 + 3 \\ y = 4x \end{array}\right\}$ $4x = x^2 + 3$; $x^2 - 4x + 3 = 0$; $(x - 3)(x - 1) = 0$; $x = 1$ or $x = 3$;
$x = 1$: $y = 4$; $x = 3$: $y = 12$; $\{(1, 4), (3, 12)\}$

2. $\left.\begin{array}{l} x^2 - y^2 = 15 \\ x - 4y = 0 \end{array}\right\}$ $x = 4y$; $(4y)^2 - y^2 = 15$; $15y^2 = 15$; $y = 1$ or $y = -1$;
$y = 1$: $x = 4$; $y = -1$: $x = -4$; $\{(-4, -1), (4, 1)\}$

3. $\left.\begin{array}{l} x^2 + y^2 = 40 \\ 3x - y = -20 \end{array}\right\}$ $y = 3x + 20$; $x^2 + (3x + 20)^2 = 40$; $10x^2 + 120x + 360 = 0$;
$10(x + 6)(x + 6) = 0$; $x = -6$; $y = 3(-6) + 20 = 2$; $\{(-6, 2)\}$

4. $\left.\begin{array}{l} x^2 + y^2 = 41 \\ x + 2y = 6 \end{array}\right\}$ $x = 6 - 2y$; $(6 - 2y)^2 + y^2 = 41$; $5y^2 - 24y - 5 = 0$;
$(5y + 1)(y - 5) = 0$; $y = -\frac{1}{5}$ or $y = 5$; $y = -\frac{1}{5}$: $x = 6 - 2\left(-\frac{1}{5}\right) = \frac{32}{5}$;
$y = 5$: $x = 6 - 2(5) = -4$; $\left\{\left(\frac{32}{5}, -\frac{1}{5}\right), (-4, 5)\right\}$

5. $\left.\begin{array}{l} x^2 - y^2 = 72 \\ 5x + y = 0 \end{array}\right\}$ $y = -5x$; $x^2 - (-5x)^2 = 72$; $-24x^2 = 72$; $x^2 = -3$; $x = i\sqrt{3}$
or $x = -i\sqrt{3}$; $x = i\sqrt{3}$: $y = -5i\sqrt{3}$; $x = -i\sqrt{3}$: $y = 5i\sqrt{3}$;
$\{(i\sqrt{3}, -5i\sqrt{3}), (-i\sqrt{3}, 5i\sqrt{3})\}$

6. $\left.\begin{array}{l} 39 - x^2 = 6y \\ 2x + y = 12 \end{array}\right\}$ $y = 12 - 2x$; $39 - x^2 = 6(12 - 2x)$; $x^2 - 12x + 33 = 0$;
$x = \dfrac{12 \pm \sqrt{144 - 4 \cdot 33}}{2} = 6 \pm \sqrt{3}$; $x = 6 + \sqrt{3}$: $y = 12 - 2(6 + \sqrt{3}) = -2\sqrt{3}$;
$x = 6 - \sqrt{3}$: $y = 12 - 2(6 - \sqrt{3}) = 2\sqrt{3}$; $\{(6 + \sqrt{3}, -2\sqrt{3}), (6 - \sqrt{3}, 2\sqrt{3})\}$

B **7.** $\left.\begin{array}{r} \sqrt{x^2 + y^2} = y + 2 \\ 2y = x - 2 \end{array}\right\}$ $x = 2y + 2$; $\sqrt{(2y + 2)^2 + y^2} = y + 2$; $\sqrt{5y^2 + 8y + 4} =$
$y + 2$; $5y^2 + 8y + 4 = y^2 + 4y + 4$; $4y^2 + 4y = 4y(y + 1) = 0$; $y = 0$ or $y = -1$;
$y = 0$: $x = 2(0) + 2 = 2$; $y = -1$: $x = 2(-1) + 2 = 0$; $\{(2, 0), (0, -1)\}$

8. $\left.\begin{array}{r} \sqrt{x^2 + y^2} = 1 - y \\ 2y + x = -1 \end{array}\right\}$ $x = -1 - 2y$; $\sqrt{(-1 - 2y)^2 + y^2} = 1 - y$; $\sqrt{5y^2 + 4y + 1} =$
$1 - y$; $5y^2 + 4y + 1 = 1 - 2y + y^2$; $4y^2 + 6y = 2y(2y + 3) = 0$; $y = 0$ or $y = -\frac{3}{2}$;
$y = 0$: $x = -1 - 2(0) = -1$; $y = -\frac{3}{2}$: $x = -1 - 2\left(-\frac{3}{2}\right) = 2$; $\left\{(-1, 0), \left(2, -\frac{3}{2}\right)\right\}$

9. $\left.\begin{array}{l} \frac{4x}{y} + \frac{y}{x} = 7 \to 4x^2 + y^2 = 7xy \\ x - y = 3 \to \qquad x = 3 + y \end{array}\right\}$ $4(3 + y)^2 + y^2 = 7(3 + y)y$;
$5y^2 + 24y + 36 = 21y + 7y^2$; $2y^2 - 3y - 36 = 0$; $y = \dfrac{3 \pm \sqrt{9 - 4(2)(-36)}}{2(2)} =$
$\dfrac{3 \pm \sqrt{297}}{4} = \dfrac{3 \pm 3\sqrt{33}}{4}$; $y = \dfrac{3 + 3\sqrt{33}}{4}$: $x = 3 + \dfrac{3 + 3\sqrt{33}}{4} = \dfrac{15 + 3\sqrt{33}}{4}$;

$$y = \frac{3 - 3\sqrt{33}}{4}: x = 3 + \frac{3 - 3\sqrt{33}}{4} = \frac{15 - 3\sqrt{33}}{4};$$
$$\left\{\left(\frac{15 + 3\sqrt{33}}{4}, \frac{3 + 3\sqrt{33}}{4}\right), \left(\frac{15 - 3\sqrt{33}}{4}, \frac{3 - 3\sqrt{33}}{4}\right)\right\}$$

10. $\left.\begin{array}{l} \frac{x}{y} + \frac{6y}{x} = \frac{-5}{xy} \rightarrow x^2 + 6y^2 = -5 \\ x - 3y = 1 \quad \rightarrow \quad x = 1 + 3y \end{array}\right\} (1 + 3y)^2 + 6y^2 = -5; 15y^2 + 6y + 6 = 0;$

$$5y^2 + 2y + 2 = 0; y = \frac{-2 \pm \sqrt{2^2 - 4(5)(2)}}{2(5)} = \frac{-2 \pm \sqrt{-36}}{10} = \frac{-1 \pm 3i}{5};$$
$$y = \frac{-1 + 3i}{5}: x = 1 + 3\left(\frac{-1 + 3i}{5}\right) = \frac{2 + 9i}{5}; y = \frac{-1 - 3i}{5}:$$
$$x = 1 + 3\left(\frac{-1 - 3i}{5}\right) = \frac{2 - 9i}{5}; \left\{\left(\frac{2}{5} + \frac{9}{5}i, -\frac{1}{5} + \frac{3}{5}i\right), \left(\frac{2}{5} - \frac{9}{5}i, -\frac{1}{5} - \frac{3}{5}i\right)\right\}$$

11. $\left.\begin{array}{r} x^2 + y^2 - 6xy = 28 \\ y + 10 = 3x \end{array}\right\} y = 3x - 10; x^2 + (3x - 10)^2 - 6x(3x - 10) = 28;$

$x^2 + (9x^2 - 60x + 100) - 18x^2 + 60x = 28; 8x^2 - 72 = 0; x^2 - 9 = 0;$
$(x + 3)(x - 3) = 0; x = \pm 3; x = 3: y = 3(3) - 10 = -1;$
$x = -3: y = 3(-3) - 10 = -19; \{(3, -1), (-3, -19)\}$

12. $\left.\begin{array}{r} y^2 + x^2 - 2xy = 1 \\ 5x - y = 5 \end{array}\right\} (5x - 5)^2 + x^2 - 2x(5x - 5) = 1;$

$(25x^2 - 50x + 25) + x^2 - 10x^2 + 10x = 1; 16x^2 - 40x + 24 = 0;$
$2x^2 - 5x + 3 = 0; (2x - 3)(x - 1) = 0;$
$$x = \frac{3}{2} \text{ or } x = 1; x = \frac{3}{2}: y = 5\left(\frac{3}{2}\right) - 5 = \frac{5}{2};$$
$$x = 1: y = 5(1) - 5 = 0; \left\{\left(\frac{3}{2}, \frac{5}{2}\right), (1, 0)\right\}$$

13. $\left.\begin{array}{l} 5\sqrt{x^2 + y^2} - \frac{17}{5\sqrt{x^2 + y^2}} = 0 \longrightarrow 25(x^2 + y^2) - 17 = 0 \\ x + y = 1 \longrightarrow \quad y = 1 - x \end{array}\right\} 25[x^2 + (1 - x)^2] -$

$17 = 0; 25(2x^2 - 2x + 1) - 17 = 0; 50x^2 - 50x + 8 = 0;$
$25x^2 - 25x + 4 = 0; (5x - 1)(5x - 4) = 0;$
$$x = \frac{1}{5} \text{ or } x = \frac{4}{5}; x = \frac{1}{5}: y = 1 - \frac{1}{5} = \frac{4}{5}; x = \frac{4}{5}: y = 1 - \frac{4}{5} = \frac{1}{5}; \left\{\left(\frac{1}{5}, \frac{4}{5}\right), \left(\frac{4}{5}, \frac{1}{5}\right)\right\}$$

14. $\left.\begin{array}{l} \frac{13}{\sqrt{x^2 - 3y^2}} - \sqrt{x^2 - 3y^2} = 0 \longrightarrow 13 - (x^2 - 3y^2) = 0 \\ 3y - x = 7 \longrightarrow \quad 3y - 7 = x \end{array}\right\} 13 - [(3y - 7)^2 - 3y^2] = 0;$

$13 - (9y^2 - 42y + 49 - 3y^2) = 0; 6y^2 - 42y + 36 = 0; y^2 - 7y + 6 = 0;$
$(y - 6)(y - 1) = 0; y = 6 \text{ or } y = 1; y = 6: x = 3(6) - 7 = 11;$
$y = 1: x = 3(1) - 7 = -4; \{(11, 6), (-4, 1)\}$

C **15.** Given the system $\left.\begin{array}{l} y = ax^2 + bx + c \\ y = mx + k \end{array}\right\}$. By substitution, we have

$mx + k = ax^2 + bx + c$ or $ax^2 + (b - m)x + (c - k) = 0$. The quadratic formula produces the solution $x = \dfrac{-(b-m) \pm \sqrt{(b-m)^2 - 4a(c-k)}}{2a}$.
If $(b - m)^2 = 4a(c - k)$, then $x = \dfrac{m-b}{2a}$, which is real, and $y = m\left(\dfrac{m-b}{2a}\right) + k$, which is also real, since a, b, m and k are real. If $(b-m)^2 > 4a(c-k)$, then x has two real solutions, $x_1 = \dfrac{-(b-m) + \sqrt{(b-m)^2 - 4a(c-k)}}{2a}$ and $x_2 = \dfrac{-(b-m) - \sqrt{(b-m)^2 - 4a(c-k)}}{2a}$; then $y_1 = mx_1 + k$ and $y_2 = mx_2 + k$, which are both real since x_1, x_2, m, and k are real. If $(b-m)^2 < 4a(c-k)$, then x has two imaginary solutions $x_1 = \dfrac{-(b-m) + i\sqrt{4a(c-k) - (b-m)^2}}{2a}$ and $x_2 = \overline{x_1}$; then $y_1 = mx_1 + k$ and $y_2 = mx_2 + k$, which are both imaginary, since x_1 and x_2 are imaginary and m and k are real. Therefore, the solution of the system must be either one ordered pair of real numbers, two ordered pairs of real numbers, or two ordered pairs of imaginary numbers.

16. Given the system $\left.\begin{aligned} ax^2 + by^2 &= c \\ y &= mx + k \end{aligned}\right\}$. By substitution, we have $ax^2 + b(mx + k)^2 = c$ or $(a + bm^2)x^2 + 2bmkx + (bk^2 - c) = 0$. The quadratic formula produces the solution $x = \dfrac{-2bmk \pm \sqrt{4b^2m^2k^2 - 4(a + bm^2)(bk^2 - c)}}{2(a + bm^2)} = \dfrac{-2bmk \pm \sqrt{D}}{2(a + bm^2)}$. If $D = 0$, then $x = \dfrac{-bmk}{a + bm^2}$ and $y = \dfrac{-bm^2k}{a + bm^2} + k$, which are both real, since a, b, m, and k are real. If $D > 0$, then $\sqrt{D}$ is real since a, b, c, k, and m are real; and $x_1 = \dfrac{-2bmk + \sqrt{D}}{2(a + bm^2)}$ and $x_2 = \dfrac{-2bmk - \sqrt{D}}{2(a + bm^2)}$ are real, so that $y_1 = mx_1 + k$ and $y_2 = mx_2 + k$ are real, since m and k are real. If $D < 0$, then $\sqrt{D}$ is imaginary and $x_1 = \dfrac{-2bmk + \sqrt{D}}{2(a + bm^2)}$ and $x_2 = \overline{x_1}$ are imaginary, so that $y_1 = mx_1 + k$ and $y_2 = mx_2 + k$ are imaginary, since m and k are real. Therefore, the solution of the system must be either one ordered pair of real numbers, two ordered pairs of real numbers, or two ordered pairs of imaginary numbers.

Pages 427–428 • PROBLEMS

A **1.** Let l = length; w = width. $lw = 72$ and $2(l + w) = 44$; $w = 22 - l$; $l(22 - l) = 72$; $l^2 - 22l + 72 = 0$; $(l - 4)(l - 18) = 0$; if $l = 4$, $w = 22 - 4 = 18$; if $l = 18$, $w = 22 - 18 = 4$; 4 cm × 18 cm

2. Let x = length of each leg; $2y$ = length of base. $2x + 2y = 36$ and $y^2 + 36 = x^2$; $x = 18 - y$; $y^2 + 36 = (18 - y)^2$; $36 = 324 - 36y$; $y = 8$, $x = 18 - 8 = 10$; 10, 10, 16

3. Let x = length of side of each square panel; y = length of side of door. $y^2 - 4x^2 = 720$ and $y = 2x + 12$; $(2x + 12)^2 - 4x^2 = 720$; $48x + 144 = 720$; $x = 12$, $y = 2(12) + 12 = 36$; door is 36 cm × 36 cm, each panel is 12 cm × 12 cm.

4. Let x = length of side of square; y = length of shorter leg of each rt. △ removed. $x^2 + y^2 = 25$ and $x + (x - 2y) + 2(25) = 84$; $y = x - 17$; $x^2 + (x - 17)^2 = 625$; $2x^2 - 34x - 336 = 0$; $x^2 - 17x - 168 = 0$; $(x - 24)(x + 7) = 0$; $x = 24$ (or $x = -7$, which is impossible); 24 cm

5. $20 = \frac{{v_2}^2 - {v_1}^2}{2(9.8)}$ and $14 = \frac{v_1 + v_2}{2}$; ${v_2}^2 - {v_1}^2 = 392$ and $v_1 + v_2 = 28$; $v_2 = 28 - v_1$; $(28 - v_1)^2 - {v_1}^2 = 392$; $784 - 56v_1 = 392$; $v_1 = 7$; $v_2 = 28 - 7 = 21$; $v_1 = 7$ m/s; $v_2 = 21$ m/s

6. Perimeter of square $EFGH = 4(x + y) = 64$; $AD = \sqrt{x^2 + y^2}$; area of square $ABCD = (AD)^2 = x^2 + y^2 = 130$: $x + y = 16$; $x^2 + (16 - x)^2 = 130$; $2x^2 - 32x + 126 = 0$; $x^2 - 16x + 63 = 0$; $(x - 9)(x - 7) = 0$; if $x = 9$, $y = 7$; if $x = 7$, $y = 9$; 7 and 9

B **7.** Perimeter $= 4x + 2y = 24$; area $= x^2 - y^2 = 21$; $2x + y = 12$; $x^2 - (12 - 2x)^2 = 21$; $-3x^2 + 48x - 144 = 21$; $x^2 - 16x + 55 = 0$; $(x - 5)(x - 11) = 0$; $x = 5$ or $x = 11$; if $x = 5$, $y = 12 - 2(5) = 2$; if $x = 11$, $y = 12 - 2(11) = -10$, which is impossible; $x = 5$ m, $y = 2$ m

8. Let w = width of original rectangle; l = length of original rectangle. $2(l + w) = 54$ and $w(l - w) = 36$; $l + w = 27$; $l = 27 - w$; $w(27 - 2w) = 36$; $2w^2 - 27w + 36 = 0$; $(2w - 3)(w - 12) = 0$; $w = \frac{3}{2}$ or $w = 12$; if $w = 1.5$, $l = 25.5$; if $w = 12$, $l = 15$; 1.5 cm × 25.5 cm or 12 cm × 15 cm

9. $3x + y = 100$ and $x^2 + \frac{1}{2}xy = 1050$; $y = 100 - 3x$; $x^2 + \frac{1}{2}x(100 - 3x) = 1050$; $2x^2 + 100x - 3x^2 = 2100$; $x^2 - 100x + 2100 = 0$; $(x - 30)(x - 70) = 0$; $x = 30$ or $x = 70$; if $x = 30$, $y = 100 - 3(30) = 10$; if $x = 70$, $y = 100 - 3(70) = -110$, which is impossible; $x = 30$ m, $y = 10$ m

C **10.** $y = 3(x - h)^2 + k$ has vertex (h, k); vertex lies on $y = 3x + 1$, so $k = 3h + 1$; the parabola passes through (1, 10), so $10 = 3(1 - h)^2 + k$; substituting gives $10 = 3(1 - h)^2 + 3h + 1$; $3h^2 - 3h - 6 = 0$; $h^2 - h - 2 = 0$; $(h - 2)(h + 1) = 0$; $h = 2$ or $h = -1$; if $h = 2$, $k = 3(2) + 1 = 7$; if $h = -1$, $k = 3(-1) + 1 = -2$; $y = 3(x - 2)^2 + 7$ or $y = 3(x + 1)^2 - 2$

11. Let x = length of each leg; $2y$ = length of the base. $2x + 2y = 50$ and $2y + \sqrt{x^2 - y^2} = 31$; $y = 25 - x$; $2(25 - x) + \sqrt{x^2 - (25 - x)^2} = 31$; $50 - 2x + \sqrt{50x - 625} = 31$; $\sqrt{50x - 625} = 2x - 19$; $50x - 625 = 4x^2 - 76x + 361$; $4x^2 - 126x + 986 = 0$; $2x^2 - 63x + 493 = 0$; $(x - 17)(2x - 29) = 0$; $x = 17$ or $x = 14.5$; if $x = 17$, $y = 8$; if $x = 14.5$, $y = 10.5$; area $= y\sqrt{x^2 - y^2} = 8\sqrt{17^2 - 8^2} = 120$ or area $= 10.5\sqrt{(14.5)^2 - (10.5)^2} = 105$; 120 cm^2 or 105 cm^2

Pages 431–432 • WRITTEN EXERCISES

A **1.** $\left.\begin{array}{l}(1)\text{: } x^2 + 16y^2 = 25 \\ (2)\text{: } x^2 - 2y^2 = 7\end{array}\right\}$ (1) − (2): $18y^2 = 18$; $y = \pm 1$; (1): $x^2 + 16(\pm 1)^2 = 25$; $x^2 = 9$; $x = \pm 3$; $\{(3, 1), (-3, 1), (3, -1), (-3, -1)\}$

2. $\left.\begin{array}{l}(1)\text{: } 2x^2 + y^2 = 33 \\ (2)\text{: } x^2 + 3y^2 = 79\end{array}\right\}$ 3(1) − (2): $5x^2 = 20$; $x^2 = 4$; $x = \pm 2$; (1): $2(\pm 2)^2 + y^2 = 33$; $y^2 = 25$; $y = \pm 5$; $\{(2, 5), (2, -5), (-2, 5), (-2, -5)\}$

3. $\left.\begin{array}{l}(1)\text{: } x^2 - 9y^2 = 7 \\ (2)\text{: } 3x^2 + 4y^2 = 52\end{array}\right\}$ (2) − 3(1): $31y^2 = 31$; $y^2 = 1$; $y = \pm 1$; (2): $x^2 - 9(\pm 1)^2 = 7$; $x^2 = 16$; $x = \pm 4$; $\{(4, 1), (4, -1), (-4, 1), (-4, -1)\}$

4. $\left.\begin{array}{l}(1)\text{: } x^2 + 4y^2 = 16 \\ (2)\text{: } 4x^2 - 5y^2 = 1\end{array}\right\}$ 4(1) − (2): $21y^2 = 63$; $y^2 = 3$; $y = \pm\sqrt{3}$; (1): $x^2 + 4(\pm\sqrt{3})^2 = 16$; $x^2 = 4$; $x = \pm 2$; $\{(2, \sqrt{3}), (2, -\sqrt{3}), (-2, \sqrt{3}), (-2, -\sqrt{3})\}$

5. $\left.\begin{array}{l}(1)\text{: } x^2 + y^2 = 4 \\ (2)\text{: } 5x^2 + 6y^2 = 8\end{array}\right\}$ 6(1) − (2): $x^2 = 16$; $x = \pm 4$; (1): $(\pm 4)^2 + y^2 = 4$; $y^2 = -12$; $y = \pm 2i\sqrt{3}$; $\{(4, 2i\sqrt{3}), (4, -2i\sqrt{3}), (-4, 2i\sqrt{3}), (-4, -2i\sqrt{3})\}$

6. $\left.\begin{array}{l}(1)\text{: } x^2 - 4y^2 = 40 \\ (2)\text{: } x^2 + 9y^2 = 66\end{array}\right\}$ (2) − (1): $13y^2 = 26$; $y^2 = 2$; $y = \pm\sqrt{2}$; (1): $x^2 - 4(\pm\sqrt{2})^2 = 40$; $x^2 = 48$; $x = \pm 4\sqrt{3}$; $\{(4\sqrt{3}, \sqrt{2}), (4\sqrt{3}, -\sqrt{2}), (-4\sqrt{3}, \sqrt{2}), (-4\sqrt{3}, -\sqrt{2})\}$

7. $\left.\begin{array}{l}(1)\text{: } y^2 - 9x^2 = 36 \\ (2)\text{: } 2x^2 + y^2 = 3\end{array}\right\}$ (2) − (1): $11x^2 = -33$; $x^2 = -3$; $x = \pm i\sqrt{3}$; (2): $2(\pm i\sqrt{3})^2 + y^2 = 3$; $2(-3) + y^2 = 3$; $y^2 = 9$; $y = \pm 3$; $\{(i\sqrt{3}, 3), (i\sqrt{3}, -3), (-i\sqrt{3}, 3), (-i\sqrt{3}, -3)\}$

8. $\left.\begin{array}{l}(1)\text{: } y^2 - 3x^2 = 23 \\ (2)\text{: } x^2 - 16y^2 = 8\end{array}\right\}$ 3(2) + (1): $-47y^2 = 47$; $y^2 = -1$; $y = \pm i$; (2): $x^2 - 16(\pm i)^2 = 8$; $x^2 + 16 = 8$; $x^2 = -8$; $x = \pm\ 2i\sqrt{2}$; $\{(2i\sqrt{2}, i), (2i\sqrt{2}, -i), (-2i\sqrt{2}, i), (-2i\sqrt{2}, -i)\}$

9. $\left.\begin{array}{l}(1)\text{: } 7x^2 + 9y^2 = 22 \\ (2)\text{: } 3x^2 - 7y^2 = -15\end{array}\right\}$ 7(1) + 9(2): $76x^2 = 19$; $x^2 = \frac{1}{4}$; $x = \pm\frac{1}{2}$; (1): $7\left(\pm\frac{1}{2}\right)^2 + 9y^2 = 22$; $9y^2 = \frac{81}{4}$; $y^2 = \frac{9}{4}$; $y = \pm\frac{3}{2}$; $\left\{\left(\frac{1}{2}, \frac{3}{2}\right), \left(\frac{1}{2}, -\frac{3}{2}\right), \left(-\frac{1}{2}, \frac{3}{2}\right), \left(-\frac{1}{2}, -\frac{3}{2}\right)\right\}$

10. $\left.\begin{array}{l}(1)\text{: } 25x^2 + 4y^2 = 99 \\ (2)\text{: } 3x^2 - 5y^2 = -21\end{array}\right\}$ 5(1) + 4(2): $137x^2 = 411$; $x^2 = 3$; $x = \pm\sqrt{3}$; (1): $25(\pm\sqrt{3})^2 + 4y^2 = 99$; $4y^2 = 24$; $y^2 = 6$; $y = \pm\sqrt{6}$; $\{(\sqrt{3}, \sqrt{6}), (\sqrt{3}, -\sqrt{6}), (-\sqrt{3}, \sqrt{6}), (-\sqrt{3}, -\sqrt{6})\}$

11. $\left.\begin{array}{l}(1)\text{: } x^2 + y^2 = 10 \\ (2)\text{: } y = x^2 - 4\end{array}\right\}$ (2): $x^2 = y + 4$; (1): $(y + 4) + y^2 = 10$; $y^2 + y - 6 = 0$; $(y + 3)(y - 2) = 0$; $y = -3$ or $y = 2$; if $y = -3$, $x^2 + (-3)^2 = 10$; $x = \pm 1$; if $y = 2$, $x^2 + 2^2 = 10$; $x = \pm\sqrt{6}$; $\{(1, -3), (-1, -3), (\sqrt{6}, 2), (-\sqrt{6}, 2)\}$

12. $\left.\begin{array}{l}(1)\text{: } 9x^2 + y^2 = 27 \\ (2)\text{: } y = 3x^2 - 3\end{array}\right\}$ (2): $3x^2 = y + 3$; $x^2 = \frac{1}{3}y + 1$; (1): $9\left(\frac{1}{3}y + 1\right) + y^2 = 27$; $y^2 + 3y - 18 = 0$; $(y + 6)(y - 3) = 0$; $y = -6$ or $y = 3$; if $y = -6$, $9x^2 + (-6)^2 = 27$; $9x^2 = -9$; $x = \pm i$; if $y = 3$, $9x^2 + 3^2 = 27$; $9x^2 = 18$; $x = \pm\sqrt{2}$; $\{(\sqrt{2}, 3), (-\sqrt{2}, 3), (i, -6), (-i, -6)\}$

B **13.** $\left.\begin{array}{l}(1)\text{: } x^2 + y^2 = 12 \\ (2)\text{: } x = 3y^2 + 16\end{array}\right\}$ (1): $y^2 = 12 - x^2$; (2): $x = 3(12 - x^2) + 16$; $3x^2 + x - 52 = 0$; $(3x + 13)(x - 4) = 0$; $x = -\frac{13}{3}$ or $x = 4$; if $x = -\frac{13}{3}$, then $\left(-\frac{13}{3}\right)^2 + y^2 = 12$; $y^2 = -\frac{61}{9}$; $y = \pm\frac{i\sqrt{61}}{3}$; if $x = 4$, then $4^2 + y^2 = 12$, $y^2 = -4$, $y = \pm 2i$; $\left\{(4, 2i), (4, -2i), \left(-\frac{13}{3}, \frac{i\sqrt{61}}{3}\right), \left(-\frac{13}{3}, -\frac{i\sqrt{61}}{3}\right)\right\}$

14. $\left.\begin{array}{l}(1)\text{: } y^2 - 8x^2 = 9 \\ (2)\text{: } y = 4x^2 - 3\end{array}\right\}$ (2): $4x^2 = y + 3$; $x^2 = \frac{1}{4}y + \frac{3}{4}$; (1): $y^2 - 8\left(\frac{1}{4}y + \frac{3}{4}\right) = 9$; $y^2 - 2y - 15 = 0$; $(y - 5)(y + 3) = 0$; $y = -3$ or $y = 5$; if $y = -3$, then $-3 = 4x^2 - 3$; $x = 0$; if $y = 5$, then $5 = 4x^2 - 3$; $4x^2 = 8$; $x = \pm\sqrt{2}$; $\{(\sqrt{2}, 5), (-\sqrt{2}, 5), (0, -3)\}$

15. $\left.\begin{array}{l}(1)\text{: } x^2 + y^2 = 15 \\ (2)\text{: } \quad xy = 6\end{array}\right\}$ (2): $x = \frac{6}{y}$; (1): $\left(\frac{6}{y}\right)^2 + y^2 = 15$; $36 + y^4 = 15y^2$; $y^4 - 15y^2 + 36 = 0$; $(y^2 - 12)(y^2 - 3) = 0$; $y^2 = 12$ or $y^2 = 3$; $y = \pm 2\sqrt{3}$ or $y = \pm\sqrt{3}$; (1): $x^2 + (\pm 2\sqrt{3})^2 = 15$; $x^2 = 3$; $x = \pm\sqrt{3}$, or $x^2 + (\pm\sqrt{3})^2 = 15$; $x^2 = 12$; $x = \pm 2\sqrt{3}$; $\{(2\sqrt{3}, \sqrt{3}), (-2\sqrt{3}, -\sqrt{3}), (\sqrt{3}, 2\sqrt{3}), (-\sqrt{3}, -2\sqrt{3})\}$
Note that other possible solutions [e.g., $(2\sqrt{3}, -\sqrt{3})$] are not solutions because the coordinates do not satisfy the equation $xy = 6$.

C **16.** By lin. comb., system becomes $\begin{array}{l} adx^2 + bdy^2 = d \\ -bcx^2 - bdy^2 = -b. \end{array}$ Adding gives $(ad - bc)x^2 = d - b$, or $x^2 = \frac{d - b}{ad - bc}$. Similarly, adding $\begin{array}{l} acx^2 + bcy^2 = c \\ -acx^2 - ady^2 = -a \end{array}$ gives $(bc - ad)y^2 = c - a$, or $y^2 = \frac{a - c}{ad - bc}$. Thus $x^2 + y^2 = \frac{a + d - (b + c)}{ad - bc}$.

Page 432 • PROBLEMS

A **1.** Let l = length; w = width. $l^2 + w^2 = 13^2$ and $lw = 60$; $l = \frac{60}{w}$; $\left(\frac{60}{w}\right)^2 + w^2 = 169$; $w^4 - 169w^2 + 3600 = 0$; $(w^2 - 25)(w^2 - 144) = 0$; $w^2 = 25$ or $w^2 = 144$; $w = \pm 5$ or $w = \pm 12$ ($w = -5, -12$ are not reasonable); if $w = 5$, $l = \frac{60}{5} = 12$; if $w = 12$, $l = \frac{60}{12} = 5$; 5 m × 12 m

2. Let l = length; w = width. $l^2 + w^2 = 50^2$ and $lw = 1200$; $l = \frac{1200}{w}$; $\left(\frac{1200}{w}\right)^2 + w^2 = 2500$; $w^4 - 2500w^2 + 1{,}440{,}000 = 0$; $(w^2 - 900)(w^2 - 1600) = 0$; $w^2 = 900$ or $w^2 = 1600$; $w = \pm 30$ or $w = \pm 40$; ($w = -30, -40$ are not reasonable); if $w = 30$, $l = \frac{1200}{30} = 40$; if $w = 40$, $l = \frac{1200}{40} = 30$; 40 cm × 30 cm

3. $\begin{matrix} (1): & \sqrt{x^2 + y^2} = 10 \longrightarrow & x^2 + y^2 = 100 \\ (2): & \sqrt{x^2 + (y+9)^2} = 17 \longrightarrow & x^2 + (y+9)^2 = 289 \end{matrix}\Bigg\}$ (2) − (1): $(y + 9)^2 - y^2 = 189$;
$18y + 81 = 189$; $18y = 108$; $y = 6$; (1): $x^2 + 6^2 = 100$; $x^2 = 64$; $x = \pm 8$;
(8, 6) and (−8, 6)

4. Let $x = AC$; $y = DC$. Note that $\angle A$ and $\angle BCD$ are right angles. $(6\sqrt{2})^2 + x^2 = y^2$ and $x^2 + y^2 = (3\sqrt{10})^2$; $y^2 - x^2 = 72$ and $x^2 + y^2 = 90$; adding the 2 equations gives $2y^2 = 162$; $y^2 = 81$; $y = \pm 9$ ($y = -9$ is not reasonable); if $y = 9$, then $x^2 + 9^2 = 90$, $x^2 = 9$, $x = \pm 3$ ($x = -3$ is not reasonable); $AC = 3$, $DC = 9$

B **5.** Let x = length of one leg; y = length of other leg. $\frac{1}{2}xy = 60$ and
$x + y + \sqrt{x^2 + y^2} = 40$; $x = \frac{120}{y}$ and $(40 - x - y)^2 = x^2 + y^2$, or
$1600 - 80x - 80y + 2xy + x^2 + y^2 = x^2 + y^2$; $800 - 40x - 40y + xy = 0$;
substitute for x to obtain $800 - 40\left(\frac{120}{y}\right) - 40y + \left(\frac{120}{y}\right)y = 0$;
$920 - \frac{4800}{y} - 40y = 0$; $23 - \frac{120}{y} - y = 0$; $y^2 - 23y + 120 = 0$;
$(y - 8)(y - 15) = 0$; $y = 8$ or $y = 15$; if $y = 8$, $x = 15$; if $y = 15$, $x = 8$;
8 cm, 15 cm, 17 cm

6. $x^2 + 60^2 = y^2$ and $2y^2 + (2x)^2 = 45{,}600$; $y^2 - x^2 = 3600$ and $y^2 + 2x^2 = 22{,}800$; subtracting the 1st equation from the 2nd equation gives $3x^2 = 19{,}200$; $x^2 = 6400$; $x = \pm 80$ ($x = -80$ is not reasonable); $y^2 - 80^2 = 3600$; $y^2 = 10{,}000$; $y = \pm 100$ ($y = -100$ is not reasonable); $x = 80$ m, $y = 100$ m

7. Let h = length of altitude; $2e$ = length of a base edge; then $2h$ = lateral edge. $e^2 + 5^2 = (2h)^2$ and $e^2 + h^2 = 5^2$; $e^2 = 4h^2 - 25$ and $e^2 + h^2 = 25$; substitute for e^2 to obtain $4h^2 - 25 + h^2 = 25$; $5h^2 = 50$; $h = \pm\sqrt{10}$ (reject $-\sqrt{10}$); if $h = \sqrt{10}$, $e^2 + 10 = 25$; $e = \pm\sqrt{15}$ (reject $-\sqrt{15}$); altitude: $\sqrt{10}$ cm; base edge: $2\sqrt{15}$ cm

Page 433 • SELF-TEST 3

1. 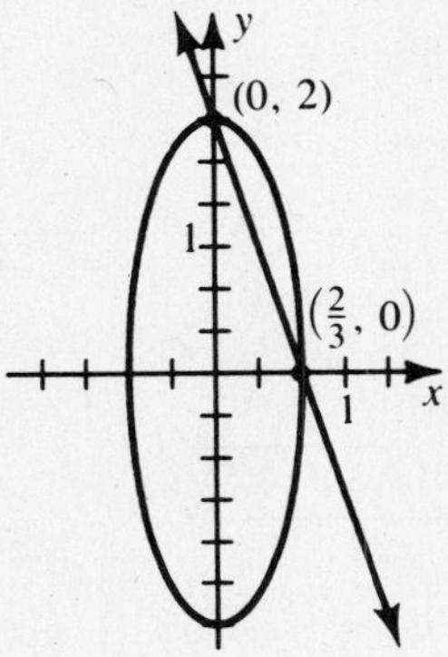 $\left\{\left(\frac{2}{3}, 0\right), (0, 2)\right\}$

y
(0, 2)
1
$(\frac{2}{3}, 0)$
1
x

2. $\begin{matrix} (1): 4x^2 + 9y^2 = 100 \\ (2): \quad x^2 - y^2 = 12 \end{matrix}\Big\}$ (1) + 9(2):
$13x^2 = 208$; $x^2 = 16$; $x = \pm 4$;
(2): $(\pm 4)^2 - y^2 = 12$; $y^2 = 4$;
$y = \pm 2$; {(4, 2), (4, −2), (−4, 2), (−4, −2)}

3. $\begin{matrix} x^2 + 3y^2 = 4 \\ x - y = 10 \end{matrix}\Big\}$ $x = y + 10$; $(y + 10)^2 + 3y^2 = 4$;
$y^2 + 20y + 100 + 3y^2 = 4$; $4y^2 + 20y + 96 = 0$; $y^2 + 5y + 24 = 0$;

$$y = \frac{-5 \pm \sqrt{25 - 4(24)}}{2} = \frac{-5 \pm i\sqrt{71}}{2};$$

$$y = \frac{-5 + i\sqrt{71}}{2}: x = \frac{-5 + i\sqrt{71}}{2} + 10 = \frac{15 + i\sqrt{71}}{2}; y = \frac{-5 - i\sqrt{71}}{2}:$$

$$x = \frac{15 - i\sqrt{71}}{2}; \left\{\left(\frac{15 + i\sqrt{71}}{2}, \frac{-5 + i\sqrt{71}}{2}\right), \left(\frac{15 - i\sqrt{71}}{2}, \frac{-5 - i\sqrt{71}}{2}\right)\right\}$$

Pages 434–435 • CHAPTER REVIEW

1. $\sqrt{(4 - 3)^2 + (-1 - 7)^2} = \sqrt{65}$; b
2. $\left(\frac{-a + a}{2}, \frac{-b + (-b)}{2}\right) = \left(\frac{0}{2}, \frac{-2b}{2}\right) = (0, -b)$; c
3. $5x + 2y = 7; y = -\frac{5}{2}x + \frac{7}{2}; m = -\frac{5}{2}$; slope of line perpendicular to $5x + 2y = 7$ is $\frac{2}{5}$; $y + 2 = \frac{2}{5}(x - 3); 5y + 10 = 2x - 6; 2x - 5y = 16$; d
4. $(x - 1)^2 + (y + 2)^2 = r^2; r = \sqrt{(1 - 5)^2 + (-2 - 1)^2} = \sqrt{(-4)^2 + (-3)^2} = \sqrt{16 + 9} = \sqrt{25} = 5; (x - 1)^2 + (y + 2)^2 = 5^2; x^2 - 2x + 1 + y^2 + 4y + 4 = 25;$ $x^2 + y^2 - 2x + 4y = 20$; d
5. The vertex of the parabola is (0, 0). Since focus is $F(0, 2)$, $p = 2$; $\therefore x^2 = 4py$, $x^2 = 4(2)y$, $x^2 = 8y$; a
6. $2a = 8, a = 4; b = 2; \frac{x^2}{16} + \frac{y^2}{4} = 1$; a
7. $36x^2 - 4y^2 = 144; \frac{x^2}{4} - \frac{y^2}{36} = 1; a = 2, b = 6$; asymptotes are $y = \pm\frac{b}{a}x$; $y = 3x$ and $y = -3x$; a
8. $x_1y_1^2 = x_2y_2^2; 3(2^2) = x_2\left(\frac{1}{4}\right)^2; x_2 = 192$; d
9. $y = x^2 + 4$ is a parabola; $8y = x^2 + y^2$ is equivalent to $x^2 + (y - 4)^2 = 16$, which is a circle. There are 2 points of intersection. c

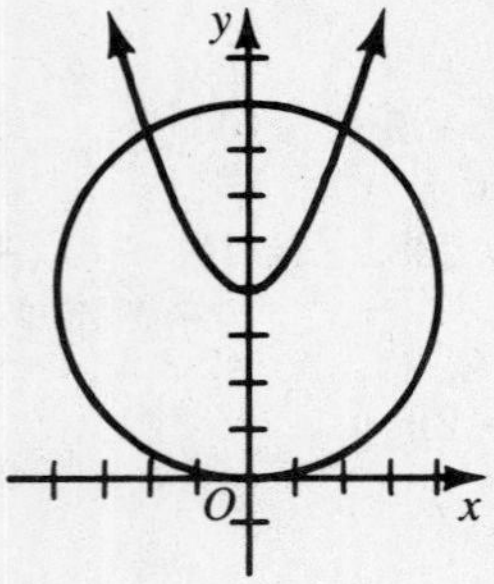

10. $\left.\begin{array}{l} y = x^2 + 5 \\ y = 4x \end{array}\right\} 4x = x^2 + 5; x^2 - 4x + 5 = 0; x = \frac{4 \pm \sqrt{16 - 4(5)}}{2};$
$x = \frac{4 \pm 2i}{2} = 2 \pm i; x = 2 + i: y = 8 + 4i; x = 2 - i: y = 8 - 4i;$
$\{(2 + i, 8 + 4i), (2 - i, 8 - 4i)\}$; a
11. $\left.\begin{array}{l} (1): x^2 + y^2 = 25 \\ (2): x^2 + 4y^2 = 46 \end{array}\right\} (2) - (1): 3y^2 = 21; y^2 = 7; y = \pm\sqrt{7}; (1): x^2 + (\pm\sqrt{7})^2 = 25;$
$x^2 = 18; x = \pm 3\sqrt{2}; \{(3\sqrt{2}, \sqrt{7}), (3\sqrt{2}, -\sqrt{7}), (-3\sqrt{2}, \sqrt{7}), (-3\sqrt{2}, -\sqrt{7})\}$; c

Pages 435–436 • CHAPTER TEST

1. $A(x, y)$: $\frac{x + 6}{2} = 4$; $\frac{y + 1}{2} = 5$; $x = 2$; $y = 9$; $(2, 9)$

2. $m_1 = \frac{6 - 4}{5 - 3} = 1$; $m_2 = -1$; $y - 3 = -1(x - 2)$; $x + y = 5$

3. $x^2 + y^2 - 6x + 2y = 6$; $(x^2 - 6x + 9) + (y^2 + 2y + 1) = 6 + 9 + 1$, $(x - 3)^2 + (y + 1)^2 = 16$; $C(3, -1)$; $r = 4$

4. $x = y^2 + 6y + 11$
$x = (y^2 + 6y + 9) + 11 - 9$
$x = (y + 3)^2 + 2$
$V(2, -3)$, axis: $y = -3$

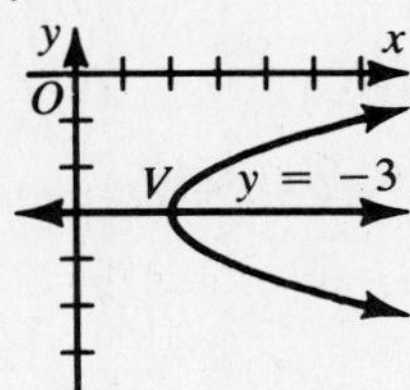

5. $\frac{(x + 2)^2}{25} + \frac{y^2}{9} = 1$
$C(-2, 0)$; $a = 5$, $b = 3$

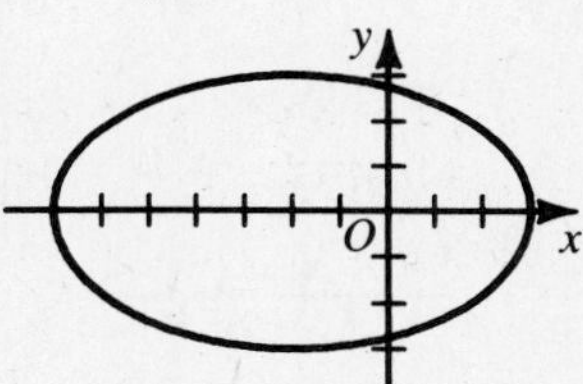

6. $9x^2 - 4y^2 = 36$
$\frac{x^2}{4} - \frac{y^2}{9} = 1$
$a = 2$, $b = 3$
$y = \frac{3}{2}x$, $y = -\frac{3}{2}x$

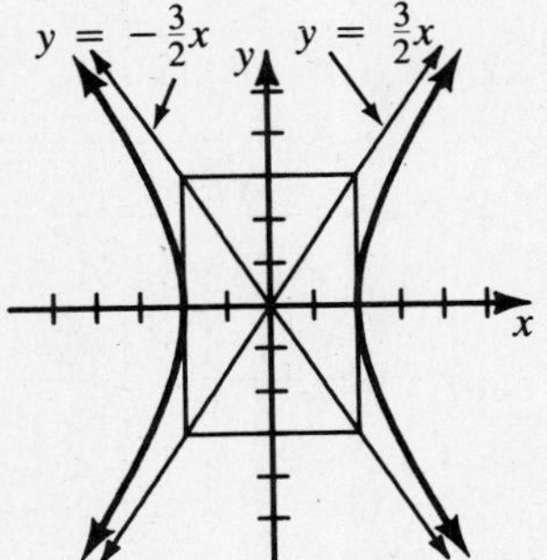

7. $\frac{xz}{y^2} = k$; $\frac{(2)(9)}{3^2} = \frac{x(8)}{2^2}$; $x = \frac{2(9)(4)}{9(8)} = 1$

8.

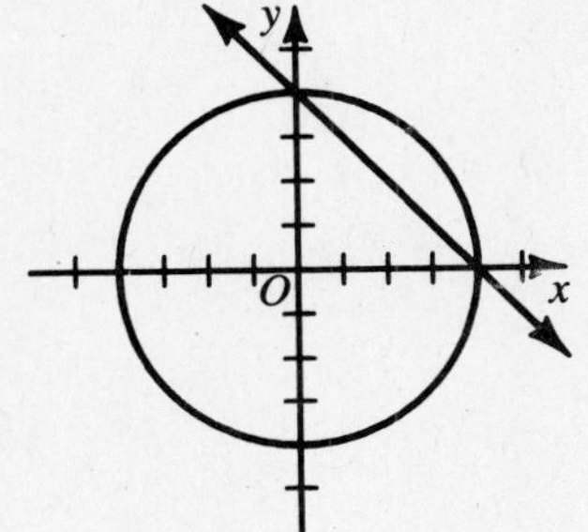

$\{(4, 0), (0, 4)\}$

9. $\left.\begin{array}{l} x^2 - 2y^2 = 7 \\ x - y \quad\;\; = 1 \end{array}\right\}$ $x = y + 1$;
$(y + 1)^2 - 2y^2 = 7$;
$y^2 + 2y + 1 - 2y^2 = 7$;
$y^2 - 2y + 6 = 0$;
$y = \frac{2 \pm \sqrt{4 - 4(6)}}{2} =$
$\frac{2 \pm 2i\sqrt{5}}{2} = 1 \pm i\sqrt{5}$;
$y = 1 + i\sqrt{5}$:
$x = 1 + 1 + i\sqrt{5} = 2 + i\sqrt{5}$;
$y = 1 - i\sqrt{5}$: $x = 2 - i\sqrt{5}$;
$\{(2 + i\sqrt{5}, 1 + i\sqrt{5}),$
$(2 - i\sqrt{5}, 1 - i\sqrt{5})\}$

10. $\left.\begin{array}{l}(1)\text{: } 4x^2 + y^2 = 52 \\ (2)\text{: } x^2 - y^2 = -7\end{array}\right\}$ $(1) + (2)$: $5x^2 = 45$; $x^2 = 9$; $x = \pm 3$; (1): $4(\pm 3)^2 + y^2 = 52$; $y^2 = 16$; $y = \pm 4$; $\{(3, 4), (3, -4), (-3, 4), (-3, -4)\}$

Page 436 • MIXED REVIEW

1. $[(5 \cdot 3)^2 + (4 \cdot 3)^2 - (2 \cdot 3)]^2 \div [3^2 \cdot 37] = (15^2 + 12^2 - 6^2) \div (9 \cdot 37) = 333 \div 333 = 1$
2. $\dfrac{(-3mn^2)^2(-2m^2n)^3}{9a^2b(-4a^2b)^2} = \dfrac{(9m^2n^4)(-8m^6n^3)}{9(a^2b)(16a^4b^2)} = \dfrac{9(-8)(m^8n^7)}{9(16)(a^6b^3)} = -\dfrac{m^8n^7}{2a^6b^3}$
3. $8\left(u - \dfrac{1}{4}v\right) - 4\left(\dfrac{1}{2}v + 2u\right) - (-v + 2) = (8u - 2v) - (2v + 8u) - (-v + 2) =$ $8u - 2v + (-2v) + (-8u) + v + (-2) = -3v - 2$
4. $\dfrac{x + 2y}{x^2 - 2xy + y^2} + \dfrac{4}{3x - 3y} = \dfrac{x + 2y}{(x - y)^2} + \dfrac{4}{3(x - y)} = \dfrac{3(x + 2y) + 4(x - y)}{3(x - y)^2} =$ $\dfrac{7x + 2y}{3(x - y)^2}$
5. $(2 + 7i) - (3 + 4i) + (7 - 6i) = (2 - 3 + 7) + (7i - 4i - 6i) = 6 - 3i$
6. $\dfrac{\sqrt{x} + 3\sqrt{y}}{\sqrt{x} - 3\sqrt{y}} \cdot \dfrac{\sqrt{x} + 3\sqrt{y}}{\sqrt{x} + 3\sqrt{y}} = \dfrac{x + 6\sqrt{xy} + 9y}{x - 9y}$
7. $-(1 - 4c) = 2(2c - 1) + 1$; $-1 + 4c = 4c - 2 + 1$; $-1 + 4c = 4c - 1$; $0 = 0$; $\mathscr{R}$
8. $|2n - 3| > 7$; $2n - 3 > 7$ or $2n - 3 < -7$; $2n > 10$ or $2n < -4$; $n > 5$ or $n < -2$; $\{n\text{: } n < -2 \text{ or } n > 5\}$
9. $6x = x^2 - 6$; $x^2 - 6x - 6 = 0$; $x = \dfrac{6 \pm \sqrt{36 - 4(-6)}}{2} = \dfrac{6 \pm \sqrt{60}}{2} = 3 \pm \sqrt{15}$; $\{3 + \sqrt{15}, 3 - \sqrt{15}\}$
10. $\dfrac{4}{a - 3} - \dfrac{2a}{9 - a^2} = \dfrac{1}{a + 3}$ $(a \neq \pm 3)$; $\dfrac{4}{a - 3} + \dfrac{2a}{a^2 - 9} = \dfrac{1}{a + 3}$; $4(a + 3) + 2a = 1(a - 3)$; $4a + 12 + 2a = a - 3$; $5a = -15$; $a = -3$; but a cannot $= -3$; $\emptyset$
11. $p \in \{\pm 1, \pm 2, \pm 4\}$, $q \in \{\pm 1\}$; $\dfrac{p}{q} \in \{\pm 1, \pm 2, \pm 4\}$; none satisfy the equation; $\therefore$ no rational roots.
12. $2x^2 = -9$; $x^2 = -\dfrac{9}{2}$; $x = \pm\sqrt{-\dfrac{9}{2}} = \pm\dfrac{3i}{\sqrt{2}} = \pm\dfrac{3i\sqrt{2}}{2}$; $\left\{\dfrac{3i\sqrt{2}}{2}, -\dfrac{3i\sqrt{2}}{2}\right\}$
13. $8 + 5t \geq 2$ or $-(8 + 5t) \geq 2$; $5t \geq -6$ or $8 + 5t \leq -2$; $t \geq -\dfrac{6}{5}$ or $5t \leq -10$; $t \geq -\dfrac{6}{5}$ or $t \leq -2$

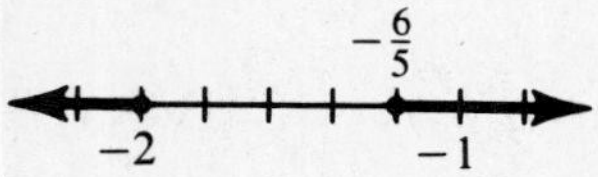

$-\frac{6}{5}$

−2 −1

14. $7x > 5y - 15$
$5y < 7x + 15$
$y < \frac{7}{5}x + 3$

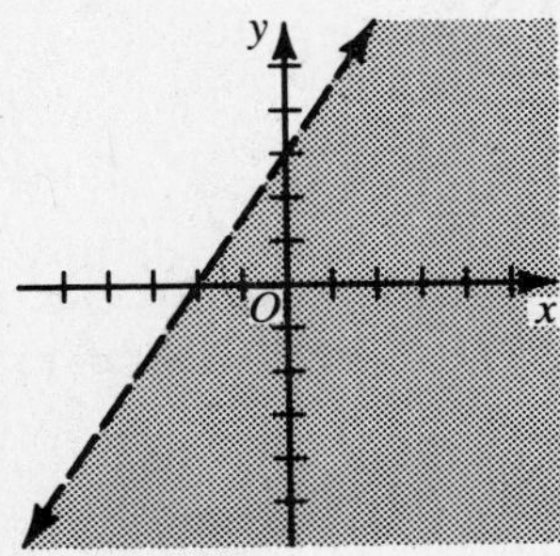

15. $16y^2 = 9x^2 + 144$
$\frac{y^2}{9} - \frac{x^2}{16} = 1$

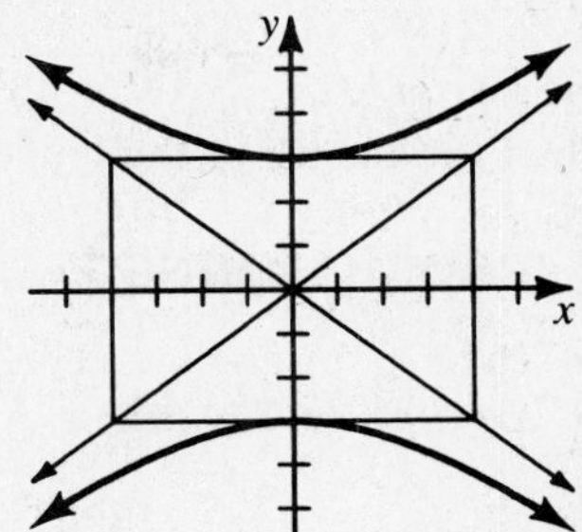

16. $y = -3|x|$

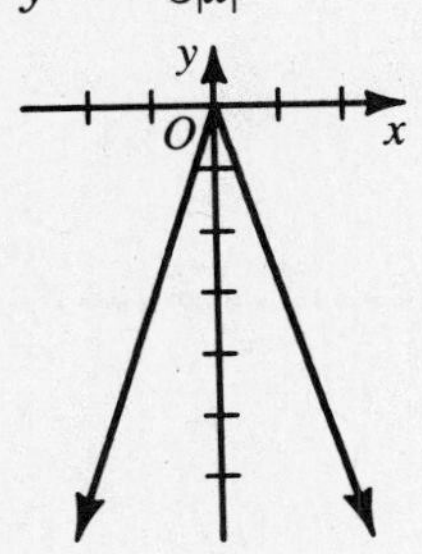

17.

Point	$2x - y$
(0, 0)	$2(0) - 0 = 0$
(0, 3)	$2(0) - 3 = -3$
(3, 2)	$2(3) - 2 = 4$
(4, 0)	$2(4) - 0 = 8$

Max.: 8; min.: -3

18. $F(x, y)$: $\frac{x + 4}{2} = 2$, $\frac{y + 2}{2} = -1$; $x = 0$, $y = -4$; $F(0, -4)$

19. $S_n = \frac{n}{2}(a_1 + a_n)$; $S_{45} = \frac{45}{2}(3 - 217) = -4815$

20. $n = 4$, $a_4 = \frac{1}{3}$, $a_1 = 9$: $\frac{1}{3} = 9r^3$; $r = \frac{1}{3}$; $a_2 = \frac{1}{3}(9) = 3$, $a_3 = \frac{1}{3}(3) = 1$; 3, 1

21. $S = \frac{a_1}{1 - r}$, $a_1 = 8$, $r = \frac{1}{2}$: $S = \frac{8}{1 - \frac{1}{2}} = 16$

22. $4x^2 + 4y^2 = 17$; $x^2 + y^2 = \frac{17}{4}$; $r^2 = \frac{17}{4}$; $r = \frac{\sqrt{17}}{2}$

23. Let x = the integer. $x^2 = 45 + 4x$; $x^2 - 4x - 45 = 0$; $(x - 9)(x + 5) = 0$; $x = -5$

24.
$$\begin{array}{r}
3x^3 + 2x^2 - x + 3 \\
2x - 2 \overline{\smash{)}\,6x^4 - 2x^3 - 6x^2 + 8x - 6} \\
\underline{6x^4 - 6x^3} \qquad\qquad\qquad \\
4x^3 - 6x^2 \qquad\qquad \\
\underline{4x^3 - 4x^2} \qquad\qquad \\
-2x^2 + 8x \qquad \\
\underline{-2x^2 + 2x} \qquad \\
6x - 6 \\
\underline{6x - 6} \\
0
\end{array}$$
$3x^3 + 2x^2 - x + 3$

25. $a_1 = 9000$, $r = 0.8$; $a_4 = a_1 r^3 = 9000(0.8)^3 = 4608$; \$4608

Page 437 • CONTEST PROBLEMS

1. The length of the diagonal of the bottom surface is $\sqrt{75^2 + 60^2} = \sqrt{5625 + 3600} = \sqrt{9225}$. The length of the diagonal through the box is $\sqrt{(\sqrt{9225})^2 + 45^2} = \sqrt{9225 + 2025} = \sqrt{11250} = 106.07$. Thus the length of the longest stick is 106 cm.

Page 437 • PROGRAMMING IN PASCAL

1.
```
{* This program analyzes the graph of an equation of the form *}
{* ax^2 + cy^2 + dx + ey + f = 0 where not both a and c are 0. *}

PROGRAM conics (INPUT, OUTPUT);

TYPE
  str9 = PACKED ARRAY[1..9] OF char;

VAR
  a, c, d, e, f : real;
  kind : str9;

{ ************************************************************ }
PROCEDURE get_coefficients;

BEGIN
  REPEAT
    write('Enter the values of a: '); readln(a);
    write('                    c: '); readln(c);
  UNTIL abs(a) + abs(c) > 0;
  write('                    d: '); readln(d);
  write('                    e: '); readln(e);
  write('                    f: '); readln(f);
  IF a < 0
     THEN BEGIN
            a := -a; c := -c; d := -d; e := -e; f := -f;
          END;
END;

{ ************************************************************ }
PROCEDURE determine_conic;

BEGIN
  IF a = c
     THEN kind := 'circle
     ELSE IF a * c = 0                    (* either a or c is 0 *)
             THEN kind := 'parabola '
             ELSE IF a * c > 0 (* a and c have the same sign *)
                     THEN kind := 'ellipse  '
                     ELSE kind := 'hyperbola';
END;
```

```
{*******************************************************************}
PROCEDURE ck_parabola;

VAR
  x_ctr, y_ctr, p : real;

BEGIN
  IF a = 0
     THEN BEGIN
            x_ctr := sqr(e)/(4 * d * c) - f/d;
            y_ctr := -e/(2 * c);
            IF d = 0
               THEN writeln('No parabola.')
               ELSE BEGIN
                      writeln('Graph is a parabola.');
                      write('Vertex: (', x_ctr:3:2, ', ');
                      writeln(y_ctr:3:2, ')');
                      p := -d/(4 * c);
                      IF -c/d > 0
                         THEN BEGIN
                                writeln('It opens to the right.');
                                write('Focus: (', (x_ctr + p):3:2);
                                writeln(', ', y_ctr:3:2, ')');
                                write('Directrix: x = ');
                                writeln((x_ctr - p):3:2);
                              END
                         ELSE BEGIN
                                writeln('It opens to the left.');
                                write('Focus: (', (x_ctr - p):3:2);
                                writeln(', ', y_ctr:3:2, ')');
                                write('Directrix: x = ');
                                writeln((x_ctr + p):3:2);
                              END;
                    END;
          END;
  IF c = 0
     THEN BEGIN
            x_ctr := -d/(2 * a);
            y_ctr := sqr(d)/(4 * a * e) - f/e;
            IF e = 0
               THEN writeln('No parabola.')
               ELSE BEGIN
                      writeln('Graph is a parabola.');
                      write('Vertex: (', x_ctr:3:2, ', ');
                      writeln(y_ctr:3:2, ')');
                      p := -e/(4 * a);
```

[*Program continued on next page*]

```
                       IF -a/e > 0
                          THEN BEGIN
                                 writeln('It opens upward.');
                                 write('Focus: (', x_ctr:3:2, ', ');
                                 writeln((y_ctr + p):3:2, ')');
                                 write('Directrix: y = ');
                                 writeln((y_ctr - p):3:2);
                               END
                          ELSE BEGIN
                                 writeln('It opens downward.');
                                 write('Focus: (', x_ctr:3:2, ', ');
                                 writeln((y_ctr - p):3:2, ')');
                                 write('Directrix: y = ');
                                 writeln((y_ctr + p):3:2);
                               END;
                     END;
          END;
END;

{ *********************************************************************** }
PROCEDURE ck_circle;

VAR
  k, x_ctr, y_ctr, r : real;

BEGIN
  x_ctr := -d/(2 * a);
  y_ctr := -e/(2 * c);
  k := sqr(x_ctr) * a + sqr(y_ctr) * c - f;
  IF k < 0
     THEN writeln('No circle.')
     ELSE BEGIN
            r := sqrt(k/a);
            write('Center of circle: (', x_ctr:3:2, ', ');
            writeln(y_ctr:3:2, ')');
            writeln('The radius is: ', r:3:2);
          END;
END;

{*********************************************************************** }
PROCEDURE ck_hyperbola;

VAR
  k, dist_to_focus,
  x_ctr, y_ctr, slope_1, slope_2 : real;
```

```
BEGIN
  x_ctr := -d/(2 * a);
  y_ctr := -e/(2 * c);
  k := sqr(x_ctr) * a + sqr(y_ctr) * c - f;
  write('Center of hyperbola: (', x_ctr:3:2, ', ');
  writeln(y_ctr:3:2, ')');
  slope_1 := sqrt(abs(k/c))/sqrt(abs(k/a));
  slope_2 := -slope_1;
  write('Asymptotes: y = ', slope_1:3:2, 'x ');
  IF -slope_1 * x_ctr + y_ctr < 0
     THEN writeln('- ', abs(-slope_1 * x_ctr + y_ctr):3:2)
     ELSE writeln('+ ', (-slope_1 * x_ctr + y_ctr):3:2);
  write('            y = ', slope_2:3:2, 'x ');
  IF -slope_2 * x_ctr + y_ctr < 0
     THEN writeln('- ', abs(-slope_2 * x_ctr + y_ctr):3:2)
     ELSE writeln('+ ', (-slope_2 * x_ctr + y_ctr):3:2);
  c := -c;
  IF k < 0
     THEN BEGIN
            k := -k;
            write('Vertices: (', x_ctr:3:2, ', ');
            writeln((y_ctr + sqrt(k/c)):3:2, ')');
            write('          (', x_ctr:3:2, ', ');
            writeln((y_ctr - sqrt(k/c)):3:2, ')');
            dist_to_focus := sqrt(k/a + k/c);
            write('Foci: (', x_ctr:3:2, ', ');
            writeln((y_ctr + dist_to_focus):3:2, ')');
            write('      (', x_ctr:3:2, ', ');
            writeln((y_ctr - dist_to_focus):3:2, ')');
          END
     ELSE BEGIN
            write('Vertices: (', (x_ctr + sqrt(k/a)):3:2, ', ');
            writeln(y_ctr:3:2, ')');
            write('          (', (x_ctr - sqrt(k/a)):3:2, ', ');
            writeln(y_ctr:3:2, ')');
            dist_to_focus := sqrt(k/c + k/a);
            write('Foci: (', (x_ctr + dist_to_focus):3:2, ', ');
            writeln(y_ctr:3:2, ')');
            write('      (', (x_ctr - dist_to_focus):3:2, ', ');
            writeln(y_ctr:3:2, ')');
          END;
END;
```

[*Program continued on next page*]

```
{***********************************************************************}
PROCEDURE ck_ellipse;

VAR
  k, dist_to_focus,
  maj_axis_len, min_axis_len,
  x_ctr, y_ctr,
  top_x_vertex, top_y_vertex,
  btm_x_vertex, btm_y_vertex,
  left_x_vertex, left_y_vertex,
  rt_x_vertex, rt_y_vertex,
  x_f1, y_f1,
  x_f2, y_f2 : real;

BEGIN
  x_ctr := -d/(2 * a);  {* found by completing the square for x *}
  y_ctr := -e/(2 * c);  {* found by completing the square for y *}
  k := sqr(x_ctr) * a + sqr(y_ctr) * c - f;
  IF k < 0
     THEN writeln('No ellipse.')
     ELSE BEGIN
            IF a < c
               THEN BEGIN
                      maj_axis_len := 2 * sqrt(k/a);
                      min_axis_len := 2 * sqrt(k/c);
                      dist_to_focus := sqrt(k * (1/a - 1/c));

                      { * the coordinates of the vertices * }
                      top_x_vertex := x_ctr;
                      top_y_vertex := y_ctr + 0.5 * min_axis_len;
                      btm_x_vertex := x_ctr;
                      btm_y_vertex := y_ctr - 0.5 * min_axis_len;
                      left_x_vertex := x_ctr - 0.5 * maj_axis_len;
                      left_y_vertex := y_ctr;
                      rt_x_vertex := x_ctr + 0.5 * maj_axis_len;
                      rt_y_vertex := y_ctr;

                      { * the coordinates of the foci * }
                      x_f1 := x_ctr - dist_to_focus;
                      y_f1 := y_ctr;
                      x_f2 := x_ctr + dist_to_focus;
                      y_f2 := y_ctr;
                    END
               ELSE BEGIN
                      maj_axis_len := 2 * sqrt(k/c);
                      min_axis_len := 2 * sqrt(k/a);
                      dist_to_focus := sqrt(k * (1/a - 1/c));
```

```
                    { * the coordinates of the vertices * }
                    top_x_vertex := x_ctr;
                    top_y_vertex := y_ctr + 0.5 * maj_axis_len;
                    btm_x_vertex := x_ctr;
                    btm_y_vertex := y_ctr - 0.5 * maj_axis_len;
                    left_x_vertex := x_ctr - 0.5 * min_axis_len;
                    left_y_vertex := y_ctr;
                    rt_x_vertex := x_ctr + 0.5 * min_axis_len;
                    rt_y_vertex := y_ctr;

                    { * the coordinates of the foci * }
                    x_f1 := x_ctr;
                    y_f1 := y_ctr + dist_to_focus;
                    x_f2 := x_ctr;
                    y_f2 := y_ctr - dist_to_focus;
                  END;
          write('Vertices: ');
          write('          (', top_x_vertex:3:2, ', ');
          writeln(top_y_vertex:3:2, ')');
          write('          (', btm_x_vertex:3:2, ', ');
          writeln(btm_y_vertex:3:2, ')');
          write('          (', left_x_vertex:3:2, ', ');
          writeln(left_y_vertex:3:2, ')');
          write('          (', rt_x_vertex:3:2, ', ');
          writeln(rt_y_vertex:3:2, ')');
          writeln('Foci: (', x_f1:3:2, ', ', y_f1:3:2, ')');
          writeln('      (', x_f2:3:2, ', ', y_f2:3:2, ')');
        END;
END;

{*******************************************************************}
BEGIN { * main * }
  get_coefficients;
  determine_conic;
  IF kind = 'ellipse '
     THEN ck_ellipse;
  IF kind = 'circle '
     THEN ck_circle;
  IF kind = 'parabola '
     THEN ck_parabola;
  IF kind = 'hyperbola'
     THEN ck_hyperbola;
END.
```

Pages 443–444 • WRITTEN EXERCISES

A **1.** $\sqrt[3]{27} = 3$ **2.** $(\sqrt[3]{8})^2 = 2^2 = 4$ **3.** $\dfrac{1}{\sqrt{36}} = \dfrac{1}{6}$ **4.** $(\sqrt[3]{64})^4 = 4^4 = 256$

5. $(\sqrt[4]{16})^3 = 2^3 = 8$ **6.** $\dfrac{1}{(\sqrt[5]{32})^3} = \dfrac{1}{2^3} = \dfrac{1}{8}$ **7.** $\left(\sqrt[3]{\dfrac{1}{27}}\right)^2 = \left(\dfrac{1}{3}\right)^2 = \dfrac{1}{9}$

8. $\left(\dfrac{64}{125}\right)^{1/3} = \sqrt[3]{\dfrac{64}{125}} = \dfrac{4}{5}$ **9.** $(\sqrt{0.25})^3 = (0.5)^3 = 0.125$, or $\dfrac{1}{8}$

10. $(\sqrt[3]{0.008})^5 = (0.2)^5 = 0.00032$ **11.** $(169)^{1/2} = \sqrt{169} = 13$

12. $(100 - 36)^{-1/2} = (64)^{-1/2} = \dfrac{1}{(64)^{1/2}} = \dfrac{1}{\sqrt{64}} = \dfrac{1}{8}$

13. $x^{9/6} = x^{6/6} \cdot x^{3/6} = x \cdot x^{1/2} = x\sqrt{x}$

14. $3^{2/4}\, y^{6/4} = 3^{1/2}\, y^{4/4} \cdot y^{2/4} = 3^{1/2}\, y \cdot y^{1/2} = y\sqrt{3y}$

15. $49^{1/4}\, n^{2/4} = 7^{2/4}\, n^{2/4} = 7^{1/2}\, n^{1/2} = \sqrt{7n}$

16. $\left(\dfrac{1}{27r^{24}}\right)^{1/6} = \dfrac{1}{27^{1/6}} \cdot \dfrac{1}{r^{24/6}} = \dfrac{1}{3^{3/6}\, r^4} = \dfrac{1}{3^{1/2}\, r^4} = \dfrac{3^{1/2}}{3^{1/2} \cdot 3^{1/2}} \cdot \dfrac{1}{r^4} = \dfrac{3^{1/2}}{3r^4} = \dfrac{\sqrt{3}}{3r^4}$

17. $81^{1/8} = 3^{4/8} = 3^{1/2} = \sqrt{3}$ **18.** $27^{1/12} = 3^{3/12} = 3^{1/4} = \sqrt[4]{3}$

19. $\sqrt[3]{125^{1/2}} = (125^{1/2})^{1/3} = 125^{1/6} = 5^{3/6} = 5^{1/2} = \sqrt{5}$

20. $\left(\sqrt{625}\right)^{1/4} = (625^{1/2})^{1/4} = 625^{1/8} = 5^{4/8} = 5^{1/2} = \sqrt{5}$

21. $\left(\dfrac{1}{1000}\right)^{1/6} = \left(\dfrac{1}{10}\right)^{3/6} = \left(\dfrac{1}{10}\right)^{1/2} \cdot \dfrac{10^{1/2}}{10^{1/2}} = \dfrac{10^{1/2}}{10} = \dfrac{\sqrt{10}}{10}$

22. $\left(\dfrac{125}{8}\right)^{1/9} = \dfrac{125^{1/9}}{8^{1/9}} = \dfrac{5^{3/9}}{2^{3/9}} = \dfrac{5^{1/3}}{2^{1/3}} \cdot \dfrac{2^{2/3}}{2^{2/3}} = \dfrac{(5 \cdot 2^2)^{1/3}}{2} = \dfrac{\sqrt[3]{20}}{2}$

23. $\left(\dfrac{64}{27}\right)^{2/3} = \dfrac{64^{2/3}}{27^{2/3}} = \dfrac{(4^3)^{2/3}}{(3^3)^{2/3}} = \dfrac{4^2}{3^2} = \dfrac{16}{9}$ **24.** $\left(\dfrac{49}{144}\right)^{-2/4} = \dfrac{49^{-1/2}}{144^{-1/2}} = \dfrac{144^{1/2}}{49^{1/2}} = \dfrac{12}{7}$

25. $36^{1/6} \cdot 36^{1/3} = 6^{2/6} \cdot 6^{2/3} = 6^{1/3} \cdot 6^{2/3} = 6$

26. $25^{1/8} \cdot 125^{1/4} = 5^{2/8} \cdot 5^{3/4} = 5^{1/4} \cdot 5^{3/4} = 5$

27. $4^{1/3} \cdot 32^{1/6} = 2^{2/3} \cdot 2^{5/6} = 2^{4/6} \cdot 2^{5/6} = 2^{9/6} = 2^{6/6} \cdot 2^{3/6} = 2 \cdot 2^{1/2} = 2\sqrt{2}$

28. $3^{1/12} \cdot 27^{1/12} = 3^{1/12} \cdot 3^{3/12} = 3^{4/12} = 3^{1/3} = \sqrt[3]{3}$

29. $\dfrac{81^{1/8}}{3^{1/6}} = \dfrac{3^{4/8}}{3^{1/6}} = \dfrac{3^{1/2}}{3^{1/6}} = \dfrac{3^{3/6}}{3^{1/6}} = 3^{2/6} = 3^{1/3} = \sqrt[3]{3}$

30. $\dfrac{27^{1/4}}{3^{1/4}} = \dfrac{3^{3/4}}{3^{1/4}} = 3^{1/2} = \sqrt{3}$

31. $\dfrac{25^{1/6}}{625^{1/3}} = \dfrac{5^{2/6}}{5^{4/3}} = \dfrac{5^{1/3}}{5^{4/3}} = \dfrac{1}{5}$

32. $\dfrac{81^{1/3}}{81^{1/12}} = \dfrac{81^{4/12}}{81^{1/12}} = 81^{3/12} = 81^{1/4} = 3$

B **33.** $\sqrt{9^{1/3}} = (9^{1/3})^{1/2} = 9^{1/6} = 3^{2/6} = 3^{1/3} = \sqrt[3]{3}$

34. $\sqrt{27^{1/2}} \cdot 3^{1/4} = (27^{1/2})^{1/2} \cdot 3^{1/4} = 27^{1/4} \cdot 3^{1/4} = 3^{3/4} \cdot 3^{1/4} = 3$

35. $\sqrt{10} \cdot \sqrt{\sqrt[3]{10}} = 10^{1/2} \cdot (10^{1/3})^{1/2} = 10^{1/2} \cdot 10^{1/6} = 10^{3/6} \cdot 10^{1/6} = 10^{4/6} = 10^{2/3} = \sqrt[3]{100}$

36. $(5^{1/4})^{1/3} \cdot (25^{1/4})^{1/3} = 5^{1/12} \cdot 25^{1/12} = 5^{1/12} \cdot 5^{2/12} = 5^{3/12} = 5^{1/4} = \sqrt[4]{5}$

37. $(7^{1/6})^{1/2} \cdot (7^{1/2})^{1/2} = 7^{1/12} \cdot 7^{1/4} = 7^{1/12} \cdot 7^{3/12} = 7^{4/12} = 7^{1/3} = \sqrt[3]{7}$

38. $(27^{1/4})^{1/3} \cdot (27^{1/2})^{1/2} = 27^{1/12} \cdot 27^{1/4} = 27^{1/12} \cdot 27^{3/12} = 27^{4/12} = 27^{1/3} = 3$

39. $z^{-4/5} = 16; (z^{-4/5})^{-5/4} = (16)^{-5/4}; z = \dfrac{1}{16^{5/4}} = \dfrac{1}{2^5} = \dfrac{1}{32}; \left\{\dfrac{1}{32}\right\}$

40. $k^{-3/4} = \dfrac{125}{8}; (k^{-3/4})^{-4/3} = \left(\dfrac{125}{8}\right)^{-4/3}; k = \left(\dfrac{8}{125}\right)^{4/3}; k = \dfrac{8^{4/3}}{125^{4/3}} = \dfrac{2^4}{5^4} = \dfrac{16}{625}; \left\{\dfrac{16}{625}\right\}$

41. $(t^{1/2} - 3)(t^{1/2} - 1) = 0; t^{1/2} = 3$ or $t^{1/2} = 1; t = 9$ or $t = 1; \{1, 9\}$

42. $(27x^{3/2} - 1)(x^{3/2} - 1) = 0; 27x^{3/2} - 1 = 0$ or $x^{3/2} - 1 = 0; 27x^{3/2} = 1$ or $x^{3/2} = 1;$ $x^{3/2} = \dfrac{1}{27}$ or $x^{3/2} = 1; (x^{3/2})^{2/3} = \left(\dfrac{1}{27}\right)^{2/3}; x = \dfrac{1}{9}$ or $x = 1; \left\{\dfrac{1}{9}, 1\right\}$

43. $(u^{2/3} - 1)(u^{2/3} - 4) = 0; u^{2/3} = 1$ or $u^{2/3} = 4; u = 1$ or $(u^{2/3})^{3/2} = 4^{3/2}; u = 2^3 = 8;$ $u = 1$ or $8; \{1, 8\}$

44. $(9x^{2/5} - 1)(x^{2/5} - 1) = 0; 9x^{2/5} = 1$ or $x^{2/5} = 1; x^{2/5} = \dfrac{1}{9}$ or $x = 1; (x^{2/5})^{5/2} = \left(\dfrac{1}{9}\right)^{5/2};$ $x = \left(\dfrac{1}{3}\right)^5 = \dfrac{1}{243}; x = 1$ or $\dfrac{1}{243}; \left\{\dfrac{1}{243}, 1\right\}$

C 45. $(b^{p/r} \cdot b^{q/s})^{rs} = b^{ps} \cdot b^{qr} = b^{ps+qr}; (b^{(ps+rq)/rs})^{rs} = b^{ps+rq}; \therefore b^{p/r} \cdot b^{q/s} = b^{(ps+rq)/rs}$

46. $((b^{p/r})^{q/s})^{rs} = b^{pq}; (b^{pq/rs})^{rs} = b^{pq}; \therefore (b^{p/r})^{q/s} = b^{pq/rs}$

47. $(a^{p/r} \cdot b^{p/r})^r = a^p \cdot b^p; ((ab)^{p/r})^r = (ab)^p = a^p \cdot b^p; \therefore a^{p/r} \cdot b^{p/r} = (ab)^{p/r}$

48. $\left(\dfrac{a^{p/r}}{b^{p/r}}\right)^r = \dfrac{a^p}{b^p}; \left(\left(\dfrac{a}{b}\right)^{p/r}\right)^r = \left(\dfrac{a}{b}\right)^p = \dfrac{a^p}{b^p}; \therefore \dfrac{a^{p/r}}{b^{p/r}} = \left(\dfrac{a}{b}\right)^{p/r}$

Page 446 • WRITTEN EXERCISES

A 1. $3^{\sqrt{2}} \cdot 3^{\sqrt{3}} = 3^{\sqrt{2}+\sqrt{3}}$ 2. $\dfrac{3^{\sqrt{3}}}{3^{\sqrt{2}}} = 3^{\sqrt{3}-\sqrt{2}}$ 3. $(3^{\sqrt{2}})^3 = 3^{3\sqrt{2}}$ 4. $\dfrac{1}{3^{\sqrt{3}}} = 3^{-\sqrt{3}}$

5. $3^{\sqrt{2}}(3^{\sqrt{3}})^2 = 3^{\sqrt{2}} \cdot 3^{2\sqrt{3}} = 3^{\sqrt{2}+2\sqrt{3}}$ 6. $(3^{\sqrt{2}})^{1/2} = 3^{\sqrt{2}/2}$

7. $\dfrac{(3^{\sqrt{2}})^3}{3^3} = \dfrac{3^{3\sqrt{2}}}{3^3} = 3^{3\sqrt{2}-3}$ 8. $\dfrac{3^2}{(3^{\sqrt{3}})^{1/3}} = \dfrac{3^2}{3^{\sqrt{3}/3}} = 3^{2-(\sqrt{3}/3)}$

9. $2^{2x} = 2^{x+3}; 2x = x + 3; x = 3; \{3\}$

10. $5^{3x-1} = 5^{2(x+4)}; 3x - 1 = 2x + 8; x = 9; \{9\}$

11. $3^{x-1} = 3^{3(2x+3)}; x - 1 = 6x + 9; 5x = -10; x = -2; \{-2\}$

12. $5^{3(2x-2)} = 5^{2(3-x)}; 6x - 6 = 6 - 2x; 8x = 12; x = \dfrac{3}{2}; \left\{\dfrac{3}{2}\right\}$

13. $6^{(-1)(x-3)} = 6^{x-1}; -x + 3 = x - 1; 2x = 4; x = 2; \{2\}$

14. $10^{2x+4} = 10^{(-2)(x-3)}; 2x + 4 = -2x + 6; 4x = 2; x = \dfrac{1}{2}; \left\{\dfrac{1}{2}\right\}$

15. $2^{3(x+3)} = 2^{(-2)(3x-6)}; 3x + 9 = -6x + 12; 9x = 3; x = \dfrac{1}{3}; \left\{\dfrac{1}{3}\right\}$

16. $2^{(-5)(x+2)} = 2^{(-3)(x-4)}; -5x - 10 = -3x + 12; 2x = -22; x = -11; \{-11\}$

B 17.

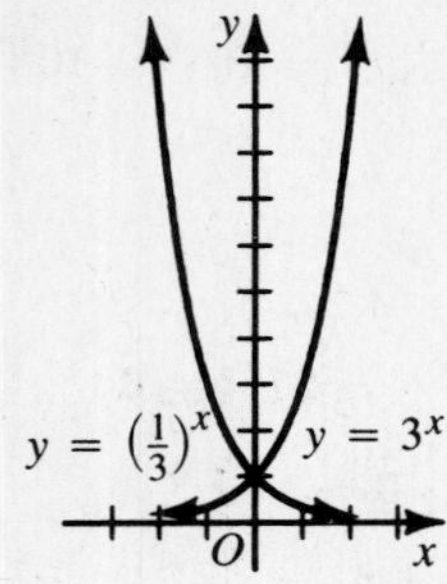

18.

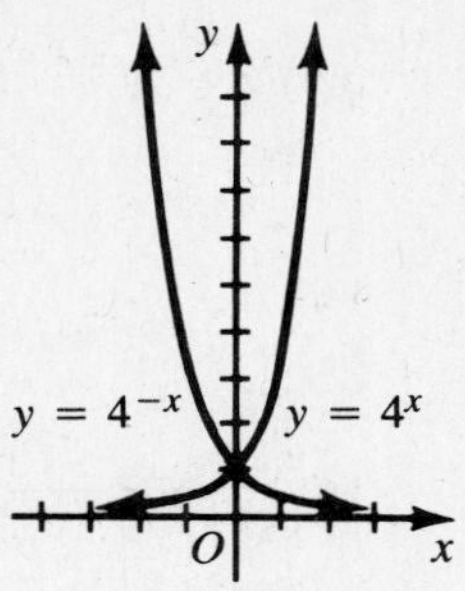

19.

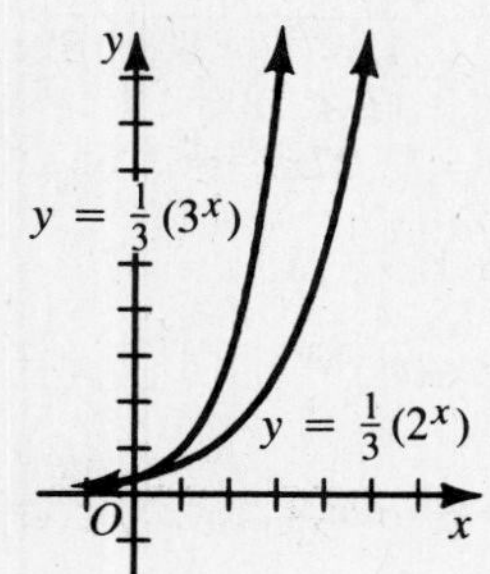

20.

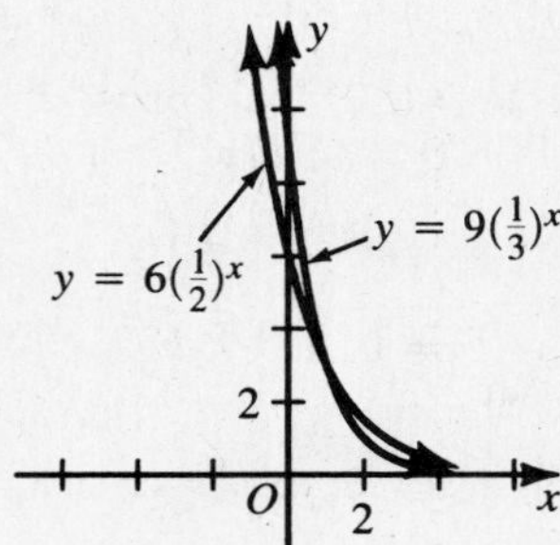

C 21. Answers may vary. For example, let $a = \sqrt{2}$ and $b = 2 - \sqrt{2}$, then $2^a \cdot 2^b = 2^2 = 4$.

22.

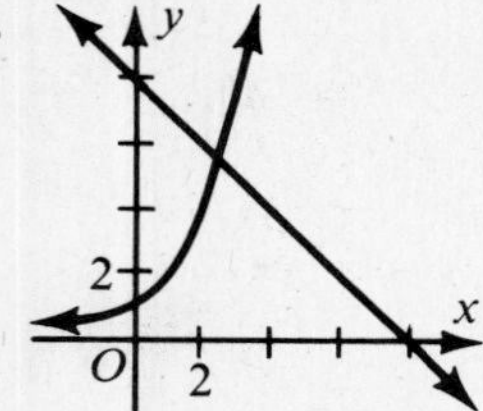

$\{(2.5, 5.5)\}$

Page 447 • SELF-TEST 1

1. $(\sqrt[4]{10{,}000})^3 = 10^3 = 1000$ 2. $\sqrt{\frac{9}{25}} = \frac{3}{5}$ 3. $\frac{1}{(\sqrt[3]{64})^2} = \frac{1}{4^2} = \frac{1}{16}$
4. $((625)^{1/3})^{1/4} = (5^4)^{1/12} = 5^{1/3} = \sqrt[3]{5}$
5. $\sqrt[8]{7^2} \cdot \sqrt[4]{7} = 7^{2/8} \cdot 7^{1/4} = 7^{1/4} \cdot 7^{1/4} = 7^{2/4} = 7^{1/2} = \sqrt{7}$
6. $\frac{36^{1/6}}{9^{1/6}} = \left(\frac{36}{9}\right)^{1/6} = 4^{1/6} = (2^2)^{1/6} = 2^{1/3} = \sqrt[3]{2}$
7. $3^{4(2x-3)} = 3^{3(7-x)}$; $8x - 12 = 21 - 3x$; $11x = 33$; $x = 3$; $\{3\}$
8. $(5^{-2})^{(2x+1)} = (5^3)^{(x+4)}$; $-4x - 2 = 3x + 12$; $7x = -14$; $x = -2$; $\{-2\}$

Page 450 • WRITTEN EXERCISES

A 1. $x = 4y - 4$; $4y = x + 4$; $y = \frac{x+4}{4}$; $f^{-1}(x) = \frac{x+4}{4}$;

$f^{-1}(f(x)) = \frac{(4x-4)+4}{4} = \frac{4x}{4} = x$; $f(f^{-1}(x)) = 4\left(\frac{x+4}{4}\right) - 4 = x + 4 - 4 = x$

2. $x = 3y + 12;\ 3y = x - 12;\ y = \dfrac{x - 12}{3};\ f^{-1}(x) = \dfrac{x - 12}{3};$

$f^{-1}(f(x)) = \dfrac{3x + 12 - 12}{3} = \dfrac{3x}{3} = x;$

$f(f^{-1}(x)) = 3\left(\dfrac{x - 12}{3}\right) + 12 = x - 12 + 12 = x$

3. $x = -\frac{1}{3}y + 7;\ 3x = -y + 21;\ y = -3x + 21;\ f^{-1}(x) = -3x + 21;\ f^{-1}(f(x)) =$
$-3(-\frac{1}{3}x + 7) + 21 = x - 21 + 21 = x;$
$f(f^{-1}(x)) = -\frac{1}{3}(-3x + 21) + 7 = x - 7 + 7 = x$

4. $x = -\frac{1}{4}y - 5;\ 4x = -y - 20;\ y = -4x - 20;\ f^{-1}(x) = -4x - 20;$
$f^{-1}(f(x)) = -4(-\frac{1}{4}x - 5) - 20 = x + 20 - 20 = x;$
$f(f^{-1}(x)) = -\frac{1}{4}(-4x - 20) - 5 = x + 5 - 5 = x$

5. $x = \dfrac{y^3}{8};\ y^3 = 8x;\ y = \sqrt[3]{8x} = 2\sqrt[3]{x};\ f^{-1}(x) = 2\sqrt[3]{x};\ f^{-1}(f(x)) = 2\sqrt[3]{\dfrac{x^3}{8}} =$

$2\left(\dfrac{x}{2}\right) = x;\ f(f^{-1}(x)) = \dfrac{(2\sqrt[3]{x})^3}{8} = \dfrac{8x}{8} = x$

6. $x = \sqrt[3]{y + 1};\ x^3 = y + 1;\ y = x^3 - 1;\ f^{-1}(x) = x^3 - 1;$
$f^{-1}(f(x)) = (\sqrt[3]{x + 1})^3 - 1 = x + 1 - 1 = x;\ f(f^{-1}(x)) = \sqrt[3]{(x^3 - 1) + 1} = \sqrt[3]{x^3} = x$

7. $x = 3y + 2;\ 3y = x - 2;$

$y = \dfrac{x - 2}{3};\ f^{-1}(x) = \dfrac{x - 2}{3}$

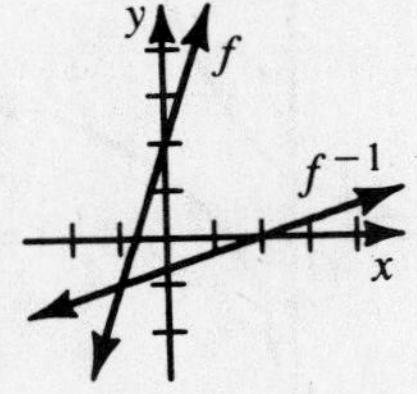

8. $x = \dfrac{6}{y};\ xy = 6;$

$y = \dfrac{6}{x};\ f^{-1}(x) = \dfrac{6}{x};\ x \neq 0$

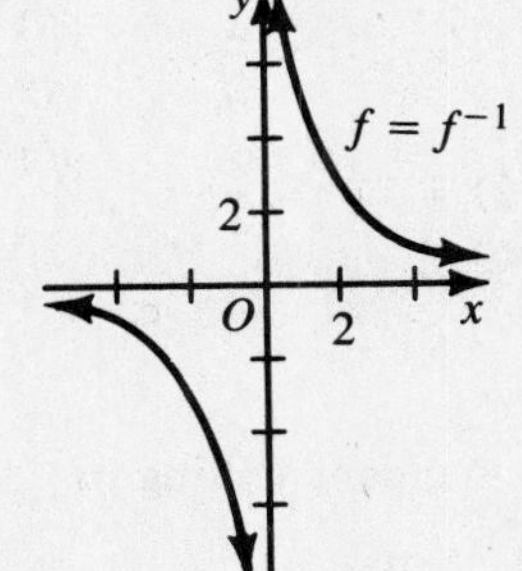

9. $x = y^3 - 1;\ y^3 = x + 1;$

$y = \sqrt[3]{x + 1};\ f^{-1}(x) = \sqrt[3]{x + 1}$

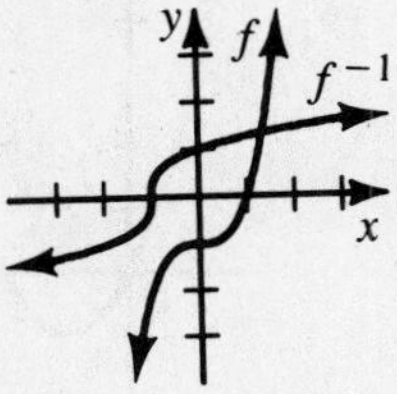

10. $x = y^2 + 3;\ y^2 = x - 3;$
$y = \sqrt{x - 3};$
$f^{-1}(x) = \sqrt{x - 3};\ x \geq 3$

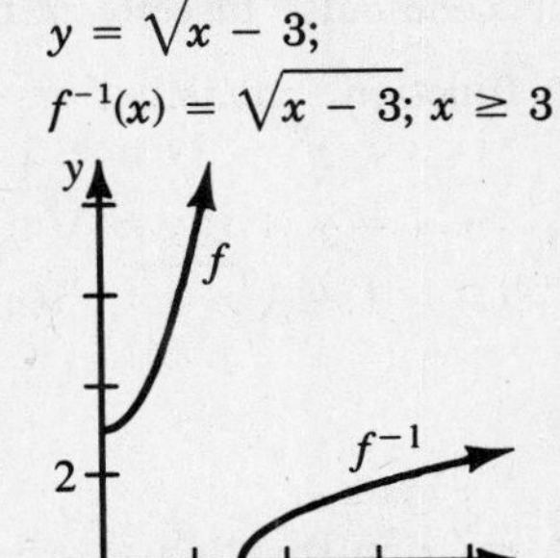

B **11.** $x = \sqrt{y^2 + 9}$; $x^2 = y^2 + 9$;
$y^2 = x^2 - 9$; $y = -\sqrt{x^2 - 9}$;
$f^{-1}(x) = -\sqrt{x^2 - 9}$; $x \geq 3$

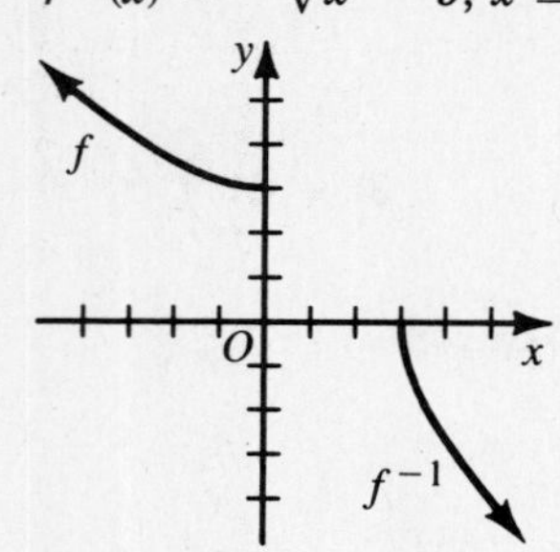

12. $x = \sqrt{y - 2}$; $x^2 = y - 2$; $y = x^2 + 2$;
$f^{-1}(x) = x^2 + 2$; $x \geq 0$

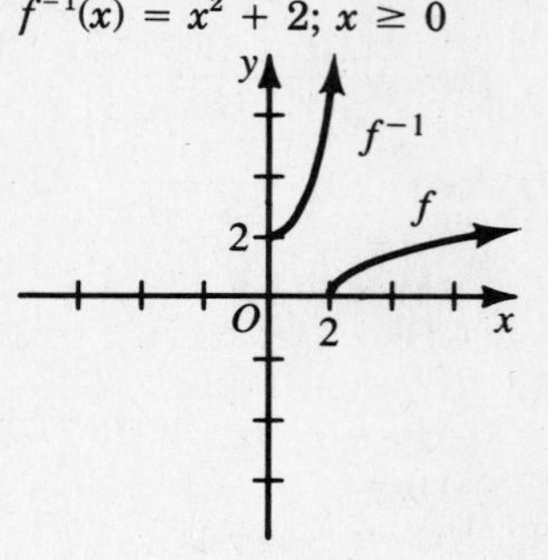

13. $x = 2\sqrt{y}$; $x^2 = 4y$; $y = \frac{x^2}{4}$; $f^{-1}(x) = \frac{x^2}{4}$; domain: $\{x: x \geq 0\}$;
$f^{-1}(f(x)) = \frac{(2\sqrt{x})^2}{4} = \frac{4x}{4} = x$ if $x \geq 0$; $f(f^{-1}(x)) = 2\sqrt{\frac{x^2}{4}} = 2\left(\frac{x}{2}\right) = x$ if $x \geq 0$

14. $x = \sqrt{y - 1}$; $x^2 = y - 1$; $y = x^2 + 1$; $f^{-1}(x) = x^2 + 1$; domain: $\{x: x \geq 0\}$;
$f^{-1}(f(x)) = (\sqrt{x - 1})^2 + 1 = x - 1 + 1 = x$ if $x \geq 1$; $f(f^{-1}(x)) = \sqrt{(x^2 + 1) - 1} =$
$\sqrt{x^2} = x$ if $x \geq 0$

15. $f(0) = 2^0 = 1$, $\therefore f^{-1}(1) = 0$; $f(1) = 2^1 = 2$, $\therefore f^{-1}(2) = 1$;
$f(2) = 2^2 = 4$, $\therefore f^{-1}(4) = 2$; $f(3) = 2^3 = 8$, $\therefore f^{-1}(8) = 3$;
$f(-1) = 2^{-1} = \frac{1}{2}$, $\therefore f^{-1}\left(\frac{1}{2}\right) = -1$;
$f(-2) = 2^{-2} = \frac{1}{4}$, $\therefore f^{-1}\left(\frac{1}{4}\right) = -2$

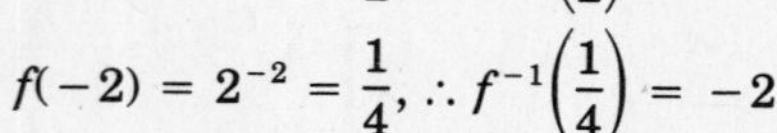

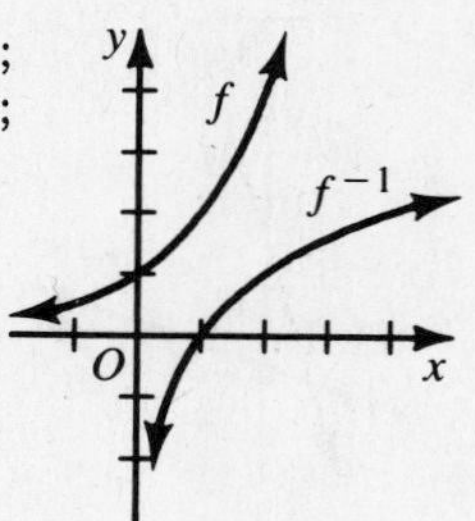

16. The segment joining (a, b) and (b, a) has slope $\frac{b - a}{a - b} = -1$ and midpoint $\left(\frac{a + b}{2}, \frac{a + b}{2}\right)$. Since $y = x$ passes through this point and has slope 1, it is the perpendicular bisector of the segment.

C **17.** A function that is not one-to-one; $\{x: x \geq 1\}$;
$f^{-1}(x)$: $x = y^2 - 2y$; $x + 1 = y^2 - 2y + 1$; $x + 1 = (y - 1)^2$;
$\sqrt{x + 1} = y - 1$; $y = \sqrt{x + 1} + 1$;
$f^{-1}(x) = 1 + \sqrt{x + 1}$; domain: $\{x: x \geq -1\}$

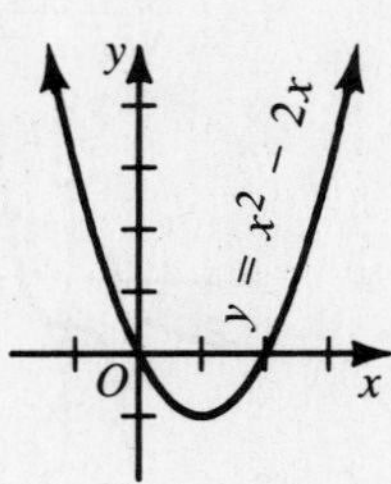

18. A relation that is not a function;
$\{(x, y): 0 \leq x \leq 3 \text{ and } 0 \leq y \leq 5\}$; $f^{-1}(x)$:
$25y^2 + 9x^2 = 225$; $25y^2 = 225 - 9x^2$; $y^2 = \frac{225 - 9x^2}{25}$;

$$y = \sqrt{\frac{225 - 9x^2}{25}} = \sqrt{\frac{9}{25}}\sqrt{25 - x^2};\ y = \frac{3}{5}\sqrt{25 - x^2};$$

$$f^{-1}(x) = \frac{3}{5}\sqrt{25 - x^2};\ \text{domain: } \{x\text{: } 0 \le x \le 5\}$$

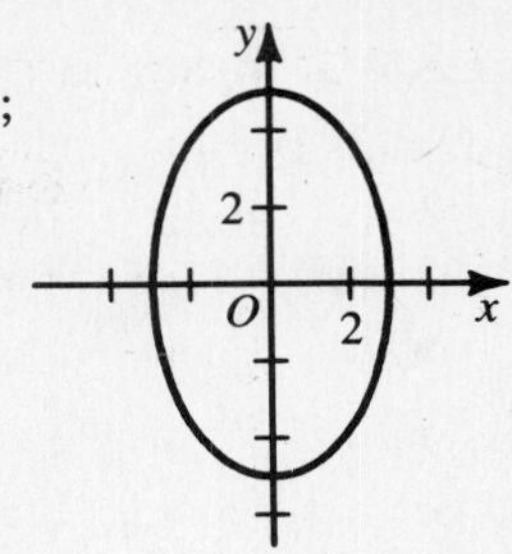

Page 453 • WRITTEN EXERCISES

A **1.** Let $\log_3 81 = x$; $3^x = 81$; $3^x = 3^4$; $x = 4$

2. Let $\log_{10} 0.0001 = x$; $10^x = 0.0001$; $10^x = 10^{-4}$; $x = -4$

3. Let $\log_2 \frac{1}{8} = x$; $2^x = \frac{1}{8}$; $2^x = \frac{1}{2^3}$; $x = -3$

4. Let $\log_7 1 = x$; $7^x = 1$; $x = 0$ **5.** Let $\log_{49} 7 = x$; $49^x = 7$; $49^x = 49^{1/2}$; $x = \frac{1}{2}$

6. Let $\log_{1/2} 128 = x$; $\left(\frac{1}{2}\right)^x = 128$; $\left(\frac{1}{2}\right)^x = 2^7$; $\left(\frac{1}{2}\right)^x = \left(\frac{1}{2}\right)^{-7}$; $x = -7$

7. Let $\log_{1/20} 8000 = x$; $\left(\frac{1}{20}\right)^x = 8000$; $\left(\frac{1}{20}\right)^x = 20^3$; $\left(\frac{1}{20}\right)^x = \left(\frac{1}{20}\right)^{-3}$; $x = -3$

8. Let $\log_{15} 15 = x$; $15^x = 15$; $x = 1$

9. Let $\log_{\sqrt{3}} 9 = x$; $(\sqrt{3})^x = 9$; $(3^{1/2})^x = 3^2$; $3^{x/2} = 3^2$; $\frac{x}{2} = 2$; $x = 4$

10. Let $\log_{25} 125 = x$; $25^x = 125$; $(5^2)^x = 5^3$; $5^{2x} = 5^3$; $2x = 3$; $x = 1.5$

11. Let $\log_{27} 9 = x$; $27^x = 9$; $(3^3)^x = 3^2$; $3x = 2$; $x = \frac{2}{3}$

12. Let $\log_{\sqrt{6}} 36\sqrt{6} = x$; $(6^{1/2})^x = 6^2 \cdot 6^{1/2}$; $6^{x/2} = 6^{5/2}$; $\frac{x}{2} = \frac{5}{2}$; $x = 5$

13. $x^6 = 64$; $x^6 = 2^6$; $x = 2$; $\{2\}$ **14.** $x^{1/2} = 16$; $(x^{1/2})^2 = 16^2$; $x = 256$; $\{256\}$

15. $8^3 = x$; $x = 512$; $\{512\}$ **16.** $1000^{-2/3} = x$; $x = \frac{1}{(10^3)^{2/3}} = \frac{1}{10^2} = \frac{1}{100} = 0.01$; $\{0.01\}$

17. $x^{4/3} = 81$; $x^{4/3} = 3^4$; $(x^{4/3})^{3/4} = (3^4)^{3/4}$; $x = 3^3 = 27$; $\{27\}$

18. $(x^{-2})^{-1/2} = (7)^{-1/2}$; $x = 7^{-1/2}$; $x = \frac{1}{\sqrt{7}} = \frac{\sqrt{7}}{7}$; $\left\{\frac{\sqrt{7}}{7}\right\}$

19. $(\sqrt{3})^x = \frac{1}{9}$; $(3^{1/2})^x = 3^{-2}$; $\frac{1}{2}x = -2$; $x = -4$; $\{-4\}$

20. $100^x = 1000$; $(10^2)^x = 10^3$; $2x = 3$; $x = 1.5$; $\{1.5\}$

21. $x^{1/4} = \sqrt{7}$; $(x^{1/4})^4 = (7^{1/2})^4$; $x = 7^2 = 49$; $\{49\}$

22. $x^6 = 125$; $(x^6)^{1/6} = (5^3)^{1/6}$; $x = 5^{1/2} = \sqrt{5}$; $\{\sqrt{5}\}$

23. $x^{-2/3} = \frac{1}{36}$; $(x^{-2/3})^{-3/2} = \left(\frac{1}{6^2}\right)^{-3/2}$; $x = 6^3 = 216$; $\{216\}$

24. $32^x = 8$; $(2^5)^x = 2^3$; $5x = 3$; $x = \frac{3}{5}$; $\left\{\frac{3}{5}\right\}$

B **25.** Let $\log_7 7^{10} = x$; $7^x = 7^{10}$; $x = 10$

26. Let $\log_3 9^7 = x$; $3^x = (3^2)^7 = 3^{14}$; $x = 14$

27. Let $\log_{1000} 10^{12} = x$; $(10^3)^x = 10^{12}$; $3x = 12$; $x = 4$

28. Let $\log_5 25^9 = x$; $5^x = (5^2)^9$; $x = 18$

29. $(4x)^2 = 36;\ 16x^2 = 36;\ x^2 = \frac{36}{16};\ x = \frac{6}{4} = \frac{3}{2};\ \left\{\frac{3}{2}\right\}$
30. $5^3 = 2x + 5;\ 2x = 125 - 5;\ 2x = 120;\ x = 60;\ \{60\}$
31. $2^{3x-1} = 32;\ 2^{3x-1} = 2^5;\ 3x - 1 = 5;\ 3x = 6;\ x = 2;\ \{2\}$
32. $3^4 = x^2 + 17;\ 81 = x^2 + 17;\ x^2 = 64;\ x = \pm 8;\ \{\pm 8\}$
33. $2^{3x+1} = 32;\ 2^{3x+1} = 2^5;\ 3x + 1 = 5;\ 3x = 4;\ x = \frac{4}{3};\ \left\{\frac{4}{3}\right\}$
34. $(x^3)^{2/3} = 64;\ x^2 = 64;\ x = \pm 8;\ \{8\}$

C **35.** Let $\log_2 16 = n;\ 2^n = 16;\ n = 4,\ \therefore \log_2 (\log_2 16) = \log_2 4$. Since $2^2 = 4$, $\log_2 4 = 2$. We have $\log_7 x = 2;\ 7^2 = x;\ x = 49;\ \{49\}$
36. Let $\log_9 3 = n;\ 9^n = 3;\ n = \frac{1}{2},\ \therefore \log_2 (\log_9 3) = \log_2 \frac{1}{2}$. Since $2^{-1} = \frac{1}{2}$, $\log_2 \frac{1}{2} = -1$. We have $\log_x 6 = -1;\ x^{-1} = 6;\ x = \frac{1}{6};\ \left\{\frac{1}{6}\right\}$
37. $\log_3 9 = 2;\ \log_2 2 = 1;\ \log_{10} 1 = 0;\ \therefore x = 0;\ \{0\}$
38. $\log_8 64 = 2;\ \log_4 2 = \frac{1}{2};\ \therefore \log_5 x = \frac{1}{2};\ 5^{1/2} = x;\ x = \sqrt{5};\ \{\sqrt{5}\}$

Page 456 • WRITTEN EXERCISES

A **1.** $\log_3 \frac{45 \cdot 6}{10} = \log_3 27 = 3$ **2.** $\log_6 4 \cdot 3^2 = \log_6 36 = 2$
3. $\log_3 \frac{144^{1/2}}{6^2} = \log_3 \frac{12}{36} = \log_3 \frac{1}{3} = -1$ **4.** $\log_2 \left(\frac{20}{5}\right)^2 = \log_2 4^2 = \log_2 (2^2)^2 = 4$
5. $\log_2 \left(\frac{12 \cdot 16}{3}\right)^{1/3} = \log_2 64^{1/3} = \log_2 (2^6)^{1/3} = 2$
6. $\log_5 \frac{16^{1/2}}{10^2} = \log_5 \frac{4}{100} = \log_5 \frac{1}{25} = -2$
7. $\log_7 x = \log_7 \frac{6^2}{4} = \log_7 9;\ x = 9;\ \{9\}$
8. $\log_{10} x = \log_{10} (8^{1/3} \cdot 81^{1/2}) = \log_{10} (2 \cdot 9) = \log_{10} 18;\ x = 18;\ \{18\}$
9. $\log_5 x = \log_5 64^{2/3} = \log_5 (2^6)^{2/3} = \log_5 2^4 = \log_5 16;\ x = 16;\ \{16\}$
10. $\log_3 x^2 = \log_3 \frac{8 \cdot 10}{5} = \log_3 16;\ x^2 = 16;\ x = \pm 4;\ \{\pm 4\}$
11. $\log_{10} x^2 = \log_{10} (12 \cdot 3) = \log_{10} 36;\ x^2 = 36;\ x = \pm 6;\ \{6\}$
12. $\log_3 x = \log_3 4 + 2 \log_3 5 = \log_3 (4 \cdot 5^2) = \log_3 100;\ x = 100;\ \{100\}$
13. $\log_2 x = \log_2 3 + \log_2 10 - \log_2 6 = \log_2 \frac{3 \cdot 10}{6} = \log_2 5;\ x = 5;\ \{5\}$
14. $\log_7 x^{1/2} = \log_7 \frac{20}{2^2 \cdot 5^2} = \log_7 \frac{20}{100} = \log_7 \frac{1}{5};\ (x^{1/2})^2 = \left(\frac{1}{5}\right)^2;\ x = \frac{1}{25};\ \left\{\frac{1}{25}\right\}$
15. $\log_b x^2 = \log_b 45 - \log_b 5 = \log_b \frac{45}{5} = \log_b 9;\ x^2 = 9;\ x = \pm 3;\ \{3\}$
16. $\log_5 x^{1/2} = \log_5 2 - \log_5 3 = \log_5 \frac{2}{3};\ (x^{1/2})^2 = \left(\frac{2}{3}\right)^2;\ x = \frac{4}{9};\ \left\{\frac{4}{9}\right\}$
17. $9^{\log_9 5} = 5$ **18.** $5^{\log_5 2 \cdot 3} = 5^{\log_5 6} = 6$
19. $(5^2)^{\log_5 7} = 5^{2 \log_5 7} = 5^{\log_5 7^2} = 5^{\log_5 49} = 49$ **20.** $2^{\log_2 27^{(1/3)}} = 2^{\log_2 3} = 3$

B **21.** $\log_b 5 - \log_b 3 = 2.2 - 1.5 = 0.7$
22. $\log_b 0.6 = \log_b \frac{3}{5} = \log_b 3 - \log_b 5 = 1.5 - 2.2 = -0.7$
23. $\log_b (3^2 \cdot 5) = 2 \log_b 3 + \log_b 5 = 2(1.5) + 2.2 = 5.2$

24. $\log_b 5^3 = 3 \log_b 5 = 3(2.2) = 6.6$

25. $\log_b 5^{1/2} = \frac{1}{2} \log_b 5 = \frac{1}{2}(2.2) = 1.1$

26. $\log_b 3^{7/3} = \frac{7}{3} \log_b 3 = \frac{7}{3}(1.5) = 3.5$

27. $\log_b 3 + 2 \log_b b = 1.5 + 2(1) = 1.5 + 2 = 3.5$

28. $\log_b b - (\log_b 3 + \log_b 5) = 1 - (1.5 + 2.2) = 1 - 3.7 = -2.7$

29. $\log_{10} x(x + 3) = 1$; $\log_{10} (x^2 + 3x) = 1$; $10^1 = x^2 + 3x$; $x^2 + 3x - 10 = 0$; $(x + 5)(x - 2) = 0$; $x = 2$ or -5; $\{2\}$

30. $\log_6 x(x + 5) = 2$; $x^2 + 5x = 6^2$; $x^2 + 5x - 36 = 0$; $(x - 4)(x + 9) = 0$; $x = 4$ or -9; $\{4\}$

31. $\log_2 \frac{x + 6}{x - 1} = 3$; $\frac{x + 6}{x - 1} = 2^3$; $x + 6 = 8x - 8$; $7x = 14$; $x = 2$; $\{2\}$

32. $\log_3 \frac{2x^2}{5x - 9} = 1$; $3^1 = \frac{2x^2}{5x - 9}$; $15x - 27 = 2x^2$; $2x^2 - 15x + 27 = 0$; $(2x - 9)(x - 3) = 0$; $x = 3$ or 4.5; $\{3, 4.5\}$

33. $\log_8 \frac{x^2 - 1}{3x + 9} = 0$; $8^0 = \frac{x^2 - 1}{3x + 9}$; $1 = \frac{x^2 - 1}{3x + 9}$; $3x + 9 = x^2 - 1$; $x^2 - 3x - 10 = 0$; $(x - 5)(x + 2) = 0$; $x = 5$ or -2; $\{5, -2\}$

34. $\log_7 \frac{x^3 + 27}{x + 3} = 2$; $7^2 = \frac{x^3 + 27}{x + 3} = x^2 - 3x + 9$; $x^2 - 3x - 40 = 0$; $(x - 8)(x + 5) = 0$; $x = 8$ or -5; $\{8\}$

C 35.

1. $x_1 = b^{\log_b x_1}$, $x_2 = b^{\log_b x_2}$	Hypothesis; def., p. 451
2. $\frac{x_1}{x_2} = \frac{b^{\log_b x_1}}{b^{\log_b x_2}}$	Subs. prin.
3. $\frac{x_1}{x_2} = b^{\log_b x_1 - \log_b x_2}$	Laws of exponents
4. $\log_b \frac{x_1}{x_2} = \log_b x_1 - \log_b x_2$	Def., p. 451

36.

1. $x_1 = b^{\log_b x_1}$	Def., p. 451
2. $x_1^n = (b^{\log_b x_1})^n$	Subs. prin.
3. $(b^{\log_b x_1})^n = b^{n \cdot \log_b x_1}$	Laws of exponents
4. $x_1^n = b^{\log_b x_1^n}$	Def., p. 451
5. $b^{\log_b x_1^n} = b^{n \cdot \log_b x_1}$	Subs. prin.
6. $\log_b x_1^n = n \cdot \log_b x_1$	Def., p. 446

Pages 456–457 • SELF-TEST 2

1. $x = \frac{1}{3}y + 1$; $\frac{1}{3}y = x - 1$; $y = 3x - 3$; $f^{-1}(x) = 3x - 3$

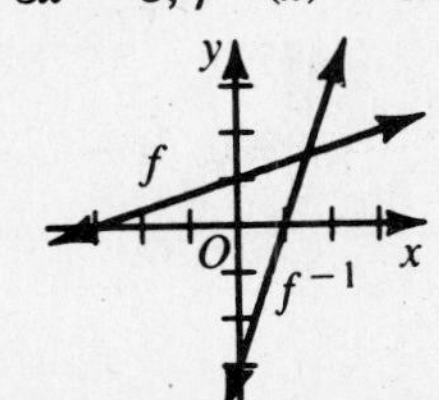

2. $x = y^2 - 2$; $y^2 = x + 2$; $y = \sqrt{x + 2}$; $f^{-1}(x) = \sqrt{x + 2}$; $x \geq -2$

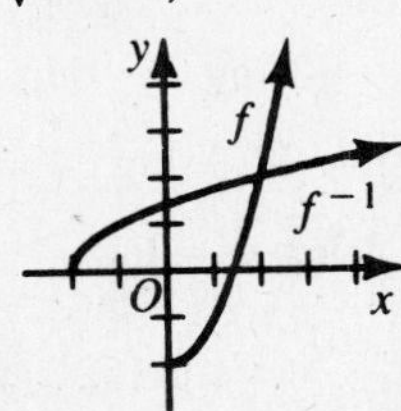

3. $\log_{1/2} 64 = -6$ **4.** $\log_{36} 6 = \frac{1}{2}$ **5.** $5^{-3} = \frac{1}{125}$ **6.** $10^3 = 1000$

7. Let $\log_4 256 = x$. $4^x = 256$; $4^x = 4^4$; $x = 4$

8. Let $\log_{27} 81 = x$; $27^x = 81$; $3^{3x} = 3^4$; $3x = 4$; $x = \frac{4}{3}$

9. $(x^{1/4})^4 = (\sqrt{2})^4$; $x = (2^{1/2})^4 = 2^2 = 4$; {4}

10. $\left(\frac{1}{125}\right)^{-4/3} = x$; $x = (125)^{4/3} = (5^3)^{4/3} = 5^4 = 625$; {625}

11. $\log_5 x = \log_5 \frac{4^3}{64^{1/3}} = \log_5 \frac{64}{4} = \log_5 16$; $x = 16$; {16}

12. $\log_{12} x^2 = \log_{12} (9^{1/2} \cdot 3^3) = \log_{12} (3 \cdot 27) = \log_{12} 81$; $x^2 = 81$; $x = \pm 9$; {9}

Pages 460–461 • WRITTEN EXERCISES

A **1.** **a.** 1.3729 **b.** 4.3729 **c.** 0.3729 − 3 or −2.6271 **d.** log 1 − log 2.36 = 0 − 0.3729 = −0.3729

2. **a.** 2.8814 **b.** 0.8814 − 2 or −1.1186 **c.** 5.8814 **d.** log 100 − log 76.1 = 2 − 1.8814 = 0.1186

3. **a.** 2560 **b.** 256,000 **c.** 0.256

4. **a.** 841 **b.** 0.0841 **c.** 0.00119

5. $\log x = \log 3.62 + \log 10{,}000 = \log (3.62)(10{,}000) = \log 36{,}200$; $x = 36{,}200$; {36,200}

6. $\log x = \log 2.69 - 1 = \log 2.69 - \log 10 = \log \frac{2.69}{10} = \log 0.269$; $x = 0.269$; {0.269}

7. $x = \log 7.520 - \log 7.52 = \log \frac{7520}{7.52} = \log 1000 = 3$; {3}

8. $x = \log \frac{831}{10^5} - \log 8.31 = \log \frac{831}{10^5 \cdot 8.31} = \log \frac{100}{10^5} = \log \frac{1}{10^3} = -3$; {−3}

9. $x = \log \frac{9.23}{0.00923} = \log 1000 = 3$; {3}

10. $x = \log 852 - \log 0.0852 = \log \frac{852}{0.0852} = \log 10{,}000 = 4$; {4}

11. $2 \log 5 + \log 3 = 2(0.6990) + 0.4771 = 1.8751$

12. $2(\log 3 - \log 2) = 2(0.4771 - 0.3010) = 0.3522$

13. $\frac{1}{2}(\log 144 - 2 \log 4) = \frac{1}{2}(2.1584 - 2(0.6021)) = 0.4771$

14. $\log 12 - \frac{1}{3} \log 8 = 1.0792 - \frac{1}{3}(0.9031) = 0.7782$

15. $\frac{1}{4} \log 81 - \log 15 = \frac{1}{4}(1.9085) - 1.1761 = -0.6990$

16. $\log 56 - \frac{1}{2} \log 64 = 1.7482 - \frac{1}{2}(1.8062) = 0.8451$

17. $\frac{1}{3}(2 \log 36 - (3 \log 3 + \log 6)) = \frac{1}{3}(2(1.5563) - (3(0.4771) + 0.7782)) = 0.3010$

18. $2 \log 7 - (\log 98 + \log 5) = 2(0.8451) - (1.9912 + 0.6990) = -1$

B **19.** $x^5 = \frac{47.9}{8.3}$; $5 \log x = \log \frac{47.9}{8.3}$; $\log x = \frac{\log 47.9 - \log 8.3}{5} = \frac{1.6803 - 0.9191}{5} = 0.1522$; $x = 1.42$; {1.42}

20. $x^4 = \frac{58}{634}$; $4 \log x = \log \frac{58}{634}$; $\log x = \frac{\log 58 - \log 634}{4} =$
$\frac{1.7634 - 2.8021}{4} = -0.2597$; $x = 0.550$; $\{0.550\}$

21. $135x^3 = 4.72$; $x^3 = \frac{4.72}{135}$; $3 \log x = \log \frac{4.72}{135}$;
$\log x = \frac{\log 4.72 - \log 135}{3} = \frac{0.6739 - 2.1303}{3} = -0.4855$;
$x = 0.327$; $\{0.327\}$

22. $\frac{2}{3} \log x = \log \frac{726}{68.1}$; $\log x = \frac{3}{2}(\log 726 - \log 68.1) = \frac{3}{2}(2.8609 - 1.8331) = 1.5417$;
$x = 34.8$; $\{34.8\}$

23. $\sqrt[3]{87}\,\sqrt[3]{x^5} = 391$; $x^{5/3} = \frac{391}{\sqrt[3]{87}}$; $\frac{5}{3} \log x = \log 391 - \frac{1}{3} \log 87$;
$\log x = \frac{3}{5}\left(\log 391 - \frac{1}{3} \log 87\right) = \frac{3}{5} \log 391 - \frac{1}{5} \log 87 =$
$1.5553 - 0.3879 = 1.1674$; $x = 14.7$; $\{14.7\}$

24. $x^{4/3}(11.6) = 2700$; $x^{4/3} = \frac{2700}{11.6}$; $\frac{4}{3} \log x = \log \left(\frac{2700}{11.6}\right)$; $\log x =$
$\frac{3}{4}(\log 2700 - \log 11.6) = \frac{3}{4}(3.4314 - 1.0645) = 1.7752$; $x = 59.6$; $\{59.6\}$

25. **a.** $\log \frac{2}{10} = \log 2 - \log 10 = 0.3010 - 1 = -0.6990$

b. $\log \frac{10}{2} = \log 10 - \log 2 = 1 - 0.3010 = 0.6990$

c. $\log \frac{1}{4} = \log 1 - \log 2^2 = \log 1 - 2 \log 2 = 0 - 2(0.3010) = -0.6020$

d. $\log \frac{10 \cdot 3}{2} = \log 10 + \log 3 - \log 2 = 1 + 0.4771 - 0.3010 = 1.1761$

26. $7^2 \approx 48 = 3 \cdot 2^4$; $\log 7^2 \approx \log (3 \cdot 2^4)$; $2 \log 7 \approx \log 3 + 4 \log 2$;
$\log 7 \approx \frac{0.4771 + 4(0.3010)}{2} \approx 0.8406$
$11^2 \approx 120 = 10 \cdot 2^2 \cdot 3$; $\log 11^2 \approx \log (10 \cdot 2^2 \cdot 3)$; $2 \log 11 \approx \log 10 + \log 2^2 + \log 3$;
$\log 11 \approx \frac{1 + 2(0.3010) + 0.4771}{2} \approx 1.0396$

27. $\log 13^3 \approx \log 3^7$; $3 \log 13 \approx 7 \log 3$; $\log 13 \approx \frac{7 \log 3}{3} \approx \frac{7(0.4771)}{3} \approx 1.1132$

28. $\log (3 \cdot 5 \cdot 17) \approx \log 2^8$; $\log 3 + \log 5 + \log 17 \approx 8 \log 2$;
$\log 17 \approx 8 \log 2 - \log 3 - \log 5 \approx 8(0.3010) - 0.4771 - 0.6990 = 1.2319$

C **29.** **a.** $10 \log \left(\frac{I_0}{I_0}\right) = 10 \log 1 = 0$; 0 decibels **b.** $10 \log \frac{10^{14} I_0}{I_0} = 10 \log 10^{14} =$
$10 \cdot 14 = 140$; 140 decibels **c.** $120 = 10 \log \frac{I}{I_0}$; $12 = \log \frac{I}{I_0}$; $10^{12} = \frac{I}{I_0}$;
$I = 10^{12} I_0$; it is 10^{12} times more intense.

30. $20 = 10 \log \left(\frac{I_w}{I_0}\right)$; $2 = \log \left(\frac{I_w}{I_0}\right)$; $10^2 = \frac{I_w}{I_0}$; $I_0 = \frac{I_w}{10^2}$
$60 = 10 \log \left(\frac{I_c}{I_0}\right)$; $6 = \log \left(\frac{I_c}{I_0}\right)$; $10^6 = \frac{I_c}{I_0}$; $I_0 = \frac{I_c}{10^6}$

$\frac{I_c}{10^6} = \frac{I_w}{10^2}$; $I_c = \frac{10^6 I_w}{10^2} = 10^4 I_w$; 10,000 times more intense

31. The intensity of one bell, I_1, is: $70 = 10 \log \frac{I_1}{I_0}$; $\log \frac{I_1}{I_0} = 7$; $10^7 = \frac{I_1}{I_0}$; $I_1 = 10^7 I_0$. The intensity of 2 bells is twice I_1 or $2 \times 10^7 I_0$, so the decibel level is: $10 \log \frac{2 \times 10^7 I_0}{I_0} = 10 \log (2 \times 10^7) = 10(\log 2 + \log 10^7) = 10 \log 2 +$ $10 \log 10^7 = 10(0.3010) + 10 \cdot 7 = 3.010 + 70 = 73.01$ or about 73 decibels.

32. The intensity, I_1, of one flute is: $41 = 10 \log \frac{I_1}{I_0}$; $4.1 = \log \frac{I_1}{I_0}$; $10^{4.1} = \frac{I_1}{I_0}$; $I_1 = 10^{4.1} I_0$; 3 flutes will have intensity $3I_1$ or $3 \times 10^{4.1} I_0$. The decibel level is $10 \log \frac{3 \times 10^{4.1} I_0}{I_0} = 10 \log (3 \times 10^{4.1}) = 10(\log 3 + \log 10^{4.1}) =$ $10(0.4771 + 4.1) = 10(4.5771) = 45.771$ or about 46 decibels.

Page 463 • WRITTEN EXERCISES

A

1. a. $x \log 1.59 = \log 4.02$; $x = \frac{\log 4.02}{\log 1.59}$ **b.** {3.00}

2. a. $-x \log 7.4 = \log 18.6$; $x = -\frac{\log 18.6}{\log 7.4}$ **b.** {−1.46}

3. a. $\frac{x}{2} \log 19 = \log 145$; $x = \frac{2 \log 145}{\log 19}$ **b.** {3.38}

4. a. $3x \log 54 = \log 19$; $x = \frac{\log 19}{3 \log 54}$ **b.** {0.246}

5. a. $4x \log 9.12 = \log 46$; $x = \frac{\log 46}{4 \log 9.12}$ **b.** {0.433}

6. a. $\frac{x}{8} \log 128 = \log 33.7$; $x = \frac{8 \log 33.7}{\log 128}$ **b.** {5.80}

7. a. $2^x = \frac{50.1}{6}$; $x \log 2 = \log 50.1 - \log 6$; $x = \frac{\log 50.1 - \log 6}{\log 2}$ **b.** {3.06}

8. a. $5^{2x} = \frac{29.3}{0.76}$; $2x \log 5 = \log 29.3 - \log 0.76$; $x = \frac{\log 29.3 - \log 0.76}{2 \log 5}$ **b.** {1.13}

9. a. $3^{-x} = \frac{4780}{8.35}$; $-x \log 3 = \log 4780 - \log 8.35$; $x = \frac{\log 8.35 - \log 4780}{\log 3}$
b. {−5.78}

10. a. $(8.2)^{x/2} = \frac{132}{4.1}$; $\frac{x}{2} \log 8.2 = \log 132 - \log 4.1$; $x = \frac{2(\log 132 - \log 4.1)}{\log 8.2}$
b. {3.30}

11. a. $6^x = \frac{122}{\sqrt[3]{25.3}}$; $x \log 6 = \log 122 - \frac{1}{3} \log 25.3$; $x = \frac{\log 122 - \frac{1}{3} \log 25.3}{\log 6}$
b. {2.08}

12. a. $7^{x/3} = \frac{96}{\sqrt[5]{203}}$; $\frac{x}{3} \log 7 = \log 96 - \frac{1}{5} \log 203$; $x = \frac{3(\log 96 - \frac{1}{5} \log 203)}{\log 7}$
b. {5.40}

13. a. $9^{-x} = \frac{18}{15^{3/4}}$; $-x \log 9 = \log 18 - \frac{3}{4}\log 15$; $x = \frac{-\log 18 + \frac{3}{4}\log 15}{\log 9}$

b. $\{-0.391\}$

14. a. $(x + 1)\log 12 = \frac{1}{4}\log 56.6$; $x + 1 = \frac{\log 56.6}{4 \log 12}$; $x = \frac{\log 56.6}{4 \log 12} - 1$

b. $\frac{1.7528}{4.3168} - 1 = -0.594$; $\{-0.594\}$

15. a. $(x - 2)\log 0.027 = \frac{1}{3}\log 94$; $x - 2 = \frac{\log 94}{3 \log 0.027}$; $x = \frac{\log 94}{3 \log 0.027} + 2$

b. $\frac{1.9731}{-4.7059} + 2 = 1.58$; $\{1.58\}$

16. $\frac{\log 11}{\log 5} \approx 1.49$ **17.** $\frac{\log 6.2}{\log 23} \approx 0.582$ **18.** $\frac{\log 85.9}{\log 7.3} \approx 2.24$

B 19. a. $x \log 3 = (x - 1)\log 5$; $x \log 3 = x \log 5 - \log 5$; $x \log 3 - x \log 5 = -\log 5$;

$x(\log 5 - \log 3) = \log 5$; $x = \frac{\log 5}{\log 5 - \log 3}$ **b.** $\{3.15\}$

20. a. $(x - 1)\log 3 = (x + 1)\log 2$; $x \log 3 - \log 3 = x \log 2 + \log 2$;

$x(\log 3 - \log 2) = \log 3 + \log 2$; $x = \frac{\log 3 + \log 2}{\log 3 - \log 2}$ **b.** $\{4.42\}$

21. a. $2x \log 6.7 = (x + 1)\log 15$; $2x \log 6.7 = x \log 15 + \log 15$;

$x(2 \log 6.7 - \log 15) = \log 15$; $x = \frac{\log 15}{2 \log 6.7 - \log 15}$ **b.** $\{2.47\}$

22. a. $(3x - 1)\log 6 = x \log 28$; $3x \log 6 - x \log 28 = \log 6$;

$x(3 \log 6 - \log 28) = \log 6$; $x = \frac{\log 6}{3 \log 6 - \log 28}$ **b.** $\{0.877\}$

23. a. $(2x - 1)\log 8 = (x + 1)\log 39$; $2x \log 8 - \log 8 = x \log 39 + \log 39$;

$x(2 \log 8 - \log 39) = \log 8 + \log 39$; $x = \frac{\log 8 + \log 39}{2 \log 8 - \log 39}$ **b.** $\{11.6\}$

24. a. $(x - 1)\log 42 = (3x - 1)\log 17$; $x \log 42 - \log 42 = 3x \log 17 - \log 17$;

$x(\log 42 - 3 \log 17) = \log 42 - \log 17$; $x = \frac{\log 42 - \log 17}{\log 42 - 3 \log 17}$ **b.** $\{-0.190\}$

C 25. $\log_{25} 2 = \frac{\log_5 2}{\log_5 25} = \frac{\log_5 2}{2 \log_5 5} = \frac{\log_5 2}{2}$

26. $\log_a 9 = \frac{\log 3^2}{\log a} = \frac{2 \log 3}{\log a}$

27. $\log_b x = \frac{\log x}{\log b} = \frac{2 \log x}{2 \log b} = \frac{2 \log x}{\log b^2} = 2 \log_{b^2} x$

28. $\log_b x = \frac{\log x}{\log b} = \frac{n \log x}{n \log b} = \frac{n \log x}{\log b^n} = n \log_{b^n} x$

29. $\log_{ab} x = \frac{\log_a x}{\log_a ab} = \frac{\log_a x}{\log_a a + \log_a b} = \frac{\log_a x}{1 + \log_a b}$

30. $\log_{a/b} x = \frac{\log_a x}{\log_a \frac{a}{b}} = \frac{\log_a x}{\log_a a - \log_a b} = \frac{\log_a x}{1 - \log_a b}$

31. $\log_x y = \frac{\log_a y}{\log_a x} = \frac{\log_b y}{\log_b x}$; $(\log_a y)(\log_b x) = (\log_a x)(\log_b y)$

Pages 466–467 • PROBLEMS

A **1.** $A = 5000\left(1 + \frac{0.08}{4}\right)^{4 \cdot 15} = 5000(1 + 0.02)^{60} = 16{,}405$; approximately \$16,400

2. $3P = P\left(1 + \frac{0.08}{2}\right)^{2t}$; $3 = (1 + 0.04)^{2t}$; $\log 3 = 2t \log 1.04$; $t = \frac{\log 3}{2 \log 1.04} = 14.0$; 14.0 yr

3. $56 = 88\left(\frac{1}{2}\right)^{3/k}$; $\frac{56}{88} = \left(\frac{1}{2}\right)^{3/k}$; $\log 56 - \log 88 = \frac{3}{k} \log \frac{1}{2}$; $k(\log 56 - \log 88) = 3 \log \frac{1}{2}$; $k = \frac{3 \log \frac{1}{2}}{\log 56 - \log 88} = 4.6$; approximately 4.6 h

4. $7.61 \times 10^5 = 3.2 \times 10^5 \cdot 2^{t/24}$; $2^{t/24} = \frac{7.61 \times 10^5}{3.2 \times 10^5} = \frac{7.61}{3.2}$; $\frac{t}{24} \log 2 = \log 7.61 - \log 3.2$; $t = \frac{24(\log 7.61 - \log 3.2)}{\log 2} = 30.0$; approximately 30 min

5. $3470 = 1000 \cdot 1.05^{t/1}$; $\frac{3470}{1000} = (1.05)^t$; $t \log 1.05 = \log 3.47$; $t = \frac{\log 3.47}{\log 1.05} = 25.5$; approximately 25.5 yr

6. $12{,}000 = 2000(b^{1/k})^{40}$; $(b^{1/k})^{40} = \frac{12{,}000}{2000} = 6$; $40 \log b^{1/k} = \log 6$; $\log b^{1/k} = \frac{\log 6}{40} = 0.01945$; $b^{1/k} = 1.0458$; $45{,}000 = 12{,}000(1.0458)^t$; $(1.0458)^t = \frac{45{,}000}{12{,}000} = 3.75$; $t \log 1.0458 = \log 3.75$; $t = \frac{\log 3.75}{\log 1.0458} = 29.5$; approximately 29.5 yr

B **7.** $0.0568\ N_0 = N_0 \cdot (0.62)^{t/1}$; $0.0568 = 0.62^t$; $\log 0.0568 = t \log 0.62$; $t = \frac{\log 0.0568}{\log 0.62} = 6$; approximately 6 s

8. $0.045\ N_0 = N_0\left(\frac{1}{2}\right)^{t/5730}$; $0.045 = \left(\frac{1}{2}\right)^{t/5730}$; $\log 0.045 = \frac{t}{5730} \log \frac{1}{2}$; $t = \frac{(\log 0.045)5730}{\log 0.5} = 25{,}600$; approximately 25,600 yr

C **9.** $653 = \frac{(0.01)(62{,}000)}{1 - (1 + 0.01)^{-n}}$; $653 - 653(1.01)^{-n} = 620$; $(1.01)^{-n} = \frac{620 - 653}{-653} = \frac{33}{653}$; $-n \log 1.01 = \log 33 - \log 653$; $n = \frac{\log 33 - \log 653}{-\log 1.01} = 300$; approximately 300 mo or 25 yr

10. $2.21 = 50\left(\frac{1}{2}\right)^{t/4.8}$; $\frac{2.21}{50} = \left(\frac{1}{2}\right)^{t/4.8}$; $\log 2.21 - \log 50 = \frac{t}{4.8} \log \frac{1}{2}$; $t = \frac{4.8(\log 2.21 - \log 50)}{\log 0.5} = 21.6$; approximately 21.6 km

Pages 467–468 • COMPUTER EXERCISES

1.
```
10 INPUT R, N
20 LET P = 1000
30 LET AM = 2 * P
40 LET I = P * (R/100) * (1/N)
50 LET M = M + 12/N
```

```
60 LET P = P + I
70 IF P < AM THEN GOTO 40
80 LET Y = M/12
90 PRINT Y;"YEARS"
100 END
```

2. 12 yr **3.** 9 yr **4.** 7.25 yr **5.** 4.7 yr

6.
```
10 INPUT P, AM, R, N
20 LET I = P * (R/100) * (1/N)
30 LET M = M + 12/N
40 LET P = P + I
50 IF P < AM THEN GOTO 20
60 LET Y = M/12
70 PRINT Y;"YEARS"
80 END
```

7. 7.75 yr **8.** 9 yr **9.** 7.75 yr

Page 471 • WRITTEN EXERCISES

A **1.** $\ln 12.18 = 2.5$ **2.** $\ln 0.135 = -2$ **3.** $\ln 1.221 = \frac{1}{5}$ **4.** $\ln 1.649 = \frac{1}{2}$

5. $e^{2.079} = 8$ **6.** $e^{-0.693} = 0.5$ **7.** $e^{-2.303} = 0.1$ **8.** $e^{-4.605} = 0.01$

9. $\ln 48 - \ln 2^4 = \ln \frac{48}{16} = \ln 3$

10. $\ln 9^{1/2} + \ln 12 - \ln 3^2 = \ln \frac{3 \cdot 12}{9} = \ln 4$

11. $\ln 45^{1/2} + \ln 5^{1/2} - \ln 3^2 = \ln \frac{\sqrt{45} \cdot \sqrt{5}}{9} = \ln \frac{15}{9} = \ln \frac{5}{3}$

12. $\ln (6 \cdot 30) - \ln 5 - \ln 2^3 = \ln \frac{6 \cdot 30}{5 \cdot 8} = \ln \frac{9}{2}$

13. $e^{\ln 6} \cdot e^{\ln 7} = 6 \cdot 7 = 42$ **14.** $e^{\ln 7^2} = e^{\ln 49} = 49$ **15.** $\frac{e^{\ln 8}}{e^{\ln 6}} = \frac{8}{6} = \frac{4}{3}$

16. $e^{\ln 3^{1/2}} = e^{\ln \sqrt{3}} = \sqrt{3}$

17. $A = 7000e^{0.08} = 7000(1.083) = 7581$; \$7581

18. $A = 4500e^{0.11} = 4500(1.116) = 5022$; \$5022

19. $A = 5500e^{0.13} = 5500(1.139) = 6264.5$; \$6264.50

20. $A = 2200e^{0.09} = 2200(1.094) = 2406.8$; \$2406.80

21. $A = 6000e^{(0.07)2} = 6000(1.150) = 6900$; \$6900

22. $A = 7500e^{(0.06)2.5} = 7500(1.162) = 8715$; \$8715

B **23.** $10 \ln x = \ln 3$; $\ln x = \frac{1.099}{10} = 0.1099$; $x \approx 1.12$; $\{1.12\}$

24. $20 \ln x = \ln 6$; $\ln x = \frac{\ln 6}{20} = \frac{1.792}{20} = 0.0896$; $x \approx 1.09$; $\{1.09\}$

25. $44 \ln x = \ln 9$; $\ln x = \frac{2.197}{44} \approx 0.050$; $x \approx 1.05$; $\{1.05\}$

26. $33 \ln x = \ln 10$; $\ln x = \frac{2.3}{33} \approx 0.070$; $x \approx 1.07$; $\{1.07\}$

27.

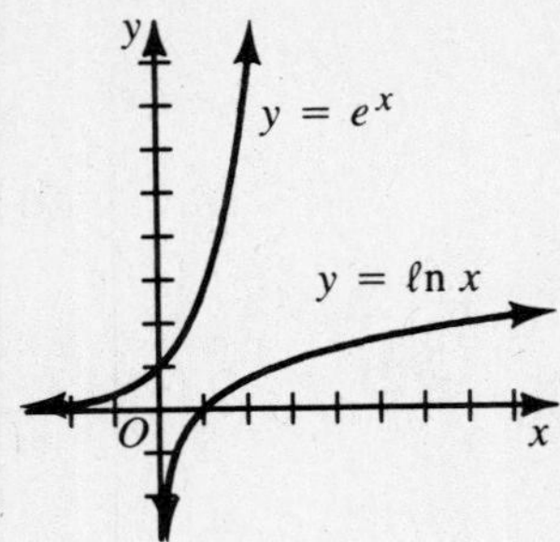

C **28.** $\lim_{k\to\infty}\left(1+\frac{1}{k}\right)^{k+1} = \lim_{k\to\infty}\left(1+\frac{1}{k}\right)^{k} \cdot \lim_{k\to\infty}\left(1+\frac{1}{k}\right)^{1} = e \cdot 1 = e$

29. $\lim_{n\to\infty}\left(1-\frac{1}{n}\right)^{n} = \lim_{n\to\infty}\left(\frac{n-1}{n}\right)^{n}$; let $k = n - 1$, then $k \to \infty$ as $n \to \infty$;

$\lim_{k\to\infty}\left(\frac{k}{k+1}\right)^{k+1} = \lim_{k\to\infty}\left(\left(\frac{k+1}{k}\right)^{k+1}\right)^{-1} = \frac{1}{\lim_{k\to\infty}\left(\frac{k+1}{k}\right)^{k+1}} = \frac{1}{e}$

Page 472 • COMPUTER EXERCISES

1.
```
10 INPUT N
30 LET D = 1
40 LET S = 1
50 FOR I = 1 TO N - 1
60 LET D = I * D
70 LET S = S + 1/D
80 NEXT I
90 PRINT S
100 END
```

2. approx. $2.708\overline{3}$ **3.** approx. 2.718254 **4.** approx. 2.7182815
5. approx. 2.7182818

Page 472 • ON THE CALCULATOR

1. Let r_1 = annual rate, r_2 = effective rate. Since $t = 1$, $Pe^{r_1} = P(1 + r_2)$; $e^{r_1} = 1 + r_2$; $r_2 = e^{r_1} - 1$. **a.** $e^{0.07} - 1 = 0.072508$; 7.2508% **b.** $e^{0.08} - 1 = 0.083287$; 8.3287% **c.** $e^{0.11} - 1 = 0.116278$; 11.6278% **d.** $e^{0.14} - 1 = 0.150274$; 15.0274%

2. Let r_1 = annual rate, r_2 = effective rate. Since $t = 1$, $Pe^{r_1} = P(1 + r_2)$; $e^{r_1} = 1 + r_2$; $r_1 \ln e = \ln (1 + r_2)$; $r_1 = \ln(1 + r_2)$. **a.** $\ln (1 + 0.08) = 0.076961$; 7.6961% **b.** $\ln (1 + 0.09) = 0.086178$; 8.6178% **c.** $\ln (1 + 0.14) = 0.131028$; 13.1028% **d.** $\ln (1 + 0.105) = 0.099845$; 9.9845%

3. 9.3083%; 9.5758%; 9.8438%; 10.1123%; 10.3813%
9.3807%; 9.6524%; 9.9248%; 10.1977%; 10.4713%
9.4174%; 9.6913%; 9.9659%; 10.2411%; 10.5171%

4. $2P = Pe^{5r}$; $2 = e^{5r}$; $\ln 2 = 5r \ln e = 5r$; $5r = 0.69315$; $r = 0.138629$; 13.8629%

5. $3P = Pe^{9r}$; $e^{9r} = 3$; $9r \ln e = \ln 3$; $9r = \ln 3$; $r = \frac{1.0986}{9} = 0.122068$ or 12.2068%

Page 473 • SELF-TEST 3

1. $0.1818 - 3$ or -2.8182

2. $\frac{1}{2}(\log 48 + \log 12 - 2 \log 30) = \frac{1}{2}(1.6812 + 1.0792 - 2(1.4771)) = -0.0969$

3. $x^3 = \frac{13.4}{5.2}$; $3 \log x = \log 13.4 - \log 5.2$; $\log x = \frac{\log 13.4 - \log 5.2}{3}$;
$\log x = \frac{1.1271 - 0.7160}{3} = \frac{0.4111}{3} = 0.1370$; $x = 1.37$; $\{1.37\}$

4. $\frac{2}{5} \log x = \log 421 - \log 26.3$; $\log x = \frac{5}{2}(\log 421 - \log 26.3)$;
$\log x = \frac{5(1.2043)}{2} = 3.0108$; $x = 1025.22 \approx 1030$; $\{1030\}$

5. **a.** $(2.9)^{2x} = \frac{\sqrt[3]{34.6}}{4}$; $2x \log 2.9 = \frac{1}{3} \log 34.6 - \log 4$; $x = \frac{\frac{1}{3}\log 34.6 - \log 4}{2 \log 2.9}$
b. $-\frac{0.08903}{0.9248} = -0.0963$; $\{-0.0963\}$

6. **a.** $(x - 1)\log 7 = 4x \log 3$; $x \log 7 - 4x \log 3 = \log 7$; $x(\log 7 - 4 \log 3) = \log 7$;
$x = \frac{\log 7}{\log 7 - 4 \log 3}$ **b.** $\frac{0.8451}{-1.0634} = -0.795$; $\{-0.795\}$

7. $3N_0 = N_0(1.05)^{t/1}$; $3 = (1.05)^t$; $t \log 1.05 = \log 3$; $t = \frac{\log 3}{\log 1.05} = 22.5$;
approximately 22.5 yr

8. $11{,}500 = 5000\left(1 + \frac{0.08}{4}\right)^{4t}$; $\frac{11{,}500}{5000} = (1.02)^{4t}$; $\log 2.3 = 4t \log 1.02$;
$t = \frac{\log 2.3}{4 \log 1.02} = 10.5$; approximately 10.5 yr

9. $\frac{e^{\ln 6}}{e^{\ln 18}} = \frac{6}{18} = \frac{1}{3}$ 10. $\ln \frac{6 \cdot \frac{2}{3}}{\frac{1}{8} \cdot 32} = \ln \frac{4}{4} = \ln 1 = 0$

11. $A = 6000e^{(0.08)(1.5)} = 6000e^{(0.12)} = 6764.98$; approximately \$6765

Pages 474–475 • CHAPTER REVIEW

1. $\sqrt[5]{4\sqrt[10]{32}} = (2^2(2^5)^{1/10})^{1/5} = (2^2 \cdot 2^{1/2})^{1/5} = (2^{5/2})^{1/5} = 2^{1/2} = \sqrt{2}$; b

2. $3^{2x+3} = \left(\frac{1}{27}\right)^{4+x}$; $3^{2x+3} = (3^{-3})^{4+x}$; $3^{2x+3} = 3^{-12-3x}$; $2x + 3 = -12 - 3x$; $5x = -15$;
$x = -3$; $\{-3\}$; d

3. 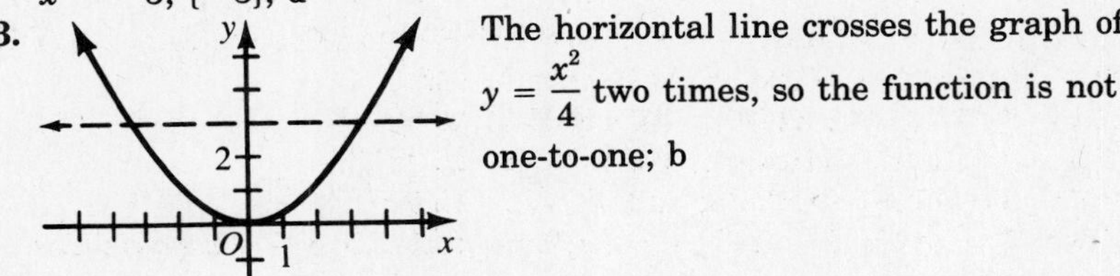

The horizontal line crosses the graph of $y = \frac{x^2}{4}$ two times, so the function is not one-to-one; b

4. Let $x = \log_{\sqrt{5}} 25$, therefore $(\sqrt{5})^x = 25$; $(5^{1/2})^x = 5^2$; $\frac{x}{2} = 2$; $x = 4$; c

5. $\log_x \frac{2}{5} = \frac{1}{3}$; $x^{1/3} = \frac{2}{5}$; $(x^{1/3})^3 = \left(\frac{2}{5}\right)^3$; $x = \frac{8}{125}$; $\left\{\frac{8}{125}\right\}$; c

6. $\frac{1}{3}\log_8 8 - 4\log_8 2 = \log_8 8^{1/3} - \log_8 2^4 = \log_8 2 - \log_8 16 = \log_8 \frac{2}{16} =$
$\log_8 \frac{1}{8} = \log_8 8^{-1} = -1$; d

7. $\log_2 (x^2) - \log_2 (x + 3) = 2$; $\log_2 \frac{x^2}{x+3} = 2$; $2^2 = \frac{x^2}{x+3}$; $4(x + 3) = x^2$;
$x^2 - 4x - 12 = 0$; $(x - 6)(x + 2) = 0$; $x = 6$ or $x = -2$; $\{6, -2\}$; a

8. $\log\left(\frac{20\sqrt{3.24}}{\sqrt[3]{1000}}\right) = \log (2 \cdot 10(3.24)^{1/2}) - \log ((10^3)^{1/3}) = \log 2 + \log 10 +$
$\log (3.24)^{1/2} - \log 10 = \log 2 + \frac{1}{2}\log 3.24 = 0.3010 + \frac{1}{2}(0.5105) = 0.55625$;
0.5563; a

9. $8^{1+x} = 5^{3-2x}$; $(1 + x)\log 8 = (3 - 2x)\log 5$; $\log 8 + x\log 8 =$
$3\log 5 - 2x\log 5$; $x\log 8 + 2x\log 5 = 3\log 5 - \log 8$; $x(\log 8 + 2\log 5) =$
$3\log 5 - \log 8$; $x = \frac{3\log 5 - \log 8}{(\log 8 + 2\log 5)} = 0.519$; c

10. $A = 2600\left(1 + \frac{0.075}{4}\right)^{4 \cdot 9} = 5074.67$; approx. \$5070; a

11. $\frac{1}{3}(\ln 48 - \frac{1}{2}\ln 36) + \ln 3 = \frac{1}{3}(\ln 48 - \ln 36^{1/2}) + \ln 3 = \frac{1}{3}(\ln 48 - \ln 6) +$
$\ln 3 = \frac{1}{3}\ln\frac{48}{6} + \ln 3 = \ln 8^{1/3} + \ln 3 = \ln 2 + \ln 3 = \ln 6$; d

Page 475 • CHAPTER TEST

1. $\frac{(5^3)^{1/15}}{(5^2)^{1/10}} = \frac{5^{1/5}}{5^{1/5}} = 1$ **2.** $(2^6)^{1/8} \cdot (2^6)^{1/12} = 2^{3/4} \cdot 2^{1/2} = 2^{5/4} = \sqrt[4]{2^5} = 2\sqrt[4]{2}$

3. $(7^{-1})^{1-3x} = (7^2)^{x+3}$; $-1 + 3x = 2x + 6$; $x = 7$; $\{7\}$

4. $x = \sqrt{3 - y}$; $x^2 = 3 - y$; $f^{-1}(x) = 3 - x^2$; $x \geq 0$
(Graph at right)

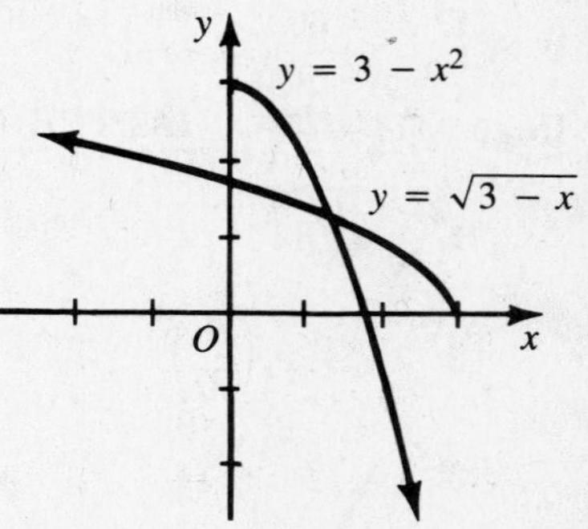

5. $9^{3/2} = x$; $x = 3^3 = 27$; $\{27\}$

6. $x^{-5} = 32$; $x^{-5} = 2^5$; $x = \frac{1}{2}$; $\left\{\frac{1}{2}\right\}$

7. $\log_3 2x = \log_3 \frac{16}{8} = \log_3 2$;
$2x = 2$; $x = 1$; $\{1\}$

8. $x(x - 3) = 4^1$; $x^2 - 3x = 4$; $x^2 - 3x - 4 = 0$; $(x - 4)(x + 1) = 0$; $x = -1, 4$; $\{4\}$

9. $x^9 = 6$; $9\log x = \log 6$; $\log x = \frac{\log 6}{9}$; $x = 1.22$; $\{1.22\}$

10. $\frac{1}{4}\log 31 + \log x^{2/4} = \log 103$; $\frac{2}{4}\log x = \log 103 - \frac{1}{4}\log 31$;

$\log x = \dfrac{\log 103 - \frac{1}{4}\log 31}{\frac{1}{2}} = 3.2800$; $x = 1910$; $\{1910\}$

11. $3x \log 2 = \log 34$; $x = \dfrac{\log 34}{3 \log 2} = 1.6958$; $\{1.70\}$

12. $(x - 1)\log 72 = x \log 19$; $x \log 72 - \log 72 = x \log 19$; $x(\log 72 - \log 19) =$ $\log 72$; $x = \dfrac{\log 72}{\log 72 - \log 19} = 3.2102$; $\{3.21\}$

13. $21{,}000 = 12{,}000(b^{1/k})^{15}$; $\dfrac{21{,}000}{12{,}000} = (b^{1/k})^{15}$; $\log 1.75 = 15 \log b^{1/k}$;

$\log b^{1/k} = \dfrac{\log 1.75}{15} = 0.0162$; $b^{1/k} = 1.038$; $36{,}750 = 21{,}000(1.038)^t$; $1.75 = (1.038)^t$;

$\log 1.75 = t \log 1.038$; $t = \dfrac{\log 1.75}{\log 1.038} = 15$; $15 + 1982 = 1997$; 1997

14. $1326 = 1200e^r$; $\dfrac{1326}{1200} = e^r$; $r \ln e = \ln 1.105$; $r = \ln 1.105 = 0.0998$; 10%

Page 478 • APPLICATION

1. $\dfrac{N}{N_0} = e^{-\lambda t}$; $-\lambda t \ln e = \ln \dfrac{N}{N_0}$; $-\lambda t = \ln N - \ln N_0$;

$t = \dfrac{\ln N - \ln N_0}{-\lambda}$ or $\dfrac{\ln N_0 - \ln N}{\lambda}$

2. $e^{-\lambda t} = b^{t/k}$; $-\lambda t \ln e = \dfrac{t}{k}\ln b$; $-\lambda t = \dfrac{t}{k}\ln b$; $-\lambda = \dfrac{\ln b}{k}$

3. $0.6\, N_0 = N_0\, e^{-\lambda t}$; $0.6 = e^{-\lambda t}$; $-\lambda t = \ln 0.6$; $t = \dfrac{0.5108}{1.21 \times 10^{-4}} = 4221$; approximately 4220 yr

4. No. The rate of decay from any living organism is 15.3 nuclei per minute for 1 g of carbon (see Example 2, p. 478). $\dfrac{13.9}{15.3} = e^{-\lambda t}$; $\ln\left(\dfrac{13.9}{15.3}\right) = -\lambda t$;

$t = -\dfrac{1}{1.21 \times 10^{-4}} \ln\left(\dfrac{13.9}{15.3}\right) \approx 793$ $\therefore$ the sandal was made less than 800 years ago.

Page 479 • PREPARING FOR COLLEGE ENTRANCE EXAMS

1. $\sqrt{(x - y)^2 + (z - z)^2} = \sqrt{(x - y)^2} = |x - y|$; C

2. The line must have slope $\frac{1}{2}$, which is the negative reciprocal of

$\dfrac{-3 + 11}{4 - 8} = -\dfrac{8}{4} = -2$, and go through the point $\left(\dfrac{4 + 8}{2}, \dfrac{-3 - 11}{2}\right)$ or $(6, -7)$.

$\dfrac{1}{2} = \dfrac{y + 7}{x - 6}$; $x - 6 = 2y + 14$; $x - 2y = 20$; A

3. Vertex is $(0, 1)$, focus is $(0, 3)$; E

4. Axis of symmetry is $x = -\dfrac{6}{2} = -3$, so I is false;

$f(x) = x^2 + 6x + 9 - 9 + 7 = (x + 3)^2 - 2$; so the vertex is $(-3, -2)$ and II is true; since $a > 0$, the graph has a minimum point and III is false; B

5. Let $x = y = 1$, $\sqrt{1^2 + 1^2} = \sqrt{2} \not> 1 + 1$, so I is false. Let $x = y = 0$, $\sqrt{0 + 0} = 0 + 0$, so II is false. $(x^2 + y^2) \le x^2 + 2xy + y^2$ is true for all nonnegative x and y, so $\sqrt{x^2 + y^2} \le \sqrt{x^2 + 2xy + y^2} = x + y$ and III is true; C
6. $a_1 = (-1)^1(3)(1) = -3$, so I is false; since $3n \to \infty$ as $n \to \infty$, II is true; because $(-1)^n$ is alternately + and −, III is false; B
7. $x(x - 3) = 10^1$; $x^2 - 3x - 10 = 0$; $(x - 5)(x + 2) = 0$; $x = 5, -2$; $\{5\}$; A
8. I is true; $\log \sqrt{x} = \frac{1}{2} \log x$, so II is false; $\log x^2 = 2 \log x$, not $2 \log \sqrt{x}$, so III is false; A

Pages 480–481 • PROGRAMMING IN PASCAL

1.
```
PROGRAM table (INPUT, OUTPUT);

VAR
  n : 1..20;

BEGIN
  writeln('n', 'exp(n)':10, 'ln(n)':18);
  writeln(n, exp(n):1:6, ln(n):1:6);
END.
```

2. a. Let $a^x = y$. Then $\ln a^x = \ln y$ and $x \ln a = \ln y$. Therefore, $e^{x \ln a} = e^{\ln y} = y$ and $a^x = e^{x \ln a}$. The only restriction is $a > 0$.

   b.
```
FUNCTION power (a, x : real) : real;

BEGIN
  power := exp(x * ln(a));
END;
```

   c.
```
PROGRAM natural (INPUT, OUTPUT);

VAR
  base, exponent : real;

{***********************************************************}
FUNCTION power (a, x : real) : real;

BEGIN
  power := exp(x * ln(a));
END;

{***********************************************************}
BEGIN {* main *}
  REPEAT
    write('Enter the value of the base: ');
    readln(base);
    IF base <= 0
```

```
        THEN BEGIN
               writeln;
               writeln('ILLEGAL BASE. RE-ENTER.');
             END;
    UNTIL base > 0;
    write('Enter the value of the exponent: ');
    readln(exponent);
    writeln;
    write(base:1:6, 'raised to the ', exponent:1:6, 'is: ');
    writeln(power(base, exponent):1:6);
  END.
```

3. {12.464458} **4. a.** When $x > 0$, $e^x \geq x^e$. In fact, the two expressions are equal only when $x = e$.

b. Program follows.

```
PROGRAM compare (INPUT, OUTPUT);

VAR
  increment, x, y, k : real;
  place : integer;
  done : boolean;

{***************************************************************}
FUNCTION power (a, x : real) : real;

BEGIN
  power := exp(x * ln(a));
END;

{***************************************************************}
BEGIN {* main *}
  increment := 10;
  place := 1;
  k := 2;

{* The value of x can't be less than 0 because x^y is not a     *}
{* continuous function and is not defined for all values of y. *}
{* The value of x can't be less than 1 since x^y would be less *}
{* than 1, which suggests that y would have to be less than 0; *}
{* again this would lead to a function, y^x, that was not      *}
{* continuous as well as defined for all values of x.          *}

  REPEAT
    write('Enter a value for x greater than 1: ');
    readln(x);
    writeln;
```

[*Program continued on next page*]

```
    IF x <= 1
      THEN BEGIN
             writeln('ILLEGAL VALUE');
             writeln('RE-ENTER');
           END;
    writeln;
  UNTIL x > 1;

{* x^y = y^x implies y ln x = x ln y, which implies y/ln y =   *}
{* x/ln x. Thus, there must exist a k such that y = kx and     *}
{* ln y = k ln x. Continuing, y = kx implies ln y = ln k +     *}
{* ln x, which implies ln y - ln x = ln k. Substitute k ln x   *}
{* for ln y to obtain k ln x - ln x = ln k. This simplifies    *}
{* to (k - 1) ln x = ln k, or ln x = ln k/(k - 1). Finally,    *}
{* we get x = k^(1/(k - 1)). When k is found, then multiply    *}
{* by x to get the value of y.                                 *}

  REPEAT
    done := FALSE;
    increment := 0.1 * increment;
    IF power(k, 1/(k - 1)) > x
      THEN REPEAT
             k := k + increment;
           UNTIL power(k, 1/(k - 1)) <= x
      ELSE IF power(k, 1/(k - 1)) < x
             THEN REPEAT
                    k := k - increment;
                    IF k = 1
                      THEN done := TRUE
                      ELSE IF power(k, 1/(k - 1)) >= x
                             THEN done := TRUE;
                  UNTIL done;
    place := place + 1;
  UNTIL place = 6;
  y := k * x;
  writeln('(', x:1:5, ', ', y:1:5, ')');
END.
```

c. `PROGRAM z_seq (INPUT, OUTPUT);`

```
VAR
  z, prev_term, term : real;
  converge : boolean;

{*********************************************************}
FUNCTION power (x, n : real) : real;

BEGIN
  power := exp(n * ln(x));
END;
```

```
{****************************************************************}
BEGIN {* main *}

  converge := FALSE;
  REPEAT
    write('Enter the value of the first term: ');
    readln(z);
    IF z <= 0
      THEN BEGIN
             writeln;
             writeln('Value of z must be greater than 0');
             writeln('RE-ENTER');
             writeln;
           END;
  UNTIL z > 0;
  prev_term := z;
  REPEAT
    term := power(z, prev_term);
    IF abs(prev_term - term) < 0.00001
      THEN converge := TRUE;
    prev_term := term;
  UNTIL converge;
  IF converge
    THEN writeln('Sequence converges at: ', prev_term:1:5);
END.
```

0.06 doesn't converge and it may cause the system to "hang" when it is entered; 0.1 converges to 0.39901; 0.25 converges to 0.5; 0.4 converges to 0.58504; 0.77 converges to 0.80934.

Pages 485–486 • WRITTEN EXERCISES

A **1.** 5×5 or 25 **2.** $8 \times 9 \times 9 = 648$ **3.** $6 \times 5 \times 4 \times 3 = 360$
4. $2 \times 2 \times 2 \times \cdots \times 2 = 2^8 = 256$
5. $76 - 3 = 52 + 45 - t$; $73 = 97 - t$; $t = 24$; 24 teach both
6. $375 = 215 + 193 - t$; $t = 33$, so $215 - 33 = 182$; 182 own a dog and no cat
7. $9 \times 8 = 72$
8. $26 \times 26 \times 10 \times 10 \times 10 \times 10 = 6{,}760{,}000$

B **9.** $26 \times 25 \times [10 \times 10 \times 10 \times 10 - 1] = 650 \times 9999 = 6{,}499{,}350$
10. $[8 \times 10 \times 10 - 3] \times 10 \times 10 \times 10 \times 10 = 797 \times 10^4 = 7{,}970{,}000$
11. They must be 1, 2 or 3-digit numbers that end in 1, 3 or 5: $7 \times 7 \times 3 = 147$
12. The hundreds' digit must be 4, 5, or 6. If 4 is the hundreds' digit, the tens' digit can be only 5, 6, 7, or 8: $1 \times 4 \times 6 + 2 \times 6 \times 6 = 24 + 72 = 96$
13. $1 \times 4 \times 3 + 2 \times 6 \times 3 = 12 + 36 = 48$

C **14.** First, consider numbers in which all three digits are the same and their sum is a multiple of 3: 111, 444, 555, 666, 888. There are 5 such numbers. Second, consider numbers composed of different sets of digits such that two of the digits are the same and their sum is a multiple of 3: 114, 441, 558, 885. The digits in each of the preceding numbers can be arranged in two other ways for a total of 3 arrangements. Therefore, there are $3 \cdot 4 = 12$ numbers in which two of the digits are identical and the sum of the digits is a multiple of 3. Finally, consider the numbers composed of different sets of digits such that all three digits are different and their sum is a multiple of 3: 156, 168, 456, 468. The digits in each of the preceding numbers can be arranged in 5 other ways for a total of 6 arrangements. Therefore, there are $6 \cdot 4 = 24$ numbers in which all three of the digits are different and their sum is a multiple of 3. To find all the specified multiples of 3 add: $5 + 12 + 24 = 41$

Page 489 • WRITTEN EXERCISES

A **1.** **a.** $9! = 362{,}880$ **b.** $8 \cdot 7 \cdot 6 \cdot 5 \cdot 1 \cdot 4 \cdot 3 \cdot 2 \cdot 1 = 8! = 40{,}320$
2. $1 \cdot 7 \cdot 6 \cdot 1 \cdot 5 \cdot 4 \cdot 3 \cdot 2 \cdot 1 = 7! = 5040$
3. $7! = 5040$ **4.** ${}_8P_3 = 8 \cdot 7 \cdot 6 = 336$ **5.** ${}_{10}P_6 = 10 \cdot 9 \cdot 8 \cdot 7 \cdot 6 \cdot 5 = 151{,}200$
6. ${}_{13}P_4 = 13 \cdot 12 \cdot 11 \cdot 10 = 17{,}160$ **7.** ${}_9P_8 = 9 \cdot 8 \cdot 7 \cdot 6 \cdot 5 \cdot 4 \cdot 3 \cdot 2 = 362{,}880$
8. ${}_9P_5 = 9 \cdot 8 \cdot 7 \cdot 6 \cdot 5 = 15{,}120$ **9.** $7! = 5040$ **10.** $8! = 40{,}320$
11. $\frac{6!}{2} = \frac{720}{2} = 360$ **12.** $\frac{7!}{2} = 2520$

B **13.** $[3 \cdot 2 \cdot 1] \cdot 5 \cdot 4 \cdot 3 \cdot 2 \cdot 1 = 720$ **14.** $4 \cdot 6 \cdot 5 \cdot 4 \cdot 3 \cdot 2 \cdot 1 \cdot 3 = 8640$
15. **a.** $2 \cdot 6! = 1440$ **b.** $7! - 2 \cdot 6! = 5040 - 1440 = 3600$
16. $7({}_6P_3) \stackrel{?}{=} {}_7P_4$; $7(6 \cdot 5 \cdot 4) \stackrel{?}{=} 7 \cdot 6 \cdot 5 \cdot 4$; $840 = 840$
17. $3({}_7P_4) \stackrel{?}{=} {}_7P_5$; $3(7 \cdot 6 \cdot 5 \cdot 4) \stackrel{?}{=} 7 \cdot 6 \cdot 5 \cdot 4 \cdot 3$; $2520 = 2520$
18. ${}_7P_3 \stackrel{?}{=} \frac{7!}{4!}$; $7 \cdot 6 \cdot 5 \stackrel{?}{=} \frac{7 \cdot 6 \cdot 5 \cdot 4 \cdot 3 \cdot 2 \cdot 1}{4 \cdot 3 \cdot 2 \cdot 1}$; $7 \cdot 6 \cdot 5 \stackrel{?}{=} 7 \cdot 6 \cdot 5$; $210 = 210$
19. $({}_7P_4)({}_3P_3) \stackrel{?}{=} {}_7P_7$; $(7 \cdot 6 \cdot 5 \cdot 4)(3 \cdot 2 \cdot 1) \stackrel{?}{=} 7 \cdot 6 \cdot 5 \cdot 4 \cdot 3 \cdot 2 \cdot 1$; $5040 = 5040$

C **20.** $n({}_{n-1}P_{r-1}) = n((n-1)(n-2) \cdots ((n-1) - (r-1) + 1)) = n(n-1)(n-2) \cdots (n - 1 - r + 1 + 1) = n(n-1)(n-2) \cdots (n - r + 1) = {}_nP_r$
21. $(n - r){}_nP_r = (n - r)n(n-1)(n-2) \cdots (n - r + 1) = n(n-1)(n-2) \cdots (n - r + 1)(n - r) = n(n-1)(n-2) \cdots (n - r + 1)(n - (r + 1) + 1) = {}_nP_{r+1}$

22. ${}_nP_{n-r} = n(n-1)(n-2) \cdots (n-(n-r)+1) = n(n-1)(n-2) \cdots$
$(n-n+r+1) = n(n-1)(n-2) \cdots (r+1) =$
$\frac{n(n-1)(n-2)\cdots(r+1)}{1} \cdot \frac{r(r-1)(r-2)\cdots 1}{r(r-1)(r-2)\cdots 1} = \frac{n!}{r!}$

23. $({}_nP_r)({}_{n-r}P_{n-r}) = n(n-1)(n-2) \cdots (n-r+1)(n-r)(n-r-1) \cdots$
$(n-r-n+r+1) = n(n-1)(n-2) \cdots (n-r+1)(n-r)(n-r-1) \cdots$
$(1) = {}_nP_n$

24. ${}_nP_r - {}_nP_{r-1} = n(n-1)(n-2) \cdots (n-r+1) - n(n-1)(n-2) \cdots (n-r+2) =$
$n(n-1)(n-2) \cdots (n-r+2)\,[(n-r+1)-1] = n(n-1)(n-2) \cdots$
$(n-r+2)(n-r) = (n-r)(n(n-1)(n-2) \cdots (n-(r-1)+1)) =$
$(n-r){}_nP_{r-1}$

25. $({}_nP_r)({}_{n-r}P_s) = n(n-1)(n-2) \cdots (n-r+1)(n-r)(n-r-1) \cdots$
$(n-r-s+1) = n(n-1)(n-2) \cdots (n-(r+s)+1) = {}_nP_{r+s}$

Page 491 • WRITTEN EXERCISES

A **1.** $\frac{6!}{2! \cdot 2!} = 180$ **2.** $\frac{8!}{2! \cdot 4!} = 840$ **3.** $\frac{7!}{3! \cdot 2!} = 420$ **4.** $\frac{7!}{3! \cdot 2! \cdot 2!} = 210$

5. $\frac{10!}{2! \cdot 2! \cdot 2!} = 453{,}600$ **6.** $\frac{11!}{2! \cdot 2! \cdot 2! \cdot 2!} = 2{,}494{,}800$ **7.** $\frac{10!}{2! \cdot 3! \cdot 2!} = 151{,}200$

8. $\frac{12!}{2! \cdot 4! \cdot 2!} = 4{,}989{,}600$ **9.** $\frac{6!}{2!} = 360$ **10.** $\frac{7!}{3! \cdot 3!} = 140$

B **11.** $\frac{10!}{3! \cdot 2!} = 302{,}400$ **12.** $\frac{10!}{2! \cdot 2! \cdot 4!} = 37{,}800$

13. $\frac{(5-1)!}{2!} = 12$ **14.** $\frac{(6-1)!}{3!} = 20$

C **15.** **a.** $5! = 120$ **b.** Pick the two w's. There are 4 ways to pick the other 3 letters to make up a 5-letter sequence. For each possible sequence, there are $\frac{5!}{2!} = 60$ distinguishable permutations. Thus, there are $4(60) = 240$ distinguishable permutations. **c.** $120 + 240 = 360$

Page 491 • SELF-TEST 1

1. $5 \cdot 5 \cdot 3 = 75$ **2.** $6! = 720$ **3.** ${}_6P_4 = 6 \cdot 5 \cdot 4 \cdot 3 = 360$ **4.** $\frac{10!}{2! \cdot 3!} = 302{,}400$

Pages 494–495 • WRITTEN EXERCISES

A **1.** ${}_{14}C_4 = \frac{14 \cdot 13 \cdot 12 \cdot 11}{4 \cdot 3 \cdot 2 \cdot 1} = 1001$ **2.** ${}_{10}C_6 = \frac{10 \cdot 9 \cdot 8 \cdot 7 \cdot 6 \cdot 5}{6 \cdot 5 \cdot 4 \cdot 3 \cdot 2 \cdot 1} = 210$

3. ${}_9C_2 = \frac{9 \cdot 8}{2 \cdot 1} = 36$ **4.** ${}_{13}C_2 = \frac{13 \cdot 12}{2 \cdot 1} = 78$ **5.** ${}_7C_2 = \frac{7 \cdot 6}{2 \cdot 1} = 21$

6. ${}_{10}C_3 = \frac{10 \cdot 9 \cdot 8}{3 \cdot 2 \cdot 1} = 120$ **7.** To score 80%, 8 answers must be correct. ${}_{10}C_8 =$
${}_{10}C_2 = \frac{10 \cdot 9}{2 \cdot 1} = 45$ **8.** ${}_{12}C_6 = \frac{12 \cdot 11 \cdot 10 \cdot 9 \cdot 8 \cdot 7}{6 \cdot 5 \cdot 4 \cdot 3 \cdot 2 \cdot 1} = 924$

B **9.** Choose the other 5 committee members from the remaining 14 club members.
${}_{14}C_5 = \frac{14 \cdot 13 \cdot 12 \cdot 11 \cdot 10}{5 \cdot 4 \cdot 3 \cdot 2 \cdot 1} = 2002$

10. Total no. of lines between all vertices: ${}_{10}C_2 = \frac{10 \cdot 9}{2 \cdot 1} = 45$. Since 10 lines are sides (not diagonals), no. of diagonals $= 45 - 10 = 35$.

11. No. of committees with president, but not v.p.: ${}_9C_4 = \frac{9 \cdot 8 \cdot 7 \cdot 6}{4 \cdot 3 \cdot 2 \cdot 1} = 126$; no. of committees with v.p., but not president: ${}_9C_4 = 126$; no. of committees with neither president nor v.p.: ${}_9C_5 = \frac{9 \cdot 8 \cdot 7 \cdot 6 \cdot 5}{5 \cdot 4 \cdot 3 \cdot 2 \cdot 1} = 126$. Total no. of committees $= 3(126) = 378$

12. No. of ways of choosing 3 vertices from the total of 10 points $= {}_{10}C_3 = \frac{10 \cdot 9 \cdot 8}{3 \cdot 2 \cdot 1} = 120$. No. of ways of choosing 3 vertices from the 6 collinear points $= {}_6C_3 = \frac{6 \cdot 5 \cdot 4}{3 \cdot 2 \cdot 1} = 20$. No. of triangles formed from the 10 points $= 120 - 20 = 100$.

13. ${}_5C_1 + {}_5C_2 + {}_5C_3 + {}_5C_4 + {}_5C_5 = 5 + 10 + 10 + 5 + 1 = 31$

14. **a.** Since ${}_nC_r = {}_nC_{n-r}$, $r = 2$, so $n - r = 98$ and $n = 100$. **b.** Since ${}_nC_4 = {}_nC_3$, $r = 4$, so $n - r = 3$, and $n - 4 = 3$; $n = 7$

15. ${}_nC_2 = 105$; $\frac{n(n-1)}{2 \cdot 1} = 105$; $n(n-1) = 210$; $n^2 - n - 210 = 0$; $(n - 15)(n + 14) = 0$; $n = 15$ or $n = -14$; 15 guests

16. ${}_nC_1 + {}_nC_2 = 21$; $\frac{n}{1} + \frac{n(n-1)}{2 \cdot 1} = 21$; $2n + n^2 - n = 42$; $n^2 + n - 42 = 0$; $(n + 7)(n - 6) = 0$; $n = 6$ or $n = -7$; 6 kinds of meat

17. ${}_nC_r = \frac{n!}{r!(n-r)!}$; ${}_nC_{n-r} = \frac{n!}{(n-r)![n-(n-r)]!} = \frac{n!}{(n-r)!r!}$; $\frac{n!}{r!(n-r)!} = \frac{n!}{(n-r)!r!}$, so ${}_nC_r = {}_nC_{n-r}$

18. ${}_nC_r = \frac{n!}{r!(n-r)!}$; $\frac{n}{r}({}_{n-1}C_{r-1}) = \frac{n}{r}\left[\frac{(n-1)!}{(r-1)!(n-1-(r-1))!}\right] = \frac{n}{r}\left[\frac{(n-1)!}{(r-1)!(n-r)!}\right] = \frac{n(n-1)!}{r(r-1)!(n-r)!} = \frac{n!}{r!(n-r)!}$; $\frac{n!}{r!(n-r)!} = \frac{n!}{r!(n-r)!}$, so ${}_nC_r = \frac{n}{r}({}_{n-1}C_{r-1})$

19. $(n-r)({}_nC_r) = \frac{(n-r)n!}{r!(n-r)!}$; $n({}_{n-1}C_r) = \frac{n(n-1)!}{r!(n-1-r)!} = \frac{n!}{r!(n-r-1)!} = \frac{n!(n-r)}{r!(n-r)(n-r-1)!} = \frac{n!(n-r)}{r!(n-r)!} = (n-r)({}_nC_r)$

C **20.** Represent the n elements of a set by n boxes: □ □ □ □ . . . To indicate a particular subset, assign T or F to the boxes to show whether or not that element is in the subset. Since there are 2^n different sequences of T or F, then there are 2^n subsets.

Page 495 • COMPUTER EXERCISES

1.
```
10 INPUT N
20 IF N < 0 THEN GOTO 10
30 LET F = 1
40 FOR I = 1 TO N
```

```
50 LET F = F * I
60 NEXT I
70 PRINT N; "! = ";F
80 END
```

2. 5040 **3.** 3,628,800 **4.** 87,178,291,200 **5.** 6,402,373,705,728,000

6.

```
10 INPUT N, R
20 IF N <= 0 OR R <= 0 THEN GOTO 10
30 IF N <> INT(N) OR R <> INT(R) THEN GOTO 10
40 IF N < R THEN GOTO 10
50 LET F1 = N
60 FOR I = N - 1 TO N - R + 1 STEP -1
70 LET F1 = F1 * I
80 NEXT I
90 LET F2 = 1
100 FOR I = 1 TO R
110 LET F2 = F2 * I
120 NEXT I
130 LET C = F1/F2
140 PRINT "C(";N;", ";R;") = ";C
150 END
```

7. 120 **8.** 1287 **9.** 2,598,960 **10.** 15,820,024,220

11.

```
10 INPUT "NUMBER OF ELEMENTS TAKEN AT A TIME ";N
20 INPUT "NUMBER OF SETS OF LIKE ELEMENTS ";S
30 FOR I = 1 TO S
40 INPUT "NUMBER IN SET ";N(I)
50 NEXT I
60 LET F = N
70 FOR I = N - 1 TO 1 STEP -1
80 LET F = F * I
90 NEXT I
100 LET D = 1
110 FOR I = 1 TO S
120 LET F(I) = N(I)
130 FOR J = N(I) - 1 TO 1 STEP -1
140 LET F(I) = F(I) * J
150 NEXT J
160 LET D = D * F(I)
170 NEXT I
180 LET P = F/D
190 PRINT "DISTINGUISHABLE PERMUTATIONS: ";P
200 END
```

12. 302,400 **13.** 1,663,200

Pages 496–497 • WRITTEN EXERCISES

A **1.** ${}_{10}C_4 = \dfrac{10 \cdot 9 \cdot 8 \cdot 7}{4 \cdot 3 \cdot 2 \cdot 1} = 210$; ${}_8C_6 = {}_8C_2 = \dfrac{8 \cdot 7}{2 \cdot 1} = 28$; $28 \times 210 = 5880$

2. ${}_{10}C_5 = \dfrac{10 \cdot 9 \cdot 8 \cdot 7 \cdot 6}{5 \cdot 4 \cdot 3 \cdot 2 \cdot 1} = 252$; ${}_8C_7 = {}_8C_1 = \dfrac{8}{1} = 8$; $8 \times 252 = 2016$

3. $_{10}C_7 = {}_{10}C_3 = \frac{10 \cdot 9 \cdot 8}{3 \cdot 2 \cdot 1} = 120$; $_8C_4 = \frac{8 \cdot 7 \cdot 6 \cdot 5}{4 \cdot 3 \cdot 2 \cdot 1} = 70$; $70 \times 120 = 8400$

4. $_{10}C_9 = {}_{10}C_1 = \frac{10}{1} = 10$; $_8C_0 = 1$; $10 \times 1 = 10$

5. $_{10}C_2 = \frac{10 \cdot 9}{2 \cdot 1} = 45$; $_8C_8 = \frac{8}{8} = 1$; $1 \times 45 = 45$

6. $_{10}C_{10} = \frac{10}{10} = 1$; $_8C_2 = \frac{8 \cdot 7}{2 \cdot 1} = 28$; $28 \times 1 = 28$

7. $_6C_2 \times {}_7C_3 = \frac{6 \cdot 5}{2 \cdot 1} \times \frac{7 \cdot 6 \cdot 5}{3 \cdot 2 \cdot 1} = 15 \times 35 = 525$

8. $_4C_2 \times {}_5C_3 \times {}_6C_4 = \frac{4 \cdot 3}{2 \cdot 1} \times \frac{5 \cdot 4}{2 \cdot 1} \times \frac{6 \cdot 5}{2 \cdot 1} = 6 \cdot 10 \cdot 15 = 900$

9. $_4C_2 \times {}_4C_3 = \frac{4 \cdot 3}{2 \cdot 1} \times \frac{4 \cdot 3 \cdot 2}{3 \cdot 2 \cdot 1} = 6 \times 4 = 24$

10. $_{13}C_2 \times {}_{13}C_3 = \frac{13 \cdot 12}{2 \cdot 1} \times \frac{13 \cdot 12 \cdot 11}{3 \cdot 2 \cdot 1} = 78 \times 286 = 22{,}308$

11. $_{12}C_4 \times {}_4C_2 = \frac{12 \cdot 11 \cdot 10 \cdot 9}{4 \cdot 3 \cdot 2 \cdot 1} \times \frac{4 \cdot 3}{2 \cdot 1} = 495 \times 6 = 2970$

12. $_{26}C_2 \times {}_{13}C_2 = \frac{26 \cdot 25}{2 \cdot 1} \times \frac{13 \cdot 12}{2 \cdot 1} = 325 \times 78 = 25{,}350$

13. $_4C_1 \times {}_4C_1 \times {}_4C_1 \times {}_4C_1 \times {}_4C_1 = 4 \times 4 \times 4 \times 4 \times 4 = 1024$

14. $_{13}C_1 \times {}_{13}C_1 \times {}_{13}C_1 \times {}_{13}C_1 = 13 \times 13 \times 13 \times 13 = 28{,}561$

15. $_4C_3 \times {}_{48}C_2 = \frac{4 \cdot 3 \cdot 2}{3 \cdot 2 \cdot 1} \times \frac{48 \cdot 47}{2 \cdot 1} = 4 \times 1128 = 4512$

16. $_4C_2 \times {}_4C_2 \times {}_{44}C_1 = \frac{4!}{2! \cdot 2!} \times \frac{4!}{2! \cdot 2!} \times \frac{44!}{1! \cdot 43!} = \left(\frac{4 \cdot 3}{2}\right)^2 (44) = 1584$

B 17. There are $_{13}C_2$ ways to choose two denominations and 2 ways to arrange them (to the pair or to the single card). There are $_4C_2$ pairs and $_4C_1$ single cards.
$2 \times {}_{13}C_2 \times {}_4C_2 \times {}_4C_1 = 2 \times 78 \times 6 \times 4 = 3744$

18. There are $_{13}C_3$ ways to choose three denominations and 3! ways to arrange them.
$_{13}C_3 \times 3! \times {}_4C_3 \times {}_4C_2 \times {}_4C_1 = 286 \times 6 \times 4 \times 6 \times 4 = 164{,}736$

19. $_3C_2 \times {}_6C_3 = \frac{3 \cdot 2}{2 \cdot 1} \times \frac{6 \cdot 5 \cdot 4}{3 \cdot 2 \cdot 1} = 60$. But there are 5!, or 120, permutations of these 60 letters. $60 \cdot 120 = 7200$ sequences

20. $_3C_2 \times {}_5C_3 = \frac{3 \cdot 2}{2 \cdot 1} \times \frac{5 \cdot 4 \cdot 3}{3 \cdot 2 \cdot 1} = 30$. But there are 5!, or 120, permutations of these 30 letters. $30 \cdot 120 = 3600$ sequences

C 21. $_{10}C_3 = \frac{10 \cdot 9 \cdot 8}{3 \cdot 2 \cdot 1} = 120$; there are 3!, or 6, permutations. $120 \cdot 6 = 720$ ways to seat 3 people in a row. There are 8 possible positions for 3 people seated next to each other, and there are 3! ways to arrange the people. $8 \times 3! = 48$ ways to seat 3 people next to each other.

22. No duplications: $_6P_5 = 720$. Both A and S are duplicated:
$_4C_1 \times \frac{5!}{2! \cdot 2!} = 4 \times 30 = 120$. Exactly one letter is duplicated:
$2 \times {}_5C_3 \times \frac{5!}{2!} = 2 \times 10 \times 60 = 1200$. Total: $720 + 120 + 1200 = 2040$

Page 497 • SELF-TEST 2

1. ${}_{10}C_4 = \frac{10 \cdot 9 \cdot 8 \cdot 7}{4 \cdot 3 \cdot 2 \cdot 1} = 210$ 2. ${}_{11}C_5 = \frac{11 \cdot 10 \cdot 9 \cdot 8 \cdot 7}{5 \cdot 4 \cdot 3 \cdot 2 \cdot 1} = 462$

3. ${}_6C_2 \times {}_6C_3 = \frac{6 \cdot 5}{2 \cdot 1} \times \frac{6 \cdot 5 \cdot 4}{3 \cdot 2 \cdot 1} = 15 \times 20 = 300$

4. ${}_6C_4 \times {}_8C_6 = \frac{6 \cdot 5}{2 \cdot 1} \times \frac{8 \cdot 7}{2 \cdot 1} = 15 \times 28 = 420$

Page 500 • WRITTEN EXERCISES

A 1. $b^5 + \frac{5}{1}b^4(2)^1 + \frac{5 \cdot 4}{2 \cdot 1}b^3(2)^2 + \frac{5 \cdot 4 \cdot 3}{3 \cdot 2 \cdot 1}b^2(2)^3 + \frac{5 \cdot 4 \cdot 3 \cdot 2}{4 \cdot 3 \cdot 2 \cdot 1}b^1(2)^4 + \frac{5 \cdot 4 \cdot 3 \cdot 2 \cdot 1}{5 \cdot 4 \cdot 3 \cdot 2 \cdot 1}(2^5) =$
$b^5 + 10b^4 + 40b^3 + 80b^2 + 80b + 32$

2. $y^4 + \frac{4}{1}y^3(-3) + \frac{4 \cdot 3}{2 \cdot 1}y^2(-3)^2 + \frac{4 \cdot 3 \cdot 2}{3 \cdot 2 \cdot 1}y(-3)^3 + \frac{4 \cdot 3 \cdot 2 \cdot 1}{4 \cdot 3 \cdot 2 \cdot 1}(-3)^4 =$
$y^4 - 12y^3 + 54y^2 - 108y + 81$

3. $5^4 + \frac{4}{1}(5^3)(-r) + \frac{4 \cdot 3}{2 \cdot 1}(5^2)(-r)^2 + \frac{4 \cdot 3 \cdot 2}{3 \cdot 2 \cdot 1}(5)(-r)^3 + \frac{4 \cdot 3 \cdot 2 \cdot 1}{4 \cdot 3 \cdot 2 \cdot 1}(-r)^4 =$
$625 - 500r + 150r^2 - 20r^3 + r^4$

4. $x^6 + \frac{6}{1}x^5(2y) + \frac{6 \cdot 5}{2 \cdot 1}x^4(2y)^2 + \frac{6 \cdot 5 \cdot 4}{3 \cdot 2 \cdot 1}x^3(2y)^3 + \frac{6 \cdot 5 \cdot 4 \cdot 3}{4 \cdot 3 \cdot 2 \cdot 1}x^2(2y)^4 +$
$\frac{6 \cdot 5 \cdot 4 \cdot 3 \cdot 2}{5 \cdot 4 \cdot 3 \cdot 2 \cdot 1}x(2y)^5 + \frac{6 \cdot 5 \cdot 4 \cdot 3 \cdot 2 \cdot 1}{6 \cdot 5 \cdot 4 \cdot 3 \cdot 2 \cdot 1}(2y)^6 =$
$x^6 + 12x^5y + 60x^4y^2 + 160x^3y^3 + 240x^2y^4 + 192xy^5 + 64y^6$

5. $(3c)^5 + \frac{5}{1}(3c)^4\left(\frac{1}{3}\right) + \frac{5 \cdot 4}{2 \cdot 1}(3c)^3\left(\frac{1}{3}\right)^2 + \frac{5 \cdot 4 \cdot 3}{3 \cdot 2 \cdot 1}(3c)^2\left(\frac{1}{3}\right)^3 + \frac{5 \cdot 4 \cdot 3 \cdot 2}{4 \cdot 3 \cdot 2 \cdot 1}(3c)\left(\frac{1}{3}\right)^4 +$
$\frac{5 \cdot 4 \cdot 3 \cdot 2 \cdot 1}{5 \cdot 4 \cdot 3 \cdot 2 \cdot 1}\left(\frac{1}{3}\right)^5 = 243c^5 + 135c^4 + 30c^3 + \frac{10}{3}c^2 + \frac{5}{27}c + \frac{1}{243}$

6. $\left(\frac{1}{2}x\right)^4 + \frac{4}{1}\left(\frac{1}{2}x\right)^3(-1) + \frac{4 \cdot 3}{2 \cdot 1}\left(\frac{1}{2}x\right)^2(-1)^2 + \frac{4 \cdot 3 \cdot 2}{3 \cdot 2 \cdot 1}\left(\frac{1}{2}x\right)(-1)^3 +$
$\frac{4 \cdot 3 \cdot 2 \cdot 1}{4 \cdot 3 \cdot 2 \cdot 1}(-1)^4 = \frac{1}{16}x^4 - \frac{1}{2}x^3 + \frac{3}{2}x^2 - 2x + 1$

7. $\left(\frac{1}{2}\right)^6 + \frac{6}{1}\left(\frac{1}{2}\right)^5(-2b) + \frac{6 \cdot 5}{2 \cdot 1}\left(\frac{1}{2}\right)^4(-2b)^2 + \frac{6 \cdot 5 \cdot 4}{3 \cdot 2 \cdot 1}\left(\frac{1}{2}\right)^3(-2b)^3 +$
$\frac{6 \cdot 5 \cdot 4 \cdot 3}{4 \cdot 3 \cdot 2 \cdot 1}\left(\frac{1}{2}\right)^2(-2b)^4 + \frac{6 \cdot 5 \cdot 4 \cdot 3 \cdot 2}{5 \cdot 4 \cdot 3 \cdot 2 \cdot 1}\left(\frac{1}{2}\right)(-2b)^5 + \frac{6 \cdot 5 \cdot 4 \cdot 3 \cdot 2 \cdot 1}{6 \cdot 5 \cdot 4 \cdot 3 \cdot 2 \cdot 1}(-2b)^6 =$
$\frac{1}{64} - \frac{3}{8}b + \frac{15}{4}b^2 - 20b^3 + 60b^4 - 96b^5 + 64b^6$

8. $(p^3)^7 + \frac{7}{1}(p^3)^6q + \frac{7 \cdot 6}{2 \cdot 1}(p^3)^5q^2 + \frac{7 \cdot 6 \cdot 5}{3 \cdot 2 \cdot 1}(p^3)^4q^3 + \frac{7 \cdot 6 \cdot 5 \cdot 4}{4 \cdot 3 \cdot 2 \cdot 1}(p^3)^3q^4 +$
$\frac{7 \cdot 6 \cdot 5 \cdot 4 \cdot 3}{5 \cdot 4 \cdot 3 \cdot 2 \cdot 1}(p^3)^2q^5 + \frac{7 \cdot 6 \cdot 5 \cdot 4 \cdot 3 \cdot 2}{6 \cdot 5 \cdot 4 \cdot 3 \cdot 2 \cdot 1}(p^3)^1q^6 + \frac{7 \cdot 6 \cdot 5 \cdot 4 \cdot 3 \cdot 2 \cdot 1}{7 \cdot 6 \cdot 5 \cdot 4 \cdot 3 \cdot 2 \cdot 1}q^7 =$
$p^{21} + 7p^{18}q + 21p^{15}q^2 + 35p^{12}q^3 + 35p^9q^4 + 21p^6q^5 + 7p^3q^6 + q^7$

9. $(y^4)^8 + \frac{8}{1}(y^4)^7(-1) + \frac{8 \cdot 7}{2 \cdot 1}(y^4)^6(-1)^2 + \frac{8 \cdot 7 \cdot 6}{3 \cdot 2 \cdot 1}(y^4)^5(-1)^3 + \frac{8 \cdot 7 \cdot 6 \cdot 5}{4 \cdot 3 \cdot 2 \cdot 1}(y^4)^4(-1)^4 +$
$\frac{8 \cdot 7 \cdot 6 \cdot 5 \cdot 4}{5 \cdot 4 \cdot 3 \cdot 2 \cdot 1}(y^4)^3(-1)^5 + \frac{8 \cdot 7 \cdot 6 \cdot 5 \cdot 4 \cdot 3}{6 \cdot 5 \cdot 4 \cdot 3 \cdot 2 \cdot 1}(y^4)^2(-1)^6 +$
$\frac{8 \cdot 7 \cdot 6 \cdot 5 \cdot 4 \cdot 3 \cdot 2}{7 \cdot 6 \cdot 5 \cdot 4 \cdot 3 \cdot 2}(y^4)^1(-1)^7 + \frac{8 \cdot 7 \cdot 6 \cdot 5 \cdot 4 \cdot 3 \cdot 2 \cdot 1}{8 \cdot 7 \cdot 6 \cdot 5 \cdot 4 \cdot 3 \cdot 2 \cdot 1}(-1)^8 =$
$y^{32} - 8y^{28} + 28y^{24} - 56y^{20} + 70y^{16} - 56y^{12} + 28y^8 - 8y^4 + 1$

10. $x^7 + \frac{7}{1}x^6(y^3) + \frac{7 \cdot 6}{2 \cdot 1}x^5(y^3)^2 + \frac{7 \cdot 6 \cdot 5}{3 \cdot 2 \cdot 1}x^4(y^3)^3 + \frac{7 \cdot 6 \cdot 5 \cdot 4}{4 \cdot 3 \cdot 2 \cdot 1}x^3(y^3)^4 +$
$\frac{7 \cdot 6 \cdot 5 \cdot 4 \cdot 3}{5 \cdot 4 \cdot 3 \cdot 2 \cdot 1}x^2(y^3)^5 + \frac{7 \cdot 6 \cdot 5 \cdot 4 \cdot 3 \cdot 2}{6 \cdot 5 \cdot 4 \cdot 3 \cdot 2 \cdot 1}x(y^3)^6 + \frac{7 \cdot 6 \cdot 5 \cdot 4 \cdot 3 \cdot 2 \cdot 1}{7 \cdot 6 \cdot 5 \cdot 4 \cdot 3 \cdot 2 \cdot 1}(y^3)^7 =$
$x^7 + 7x^6y^3 + 21x^5y^6 + 35x^4y^9 + 35x^3y^{12} + 21x^2y^{15} + 7xy^{18} + y^{21}$

11. $3^7 + \frac{7}{1}(3)^6(2b) + \frac{7 \cdot 6}{2 \cdot 1}(3)^5(2b)^2 + \frac{7 \cdot 6 \cdot 5}{3 \cdot 2 \cdot 1}(3)^4(2b)^3 + \frac{7 \cdot 6 \cdot 5 \cdot 4}{4 \cdot 3 \cdot 2 \cdot 1}(3)^3(2b)^4 +$
$\frac{7 \cdot 6 \cdot 5 \cdot 4 \cdot 3}{5 \cdot 4 \cdot 3 \cdot 2 \cdot 1}(3)^2(2b)^5 + \frac{7 \cdot 6 \cdot 5 \cdot 4 \cdot 3 \cdot 2}{6 \cdot 5 \cdot 4 \cdot 3 \cdot 2 \cdot 1}(3)(2b)^6 + \frac{7 \cdot 6 \cdot 5 \cdot 4 \cdot 3 \cdot 2 \cdot 1}{7 \cdot 6 \cdot 5 \cdot 4 \cdot 3 \cdot 2 \cdot 1}(2b)^7 =$
$2187 + 10{,}206b + 20{,}412b^2 + 22{,}680b^3 + 15{,}120b^4 + 6048b^5 + 1344b^6 + 128b^7$

12. $(c^4)^8 + \frac{8}{1}(c^4)^7(d^3) + \frac{8 \cdot 7}{2 \cdot 1}(c^4)^6(d^3)^2 + \frac{8 \cdot 7 \cdot 6}{3 \cdot 2 \cdot 1}(c^4)^5(d^3)^3 + \frac{8 \cdot 7 \cdot 6 \cdot 5}{4 \cdot 3 \cdot 2 \cdot 1}(c^4)^4(d^3)^4 +$
$\frac{8 \cdot 7 \cdot 6 \cdot 5 \cdot 4}{5 \cdot 4 \cdot 3 \cdot 2 \cdot 1}(c^4)^3(d^3)^5 + \frac{8 \cdot 7 \cdot 6 \cdot 5 \cdot 4 \cdot 3}{6 \cdot 5 \cdot 4 \cdot 3 \cdot 2 \cdot 1}(c^4)^2(d^3)^6 + \frac{8 \cdot 7 \cdot 6 \cdot 5 \cdot 4 \cdot 3 \cdot 2}{7 \cdot 6 \cdot 5 \cdot 4 \cdot 3 \cdot 2 \cdot 1}c^4(d^3)^7 +$
$\frac{8 \cdot 7 \cdot 6 \cdot 5 \cdot 4 \cdot 3 \cdot 2 \cdot 1}{8 \cdot 7 \cdot 6 \cdot 5 \cdot 4 \cdot 3 \cdot 2 \cdot 1}(d^3)^8 = c^{32} + 8c^{28}d^3 + 28c^{24}d^6 + 56c^{20}d^9 + 70c^{16}d^{12} +$
$56c^{12}d^{15} + 28c^8d^{18} + 8c^4d^{21} + d^{24}$

13. $\frac{10 \cdot 9 \cdot 8 \cdot 7}{4 \cdot 3 \cdot 2 \cdot 1}a^6b^4 = 210a^6b^4$ **14.** $\frac{12 \cdot 11}{2 \cdot 1}(x^2)^{10}\,(-1)^2 = 66x^{20}$

15. $\frac{9 \cdot 8 \cdot 7}{3 \cdot 2 \cdot 1}p^6(2q)^3 = 672p^6q^3$ **16.** $\frac{11 \cdot 10 \cdot 9 \cdot 8 \cdot 7}{5 \cdot 4 \cdot 3 \cdot 2 \cdot 1}(n^3)^6(-3)^5 = -112{,}266n^{18}$

17. $\frac{13 \cdot 12 \cdot 11 \cdot 10 \cdot 9 \cdot 8 \cdot 7 \cdot 6}{8 \cdot 7 \cdot 6 \cdot 5 \cdot 4 \cdot 3 \cdot 2 \cdot 1}(2c)^5\left(-\frac{d}{2}\right)^8 = \frac{1287}{8}c^5d^8$

18. $\frac{14 \cdot 13 \cdot 12 \cdot 11 \cdot 10}{5 \cdot 4 \cdot 3 \cdot 2 \cdot 1}(1)^9(-u^2)^5 = -2002u^{10}$

B **19.** $(x^2)^{15} + \frac{15}{1}(x^2)^{14}(2) + \frac{15 \cdot 14}{2 \cdot 1}(x^2)^{13}(2)^2 + \frac{15 \cdot 14 \cdot 13}{3 \cdot 2 \cdot 1}(x^2)^{12}(2)^3 =$
$x^{30} + 30x^{28} + 420x^{26} + 3640x^{24}$

20. $(2z^2)^{10} + \frac{10}{1}(2z^2)^9(-1) + \frac{10 \cdot 9}{2 \cdot 1}(2z^2)^8(-1)^2 + \frac{10 \cdot 9 \cdot 8}{3 \cdot 2 \cdot 1}(2z^2)^7(-1)^3 =$
$1024z^{20} - 5120z^{18} + 11{,}520z^{16} - 15{,}360z^{14}$

21. $(x^2)^{24} + \frac{24}{1}(x^2)^{23}\left(-\frac{1}{2}y^3\right) + \frac{24 \cdot 23}{2 \cdot 1}(x^2)^{22}\left(-\frac{1}{2}y^3\right)^2 + \frac{24 \cdot 23 \cdot 22}{3 \cdot 2 \cdot 1}(x^2)^{21}\left(-\frac{1}{2}y^3\right)^3 =$
$x^{48} - 12x^{46}y^3 + 69x^{44}y^6 - 253x^{42}y^9$

22. $x^{40} + \frac{40}{1}x^{39}(x^{3/2}) + \frac{40 \cdot 39}{2 \cdot 1}x^{38}(x^{3/2})^2 + \frac{40 \cdot 39 \cdot 38}{3 \cdot 2 \cdot 1}x^{37}(x^{3/2})^3 =$
$x^{40} + 40x^{39}x^{3/2} + 780x^{41} + 9880x^{37}x^{9/2}$, or $x^{40} + 40x^{81/2} + 780x^{41} + 9880x^{83/2}$

C **23.** $64^{1/2} + \frac{1}{2}(64)^{-1/2}(1)^1 + \frac{\frac{1}{2}\left(-\frac{1}{2}\right)}{2 \cdot 1}(64)^{-3/2}(1)^2 = 8 + \frac{1}{2}\left(\frac{1}{8}\right) + \left(-\frac{1}{8}\right)\left(\frac{1}{8^3}\right) = 8 + \frac{1}{16} - \frac{1}{4096}$

24. $81^{1/2} + \frac{1}{2}(81)^{-1/2}(-1)^1 + \frac{\frac{1}{2}\left(-\frac{1}{2}\right)}{2 \cdot 1}(81)^{-3/2}(-1)^2 = 9 + \frac{1}{2}\left(\frac{1}{9}\right)(-1) + \left(-\frac{1}{8}\right)\left(\frac{1}{9^3}\right) = 9 - \frac{1}{18} - \frac{1}{5832}$

25. $16^{1/2} + \frac{1}{2}(16)^{-1/2}(-r)^1 + \frac{\frac{1}{2}\left(-\frac{1}{2}\right)}{2 \cdot 1}(16)^{-3/2}(-r)^2 = 4 + \frac{1}{2}\left(\frac{1}{4}\right)(-r) + \left(-\frac{1}{8}\right)\left(\frac{1}{4^3}\right)r^2 = 4 - \frac{1}{8}r - \frac{1}{512}r^2$

26. $(27a^3)^{2/3} + \frac{2}{3}(27a^3)^{-1/3}(1)^1 + \frac{\frac{2}{3}\left(-\frac{1}{3}\right)}{2 \cdot 1}(27a^3)^{-4/3}(1)^2 = 9a^2 + \frac{2}{3}\left(\frac{1}{3a}\right) + \left(-\frac{1}{9}\right)\left(\frac{1}{81a^4}\right) = 9a^2 + \frac{2}{9a} - \frac{1}{729a^4}$

Page 500 • COMPUTER EXERCISES

1.
```
10 INPUT N
20 DIM C(N)
30 LET C(0) = 1
40 FOR K = 1 TO N
50 LET C(K) = C(K - 1) * (N - K + 1)/K
60 NEXT K
70 LET EA = N : LET EB = 0
80 FOR I = 0 TO N
90 IF I = 0 THEN GOTO 110
100 PRINT " + ";
110 IF C(I) = 1 THEN GOTO 130
120 PRINT INT(C(I) + 0.5);
130 IF EA <> 0 THEN PRINT "A";
140 IF EA <> 1 AND EA <> 0 THEN PRINT "^";EA;
150 IF EB <> 0 THEN PRINT "B";
160 IF EB <> 0 AND EB <> 1 THEN PRINT "^";EB;
170 LET EA = EA - 1 : LET EB = EB + 1
180 NEXT I
190 END
```

2. $a^7 + 7a^6b + 21a^5b^2 + 35a^4b^3 + 35a^3b^4 + 21a^2b^5 + 7ab^6 + b^7$

3. $a^{10} + 10a^9b + 45a^8b^2 + 120a^7b^3 + 210a^6b^4 + 252a^5b^5 + 210a^4b^6 + 120a^3b^7 + 45a^2b^8 + 10ab^9 + b^{10}$

4. $a^{12} + 12a^{11}b + 66a^{10}b^2 + 220a^9b^3 + 495a^8b^4 + 792a^7b^5 + 924a^6b^6 + 792a^5b^7 + 495a^4b^8 + 220a^3b^9 + 66a^2b^{10} + 12ab^{11} + b^{12}$

5. $a^{20} + 20a^{19}b + 190a^{18}b^2 + 1140a^{17}b^3 + 4845a^{16}b^4 + 15{,}504a^{15}b^5 + 38{,}760a^{14}b^6 + 77{,}520a^{13}b^7 + 125{,}970a^{12}b^8 + 167{,}960a^{11}b^9 + 184{,}756a^{10}b^{10} + 167{,}960a^9b^{11} + 125{,}970a^8b^{12} + 77{,}520a^7b^{13} + 38{,}760a^6b^{14} + 15{,}504a^5b^{15} + 4845a^4b^{16} + 1140a^3b^{17} + 190a^2b^{18} + 20ab^{19} + b^{20}$

6.
```
10 INPUT A,B,N
20 DIM C(N)
30 LET C(0) = 1
40 FOR K = 1 TO N
50 LET C(K) = C(K - 1) * (N - K + 1)/K
60 NEXT K
70 LET EA = N : LET EB = 0
80 FOR I = 0 TO N
90 LET V = V + C(I) * A^EA * B^EB
100 LET EA = EA - 1 : LET EB = EB + 1
110 NEXT I
120 PRINT V
130 END
```

7. 2.01219647 8. 11.390625 9. 0.472161363

Page 504 • WRITTEN EXERCISES

A

1. Seventh row: 1, 6, 15, 20, 15, 6, 1. $(1)c^6 + 6c^5(1) + 15c^4(1)^2 + 20c^3(1)^3 + 15c^2(1)^4 + 6c(1)^5 + 1(1)^6 = c^6 + 6c^5 + 15c^4 + 20c^3 + 15c^2 + 6c + 1$
2. Eighth row: 1, 7, 21, 35, 35, 21, 7, 1. $(1)r^7 + (7)r^6(-2) + (21)r^5(-2)^2 + (35)r^4(-2)^3 + (35)r^3(-2)^4 + (21)r^2(-2)^5 + (7)r(-2)^6 + 1(-2)^7 = r^7 - 14r^6 + 84r^5 - 280r^4 + 560r^3 - 672r^2 + 448r - 128$
3. Sixth row: 1, 5, 10, 10, 5, 1. $(1)(3x)^5 + (5)(3x)^4y + (10)(3x)^3y^2 + (10)(3x)^2y^3 + (5)(3x)y^4 + (1)y^5 = 243x^5 + 405x^4y + 270x^3y^2 + 90x^2y^3 + 15xy^4 + y^5$
4. Seventh row: 1, 6, 15, 20, 15, 6, 1. $(1)\left(\frac{a}{2}\right)^6 + (6)\left(\frac{a}{2}\right)^5 b + (15)\left(\frac{a}{2}\right)^4 b^2 + (20)\left(\frac{a}{2}\right)^3 b^3 + (15)\left(\frac{a}{2}\right)^2 b^4 + (6)\left(\frac{a}{2}\right)b^5 + (1)b^6 = \frac{1}{64}a^6 + \frac{3}{16}a^5b + \frac{15}{16}a^4b^2 + \frac{5}{2}a^3b^3 + \frac{15}{4}a^2b^4 + 3ab^5 + b^6$
5. Eighth row: 1, 7, 21, 35, 35, 21, 7, 1. $(1)(1)^7 + (7)(1)^6(-y^3) + (21)(1)^5(-y^3)^2 + 35(1)^4(-y^3)^3 + 35(1)^3(-y^3)^4 + 21(1)^2(-y^3)^5 + 7(1)(-y^3)^6 + 1(-y^3)^7 = 1 - 7y^3 + 21y^6 - 35y^9 + 35y^{12} - 21y^{15} + 7y^{18} - y^{21}$
6. Sixth row: 1, 5, 10, 10, 5, 1. $(1)\left(\frac{1}{3}a\right)^5 + (5)\left(\frac{1}{3}a\right)^4(3) + (10)\left(\frac{1}{3}a\right)^3(3)^2 + (10)\left(\frac{1}{3}a\right)^2(3)^3 + (5)\left(\frac{1}{3}a\right)(3)^4 + (1)(3)^5 = \frac{1}{243}a^5 + \frac{5}{27}a^4 + \frac{10}{3}a^3 + 30a^2 + 135a + 243$
7. Seventh row: 1, 6, 15, 20, 15, 6, 1. $(1)(r^3)^6 + (6)(r^3)^5(-t^2) + (15)(r^3)^4(-t^2)^2 + (20)(r^3)^3(-t^2)^3 + (15)(r^3)^2(-t^2)^4 + (6)(r^3)(-t^2)^5 + (1)(-t^2)^6 = r^{18} - 6r^{15}t^2 + 15r^{12}t^4 - 20r^9t^6 + 15r^6t^8 - 6r^3t^{10} + t^{12}$
8. Ninth row: 1, 8, 28, 56, 70, 56, 28, 8, 1. $(1)(k^2)^8 + (8)(k^2)^7(2n) + (28)(k^2)^6(2n)^2 + (56)(k^2)^5(2n)^3 + (70)(k^2)^4(2n)^4 + (56)(k^2)^3(2n)^5 + (28)(k^2)^2(2n)^6 + (8)(k^2)(2n)^7 + (1)(2n)^8 = k^{16} + 16k^{14}n + 112k^{12}n^2 + 448k^{10}n^3 + 1120k^8n^4 + 1792k^6n^5 + 1792k^4n^6 + 1024k^2n^7 + 256n^8$

B **9.** $r - 1 = 4$, $r = 5$; find the fifth term: ${}_8C_4(1)^4(m)^4 = \frac{8 \cdot 7 \cdot 6 \cdot 5}{4 \cdot 3 \cdot 2 \cdot 1}m^4 = 70m^4$

10. $n - r + 1 = 6$ and $n = 9$, $10 - r = 6$, $r = 4$; find the fourth term: ${}_9C_3(x)^6\left(\frac{1}{2}\right)^3 =$
$\frac{9 \cdot 8 \cdot 7}{3 \cdot 2 \cdot 1}x^6\left(\frac{1}{8}\right) = \frac{21}{2}x^6$

11. $r - 1 = 7$, $r = 8$; find the eighth term: ${}_{10}C_7\left(\frac{p}{2}\right)^3(q)^7 = \frac{10 \cdot 9 \cdot 8}{3 \cdot 2 \cdot 1}\left(\frac{p^3}{8}\right)q^7 = 15p^3q^7$

12. $n - r + 1 = 7$ and $n = 11$, $11 - r + 1 = 7$, $r = 5$; find the fifth term:
${}_{11}C_4(a)^7\left(\frac{1}{3}b\right)^4 = \frac{11 \cdot 10 \cdot 9 \cdot 8}{4 \cdot 3 \cdot 2 \cdot 1}a^7\left(\frac{b^4}{81}\right) = \frac{110}{27}a^7b^4$

13. $r - 1 = 4$, $r = 5$; find the fifth term: ${}_{12}C_4(1)^8(-2x)^4 = \frac{12 \cdot 11 \cdot 10 \cdot 9}{4 \cdot 3 \cdot 2 \cdot 1}(16x^4) =$
$7920x^4$

14. $r - 1 = 5$, $r = 6$; find the sixth term: ${}_{14}C_5(r^2)^9(-t^3)^5 =$
$\frac{14 \cdot 13 \cdot 12 \cdot 11 \cdot 10}{5 \cdot 4 \cdot 3 \cdot 2 \cdot 1}r^{18}(-t^{15}) = -2002r^{18}t^{15}$

15. **a.** ${}_{15}C_8(x^7)(-y)^8 = 6435x^7y^8$ **b.** ${}_{16}C_8(x^8)(-y)^8 = 12{,}870x^8y^8$

16. **a.** ${}_{30}C_{16}(z^{14})(1^{16}) = 145{,}422{,}675z^{14}$ **b.** ${}_{30}C_{15}(z^{15})(1^{15}) = 155{,}117{,}520z^{15}$

C **17.** ${}_{n-1}C_2 + {}_{n-1}C_3 = \frac{(n-1)(n-2)}{2 \cdot 1} + \frac{(n-1)(n-2)(n-3)}{3 \cdot 2 \cdot 1} =$
$\frac{3(n-1)(n-2) + (n-1)(n-2)(n-3)}{3 \cdot 2 \cdot 1} = \frac{(n-1)(n-2)[3 + (n-3)]}{3 \cdot 2 \cdot 1} =$
$\frac{(n-1)(n-2)n}{3 \cdot 2 \cdot 1} = \frac{n(n-1)(n-2)}{3 \cdot 2 \cdot 1} = {}_nC_3$

18. ${}_{n-1}C_{r-1} + {}_{n-1}C_r = \frac{(n-1)(n-2) \cdots (n - 1 - (r-1) + 1)}{(r-1)(r-2) \cdots 1} +$
$\frac{(n-1)(n-2) \cdots (n - 1 - r + 1)}{r(r-1)(r-2) \cdots 1} =$
$\frac{r(n-1)(n-2) \cdots (n-r+1) + (n-1)(n-2) \cdots (n-r)}{r(r-1)(r-2) \cdots 1}$; factoring the numerator yields $\frac{(n-1)(n-2) \cdots (n-r+1)[r + (n-r)]}{r!} =$
$\frac{(n-1)(n-2) \cdots (n-r+1)(n)}{r!} = \frac{n(n-1) \cdots (n-r+1)}{r!} = {}_nC_r$

19. Row 1: 1; row 2: 2; row 3: 4; row 4: 8; row 5: 16; conjecture: $f(n) = 2^{n-1}$. For $n = 8$, the sum is $1 + 7 + 21 + 35 + 35 + 21 + 7 + 1 = 128$ and $f(8) = 2^7 = 128$.

Page 504 • SELF-TEST 3

1. $(a^2)^5 + \frac{5}{1}(a^2)^4(-2) + \frac{5 \cdot 4}{2 \cdot 1}(a^2)^3(-2)^2 + \frac{5 \cdot 4 \cdot 3}{3 \cdot 2 \cdot 1}(a^2)^2(-2)^3 + \frac{5 \cdot 4 \cdot 3 \cdot 2}{4 \cdot 3 \cdot 2 \cdot 1}a^2(-2)^4 +$
$\frac{5 \cdot 4 \cdot 3 \cdot 2 \cdot 1}{5 \cdot 4 \cdot 3 \cdot 2 \cdot 1}(-2)^5 = a^{10} - 10a^8 + 40a^6 - 80a^4 + 80a^2 - 32$

2. $\frac{10 \cdot 9 \cdot 8 \cdot 7 \cdot 6 \cdot 5}{6 \cdot 5 \cdot 4 \cdot 3 \cdot 2 \cdot 1}(2p)^4(q)^6 = 3360p^4q^6$

3. Seventh row: 1, 6, 15, 20, 15, 6, 1. $(1)x^6 + (6)x^5\left(-\frac{y}{2}\right) + (15)x^4\left(-\frac{y}{2}\right)^2 +$
$20x^3\left(-\frac{y}{2}\right)^3 + 15x^2\left(-\frac{y}{2}\right)^4 + 6x\left(-\frac{y}{2}\right)^5 + \left(-\frac{y}{2}\right)^6 = x^6 - 3x^5y + \frac{15}{4}x^4y^2 - \frac{5}{2}x^3y^3 +$
$\frac{15}{16}x^2y^4 - \frac{3}{16}xy^5 + \frac{1}{64}y^6$

Pages 506–507 • WRITTEN EXERCISES

A **1. a.** {1, 2, 3, 4, 5, 6, 7, 8, 9, 10, 11, 12, 13, 14} **b.** {4, 8, 12}
2. a. {W1, W2, B1, B2, B3, B4, B5} **b.** {B1, B2, B3, B4, B5}
3. a. {(1, 1), (1, 2), (1, 3), (2, 1), (2, 2), (2, 3), (3, 1), (3, 2), (3, 3)}
b. {(1, 1), (2, 2), (3, 3)}
4. a. {(1, 2), (1, 3), (2, 1), (2, 3), (3, 1), (3, 2)} **b.** {(1, 3), (3, 1)}
5. a. {(D, H), (D, C), (D, S), (C, H), (C, D), (C, S), (H, C), (H, D), (H, S), (S, H), (S, C), (S, D)} **b.** {(D, H), (H, D)}
6. a. {(H, H), (H, T), (T, H), (T, T)} **b.** {(H, H), (H, T), (T, H)}
7. a. {(A, B), (A, C), (A, E), (B, A), (B, C), (B, E), (C, A), (C, B), (C, E), (E, A), (E, B), (E, C)} **b.** {(A, B), (A, C), (A, E), (B, A), (B, E), (C, A), (C, E), (E, A), (E, B), (E, C)}
8. a. {(A, B), (A, C), (A, E), (B, A), (B, C), (B, E), (C, A), (C, B), (C, E), (E, A), (E, B), (E, C)} **b.** {(A, B), (A, C), (B, A), (B, C), (B, E), (C, A), (C, B), (C, E), (E, B), (E, C)}

B **9.** ${}_{52}C_1 = 52$; ${}_{13}C_1 = 13$ **10.** ${}_{52}C_2 = \frac{52 \cdot 51}{2 \cdot 1} = 1326$; ${}_4C_2 = 6$
11. ${}_{52}C_2 = 1326$; ${}_{26}C_2 = \frac{26 \cdot 25}{2 \cdot 1} = 325$ **12.** ${}_{52}C_2 = 1326$; ${}_{12}C_2 = \frac{12 \cdot 11}{2 \cdot 1} = 66$
13. ${}_{52}C_2 = 1326$; ${}_{13}C_2 = \frac{13 \cdot 12}{2 \cdot 1} = 78$
14. ${}_{52}C_3 = \frac{52 \cdot 51 \cdot 50}{3 \cdot 2 \cdot 1} = 22{,}100$; ${}_{13}C_3 = \frac{13 \cdot 12 \cdot 11}{3 \cdot 2 \cdot 1} = 286$

C **15.** ${}_{52}C_5 = \frac{52 \cdot 51 \cdot 50 \cdot 49 \cdot 48}{5 \cdot 4 \cdot 3 \cdot 2 \cdot 1} = 2{,}598{,}960$; ${}_{13}C_3 \cdot {}_{13}C_2 = 286 \cdot 78 = 22{,}308$
16. ${}_{52}C_5 = 2{,}598{,}960$; there are 4 ways to choose the suit that will be drawn twice; $4 \cdot {}_{13}C_2 \cdot {}_{13}C_1 \cdot {}_{13}C_1 \cdot {}_{13}C_1 = 4 \cdot 78 \cdot 13^3 = 685{,}464$

Page 507 • ON THE CALCULATOR

1. ${}_{52}C_4 = 270{,}725$; ${}_{26}C_4 = 14{,}950$ **2.** ${}_{52}C_5 = 2{,}598{,}960$; ${}_{26}C_5 = 1287$
3. ${}_{52}C_5 = 2{,}598{,}960$; ${}_{24}C_5 = 42{,}504$
4. ${}_{52}C_6 = 20{,}358{,}520$; ${}_{13}C_4 \cdot {}_{13}C_2 = 715 \cdot 78 = 55{,}770$
5. ${}_{52}C_{15} = 4.4814 \times 10^{12}$; ${}_{26}C_{15} = 7{,}726{,}160$
6. ${}_{52}C_{12} = 2.0638 \times 10^{11}$; ${}_{12}C_8 \cdot {}_{12}C_4 = 495 \cdot 495 = 245{,}025$

Pages 510–511 • WRITTEN EXERCISES

A **1. a.** $\frac{3}{9} = \frac{1}{3}$ **b.** $\frac{6}{9} = \frac{2}{3}$ **c.** $\frac{4}{9}$ **d.** $\frac{5}{9}$
2. a. $\frac{4}{36} = \frac{1}{9}$ **b.** $\frac{6}{36} = \frac{1}{6}$ **c.** $\frac{30}{36} = \frac{5}{6}$ **d.** $\frac{2 + 6}{36} = \frac{8}{36} = \frac{2}{9}$

3. a. $\frac{1}{2}$ **b.** $\frac{2}{1}$ **c.** $\frac{4}{5}$ **d.** $\frac{5}{4}$

4. a. $\frac{1}{8}$ **b.** $\frac{1}{5}$ **c.** $\frac{5}{1}$ **d.** $\frac{2}{7}$

5. a. $\frac{{}_4C_2}{{}_{16}C_2} = \frac{6}{120} = \frac{1}{20}$ **b.** $\frac{{}_5C_2}{{}_{16}C_2} = \frac{10}{120} = \frac{1}{12}$ **c.** $\frac{{}_7C_2}{{}_{16}C_2} = \frac{21}{120} = \frac{7}{40}$

d. $\frac{{}_{12}C_2}{{}_{16}C_2} = \frac{66}{120} = \frac{11}{20}$ **e.** $\frac{{}_{11}C_2}{{}_{16}C_2} = \frac{55}{120} = \frac{11}{24}$ **f.** $\frac{{}_9C_2}{{}_{16}C_2} = \frac{36}{120} = \frac{3}{10}$

6. no; $\frac{1}{20} + \frac{11}{20} \neq 1$

7. a. $\frac{1}{2^5} = \frac{1}{32}$ **b.** $\frac{{}_5C_1}{2^5} = \frac{5}{32}$ **c.** $\frac{{}_5C_3}{2^5} = \frac{10}{32} = \frac{5}{16}$ **d.** $\frac{{}_5C_2}{2^5} = \frac{10}{32} = \frac{5}{16}$

e. The complement of "at least one head" is "all tails". Probability of "all tails" is $\frac{1}{2^5}$, or $\frac{1}{32}$; $1 - \frac{1}{32} = \frac{31}{32}$

8. a. $2^{10} = 1024$ **b.** $\frac{{}_{10}C_9}{2^{10}} = \frac{10}{1024} = \frac{5}{512}$ **c.** $\frac{{}_{10}C_7}{2^{10}} = \frac{120}{1024} = \frac{15}{128}$

B

9. a. Units' digit must be odd; $\frac{3}{6} = \frac{1}{2}$ **b.** Units' digit must be a 5; $\frac{1}{6}$

c. Hundreds' digit must be a 1, 2, or 3; $\frac{3}{6} = \frac{1}{2}$

d. If the no. is between 200 and 500, hundreds' digit must be a 2, 3, or 4. If the no. is between 500 and 550, hundreds' digit must be a 5 and tens' digit must be a 1, 2, 3, or 4.

$P(E) = \frac{3 \cdot {}_5P_2 + 1 \cdot {}_4P_1 \cdot {}_4P_1}{{}_6P_3} = \frac{60 + 16}{120} = \frac{76}{120} = \frac{19}{30}$

10. a. $\frac{{}_{10}C_4}{{}_{13}C_4} = \frac{210}{715} = \frac{42}{143}$ **b.** $\frac{{}_1C_1 \cdot {}_1C_1 \cdot {}_{11}C_2}{{}_{13}C_4} = \frac{1 \cdot 1 \cdot 55}{715} = \frac{1}{13}$ **c.** $\frac{{}_1C_1 \cdot {}_{11}C_3}{{}_{13}C_4} = \frac{1 \cdot 165}{715} = \frac{3}{13}$

d. P(2 face cards) + P(3 face cards) = $\frac{{}_3C_2 \cdot {}_{10}C_2}{{}_{13}C_4} + \frac{{}_3C_3 \cdot {}_{10}C_1}{{}_{13}C_4} =$

$\frac{3 \cdot 45}{715} + \frac{1 \cdot 10}{715} = \frac{135 + 10}{715} = \frac{145}{715} = \frac{29}{143}$

11. a. $\frac{{}_1C_1 \cdot {}_1C_1 \cdot {}_9C_3}{{}_{11}C_5} = \frac{1 \cdot 1 \cdot 84}{462} = \frac{2}{11}$ **b.** $\frac{{}_9C_5}{{}_{11}C_5} = \frac{126}{462} = \frac{3}{11}$

c. $\frac{{}_1C_1 \cdot {}_9C_4}{{}_{11}C_5} = \frac{1 \cdot 126}{462} = \frac{3}{11}$

d. P(at least one) = 1 − P(neither) = $1 - \frac{3}{11} = \frac{8}{11}$

12. a. There are 4 ways to place QU in the sequence; there are ${}_3P_3$ ways to arrange the other 3 letters; $\frac{4 \cdot {}_3P_3}{{}_5P_5} = \frac{4 \cdot 6}{120} = \frac{1}{5}$ **b.** $\frac{4 \cdot {}_2P_2 \cdot {}_3P_3}{{}_5P_5} = \frac{4 \cdot 2 \cdot 6}{120} = \frac{2}{5}$

13. a. $\frac{{}_4C_2}{{}_9C_2} = \frac{6}{36} = \frac{1}{6}$ **b.** $\frac{{}_5C_2}{{}_9C_2} = \frac{10}{36} = \frac{5}{18}$

14. $\frac{{}_7C_3 \cdot {}_4C_2}{{}_{11}C_5} = \frac{35 \cdot 6}{462} = \frac{210}{462} = \frac{5}{11}$

15. There is 1 possible seating arrangement: G, B, G, B, G, B, G, B, G. There are ${}_5P_5$ ways of arranging the 5 girls and ${}_4P_4$ ways of arranging the 4 boys.

$P(E) = \frac{1 \cdot {}_5P_5 \cdot {}_4P_4}{{}_9P_9} = \frac{120 \cdot 24}{362{,}880} = \frac{2880}{362{,}880} = \frac{1}{126}$

C **16.** There are 6 possible positions for the 4 adjacent seats and ${}_4P_4$ ways to arrange the boys in the seats. There are ${}_5P_5$ ways of arranging the girls in the remaining 5 seats. $P(E) = \frac{6 \cdot {}_4P_4 \cdot {}_5P_5}{{}_9P_9} = \frac{6 \cdot 120 \cdot 24}{362{,}880} = \frac{17{,}280}{362{,}880} = \frac{1}{21}$

Page 513 • WRITTEN EXERCISES

A **1. a.** $P(5\text{ H}) + P(6\text{ H}) = \frac{{}_6C_5}{2^6} + \frac{{}_6C_6}{2^6} = \frac{6}{64} + \frac{1}{64} = \frac{7}{64}$

b. $P(4\text{ H}) + P(5\text{ H}) + P(6\text{ H}) = \frac{{}_6C_4}{64} + \frac{6}{64} + \frac{1}{64} = \frac{15}{64} + \frac{6}{64} + \frac{1}{64} = \frac{22}{64} = \frac{11}{32}$

c. $P(3\text{ H}) + P(4\text{ H}) + P(5\text{ H}) + P(6\text{ H}) = \frac{{}_6C_3}{64} + \frac{15}{64} + \frac{6}{64} + \frac{1}{64} = \frac{20}{64} + \frac{15}{64} + \frac{6}{64} + \frac{1}{64} = \frac{42}{64} = \frac{21}{32}$

d. $P(\text{no H}) = \frac{{}_6C_0}{64} = \frac{1}{64}$

e. $P(\text{at least 4 H}) + P(\text{at least 3 T}) = \frac{11}{32} + \frac{21}{32} = \frac{32}{32} = 1$

2. a. $P(1\text{ R}) + P(2\text{ R}) = \frac{{}_3C_1 \cdot {}_{11}C_1}{{}_{14}C_2} + \frac{{}_3C_2}{{}_{14}C_2} = \frac{3 \cdot 11}{91} + \frac{3}{91} = \frac{36}{91}$

b. $P(1\text{ G}) + P(2\text{ G}) = \frac{{}_6C_1 \cdot {}_8C_1}{91} + \frac{{}_6C_2}{91} = \frac{6 \cdot 8}{91} + \frac{15}{91} = \frac{63}{91} = \frac{9}{13}$

c. $P(2\text{ R}) + P(2\text{ B}) + P(2\text{ G}) = \frac{3}{91} + \frac{{}_5C_2}{91} + \frac{15}{91} = \frac{3}{91} + \frac{10}{91} + \frac{15}{91} = \frac{28}{91} = \frac{4}{13}$

d. $1 - P(2\text{ same color}) = 1 - \frac{4}{13} = \frac{9}{13}$

3. $P(8) + P(9) + P(10) = \frac{{}_{10}C_8}{2^{10}} + \frac{{}_{10}C_9}{2^{10}} + \frac{{}_{10}C_{10}}{2^{10}} = \frac{45}{1024} + \frac{10}{1024} + \frac{1}{1024} = \frac{56}{1024} = \frac{7}{128}$

4. a. $P(2\text{ S}) = \frac{{}_{13}C_2}{{}_{52}C_2} = \frac{78}{1326} = \frac{1}{17}$

b. $P(\text{same suit}) = P(2\text{ S}) + P(2\text{ C}) + P(2\text{ D}) + P(2\text{ H}) = \frac{1}{17} + \frac{1}{17} + \frac{1}{17} + \frac{1}{17} = \frac{4}{17}$

c. $P(2\text{ J}) = \frac{{}_4C_2}{1326} = \frac{6}{1326} = \frac{1}{221}$

d. $P(\text{pair}) = P(\text{two 2's}) + P(\text{two 3's}) + \cdots + P(2\text{ K}) + P(2\text{ A}) = 13 \cdot P(2\text{ J}) = 13 \cdot \frac{1}{221} = \frac{1}{17}$

e. $P(2\text{ S}) + P(2\text{ J}) = \frac{1}{17} + \frac{1}{221} = \frac{14}{221}$

f. $P(\text{same suit}) + P(\text{pair}) = \frac{4}{17} + \frac{1}{17} = \frac{5}{17}$

5. a. $P(\text{E or O, not both}) = P(\text{E, no O}) + P(\text{O, no E}) = \frac{1 \cdot {}_4C_2}{{}_6C_3} + \frac{1 \cdot {}_4C_2}{{}_6C_3} = \frac{6}{20} + \frac{6}{20} = \frac{3}{5}$

b. $P(\text{E or O, not both}) + P(\text{both}) = \frac{3}{5} + \frac{{}_2C_2 \cdot {}_4C_1}{20} = \frac{3}{5} + \frac{4}{20} = \frac{4}{5}$

B **6. a.** $P(\text{same no. on 3}) = \frac{{}_6C_1}{6^3} = \frac{6}{216} = \frac{1}{36}$

b. $P(\text{same no. on 2}) = \frac{{}_6C_1 \cdot {}_5C_1}{216} \cdot {}_3P_2 = \frac{6 \cdot 5}{216} \cdot 3 = \frac{90}{216} = \frac{5}{12}$

c. $P(\text{same no. on 2}) + P(\text{same no. on 3}) = \frac{5}{12} + \frac{1}{36} = \frac{16}{36} = \frac{4}{9}$

7. $P(4\ \text{E}) + P(\text{no E}) = \frac{{}_3C_2}{{}_6C_2} \cdot \frac{{}_2C_2}{{}_5C_2} + \frac{{}_3C_2}{{}_6C_2} \cdot \frac{{}_3C_2}{{}_5C_2} = \frac{3}{15} \cdot \frac{1}{10} + \frac{3}{15} \cdot \frac{3}{10} = \frac{3}{150} + \frac{9}{150} = \frac{12}{150} = \frac{2}{25}$

8. $P(\text{A or B, not both}) = \frac{{}_2C_1 \cdot {}_8C_4}{{}_{10}C_5} = \frac{2 \cdot 70}{252} = \frac{140}{252} = \frac{5}{9}$

C **9.** $P(\text{2 B or 2 W}) = P(\text{2 B}) + P(\text{2 W}) - P(\text{2 B and 2 W at same time}) = \frac{{}_4C_2 \cdot {}_6C_2}{{}_{10}C_4} + \frac{{}_4C_2 \cdot {}_6C_2}{{}_{10}C_4} - \frac{{}_4C_2 \cdot {}_4C_2}{{}_{10}C_4} = \frac{6 \cdot 15}{210} + \frac{6 \cdot 15}{210} - \frac{6 \cdot 6}{210} = \frac{90}{210} + \frac{90}{210} - \frac{36}{210} = \frac{144}{210} = \frac{24}{35}$

10. $P(\text{A or E or both, not I}) = P(\text{A, not E, not I}) + P(\text{E, not A, not I}) + P(\text{A and E, not I}) = \frac{{}_1C_1 \cdot {}_7C_4}{{}_{10}C_5} + \frac{{}_1C_1 \cdot {}_7C_4}{{}_{10}C_5} + \frac{{}_2C_2 \cdot {}_7C_3}{{}_{10}C_5} = \frac{35}{252} + \frac{35}{252} + \frac{35}{252} = \frac{105}{252} = \frac{5}{12}$

Pages 517–518 • WRITTEN EXERCISES

A **1.** $P(C \cap D) = P(C) \cdot P(D|C) = 0.4(0.55) = 0.22$

2. $P(J \cap K) = P(K) \cdot P(J|K); \frac{1}{12} = \frac{1}{4} \cdot P(J|K); P(J|K) = \frac{1}{3}$

3. a. $P(\text{T, T}) \cdot P(6) = \frac{1}{4} \cdot \frac{1}{6} = \frac{1}{24}$ **b.** $P(\text{1 H}) \cdot P(\text{3 or 4}) = \frac{2}{4} \cdot \frac{2}{6} = \frac{1}{6}$

c. $P(\text{at least 1 H}) \cdot P(\text{not 5}) = \frac{3}{4} \cdot \frac{5}{6} = \frac{5}{8}$ **d.** $P(\text{at most 1 H}) \cdot P(< 4) = \frac{3}{4} \cdot \frac{3}{6} = \frac{3}{8}$

4. a. $P(\text{green, then red}) + P(\text{red, then green}) = \frac{2}{11} \cdot \frac{6}{10} + \frac{6}{11} \cdot \frac{2}{10} = \frac{24}{110} = \frac{12}{55}$

b. $P(\text{orange, then green}) + P(\text{green, then orange}) = \frac{3}{11} \cdot \frac{2}{10} + \frac{2}{11} \cdot \frac{3}{10} = \frac{12}{110} = \frac{6}{55}$

c. $P(\text{red, then green}) + P(\text{red, then red}) + P(\text{green, then red}) = \frac{6}{11} \cdot \frac{2}{10} + \frac{6}{11} \cdot \frac{5}{10} + \frac{2}{11} \cdot \frac{6}{10} = \frac{12}{110} + \frac{30}{110} + \frac{12}{110} = \frac{54}{110} = \frac{27}{55}$

d. $P(\text{orange, then orange}) + P(\text{orange, then green}) + P(\text{green, then orange}) = \frac{3}{11} \cdot \frac{2}{10} + \frac{3}{11} \cdot \frac{2}{10} + \frac{2}{11} \cdot \frac{3}{10} = \frac{18}{110} = \frac{9}{55}$

B **5. a.** $\frac{1}{13} \cdot \frac{3}{13} \cdot \frac{2}{12} = \frac{6}{2028} = \frac{1}{338}$ **b.** $\frac{3}{13} \cdot \frac{3}{13} \cdot \frac{2}{12} = \frac{18}{2028} = \frac{3}{338}$

6. a. $[P(\text{A, M}) + P(\text{M, A})][6 \cdot P(1, 1, 1)] = \left[\frac{1}{7} \cdot \frac{1}{6} + \frac{1}{7} \cdot \frac{1}{6}\right]\left[6 \cdot \frac{1}{6} \cdot \frac{1}{6} \cdot \frac{1}{6}\right] = \frac{2}{42} \cdot \frac{1}{36} = \frac{1}{756}$ **b.** $\left(\frac{3}{7} \cdot \frac{2}{6}\right)\left(\frac{3}{6} \cdot \frac{3}{6} \cdot \frac{3}{6}\right) = \frac{1}{56}$ **c.** $\left(\frac{4}{7} \cdot \frac{3}{6}\right)\left(\frac{5}{6} \cdot \frac{5}{6} \cdot \frac{5}{6}\right) = \frac{500}{1512} = \frac{125}{756}$

7. $P(A) = \frac{6}{36} = \frac{1}{6}$; $P(B) = \frac{1}{6}\cdot\frac{5}{6} + \frac{1}{6}\cdot\frac{6}{6} = \frac{5}{36} + \frac{6}{36} = \frac{11}{36}$; $P(A|B) = \frac{2}{11}$;
$P(B|A) = \frac{2}{6} = \frac{1}{3}$; $\frac{1}{6}\cdot\frac{1}{3} \stackrel{?}{=} \frac{11}{36}\cdot\frac{2}{11}$; $\frac{1}{18} = \frac{1}{18}$

8. $P(A) = P(\text{both black}) + P(\text{both red}) = \frac{26}{52}\cdot\frac{25}{51} + \frac{26}{52}\cdot\frac{25}{51} = \frac{25}{102} + \frac{25}{102} =$
$\frac{50}{102} = \frac{25}{51}$; $P(B) = 1 - P(A) = 1 - \frac{25}{51} = \frac{26}{51}$

9. $P(A) = \frac{2}{6} = \frac{1}{3}$; $P(B) = \frac{2}{6} = \frac{1}{3}$; $P(C) = 1 - P(\text{top or bottom is green}) =$
$1 - \left(\frac{2}{6} + \frac{2}{6}\right) = \frac{1}{3}$; $P(D) = P(\text{top or bottom is red}) = \frac{2}{6} + \frac{2}{6} = \frac{2}{3}$

10. No; $P(A|B) = \frac{1}{2} \neq \frac{1}{3} = P(A)$

11. No; $P(B|C) = \frac{1}{2} \neq \frac{1}{3} = P(B)$

12. No; $P(B|D) = \frac{1}{2} \neq \frac{1}{3} = P(B)$

13. A and C

14. $P(2\text{ G}, 2\text{ B}) + P(3\text{ G}, 1\text{ B}) = \left(\frac{3}{10}\cdot\frac{2}{9}\right)\left(\frac{3}{8}\cdot\frac{2}{7}\right)\frac{4!}{2!\cdot 2!} + \left(\frac{3}{10}\cdot\frac{2}{9}\cdot\frac{1}{8}\right)\left(\frac{3}{7}\right)\frac{4!}{3!} =$
$\frac{3}{70} + \frac{1}{70} = \frac{4}{70} = \frac{2}{35}$

C **15. a.** $\frac{1}{2}\left(\frac{p}{p+q}\right) = \frac{p}{2(p+q)}$

b. $\frac{1}{2}\left(\frac{p}{p+q}\right) + \frac{1}{2}\left(\frac{r}{r}\right) = \frac{p}{2(p+q)} + \frac{1}{2} = \frac{p+p+q}{2(p+q)} = \frac{2p+q}{2(p+q)}$

16. From Ex. 15b., $P(\text{red}) = \frac{2(4)+4}{2(4+4)} = \frac{3}{4}$;

$$P(\text{Box Y}|\text{red}) = \frac{P(\text{Box Y and red})}{P(\text{red})} = \frac{\frac{1}{2}}{\frac{3}{4}} = \frac{2}{3}$$

Page 518 • SELF-TEST 4

1. {(A, 1), (A, 2), (A, 3), (B, 1), (B, 2), (B, 3), (C, 1), (C, 2), (C, 3)};
{(A, 1), (A, 3), (B, 1), (B, 3)}

2. a. $\frac{2}{8} = \frac{1}{4}$ **b.** $\frac{6}{8} = \frac{3}{4}$

3. $P(7\text{ or }8) + P(3) + P(5) = \frac{2}{8} + \frac{1}{8} + \frac{1}{8} = \frac{4}{8} = \frac{1}{2}$

4. No; if the sum of the numbers equals 8, it is impossible for one die to show a 1.

Pages 519–520 • CHAPTER REVIEW

1. d **2.** $4 + 3 - 2 = 5$; b **3.** $4! = 24$; a **4.** $(5 - 1)! = 4! = 24$; c

5. $\frac{11!}{4!\cdot 4!\cdot 2!} = 34{,}650$; d **6.** ${}_{36}C_{33} = {}_{36}C_3 = \frac{36\cdot 35\cdot 34}{3\cdot 2\cdot 1} = 7140$; b

7. ${}_7C_3 = \frac{7 \cdot 6 \cdot 5}{3 \cdot 2 \cdot 1} = 35$; a **8.** ${}_6C_3 \times {}_5C_4 = \frac{6 \cdot 5 \cdot 4}{3 \cdot 2 \cdot 1} \cdot 5 = 100$; d

9. $\frac{7 \cdot 6 \cdot 5}{3 \cdot 2 \cdot 1}(3x)^4(2)^3 = 22{,}680x^4$; c

10. Sixth row: 1, 5, 10, 10, 5, 1. 3rd term: $10\left(\frac{x}{2}\right)^3(3)^2 = \frac{45}{4}x^3$; a

11. $2 \cdot 2 \cdot 2 \cdot 2 = 16$; b **12.** c **13.** $\frac{4}{6} = \frac{2}{3}$; b

14. $P(\text{at least 3 R}) = P(3\text{ R}) + P(4\text{ R}) + P(5\text{ R})$; d

15. $P(3) \cdot P(\text{H}) = \left(\frac{1}{6}\right)\left(\frac{1}{2}\right) = \frac{1}{12}$; d

Page 521 • CHAPTER TEST

1. $7 \cdot 7 \cdot 7 = 343$ **2.** ${}_{10}P_4 = 10 \cdot 9 \cdot 8 \cdot 7 = 5040$ **3.** $\frac{11!}{2! \cdot 2! \cdot 2! \cdot 2!} = 2{,}494{,}800$

4. ${}_{10}C_4 = \frac{10 \cdot 9 \cdot 8 \cdot 7}{4 \cdot 3 \cdot 2 \cdot 1} = 210$ **5.** ${}_4C_2 \times {}_4C_3 = \frac{4 \cdot 3}{2 \cdot 1} \times 4 = 24$

6. $\frac{13 \cdot 12 \cdot 11 \cdot 10 \cdot 9 \cdot 8 \cdot 7}{7 \cdot 6 \cdot 5 \cdot 4 \cdot 3 \cdot 2 \cdot 1}\left(\frac{3}{4}x\right)^6(-1)^7 = -\frac{1250964}{4096}x^6 = -\frac{312741}{1024}x^6$

7. Sixth row: 1, 5, 10, 10, 5, 1. $1(y)^5 + 5(y^4)(-3) + 10(y^3)(-3)^2 + 10(y^2)(-3)^3 + 5(y)(-3)^4 + 1(-3)^5 = y^5 - 15y^4 + 90y^3 - 270y^2 + 405y - 243$

8. {(H, 1), (H, 2), (H, 3), (H, 4), (H, 5), (H, 6), (T, 1), (T, 2), (T, 3), (T, 4), (T, 5), (T, 6)}

9. $\frac{{}_9C_3}{{}_{16}C_3} = \frac{84}{560} = \frac{3}{20}$

10. $P(\text{at least 2 B}) = P(2\text{ B}) + P(3\text{ B}) = \frac{{}_4C_2 \cdot {}_{12}C_1}{{}_{16}C_3} + \frac{{}_4C_3}{{}_{16}C_3} = \frac{6 \cdot 12}{560} + \frac{4}{560} = \frac{76}{560} = \frac{19}{140}$

11. $P(2\text{ B}) = \frac{7}{16} \cdot \frac{6}{15} = \frac{7}{40}$

Pages 521–523 • CUMULATIVE REVIEW

1. $x = \frac{-4 \pm \sqrt{16 - 4(20)}}{2} = \frac{-4 \pm \sqrt{-64}}{2} = \frac{-4 \pm 8i}{2} = -2 \pm 4i$; $\{-2 + 4i, -2 - 4i\}$

2. $x = \frac{2 \pm \sqrt{4 - 4(4)}}{2} = \frac{2 \pm \sqrt{-12}}{2} = \frac{2 \pm 2i\sqrt{3}}{2} = 1 \pm i\sqrt{3}$; $\{1 + i\sqrt{3}, 1 - i\sqrt{3}\}$

3. $D = b^2 - 4ac = 0$; $144 - 4(4)(3k) = 0$; $48k = 144$; $k = 3$

4. $r_1 + r_2 = -2 + i\sqrt{2} + (-2 - i\sqrt{2}) = -4 = -\frac{b}{a}$;

$r_1r_2 = (-2 + i\sqrt{2})(-2 - i\sqrt{2}) = 4 - 2i^2 = 6 = \frac{c}{a}$; $x^2 + 4x + 6 = 0$

5. $x = \frac{3}{2}$

6. $y = a(x - 4)^2 + 1$; $0 = a(3 - 4)^2 + 1$; $a = -1$; $y = -(x - 4)^2 + 1$

7. $y = ax^2 + bx + c$

(1): $-1 = a + b + c$ (1): $-6 = a + b$

(2): $5 = c$ (3): $10 = a - b$

(3): $15 = a - b + c$ $4 = 2a$; $a = 2, b = -8$; $y = 2x^2 - 8x + 5$

8. $\begin{array}{r|rrrr} -2 & 12 & 7 & -14 & 3 \\ & & -24 & 34 & -40 \\ \hline & 12 & -17 & 20 & -37 \end{array}$ $Q(-2) = -37$

9. possible roots are $\pm 1, \pm 3$; $\begin{array}{r|rrr|r} 1 & 1 & 3 & -1 & -3 \\ & & 1 & 4 & 3 \\ \hline & 1 & 4 & 3 & 0 \end{array}$

$x^2 + 4x + 3 = 0$; $(x + 3)(x + 1) = 0$; $x = -3$ or $x = -1$; $\{1, -1, -3\}$

10. variations of sign in $x^3 - 2x^2 - x + 7 = 0$ are 2; thus, 2 or 0 positive roots; variation of sign in $-x^3 - 2x^2 + x + 7 = 0$ is 1; thus, 1 negative root

11. Midpt. is $(3, -1)$; slope $= \frac{-2}{4} = -\frac{1}{2}$; $y + 1 = 2(x - 3)$; $y = 2x - 7$

12. $(x - 0)^2 + (y - 0)^2 = r^2$; $(-2)^2 + (-3)^2 = r^2$; $13 = r^2$; $x^2 + y^2 = 13$

13. $|y - 3| = \sqrt{(x - 0)^2 + (y + 3)^2}$; $(y - 3)^2 = x^2 + (y + 3)^2$;

$y^2 - 6y + 9 = x^2 + y^2 + 6y + 9$; $-12y = x^2$; $y = -\frac{1}{12}x^2$

14. $2a = 12, a = 6$; $b = 3$; $\frac{x^2}{36} + \frac{y^2}{9} = 1$

15. $F_1(0, 5), F_2(0, -5), V_1(0, -3)$: $F_1V_1 = 8, F_2V_1 = 2$; difference of focal radii $= 2a = 8 - 2, a = 3$; $b^2 = 5^2 - 3^2 = 25 - 9 = 16, b = 4$; $\frac{y^2}{16} - \frac{x^2}{9} = 1$

16. $y_1x_1^2 = y_2x_2^2$; $\frac{1}{2}(3\sqrt{2})^2 = y\left(\frac{2}{3}\right)^2$; $9 = \frac{4}{9}y$; $y = \frac{81}{4}$

17. $y = -4x$; $2x^2 - (-4x)^2 = 28$; $-14x^2 = 28$; $x^2 = -2$; $x = \pm i\sqrt{2}$;

$y = -4(\pm i\sqrt{2}) = \pm 4i\sqrt{2}$; $\{(i\sqrt{2}, -4i\sqrt{2}), (-i\sqrt{2}, 4i\sqrt{2})\}$

18. $y = -x^2$; $2x - (-x^2) - 3 = 0$; $x^2 + 2x - 3 = 0$; $(x + 3)(x - 1) = 0$; $x = -3$ or $x = 1$; $\{(-3, -9), (1, -1)\}$

19. $x = -y - 2\sqrt{3}$; $(-y - 2\sqrt{3})^2 + y^2 = 6$; $y^2 + 4y\sqrt{3} + 12 + y^2 = 6$;

$y^2 + 2y\sqrt{3} + 3 = 0$; $y = \frac{-2\sqrt{3} \pm \sqrt{12 - 4(3)}}{2} = -\sqrt{3}$; $x = \sqrt{3} - 2\sqrt{3} = -\sqrt{3}$; $\{(-\sqrt{3}, -\sqrt{3})\}$

20. $x^2 = \frac{5y^2 - 24}{2}$; $3\left(\frac{5y^2 - 24}{2}\right) + 2y^2 = 21$; $15y^2 - 72 + 4y^2 = 42$; $19y^2 = 114$;

$y^2 = 6$; $y = \pm\sqrt{6}$; $x^2 = \frac{30 - 24}{2} = 3$; $x = \pm\sqrt{3}$; $\{(\sqrt{3}, \sqrt{6}), (-\sqrt{3}, \sqrt{6}), (\sqrt{3}, -\sqrt{6}), (-\sqrt{3}, -\sqrt{6})\}$

21. Let y = length of altitude; x = length of base. $\frac{1}{2}xy = 30$ and

$x^2 + y^2 = (30 - x - y)^2$; $y = \frac{60}{x}$ and $x^2 + \left(\frac{60}{x}\right)^2 = \left(30 - x - \frac{60}{x}\right)^2$;

$x^2 + \frac{3600}{x^2} = 1020 - 60x - \frac{3600}{x} + x^2 + \frac{3600}{x^2}$; $0 = 1020 - 60x - \frac{3600}{x}$;

$60x^2 - 1020x + 3600 = 0$; $x^2 - 17x + 60 = 0$; $(x - 12)(x - 5) = 0$; $x = 12$ or $x = 5$; $y = 5$ or $y = 12$; $\sqrt{5^2 + 12^2} = 13$; 5 m, 12 m, 13 m

22. $\sqrt[4]{4} = \sqrt[2\cdot 2]{4} = \sqrt{2}$

23. $(2^3)^{2x-2} = (2^2)^{2-x}$; $2^{6x-6} = 2^{4x-2}$; $6x - 6 = 4x - 2$; $x = \frac{5}{4}$; $\left\{\frac{5}{4}\right\}$

24. $25^{3/2} = 5^3 = 125$; $\{125\}$

25. $\log_3 \frac{63}{7x} = \log_3 2$; $\frac{63}{7x} = 2$; $14x = 63$; $x = \frac{9}{2}$; $\left\{\frac{9}{2}\right\}$

26. $(5^3)^{x-6} = (5^2)^{2x-3}$; $5^{3x-18} = 5^{4x-6}$; $3x - 18 = 4x - 6$; $x = -12$; $\{-12\}$

27. $2^{3x-2} = 2^6$; $3x - 2 = 6$; $3x = 8$; $x = \frac{8}{3} \approx 2.67$; $\{2.67\}$

28. $A = 3000\left(1 + \frac{0.12}{4}\right)^{44} = 3000(1.03)^{44} \approx 11{,}014.36$; \$11,014.36

29. $e^7 = 1097$; $7 \ln e = \ln 1097$; $7 = \ln 1097$

30. $4 \cdot 4 = 16$ **31.** $6! = 720$ **32.** $\frac{8!}{2! \cdot 3! \cdot 3!} = 560$ **33.** ${}_7P_2 = 7 \cdot 6 = 42$

34. ${}_9P_9 = 9! = 362{,}880$ **35.** ${}_7C_2 = \frac{7 \cdot 6}{2 \cdot 1} = 21$ **36.** ${}_{12}C_0 = 1$

37. ${}_9C_4 = \frac{9 \cdot 8 \cdot 7 \cdot 6}{4 \cdot 3 \cdot 2 \cdot 1} = 126$ **38.** ${}_4C_3 \cdot {}_4C_3 = 4 \cdot 4 = 16$

39. $(2x)^4 + 4(2x)^3(3b) + \frac{4 \cdot 3}{2 \cdot 1}(2x)^2(3b)^2 + \frac{4 \cdot 3 \cdot 2}{3 \cdot 2 \cdot 1}(2x)(3b)^3 + \frac{4 \cdot 3 \cdot 2 \cdot 1}{4 \cdot 3 \cdot 2 \cdot 1}(3b)^4 =$
$16x^4 + 96x^3b + 216x^2b^2 + 216xb^3 + 81b^4$

40. $\frac{11 \cdot 10 \cdot 9 \cdot 8}{4 \cdot 3 \cdot 2 \cdot 1}(3a)^7(-b)^4 = 721{,}710a^7b^4$

41. $\frac{{}_5C_3}{2^5} = \frac{10}{32} = \frac{5}{16}$

42. $P(7, \text{no } 8) + P(8, \text{no } 7) = \frac{{}_1C_1 \cdot {}_8C_4}{{}_{10}C_5} + \frac{{}_1C_1 \cdot {}_8C_4}{{}_{10}C_5} = \frac{70}{252} + \frac{70}{252} = \frac{140}{252} = \frac{5}{9}$

Page 523 • CONTEST PROBLEMS

1. *Step 1*: Put the coins into three groups of 3; call the groups A, B, and C. Place groups A and B on opposite sides of the scale. *Case I*: A and B have the same weight. Then C contains the counterfeit coin. Replace B with C. If C is lighter than A, then the counterfeit coin is lighter than the others. If C is heavier, then the counterfeit coin is heavier than the others. *Case II*: A and B have different weights. Replace the lighter group with C. If C and the heavier group have the same weight, then the lighter group contains the counterfeit coin and the counterfeit coin is lighter than the others. If C and the heavier group have different weights (i.e., C is lighter), then the heavier group contains the counterfeit coin and the counterfeit coin is heavier than the others.

   The two weighings in Step 1 have determined which group contains the counterfeit coin and whether that coin is heavier or lighter than all of the others.

   Step 2: Clear the scales. Select 2 coins from the group containing the counterfeit coin. Place one coin on each side of the scale. *Case I*: The 2 coins have the same weight. Then the third coin is the counterfeit coin. *Case II*: The 2 coins have different weights. Using the information from Step 1, it can be determined which of the 2 coins (the lighter or heavier one) is the counterfeit coin.

2. Let w = width of river; x = rate of boat leaving A; y = rate of boat leaving B. Using $t = \frac{d}{r}$ and equating the time results in (*1*): $\frac{720}{x} = \frac{w - 720}{y}$ and (*2*): $\frac{w}{x} + \frac{1}{3} + \frac{400}{x} = \frac{w}{y} + \frac{1}{3} + \frac{w - 400}{y}$. Simplify each equation and solve for w.
(*1*): $720y = wx - 720x$, $w = \frac{720x + 720y}{x}$; (*2*): $\frac{w + 400}{x} = \frac{2w - 400}{y}$,
$wy + 400y = 2wx - 400x$, $2wx - wy = 400x + 400y$, $w = \frac{400x + 400y}{2x - y}$.
Set the two equations equal to each other: $\frac{720x + 720y}{x} = \frac{400x + 400y}{2x - y}$,
$1440x^2 + 720xy - 720y^2 = 400x^2 + 400xy$, $1040x^2 + 320xy - 720y^2 = 0$,
$13x^2 + 4xy - 9y^2 = 0$, $(13x - 9y)(x + y) = 0$, $x = \frac{9}{13}y$ or $x = -y$. Eliminate $x = -y$ since one of the rates would be negative. Substitute $x = \frac{9}{13}y$ into (*1*):
$\frac{720}{\frac{9}{13}y} = \frac{w - 720}{y}$, $720y = \frac{9}{13}y(w - 720)$, $720 = \frac{9}{13}w - \frac{6480}{13}$, $\frac{9}{13}w = \frac{15{,}840}{13}$,
$w = 1760$; 1760 ft

3. The pattern is consecutive two-digit prime numbers with the digits reversed. The next three prime numbers after 43 (i.e., 34, the last number shown in the set) are 47, 53, and 59. Thus the next three numbers in the pattern are 74, 35, and 95.

Pages 524–525 • PROGRAMMING IN PASCAL

1. a.
```
FUNCTION factorial (n : integer) : real;

VAR
  j : integer;
  prod : real;

BEGIN
  IF n = 0
    THEN factorial := 1
    ELSE BEGIN
           prod := 1;
           FOR j := n DOWNTO 1 DO
             prod := prod * j;
           factorial := prod;
         END;
END;
```

b. Answers may vary depending on the system used.

c.

n	n!
1	1
2	2
3	6
4	24
5	120
6	720
7	5040
8	40320
9	362880
10	3628800
11	39916800
12	479001600
13	6227020800
14	87178291200
15	1307674368000
16	20922789888000
17	355687428100000
18	6402373705700000
19	121645100410000000
20	2432902008200000000
21	51090942172000000000
22	1124000727800000000000
23	25852016739000000000000
24	620448401730000000000000
25	15511210043000000000000000
26	403291461120000000000000000
27	10888869450000000000000000000
28	304888344610000000000000000000
29	8841761993700000000000000000000
30	265252859810000000000000000000000

2. a. See the function *permutations* in Exercise 3b.
b. See the program in Exercise 3b.

3. a. See the function *combinations* in Exercise 3b.
b. $n \geq 0$; $0 \leq r \leq n$; ${}_nP_r$ and ${}_nC_r$ are accurate to 11 significant digits.

```
PROGRAM evaluate (INPUT, OUTPUT);

VAR
  n, r : integer;
  ans : char;

{ ********************************************************** }
FUNCTION factorial (n : integer) : real;

VAR
  j : integer;
  prod : real;
```

[*Program continued on next page*]

```
BEGIN
  IF n = 0
     THEN factorial := 1
     ELSE BEGIN
            prod := 1;
            FOR j := n DOWNTO 1 DO
              prod := prod * j;
            factorial := prod;
          END;
END;

{ ************************************************************ }
FUNCTION permutations (n, r : integer) : real;

BEGIN
  permutations := factorial(n)/factorial(n - r);
END;

{ ************************************************************ }
FUNCTION combinations (n, r : integer) : real;

BEGIN
  combinations := permutations(n, r)/factorial(r);
END;

{ ************************************************************ }
BEGIN {* main *}
  REPEAT
    write('Enter how many objects there are: ');
    readln(n);
  UNTIL n >= 0;
  REPEAT
    write('Enter how many are taken at a time: ');
    readln(r);
  UNTIL (r >= 0) AND (r <= n);
  writeln;
  REPEAT
    write('Do you want to find <p>ermutations or ');
    write('<c>ombinations: ');
    readln(ans);
  UNTIL ans IN ['P', 'p', 'C', 'c'];
  writeln;
  IF ans IN ['P', 'p']
     THEN BEGIN
            writeln('The number of permutations of');
            writeln(n:1, ' items taken ', r:1, ' at a time');
            writeln('is ', permutations(n, r):1:0, '.');
          END
```

```
    ELSE BEGIN
           writeln('The number of combinations of');
           writeln(n:1, ' items taken ', r:1, ' at a time');
           writeln('is ', combinations(n, r):1:0, '.');
         END;
END.
```

c. Modify the function *combinations* in the following way.

```
BEGIN
  IF r <= n/2
    THEN combinations := permutations(n, r)/factorial(r)
    ELSE combinations := permutations(n, n - r)/factorial(r);
END;
```

4. a. See the program in Exercise 4b.

b.
```
PROGRAM pascal_triangle (INPUT, OUTPUT);

VAR
  n, r, k : integer;
  ans : 1..2;

{*********************************************************}
FUNCTION factorial (n : integer) : real;

VAR
  j : integer;
  prod : real;

BEGIN
  IF n = 0
     THEN factorial := 1
     ELSE BEGIN
            prod := 1;
            FOR j := n DOWNTO 1 DO
                prod := prod * j;
            factorial := prod;
          END;
END;

{*********************************************************}
FUNCTION permutations (n, r : integer) : real;

BEGIN
  permutations := factorial(n)/factorial(n - r);
END;
```

[*Program continued on next page*]

```
{**********************************************************}
FUNCTION combinations (n, r : integer) : real;

BEGIN
  combinations := permutations(n, r)/factorial(r);
END;

{**********************************************************}
BEGIN {* main *}
  REPEAT
    write('Enter <1> for the first n rows or ');
    write('<2> for just the nth row: ');
    readln(ans);
    writeln;
  UNTIL ans IN [1, 2];
  write('Enter the value of n: ');
  readln(k);
  IF ans = 1
     THEN FOR n := 0 TO k - 1 DO
              BEGIN
                FOR r := 0 TO n DO
                    write(combinations(n, r):1:0, ' ');
                writeln;
              END
     ELSE FOR r := 0 TO k - 1 DO
              write(combinations(k - 1, r):1:0, ' ');
END.
```

Page 531 • WRITTEN EXERCISES

A **1.** $\begin{bmatrix} -4 \\ -1 \end{bmatrix}$ **2.** $[-2 \quad -3]$ **3.** $\begin{bmatrix} 4 & -1 \\ 12 & 1 \end{bmatrix}$ **4.** $\begin{bmatrix} 8 & -7 \\ -4 & 6 \end{bmatrix}$ **5.** $\begin{bmatrix} 2 & 8 & -8 \\ -4 & 2 & 6 \end{bmatrix}$

6. $[7 \quad -1 \quad 3]$ **7.** $\begin{bmatrix} -9 & 5 \\ 1 & -7 \\ 7 & -6 \end{bmatrix}$ **8.** $\begin{bmatrix} 20 & 22 & -11 \\ 4 & -2 & -3 \\ 10 & 0 & 16 \end{bmatrix}$

9. $w + 3 = -2$, $w = -5$; $2 - x = 5$, $x = -3$; $z + 1 = 6$, $z = 5$; $2y = 10$, $y = 5$

10. $w - 6 = 2$, $w = 8$; $3x - 1 = 11$, $x = 4$; $y + 7 = -2$, $y = -9$; $2z + 3 = -5$, $2z = -8$, $z = -4$

11. $x + 3 = -2$, $x = -5$; $x - y = 3$, $-5 - y = 3$, $y = -8$

12. $$\begin{array}{r} x - y = 17 \\ \underline{x + y = 7} \\ 2x = 24; x = 12; 12 + y = 7, y = -5 \end{array}$$

13. (1): $w - x = 5$ (2): $x - y = -1$ (3): $w + y = 2$;
from (2): $x = y - 1$; (1): $w - (y - 1) = 5$, $w - y = 4$
$$\begin{array}{rl} & w - y = 4 \\ (3): & \underline{w + y = 2} \\ & 2w = 6;\ w = 3;\ 3 + y = 2,\ y = -1;\ 3 - x = 5,\ x = -2 \end{array}$$

14. (1): $w + x = 7$ (2): $w - y = -1$ (3): $y + 2 = 0$ (4): $z + y = 2$; from (3): $y = -2$; $w - (-2) = -1$, $w = -3$; $-3 + x = 7$, $x = 10$; $z - 2 = 2$, $z = 4$

15. (1): $w - x = 11$ (2): $y + w = 6$ (3): $x - y = -5$; from (2): $w = 6 - y$;
(1): $6 - y - x = 11$, $x + y = -5$
$$\begin{array}{rl} & x + y = -5 \\ (3): & \underline{x - y = -5} \\ & 2x = -10;\ x = -5;\ -5 - y = -5,\ y = 0;\ 0 + w = 6,\ w = 6 \end{array}$$

16. (1): $w + 3 + x - 2 = 5$ (2): $y - 4 + 3 = 1$ (3): $x - z = 2$

(4): $z + 3z = -2$ or (1): $w + x = 4$ (2): $y = 2$ (3): $x - z = 2$ (4): $z = -\frac{1}{2}$;
thus $x - (-\frac{1}{2}) = 2$, $x = \frac{3}{2}$; $w + \frac{3}{2} = 4$, $w = \frac{5}{2}$

B **17.** $|A| = 2(6) - 3(-1) = 15$; $|B| = -4(-7) - 2(5) = 18$; $A + B = \begin{bmatrix} -2 & 8 \\ 1 & -1 \end{bmatrix}$;
$|A + B| = -2(-1) - 1(8) = -6$; $|A| + |B| = 33 \neq |A + B|$

18. $|A| = ad - bc$; $|B| = rctb - rdta$; $A + B = \begin{bmatrix} a & b \\ c & d \end{bmatrix} + \begin{bmatrix} rc & rd \\ ta & tb \end{bmatrix} =$
$\begin{bmatrix} a + rc & b + rd \\ c + ta & d + tb \end{bmatrix}$; $|A + B| = (a + rc)(d + tb) - (b + rd)(c + ta) =$
$ad + rcd + atb + rctb - bc - rdc - atb - rdta = (ad - bc) + (rctb - rdta) =$
$|A| + |B|$

C **19.** $|A| = ad - bc$; $|B| = wz - xy$; $A + B = \begin{bmatrix} a & b \\ c & d \end{bmatrix} + \begin{bmatrix} w & x \\ y & z \end{bmatrix} = \begin{bmatrix} a + w & b + x \\ c + y & d + z \end{bmatrix}$;
$|A + B| = (a + w)(d + z) - (b + x)(c + y) = ad + dw + az + wz - bc -$
$cx - by - xy$; $|A| + |B| + \begin{vmatrix} a & b \\ y & z \end{vmatrix} + \begin{vmatrix} w & x \\ c & d \end{vmatrix} = ad - bc + wz - xy + az - by +$
$wd - cx = |A + B|$

Page 534 • WRITTEN EXERCISES

A 1. $X + \begin{bmatrix} 2 & 0 \\ -3 & 1 \end{bmatrix} + \begin{bmatrix} -2 & 0 \\ 3 & -1 \end{bmatrix} = \begin{bmatrix} 7 & -4 \\ 2 & -5 \end{bmatrix} + \begin{bmatrix} -2 & 0 \\ 3 & -1 \end{bmatrix}$; $X + \begin{bmatrix} 0 & 0 \\ 0 & 0 \end{bmatrix} =$ $\begin{bmatrix} 5 & -4 \\ 5 & -6 \end{bmatrix}$; $X = \begin{bmatrix} 5 & -4 \\ 5 & -6 \end{bmatrix}$

2. $X + \begin{bmatrix} -1 & 7 \\ 5 & 8 \end{bmatrix} + \begin{bmatrix} 1 & -7 \\ -5 & -8 \end{bmatrix} = \begin{bmatrix} 8 & -6 \\ -2 & 0 \end{bmatrix} + \begin{bmatrix} 1 & -7 \\ -5 & -8 \end{bmatrix}$; $X + \begin{bmatrix} 0 & 0 \\ 0 & 0 \end{bmatrix} =$ $\begin{bmatrix} 9 & -13 \\ -7 & -8 \end{bmatrix}$; $X = \begin{bmatrix} 9 & -13 \\ -7 & -8 \end{bmatrix}$

3. $X - \begin{bmatrix} 2 & 1 \\ -3 & -9 \end{bmatrix} + \begin{bmatrix} 2 & 1 \\ -3 & -9 \end{bmatrix} = \begin{bmatrix} 6 & -2 \\ 3 & -4 \end{bmatrix} + \begin{bmatrix} 2 & 1 \\ -3 & -9 \end{bmatrix}$; $X + \begin{bmatrix} 0 & 0 \\ 0 & 0 \end{bmatrix} =$ $\begin{bmatrix} 8 & -1 \\ 0 & -13 \end{bmatrix}$; $X = \begin{bmatrix} 8 & -1 \\ 0 & -13 \end{bmatrix}$

4. $X - \begin{bmatrix} 8 & -3 \\ 5 & -4 \end{bmatrix} + \begin{bmatrix} 8 & -3 \\ 5 & -4 \end{bmatrix} = \begin{bmatrix} 1 & 4 \\ -11 & 0 \end{bmatrix} + \begin{bmatrix} 8 & -3 \\ 5 & -4 \end{bmatrix}$; $X + \begin{bmatrix} 0 & 0 \\ 0 & 0 \end{bmatrix} =$ $\begin{bmatrix} 9 & 1 \\ -6 & -4 \end{bmatrix}$; $X = \begin{bmatrix} 9 & 1 \\ -6 & -4 \end{bmatrix}$

5. $\begin{bmatrix} 3 & 8 \\ -6 & 7 \end{bmatrix} - X + \begin{bmatrix} -3 & -8 \\ 6 & -7 \end{bmatrix} = \begin{bmatrix} 4 & 5 \\ 2 & -1 \end{bmatrix} + \begin{bmatrix} -3 & -8 \\ 6 & -7 \end{bmatrix}$; $-X + \begin{bmatrix} 0 & 0 \\ 0 & 0 \end{bmatrix} =$ $\begin{bmatrix} 1 & -3 \\ 8 & -8 \end{bmatrix}$; $-X = \begin{bmatrix} 1 & -3 \\ 8 & -8 \end{bmatrix}$; $X = \begin{bmatrix} -1 & 3 \\ -8 & 8 \end{bmatrix}$

6. $\begin{bmatrix} -4 & 0 \\ 2 & 5 \end{bmatrix} - X + \begin{bmatrix} 4 & 0 \\ -2 & -5 \end{bmatrix} = \begin{bmatrix} -7 & 3 \\ -2 & 1 \end{bmatrix} + \begin{bmatrix} 4 & 0 \\ -2 & -5 \end{bmatrix}$; $-X + \begin{bmatrix} 0 & 0 \\ 0 & 0 \end{bmatrix} =$ $\begin{bmatrix} -3 & 3 \\ -4 & -4 \end{bmatrix}$; $-X = \begin{bmatrix} -3 & 3 \\ -4 & -4 \end{bmatrix}$; $X = \begin{bmatrix} 3 & -3 \\ 4 & 4 \end{bmatrix}$

7. $\begin{bmatrix} 4 & -3 & 9 \\ 2 & 5 & -6 \end{bmatrix} + X + \begin{bmatrix} -4 & 3 & -9 \\ -2 & -5 & 6 \end{bmatrix} = \begin{bmatrix} -2 & 0 & 6 \\ 2 & -3 & 0 \end{bmatrix} + \begin{bmatrix} -4 & 3 & -9 \\ -2 & -5 & 6 \end{bmatrix}$; $X + \begin{bmatrix} 0 & 0 & 0 \\ 0 & 0 & 0 \end{bmatrix} = \begin{bmatrix} -6 & 3 & -3 \\ 0 & -8 & 6 \end{bmatrix}$; $X = \begin{bmatrix} -6 & 3 & -3 \\ 0 & -8 & 6 \end{bmatrix}$

8. $\begin{bmatrix} 6 & -10 & 0 \\ -9 & 7 & 4 \end{bmatrix} - X + \begin{bmatrix} -6 & 10 & 0 \\ 9 & -7 & -4 \end{bmatrix} = \begin{bmatrix} 4 & -6 & 2 \\ 1 & 0 & -3 \end{bmatrix} + \begin{bmatrix} -6 & 10 & 0 \\ 9 & -7 & -4 \end{bmatrix}$; $-X + \begin{bmatrix} 0 & 0 & 0 \\ 0 & 0 & 0 \end{bmatrix} = \begin{bmatrix} -2 & 4 & 2 \\ 10 & -7 & -7 \end{bmatrix}$; $-X = \begin{bmatrix} -2 & 4 & 2 \\ 10 & -7 & -7 \end{bmatrix}$; $X = \begin{bmatrix} 2 & -4 & -2 \\ -10 & 7 & 7 \end{bmatrix}$

B 9.

$-X + [a \quad b] = [c \quad d]$	Comm. prop.
$-X + ([a \quad b] + [-a \quad -b]) = [c \quad d] + [-a \quad -b]$	Add. prop. of =; assoc. prop.
$-X + [0 \quad 0] = [c - a \quad d - b]$	Add. inv. prop.
$-X = [c - a \quad d - b]$	Iden. prop. for add.
$X = [a - c \quad b - d]$	Def. of opp. of a matrix

10. $X + \begin{bmatrix} a \\ b \end{bmatrix} = \begin{bmatrix} c \\ d \end{bmatrix}$ Comm. prop.

$X + \left(\begin{bmatrix} a \\ b \end{bmatrix} + \begin{bmatrix} -a \\ -b \end{bmatrix}\right) = \begin{bmatrix} c \\ d \end{bmatrix} + \begin{bmatrix} -a \\ -b \end{bmatrix}$ Add. prop. of =; assoc. prop.

$X + \begin{bmatrix} 0 \\ 0 \end{bmatrix} = \begin{bmatrix} c - a \\ d - b \end{bmatrix}$ Add. inv. prop.

$X = \begin{bmatrix} c - a \\ d - b \end{bmatrix}$ Iden. prop. for add.

In Exs. 11–16, $A = \begin{bmatrix} a_1 & b_1 \\ a_2 & b_2 \end{bmatrix}$, $B = \begin{bmatrix} c_1 & d_1 \\ c_2 & d_2 \end{bmatrix}$ and $C = \begin{bmatrix} e_1 & f_1 \\ e_2 & f_2 \end{bmatrix}$.

11. $(A + B) + C = \begin{bmatrix} a_1 + c_1 & b_1 + d_1 \\ a_2 + c_2 & b_2 + d_2 \end{bmatrix} + C = \begin{bmatrix} (a_1 + c_1) + e_1 & (b_1 + d_1) + f_1 \\ (a_2 + c_2) + e_2 & (b_2 + d_2) + f_2 \end{bmatrix} =$

$\begin{bmatrix} a_1 + (c_1 + e_1) & b_1 + (d_1 + f_1) \\ a_2 + (c_2 + e_2) & b_2 + (d_2 + f_2) \end{bmatrix} = A + \begin{bmatrix} c_1 + e_1 & d_1 + f_1 \\ c_2 + e_2 & d_2 + f_2 \end{bmatrix} = A + (B + C)$

12. $A + O = \begin{bmatrix} a_1 & b_1 \\ a_2 & b_2 \end{bmatrix} + \begin{bmatrix} 0 & 0 \\ 0 & 0 \end{bmatrix} = \begin{bmatrix} a_1 + 0 & b_1 + 0 \\ a_2 + 0 & b_2 + 0 \end{bmatrix} = \begin{bmatrix} a_1 & b_1 \\ a_2 & b_2 \end{bmatrix} = A$

13. $A + (-A) = \begin{bmatrix} a_1 & b_1 \\ a_2 & b_2 \end{bmatrix} + \begin{bmatrix} -a_1 & -b_1 \\ -a_2 & -b_2 \end{bmatrix} = \begin{bmatrix} 0 & 0 \\ 0 & 0 \end{bmatrix} = O$

14. If $A = -A$, then $\begin{bmatrix} a_1 & b_1 \\ a_2 & b_2 \end{bmatrix} = \begin{bmatrix} -a_1 & -b_1 \\ -a_2 & -b_2 \end{bmatrix}$. By the definition of equality of matrices, $a_1 = -a_1$, $b_1 = -b_1$, $a_2 = -a_2$, and $b_2 = -b_2$.
$\therefore a_1 = b_1 = a_2 = b_2 = 0$, and $A = O$.

15. $(A + B) - A = \begin{bmatrix} a_1 + c_1 & b_1 + d_1 \\ a_2 + c_2 & b_2 + d_2 \end{bmatrix} - \begin{bmatrix} a_1 & b_1 \\ a_2 & b_2 \end{bmatrix} =$

$\begin{bmatrix} a_1 + c_1 - a_1 & b_1 + d_1 - b_1 \\ a_2 + c_2 - a_2 & b_2 + d_2 - b_2 \end{bmatrix} = \begin{bmatrix} c_1 & d_1 \\ c_2 & d_2 \end{bmatrix} = B$

16. If $A + B = A$, then $\begin{bmatrix} a_1 + c_1 & b_1 + d_1 \\ a_2 + c_2 & b_2 + d_2 \end{bmatrix} = \begin{bmatrix} a_1 & b_1 \\ a_2 & b_2 \end{bmatrix}$ and $a_1 + c_1 = a_1$, $b_1 + d_1 = b_1$, $a_2 + c_2 = a_2$, $b_2 + d_2 = b_2$. $\therefore c_1 = d_1 = c_2 = d_2 = 0$ and $B = O$.

C 17. $3x + 9 = 0$; $3x = -9$; $x = -3$; $y^2 - 3y - 10 = 0$; $(y - 5)(y + 2) = 0$; $y = 5$ or $y = -2$; $6z + (8 + z^2) = 0$; $z^2 + 6z + 8 = 0$; $(z + 4)(z + 2) = 0$; $z = -2$ or $z = -4$; $\{(-3, 5, -2), (-3, 5, -4), (-3, -2, -2), (-3, -2, -4)\}$

Pages 537–538 • WRITTEN EXERCISES

A 1. $X = 3\begin{bmatrix} 9 & -3 \\ -21 & 0 \end{bmatrix} = \begin{bmatrix} 27 & -9 \\ -63 & 0 \end{bmatrix}$ **2.** $X = -4\begin{bmatrix} 11 & -7 \\ 1 & 10 \end{bmatrix} = \begin{bmatrix} -44 & 28 \\ -4 & -40 \end{bmatrix}$

3. $X = 2\begin{bmatrix} 9 & -3 \\ -21 & 0 \end{bmatrix} + \begin{bmatrix} 11 & -7 \\ 1 & 10 \end{bmatrix} = \begin{bmatrix} 18 & -6 \\ -42 & 0 \end{bmatrix} + \begin{bmatrix} 11 & -7 \\ 1 & 10 \end{bmatrix} = \begin{bmatrix} 29 & -13 \\ -41 & 10 \end{bmatrix}$

4. $X = \begin{bmatrix} 9 & -3 \\ -21 & 0 \end{bmatrix} - 2\begin{bmatrix} 11 & -7 \\ 1 & 10 \end{bmatrix} = \begin{bmatrix} 9 & -3 \\ -21 & 0 \end{bmatrix} - \begin{bmatrix} 22 & -14 \\ 2 & 20 \end{bmatrix} = \begin{bmatrix} -13 & 11 \\ -23 & -20 \end{bmatrix}$

5. $3X = \begin{bmatrix} 9 & -3 \\ -21 & 0 \end{bmatrix}$; $X = \frac{1}{3}\begin{bmatrix} 9 & -3 \\ -21 & 0 \end{bmatrix} = \begin{bmatrix} 3 & -1 \\ -7 & 0 \end{bmatrix}$

6. $\frac{1}{2}X = \begin{bmatrix} 11 & -7 \\ 1 & 10 \end{bmatrix}$; $X = 2\begin{bmatrix} 11 & -7 \\ 1 & 10 \end{bmatrix} = \begin{bmatrix} 22 & -14 \\ 2 & 20 \end{bmatrix}$

7. $X + \begin{bmatrix} 9 & -3 \\ -21 & 0 \end{bmatrix} = 2\begin{bmatrix} 11 & -7 \\ 1 & 10 \end{bmatrix}$; $X + \begin{bmatrix} 9 & -3 \\ -21 & 0 \end{bmatrix} + \begin{bmatrix} -9 & 3 \\ 21 & 0 \end{bmatrix} =$
$\begin{bmatrix} 22 & -14 \\ 2 & 20 \end{bmatrix} + \begin{bmatrix} -9 & 3 \\ 21 & 0 \end{bmatrix}$; $X = \begin{bmatrix} 13 & -11 \\ 23 & 20 \end{bmatrix}$

8. $X - \begin{bmatrix} 11 & -7 \\ 1 & 10 \end{bmatrix} = 3\begin{bmatrix} 9 & -3 \\ -21 & 0 \end{bmatrix}$; $X - \begin{bmatrix} 11 & -7 \\ 1 & 10 \end{bmatrix} + \begin{bmatrix} 11 & -7 \\ 1 & 10 \end{bmatrix} =$
$\begin{bmatrix} 27 & -9 \\ -63 & 0 \end{bmatrix} + \begin{bmatrix} 11 & -7 \\ 1 & 10 \end{bmatrix}$; $X = \begin{bmatrix} 38 & -16 \\ -62 & 10 \end{bmatrix}$

9. $\frac{1}{3}X = \begin{bmatrix} 11 & -7 \\ 1 & 10 \end{bmatrix} - \begin{bmatrix} 9 & -3 \\ -21 & 0 \end{bmatrix}$; $\frac{1}{3}X = \begin{bmatrix} 2 & -4 \\ 22 & 10 \end{bmatrix}$; $X = 3\begin{bmatrix} 2 & -4 \\ 22 & 10 \end{bmatrix} =$
$\begin{bmatrix} 6 & -12 \\ 66 & 30 \end{bmatrix}$

10. $3X = \begin{bmatrix} 9 & -3 \\ -21 & 0 \end{bmatrix} - 3\begin{bmatrix} 11 & -7 \\ 1 & 10 \end{bmatrix}$; $3X = \begin{bmatrix} 9 & -3 \\ -21 & 0 \end{bmatrix} - \begin{bmatrix} 33 & -21 \\ 3 & 30 \end{bmatrix}$;
$3X = \begin{bmatrix} -24 & 18 \\ -24 & -30 \end{bmatrix}$; $X = \frac{1}{3}\begin{bmatrix} -24 & 18 \\ -24 & -30 \end{bmatrix} = \begin{bmatrix} -8 & 6 \\ -8 & -10 \end{bmatrix}$

11. $5X = 3\begin{bmatrix} 9 & -3 \\ -21 & 0 \end{bmatrix} - 2\begin{bmatrix} 11 & -7 \\ 1 & 10 \end{bmatrix}$; $5X = \begin{bmatrix} 27 & -9 \\ -63 & 0 \end{bmatrix} - \begin{bmatrix} 22 & -14 \\ 2 & 20 \end{bmatrix}$;
$5X = \begin{bmatrix} 5 & 5 \\ -65 & -20 \end{bmatrix}$; $X = \frac{1}{5}\begin{bmatrix} 5 & 5 \\ -65 & -20 \end{bmatrix} = \begin{bmatrix} 1 & 1 \\ -13 & -4 \end{bmatrix}$

12. $2X - 3\begin{bmatrix} 9 & -3 \\ -21 & 0 \end{bmatrix} = \begin{bmatrix} 11 & -7 \\ 1 & 10 \end{bmatrix}$; $2X - \begin{bmatrix} 27 & -9 \\ -63 & 0 \end{bmatrix} = \begin{bmatrix} 11 & -7 \\ 1 & 10 \end{bmatrix}$;
$2X - \begin{bmatrix} 27 & -9 \\ -63 & 0 \end{bmatrix} + \begin{bmatrix} 27 & -9 \\ -63 & 0 \end{bmatrix} = \begin{bmatrix} 11 & -7 \\ 1 & 10 \end{bmatrix} + \begin{bmatrix} 27 & -9 \\ -63 & 0 \end{bmatrix}$;
$2X = \begin{bmatrix} 38 & -16 \\ -62 & 10 \end{bmatrix}$; $X = \frac{1}{2}\begin{bmatrix} 38 & -16 \\ -62 & 10 \end{bmatrix} = \begin{bmatrix} 19 & -8 \\ -31 & 5 \end{bmatrix}$

13. $X + 2\begin{bmatrix} 1 & -2 \\ 4 & -5 \end{bmatrix} = \begin{bmatrix} -13 & -4 \\ 8 & 5 \end{bmatrix}$; $X + \begin{bmatrix} 2 & -4 \\ 8 & -10 \end{bmatrix} = \begin{bmatrix} -13 & -4 \\ 8 & 5 \end{bmatrix}$;
$X + \begin{bmatrix} 2 & -4 \\ 8 & -10 \end{bmatrix} + \begin{bmatrix} -2 & 4 \\ -8 & 10 \end{bmatrix} = \begin{bmatrix} -13 & -4 \\ 8 & 5 \end{bmatrix} + \begin{bmatrix} -2 & 4 \\ -8 & 10 \end{bmatrix}$; $X = \begin{bmatrix} -15 & 0 \\ 0 & 15 \end{bmatrix}$

14. $4\begin{bmatrix} 1 & -2 \\ 4 & -5 \end{bmatrix} - X = 2\begin{bmatrix} -13 & -4 \\ 8 & 5 \end{bmatrix}$; $\begin{bmatrix} 4 & -8 \\ 16 & -20 \end{bmatrix} - X = \begin{bmatrix} -26 & -8 \\ 16 & 10 \end{bmatrix}$;
$\begin{bmatrix} 4 & -8 \\ 16 & -20 \end{bmatrix} + \begin{bmatrix} -4 & 8 \\ -16 & 20 \end{bmatrix} - X = \begin{bmatrix} -26 & -8 \\ 16 & 10 \end{bmatrix} + \begin{bmatrix} -4 & 8 \\ -16 & 20 \end{bmatrix}$;
$-X = \begin{bmatrix} -30 & 0 \\ 0 & 30 \end{bmatrix}$; $X = \begin{bmatrix} 30 & 0 \\ 0 & -30 \end{bmatrix}$

15. $\frac{1}{2}X + 5\begin{bmatrix} 1 & -2 \\ 4 & -5 \end{bmatrix} = \begin{bmatrix} -13 & -4 \\ 8 & 5 \end{bmatrix}$; $\frac{1}{2}X + \begin{bmatrix} 5 & -10 \\ 20 & -25 \end{bmatrix} = \begin{bmatrix} -13 & -4 \\ 8 & 5 \end{bmatrix}$;
$\frac{1}{2}X + \begin{bmatrix} 5 & -10 \\ 20 & -25 \end{bmatrix} + \begin{bmatrix} -5 & 10 \\ -20 & 25 \end{bmatrix} = \begin{bmatrix} -13 & -4 \\ 8 & 5 \end{bmatrix} + \begin{bmatrix} -5 & 10 \\ -20 & 25 \end{bmatrix}$;

$$\frac{1}{2}X = \begin{bmatrix} -18 & 6 \\ -12 & 30 \end{bmatrix}; X = 2\begin{bmatrix} -18 & 6 \\ -12 & 30 \end{bmatrix} = \begin{bmatrix} -36 & 12 \\ -24 & 60 \end{bmatrix}$$

16. $3X - 5\begin{bmatrix} 1 & -2 \\ 4 & -5 \end{bmatrix} = 2\begin{bmatrix} -13 & -4 \\ 8 & 5 \end{bmatrix}; 3X - \begin{bmatrix} 5 & -10 \\ 20 & -25 \end{bmatrix} = \begin{bmatrix} -26 & -8 \\ 16 & 10 \end{bmatrix};$

$3X - \begin{bmatrix} 5 & -10 \\ 20 & -25 \end{bmatrix} + \begin{bmatrix} 5 & -10 \\ 20 & -25 \end{bmatrix} = \begin{bmatrix} -26 & -8 \\ 16 & 10 \end{bmatrix} + \begin{bmatrix} 5 & -10 \\ 20 & -25 \end{bmatrix};$

$3X = \begin{bmatrix} -21 & -18 \\ 36 & -15 \end{bmatrix}; X = \frac{1}{3}\begin{bmatrix} -21 & -18 \\ 36 & -15 \end{bmatrix} = \begin{bmatrix} -7 & -6 \\ 12 & -5 \end{bmatrix}$

17. $4\begin{bmatrix} 1 & -2 \\ 4 & -5 \end{bmatrix} - 5X = -3\begin{bmatrix} -13 & -4 \\ 8 & 5 \end{bmatrix}; \begin{bmatrix} 4 & -8 \\ 16 & -20 \end{bmatrix} - 5X = \begin{bmatrix} 39 & 12 \\ -24 & -15 \end{bmatrix};$

$\begin{bmatrix} 4 & -8 \\ 16 & -20 \end{bmatrix} + \begin{bmatrix} -4 & 8 \\ -16 & 20 \end{bmatrix} - 5X = \begin{bmatrix} 39 & 12 \\ -24 & -15 \end{bmatrix} + \begin{bmatrix} -4 & 8 \\ -16 & 20 \end{bmatrix};$

$-5X = \begin{bmatrix} 35 & 20 \\ -40 & 5 \end{bmatrix}; X = -\frac{1}{5}\begin{bmatrix} 35 & 20 \\ -40 & 5 \end{bmatrix} = \begin{bmatrix} -7 & -4 \\ 8 & -1 \end{bmatrix}$

18. $2X + 3\begin{bmatrix} -13 & -4 \\ 8 & 5 \end{bmatrix} = 5\begin{bmatrix} 1 & -2 \\ 4 & -5 \end{bmatrix}; 2X + \begin{bmatrix} -39 & -12 \\ 24 & 15 \end{bmatrix} = \begin{bmatrix} 5 & -10 \\ 20 & -25 \end{bmatrix};$

$2X + \begin{bmatrix} -39 & -12 \\ 24 & 15 \end{bmatrix} + \begin{bmatrix} 39 & 12 \\ -24 & -15 \end{bmatrix} = \begin{bmatrix} 5 & -10 \\ 20 & -25 \end{bmatrix} + \begin{bmatrix} 39 & 12 \\ -24 & -15 \end{bmatrix};$

$2X = \begin{bmatrix} 44 & 2 \\ -4 & -40 \end{bmatrix}; X = \frac{1}{2}\begin{bmatrix} 44 & 2 \\ -4 & -40 \end{bmatrix} = \begin{bmatrix} 22 & 1 \\ -2 & -20 \end{bmatrix}$

B **19.** Let $A = \begin{bmatrix} a_1 & b_1 \\ a_2 & b_2 \end{bmatrix}$. Then $cA = c\begin{bmatrix} a_1 & b_1 \\ a_2 & b_2 \end{bmatrix} = \begin{bmatrix} ca_1 & cb_1 \\ ca_2 & cb_2 \end{bmatrix}$, by the definition of the product of a scalar and a matrix; so $cA \in S_{2\times 2}$ because the product of two real numbers is a real number.

20. Let $A = \begin{bmatrix} a_1 & b_1 \\ a_2 & b_2 \end{bmatrix}$. $c(dA) = c\begin{bmatrix} da_1 & db_1 \\ da_2 & db_2 \end{bmatrix} = \begin{bmatrix} cda_1 & cdb_1 \\ cda_2 & cdb_2 \end{bmatrix};$

$(cd)A = cd\begin{bmatrix} a_1 & b_1 \\ a_2 & b_2 \end{bmatrix} = \begin{bmatrix} cda_1 & cdb_1 \\ cda_2 & cdb_2 \end{bmatrix}; \therefore c(dA) = (cd)A$

21. Let $A = \begin{bmatrix} a_1 & b_1 \\ a_2 & b_2 \end{bmatrix}$. $(c + d)A = \begin{bmatrix} (c+d)a_1 & (c+d)b_1 \\ (c+d)a_2 & (c+d)b_2 \end{bmatrix} =$

$\begin{bmatrix} ca_1 + da_1 & cb_1 + db_1 \\ ca_2 + da_2 & cb_2 + db_2 \end{bmatrix}; cA + dA = c\begin{bmatrix} a_1 & b_1 \\ a_2 & b_2 \end{bmatrix} + d\begin{bmatrix} a_1 & b_1 \\ a_2 & b_2 \end{bmatrix} = \begin{bmatrix} ca_1 & cb_1 \\ ca_2 & cb_2 \end{bmatrix} +$

$\begin{bmatrix} da_1 & db_1 \\ da_2 & db_2 \end{bmatrix} = \begin{bmatrix} ca_1 + da_1 & cb_1 + db_1 \\ ca_2 + da_2 & cb_2 + db_2 \end{bmatrix}; \therefore (c + d)A = cA + dA$

22. Let $A = \begin{bmatrix} a_1 & b_1 \\ a_2 & b_2 \end{bmatrix}$ and $B = \begin{bmatrix} c_1 & d_1 \\ c_2 & d_2 \end{bmatrix}$. $c(A + B) = c\begin{bmatrix} a_1 + c_1 & b_1 + d_1 \\ a_2 + c_2 & b_2 + d_2 \end{bmatrix} =$

$\begin{bmatrix} c(a_1 + c_1) & c(b_1 + d_1) \\ c(a_2 + c_2) & c(b_2 + d_2) \end{bmatrix} = \begin{bmatrix} ca_1 + cc_1 & cb_1 + cd_1 \\ ca_2 + cc_2 & cb_2 + cd_2 \end{bmatrix}; cA + cB = c\begin{bmatrix} a_1 & b_1 \\ a_2 & b_2 \end{bmatrix} +$

$c\begin{bmatrix} c_1 & d_1 \\ c_2 & d_2 \end{bmatrix} = \begin{bmatrix} ca_1 & cb_1 \\ ca_2 & cb_2 \end{bmatrix} + \begin{bmatrix} cc_1 & cd_1 \\ cc_2 & cd_2 \end{bmatrix} = \begin{bmatrix} ca_1 + cc_1 & cb_1 + cd_1 \\ ca_2 + cc_2 & cb_2 + cd_2 \end{bmatrix};$

$\therefore c(A + B) = cA + cB$

23. Let $A = \begin{bmatrix} a_1 & b_1 \\ a_2 & b_2 \end{bmatrix}$. $1 \cdot A = 1\begin{bmatrix} a_1 & b_1 \\ a_2 & b_2 \end{bmatrix} = \begin{bmatrix} 1 \cdot a_1 & 1 \cdot b_1 \\ 1 \cdot a_2 & 1 \cdot b_2 \end{bmatrix} = \begin{bmatrix} a_1 & b_1 \\ a_2 & b_2 \end{bmatrix} = A$

24. Let $A = \begin{bmatrix} a_1 & b_1 \\ a_2 & b_2 \end{bmatrix}$. $0 \cdot A = 0\begin{bmatrix} a_1 & b_1 \\ a_2 & b_2 \end{bmatrix} = \begin{bmatrix} 0 \cdot a_1 & 0 \cdot b_1 \\ 0 \cdot a_2 & 0 \cdot b_2 \end{bmatrix} = \begin{bmatrix} 0 & 0 \\ 0 & 0 \end{bmatrix} = O_{2\times 2}$

25. $O_{2\times 2} = \begin{bmatrix} 0 & 0 \\ 0 & 0 \end{bmatrix}$. $cO_{2\times 2} = \begin{bmatrix} c \cdot 0 & c \cdot 0 \\ c \cdot 0 & c \cdot 0 \end{bmatrix} = \begin{bmatrix} 0 & 0 \\ 0 & 0 \end{bmatrix} = O_{2\times 2}$

26. Let $B = \begin{bmatrix} c_1 & d_1 \\ c_2 & d_2 \end{bmatrix}$ and $X = \begin{bmatrix} x_1 & y_1 \\ x_2 & y_2 \end{bmatrix}$. Since $cX = B$, $\begin{bmatrix} cx_1 & cy_1 \\ cx_2 & cy_2 \end{bmatrix} = \begin{bmatrix} c_1 & d_1 \\ c_2 & d_2 \end{bmatrix}$ and $cx_1 = c_1, cy_1 = d_1, cx_2 = c_2, cy_2 = d_2. \therefore\ x_1 = \frac{1}{c}c_1, y_1 = \frac{1}{c}d_1, x_2 = \frac{1}{c}c_2, y_2 = \frac{1}{c}d_2,$

and $X = \begin{bmatrix} \frac{1}{c}c_1 & \frac{1}{c}d_1 \\ \frac{1}{c}c_2 & \frac{1}{c}d_2 \end{bmatrix} = \frac{1}{c}B.$

27. If $cX + B = A$, then $cX + B + (-B) = A + (-B)$ and $cX = A - B$. From Exercise 26, $X = \frac{1}{c}(A - B)$.

C **28.** $x^2 - 4x - 8 = 0$ and $y^2 - 5y - 24 = 0$; $x = \frac{4 \pm \sqrt{16 + 32}}{2} = \frac{4 \pm 4\sqrt{3}}{2} =$ $2 \pm 2\sqrt{3}$; $(y - 8)(y + 3) = 0$; $y = 8$ or $y = -3$; $(2 + 2\sqrt{3}, 8)$, $(2 + 2\sqrt{3}, -3)$, $(2 - 2\sqrt{3}, 8)$, $(2 - 2\sqrt{3}, -3)$

Pages 541–542 • WRITTEN EXERCISES

A **1.** $\begin{bmatrix} 15 - 4 \\ -10 - 1 \end{bmatrix} = \begin{bmatrix} 11 \\ -11 \end{bmatrix}$ **2.** $\begin{bmatrix} 6 + 11 \\ -12 - 6 \end{bmatrix} = \begin{bmatrix} 17 \\ -18 \end{bmatrix}$

3. $\begin{bmatrix} 12 + 0 - 12 \\ -14 + 5 + 6 \end{bmatrix} = \begin{bmatrix} 0 \\ -3 \end{bmatrix}$ **4.** $\begin{bmatrix} -16 + 18 \\ -10 + 0 \\ 12 - 9 \end{bmatrix} = \begin{bmatrix} 2 \\ -10 \\ 3 \end{bmatrix}$

5. $\begin{bmatrix} -2 + 0 & 2 + 5 \\ 6 + 0 & -6 + 2 \end{bmatrix} = \begin{bmatrix} -2 & 7 \\ 6 & -4 \end{bmatrix}$

6. $\begin{bmatrix} 0 - 6 & -15 + 21 \\ 0 - 4 & 12 + 14 \end{bmatrix} = \begin{bmatrix} -6 & 6 \\ -4 & 26 \end{bmatrix}$ **7.** $\begin{bmatrix} -20 + 21 & -12 + 0 \\ 35 - 35 & 21 + 0 \end{bmatrix} = \begin{bmatrix} -1 & -12 \\ 0 & 21 \end{bmatrix}$

8. 1 col. in 1st matrix; 2 rows in 2nd matrix; undefined

9. 1 col. in 1st matrix; 2 rows in 2nd matrix; undefined

10. $\begin{bmatrix} -5 + 0 & 2 + 0 \\ 0 + 3 & 0 + 6 \end{bmatrix} = \begin{bmatrix} -5 & 2 \\ 3 & 6 \end{bmatrix}$

11. $\begin{bmatrix} 7 - 12 & 0 + 6 & 4 + 6 \\ 14 + 8 & 0 - 4 & 8 - 4 \\ 0 - 6 & 0 + 3 & 0 + 3 \end{bmatrix} = \begin{bmatrix} -5 & 6 & 10 \\ 22 & -4 & 4 \\ -6 & 3 & 3 \end{bmatrix}$

12. $\begin{bmatrix} 14 + 0 + 15 & -7 - 12 + 25 & 0 + 6 + 10 \\ 8 + 0 - 18 & -4 - 6 - 30 & 0 + 3 - 12 \end{bmatrix} = \begin{bmatrix} 29 & 6 & 16 \\ -10 & -40 & -9 \end{bmatrix}$

13. $\begin{bmatrix} 5 & -3 \\ 4 & 1 \end{bmatrix}\begin{bmatrix} -2 & 3 \\ -1 & 2 \end{bmatrix} = \begin{bmatrix} -10 + 3 & 15 - 6 \\ -8 - 1 & 12 + 2 \end{bmatrix} = \begin{bmatrix} -7 & 9 \\ -9 & 14 \end{bmatrix}$

14. $\begin{bmatrix} -2 & 3 \\ -1 & 2 \end{bmatrix}\begin{bmatrix} 5 & -3 \\ 4 & 1 \end{bmatrix} = \begin{bmatrix} -10 + 12 & 6 + 3 \\ -5 + 8 & 3 + 2 \end{bmatrix} = \begin{bmatrix} 2 & 9 \\ 3 & 5 \end{bmatrix}$

15. $\begin{bmatrix} -2 & 3 \\ -1 & 2 \end{bmatrix}\begin{bmatrix} -2 & 3 \\ -1 & 2 \end{bmatrix} = \begin{bmatrix} 4-3 & -6+6 \\ 2-2 & -3+4 \end{bmatrix} = \begin{bmatrix} 1 & 0 \\ 0 & 1 \end{bmatrix}$

16. $\begin{bmatrix} 5 & -3 \\ 4 & 1 \end{bmatrix}\begin{bmatrix} 5 & -3 \\ 4 & 1 \end{bmatrix} = \begin{bmatrix} 25-12 & -15-3 \\ 20+4 & -12+1 \end{bmatrix} = \begin{bmatrix} 13 & -18 \\ 24 & -11 \end{bmatrix}$

17. $\left(\begin{bmatrix} 5 & -3 \\ 4 & 1 \end{bmatrix} + \begin{bmatrix} -2 & 3 \\ -1 & 2 \end{bmatrix}\right)^2 = \begin{bmatrix} 3 & 0 \\ 3 & 3 \end{bmatrix}^2 = \begin{bmatrix} 9+0 & 0+0 \\ 9+9 & 0+9 \end{bmatrix} = \begin{bmatrix} 9 & 0 \\ 18 & 9 \end{bmatrix}$

18. $\left(\begin{bmatrix} 5 & -3 \\ 4 & 1 \end{bmatrix} + \begin{bmatrix} -2 & 3 \\ -1 & 2 \end{bmatrix}\right)\left(\begin{bmatrix} 5 & -3 \\ 4 & 1 \end{bmatrix} - \begin{bmatrix} -2 & 3 \\ -1 & 2 \end{bmatrix}\right) = \begin{bmatrix} 3 & 0 \\ 3 & 3 \end{bmatrix}\begin{bmatrix} 7 & -6 \\ 5 & -1 \end{bmatrix} =$
$\begin{bmatrix} 21+0 & -18+0 \\ 21+15 & -18-3 \end{bmatrix} = \begin{bmatrix} 21 & -18 \\ 36 & -21 \end{bmatrix}$

19. $AB = \begin{bmatrix} ax+0z & ay+0w \\ 0x+az & 0y+aw \end{bmatrix} = \begin{bmatrix} ax & ay \\ az & aw \end{bmatrix}$; $BA = \begin{bmatrix} ax+0y & 0x+ay \\ az+0w & 0z+aw \end{bmatrix} =$
$\begin{bmatrix} ax & ay \\ az & aw \end{bmatrix} = AB$

20. $AB = \begin{bmatrix} ax+0 & 0a+0y \\ 0x+0b & 0+by \end{bmatrix} = \begin{bmatrix} ax & 0 \\ 0 & by \end{bmatrix}$; $BA = \begin{bmatrix} ax+0 & 0x+0b \\ 0a+0y & 0+by \end{bmatrix} =$
$\begin{bmatrix} ax & 0 \\ 0 & by \end{bmatrix} = AB$

B 21. $\begin{bmatrix} 3x+0y \\ -3x+y \end{bmatrix} = \begin{bmatrix} 6 \\ -7 \end{bmatrix}$ $3x = 6$ and $-3x + y = -7$; $x = 2$; $-3(2) + y = -7$; $y = -1$

22. $\begin{bmatrix} 4x-7y \\ 2x-3y \end{bmatrix} = \begin{bmatrix} -3 \\ 1 \end{bmatrix}$ $\begin{aligned} 4x - 7y &= -3 \\ 2x - 3y &= 1 \end{aligned}$ $\begin{aligned} 4x - 7y &= -3 \\ -4x + 6y &= -2 \\ \hline -y &= -5 \end{aligned}$; $y = 5$; $2x - 3(5) = 1$, $x = 8$

23. $\begin{bmatrix} -5x-2y \\ 4x+3y \end{bmatrix} = \begin{bmatrix} -8 \\ 12 \end{bmatrix}$ $\begin{aligned} -5x - 2y &= -8 \\ 4x + 3y &= 12 \end{aligned}$ $\begin{aligned} -15x - 6y &= -24 \\ 8x + 6y &= 24 \\ \hline -7x &= 0 \end{aligned}$, $x = 0$;
$4(0) + 3y = 12$, $y = 4$

24. $[4x - 2y \quad -5x + 6y] = [2 \quad 15]$; $4x - 2y = 2$ and $-5x + 6y = 15$;
$\begin{aligned} 12x - 6y &= 6 \\ -5x + 6y &= 15 \\ \hline 7x &= 21 \end{aligned}$; $x = 3$; $-5(3) + 6y = 15$, $y = 5$

25. $\begin{bmatrix} -x+0y+z \\ x-y+0z \\ 0x+y+z \end{bmatrix} = \begin{bmatrix} -4 \\ 2 \\ 4 \end{bmatrix}$ $\begin{aligned} (1)&: -x + z = -4 \\ (2)&: x - y = 2 \\ (3)&: y + z = 4 \end{aligned}$ $\begin{aligned} (1)+(2)&: -y + z = -2 \\ (3)&: y + z = 4 \\ \hline & 2z = 2 \end{aligned}$;
$z = 1$; $-x + 1 = -4$, $x = 5$; $y + 1 = 4$, $y = 3$

26. $\begin{bmatrix} 2x+0y-z \\ 0x+3y+0z \\ x+y+0z \end{bmatrix} = \begin{bmatrix} 6 \\ 9 \\ 5 \end{bmatrix}$ $\begin{aligned} 2x - z &= 6 \\ 3y &= 9 \\ x + y &= 5 \end{aligned}$ $3y = 9$, $y = 3$; $x + 3 = 5$, $x = 2$; $2(2) - z = 6$, $z = -2$

27. $AB = \begin{bmatrix} ad-bc & -ab+ab \\ cd-cd & -bc+ad \end{bmatrix} = \begin{bmatrix} ad-bc & 0 \\ 0 & ad-bc \end{bmatrix}$, which is in the form
$\begin{bmatrix} x & 0 \\ 0 & x \end{bmatrix}$.

28. See Exercise 27 for AB. $BA = \begin{bmatrix} d & -b \\ -c & a \end{bmatrix}\begin{bmatrix} a & b \\ c & d \end{bmatrix} = \begin{bmatrix} ad - bc & bd - bd \\ -ac + ac & -bc + ad \end{bmatrix} = \begin{bmatrix} ad - bc & 0 \\ 0 & ad - bc \end{bmatrix}$; $\therefore AB = BA$

C **29.** $AB = \begin{bmatrix} 17 & 21 \\ 2 & 4 \end{bmatrix}\begin{bmatrix} 20 & 15 & 10 \\ 30 & 40 & 50 \end{bmatrix} = \begin{bmatrix} 340 + 630 & 225 + 840 & 170 + 1050 \\ 40 + 120 & 30 + 160 & 20 + 200 \end{bmatrix} = \begin{bmatrix} 970 & 1065 & 1220 \\ 160 & 190 & 220 \end{bmatrix}$. The entries in the top row represent the number of jewels used in producing both models on Monday, Tuesday, and Wednesday, respectively. The entries in the second row represent the number of straps used in producing both models on Monday, Tuesday, and Wednesday, respectively.

30. $ABC = [125 \quad 15]\begin{bmatrix} 4 & 0.6 \\ 8 & 1.0 \end{bmatrix}\begin{bmatrix} 12.50 \\ 14.00 \end{bmatrix} = [500 + 120 \quad 75 + 15]\begin{bmatrix} 12.50 \\ 14.00 \end{bmatrix} = [620 \quad 90]\begin{bmatrix} 12.50 \\ 14.00 \end{bmatrix} = [620(12.50) + 90(14.00)] = [9010]$. The entry in ABC represents the total labor cost of producing the tennis rackets.

Page 542 • COMPUTER EXERCISES

1.

```
10  INPUT "NUMBER OF ROWS IN MATRIX";R
20  FOR I = 1 TO R
30  FOR J = 1 TO R
40  INPUT A(I,J)
50  LET B(I,J) = A(I,J)
60  NEXT J
70  NEXT I
80  FOR C = 2 TO 4
90  FOR I = 1 TO R
100  FOR J = 1 TO R
110  FOR K = 1 TO R
120  LET C(I,J) = C(I,J) + A(I,K) * B(K,J)
130  NEXT K
140  PRINT C(I,J);" ";
150  NEXT J
160  PRINT
170  NEXT I
180  PRINT : PRINT : PRINT
190  FOR I = 1 TO R
200  FOR J = 1 TO R
210  LET A(I,J) = C(I,J)
220  LET C(I,J) = 0
230  NEXT J
240  NEXT I
250  NEXT C
260  END
```

2. $\begin{bmatrix} 7 & -6 \\ 3 & -2 \end{bmatrix}$; $\begin{bmatrix} 15 & -14 \\ 7 & -6 \end{bmatrix}$; $\begin{bmatrix} 31 & -30 \\ 15 & -14 \end{bmatrix}$

3. $\begin{bmatrix} -19 & -28 \\ 35 & 44 \end{bmatrix}$; $\begin{bmatrix} -121 & -148 \\ 185 & 212 \end{bmatrix}$; $\begin{bmatrix} -619 & -700 \\ 875 & 956 \end{bmatrix}$

4. $\begin{bmatrix} -3 & -15 & -1 \\ -21 & 1 & 11 \\ -17 & 5 & -17 \end{bmatrix}$; $\begin{bmatrix} 69 & -31 & 39 \\ 7 & 67 & 61 \\ -93 & -41 & 41 \end{bmatrix}$; $\begin{bmatrix} 341 & 25 & -145 \\ -145 & 371 & -155 \\ 235 & 175 & 361 \end{bmatrix}$

5. $\begin{bmatrix} 1 & 0 & 0 \\ 0 & 1 & 0 \\ 0 & 0 & 1 \end{bmatrix}$; $\begin{bmatrix} 1 & 0 & 0 \\ 0 & 1 & 0 \\ 0 & 0 & 1 \end{bmatrix}$; $\begin{bmatrix} 1 & 0 & 0 \\ 0 & 1 & 0 \\ 0 & 0 & 1 \end{bmatrix}$

Pages 545–546 • WRITTEN EXERCISES

A 1. $AB = \begin{bmatrix} 4 & 2 \\ -8 & -4 \end{bmatrix}\begin{bmatrix} -1 & 3 \\ 2 & -6 \end{bmatrix} = \begin{bmatrix} 0 & 0 \\ 0 & 0 \end{bmatrix}$; $BA = \begin{bmatrix} -1 & 3 \\ 2 & -6 \end{bmatrix}\begin{bmatrix} 4 & 2 \\ -8 & -4 \end{bmatrix} =$
$\begin{bmatrix} -28 & -14 \\ 56 & 28 \end{bmatrix}$; $AB \neq BA$

2. $AC = \begin{bmatrix} 4 & 2 \\ -8 & -4 \end{bmatrix}\begin{bmatrix} 8 & 2 \\ 4 & 6 \end{bmatrix} = \begin{bmatrix} 40 & 20 \\ -80 & -40 \end{bmatrix}$; $CA = \begin{bmatrix} 8 & 2 \\ 4 & 6 \end{bmatrix}\begin{bmatrix} 4 & 2 \\ -8 & -4 \end{bmatrix} =$
$\begin{bmatrix} 16 & 8 \\ -32 & -16 \end{bmatrix}$; $AC \neq CA$

3. $AD = \begin{bmatrix} 4 & 2 \\ -8 & -4 \end{bmatrix}\begin{bmatrix} 3 & -1 \\ -2 & 4 \end{bmatrix} = \begin{bmatrix} 8 & 4 \\ -16 & -8 \end{bmatrix}$; $DA = \begin{bmatrix} 3 & -1 \\ -2 & 4 \end{bmatrix}\begin{bmatrix} 4 & 2 \\ -8 & -4 \end{bmatrix} =$
$\begin{bmatrix} 20 & 10 \\ -40 & -20 \end{bmatrix}$; $AD \neq DA$

4. $BC = \begin{bmatrix} -1 & 3 \\ 2 & -6 \end{bmatrix}\begin{bmatrix} 8 & 2 \\ 4 & 6 \end{bmatrix} = \begin{bmatrix} 4 & 16 \\ -8 & -32 \end{bmatrix}$; $CB = \begin{bmatrix} 8 & 2 \\ 4 & 6 \end{bmatrix}\begin{bmatrix} -1 & 3 \\ 2 & -6 \end{bmatrix} =$
$\begin{bmatrix} -4 & 12 \\ 8 & -24 \end{bmatrix}$; $BC \neq CB$

5. $CD = \begin{bmatrix} 8 & 2 \\ 4 & 6 \end{bmatrix}\begin{bmatrix} 3 & -1 \\ -2 & 4 \end{bmatrix} = \begin{bmatrix} 20 & 0 \\ 0 & 20 \end{bmatrix}$; $DC = \begin{bmatrix} 3 & -1 \\ -2 & 4 \end{bmatrix}\begin{bmatrix} 8 & 2 \\ 4 & 6 \end{bmatrix} = \begin{bmatrix} 20 & 0 \\ 0 & 20 \end{bmatrix}$;
$CD = DC$

6. $BD = \begin{bmatrix} -1 & 3 \\ 2 & -6 \end{bmatrix}\begin{bmatrix} 3 & -1 \\ -2 & 4 \end{bmatrix} = \begin{bmatrix} -9 & 13 \\ 18 & -26 \end{bmatrix}$; $DB = \begin{bmatrix} 3 & -1 \\ -2 & 4 \end{bmatrix}\begin{bmatrix} -1 & 3 \\ 2 & -6 \end{bmatrix} =$
$\begin{bmatrix} -5 & 15 \\ 10 & -30 \end{bmatrix}$; $BD \neq DB$

7. $A(B + C) = A\left(\begin{bmatrix} -1 & 3 \\ 2 & -6 \end{bmatrix} + \begin{bmatrix} 8 & 2 \\ 4 & 6 \end{bmatrix}\right) = \begin{bmatrix} 4 & 2 \\ -8 & -4 \end{bmatrix}\begin{bmatrix} 7 & 5 \\ 6 & 0 \end{bmatrix} = \begin{bmatrix} 40 & 20 \\ -80 & -40 \end{bmatrix}$;
$AB = \begin{bmatrix} 0 & 0 \\ 0 & 0 \end{bmatrix}$ (See Ex. 1.); $AC = \begin{bmatrix} 40 & 20 \\ -80 & -40 \end{bmatrix}$ (See Ex. 2.);
$AB + AC = \begin{bmatrix} 40 & 20 \\ -80 & -40 \end{bmatrix}$; $A(B + C) = AB + AC$

8. $BC = \begin{bmatrix} 4 & 16 \\ -8 & -32 \end{bmatrix}$ (See Ex. 4.); $A(BC) = \begin{bmatrix} 4 & 2 \\ -8 & -4 \end{bmatrix}\begin{bmatrix} 4 & 16 \\ -8 & -32 \end{bmatrix} = \begin{bmatrix} 0 & 0 \\ 0 & 0 \end{bmatrix}$;
$AB = \begin{bmatrix} 0 & 0 \\ 0 & 0 \end{bmatrix}$ (See Ex. 1.); $(AB)C = \begin{bmatrix} 0 & 0 \\ 0 & 0 \end{bmatrix}\begin{bmatrix} 8 & 2 \\ 4 & 6 \end{bmatrix} = \begin{bmatrix} 0 & 0 \\ 0 & 0 \end{bmatrix}$; $A(BC) = (AB)C$

9. $AC = \begin{bmatrix} 40 & 20 \\ -80 & -40 \end{bmatrix}$ (See Ex. 2.); $B(AC) = \begin{bmatrix} -1 & 3 \\ 2 & -6 \end{bmatrix}\begin{bmatrix} 40 & 20 \\ -80 & -40 \end{bmatrix} =$

$\begin{bmatrix} -280 & -140 \\ 560 & 280 \end{bmatrix}$; $BA = \begin{bmatrix} -28 & -14 \\ 56 & 28 \end{bmatrix}$ (See Ex. 1.);

$(BA)C = \begin{bmatrix} -28 & -14 \\ 56 & 28 \end{bmatrix}\begin{bmatrix} 8 & 2 \\ 4 & 6 \end{bmatrix} = \begin{bmatrix} -280 & -140 \\ 560 & 280 \end{bmatrix}$; $B(AC) = (BA)C$

10. $B(A + C) = B\left(\begin{bmatrix} 4 & 2 \\ -8 & -4 \end{bmatrix} + \begin{bmatrix} 8 & 2 \\ 4 & 6 \end{bmatrix}\right) = \begin{bmatrix} -1 & 3 \\ 2 & -6 \end{bmatrix}\begin{bmatrix} 12 & 4 \\ -4 & 2 \end{bmatrix} = \begin{bmatrix} -24 & 2 \\ 48 & -4 \end{bmatrix}$;

$BA = \begin{bmatrix} -28 & -14 \\ 56 & 28 \end{bmatrix}$ (See Ex. 1.); $BC = \begin{bmatrix} 4 & 16 \\ -8 & -32 \end{bmatrix}$ (See Ex. 4.);

$BA + BC = \begin{bmatrix} -24 & 2 \\ 48 & -4 \end{bmatrix}$; $B(A + C) = BA + BC$

11. $BC = \begin{bmatrix} 4 & 16 \\ -8 & -32 \end{bmatrix}$ (See Ex. 4.); $A(BC) = \begin{bmatrix} 4 & 2 \\ -8 & -4 \end{bmatrix}\begin{bmatrix} 4 & 16 \\ -8 & -32 \end{bmatrix} = \begin{bmatrix} 0 & 0 \\ 0 & 0 \end{bmatrix}$;

$(BC)A = \begin{bmatrix} 4 & 16 \\ -8 & -32 \end{bmatrix}\begin{bmatrix} 4 & 2 \\ -8 & -4 \end{bmatrix} = \begin{bmatrix} -112 & -56 \\ 224 & 112 \end{bmatrix}$; $A(BC) \neq (BC)A$

12. $CD = \begin{bmatrix} 20 & 0 \\ 0 & 20 \end{bmatrix}$ (See Ex. 5.); $A(CD) = \begin{bmatrix} 4 & 2 \\ -8 & -4 \end{bmatrix}\begin{bmatrix} 20 & 0 \\ 0 & 20 \end{bmatrix} = \begin{bmatrix} 80 & 40 \\ -160 & -80 \end{bmatrix}$;

$(CD)A = \begin{bmatrix} 20 & 0 \\ 0 & 20 \end{bmatrix}\begin{bmatrix} 4 & 2 \\ -8 & -4 \end{bmatrix} = \begin{bmatrix} 80 & 40 \\ -160 & -80 \end{bmatrix}$; $A(CD) = (CD)A$

13. $C(A + B) = C\left(\begin{bmatrix} 4 & 2 \\ -8 & -4 \end{bmatrix} + \begin{bmatrix} -1 & 3 \\ 2 & -6 \end{bmatrix}\right) = \begin{bmatrix} 8 & 2 \\ 4 & 6 \end{bmatrix}\begin{bmatrix} 3 & 5 \\ -6 & -10 \end{bmatrix} =$

$\begin{bmatrix} 12 & 20 \\ -24 & -40 \end{bmatrix}$; $(A + B)C = \begin{bmatrix} 3 & 5 \\ -6 & -10 \end{bmatrix}\begin{bmatrix} 8 & 2 \\ 4 & 6 \end{bmatrix} = \begin{bmatrix} 44 & 36 \\ -88 & -72 \end{bmatrix}$;
$C(A + B) \neq (A + B)C$

14. $AB = \begin{bmatrix} 0 & 0 \\ 0 & 0 \end{bmatrix}$ (See Ex. 1.); $CD = \begin{bmatrix} 20 & 0 \\ 0 & 20 \end{bmatrix}$ (See Ex. 5.);

$(AB)(CD) = \begin{bmatrix} 0 & 0 \\ 0 & 0 \end{bmatrix}\begin{bmatrix} 20 & 0 \\ 0 & 20 \end{bmatrix} = \begin{bmatrix} 0 & 0 \\ 0 & 0 \end{bmatrix}$; $(CD)(AB) = \begin{bmatrix} 20 & 0 \\ 0 & 20 \end{bmatrix}\begin{bmatrix} 0 & 0 \\ 0 & 0 \end{bmatrix} =$

$\begin{bmatrix} 0 & 0 \\ 0 & 0 \end{bmatrix}$; $(AB)(CD) = (CD)(AB)$

B **15. a.** $A + B = \begin{bmatrix} 3 & 5 \\ -6 & -10 \end{bmatrix}$ (See Ex. 13.); $(A + B)^2 = \begin{bmatrix} 3 & 5 \\ -6 & -10 \end{bmatrix}\begin{bmatrix} 3 & 5 \\ -6 & -10 \end{bmatrix} =$

$\begin{bmatrix} -21 & -35 \\ 42 & 70 \end{bmatrix}$

b. $AB = \begin{bmatrix} 0 & 0 \\ 0 & 0 \end{bmatrix}$ (See Ex. 1.); $A^2 + 2AB + B^2 = \begin{bmatrix} 4 & 2 \\ -8 & -4 \end{bmatrix}\begin{bmatrix} 4 & 2 \\ -8 & -4 \end{bmatrix} +$

$2\begin{bmatrix} 0 & 0 \\ 0 & 0 \end{bmatrix} + \begin{bmatrix} -1 & 3 \\ 2 & -6 \end{bmatrix}\begin{bmatrix} -1 & 3 \\ 2 & -6 \end{bmatrix} = \begin{bmatrix} 0 & 0 \\ 0 & 0 \end{bmatrix} + \begin{bmatrix} 0 & 0 \\ 0 & 0 \end{bmatrix} + \begin{bmatrix} 7 & -21 \\ -14 & 42 \end{bmatrix} =$

$\begin{bmatrix} 7 & -21 \\ -14 & 42 \end{bmatrix}$

c. $A^2 = \begin{bmatrix} 0 & 0 \\ 0 & 0 \end{bmatrix}$ (See Ex. 15b.); $AB = \begin{bmatrix} 0 & 0 \\ 0 & 0 \end{bmatrix}$ (See Ex. 1.);

$BA = \begin{bmatrix} -28 & -14 \\ 56 & 28 \end{bmatrix}$ (See Ex. 1.); $B^2 = \begin{bmatrix} 7 & -21 \\ -14 & 42 \end{bmatrix}$ (See Ex. 15b.);

$A^2 + AB + BA + B^2 = \begin{bmatrix} 0 & 0 \\ 0 & 0 \end{bmatrix} + \begin{bmatrix} 0 & 0 \\ 0 & 0 \end{bmatrix} + \begin{bmatrix} -28 & -14 \\ 56 & 28 \end{bmatrix} +$

$\begin{bmatrix} 7 & -21 \\ -14 & 42 \end{bmatrix} = \begin{bmatrix} -21 & -35 \\ 42 & 70 \end{bmatrix}$; parts a and c are equal.

16. a. $(A - B)^2 = \left(\begin{bmatrix} 4 & 2 \\ -8 & -4 \end{bmatrix} - \begin{bmatrix} -1 & 3 \\ 2 & -6 \end{bmatrix}\right)^2 = \begin{bmatrix} 5 & -1 \\ -10 & 2 \end{bmatrix}\begin{bmatrix} 5 & -1 \\ -10 & 2 \end{bmatrix} =$

$\begin{bmatrix} 35 & -7 \\ -70 & 14 \end{bmatrix}$

b. $A^2 = \begin{bmatrix} 0 & 0 \\ 0 & 0 \end{bmatrix}$ (See Ex. 15b.); $AB = \begin{bmatrix} 0 & 0 \\ 0 & 0 \end{bmatrix}$ (See Ex. 1.);

$BA = \begin{bmatrix} -28 & -14 \\ 56 & 28 \end{bmatrix}$ (See Ex. 1.); $B^2 = \begin{bmatrix} 7 & -21 \\ -14 & 42 \end{bmatrix}$ (See Ex. 15b.);

$A^2 - AB - BA + B^2 = \begin{bmatrix} 0 & 0 \\ 0 & 0 \end{bmatrix} - \begin{bmatrix} 0 & 0 \\ 0 & 0 \end{bmatrix} - \begin{bmatrix} -28 & -14 \\ 56 & 28 \end{bmatrix} +$

$\begin{bmatrix} 7 & -21 \\ -14 & 42 \end{bmatrix} = \begin{bmatrix} 35 & -7 \\ -70 & 14 \end{bmatrix}$

c. $A^2 = \begin{bmatrix} 0 & 0 \\ 0 & 0 \end{bmatrix}$ (See Ex. 15b.); $AB = \begin{bmatrix} 0 & 0 \\ 0 & 0 \end{bmatrix}$ (See Ex. 1.); $B^2 = \begin{bmatrix} 7 & -21 \\ -14 & 42 \end{bmatrix}$

(See Ex. 15b.); $A^2 - 2AB + B^2 = \begin{bmatrix} 0 & 0 \\ 0 & 0 \end{bmatrix} - 2\begin{bmatrix} 0 & 0 \\ 0 & 0 \end{bmatrix} + \begin{bmatrix} 7 & -21 \\ -14 & 42 \end{bmatrix} =$

$\begin{bmatrix} 7 & -21 \\ -14 & 42 \end{bmatrix}$; parts a and b are equal.

17. a. $A + B = \begin{bmatrix} 3 & 5 \\ -6 & -10 \end{bmatrix}$ (See Ex. 13.); $(A + B)(C + D) =$

$\begin{bmatrix} 3 & 5 \\ -6 & -10 \end{bmatrix}\left(\begin{bmatrix} 8 & 2 \\ 4 & 6 \end{bmatrix} + \begin{bmatrix} 3 & -1 \\ -2 & 4 \end{bmatrix}\right) = \begin{bmatrix} 3 & 5 \\ -6 & -10 \end{bmatrix}\begin{bmatrix} 11 & 1 \\ 2 & 10 \end{bmatrix} =$

$\begin{bmatrix} 43 & 53 \\ -86 & -106 \end{bmatrix}$

b. $AC = \begin{bmatrix} 40 & 20 \\ -80 & -40 \end{bmatrix}$ (See Ex. 2.); $BC = \begin{bmatrix} 4 & 16 \\ -8 & -32 \end{bmatrix}$ (See Ex. 4.);

$AD = \begin{bmatrix} 8 & 4 \\ -16 & -8 \end{bmatrix}$ (See Ex. 3.); $BD = \begin{bmatrix} -9 & 13 \\ 18 & -26 \end{bmatrix}$ (See Ex. 6.);

$AC + BC + AD + BD = \begin{bmatrix} 40 & 20 \\ -80 & -40 \end{bmatrix} + \begin{bmatrix} 4 & 16 \\ -8 & -32 \end{bmatrix} + \begin{bmatrix} 8 & 4 \\ -16 & -8 \end{bmatrix} +$

$\begin{bmatrix} -9 & 13 \\ 18 & -26 \end{bmatrix} = \begin{bmatrix} 43 & 53 \\ -86 & -106 \end{bmatrix}$

c. $A + B = \begin{bmatrix} 3 & 5 \\ -6 & -10 \end{bmatrix}$ (See Ex. 13.); $C(A + B) + D(A + B) =$

$\begin{bmatrix} 8 & 2 \\ 4 & 6 \end{bmatrix}\begin{bmatrix} 3 & 5 \\ -6 & -10 \end{bmatrix} + \begin{bmatrix} 3 & -1 \\ -2 & 4 \end{bmatrix}\begin{bmatrix} 3 & 5 \\ -6 & -10 \end{bmatrix} = \begin{bmatrix} 12 & 20 \\ -24 & -40 \end{bmatrix} +$

$\begin{bmatrix} 15 & 25 \\ -30 & -50 \end{bmatrix} = \begin{bmatrix} 27 & 45 \\ -54 & -90 \end{bmatrix}$; parts a and b are equal.

18. a. $A - B = \begin{bmatrix} 5 & -1 \\ -10 & 2 \end{bmatrix}$ (See Ex. 16a.); $C + D = \begin{bmatrix} 11 & 1 \\ 2 & 10 \end{bmatrix}$ (See Ex. 17a.);

$(A - B)(C + D) = \begin{bmatrix} 5 & -1 \\ -10 & 2 \end{bmatrix}\begin{bmatrix} 11 & 1 \\ 2 & 10 \end{bmatrix} = \begin{bmatrix} 53 & -5 \\ -106 & 10 \end{bmatrix}$

b. $AC = \begin{bmatrix} 40 & 20 \\ -80 & -40 \end{bmatrix}$ (See Ex. 2.); $BC = \begin{bmatrix} 4 & 16 \\ -8 & -32 \end{bmatrix}$ (See Ex. 4.);

$AD = \begin{bmatrix} 8 & 4 \\ -16 & -8 \end{bmatrix}$ (See Ex. 3.); $BD = \begin{bmatrix} -9 & 13 \\ 18 & -26 \end{bmatrix}$ (See Ex. 6.);

$AC - BC - BD + AD = \begin{bmatrix} 40 & 20 \\ -80 & -40 \end{bmatrix} - \begin{bmatrix} 4 & 16 \\ -8 & -32 \end{bmatrix} - \begin{bmatrix} -9 & 13 \\ 18 & -26 \end{bmatrix} +$

$\begin{bmatrix} 8 & 4 \\ -16 & -8 \end{bmatrix} = \begin{bmatrix} 53 & -5 \\ -106 & 10 \end{bmatrix}$

c. $C + D = \begin{bmatrix} 11 & 1 \\ 2 & 10 \end{bmatrix}$ (See Ex. 17a.); $A(C + D) - B(C + D) =$

$\begin{bmatrix} 4 & 2 \\ -8 & -4 \end{bmatrix}\begin{bmatrix} 11 & 1 \\ 2 & 10 \end{bmatrix} - \begin{bmatrix} -1 & 3 \\ 2 & -6 \end{bmatrix}\begin{bmatrix} 11 & 1 \\ 2 & 10 \end{bmatrix} = \begin{bmatrix} 48 & 24 \\ -96 & -48 \end{bmatrix} -$

$\begin{bmatrix} -5 & 29 \\ 10 & -58 \end{bmatrix} = \begin{bmatrix} 53 & -5 \\ -106 & 10 \end{bmatrix}$; parts a, b, and c are equal.

In Exs. 19–27, let $A = \begin{bmatrix} a_1 & b_1 \\ a_2 & b_2 \end{bmatrix}$, $B = \begin{bmatrix} c_1 & d_1 \\ c_2 & d_2 \end{bmatrix}$, and $C = \begin{bmatrix} e_1 & f_1 \\ e_2 & f_2 \end{bmatrix}$.

19. $AB = \begin{bmatrix} a_1 & b_1 \\ a_2 & b_2 \end{bmatrix}\begin{bmatrix} c_1 & d_1 \\ c_2 & d_2 \end{bmatrix} = \begin{bmatrix} a_1c_1 + b_1c_2 & a_1d_1 + b_1d_2 \\ a_2c_1 + b_2c_2 & a_2d_1 + b_2d_2 \end{bmatrix}$; $a_1c_1 + b_1c_2 \in \mathcal{R}$,
$a_1d_1 + b_1d_2 \in \mathcal{R}$, $a_2c_1 + b_2c_2 \in \mathcal{R}$, $a_2d_1 + b_2d_2 \in \mathcal{R}$; $\therefore AB \in S_{2\times 2}$

20. $(AB)C = \left(\begin{bmatrix} a_1 & b_1 \\ a_2 & b_2 \end{bmatrix}\begin{bmatrix} c_1 & d_1 \\ c_2 & d_2 \end{bmatrix}\right)\begin{bmatrix} e_1 & f_1 \\ e_2 & f_2 \end{bmatrix} =$

$\begin{bmatrix} a_1c_1 + b_1c_2 & a_1d_1 + b_1d_2 \\ a_2c_1 + b_2c_2 & a_2d_1 + b_2d_2 \end{bmatrix}\begin{bmatrix} e_1 & f_1 \\ e_2 & f_2 \end{bmatrix} =$

$\begin{bmatrix} e_1(a_1c_1 + b_1c_2) + e_2(a_1d_1 + b_1d_2) & f_1(a_1c_1 + b_1c_2) + f_2(a_1d_1 + b_1d_2) \\ e_1(a_2c_1 + b_2c_2) + e_2(a_2d_1 + b_2d_2) & f_1(a_2c_1 + b_2c_2) + f_2(a_2d_1 + b_2d_2) \end{bmatrix} =$

$\begin{bmatrix} a_1c_1e_1 + b_1c_2e_1 + a_1d_1e_2 + b_1d_2e_2 & a_1c_1f_1 + b_1c_2f_1 + a_1d_1f_2 + b_1d_2f_2 \\ a_2c_1e_1 + b_2c_2e_1 + a_2d_1e_2 + b_2d_2e_2 & a_2c_1f_1 + b_2c_2f_1 + a_2d_1f_2 + b_2d_2f_2 \end{bmatrix}$;

$A(BC) = \begin{bmatrix} a_1 & b_1 \\ a_2 & b_2 \end{bmatrix}\left(\begin{bmatrix} c_1 & d_1 \\ c_2 & d_2 \end{bmatrix}\begin{bmatrix} e_1 & f_1 \\ e_2 & f_2 \end{bmatrix}\right) =$

$\begin{bmatrix} a_1 & b_1 \\ a_2 & b_2 \end{bmatrix}\begin{bmatrix} c_1e_1 + d_1e_2 & c_1f_1 + d_1f_2 \\ c_2e_1 + d_2e_2 & c_2f_1 + d_2f_2 \end{bmatrix} =$

$\begin{bmatrix} a_1(c_1e_1 + d_1e_2) + b_1(c_2e_1 + d_2e_2) & a_1(c_1f_1 + d_1f_2) + b_1(c_2f_1 + d_2f_2) \\ a_2(c_1e_1 + d_1e_2) + b_2(c_2e_1 + d_2e_2) & a_2(c_1f_1 + d_1f_2) + b_2(c_2f_1 + d_2f_2) \end{bmatrix} =$

$$\begin{bmatrix} a_1c_1e_1 + a_1d_1e_2 + b_1c_2e_1 + b_1d_2e_2 & a_1c_1f_1 + a_1d_1f_2 + b_1c_2f_1 + b_1d_2f_2 \\ a_2c_1e_1 + a_2d_1e_2 + b_2c_2e_1 + b_2d_2e_2 & a_2c_1f_1 + a_2d_1f_2 + b_2c_2f_1 + b_2d_2f_2 \end{bmatrix};$$

$\therefore (AB)C = A(BC)$

21. $A(B + C) = \begin{bmatrix} a_1 & b_1 \\ a_2 & b_2 \end{bmatrix}\begin{bmatrix} c_1 + e_1 & d_1 + f_1 \\ c_2 + e_2 & d_2 + f_2 \end{bmatrix} =$

$$\begin{bmatrix} a_1c_1 + a_1e_1 + b_1c_2 + b_1e_2 & a_1d_1 + a_1f_1 + b_1d_2 + b_1f_2 \\ a_2c_1 + a_2e_1 + b_2c_2 + b_2e_2 & a_2d_1 + a_2f_1 + b_2d_2 + b_2f_2 \end{bmatrix};$$

$AB + AC = \begin{bmatrix} a_1 & b_1 \\ a_2 & b_2 \end{bmatrix}\begin{bmatrix} c_1 & d_1 \\ c_2 & d_2 \end{bmatrix} + \begin{bmatrix} a_1 & b_1 \\ a_2 & b_2 \end{bmatrix}\begin{bmatrix} e_1 & f_1 \\ e_2 & f_2 \end{bmatrix} =$

$$\begin{bmatrix} a_1c_1 + b_1c_2 & a_1d_1 + b_1d_2 \\ a_2c_1 + b_2c_2 & a_2d_1 + b_2d_2 \end{bmatrix} + \begin{bmatrix} a_1e_1 + b_1e_2 & a_1f_1 + b_1f_2 \\ a_2e_1 + b_2e_2 & a_2f_1 + b_2f_2 \end{bmatrix} =$$

$$\begin{bmatrix} a_1c_1 + b_1c_2 + a_1e_1 + b_1e_2 & a_1d_1 + b_1d_2 + a_1f_1 + b_1f_2 \\ a_2c_1 + b_2c_2 + a_2e_1 + b_2e_2 & a_2d_1 + b_2d_2 + a_2f_1 + b_2f_2 \end{bmatrix};$$

$\therefore A(B + C) = AB + AC$

22. $(B + C)A = \begin{bmatrix} c_1 + e_1 & d_1 + f_1 \\ c_2 + e_2 & d_2 + f_2 \end{bmatrix}\begin{bmatrix} a_1 & b_1 \\ a_2 & b_2 \end{bmatrix} =$

$$\begin{bmatrix} a_1c_1 + a_1e_1 + a_2d_1 + a_2f_1 & b_1c_1 + b_1e_1 + b_2d_1 + b_2f_1 \\ a_1c_2 + a_1e_2 + a_2d_2 + a_2f_2 & b_1c_2 + b_1e_2 + b_2d_2 + b_2f_2 \end{bmatrix};$$

$BA + CA = \begin{bmatrix} c_1 & d_1 \\ c_2 & d_2 \end{bmatrix}\begin{bmatrix} a_1 & b_1 \\ a_2 & b_2 \end{bmatrix} + \begin{bmatrix} e_1 & f_1 \\ e_2 & f_2 \end{bmatrix}\begin{bmatrix} a_1 & b_1 \\ a_2 & b_2 \end{bmatrix} =$

$$\begin{bmatrix} a_1c_1 + a_2d_1 & b_1c_1 + b_2d_1 \\ a_1c_2 + a_2d_2 & b_1c_2 + b_2d_2 \end{bmatrix} + \begin{bmatrix} a_1e_1 + a_2f_1 & b_1e_1 + b_2f_1 \\ a_1e_2 + a_2f_2 & b_1e_2 + b_2f_2 \end{bmatrix} =$$

$$\begin{bmatrix} a_1c_1 + a_2d_1 + a_1e_1 + a_2f_1 & b_1c_1 + b_2d_1 + b_1e_1 + b_2f_1 \\ a_1c_2 + a_2d_2 + a_1e_2 + a_2f_2 & b_1c_2 + b_2d_2 + b_1e_2 + b_2f_2 \end{bmatrix};$$

$\therefore (B + C)A = BA + CA$

23. $a(AB) = a\begin{bmatrix} a_1c_1 + b_1c_2 & a_1d_1 + b_1b_2 \\ a_2c_1 + b_2c_2 & a_2d_1 + b_2d_2 \end{bmatrix} = \begin{bmatrix} aa_1c_1 + ab_1c_2 & aa_1d_1 + ab_1b_2 \\ aa_2c_1 + ab_2c_2 & aa_2d_1 + ab_2d_2 \end{bmatrix};$

$(aA)B = \begin{bmatrix} aa_1 & ab_1 \\ aa_2 & ab_2 \end{bmatrix}\begin{bmatrix} c_1 & d_1 \\ c_2 & d_2 \end{bmatrix} = \begin{bmatrix} aa_1c_1 + ab_1c_2 & aa_1d_1 + ab_1d_2 \\ aa_2c_1 + ab_2c_2 & aa_2d_1 + ab_2d_2 \end{bmatrix};$

$\therefore a(AB) = (aA)B$

24. $a(AB) = \begin{bmatrix} aa_1c_1 + ab_1c_2 & aa_1d_1 + ab_1d_2 \\ aa_2c_1 + ab_2c_2 & aa_2d_1 + ab_2d_2 \end{bmatrix};$

$A(aB) = \begin{bmatrix} a_1 & b_1 \\ a_2 & b_2 \end{bmatrix}\begin{bmatrix} ac_1 & ad_1 \\ ac_2 & ad_2 \end{bmatrix} = \begin{bmatrix} aa_1c_1 + ab_1c_2 & aa_1d_1 + ab_1d_2 \\ aa_2c_1 + ab_2c_2 & aa_2d_1 + ab_2d_2 \end{bmatrix};$

$\therefore a(AB) = A(aB)$

25. $O_{2\times 2} = \begin{bmatrix} 0 & 0 \\ 0 & 0 \end{bmatrix}; O_{2\times 2}A = \begin{bmatrix} 0 & 0 \\ 0 & 0 \end{bmatrix}\begin{bmatrix} a_1 & b_1 \\ a_2 & b_2 \end{bmatrix} = \begin{bmatrix} 0 & 0 \\ 0 & 0 \end{bmatrix} = O_{2\times 2}$

26. $O_{2\times 2} = \begin{bmatrix} 0 & 0 \\ 0 & 0 \end{bmatrix}; AO_{2\times 2} = \begin{bmatrix} a_1 & b_1 \\ a_2 & b_2 \end{bmatrix}\begin{bmatrix} 0 & 0 \\ 0 & 0 \end{bmatrix} = \begin{bmatrix} 0 & 0 \\ 0 & 0 \end{bmatrix} = O_{2\times 2}$

C 27. a. $|A| = \begin{vmatrix} a_1 & b_1 \\ a_2 & b_2 \end{vmatrix} = a_1b_2 - a_2b_1; |cA| = \left| c\begin{bmatrix} a_1 & b_1 \\ a_2 & b_2 \end{bmatrix} \right| = \begin{vmatrix} ca_1 & cb_1 \\ ca_2 & cb_2 \end{vmatrix} =$
$ca_1 \cdot cb_2 - (ca_2 \cdot cb_1) = c^2a_1b_2 - c^2a_2b_1 = c^2(a_1b_2 - a_2b_1) = c^2|A|$

b. $|A| = \begin{vmatrix} a_1 & b_1 \\ a_2 & b_2 \end{vmatrix}, |B| = \begin{vmatrix} c_1 & d_1 \\ c_2 & d_2 \end{vmatrix}, |A| = a_1b_2 - a_2b_1, |B| = c_1d_2 - c_2d_1;$

$|AB| = \left| \begin{bmatrix} a_1 & b_1 \\ a_2 & b_2 \end{bmatrix} \begin{bmatrix} c_1 & d_1 \\ c_2 & d_2 \end{bmatrix} \right| = \begin{vmatrix} a_1c_1 + b_1c_2 & a_1d_1 + b_1d_2 \\ a_2c_1 + b_2c_2 & a_2d_1 + b_2d_2 \end{vmatrix} =$
$(a_1c_1 + b_1c_2)(a_2d_1 + b_2d_2) - (a_2c_1 + b_2c_2)(a_1d_1 + b_1d_2) =$
$a_1c_1a_2d_1 + a_1c_1b_2d_2 + b_1c_2a_2d_1 + b_1c_2b_2d_2 -$
$(a_2c_1a_1d_1 + a_2c_1b_1d_2 + b_2c_2a_1d_1 + b_2c_2b_1d_2) = a_1b_2(c_1d_2 - c_2d_1) -$
$a_2b_1(c_1d_2 - c_2d_1) + a_1a_2c_1d_1 + b_1b_2c_2d_2 - (a_1a_2c_1d_1 + b_1b_2c_2d_2) =$
$(a_1b_2 - a_2b_1)(c_1d_2 - c_2d_1) = |A| \cdot |B|$

Pages 546–547 • SELF-TEST 1

1. $\begin{bmatrix} 9 & 13 \\ 3 & 7 \end{bmatrix}$ **2.** $\begin{bmatrix} -7 & 7 & -4 \\ -6 & 6 & 11 \end{bmatrix}$

3. $5\begin{bmatrix} -4 & -1 \\ 8 & 5 \end{bmatrix} - 4\begin{bmatrix} -4 & 2 \\ 4 & 7 \end{bmatrix} = \begin{bmatrix} -20 & -5 \\ 40 & 25 \end{bmatrix} - \begin{bmatrix} -16 & 8 \\ 16 & 28 \end{bmatrix} = \begin{bmatrix} -4 & -13 \\ 24 & -3 \end{bmatrix}$

4. $X - \begin{bmatrix} -4 & 2 \\ 4 & 7 \end{bmatrix} = \begin{bmatrix} -4 & -1 \\ 8 & 5 \end{bmatrix}; X - \begin{bmatrix} -4 & 2 \\ 4 & 7 \end{bmatrix} + \begin{bmatrix} -4 & 2 \\ 4 & 7 \end{bmatrix} = \begin{bmatrix} -4 & -1 \\ 8 & 5 \end{bmatrix} +$
$\begin{bmatrix} -4 & 2 \\ 4 & 7 \end{bmatrix}; X = \begin{bmatrix} -8 & 1 \\ 12 & 12 \end{bmatrix}$

5. $2X + 5\begin{bmatrix} -4 & 2 \\ 4 & 7 \end{bmatrix} = 3\begin{bmatrix} -4 & -1 \\ 8 & 5 \end{bmatrix}; 2X + \begin{bmatrix} -20 & 10 \\ 20 & 35 \end{bmatrix} = \begin{bmatrix} -12 & -3 \\ 24 & 15 \end{bmatrix};$
$2X + \begin{bmatrix} -20 & 10 \\ 20 & 35 \end{bmatrix} + \begin{bmatrix} 20 & -10 \\ -20 & -35 \end{bmatrix} = \begin{bmatrix} -12 & -3 \\ 24 & 15 \end{bmatrix} + \begin{bmatrix} 20 & -10 \\ -20 & -35 \end{bmatrix};$
$2X = \begin{bmatrix} 8 & -13 \\ 4 & -20 \end{bmatrix}; X = \frac{1}{2}\begin{bmatrix} 8 & -13 \\ 4 & -20 \end{bmatrix} = \begin{bmatrix} 4 & -\frac{13}{2} \\ 2 & -10 \end{bmatrix}$

6. $\begin{bmatrix} 9 - 5 & 2 - 15 \\ -54 + 0 & -12 + 0 \end{bmatrix} = \begin{bmatrix} 4 & -13 \\ -54 & -12 \end{bmatrix}$

7. $\begin{bmatrix} 5 + 0 - 4 \\ 30 + 0 + 3 \end{bmatrix} = \begin{bmatrix} 1 \\ 33 \end{bmatrix}$

8. $AB = \begin{bmatrix} -4 & -1 \\ 8 & 5 \end{bmatrix}\begin{bmatrix} -4 & 2 \\ 4 & 7 \end{bmatrix} = \begin{bmatrix} 12 & -15 \\ -12 & 51 \end{bmatrix}; BA = \begin{bmatrix} -4 & 2 \\ 4 & 7 \end{bmatrix}\begin{bmatrix} -4 & -1 \\ 8 & 5 \end{bmatrix} =$
$\begin{bmatrix} 32 & 14 \\ 40 & 31 \end{bmatrix}; AB \neq BA$

Pages 551–552 • WRITTEN EXERCISES

A **1.** $\begin{bmatrix} 5 & 2 \\ 7 & 3 \end{bmatrix}^{-1} = \frac{1}{1}\begin{bmatrix} 3 & -2 \\ -7 & 5 \end{bmatrix} = \begin{bmatrix} 3 & -2 \\ -7 & 5 \end{bmatrix}$

2. $\begin{bmatrix} 9 & 2 \\ 5 & 1 \end{bmatrix}^{-1} = \frac{1}{-1}\begin{bmatrix} 1 & -2 \\ -5 & 9 \end{bmatrix} = \begin{bmatrix} -1 & 2 \\ 5 & -9 \end{bmatrix}$

3. Since $8 \cdot 3 - 6 \cdot 4 = 0$, the matrix is singular.

4. $\begin{bmatrix} 0 & 2 \\ 2 & 0 \end{bmatrix}^{-1} = \frac{1}{-4}\begin{bmatrix} 0 & -2 \\ -2 & 0 \end{bmatrix} = \begin{bmatrix} 0 & \frac{1}{2} \\ \frac{1}{2} & 0 \end{bmatrix}$

5. $\begin{bmatrix} 3 & 0 \\ 0 & 3 \end{bmatrix}^{-1} = \frac{1}{9}\begin{bmatrix} 3 & 0 \\ 0 & 3 \end{bmatrix} = \begin{bmatrix} \frac{1}{3} & 0 \\ 0 & \frac{1}{3} \end{bmatrix}$

6. $\begin{bmatrix} 4 & -7 \\ 3 & -5 \end{bmatrix}^{-1} = \frac{1}{1}\begin{bmatrix} -5 & 7 \\ -3 & 4 \end{bmatrix} = \begin{bmatrix} -5 & 7 \\ -3 & 4 \end{bmatrix}$

7. $\begin{bmatrix} 4 & 3 \\ 6 & 5 \end{bmatrix}^{-1} = \frac{1}{2}\begin{bmatrix} 5 & -3 \\ -6 & 4 \end{bmatrix} = \begin{bmatrix} \frac{5}{2} & -\frac{3}{2} \\ -3 & 2 \end{bmatrix}$

8. Since $9 \cdot 2 - (-6)(-3) = 0$, the matrix is singular.

9. $\begin{bmatrix} 9 & 4 \\ -6 & -3 \end{bmatrix}^{-1} = \frac{1}{-3}\begin{bmatrix} -3 & -4 \\ 6 & 9 \end{bmatrix} = \begin{bmatrix} 1 & \frac{4}{3} \\ -2 & -3 \end{bmatrix}$

10. $\begin{bmatrix} -11 & 5 \\ -4 & 2 \end{bmatrix}^{-1} = \frac{1}{-2}\begin{bmatrix} 2 & -5 \\ 4 & -11 \end{bmatrix} = \begin{bmatrix} -1 & \frac{5}{2} \\ -2 & \frac{11}{2} \end{bmatrix}$

11. $A^{-1}AX = A^{-1}B$; $X = A^{-1}B$; $X = \begin{bmatrix} 3 & 4 \\ 2 & 3 \end{bmatrix}^{-1}\begin{bmatrix} 2 & 1 \\ 5 & 3 \end{bmatrix}$; $X = \frac{1}{1}\begin{bmatrix} 3 & -4 \\ -2 & 3 \end{bmatrix}\begin{bmatrix} 2 & 1 \\ 5 & 3 \end{bmatrix} =$

$\begin{bmatrix} 3 & -4 \\ -2 & 3 \end{bmatrix}\begin{bmatrix} 2 & 1 \\ 5 & 3 \end{bmatrix} = \begin{bmatrix} -14 & -9 \\ 11 & 7 \end{bmatrix}$

12. $A^{-1}AX = A^{-1}C$; $X = A^{-1}C = \begin{bmatrix} 3 & 4 \\ 2 & 3 \end{bmatrix}^{-1}\begin{bmatrix} 2 & 0 \\ 0 & 2 \end{bmatrix} = \frac{1}{1}\begin{bmatrix} 3 & -4 \\ -2 & 3 \end{bmatrix}\begin{bmatrix} 2 & 0 \\ 0 & 2 \end{bmatrix} =$

$\begin{bmatrix} 3 & -4 \\ -2 & 3 \end{bmatrix}\begin{bmatrix} 2 & 0 \\ 0 & 2 \end{bmatrix} = \begin{bmatrix} 6 & -8 \\ -4 & 6 \end{bmatrix}$

13. $B^{-1}BX = B^{-1}A$; $X = B^{-1}A = \begin{bmatrix} 2 & 1 \\ 5 & 3 \end{bmatrix}^{-1}\begin{bmatrix} 3 & 4 \\ 2 & 3 \end{bmatrix} = \frac{1}{1}\begin{bmatrix} 3 & -1 \\ -5 & 2 \end{bmatrix}\begin{bmatrix} 3 & 4 \\ 2 & 3 \end{bmatrix} =$

$\begin{bmatrix} 3 & -1 \\ -5 & 2 \end{bmatrix}\begin{bmatrix} 3 & 4 \\ 2 & 3 \end{bmatrix} = \begin{bmatrix} 7 & 9 \\ -11 & -14 \end{bmatrix}$

14. $B^{-1}BX = B^{-1}D$; $X = B^{-1}D = \begin{bmatrix} 2 & 1 \\ 5 & 3 \end{bmatrix}^{-1}\begin{bmatrix} 5 & 4 \\ -2 & -2 \end{bmatrix} =$

$\frac{1}{1}\begin{bmatrix} 3 & -1 \\ -5 & 2 \end{bmatrix}\begin{bmatrix} 5 & 4 \\ -2 & -2 \end{bmatrix} = \begin{bmatrix} 3 & -1 \\ -5 & 2 \end{bmatrix}\begin{bmatrix} 5 & 4 \\ -2 & -2 \end{bmatrix} = \begin{bmatrix} 17 & 14 \\ -29 & -24 \end{bmatrix}$

15. $C^{-1}CX = C^{-1}D$; $X = C^{-1}D = \begin{bmatrix} 2 & 0 \\ 0 & 2 \end{bmatrix}^{-1}\begin{bmatrix} 5 & 4 \\ -2 & -2 \end{bmatrix} = \frac{1}{4}\begin{bmatrix} 2 & 0 \\ 0 & 2 \end{bmatrix}\begin{bmatrix} 5 & 4 \\ -2 & -2 \end{bmatrix} =$

$\begin{bmatrix} \frac{1}{2} & 0 \\ 0 & \frac{1}{2} \end{bmatrix}\begin{bmatrix} 5 & 4 \\ -2 & -2 \end{bmatrix} = \begin{bmatrix} \frac{5}{2} & 2 \\ -1 & -1 \end{bmatrix}$

16. $C^{-1}CX = C^{-1}A$; $X = C^{-1}A = \begin{bmatrix} 2 & 0 \\ 0 & 2 \end{bmatrix}^{-1} \begin{bmatrix} 3 & 4 \\ 2 & 3 \end{bmatrix} = \frac{1}{4}\begin{bmatrix} 2 & 0 \\ 0 & 2 \end{bmatrix}\begin{bmatrix} 3 & 4 \\ 2 & 3 \end{bmatrix} =$

$\begin{bmatrix} \frac{1}{2} & 0 \\ 0 & \frac{1}{2} \end{bmatrix}\begin{bmatrix} 3 & 4 \\ 2 & 3 \end{bmatrix} = \begin{bmatrix} \frac{3}{2} & 2 \\ 1 & \frac{3}{2} \end{bmatrix}$

17. $D^{-1}DX = D^{-1}C$; $X = D^{-1}C = \begin{bmatrix} 5 & 4 \\ -2 & -2 \end{bmatrix}^{-1} \begin{bmatrix} 2 & 0 \\ 0 & 2 \end{bmatrix} =$

$\frac{1}{-2}\begin{bmatrix} -2 & -4 \\ 2 & 5 \end{bmatrix}\begin{bmatrix} 2 & 0 \\ 0 & 2 \end{bmatrix} = \begin{bmatrix} 1 & 2 \\ -1 & -\frac{5}{2} \end{bmatrix}\begin{bmatrix} 2 & 0 \\ 0 & 2 \end{bmatrix} = \begin{bmatrix} 2 & 4 \\ -2 & -5 \end{bmatrix}$

18. $D^{-1}DX = D^{-1}A$; $X = D^{-1}A = \begin{bmatrix} 5 & 4 \\ -2 & -2 \end{bmatrix}^{-1} \begin{bmatrix} 3 & 4 \\ 2 & 3 \end{bmatrix} =$

$\frac{1}{-2}\begin{bmatrix} -2 & -4 \\ 2 & 5 \end{bmatrix}\begin{bmatrix} 3 & 4 \\ 2 & 3 \end{bmatrix} = \begin{bmatrix} 1 & 2 \\ -1 & -\frac{5}{2} \end{bmatrix}\begin{bmatrix} 3 & 4 \\ 2 & 3 \end{bmatrix} = \begin{bmatrix} 7 & 10 \\ -8 & -\frac{23}{2} \end{bmatrix}$

B 19. $\begin{bmatrix} 1 & 3 \\ 1 & 2 \end{bmatrix}\begin{bmatrix} x \\ y \end{bmatrix} = \begin{bmatrix} 0 \\ 1 \end{bmatrix}$; $\begin{bmatrix} 1 & 3 \\ 1 & 2 \end{bmatrix}^{-1} = \frac{1}{-1}\begin{bmatrix} 2 & -3 \\ -1 & 1 \end{bmatrix} = \begin{bmatrix} -2 & 3 \\ 1 & -1 \end{bmatrix}$;

$\begin{bmatrix} x \\ y \end{bmatrix} = \begin{bmatrix} -2 & 3 \\ 1 & -1 \end{bmatrix}\begin{bmatrix} 0 \\ 1 \end{bmatrix} = \begin{bmatrix} 3 \\ -1 \end{bmatrix}$; $\{(3, -1)\}$

20. $\begin{bmatrix} 3 & 2 \\ 7 & 4 \end{bmatrix}\begin{bmatrix} x \\ y \end{bmatrix} = \begin{bmatrix} -3 \\ 1 \end{bmatrix}$; $\begin{bmatrix} 3 & 2 \\ 7 & 4 \end{bmatrix}^{-1} = \frac{1}{-2}\begin{bmatrix} 4 & -2 \\ -7 & 3 \end{bmatrix} = \begin{bmatrix} -2 & 1 \\ \frac{7}{2} & -\frac{3}{2} \end{bmatrix}$;

$\begin{bmatrix} x \\ y \end{bmatrix} = \begin{bmatrix} -2 & 1 \\ \frac{7}{2} & -\frac{3}{2} \end{bmatrix}\begin{bmatrix} -3 \\ 1 \end{bmatrix} = \begin{bmatrix} 7 \\ -12 \end{bmatrix}$; $\{(7, -12)\}$

21. $\begin{bmatrix} 4 & 3 \\ -5 & -4 \end{bmatrix}\begin{bmatrix} x \\ y \end{bmatrix} = \begin{bmatrix} -5 \\ 6 \end{bmatrix}$; $\begin{bmatrix} 4 & 3 \\ -5 & -4 \end{bmatrix}^{-1} = \frac{1}{-1}\begin{bmatrix} -4 & -3 \\ 5 & 4 \end{bmatrix} = \begin{bmatrix} 4 & 3 \\ -5 & -4 \end{bmatrix}$;

$\begin{bmatrix} x \\ y \end{bmatrix} = \begin{bmatrix} 4 & 3 \\ -5 & -4 \end{bmatrix}\begin{bmatrix} -5 \\ 6 \end{bmatrix} = \begin{bmatrix} -2 \\ 1 \end{bmatrix}$; $\{(-2, 1)\}$

22. $\begin{bmatrix} 4 & -2 \\ 5 & -2 \end{bmatrix}\begin{bmatrix} x \\ y \end{bmatrix} = \begin{bmatrix} 8 \\ 11 \end{bmatrix}$; $\begin{bmatrix} 4 & -2 \\ 5 & -2 \end{bmatrix}^{-1} = \frac{1}{2}\begin{bmatrix} -2 & 2 \\ -5 & 4 \end{bmatrix} = \begin{bmatrix} -1 & 1 \\ -\frac{5}{2} & 2 \end{bmatrix}$;

$\begin{bmatrix} x \\ y \end{bmatrix} = \begin{bmatrix} -1 & 1 \\ -\frac{5}{2} & 2 \end{bmatrix}\begin{bmatrix} 8 \\ 11 \end{bmatrix} = \begin{bmatrix} 3 \\ 2 \end{bmatrix}$; $\{(3, 2)\}$

23. $\begin{bmatrix} 6 & -2 \\ 10 & -2 \end{bmatrix}\begin{bmatrix} x \\ y \end{bmatrix} = \begin{bmatrix} 3 \\ 5 \end{bmatrix}$; $\begin{bmatrix} 6 & -2 \\ 10 & -2 \end{bmatrix}^{-1} = \frac{1}{8}\begin{bmatrix} -2 & 2 \\ -10 & 6 \end{bmatrix} = \begin{bmatrix} -\frac{1}{4} & \frac{1}{4} \\ -\frac{5}{4} & \frac{3}{4} \end{bmatrix}$;

$\begin{bmatrix} x \\ y \end{bmatrix} = \begin{bmatrix} -\frac{1}{4} & \frac{1}{4} \\ -\frac{5}{4} & \frac{3}{4} \end{bmatrix}\begin{bmatrix} 3 \\ 5 \end{bmatrix} = \begin{bmatrix} \frac{1}{2} \\ 0 \end{bmatrix}$; $\left\{\left(\frac{1}{2}, 0\right)\right\}$

24. $\begin{bmatrix} 3 & 2 \\ 6 & 5 \end{bmatrix}\begin{bmatrix} x \\ y \end{bmatrix} = \begin{bmatrix} -6 \\ -9 \end{bmatrix}$; $\begin{bmatrix} 3 & 2 \\ 6 & 5 \end{bmatrix}^{-1} = \frac{1}{3}\begin{bmatrix} 5 & -2 \\ -6 & 3 \end{bmatrix} = \begin{bmatrix} \frac{5}{3} & -\frac{2}{3} \\ -2 & 1 \end{bmatrix}$;

$\begin{bmatrix} x \\ y \end{bmatrix} = \begin{bmatrix} \frac{5}{3} & -\frac{2}{3} \\ -2 & 1 \end{bmatrix}\begin{bmatrix} -6 \\ -9 \end{bmatrix} = \begin{bmatrix} -4 \\ 3 \end{bmatrix}$; $\{(-4, 3)\}$

C 25. $\det A = ad - bc$; $A \cdot \frac{1}{ad - bc}\begin{bmatrix} d & -b \\ -c & a \end{bmatrix} = A\begin{bmatrix} \frac{d}{ad - bc} & \frac{-b}{ad - bc} \\ \frac{-c}{ad - bc} & \frac{a}{ad - bc} \end{bmatrix} =$

$\begin{bmatrix} a & b \\ c & d \end{bmatrix}\begin{bmatrix} \frac{d}{ad - bc} & \frac{-b}{ad - bc} \\ \frac{-c}{ad - bc} & \frac{a}{ad - bc} \end{bmatrix} = \begin{bmatrix} \frac{ad - bc}{ad - bc} & \frac{-ab + ab}{ad + bc} \\ \frac{cd - cd}{ad - bc} & \frac{-cb + ad}{ad - bc} \end{bmatrix} = \begin{bmatrix} 1 & 0 \\ 0 & 1 \end{bmatrix}$

26. $\det A = ad - bc$; $\frac{1}{ad - bc}\begin{bmatrix} d & -b \\ -c & a \end{bmatrix} \cdot A = \begin{bmatrix} \frac{d}{ad - bc} & \frac{-b}{ad - bc} \\ \frac{-c}{ad - bc} & \frac{a}{ad - bc} \end{bmatrix} \cdot A =$

$\begin{bmatrix} \frac{d}{ad - bc} & \frac{-b}{ad - bc} \\ \frac{-c}{ad - bc} & \frac{a}{ad - bc} \end{bmatrix}\begin{bmatrix} a & b \\ c & d \end{bmatrix} = \begin{bmatrix} \frac{ad - bc}{ad - bc} & \frac{bd - bd}{ad - bc} \\ \frac{-ac + ac}{ad - bc} & \frac{-bc + ad}{ad - bc} \end{bmatrix} = \begin{bmatrix} 1 & 0 \\ 0 & 1 \end{bmatrix}$

27. Assume A^{-1} exists. Then $AA^{-1} = I$ and $\det AA^{-1} = \det I = 1$. By Exercise 27(b) on page 546, $\det AA^{-1} = \det A \det A^{-1}$. But if $\det A = 0$, $\det A \cdot \det A^{-1} = 0$, and $0 \neq 1$ so A^{-1} cannot exist.

Page 552 • COMPUTER EXERCISES

1.
```
10  INPUT A,B,C,D
20  LET Z = A * D - B * C
30  IF Z = 0 THEN PRINT "MATRIX HAS NO INVERSE." : END
40  PRINT D/Z;"    ";-B/Z
50  PRINT -C/Z;"    ";A/Z
60  END
```

2. $\begin{bmatrix} 0.25 & -1.25 \\ 0 & 1 \end{bmatrix}$ 3. no inverse 4. $\begin{bmatrix} 0.5 & 0 \\ 0 & 0.5 \end{bmatrix}$ 5. $\begin{bmatrix} -1.5 & -2 \\ 2.5 & 3 \end{bmatrix}$

6. no inverse

Page 552 • SELF-TEST 2

1. $\begin{bmatrix} -2 & 1 \\ 5 & 7 \end{bmatrix}^{-1} = \frac{1}{-19}\begin{bmatrix} 7 & -1 \\ -5 & -2 \end{bmatrix} = \begin{bmatrix} -\frac{7}{19} & \frac{1}{19} \\ \frac{5}{19} & \frac{2}{19} \end{bmatrix}$

2. $\begin{bmatrix} 4 & 9 \\ 2 & 3 \end{bmatrix}^{-1}\begin{bmatrix} 4 & 9 \\ 2 & 3 \end{bmatrix}X = \begin{bmatrix} 4 & 9 \\ 2 & 3 \end{bmatrix}^{-1}\begin{bmatrix} -4 & 5 \\ -2 & 7 \end{bmatrix}$; $X = \frac{1}{-6}\begin{bmatrix} 3 & -9 \\ -2 & 4 \end{bmatrix}\begin{bmatrix} -4 & 5 \\ -2 & 7 \end{bmatrix} =$

$\begin{bmatrix} -\frac{1}{2} & \frac{3}{2} \\ \frac{1}{3} & -\frac{2}{3} \end{bmatrix}\begin{bmatrix} -4 & 5 \\ -2 & 7 \end{bmatrix} = \begin{bmatrix} -1 & 8 \\ 0 & -3 \end{bmatrix}$

3. $\begin{bmatrix} 2 & -1 \\ 7 & 8 \end{bmatrix}\begin{bmatrix} x \\ y \end{bmatrix} = \begin{bmatrix} -3 \\ 1 \end{bmatrix}$

4. $\begin{bmatrix} 4 & 3 \\ 2 & -4 \end{bmatrix}\begin{bmatrix} x \\ y \end{bmatrix} = \begin{bmatrix} -10 \\ 17 \end{bmatrix}$; $\begin{bmatrix} 4 & 3 \\ 2 & -4 \end{bmatrix}^{-1} = \frac{1}{-22}\begin{bmatrix} -4 & -3 \\ -2 & 4 \end{bmatrix} = \begin{bmatrix} \frac{2}{11} & \frac{3}{22} \\ \frac{1}{11} & -\frac{2}{11} \end{bmatrix}$;

$\begin{bmatrix} x \\ y \end{bmatrix} = \begin{bmatrix} \frac{2}{11} & \frac{3}{22} \\ \frac{1}{11} & -\frac{2}{11} \end{bmatrix}\begin{bmatrix} -10 \\ 17 \end{bmatrix} = \begin{bmatrix} \frac{1}{2} \\ -4 \end{bmatrix}$; $\left\{\left(\frac{1}{2}, -4\right)\right\}$

Page 556 • WRITTEN EXERCISES

A 1. $\begin{bmatrix} x' \\ y' \end{bmatrix} = \begin{bmatrix} -6 \\ 1 \end{bmatrix} + \begin{bmatrix} 5 \\ 2 \end{bmatrix} = \begin{bmatrix} -1 \\ 3 \end{bmatrix}$; $(-1, 3)$

2. $\begin{bmatrix} x' \\ y' \end{bmatrix} = \begin{bmatrix} -4 \\ 9 \end{bmatrix} + \begin{bmatrix} 7 \\ -6 \end{bmatrix} = \begin{bmatrix} 3 \\ 3 \end{bmatrix}$; $(3, 3)$

3. $\begin{bmatrix} x' \\ y' \end{bmatrix} = \begin{bmatrix} -3 \\ -8 \end{bmatrix} + \begin{bmatrix} -1 \\ 11 \end{bmatrix} = \begin{bmatrix} -4 \\ 3 \end{bmatrix}$; $(-4, 3)$

4. $\begin{bmatrix} x' \\ y' \end{bmatrix} = \begin{bmatrix} 0 \\ 10 \end{bmatrix} + \begin{bmatrix} -8 \\ -1 \end{bmatrix} = \begin{bmatrix} -8 \\ 9 \end{bmatrix}$; $(-8, 9)$

5. $\begin{bmatrix} x' \\ y' \end{bmatrix} = \begin{bmatrix} a \\ b \end{bmatrix} + \begin{bmatrix} a \\ -b \end{bmatrix} = \begin{bmatrix} 2a \\ 0 \end{bmatrix}$; $(2a, 0)$

6. $\begin{bmatrix} x' \\ y' \end{bmatrix} = \begin{bmatrix} a \\ a - b \end{bmatrix} + \begin{bmatrix} a \\ a + b \end{bmatrix} = \begin{bmatrix} 2a \\ 2a \end{bmatrix}$; $(2a, 2a)$

7. $\begin{bmatrix} 2 \\ -7 \end{bmatrix} = \begin{bmatrix} x \\ y \end{bmatrix} + \begin{bmatrix} 4 \\ -5 \end{bmatrix}$; $\begin{bmatrix} 2 \\ -7 \end{bmatrix} - \begin{bmatrix} 4 \\ -5 \end{bmatrix} = \begin{bmatrix} x \\ y \end{bmatrix}$; $\begin{bmatrix} x \\ y \end{bmatrix} = \begin{bmatrix} -2 \\ -2 \end{bmatrix}$; $(-2, -2)$

8. $\begin{bmatrix} 0 \\ 4 \end{bmatrix} = \begin{bmatrix} x \\ y \end{bmatrix} + \begin{bmatrix} -3 \\ 9 \end{bmatrix}$; $\begin{bmatrix} 0 \\ 4 \end{bmatrix} - \begin{bmatrix} -3 \\ 9 \end{bmatrix} = \begin{bmatrix} x \\ y \end{bmatrix}$; $\begin{bmatrix} x \\ y \end{bmatrix} = \begin{bmatrix} 3 \\ -5 \end{bmatrix}$; $(3, -5)$

9. $\begin{bmatrix} -1 \\ -9 \end{bmatrix} = \begin{bmatrix} x \\ y \end{bmatrix} + \begin{bmatrix} 3 \\ -4 \end{bmatrix}$; $\begin{bmatrix} -1 \\ -9 \end{bmatrix} - \begin{bmatrix} 3 \\ -4 \end{bmatrix} = \begin{bmatrix} x \\ y \end{bmatrix}$; $\begin{bmatrix} x \\ y \end{bmatrix} = \begin{bmatrix} -4 \\ -5 \end{bmatrix}$; $(-4, -5)$

10. $\begin{bmatrix} 6 \\ -2 \end{bmatrix} = \begin{bmatrix} x \\ y \end{bmatrix} + \begin{bmatrix} -5 \\ 8 \end{bmatrix}$; $\begin{bmatrix} 6 \\ -2 \end{bmatrix} - \begin{bmatrix} -5 \\ 8 \end{bmatrix} = \begin{bmatrix} x \\ y \end{bmatrix}$; $\begin{bmatrix} x \\ y \end{bmatrix} = \begin{bmatrix} 11 \\ -10 \end{bmatrix}$; $(11, -10)$

11. $\begin{bmatrix} 0 \\ 0 \end{bmatrix} = \begin{bmatrix} x \\ y \end{bmatrix} + \begin{bmatrix} a \\ -b \end{bmatrix}$; $\begin{bmatrix} 0 \\ 0 \end{bmatrix} - \begin{bmatrix} a \\ -b \end{bmatrix} = \begin{bmatrix} x \\ y \end{bmatrix}$; $\begin{bmatrix} x \\ y \end{bmatrix} = \begin{bmatrix} -a \\ b \end{bmatrix}$; $(-a, b)$

12. $\begin{bmatrix} a \\ -b \end{bmatrix} = \begin{bmatrix} x \\ y \end{bmatrix} + \begin{bmatrix} a - b \\ a + b \end{bmatrix}$; $\begin{bmatrix} a \\ -b \end{bmatrix} - \begin{bmatrix} a - b \\ a + b \end{bmatrix} = \begin{bmatrix} x \\ y \end{bmatrix}$; $\begin{bmatrix} x \\ y \end{bmatrix} = \begin{bmatrix} b \\ -a - 2b \end{bmatrix}$;
$(b, -a - 2b)$

13. $\begin{bmatrix} 3 \\ 4 \end{bmatrix} = \begin{bmatrix} 2 \\ 8 \end{bmatrix} + \begin{bmatrix} h \\ k \end{bmatrix}$; $\begin{bmatrix} h \\ k \end{bmatrix} = \begin{bmatrix} 3 \\ 4 \end{bmatrix} - \begin{bmatrix} 2 \\ 8 \end{bmatrix} = \begin{bmatrix} 1 \\ -4 \end{bmatrix}$; $\begin{bmatrix} x' \\ y' \end{bmatrix} = \begin{bmatrix} x \\ y \end{bmatrix} + \begin{bmatrix} 1 \\ -4 \end{bmatrix}$

14. $\begin{bmatrix} 4 \\ -1 \end{bmatrix} = \begin{bmatrix} -3 \\ 7 \end{bmatrix} + \begin{bmatrix} h \\ k \end{bmatrix}$; $\begin{bmatrix} h \\ k \end{bmatrix} = \begin{bmatrix} 4 \\ -1 \end{bmatrix} - \begin{bmatrix} -3 \\ 7 \end{bmatrix} = \begin{bmatrix} 7 \\ -8 \end{bmatrix}$; $\begin{bmatrix} x' \\ y' \end{bmatrix} = \begin{bmatrix} x \\ y \end{bmatrix} + \begin{bmatrix} 7 \\ -8 \end{bmatrix}$

15. $\begin{bmatrix} 7 \\ -1 \end{bmatrix} = \begin{bmatrix} 2 \\ -4 \end{bmatrix} + \begin{bmatrix} h \\ k \end{bmatrix}$; $\begin{bmatrix} h \\ k \end{bmatrix} = \begin{bmatrix} 7 \\ -1 \end{bmatrix} - \begin{bmatrix} 2 \\ -4 \end{bmatrix} = \begin{bmatrix} 5 \\ 3 \end{bmatrix}$; $\begin{bmatrix} x' \\ y' \end{bmatrix} = \begin{bmatrix} x \\ y \end{bmatrix} + \begin{bmatrix} 5 \\ 3 \end{bmatrix}$

16. $\begin{bmatrix}-7\\2\end{bmatrix} = \begin{bmatrix}-1\\-8\end{bmatrix} + \begin{bmatrix}h\\k\end{bmatrix}; \begin{bmatrix}h\\k\end{bmatrix} = \begin{bmatrix}-7\\2\end{bmatrix} - \begin{bmatrix}-1\\-8\end{bmatrix} = \begin{bmatrix}-6\\10\end{bmatrix}; \begin{bmatrix}x'\\y'\end{bmatrix} = \begin{bmatrix}x\\y\end{bmatrix} + \begin{bmatrix}-6\\10\end{bmatrix}$

17. $\begin{bmatrix}c\\d\end{bmatrix} = \begin{bmatrix}a\\b\end{bmatrix} + \begin{bmatrix}h\\k\end{bmatrix}; \begin{bmatrix}h\\k\end{bmatrix} = \begin{bmatrix}c\\d\end{bmatrix} - \begin{bmatrix}a\\b\end{bmatrix} = \begin{bmatrix}c-a\\d-b\end{bmatrix}; \begin{bmatrix}x'\\y'\end{bmatrix} = \begin{bmatrix}x\\y\end{bmatrix} + \begin{bmatrix}c-a\\d-b\end{bmatrix}$

18. $\begin{bmatrix}a\\b\end{bmatrix} = \begin{bmatrix}a-b\\a+b\end{bmatrix} + \begin{bmatrix}h\\k\end{bmatrix}; \begin{bmatrix}h\\k\end{bmatrix} = \begin{bmatrix}a\\b\end{bmatrix} - \begin{bmatrix}a-b\\a+b\end{bmatrix} = \begin{bmatrix}b\\-a\end{bmatrix}; \begin{bmatrix}x'\\y'\end{bmatrix} = \begin{bmatrix}x\\y\end{bmatrix} + \begin{bmatrix}b\\-a\end{bmatrix}$

B 19. $\begin{bmatrix}7\\-8\end{bmatrix} = \begin{bmatrix}0\\-3\end{bmatrix} + \begin{bmatrix}h\\k\end{bmatrix}; \begin{bmatrix}h\\k\end{bmatrix} = \begin{bmatrix}7\\-8\end{bmatrix} - \begin{bmatrix}0\\-3\end{bmatrix} = \begin{bmatrix}7\\-5\end{bmatrix}; \begin{bmatrix}x'\\y'\end{bmatrix} = \begin{bmatrix}x\\y\end{bmatrix} +$
$\begin{bmatrix}7\\-5\end{bmatrix} = \begin{bmatrix}6\\1\end{bmatrix} + \begin{bmatrix}7\\-5\end{bmatrix} = \begin{bmatrix}13\\-4\end{bmatrix}$; (13, −4)

20. $\begin{bmatrix}9\\-4\end{bmatrix} = \begin{bmatrix}-2\\1\end{bmatrix} + \begin{bmatrix}h\\k\end{bmatrix}; \begin{bmatrix}h\\k\end{bmatrix} = \begin{bmatrix}9\\-4\end{bmatrix} - \begin{bmatrix}-2\\1\end{bmatrix} = \begin{bmatrix}11\\-5\end{bmatrix}; \begin{bmatrix}x'\\y'\end{bmatrix} = \begin{bmatrix}x\\y\end{bmatrix} +$
$\begin{bmatrix}11\\-5\end{bmatrix} = \begin{bmatrix}-5\\3\end{bmatrix} + \begin{bmatrix}11\\-5\end{bmatrix} = \begin{bmatrix}6\\-2\end{bmatrix}$; (6, −2)

21. $\begin{bmatrix}-6\\4\end{bmatrix} = \begin{bmatrix}-6\\-7\end{bmatrix} + \begin{bmatrix}h\\k\end{bmatrix}; \begin{bmatrix}h\\k\end{bmatrix} = \begin{bmatrix}-6\\4\end{bmatrix} - \begin{bmatrix}-6\\-7\end{bmatrix} = \begin{bmatrix}0\\11\end{bmatrix}; \begin{bmatrix}x'\\y'\end{bmatrix} = \begin{bmatrix}x\\y\end{bmatrix} +$
$\begin{bmatrix}0\\11\end{bmatrix} = \begin{bmatrix}8\\3\end{bmatrix} + \begin{bmatrix}0\\11\end{bmatrix} = \begin{bmatrix}8\\14\end{bmatrix}$; (8, 14)

22. $\begin{bmatrix}-a\\b\end{bmatrix} = \begin{bmatrix}c\\d\end{bmatrix} + \begin{bmatrix}h\\k\end{bmatrix}; \begin{bmatrix}h\\k\end{bmatrix} = \begin{bmatrix}-a\\b\end{bmatrix} - \begin{bmatrix}c\\d\end{bmatrix} = \begin{bmatrix}-a-c\\b-d\end{bmatrix}; \begin{bmatrix}x'\\y'\end{bmatrix} = \begin{bmatrix}x\\y\end{bmatrix} +$
$\begin{bmatrix}-a-c\\b-d\end{bmatrix} = \begin{bmatrix}a\\b\end{bmatrix} + \begin{bmatrix}-a-c\\b-d\end{bmatrix} = \begin{bmatrix}-c\\2b-d\end{bmatrix}$; $(-c, 2b-d)$

23. $\begin{bmatrix}9\\-8\end{bmatrix} = \begin{bmatrix}4\\0\end{bmatrix} + \begin{bmatrix}h\\k\end{bmatrix}; \begin{bmatrix}h\\k\end{bmatrix} = \begin{bmatrix}9\\-8\end{bmatrix} - \begin{bmatrix}4\\0\end{bmatrix} = \begin{bmatrix}5\\-8\end{bmatrix}; \begin{bmatrix}x'\\y'\end{bmatrix} = \begin{bmatrix}x\\y\end{bmatrix} + \begin{bmatrix}5\\-8\end{bmatrix};$
$\begin{bmatrix}6\\-2\end{bmatrix} = \begin{bmatrix}x\\y\end{bmatrix} + \begin{bmatrix}5\\-8\end{bmatrix}; \begin{bmatrix}x\\y\end{bmatrix} = \begin{bmatrix}6\\-2\end{bmatrix} - \begin{bmatrix}5\\-8\end{bmatrix} = \begin{bmatrix}1\\6\end{bmatrix}$; (1, 6)

24. $\begin{bmatrix}-3\\7\end{bmatrix} = \begin{bmatrix}-1\\-6\end{bmatrix} + \begin{bmatrix}h\\k\end{bmatrix}; \begin{bmatrix}h\\k\end{bmatrix} = \begin{bmatrix}-3\\7\end{bmatrix} - \begin{bmatrix}-1\\-6\end{bmatrix} = \begin{bmatrix}-2\\13\end{bmatrix}; \begin{bmatrix}x'\\y'\end{bmatrix} = \begin{bmatrix}x\\y\end{bmatrix} + \begin{bmatrix}-2\\13\end{bmatrix};$
$\begin{bmatrix}8\\2\end{bmatrix} = \begin{bmatrix}x\\y\end{bmatrix} + \begin{bmatrix}-2\\13\end{bmatrix}; \begin{bmatrix}x\\y\end{bmatrix} = \begin{bmatrix}8\\2\end{bmatrix} - \begin{bmatrix}-2\\13\end{bmatrix} = \begin{bmatrix}10\\-11\end{bmatrix}$; (10, −11)

25. $\begin{bmatrix}1\\-2\end{bmatrix} = \begin{bmatrix}5\\9\end{bmatrix} + \begin{bmatrix}h\\k\end{bmatrix}; \begin{bmatrix}h\\k\end{bmatrix} = \begin{bmatrix}1\\-2\end{bmatrix} - \begin{bmatrix}5\\9\end{bmatrix} = \begin{bmatrix}-4\\-11\end{bmatrix}; \begin{bmatrix}x'\\y'\end{bmatrix} = \begin{bmatrix}x\\y\end{bmatrix} + \begin{bmatrix}-4\\-11\end{bmatrix};$
$\begin{bmatrix}-1\\-6\end{bmatrix} = \begin{bmatrix}x\\y\end{bmatrix} + \begin{bmatrix}-4\\-11\end{bmatrix}; \begin{bmatrix}x\\y\end{bmatrix} = \begin{bmatrix}-1\\-6\end{bmatrix} - \begin{bmatrix}-4\\-11\end{bmatrix} = \begin{bmatrix}3\\5\end{bmatrix}$; (3, 5)

26. $\begin{bmatrix}2a\\b\end{bmatrix} = \begin{bmatrix}c\\d\end{bmatrix} + \begin{bmatrix}h\\k\end{bmatrix}; \begin{bmatrix}h\\k\end{bmatrix} = \begin{bmatrix}2a\\b\end{bmatrix} - \begin{bmatrix}c\\d\end{bmatrix} = \begin{bmatrix}2a-c\\b-d\end{bmatrix}; \begin{bmatrix}x'\\y'\end{bmatrix} = \begin{bmatrix}x\\y\end{bmatrix} +$
$\begin{bmatrix}2a-c\\b-d\end{bmatrix}; \begin{bmatrix}a\\b\end{bmatrix} = \begin{bmatrix}x\\y\end{bmatrix} + \begin{bmatrix}2a-c\\b-d\end{bmatrix}; \begin{bmatrix}x\\y\end{bmatrix} = \begin{bmatrix}a\\b\end{bmatrix} - \begin{bmatrix}2a-c\\b-d\end{bmatrix} = \begin{bmatrix}-a+c\\d\end{bmatrix};$
$(c-a, d)$

C 27. $\begin{bmatrix}x'\\y'\end{bmatrix} = \begin{bmatrix}a\\b\end{bmatrix} + \begin{bmatrix}h_1\\k_1\end{bmatrix} = \begin{bmatrix}a+h_1\\b+k_1\end{bmatrix}; \begin{bmatrix}x'\\y'\end{bmatrix} = \begin{bmatrix}a+h_1\\b+k_1\end{bmatrix} + \begin{bmatrix}h_2\\k_2\end{bmatrix} =$
$\begin{bmatrix}a+h_1+h_2\\b+k_1+k_2\end{bmatrix}$; $(a+h_1+h_2, b+k_1+k_2)$

28. $\begin{bmatrix} x' \\ y' \end{bmatrix} = \begin{bmatrix} a \\ b \end{bmatrix} + \begin{bmatrix} h_2 \\ k_2 \end{bmatrix} = \begin{bmatrix} a + h_2 \\ b + k_2 \end{bmatrix}$; $\begin{bmatrix} x' \\ y' \end{bmatrix} = \begin{bmatrix} a + h_2 \\ b + k_2 \end{bmatrix} + \begin{bmatrix} h_1 \\ k_1 \end{bmatrix} =$
$\begin{bmatrix} a + h_2 + h_1 \\ b + k_2 + k_1 \end{bmatrix}$; $(a + h_2 + h_1, b + k_2 + k_1)$

29. yes

Pages 562–563 • WRITTEN EXERCISES

A

1. $\begin{bmatrix} x' \\ y' \end{bmatrix} = \begin{bmatrix} 5 & 0 \\ 0 & 5 \end{bmatrix}\begin{bmatrix} 2 \\ 1 \end{bmatrix} = \begin{bmatrix} 10 \\ 5 \end{bmatrix}$; (10, 5)

2. $\begin{bmatrix} x' \\ y' \end{bmatrix} = \begin{bmatrix} 0 & 4 \\ 4 & 0 \end{bmatrix}\begin{bmatrix} -3 \\ 2 \end{bmatrix} = \begin{bmatrix} 8 \\ -12 \end{bmatrix}$; (8, −12)

3. $\begin{bmatrix} x' \\ y' \end{bmatrix} = \begin{bmatrix} 3 & 5 \\ 0 & 2 \end{bmatrix}\begin{bmatrix} 7 \\ -2 \end{bmatrix} = \begin{bmatrix} 11 \\ -4 \end{bmatrix}$; (11, −4)

4. $\begin{bmatrix} x' \\ y' \end{bmatrix} = \begin{bmatrix} 0 & 3 \\ -3 & 0 \end{bmatrix}\begin{bmatrix} -1 \\ 2 \end{bmatrix} = \begin{bmatrix} 6 \\ 3 \end{bmatrix}$; (6, 3)

5. $\begin{bmatrix} x' \\ y' \end{bmatrix} = \begin{bmatrix} 4 & -5 \\ -1 & 0 \end{bmatrix}\begin{bmatrix} -3 \\ -1 \end{bmatrix} = \begin{bmatrix} -7 \\ 3 \end{bmatrix}$; (−7, 3)

6. $\begin{bmatrix} x' \\ y' \end{bmatrix} = \begin{bmatrix} -1 & 2 \\ 5 & -3 \end{bmatrix}\begin{bmatrix} 4 \\ 3 \end{bmatrix} = \begin{bmatrix} 2 \\ 11 \end{bmatrix}$; (2, 11)

7. $\begin{bmatrix} 6 \\ 10 \end{bmatrix} = \begin{bmatrix} -2 & 0 \\ 0 & -2 \end{bmatrix}\begin{bmatrix} x \\ y \end{bmatrix}$; $\begin{bmatrix} -2 & 0 \\ 0 & -2 \end{bmatrix}^{-1} = \frac{1}{4}\begin{bmatrix} -2 & 0 \\ 0 & -2 \end{bmatrix} = \begin{bmatrix} -\frac{1}{2} & 0 \\ 0 & -\frac{1}{2} \end{bmatrix}$;
$\begin{bmatrix} x \\ y \end{bmatrix} = \begin{bmatrix} -\frac{1}{2} & 0 \\ 0 & -\frac{1}{2} \end{bmatrix}\begin{bmatrix} 6 \\ 10 \end{bmatrix} = \begin{bmatrix} -3 \\ -5 \end{bmatrix}$; (−3, −5)

8. $\begin{bmatrix} 9 \\ -3 \end{bmatrix} = \begin{bmatrix} 0 & 3 \\ -3 & 0 \end{bmatrix}\begin{bmatrix} x \\ y \end{bmatrix}$; $\begin{bmatrix} 0 & 3 \\ -3 & 0 \end{bmatrix}^{-1} = \frac{1}{9}\begin{bmatrix} 0 & -3 \\ 3 & 0 \end{bmatrix} = \begin{bmatrix} 0 & -\frac{1}{3} \\ \frac{1}{3} & 0 \end{bmatrix}$;
$\begin{bmatrix} x \\ y \end{bmatrix} = \begin{bmatrix} 0 & -\frac{1}{3} \\ \frac{1}{3} & 0 \end{bmatrix}\begin{bmatrix} 9 \\ -3 \end{bmatrix} = \begin{bmatrix} 1 \\ 3 \end{bmatrix}$; (1, 3)

9. $\begin{bmatrix} -1 \\ 3 \end{bmatrix} = \begin{bmatrix} 3 & 4 \\ 2 & 3 \end{bmatrix}\begin{bmatrix} x \\ y \end{bmatrix}$; $\begin{bmatrix} 3 & 4 \\ 2 & 3 \end{bmatrix}^{-1} = \frac{1}{1}\begin{bmatrix} 3 & -4 \\ -2 & 3 \end{bmatrix} = \begin{bmatrix} 3 & -4 \\ -2 & 3 \end{bmatrix}$;
$\begin{bmatrix} x \\ y \end{bmatrix} = \begin{bmatrix} 3 & -4 \\ -2 & 3 \end{bmatrix}\begin{bmatrix} -1 \\ 3 \end{bmatrix} = \begin{bmatrix} -15 \\ 11 \end{bmatrix}$; (−15, 11)

10. $\begin{bmatrix} -3 \\ -1 \end{bmatrix} = \begin{bmatrix} 4 & 3 \\ 3 & 2 \end{bmatrix}\begin{bmatrix} x \\ y \end{bmatrix}$; $\begin{bmatrix} 4 & 3 \\ 3 & 2 \end{bmatrix}^{-1} = \frac{1}{-1}\begin{bmatrix} 2 & -3 \\ -3 & 4 \end{bmatrix} = \begin{bmatrix} -2 & 3 \\ 3 & -4 \end{bmatrix}$;
$\begin{bmatrix} x \\ y \end{bmatrix} = \begin{bmatrix} -2 & 3 \\ 3 & -4 \end{bmatrix}\begin{bmatrix} -3 \\ -1 \end{bmatrix} = \begin{bmatrix} 3 \\ -5 \end{bmatrix}$; (3, −5)

11. $\begin{bmatrix} -5 \\ 9 \end{bmatrix} = \begin{bmatrix} 7 & 4 \\ 5 & 6 \end{bmatrix}\begin{bmatrix} x \\ y \end{bmatrix}$; $\begin{bmatrix} 7 & 4 \\ 5 & 6 \end{bmatrix}^{-1} = \frac{1}{22}\begin{bmatrix} 6 & -4 \\ -5 & 7 \end{bmatrix} = \begin{bmatrix} \frac{3}{11} & -\frac{2}{11} \\ -\frac{5}{22} & \frac{7}{22} \end{bmatrix}$;

$\begin{bmatrix} x \\ y \end{bmatrix} = \begin{bmatrix} \frac{3}{11} & -\frac{2}{11} \\ -\frac{5}{22} & \frac{7}{22} \end{bmatrix}\begin{bmatrix} -5 \\ 9 \end{bmatrix} = \begin{bmatrix} -3 \\ 4 \end{bmatrix}$; $(-3, 4)$

12. $\begin{bmatrix} 4 \\ -3 \end{bmatrix} = \begin{bmatrix} 2 & 5 \\ 3 & 9 \end{bmatrix}\begin{bmatrix} x \\ y \end{bmatrix}$; $\begin{bmatrix} 2 & 5 \\ 3 & 9 \end{bmatrix}^{-1} = \frac{1}{3}\begin{bmatrix} 9 & -5 \\ -3 & 2 \end{bmatrix} = \begin{bmatrix} 3 & -\frac{5}{3} \\ -1 & \frac{2}{3} \end{bmatrix}$;

$\begin{bmatrix} x \\ y \end{bmatrix} = \begin{bmatrix} 3 & -\frac{5}{3} \\ -1 & \frac{2}{3} \end{bmatrix}\begin{bmatrix} 4 \\ -3 \end{bmatrix} = \begin{bmatrix} 17 \\ -6 \end{bmatrix}$; $(17, -6)$

B 13. $\begin{bmatrix} x' \\ y' \end{bmatrix} = \begin{bmatrix} 2 & 3 \\ 4 & 6 \end{bmatrix}\begin{bmatrix} x \\ y \end{bmatrix} = \begin{bmatrix} 2x + 3y \\ 4x + 6y \end{bmatrix}$; $x' = 2x + 3y$; $y' = 4x + 6y = 2(2x + 3y)$, $y' = 2x'$; $\therefore m = 2$

14. $\begin{bmatrix} x' \\ y' \end{bmatrix} = \begin{bmatrix} 3 & 1 \\ -6 & -2 \end{bmatrix}\begin{bmatrix} x \\ y \end{bmatrix} = \begin{bmatrix} 3x + y \\ -6x - 2y \end{bmatrix}$; $x' = 3x + y$; $y' = -6x - 2y = -2(3x + y)$, $y' = -2x'$; $\therefore m = -2$

15. $\begin{bmatrix} x' \\ y' \end{bmatrix} = \begin{bmatrix} -10 & -12 \\ -5 & -6 \end{bmatrix}\begin{bmatrix} x \\ y \end{bmatrix} = \begin{bmatrix} -10x - 12y \\ -5x - 6y \end{bmatrix}$; $x' = -10x - 12y$;

$y' = -5x - 6y = \frac{1}{2}(-10x - 12y)$, $y' = \frac{1}{2}x'$; $\therefore m = \frac{1}{2}$

16. $\begin{bmatrix} x' \\ y' \end{bmatrix} = \begin{bmatrix} 6 & 15 \\ 4 & 10 \end{bmatrix}\begin{bmatrix} x \\ y \end{bmatrix} = \begin{bmatrix} 6x + 15y \\ 4x + 10y \end{bmatrix}$; $x' = 6x + 15y$; $y' = 4x + 10y = \frac{2}{3}(6x + 15y)$, $y' = \frac{2}{3}x'$; $\therefore m = \frac{2}{3}$

17. $\begin{bmatrix} x' \\ y' \end{bmatrix} = \begin{bmatrix} 3 & 0 \\ 0 & 3 \end{bmatrix}\begin{bmatrix} 0 \\ 0 \end{bmatrix} = \begin{bmatrix} 0 \\ 0 \end{bmatrix}$, (0, 0); $\begin{bmatrix} x' \\ y' \end{bmatrix} = \begin{bmatrix} 3 & 0 \\ 0 & 3 \end{bmatrix}\begin{bmatrix} 1 \\ 0 \end{bmatrix} = \begin{bmatrix} 3 \\ 0 \end{bmatrix}$, (3, 0);

$\begin{bmatrix} x' \\ y' \end{bmatrix} = \begin{bmatrix} 3 & 0 \\ 0 & 3 \end{bmatrix}\begin{bmatrix} 1 \\ 1 \end{bmatrix} = \begin{bmatrix} 3 \\ 3 \end{bmatrix}$, (3, 3); $\begin{bmatrix} x' \\ y' \end{bmatrix} = \begin{bmatrix} 3 & 0 \\ 0 & 3 \end{bmatrix}\begin{bmatrix} 0 \\ 1 \end{bmatrix} = \begin{bmatrix} 0 \\ 3 \end{bmatrix}$, (0, 3);

a square with vertices (0, 0), (3, 0), (3, 3), and (0, 3)

18. $\begin{bmatrix} x' \\ y' \end{bmatrix} = \begin{bmatrix} 4 & 0 \\ 0 & -4 \end{bmatrix}\begin{bmatrix} 0 \\ 0 \end{bmatrix} = \begin{bmatrix} 0 \\ 0 \end{bmatrix}$, (0, 0); $\begin{bmatrix} x' \\ y' \end{bmatrix} = \begin{bmatrix} 4 & 0 \\ 0 & -4 \end{bmatrix}\begin{bmatrix} 1 \\ 0 \end{bmatrix} = \begin{bmatrix} 4 \\ 0 \end{bmatrix}$, (4, 0);

$\begin{bmatrix} x' \\ y' \end{bmatrix} = \begin{bmatrix} 4 & 0 \\ 0 & -4 \end{bmatrix}\begin{bmatrix} 1 \\ 1 \end{bmatrix} = \begin{bmatrix} 4 \\ -4 \end{bmatrix}$, (4, −4); $\begin{bmatrix} x' \\ y' \end{bmatrix} = \begin{bmatrix} 4 & 0 \\ 0 & -4 \end{bmatrix}\begin{bmatrix} 0 \\ 1 \end{bmatrix} = \begin{bmatrix} 0 \\ -4 \end{bmatrix}$, (0, −4);

a square with vertices (0, 0), (4, 0), (4, −4), and (0, −4)

19. $\begin{bmatrix} x' \\ y' \end{bmatrix} = \begin{bmatrix} 0 & -1 \\ -1 & 0 \end{bmatrix}\begin{bmatrix} 0 \\ 0 \end{bmatrix} = \begin{bmatrix} 0 \\ 0 \end{bmatrix}$, (0, 0); $\begin{bmatrix} x' \\ y' \end{bmatrix} = \begin{bmatrix} 0 & -1 \\ -1 & 0 \end{bmatrix}\begin{bmatrix} 1 \\ 0 \end{bmatrix} = \begin{bmatrix} 0 \\ -1 \end{bmatrix}$, (0, −1);

$\begin{bmatrix} x' \\ y' \end{bmatrix} = \begin{bmatrix} 0 & -1 \\ -1 & 0 \end{bmatrix}\begin{bmatrix} 1 \\ 1 \end{bmatrix} = \begin{bmatrix} -1 \\ -1 \end{bmatrix}$, (−1, −1); $\begin{bmatrix} x' \\ y' \end{bmatrix} = \begin{bmatrix} 0 & -1 \\ -1 & 0 \end{bmatrix}\begin{bmatrix} 0 \\ 1 \end{bmatrix} = \begin{bmatrix} -1 \\ 0 \end{bmatrix}$,

(−1, 0); a square with vertices (0, 0), (0, −1), (−1, −1), and (−1, 0)

20. $\begin{bmatrix} x' \\ y' \end{bmatrix} = \begin{bmatrix} 1 & 0 \\ 2 & 1 \end{bmatrix}\begin{bmatrix} 0 \\ 0 \end{bmatrix} = \begin{bmatrix} 0 \\ 0 \end{bmatrix}$, (0, 0); $\begin{bmatrix} x' \\ y' \end{bmatrix} = \begin{bmatrix} 1 & 0 \\ 2 & 1 \end{bmatrix}\begin{bmatrix} 1 \\ 0 \end{bmatrix} = \begin{bmatrix} 1 \\ 2 \end{bmatrix}$, (1, 2);
$\begin{bmatrix} x' \\ y' \end{bmatrix} = \begin{bmatrix} 1 & 0 \\ 2 & 1 \end{bmatrix}\begin{bmatrix} 1 \\ 1 \end{bmatrix} = \begin{bmatrix} 1 \\ 3 \end{bmatrix}$, (1, 3); $\begin{bmatrix} x' \\ y' \end{bmatrix} = \begin{bmatrix} 1 & 0 \\ 2 & 1 \end{bmatrix}\begin{bmatrix} 0 \\ 1 \end{bmatrix} = \begin{bmatrix} 0 \\ 1 \end{bmatrix}$, (0, 1);
parallelogram with vertices (0, 0), (1, 2), (1, 3), and (0, 1)

21. $\begin{bmatrix} x' \\ y' \end{bmatrix} = \begin{bmatrix} 3 & 0 \\ 0 & 3 \end{bmatrix}\begin{bmatrix} x \\ y \end{bmatrix} = \begin{bmatrix} 3x \\ 3y \end{bmatrix}$; $x' = 3x$, $y' = 3y$; $(x', y') = (3x, 3y)$;
an expansion by a factor of 3

22. $\begin{bmatrix} x' \\ y' \end{bmatrix} = \begin{bmatrix} 4 & 0 \\ 0 & -4 \end{bmatrix}\begin{bmatrix} x \\ y \end{bmatrix} = \begin{bmatrix} 4x \\ -4y \end{bmatrix}$; $x' = 4x$, $y' = -4y$; $(x', y') = (4x, -4y)$;
an expansion by a factor of 4 and a reflection in the x-axis

23. $\begin{bmatrix} x' \\ y' \end{bmatrix} = \begin{bmatrix} 0 & -1 \\ -1 & 0 \end{bmatrix}\begin{bmatrix} x \\ y \end{bmatrix} = \begin{bmatrix} -y \\ -x \end{bmatrix}$; $x' = -y$, $y' = -x$; $(x', y') = (-y, -x)$;
a reflection in the line $y = -x$

24. $\begin{bmatrix} x' \\ y' \end{bmatrix} = \begin{bmatrix} 1 & 0 \\ 2 & 1 \end{bmatrix}\begin{bmatrix} x \\ y \end{bmatrix} = \begin{bmatrix} x \\ 2x + y \end{bmatrix}$; $x' = x$, $y' = 2x + y$; $(x', y') = (x, 2x + y)$;
the image of each point has the same x-coordinate, but the y-coordinate is translated by an amount equal to twice the x-coordinate.

25. $\begin{bmatrix} x' \\ y' \end{bmatrix} = \begin{bmatrix} 0 & 0 \\ 0 & 1 \end{bmatrix}\begin{bmatrix} x \\ y \end{bmatrix} = \begin{bmatrix} 0 \\ y \end{bmatrix}$; $x' = 0$, $y' = y$; $(x', y') = (0, y)$;
a projection onto the y-axis

26. $\begin{bmatrix} x' \\ y' \end{bmatrix} = \begin{bmatrix} 1 & 0 \\ 1 & 0 \end{bmatrix}\begin{bmatrix} x \\ y \end{bmatrix} = \begin{bmatrix} x \\ x \end{bmatrix}$; $x' = x$, $y' = x$; $(x', y') = (x, x)$;
a projection onto the line $y = x$

27. $\begin{bmatrix} x' \\ y' \end{bmatrix} = \begin{bmatrix} 0 & 0 \\ 3 & 0 \end{bmatrix}\begin{bmatrix} x \\ y \end{bmatrix} = \begin{bmatrix} 0 \\ 3x \end{bmatrix}$; $x' = 0$, $y' = 3x$; $(x', y') = (0, 3x)$;
a projection onto the y-axis in such a way that the y-coordinate of the image is three times the x-coordinate of the preimage

28. $\begin{bmatrix} x' \\ y' \end{bmatrix} = \begin{bmatrix} 0 & 0 \\ 0 & -1 \end{bmatrix}\begin{bmatrix} x \\ y \end{bmatrix} = \begin{bmatrix} 0 \\ -y \end{bmatrix}$; $x' = 0$, $y' = -y$; $(x', y') = (0, -y)$;
a projection onto the y-axis, followed by a reflection in the x-axis

C 29. Let $(x_1, y_1) = (0, 0)$; $(x_2, y_2) = (1, 0)$; $(x_3, y_3) = (1, 1)$; $(x_4, y_4) = (0, 1)$.
$\begin{bmatrix} x_1' \\ y_1' \end{bmatrix} = \begin{bmatrix} a & b \\ c & d \end{bmatrix}\begin{bmatrix} 0 \\ 0 \end{bmatrix} = \begin{bmatrix} 0 \\ 0 \end{bmatrix}$; $\begin{bmatrix} x_2' \\ y_2' \end{bmatrix} = \begin{bmatrix} a & b \\ c & d \end{bmatrix}\begin{bmatrix} 1 \\ 0 \end{bmatrix} = \begin{bmatrix} a \\ c \end{bmatrix}$; $\begin{bmatrix} x_3' \\ y_3' \end{bmatrix} =$
$\begin{bmatrix} a & b \\ c & d \end{bmatrix}\begin{bmatrix} 1 \\ 1 \end{bmatrix} = \begin{bmatrix} a + b \\ c + d \end{bmatrix}$; $\begin{bmatrix} x_4' \\ y_4' \end{bmatrix} = \begin{bmatrix} a & b \\ c & d \end{bmatrix}\begin{bmatrix} 0 \\ 1 \end{bmatrix} = \begin{bmatrix} b \\ d \end{bmatrix}$. The line through (0, 0) and (a, c) is parallel to the line through (b, d) and $(a + b, c + d)$; both have slope $\frac{c}{a}$. The line through (0, 0) and (b, d) is parallel to the line through (a, c) and $(a + b, c + d)$; both have slope $\frac{d}{b}$. Thus the points (0, 0), (a, c), $(a + b, c + d)$, and (b, d) are the vertices of a parallelogram.

30. $A(BX) = \begin{bmatrix} a & b \\ c & d \end{bmatrix}\left(\begin{bmatrix} e & f \\ g & h \end{bmatrix}\begin{bmatrix} x \\ y \end{bmatrix}\right) = \begin{bmatrix} a & b \\ c & d \end{bmatrix}\begin{bmatrix} ex + fy \\ gx + hy \end{bmatrix} =$
$\begin{bmatrix} a(ex + fy) + b(gx + hy) \\ c(ex + fy) + d(gx + hy) \end{bmatrix} = \begin{bmatrix} aex + afy + bgx + bhy \\ cex + cfy + dgx + dhy \end{bmatrix}$; $(AB)X =$

$$\left(\begin{bmatrix} a & b \\ c & d \end{bmatrix}\begin{bmatrix} e & f \\ g & h \end{bmatrix}\right)\begin{bmatrix} x \\ y \end{bmatrix} = \begin{bmatrix} ae + bg & af + bh \\ ce + dg & cf + dh \end{bmatrix}\begin{bmatrix} x \\ y \end{bmatrix} =$$

$$\begin{bmatrix} (ae + bg)x + (af + bh)y \\ (ce + dg)x + (cf + dh)y \end{bmatrix} = \begin{bmatrix} aex + afy + bgx + bhy \\ cex + cfy + dgx + dhy \end{bmatrix}; \therefore A(BX) = (AB)X$$

Page 563 • SELF-TEST 3

1. $\begin{bmatrix} 4 \\ 6 \end{bmatrix} = \begin{bmatrix} 3 \\ -1 \end{bmatrix} + \begin{bmatrix} h \\ k \end{bmatrix}; \begin{bmatrix} h \\ k \end{bmatrix} = \begin{bmatrix} 4 \\ 6 \end{bmatrix} - \begin{bmatrix} 3 \\ -1 \end{bmatrix} = \begin{bmatrix} 1 \\ 7 \end{bmatrix}; \begin{bmatrix} x' \\ y' \end{bmatrix} = \begin{bmatrix} x \\ y \end{bmatrix} + \begin{bmatrix} 1 \\ 7 \end{bmatrix}$

2. $\begin{bmatrix} x' \\ y' \end{bmatrix} = \begin{bmatrix} -3 \\ 5 \end{bmatrix} + \begin{bmatrix} -2 \\ 2 \end{bmatrix} = \begin{bmatrix} -5 \\ 7 \end{bmatrix}$; $(-5, 7)$

3. $\begin{bmatrix} -5 \\ 8 \end{bmatrix} = \begin{bmatrix} x \\ y \end{bmatrix} + \begin{bmatrix} -2 \\ 2 \end{bmatrix}; \begin{bmatrix} x \\ y \end{bmatrix} = \begin{bmatrix} -5 \\ 8 \end{bmatrix} - \begin{bmatrix} -2 \\ 2 \end{bmatrix} = \begin{bmatrix} -3 \\ 6 \end{bmatrix}$; $(-3, 6)$

4. $\begin{bmatrix} x' \\ y' \end{bmatrix} = \begin{bmatrix} 6 & 1 \\ 4 & 5 \end{bmatrix}\begin{bmatrix} 7 \\ -2 \end{bmatrix} = \begin{bmatrix} 40 \\ 18 \end{bmatrix}$; $(40, 18)$

5. $\begin{bmatrix} 8 \\ 3 \end{bmatrix} = \begin{bmatrix} 6 & 1 \\ 4 & 5 \end{bmatrix}\begin{bmatrix} x \\ y \end{bmatrix}; \begin{bmatrix} 6 & 1 \\ 4 & 5 \end{bmatrix}^{-1} = \frac{1}{26}\begin{bmatrix} 5 & -1 \\ -4 & 6 \end{bmatrix} = \begin{bmatrix} \frac{5}{26} & -\frac{1}{26} \\ -\frac{2}{13} & \frac{3}{13} \end{bmatrix}$;

$\begin{bmatrix} x \\ y \end{bmatrix} = \begin{bmatrix} \frac{5}{26} & -\frac{1}{26} \\ -\frac{2}{13} & \frac{3}{13} \end{bmatrix}\begin{bmatrix} 8 \\ 3 \end{bmatrix} = \begin{bmatrix} \frac{37}{26} \\ -\frac{7}{13} \end{bmatrix}; \left(\frac{37}{26}, -\frac{7}{13}\right)$

Pages 565–566 • CHAPTER REVIEW

1. a **2.** $\begin{bmatrix} 0 & 5 \\ 2 & 8 \end{bmatrix}$; a

3. $X + \begin{bmatrix} 3 & 0 \\ 4 & 7 \end{bmatrix} + \begin{bmatrix} -3 & 0 \\ -4 & -7 \end{bmatrix} = \begin{bmatrix} 4 & 1 \\ 4 & 9 \end{bmatrix} + \begin{bmatrix} -3 & 0 \\ -4 & -7 \end{bmatrix}; X = \begin{bmatrix} 1 & 1 \\ 0 & 2 \end{bmatrix}$; c

4. $X - \begin{bmatrix} 2 & 6 \\ 0 & 3 \end{bmatrix} + \begin{bmatrix} 2 & 6 \\ 0 & 3 \end{bmatrix} = \begin{bmatrix} 1 & -8 \\ 2 & 7 \end{bmatrix} + \begin{bmatrix} 2 & 6 \\ 0 & 3 \end{bmatrix}; X = \begin{bmatrix} 3 & -2 \\ 2 & 10 \end{bmatrix}$; d

5. $X = \begin{bmatrix} 8 & -6 \\ 7 & -2 \end{bmatrix} - 2\begin{bmatrix} -3 & -4 \\ 1 & 2 \end{bmatrix}; X = \begin{bmatrix} 8 & -6 \\ 7 & -2 \end{bmatrix} - \begin{bmatrix} -6 & -8 \\ 2 & 4 \end{bmatrix} = \begin{bmatrix} 14 & 2 \\ 5 & -6 \end{bmatrix}$; d

6. $X - \begin{bmatrix} -3 & -4 \\ 1 & 2 \end{bmatrix} = \begin{bmatrix} 8 & -6 \\ 7 & -2 \end{bmatrix}; X - \begin{bmatrix} -3 & -4 \\ 1 & 2 \end{bmatrix} + \begin{bmatrix} -3 & -4 \\ 1 & 2 \end{bmatrix} = \begin{bmatrix} 8 & -6 \\ 7 & -2 \end{bmatrix} +$

$\begin{bmatrix} -3 & -4 \\ 1 & 2 \end{bmatrix}; X = \begin{bmatrix} 5 & -10 \\ 8 & 0 \end{bmatrix}$; a

7. $\begin{bmatrix} -4 - 30 \\ -2 + 0 \end{bmatrix} = \begin{bmatrix} -34 \\ -2 \end{bmatrix}$; c **8.** $\begin{bmatrix} -5 - 4 & 0 + 8 \\ 0 + 6 & 0 - 12 \end{bmatrix} = \begin{bmatrix} -9 & 8 \\ 6 & -12 \end{bmatrix}$; c

9. b (matrix multiplication is associative.)

10. $(B + A)(B - A) = B(B - A) + A(B - A) = B^2 - BA + AB - A^2$; b

11. $\begin{bmatrix} 5 & 9 \\ 4 & 8 \end{bmatrix}^{-1} = \frac{1}{4}\begin{bmatrix} 8 & -9 \\ -4 & 5 \end{bmatrix} = \begin{bmatrix} 2 & -\frac{9}{4} \\ -1 & \frac{5}{4} \end{bmatrix}$; d **12.** b

13. $\begin{bmatrix} x' \\ y' \end{bmatrix} = \begin{bmatrix} -6 \\ 1 \end{bmatrix} + \begin{bmatrix} 3 \\ -5 \end{bmatrix} = \begin{bmatrix} -3 \\ -4 \end{bmatrix}$; $P'(-3, -4)$; a

14. $\begin{bmatrix} -2 \\ 4 \end{bmatrix} = \begin{bmatrix} x \\ y \end{bmatrix} + \begin{bmatrix} 3 \\ -5 \end{bmatrix}$; $\begin{bmatrix} x \\ y \end{bmatrix} = \begin{bmatrix} -2 \\ 4 \end{bmatrix} - \begin{bmatrix} 3 \\ -5 \end{bmatrix} = \begin{bmatrix} -5 \\ 9 \end{bmatrix}$; $Q(-5, 9)$; a

15. $\begin{bmatrix} x' \\ y' \end{bmatrix} = \begin{bmatrix} 2 & 2 \\ 1 & -3 \end{bmatrix}\begin{bmatrix} 3 \\ -4 \end{bmatrix} = \begin{bmatrix} -2 \\ 15 \end{bmatrix}$; $R'(-2, 15)$; d

16. $\begin{bmatrix} -8 \\ -16 \end{bmatrix} = \begin{bmatrix} 2 & 2 \\ 1 & -3 \end{bmatrix}\begin{bmatrix} x \\ y \end{bmatrix}$; $\begin{bmatrix} 2 & 2 \\ 1 & -3 \end{bmatrix}^{-1} = \frac{1}{-8}\begin{bmatrix} -3 & -2 \\ -1 & 2 \end{bmatrix} = \begin{bmatrix} \frac{3}{8} & \frac{1}{4} \\ \frac{1}{8} & -\frac{1}{4} \end{bmatrix}$;

$\begin{bmatrix} x \\ y \end{bmatrix} = \begin{bmatrix} \frac{3}{8} & \frac{1}{4} \\ \frac{1}{8} & -\frac{1}{4} \end{bmatrix}\begin{bmatrix} -8 \\ -16 \end{bmatrix} = \begin{bmatrix} -7 \\ 3 \end{bmatrix}$; $S(-7, 3)$; b

Page 567 • CHAPTER TEST

1. (*1*): $2x - 3y = 5$ (*2*): $3y + x = -11$; from (*2*): $x = -3y - 11$;
(*1*): $2(-3y - 11) - 3y = 5$, $-9y - 22 = 5$, $y = -3$; $x = -3(-3) - 11 = -2$

2. $X + \begin{bmatrix} 2 & 0 & 1 \\ -1 & 6 & 4 \end{bmatrix} + \begin{bmatrix} -2 & 0 & -1 \\ 1 & -6 & -4 \end{bmatrix} = \begin{bmatrix} -3 & 9 & 8 \\ 2 & 4 & 0 \end{bmatrix} + \begin{bmatrix} -2 & 0 & -1 \\ 1 & -6 & -4 \end{bmatrix}$;

$X = \begin{bmatrix} -5 & 9 & 7 \\ 3 & -2 & -4 \end{bmatrix}$

3. $2X - \begin{bmatrix} 0 & -12 \\ 14 & 2 \end{bmatrix} = \begin{bmatrix} 12 & 0 \\ -6 & 6 \end{bmatrix}$; $2X = \begin{bmatrix} 12 & 0 \\ -6 & 6 \end{bmatrix} + \begin{bmatrix} 0 & -12 \\ 14 & 2 \end{bmatrix} = \begin{bmatrix} 12 & -12 \\ 8 & 8 \end{bmatrix}$;

$X = \frac{1}{2}\begin{bmatrix} 12 & -12 \\ 8 & 8 \end{bmatrix} = \begin{bmatrix} 6 & -6 \\ 4 & 4 \end{bmatrix}$

4. $\begin{bmatrix} -1 + 0 + 14 & 8 + 0 + 0 \\ 0 - 9 + 7 & 0 - 3 + 0 \end{bmatrix} = \begin{bmatrix} 13 & 8 \\ -2 & -3 \end{bmatrix}$

5. $(2A)B = \left(2\begin{bmatrix} -3 & 0 \\ 2 & 4 \end{bmatrix}\right)\begin{bmatrix} 5 & 0 \\ 1 & 1 \end{bmatrix} = \begin{bmatrix} -6 & 0 \\ 4 & 8 \end{bmatrix}\begin{bmatrix} 5 & 0 \\ 1 & 1 \end{bmatrix} = \begin{bmatrix} -30 & 0 \\ 28 & 8 \end{bmatrix}$;

$A(2B) = \begin{bmatrix} -3 & 0 \\ 2 & 4 \end{bmatrix}\left(2\begin{bmatrix} 5 & 0 \\ 1 & 1 \end{bmatrix}\right) = \begin{bmatrix} -3 & 0 \\ 2 & 4 \end{bmatrix}\begin{bmatrix} 10 & 0 \\ 2 & 2 \end{bmatrix} = \begin{bmatrix} -30 & 0 \\ 28 & 8 \end{bmatrix}$;

$(2A)B = A(2B)$

6. $X = \begin{bmatrix} 5 & -4 \\ 1 & -1 \end{bmatrix}^{-1}\begin{bmatrix} 3 & 0 \\ 1 & -12 \end{bmatrix}$; $\begin{bmatrix} 5 & -4 \\ 1 & -1 \end{bmatrix}^{-1} = \frac{1}{-1}\begin{bmatrix} -1 & 4 \\ -1 & 5 \end{bmatrix} = \begin{bmatrix} 1 & -4 \\ 1 & -5 \end{bmatrix}$;

$X = \begin{bmatrix} 1 & -4 \\ 1 & -5 \end{bmatrix}\begin{bmatrix} 3 & 0 \\ 1 & -12 \end{bmatrix} = \begin{bmatrix} -1 & 48 \\ -2 & 60 \end{bmatrix}$

7. $\begin{bmatrix} 4 \\ -2 \end{bmatrix} = \begin{bmatrix} -1 \\ 7 \end{bmatrix} + \begin{bmatrix} h \\ k \end{bmatrix}$; $\begin{bmatrix} h \\ k \end{bmatrix} = \begin{bmatrix} 4 \\ -2 \end{bmatrix} - \begin{bmatrix} -1 \\ 7 \end{bmatrix} = \begin{bmatrix} 5 \\ -9 \end{bmatrix}$;

$\begin{bmatrix} x' \\ y' \end{bmatrix} = \begin{bmatrix} x \\ y \end{bmatrix} + \begin{bmatrix} 5 \\ -9 \end{bmatrix}$

8. $\begin{bmatrix} x' \\ y' \end{bmatrix} = \begin{bmatrix} 9 \\ -6 \end{bmatrix} + \begin{bmatrix} 5 \\ -9 \end{bmatrix} = \begin{bmatrix} 14 \\ -15 \end{bmatrix}$; $(14, -15)$

9. $\begin{bmatrix} x' \\ y' \end{bmatrix} = \begin{bmatrix} 6 & -2 \\ 0 & 1 \end{bmatrix}\begin{bmatrix} 3 \\ -4 \end{bmatrix} = \begin{bmatrix} 26 \\ -4 \end{bmatrix}$; (26, −4)

10. $\begin{bmatrix} -5 \\ 2 \end{bmatrix} = \begin{bmatrix} 1 & 3 \\ 2 & 7 \end{bmatrix}\begin{bmatrix} x \\ y \end{bmatrix}$; $\begin{bmatrix} 1 & 3 \\ 2 & 7 \end{bmatrix}^{-1} = \frac{1}{1}\begin{bmatrix} 7 & -3 \\ -2 & 1 \end{bmatrix} = \begin{bmatrix} 7 & -3 \\ -2 & 1 \end{bmatrix}$;
$\begin{bmatrix} x \\ y \end{bmatrix} = \begin{bmatrix} 7 & -3 \\ -2 & 1 \end{bmatrix}\begin{bmatrix} -5 \\ 2 \end{bmatrix} = \begin{bmatrix} -41 \\ 12 \end{bmatrix}$; (−41, 12)

Page 569 • APPLICATION

1. $\begin{array}{c} \\ 1 \\ 2 \\ 3 \end{array}\begin{array}{c} \begin{array}{ccc} 1 & 2 & 3 \end{array} \\ \begin{bmatrix} 0 & 1 & 0 \\ 1 & 0 & 1 \\ 1 & 0 & 0 \end{bmatrix} \end{array}$
2. $\begin{array}{c} \\ 1 \\ 2 \\ 3 \\ 4 \end{array}\begin{array}{c} \begin{array}{cccc} 1 & 2 & 3 & 4 \end{array} \\ \begin{bmatrix} 0 & 1 & 0 & 0 \\ 1 & 0 & 1 & 1 \\ 0 & 0 & 0 & 1 \\ 1 & 0 & 0 & 0 \end{bmatrix} \end{array}$
3. $\begin{array}{c} \\ 1 \\ 2 \\ 3 \\ 4 \end{array}\begin{array}{c} \begin{array}{cccc} 1 & 2 & 3 & 4 \end{array} \\ \begin{bmatrix} 0 & 1 & 0 & 0 \\ 1 & 0 & 1 & 0 \\ 0 & 0 & 0 & 1 \\ 0 & 1 & 0 & 0 \end{bmatrix} \end{array}$

4.

5. 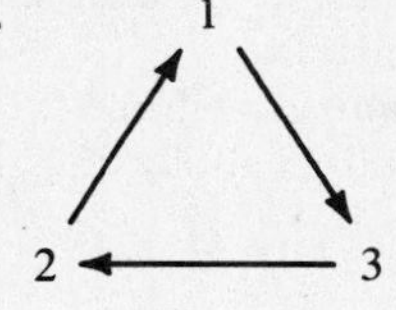

6. 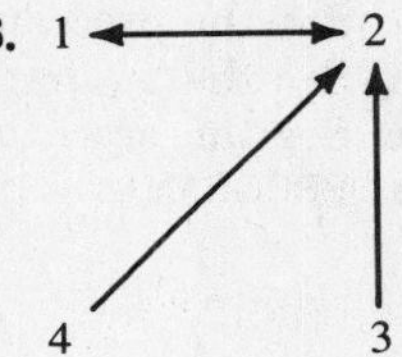

7. Dot 8. Pat, Dot

9. $\begin{bmatrix} 0 & 1 & 1 & 0 \\ 0 & 0 & 1 & 1 \\ 0 & 0 & 0 & 0 \\ 1 & 0 & 1 & 0 \end{bmatrix}\begin{bmatrix} 0 & 1 & 1 & 0 \\ 0 & 0 & 1 & 1 \\ 0 & 0 & 0 & 0 \\ 1 & 0 & 1 & 0 \end{bmatrix} = \begin{array}{c} \\ \text{Ed} \\ \text{Tod} \\ \text{Pat} \\ \text{Dot} \end{array}\begin{array}{c} \begin{array}{cccc} \text{Ed} & \text{Tod} & \text{Pat} & \text{Dot} \end{array} \\ \begin{bmatrix} 0 & 0 & 1 & 1 \\ 1 & 0 & 1 & 0 \\ 1 & 0 & 0 & 0 \\ 0 & 1 & 1 & 0 \end{bmatrix} \end{array}$; Dot

10. Ed, Tod, and Dot

Page 570 • PREPARING FOR COLLEGE ENTRANCE EXAMS

1. B contains 7 elements; $A \times B$ contains 42 elements; $B \not\subset A$; E

2. There are (8 − 1)! circular permutations of 8 people. (8 − 1)! = 7! = 5040; C
3. $\frac{12 \cdot 11 \cdot 10 \cdot 9 \cdot 8}{5 \cdot 4 \cdot 3 \cdot 2 \cdot 1}(x^2)^7(-1)^5 = -792x^{14}$; C
4. $x - 1 = 2\left(y^2 + 3y + \frac{9}{4}\right) - \frac{9}{2}$; $x = 2\left(y + \frac{3}{2}\right)^2 - \frac{7}{2}$; $V\left(-\frac{7}{2}, -\frac{3}{2}\right)$; A
5. $(x^2 - 6x + 9) + (y^2 + 4y + 4) = -4 + 9 + 4$; $(x - 3)^2 + (y + 2)^2 = 3^2$; center is $(3, -2)$; diameter is 2(3), or 6; (0, 2) satisfies the equation of the circle; C
6. $\begin{array}{r|rrrrr} -2i & 1 & -1 & 2 & -4 & -8 \\ & & 0 - 2i & -4 + 2i & 4 + 4i & 8 \\ \hline & 1 & -1 - 2i & -2 + 2i & 0 + 4i & 0 \end{array}$

   $P(-2i) = 0$; $P(i) = -9 - 3i$; $P\left(\frac{1}{2}\right) = -9\frac{9}{16}$; $P(i\sqrt{2}) = -8 - 2i\sqrt{2}$; A

7. $a - b = 16$
$\underline{a + b = \ \ 6}$
$2a = 22; a = 11; 11 + b = 6, b = -5$; A

8. $(A + B)^2 = (A + B)(A + B) = A(A + B) + B(A + B) = A^2 + AB + BA + B^2$, so I and III are equivalent. II will be equivalent to III if $AB = BA$, since then $2AB = AB + BA$ and $A^2 + AB + BA + B^2 = A^2 + 2AB + B^2$;
$AB = \begin{bmatrix} 2 & 1 \\ -4 & -2 \end{bmatrix}\begin{bmatrix} -1 & -3 \\ 2 & 6 \end{bmatrix} = \begin{bmatrix} 0 & 0 \\ 0 & 0 \end{bmatrix}$; $BA = \begin{bmatrix} -1 & -3 \\ 2 & 6 \end{bmatrix}\begin{bmatrix} 2 & 1 \\ -4 & -2 \end{bmatrix} =$
$\begin{bmatrix} 10 & 5 \\ -20 & -10 \end{bmatrix}$; $AB \neq BA$ so II and III are not equivalent; B

Pages 572–573 • PROGRAMMING IN PASCAL

1. a. See the procedure *get_matrix* in Exercise 2b.
b. See the procedure *print_matrix* in Exercise 2b.
c. See the program in Exercise 2b.

2. a. See the procedure *mult_matrix* in Exercise 2b.
b.

```
PROGRAM matrices (INPUT, OUTPUT);

CONST
  dim = 2;

TYPE
  scalar = real;
  vector = ARRAY[1..dim] OF real;
  matrix = ARRAY[1..dim, 1..dim] OF real;

VAR
  a, b : scalar;
  u, v : vector;
  m, n, mat : matrix;

(**************************************************************)
PROCEDURE get_matrix (VAR m : matrix);

VAR
  row, col : 1..dim;

BEGIN
  FOR row := 1 TO dim DO
      FOR col := 1 TO dim DO
         BEGIN
           write('Enter the item in row ', row:1);
           write(' and col ', col:1, ': ');
           readln(m[row, col]);
         END;
END;
```

```
(**************************************************************)
PROCEDURE print_matrix (m : matrix);

VAR
  row, col : 1..dim;

BEGIN
  FOR row := 1 TO dim DO
      BEGIN
        FOR col := 1 TO dim DO
            write(m[row, col]:10:3);
        writeln;
      END;
END;

(**************************************************************)
PROCEDURE add_matrices (m, n : matrix; VAR sum : matrix);

VAR
  row, col : 1..dim;

BEGIN
  FOR row := 1 TO dim DO
      FOR col := 1 TO dim DO
          sum[row, col] := m[row, col] + n[row, col];
END;

(**************************************************************)
PROCEDURE mult_matrices (m, n : matrix; VAR prod : matrix);

VAR
  row, col, j : 1..dim;

BEGIN
  FOR row := 1 TO dim DO
      FOR col := 1 TO dim DO
          BEGIN
            prod[row, col] := 0;
            FOR j := 1 TO dim DO
                prod[row, col] := prod[row, col] + m[row, j] *
                                                      n[j, col];
          END;
END;
```

[*Program continued on next page*]

```
(*************************************************************)
BEGIN (* main *)
  writeln('THE FIRST MATRIX');
  get_matrix(m);
  writeln;
  writeln('THE SECOND MATRIX');
  get_matrix(n);
  writeln;
  add_matrices(m, n, mat);
  print_matrix(mat);
  writeln;
  writeln;
  mult_matrices(m, n, mat);
  print_matrix(mat);
END.
```

3. a. See the procedure *inv* in Exercise 3b.

b.

```
PROGRAM solve (INPUT, OUTPUT);

CONST
  dim = 2;

TYPE
  scalar = real;
  vector = ARRAY[1..dim] OF real;
  matrix = ARRAY[1..dim, 1..dim] OF real;

VAR
  u, v : vector;
  m, n : matrix;

(*************************************************************)
PROCEDURE get_matrix (VAR m : matrix);

VAR
  row, col : 1..dim;

BEGIN
  FOR row := 1 TO dim DO
      FOR col := 1 TO dim DO
          BEGIN
            write('Enter the item in row ', row:1);
            write(' and col ', col:1, ': ');
            readln(m[row, col]);
          END;
END;
```

```
(**********************************************************)
PROCEDURE get_vector (VAR u : vector);

VAR
  row : 1..dim;

BEGIN
  FOR row := 1 TO dim DO
      BEGIN
        write('Enter the item on row ', row:1, ': ');
        readln(u[row]);
      END;
END;

(**********************************************************)
PROCEDURE inv (a : matrix; VAR b : matrix);

VAR
  det : scalar;

BEGIN
  det := a[1, 1] * a[2, 2] - a[1, 2] * a[2, 1];
  det := 1/det;
  b[1, 1] := det * a[2, 2];
  b[1, 2] := det * -a[1, 2];
  b[2, 1] := det * -a[2, 1];
  b[2, 2] := det * a[1, 1];
END;

(**********************************************************)
PROCEDURE multiply (n : matrix; u : vector; VAR v : vector);

VAR
  row, j : 1..dim;

BEGIN
  FOR row := 1 TO dim DO
      BEGIN
        v[row] := 0;
        FOR j := 1 TO dim DO
            v[row] := v[row] + n[row, j] * u[j];
      END;
END;
```

[*Program continued on next page*]

```
(*************************************************************)
PROCEDURE print_result (v : vector);

BEGIN
  writeln('x = ', v[1]:6:3);
  writeln('y = ', v[2]:6:3);
END;

(*************************************************************)
BEGIN (* main *)
  writeln('THE COEFFICIENTS MATRIX.');
  get_matrix(m);
  writeln;
  writeln('THE CONSTANTS.');
  get_vector(u);
  writeln;
  inv(m, n);
  multiply(n, u, v);
  print_result(v);
END.
```

4. PROCEDURE transpose (a : matrix; VAR t : matrix);

```
VAR
  row, col : 1..dim;

BEGIN
  FOR row := 1 TO dim DO
      FOR col := 1 TO dim DO
          t[col, row] := a[row, col];

END;
```

Page 578 • WRITTEN EXERCISES

Note: Answers are given to 3 significant digits because a 3-significant-digit approximation has been used for π.

A 1. $C \approx 3.14(25) = 78.5$; $s \approx 78.5(1.6) \approx 126$; 126 cm
2. $C \approx 3.14(20) = 62.8$; $s \approx 62.8(3.5) \approx 220$; 220 m
3. $C \approx 3.14(12) = 37.68$; $s \approx 37.68(6.9) \approx 260$; 260 m
4. $C \approx 3.14(10) = 31.4$; $s \approx 31.4(0.6) \approx 18.8$; 18.8 m
5. $r = \frac{s}{2\pi k}$, where k = number of revolutions; $r \approx \frac{942}{2(3.14)(4)} = 37.5$; 37.5 cm
6. $r \approx \frac{15.7}{2(3.14)(8)} \approx 0.313$; 0.313 m 7. $r \approx \frac{62.8}{2(3.14)(6)} \approx 1.67$; 1.67 m
8. $r \approx \frac{25.12}{2(3.14)(6)} \approx 0.667$; 0.667 m

9. 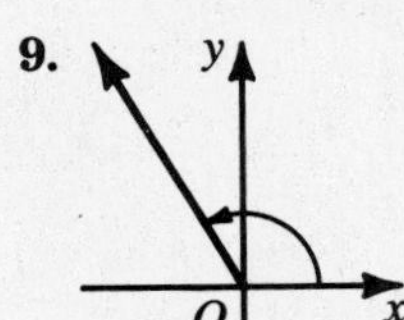

10. 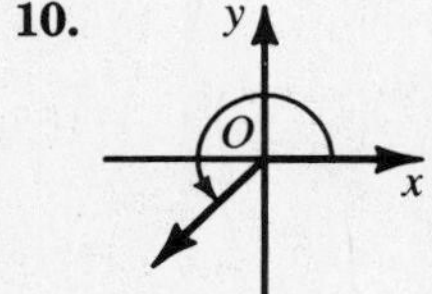

11. 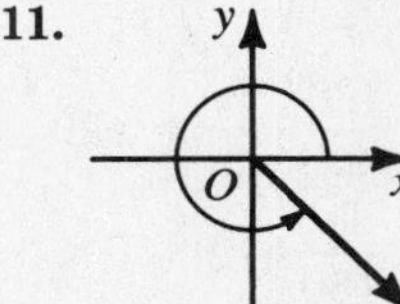

12. 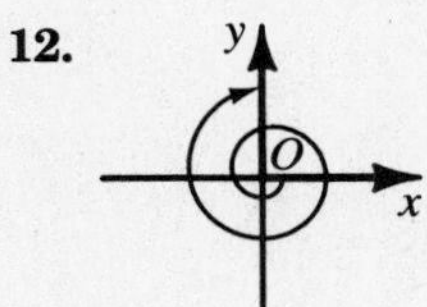

13. 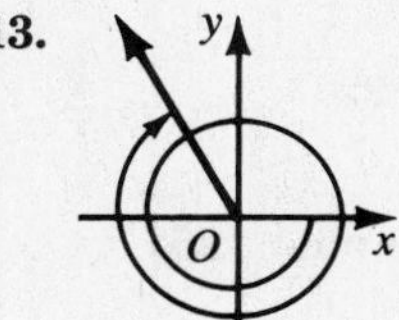

14. 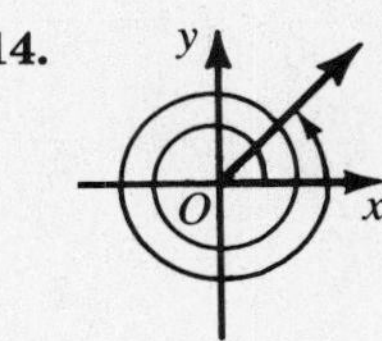

15. $\frac{2}{3}$ revolution clockwise 16. $\frac{3}{8}$ revolution clockwise 17. $\frac{1}{8}$ revolution clockwise
18. $\frac{1}{4}$ revolution counterclockwise 19. $\frac{1}{3}$ revolution counterclockwise
20. $\frac{7}{8}$ revolution clockwise

B 21. $s = 2\pi rk$, where k is the number of revolutions; $s \approx 2(3.14)(42)\left(\frac{5}{60}\right) \approx 22.0$; 22.0 cm
22. $s = 2\pi rk$; $k = \left(\frac{1}{12}\right)\left(\frac{5}{60}\right)$; $s \approx 2(3.14)(7.5)\left(\frac{1}{12}\right)\left(\frac{5}{60}\right) \approx 0.327$; 0.327 cm
23. $C = 2\pi r \approx 2(3.14)(6800) \approx 42{,}700$; speed $\approx \frac{42{,}700}{6} \approx 7120$; 7120 km/h

C 24. To find the number of revolutions a wheel makes as it rolls, find the distance traveled by its center and divide this distance by the circumference of the wheel. P travels in a circle with radius $5AP$. $\therefore$ number of revolutions $= \frac{2\pi(5AP)}{2\pi(AP)} = 5$.
25. (See the explanation for Ex. 24.) Now P travels in a circle with radius $3AP$, so the number of revolutions $= \frac{2\pi(3AP)}{2\pi(AP)} = 3$.

Pages 582–583 • WRITTEN EXERCISES

A 1. $\frac{\pi}{180} \cdot 240 = \frac{4\pi}{3}; \frac{4\pi}{3}^R$ 2. $\frac{\pi}{180}(-225) = -\frac{5\pi}{4}; -\frac{5\pi}{4}^R$ 3. $\frac{\pi}{180} \cdot 150 = \frac{5\pi}{6}; \frac{5\pi}{6}^R$

4. $\frac{\pi}{180}(-60) = -\frac{\pi}{3}; -\frac{\pi}{3}^R$ 5. $\frac{\pi}{180} \cdot 330 = \frac{11\pi}{6}; \frac{11\pi}{6}^R$ 6. $\frac{\pi}{180}(-300) = -\frac{5\pi}{3}; -\frac{5\pi}{3}^R$

7. $\frac{\pi}{180} \cdot 315 = \frac{7\pi}{4}; \frac{7\pi}{4}^R$ 8. $\frac{\pi}{180} \cdot 270 = \frac{3\pi}{2}; \frac{3\pi}{2}^R$ 9. $\frac{\pi}{180}(-108) = -\frac{3\pi}{5}; -\frac{3\pi}{5}^R$

10. $\frac{\pi}{180} \cdot 144 = \frac{4\pi}{5}; \frac{4\pi}{5}^R$ 11. $\frac{180}{\pi} \cdot \frac{3\pi}{4} = 135; 135°$ 12. $\frac{180}{\pi} \cdot \frac{5\pi}{3} = 300; 300°$

13. $\frac{180}{\pi}\left(-\frac{7\pi}{4}\right) = -315; -315°$ 14. $\frac{180}{\pi} \cdot \frac{3\pi}{2} = 270; 270°$ 15. $\frac{180}{\pi} \cdot \frac{5\pi}{6} = 150; 150°$

16. $\frac{180}{\pi}\left(-\frac{3\pi}{2}\right) = -270; -270°$ 17. $\frac{180}{\pi}\left(-\frac{11\pi}{6}\right) = -330; -330°$

18. $\frac{180}{\pi} \cdot \frac{7\pi}{3} = 420; 420°$ 19. $\frac{180}{\pi}\left(-\frac{8\pi}{9}\right) = -160; -160°$ 20. $\frac{180}{\pi} \cdot \frac{7\pi}{12} = 105; 105°$

21. $\frac{\pi}{180} \cdot 72 = \frac{2\pi}{5}; s = r \cdot m(\alpha); s \approx 35 \cdot \frac{2}{5} \cdot \frac{22}{7} = 44$; 44 cm

22. $\frac{\pi}{180} \cdot 330 = \frac{11\pi}{6}; s \approx 2.8 \cdot \frac{11}{6} \cdot \frac{22}{7} \approx 16.1$; 16.1 cm

23. $\frac{\pi}{180} \cdot 150 = \frac{5\pi}{6}; s \approx 105 \cdot \frac{5}{6} \cdot \frac{22}{7} = 275$; 275 cm

24. $s \approx 630 \cdot \frac{5}{6} \cdot \frac{22}{7} = 1650$; 1650 mm 25. $s \approx 56 \cdot \frac{1}{8} \cdot \frac{22}{7} = 22$; 22 cm

26. $s \approx 0.42 \cdot \frac{9}{2} \cdot \frac{22}{7} = 5.94$; 5.94 cm

27. $\frac{\pi}{180} \cdot 150 = \frac{5\pi}{6}; s = r \cdot m(\alpha); 30\pi = r\left(\frac{5\pi}{6}\right); r = 36$

28. $\frac{\pi}{180} \cdot 315 = \frac{7\pi}{4}; 35\pi = r\left(\frac{7\pi}{4}\right); r = 20$

29. $\frac{\pi}{180} \cdot 210 = \frac{7\pi}{6}; \frac{7\pi}{4} = r \cdot \frac{7\pi}{6}; r = \frac{3}{2}$

30. $\frac{\pi}{180} \cdot 300 = \frac{5\pi}{3}; \frac{25\pi}{9} = r \cdot \frac{5\pi}{3}; r = \frac{5}{3}$

31. $\frac{\pi}{180} \cdot 225 = \frac{5\pi}{4}; \frac{15\pi}{8} = r \cdot \frac{5\pi}{4}; r = \frac{3}{2}$

32. $\frac{\pi}{180} \cdot c = \frac{\pi c}{180}; \frac{j\pi}{k} = r \cdot \frac{\pi c}{180}; r = \frac{j\pi}{k} \cdot \frac{180}{\pi c} = \frac{180j}{kc}$

B 33. Let s = length of arc intercepted by α; $\frac{m(\alpha)}{2\pi} = \frac{s}{2\pi r}; 2\pi s = m(\alpha) \cdot 2\pi r;$
$s = r \cdot m(\alpha)$

C 34. Any angle coterminal with $\frac{\pi}{2}$ must differ from $\frac{\pi}{2}$ by a whole number of rotations, or $2\pi \cdot n$ radians, where n is an integer. Thus we have a sequence with numbers such as $\frac{\pi}{2} + 2\pi, \frac{\pi}{2} + 4\pi, \frac{\pi}{2} + 6\pi, \ldots$ and $\frac{\pi}{2} - 2\pi, \frac{\pi}{2} - 4\pi, \frac{\pi}{2} - 6\pi, \ldots$, or
$\ldots, -\frac{11\pi}{2}, -\frac{7\pi}{2}, -\frac{3\pi}{2}, \frac{\pi}{2}, \frac{5\pi}{2}, \frac{9\pi}{2}, \frac{13\pi}{2}, \ldots$

Page 583 • COMPUTER EXERCISES

1.
```
10  INPUT R
20  LET PI = 3.14159
30  LET D = R * 180/PI
40  LET M = (D - INT(D)) * 60
50  LET S = INT((M - INT(M)) * 60 + .5)
60  LET D = INT(D)
70  LET M = INT(M)
80  PRINT D;" DEGREES ";M;" MINUTES ";S;" SECONDS"
90  END
```

2. 38°57′40″ **3.** 65°19′2″ **4.** 146°40′38″ **5.** 207°59′2″

Page 583 • ON THE CALCULATOR

1. 0.9774 **2.** 0.5725 **3.** −0.4206 **4.** −8.3776 **5.** 0.2342 **6.** 0.7959
7. 0.4931 **8.** −0.0475 **9.** 25.9° **10.** 21.0° **11.** 41.8° **12.** −49.4°
13. 138.7° **14.** 360.0° **15.** −259.2° **16.** 76.0°

Page 584 • SELF-TEST 1

1. $C = 2\pi r \approx 2 \cdot \frac{22}{7} \cdot 28 = 176$; $3.2(176) = 563.2$; 563.2 cm

2. $\frac{\pi}{180} \cdot 480 = \frac{8\pi}{3}; \frac{8\pi^{R}}{3}$ **3.** $\frac{180}{\pi} \cdot \frac{11\pi}{6} = 330$; 330°

Page 588 • WRITTEN EXERCISES

A **1.**

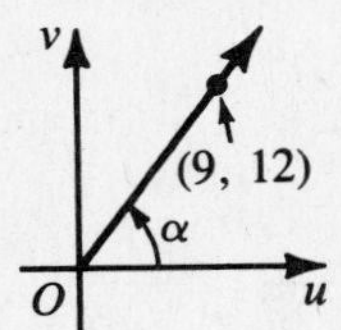

$r = \sqrt{9^2 + 12^2} = 15$;
$\sin \alpha = \frac{12}{15} = \frac{4}{5}$; $\cos \alpha = \frac{9}{15} = \frac{3}{5}$

2.

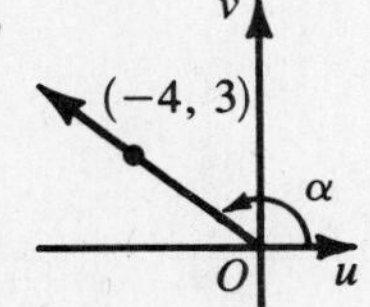

$r = \sqrt{(-4)^2 + 3^2} = \sqrt{25} = 5$;
$\sin \alpha = \frac{3}{5}$; $\cos \alpha = -\frac{4}{5}$

3.

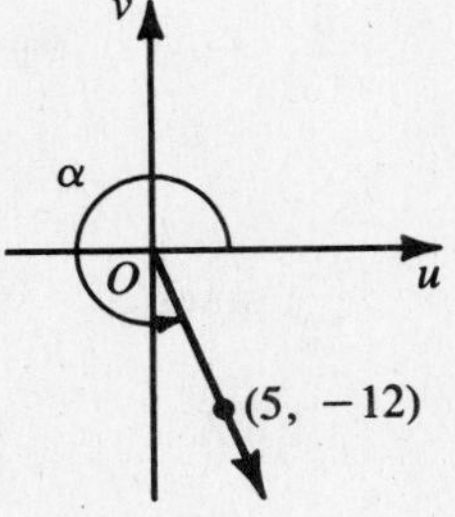

$r = \sqrt{5^2 + (-12)^2} = \sqrt{169} = 13$;
$\sin \alpha = -\frac{12}{13}$; $\cos \alpha = \frac{5}{13}$

4.

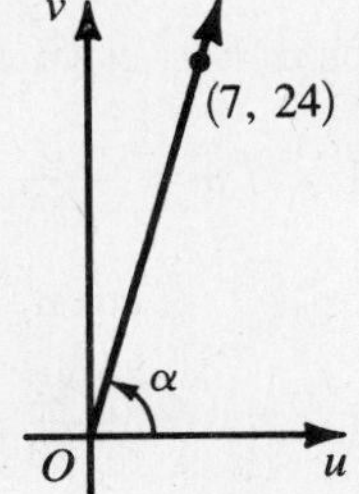

$r = \sqrt{7^2 + 24^2} = \sqrt{625} = 25$;
$\sin \alpha = \frac{24}{25}$; $\cos \alpha = \frac{7}{25}$

5. 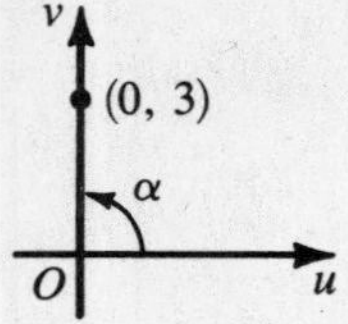

$r = \sqrt{0^2 + 3^2} = 3;$

$\sin \alpha = \frac{3}{3} = 1;\ \cos \alpha = \frac{0}{3} = 0$

6. 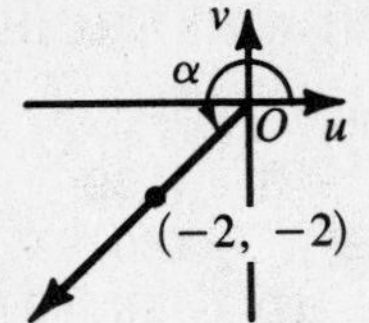

$r = \sqrt{(-2)^2 + (-2)^2} = \sqrt{8} = 2\sqrt{2};$

$\sin \alpha = -\frac{2}{2\sqrt{2}} = -\frac{1}{\sqrt{2}};$

$\cos \alpha = -\frac{2}{2\sqrt{2}} = -\frac{1}{\sqrt{2}}$

7. 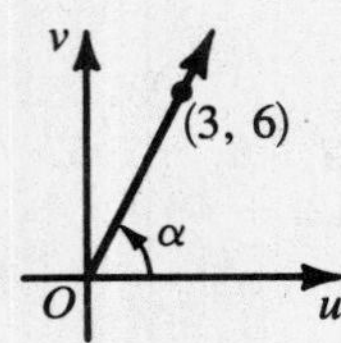

$r = \sqrt{3^2 + 6^2} = \sqrt{45} = 3\sqrt{5};$

$\sin \alpha = \frac{6}{3\sqrt{5}} = \frac{2}{\sqrt{5}};$

$\cos \alpha = \frac{3}{3\sqrt{5}} = \frac{1}{\sqrt{5}}$

8. 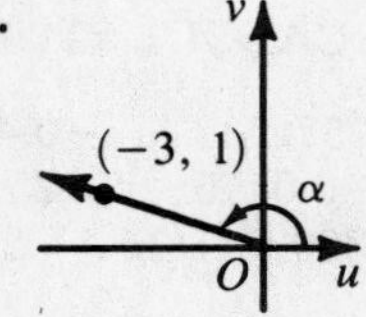

$r = \sqrt{(-3)^2 + 1^2} = \sqrt{10};$

$\sin \alpha = \frac{1}{\sqrt{10}};\ \cos \alpha = -\frac{3}{\sqrt{10}}$

9. 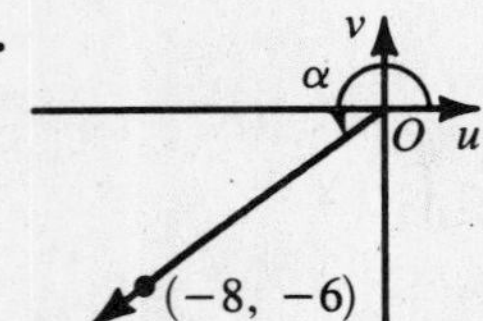

$r = \sqrt{(-8)^2 + (-6)^2} = \sqrt{100} = 10;$

$\sin \alpha = -\frac{6}{10} = -\frac{3}{5};$

$\cos \alpha = -\frac{8}{10} = -\frac{4}{5}$

10. 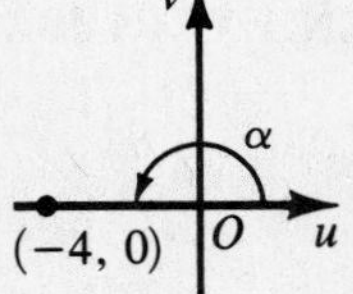

$r = \sqrt{(-4)^2 + 0^2} = \sqrt{16} = 4;$

$\sin \alpha = \frac{0}{4} = 0;\ \cos \alpha = -\frac{4}{4} = -1$

11. $\left(-\frac{12}{13}\right)^2 + \cos^2\alpha = 1;\ \cos^2\alpha = 1 - \frac{144}{169} = \frac{25}{169};\ \cos \alpha = \pm\frac{5}{13};$ since α is in Quadrant III, $\cos \alpha = -\frac{5}{13}$.

12. $\left(-\frac{5}{13}\right)^2 + \cos^2\alpha = 1;\ \cos^2\alpha = 1 - \frac{25}{169} = \frac{144}{169};\ \cos \alpha = \pm\frac{12}{13};$ since α is in Quadrant III, $\cos \alpha = -\frac{12}{13}$.

13. $\sin^2\alpha + \left(\frac{15}{17}\right)^2 = 1;\ \sin^2\alpha = 1 - \frac{225}{289} = \frac{64}{289};\ \sin \alpha = \pm\frac{8}{17};$ since α is in Quadrant I, $\sin \alpha = \frac{8}{17}$.

14. $\sin^2\alpha + \left(-\frac{21}{25}\right)^2 = 1$; $\sin^2\alpha = 1 - \frac{441}{625} = \frac{184}{625}$; $\sin\alpha = \pm\frac{2\sqrt{46}}{25}$; since α is in Quadrant II, $\sin\alpha = \frac{2\sqrt{46}}{25}$.

15. $\sin^2\alpha + \left(\frac{1}{4}\right)^2 = 1$; $\sin^2\alpha = 1 - \frac{1}{16} = \frac{15}{16}$; $\sin\alpha = \pm\frac{\sqrt{15}}{4}$; since α is in Quadrant IV, $\sin\alpha = -\frac{\sqrt{15}}{4}$.

16. $\left(-\frac{\sqrt{3}}{2}\right)^2 + \cos^2\alpha = 1$; $\cos^2\alpha = 1 - \frac{3}{4} = \frac{1}{4}$; $\cos\alpha = \pm\frac{1}{2}$; since α is in Quadrant IV, $\cos\alpha = \frac{1}{2}$.

17. $\left(\frac{2\sqrt{6}}{5}\right)^2 + \cos^2\alpha = 1$; $\cos^2\alpha = 1 - \frac{24}{25} = \frac{1}{25}$; $\cos\alpha = \pm\frac{1}{5}$; since α is in Quadrant II, $\cos\alpha = -\frac{1}{5}$.

18. $\sin^2\alpha + \left(\frac{5}{7}\right)^2 = 1$; $\sin^2\alpha = 1 - \frac{25}{49} = \frac{24}{49}$; $\sin\alpha = \pm\frac{2\sqrt{6}}{7}$; since α is in Quadrant I, $\sin\alpha = \frac{2\sqrt{6}}{7}$.

B **19.** $u^2 + v^2 = 1$; $u^2 + (2u)^2 = 1$; $5u^2 = 1$; $u^2 = \frac{1}{5}$; $u = \pm\frac{1}{\sqrt{5}}$; since $u < 0$, $\cos\alpha = u = -\frac{1}{\sqrt{5}}$ and $\sin\alpha = v = 2u = -\frac{2}{\sqrt{5}}$.

20. $u^2 + \left(-\frac{4u}{3}\right)^2 = 1$; $\frac{25u^2}{9} = 1$; $u^2 = \frac{9}{25}$; $u = \pm\frac{3}{5}$; since $u > 0$, $\cos\alpha = u = \frac{3}{5}$ and $\sin\alpha = v = \left(-\frac{4}{3}\right)\left(\frac{3}{5}\right) = -\frac{4}{5}$.

21. $u^2 + \left(-\frac{u}{3}\right)^2 = 1$; $\frac{10u^2}{9} = 1$; $u^2 = \frac{9}{10}$; $u = \pm\frac{3}{\sqrt{10}}$; since $u < 0$, $\cos\alpha = u = -\frac{3}{\sqrt{10}}$ and $\sin\alpha = v = \left(-\frac{1}{3}\right)\left(-\frac{3}{\sqrt{10}}\right) = \frac{1}{\sqrt{10}}$.

22. $u^2 + (u\sqrt{3})^2 = 1$; $u^2 + 3u^2 = 1$; $4u^2 = 1$; $u^2 = \frac{1}{4}$; $u = \pm\frac{1}{2}$; since $u < 0$, $\cos\alpha = u = -\frac{1}{2}$ and $\sin\alpha = v = -\frac{\sqrt{3}}{2}$.

C **23.** $\frac{v+2}{2} = u^2$; $\frac{v+2}{2} + v^2 = 1$; $v + 2 + 2v^2 = 2$; $2v^2 + v = 0$; $v(2v+1) = 0$; $v = 0$ or $2v + 1 = 0$; $v = 0$ or $v = -\frac{1}{2}$; since $v < 0$, $\sin\alpha = v = -\frac{1}{2}$; $\therefore \left(-\frac{1}{2}\right) = 2u^2 - 2$; $u^2 = \frac{3}{4}$; since $u < 0$, $\cos\alpha = u = -\frac{\sqrt{3}}{2}$.

24. $v^2 = \frac{9u}{20}$; $u^2 + \frac{9u}{20} = 1$; $20u^2 + 9u = 20$; $20u^2 + 9u - 20 = 0$;

$(5u - 4)(4u + 5) = 0$; $5u = 4$ or $4u = -5$; $u = \frac{4}{5}$ or $u = -\frac{5}{4}$ (impossible,

since $-1 \le u \le 1$); thus $\cos \alpha = u = \frac{4}{5}$; $v^2 = \left(\frac{9}{20}\right)\left(\frac{4}{5}\right) = \frac{9}{25}$; $v = \pm\frac{3}{5}$;

since $v > 0$, $\sin \alpha = v = \frac{3}{5}$.

25. $x' = x \cos \alpha + y \sin \alpha$ and $y' = -x \sin \alpha + y \cos \alpha$. If $P(a, b)$ is on the terminal side of α, then $\sin \alpha = \frac{b}{\sqrt{a^2 + b^2}}$ and $\cos \alpha = \frac{a}{\sqrt{a^2 + b^2}}$.

The image of $P(a, b)$ under the given transformation is $P'(a', b')$, where:

$a' = a \cos \alpha + b \sin \alpha = a \cdot \frac{a}{\sqrt{a^2 + b^2}} + b \cdot \frac{b}{\sqrt{a^2 + b^2}} = \frac{a^2 + b^2}{\sqrt{a^2 + b^2}}$ and

$b' = -a \sin \alpha + b \cos \alpha = -a \cdot \frac{b}{\sqrt{a^2 + b^2}} + b \cdot \frac{a}{\sqrt{a^2 + b^2}} = 0$

Since $b' = 0$, $P'(a', b')$ is on the x-axis. Thus the distance from P' to the origin is a', or $\sqrt{a^2 + b^2}$, which is also the distance from P to the origin.

Pages 591–592 • WRITTEN EXERCISES

A **1.** $\cos 495° = \cos(135° + 360°) = \cos 135° = -\frac{1}{\sqrt{2}}$

2. $\sin(-210°) = \sin[150° + (-1)360°] = \sin 150° = \frac{1}{2}$

3. $\cos 765° = \cos(45° + 2 \cdot 360°) = \cos 45° = \frac{1}{\sqrt{2}}$

4. $\sin 600° = \sin(240° + 360°) = \sin 240° = -\frac{\sqrt{3}}{2}$

5. $\sin \frac{9\pi^R}{4} = \sin\left(\frac{\pi^R}{4} + 2\pi^R\right) = \sin \frac{\pi^R}{4} = \frac{1}{\sqrt{2}}$

6. $\sin \frac{5\pi^R}{2} = \sin\left(\frac{\pi^R}{2} + 2\pi^R\right) = \sin \frac{\pi^R}{2} = 1$

7. $\cos\left(-\frac{7\pi^R}{3}\right) = \cos\left[\frac{5\pi^R}{3} - 2(2\pi^R)\right] = \cos \frac{5\pi^R}{3} = \frac{1}{2}$

8. $\cos\left(-\frac{7\pi^R}{6}\right) = \cos\left[\frac{5\pi^R}{6} - 2\pi^R\right] = \cos \frac{5\pi^R}{6} = -\frac{\sqrt{3}}{2}$

9. $\sin\left(-\frac{4\pi^R}{3}\right) = \sin\left[\frac{2\pi^R}{3} - 2\pi^R\right] = \sin \frac{2\pi^R}{3} = \frac{\sqrt{3}}{2}$

10. $\cos \frac{19\pi^R}{6} = \cos\left(\frac{7\pi^R}{6} + 2\pi^R\right) = \cos \frac{7\pi^R}{6} = -\frac{\sqrt{3}}{2}$

11. $\cos(-5\pi^R) = \cos[\pi^R - 3(2\pi^R)] = \cos \pi^R = -1$

12. $\sin \frac{15\pi^R}{4} = \sin\left(\frac{7\pi^R}{4} + 2\pi^R\right) = \sin \frac{7\pi^R}{4} = -\frac{1}{\sqrt{2}}$

13. $u = 1, v = \sqrt{3}; r = \sqrt{1^2 + (\sqrt{3})^2} = \sqrt{4} = 2; \sin \alpha = \frac{\sqrt{3}}{2}, \cos \alpha = \frac{1}{2}$

a. $60° + k \cdot 360°$ **b.** $\frac{\pi^R}{3} + 2k\pi^R$

14. $u = 5, v = -5; r = \sqrt{5^2 + (-5)^2} = \sqrt{50} = 5\sqrt{2}; \sin \alpha = \frac{-5}{5\sqrt{2}} = -\frac{1}{\sqrt{2}};$

$\cos \alpha = \frac{5}{5\sqrt{2}} = \frac{1}{\sqrt{2}}$ **a.** $315° + k \cdot 360°$ **b.** $\frac{7\pi^R}{4} + 2k\pi^R$

15. $u = -2\sqrt{3}, v = -2; r = \sqrt{(-2\sqrt{3})^2 + (-2)^2} = \sqrt{16} = 4; \sin \alpha = \frac{-2}{4} = -\frac{1}{2};$

$\cos \alpha = \frac{-2\sqrt{3}}{4} = -\frac{\sqrt{3}}{2}$ **a.** $210° + k \cdot 360°$ **b.** $\frac{7\pi^R}{6} + 2k\pi^R$

16. $u = 9, v = 9; r = \sqrt{9^2 + 9^2} = \sqrt{162} = 9\sqrt{2}; \sin \alpha = \frac{9}{9\sqrt{2}} = \frac{1}{\sqrt{2}};$

$\cos \alpha = \frac{9}{9\sqrt{2}} = \frac{1}{\sqrt{2}}$ **a.** $45° + k \cdot 360°$ **b.** $\frac{\pi^R}{4} + 2k\pi^R$

17. $u = -3, v = -3; r = \sqrt{(-3)^2 + (-3)^2} = \sqrt{18} = 3\sqrt{2}; \sin \alpha = \frac{-3}{3\sqrt{2}} = -\frac{1}{\sqrt{2}};$

$\cos \alpha = \frac{-3}{3\sqrt{2}} = -\frac{1}{\sqrt{2}}$ **a.** $225° + k \cdot 360°$ **b.** $\frac{5\pi^R}{4} + 2k\pi^R$

18. $u = \sqrt{5}, v = -\sqrt{15}; r = \sqrt{(\sqrt{5})^2 + (-\sqrt{15})^2} = \sqrt{20} = 2\sqrt{5};$

$\sin \alpha = \frac{-\sqrt{15}}{2\sqrt{5}} = -\frac{\sqrt{3}}{2}; \cos \alpha = \frac{\sqrt{5}}{2\sqrt{5}} = \frac{1}{2}$ **a.** $300° + k \cdot 360°$ **b.** $\frac{5\pi^R}{3} + 2k\pi^R$

19. $u = -\sqrt{3}, v = -\sqrt{3}; r = \sqrt{(-\sqrt{3})^2 + (-\sqrt{3})^2} = \sqrt{6}; \sin \alpha = \frac{-\sqrt{3}}{\sqrt{6}} = -\frac{1}{\sqrt{2}};$

$\cos \alpha = \frac{-\sqrt{3}}{\sqrt{6}} = -\frac{1}{\sqrt{2}}$ **a.** $225° + k \cdot 360°$ **b.** $\frac{5\pi^R}{4} + 2k\pi^R$

20. $u = -\frac{1}{3}, v = \frac{1}{3}\sqrt{3}; r = \sqrt{\left(-\frac{1}{3}\right)^2 + \left(\frac{1}{3}\sqrt{3}\right)^2} = \sqrt{\frac{4}{9}} = \frac{2}{3}; \sin \alpha = \frac{\frac{1}{3}\sqrt{3}}{\frac{2}{3}} = \frac{\sqrt{3}}{2};$

$\cos \alpha = \frac{-\frac{1}{3}}{\frac{2}{3}} = -\frac{1}{2}$ **a.** $120° + k \cdot 360°$ **b.** $\frac{2\pi^R}{3} + 2k\pi^R$

B 21. The line defined by $y = \sqrt{3}x$ goes through $(0, 0)$ and $(1, \sqrt{3})$. Since $\sqrt{(1)^2 + (\sqrt{3})^2} = \sqrt{4} = 2$, the line makes an angle α with the x-axis such that $\sin \alpha = \frac{\sqrt{3}}{2}$ and $\cos \alpha = \frac{1}{2}$. From the table on page 590, $m(\alpha) = 60°$.

22. The line $y = -\frac{x}{\sqrt{3}}$ goes through (0, 0), $(\sqrt{3}, -1)$, and $(-\sqrt{3}, 1)$. Let α be the angle that the part of the line through (0, 0) and $(\sqrt{3}, -1)$ makes with the positive x-axis. Then $\sqrt{(\sqrt{3})^2 + (-1)^2} = \sqrt{4} = 2$, and $\cos \alpha = \frac{\sqrt{3}}{2}$, $\sin \alpha = -\frac{1}{2}$. Thus $m(\alpha) = 330°$. The part of the line through (0, 0) and $(-\sqrt{3}, 1)$ makes an angle of 150° with the positive x-axis.

C **23.** Since the fundamental period of sine is 2π,

$$\sin nx = \sin(nx + 2\pi) = \sin n\left(x + \frac{2\pi}{n}\right)$$

$$\text{and } \sin nx = \sin(nx - 2\pi) = \sin n\left(x - \frac{2\pi}{n}\right).$$

Thus $\frac{2\pi}{n}$ is a period of $y = \sin nx$. It is the fundamental period because if there were a smaller period p, then np would be a period of $y = \sin x$ smaller than the fundamental period 2π, since $p < \frac{2\pi}{n}$ implies $np < 2\pi$.

Page 592 • COMPUTER EXERCISES

1.
```
10  INPUT N,X
20  LET S = 0 : LET K = 0
30  FOR I = 1 TO 2 * N - 1 STEP 2
40  LET P = 1
50  FOR J = I TO 1 STEP -1
60  LET P = P * J
70  NEXT J
80  LET S = S + ((-1)^K * X^I)/P
90  LET K = K + 1
100  NEXT I
110  PRINT S
120  END
```

2. 0.1986667 **3.** 0.9949261 **4.** 0.9949244

Pages 594–595 • WRITTEN EXERCISES

A **1. a.** $m(\theta) = 180° - m(\alpha)$
$= 180° - 105° = 75°$

b.

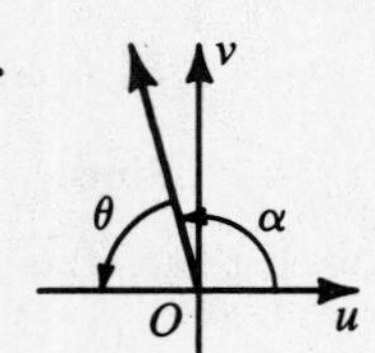

c. $\sin 105° = \sin 75° = 0.9659$

2. a. $m(\theta) = 235° - 180° = 55°$

b.

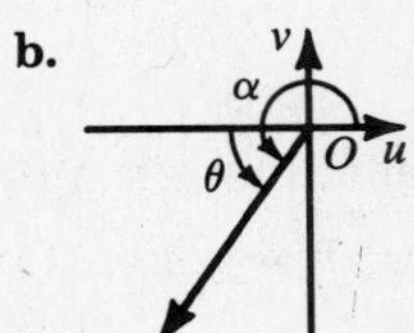

c. $\cos 235° = -\cos 55° = -0.5736$

3. a. $m(\theta) = 360° - 348° = 12°$

b.

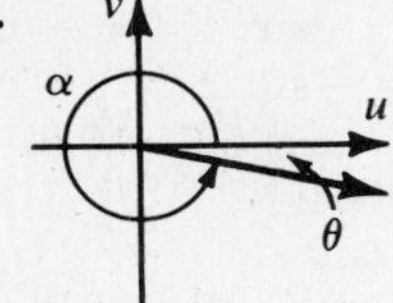

c. $\cos 348° = \cos 12° = 0.9781$

4. a. $m(\theta) = 260° - 180° = 80°$

b.

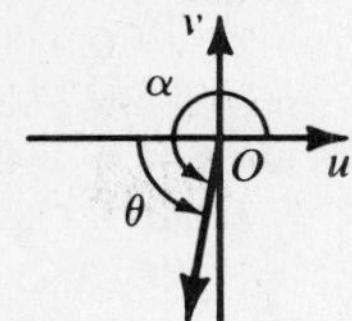

c. $\sin 260° = -\sin 80° = -0.9848$

5. a. $m(\theta) = 195° - 180° = 15°$

b.

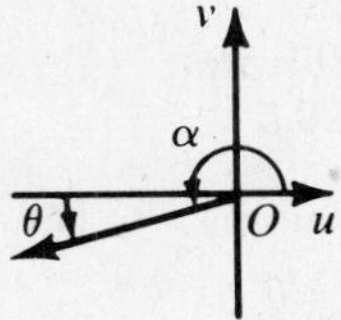

c. $\cos 195° = -\cos 15° = -0.9659$

6. a. $m(\theta) = 180° - 157° = 23°$

b.

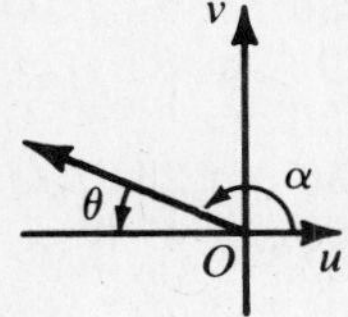

c. $\sin 157° = \sin 23° = 0.3907$

7. a. $m(\theta) = 360° - 295.8° = 64.2°$

b.

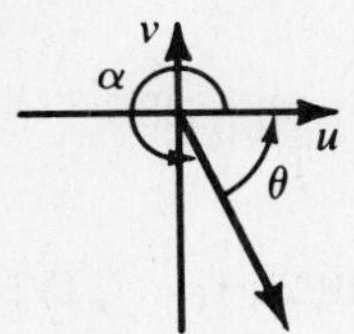

c. $\sin 295.8° = -\sin 64.2° = -0.9003$

8. a. $m(\theta) = 216.5° - 180° = 36.5°$

b.

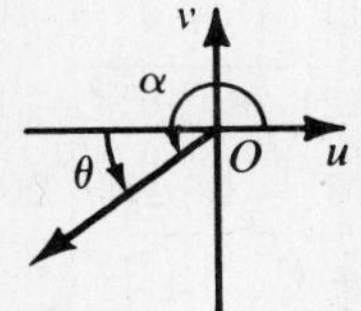

c. $\cos 216.5° = -\cos 36.5° = -0.8039$

9. a. $m(\theta) \approx 4.23^R - 3.14^R = 1.09^R$

b.

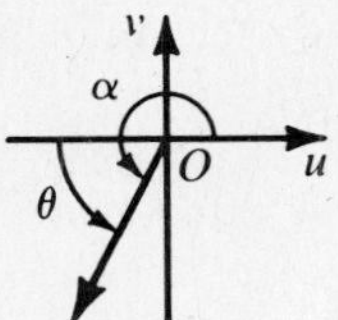

c. $\cos 4.23^R \approx -\cos 1.09^R = -0.4625$

10. a. $m(\theta) \approx 6.28^R - 5.78^R = 0.50^R$

b.

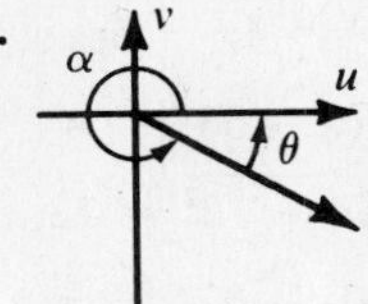

c. $\sin 5.78^R \approx -\sin 0.5^R = -0.4794$

11. a. $m(\theta) \approx 3.14^R - 1.86^R = 1.28^R$

b.

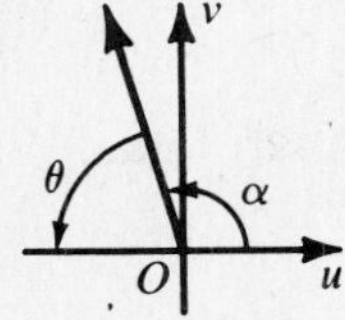

c. $\cos 1.86^R \approx -\cos 1.28^R = -0.2867$

12. a. $m(\theta) \approx 3.14^R - 2.69^R = 0.45^R$

b.

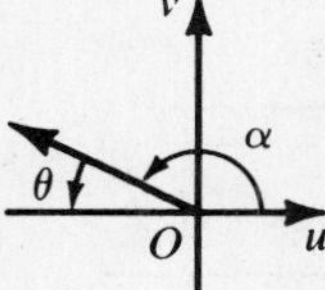

c. $\sin 2.69^R \approx \sin 0.45^R = 0.4350$

13. $\sin 136.7° = \sin 43.3° = 0.6858$

14. $\sin 314.8° = -\sin 45.2° = -0.7096$

15. $\cos 226°40' = -\cos 46°40' = -0.6862$

16. $\cos 117°10' = -\cos 62°50' = -0.4566$
17. $\sin(-128°) = -\sin 52° = -0.7880$
18. $\cos(-251°) = -\cos 71° = -0.3256$
19. $\cos 339° = \cos 21° = 0.9336$
20. $\sin 675° = -\sin 45° = -0.7071$
21. $\cos 812° = -\cos 88° = -0.0349$
22. $\sin 542° = -\sin 2° = -0.0349$
23. $\sin 963° = -\sin 63° = -0.8910$
24. $\cos 1022° = \cos 58° = 0.5299$
25. $\cos(-3.4^R) \approx -\cos(3.4^R - 3.14^R) = -\cos 0.26^R = -0.9664$
26. $\sin(-2.7^R) \approx -\sin(3.14^R - 2.7^R) = -\sin 0.44^R = -0.4259$
27. $\sin 8.7^R \approx \sin(8.7^R - 6.28^R) = \sin 2.42^R \approx \sin(3.14^R - 2.42^R) = \sin 0.72^R = 0.6594$
28. $\cos 11.6^R \approx \cos(11.6^R - 6.28^R) = \cos 5.32^R \approx \cos(6.28^R - 5.32^R) =$ $\cos 0.96^R = 0.5735$
29. $m(\theta) = 12°$; since $\sin \alpha < 0$ and $\cos \alpha < 0$, α is in Quad. III; $m(\alpha) = 192°$
30. $m(\theta) = 22°30'$, or $22.5°$; Quad. IV; $m(\alpha) = 337°30'$, or $337.5°$
31. $m(\theta) = 83°30'$, or $83.5°$; Quad. III; $m(\alpha) = 263°30'$, or $263.5°$
32. $m(\theta) = 50°30'$, or $50.5°$; Quad. II; $m(\alpha) = 129°30'$, or $129.5°$
33. $m(\theta) \approx 23°50'$, or $23.8°$; Quad. II; $m(\alpha) \approx 156°10'$, or $156.2°$
34. $m(\theta) \approx 47°50'$, or $47.8°$; Quad. III; $m(\alpha) \approx 227°50'$, or $227.8°$
35. $m(\theta) \approx 62°20'$, or $62.3°$; Quad. II; $m(\alpha) \approx 117°40'$, or $117.7°$
36. $m(\theta) \approx 75°20'$, or $75.3°$; Quad. III; $m(\alpha) \approx 255°20'$, or $255.3°$

B 37. $r = \sqrt{(-20)^2 + 21^2} = 29$; $\sin \alpha = \frac{21}{29} \approx 0.7241$; $m(\theta) \approx 46.4°$; Quad. II; $m(\alpha) \approx 180° - 46.4° = 133.6°$

38. $r = \sqrt{(-15)^2 + (-8)^2} = 17$; $\sin \alpha = -\frac{8}{17} \approx -0.4706$; $m(\theta) \approx 28.1°$; Quad. III; $m(\alpha) \approx 180° + 28.1° = 208.1°$

39. $r = \sqrt{2^2 + (-\sqrt{5})^2} = 3$; $\cos \alpha = \frac{2}{3} \approx 0.6667$; $m(\theta) \approx 48.2°$; Quad. IV; $m(\alpha) \approx 360° - 48.2° = 311.8°$

40. $r = \sqrt{(-7)^2 + (6\sqrt{2})^2} = 11$; $\cos \alpha = -\frac{7}{11} \approx -0.6364$; $m(\theta) \approx 50.5°$; Quad. II; $m(\alpha) = 180° - 50.5° = 129.5°$

41. $r = \sqrt{3^2 + (-4)^2} = 5$; $\cos \alpha = \frac{3}{5} = 0.6$; $m(\theta) \approx 53°10'$; Quad. IV; $m(\alpha) \approx 360° - 53°10' = 306°50'$

42. $r = \sqrt{(-5)^2 + 12^2} = 13$; $\sin \alpha = \frac{12}{13} \approx 0.9231$; $m(\theta) \approx 67°20'$; Quad. II; $m(\alpha) \approx 180° - 67°20' = 112°40'$

43. $r = \sqrt{(-9)^2 + (-40)^2} = 41$; $\sin \alpha = -\frac{40}{41} \approx -0.9756$; $m(\theta) \approx 77°20'$; Quad. III; $m(\alpha) \approx 180° + 77°20' = 257°20'$

44. $r = \sqrt{3^2 + (-2\sqrt{10})^2} = 7$; $\cos \alpha = \frac{3}{7} \approx 0.4286$; $m(\theta) \approx 64°40'$; Quad. IV; $m(\alpha) \approx 360° - 64°40' = 295°20'$

C 45. $r = \sqrt{15^2 + (-8)^2} = 17$; $\cos \alpha = \frac{15}{17} \approx 0.8824$; $m(\theta) \approx 0.49^R$; Quad. IV; $m(\alpha) \approx 6.28^R - 0.49^R = 5.79^R$

46. $r = \sqrt{(-\sqrt{7})^2 + 3^2} = 4$; $\sin \alpha = \frac{3}{4} = 0.75$; $m(\theta) \approx 0.85^R$; Quad. II; $m(\alpha) \approx 3.14^R - 0.85^R = 2.29^R$

47. $r = \sqrt{(-2)^2 + (-3\sqrt{5})^2} = 7$; $\cos \alpha = -\dfrac{2}{7} \approx -0.2857$; $m(\theta) \approx 1.28^R$; Quad. III;
$m(\alpha) \approx 3.14^R + 1.28^R = 4.42^R$

48. $r = \sqrt{7^2 + (-4\sqrt{2})^2} = 9$; $\cos \alpha = \dfrac{7}{9} \approx 0.7778$; $m(\theta) \approx 0.68^R$; Quad. IV;
$m(\alpha) \approx 6.28^R - 0.68^R = 5.60^R$

Page 595 • COMPUTER EXERCISES

1.
```
10  INPUT D
20  IF D < 0 THEN LET D = D + 360 : GOTO 20
30  IF D > 360 THEN LET D = D - 360 : GOTO 30
40  IF D > 0 AND D <= 90 THEN LET R = D
50  IF D > 90 AND D <= 180 THEN LET R = 180 - D
60  IF D > 180 AND D <= 270 THEN LET R = D - 180
70  IF D > 270 AND D <= 360 THEN LET R = 360 - D
80  PRINT "REFERENCE ANGLE IS ";R
90  END
```

2. 75° **3.** 37° **4.** 20° **5.** 79°

Pages 598–599 • WRITTEN EXERCISES

A 1.

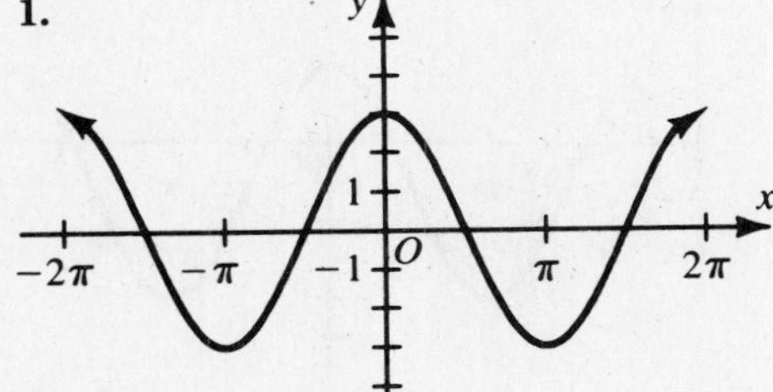

2.

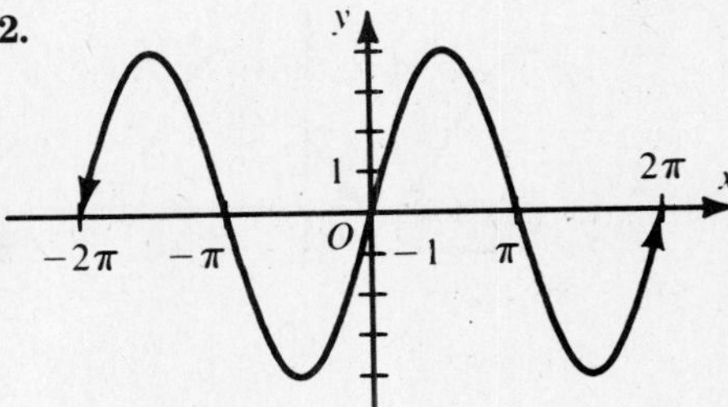

3.

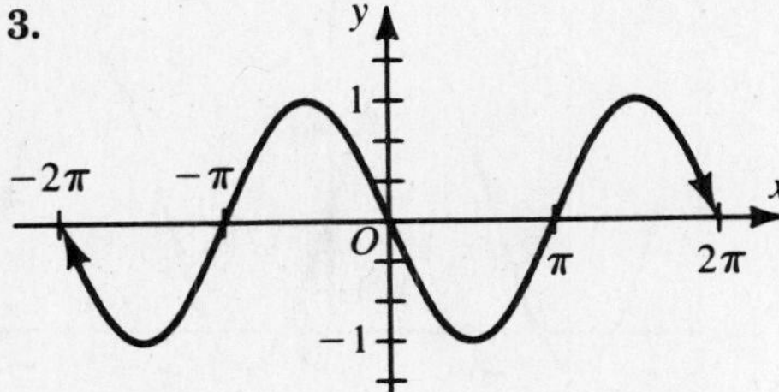

4.

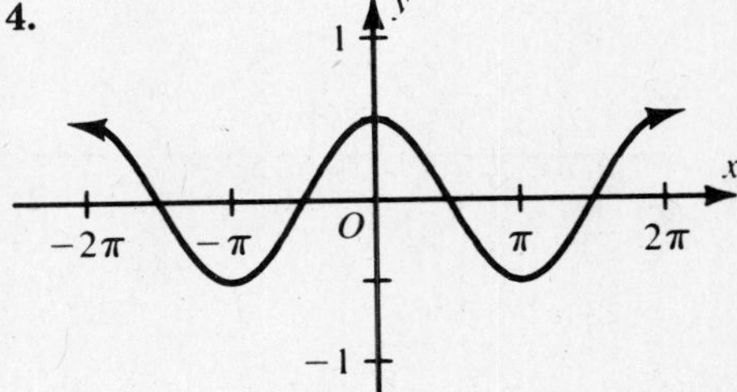

5.

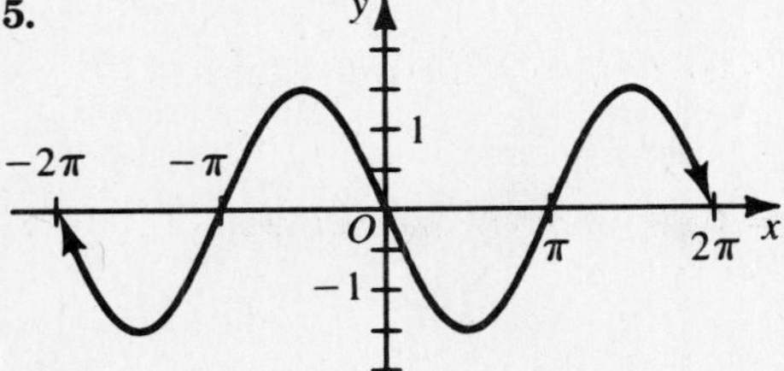

6.

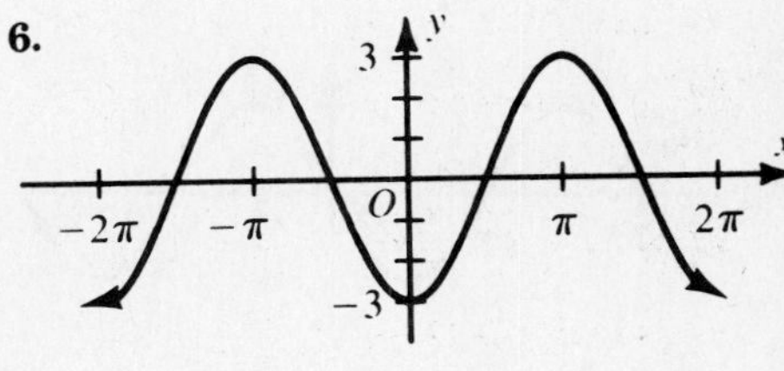

7.

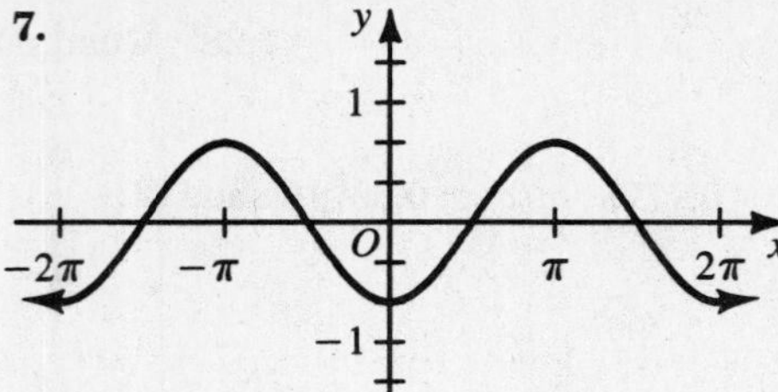

8.

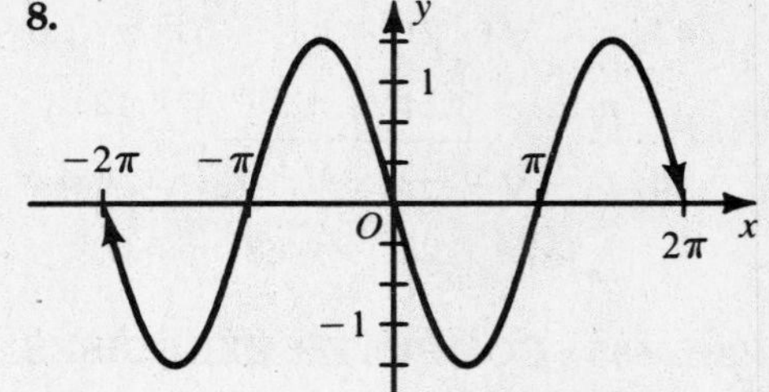

9.

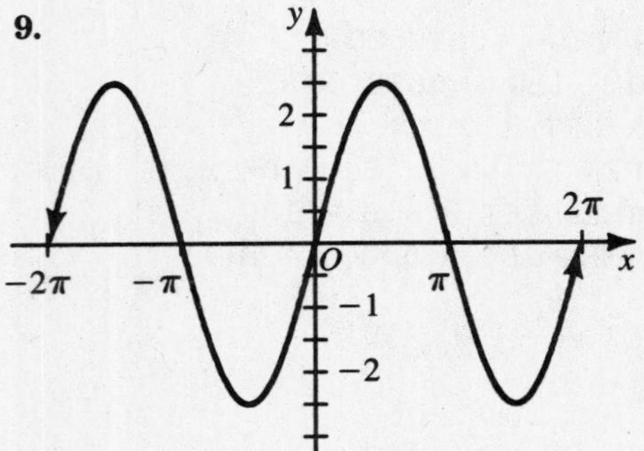

10. max.: 2 + 2, or 4; min.: 2(−1) + 2, or 0; amplitude: 2

11. max.: 3 − 1, or 2; min.: 3(−1) − 1, or −4; amplitude: 3

12. max.: $\frac{1}{2} + 1$, or $\frac{3}{2}$; min.: $\frac{1}{2}(-1) + 1$, or $\frac{1}{2}$; amplitude: $\frac{1}{2}$

13. max.: −4(−1) + 3, or 7; min.: −4 + 3, or −1; amplitude: 4

B 14.

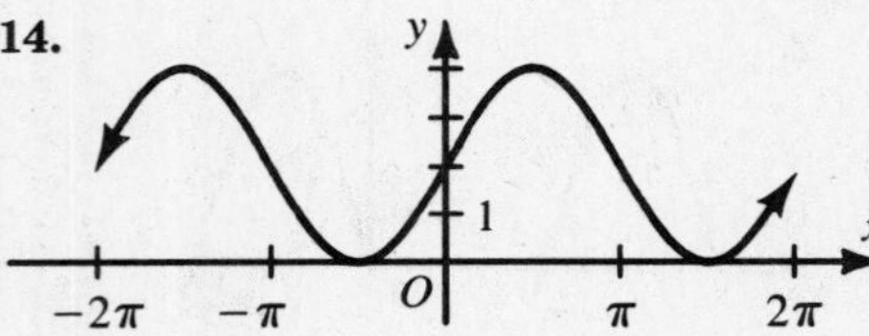

15.

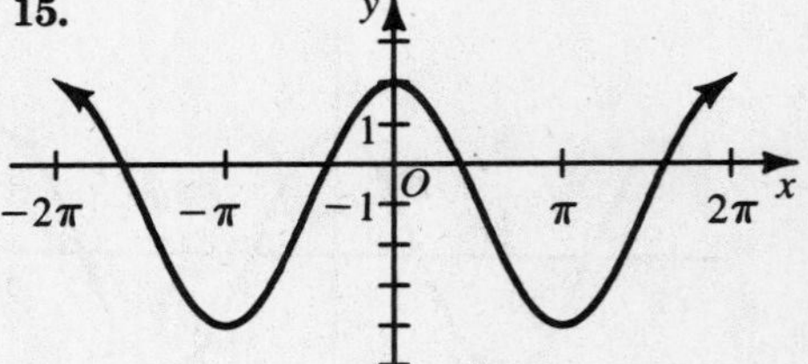

16.

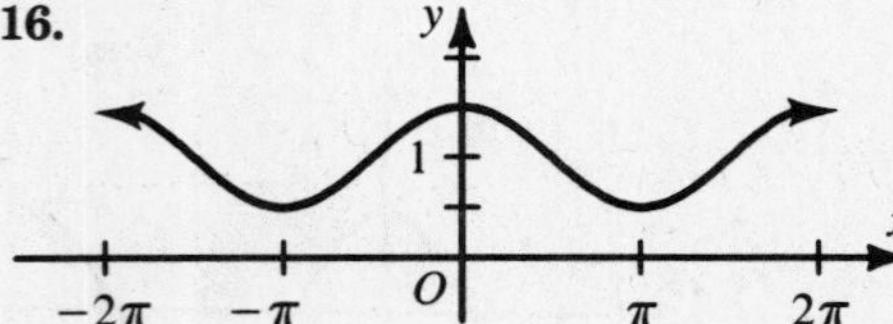

17.

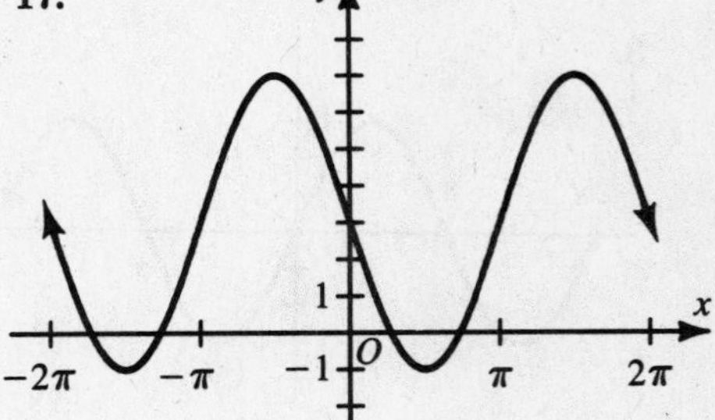

C 18.

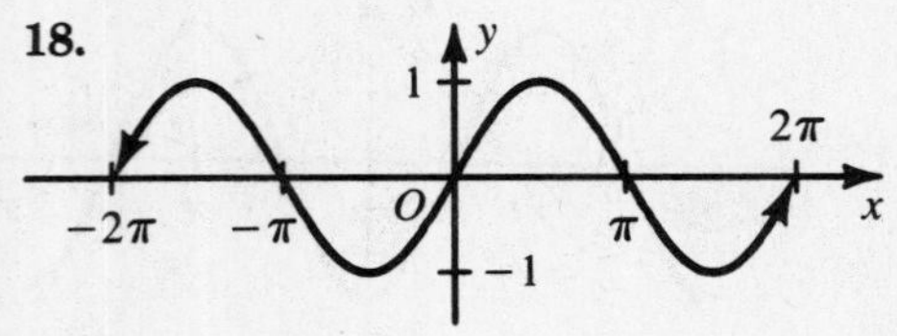

yes; $y = \sin x$

19.

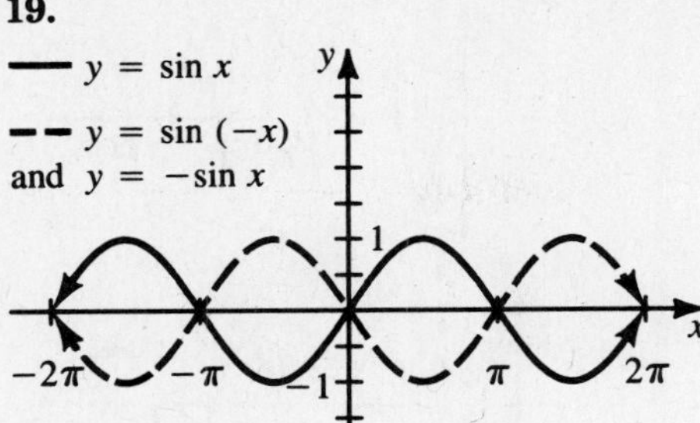

20.

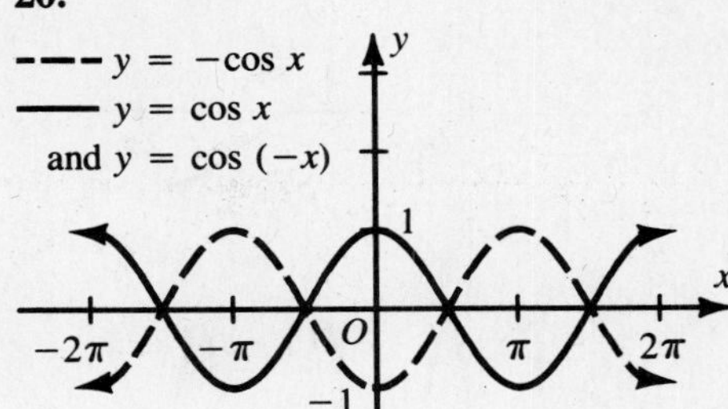

Pages 600–601 • WRITTEN EXERCISES

A **1.**

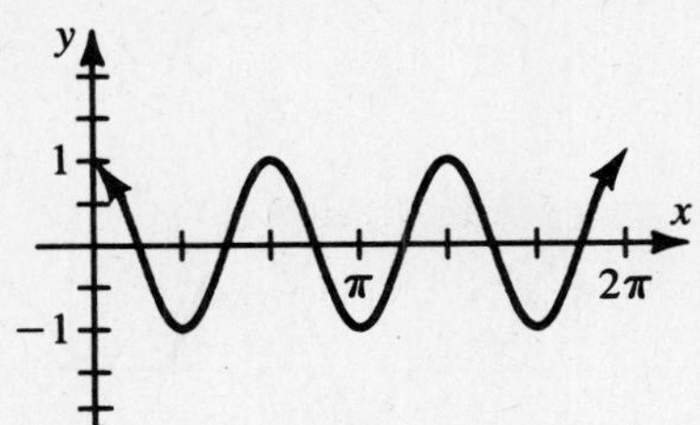

2.

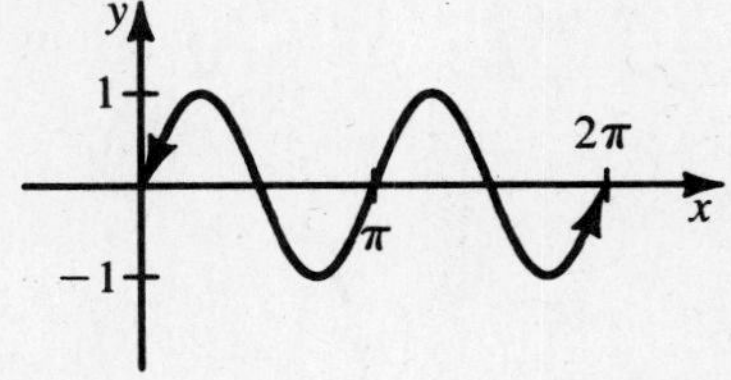

3.

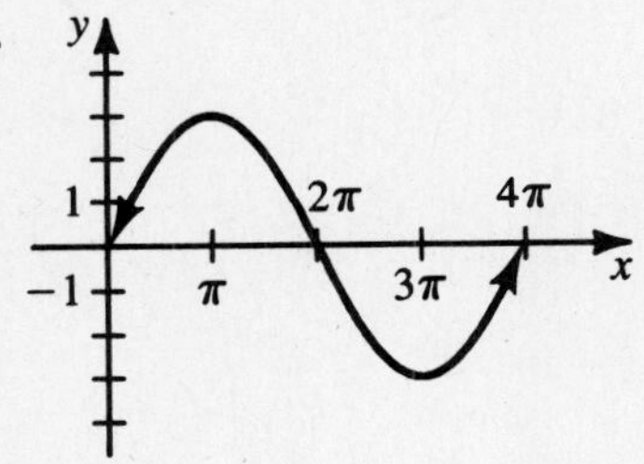

4.

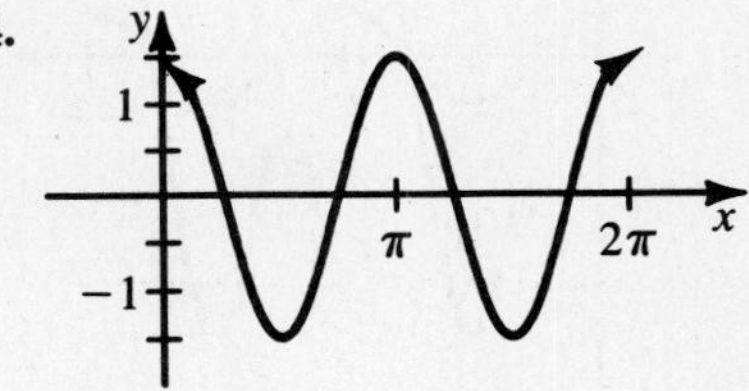

5.

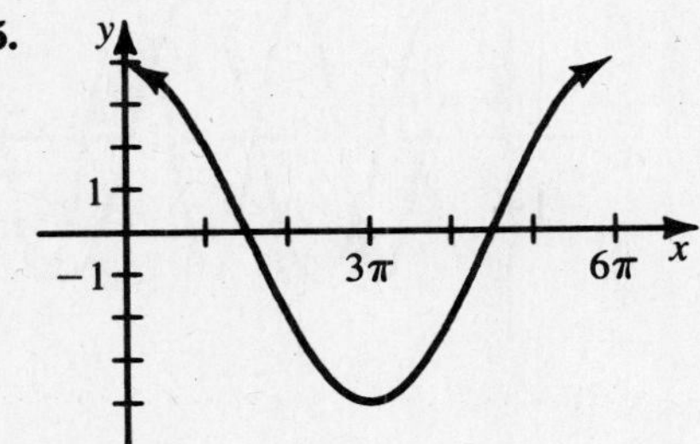

6.

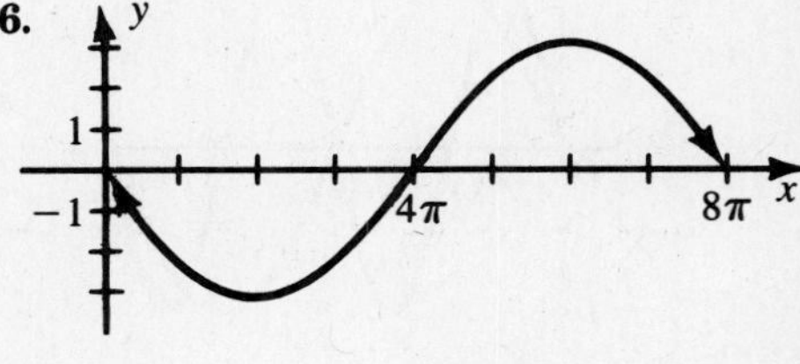

7.

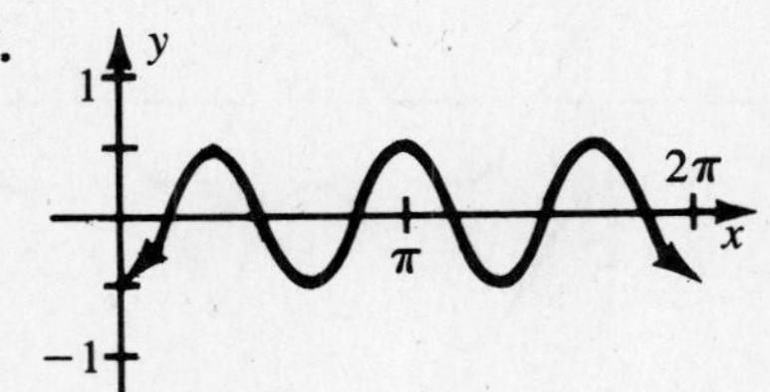

8.

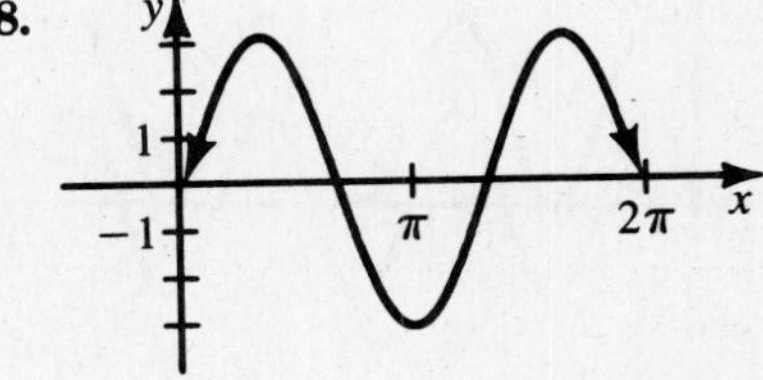

9.

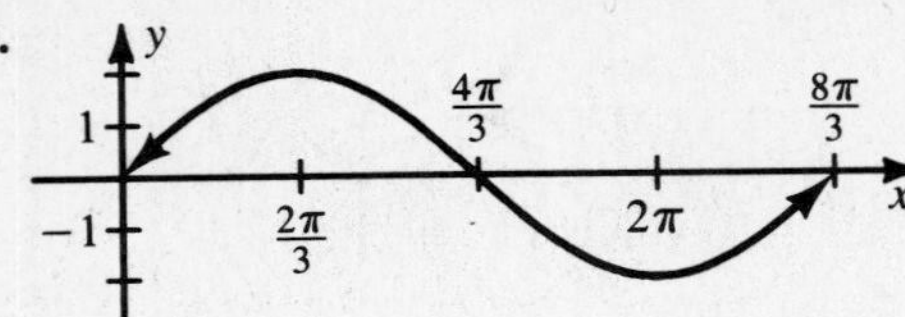

10.

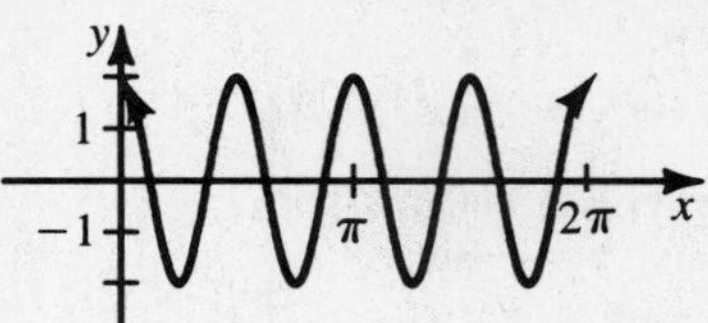

11.

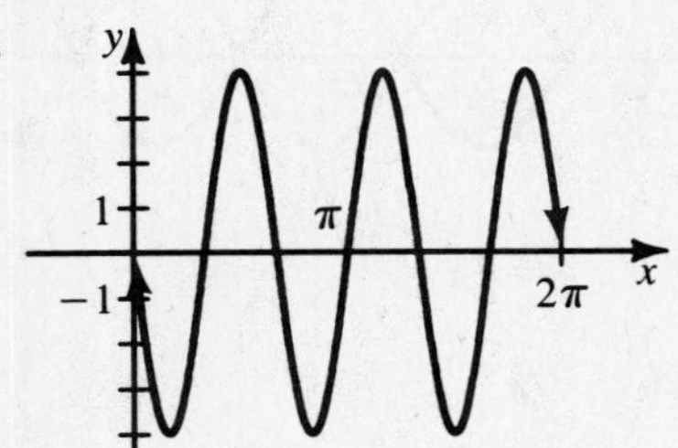

12.

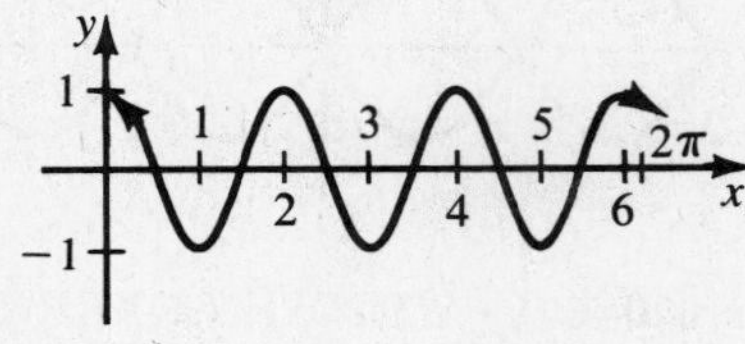

B 13.

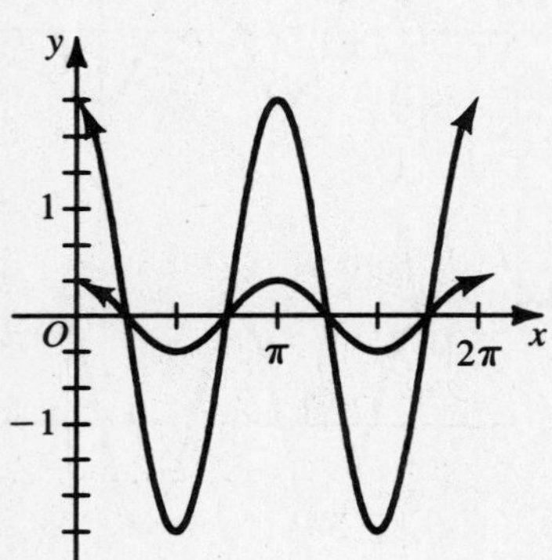

14.

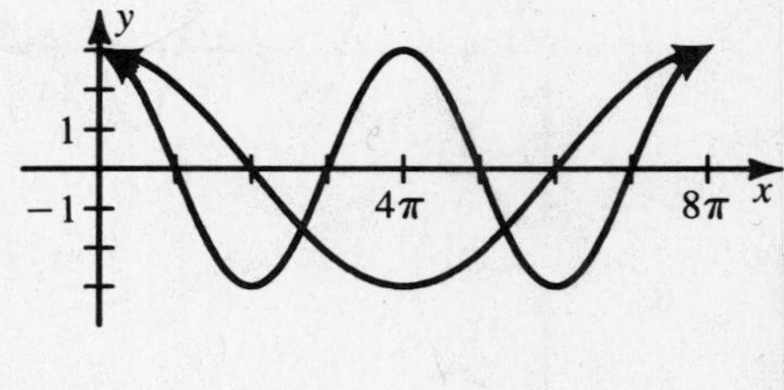

15.

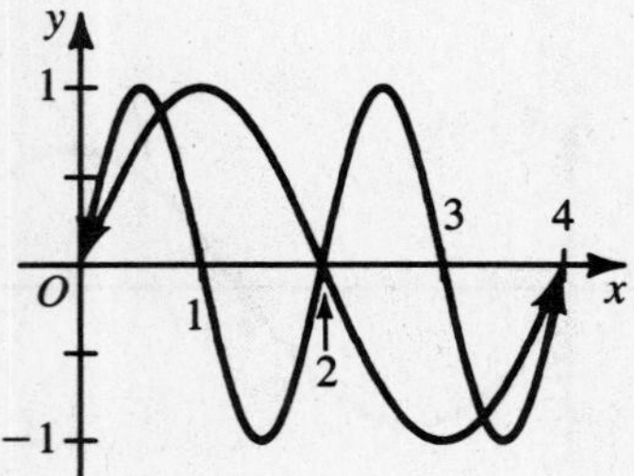

16.

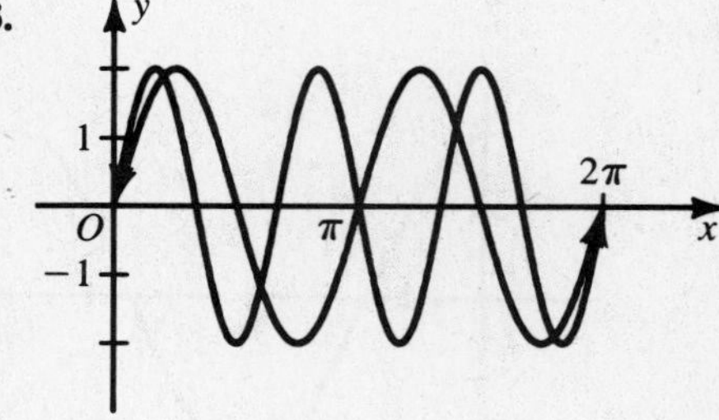

C 17.

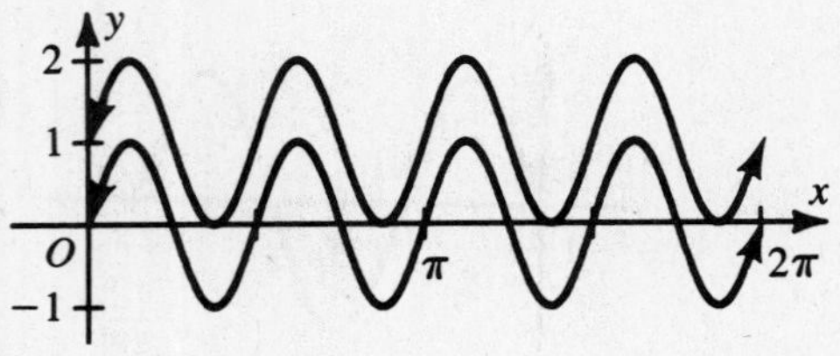

18.

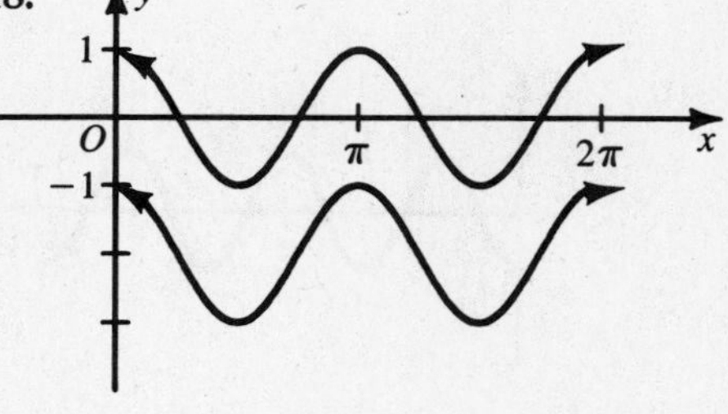

19.

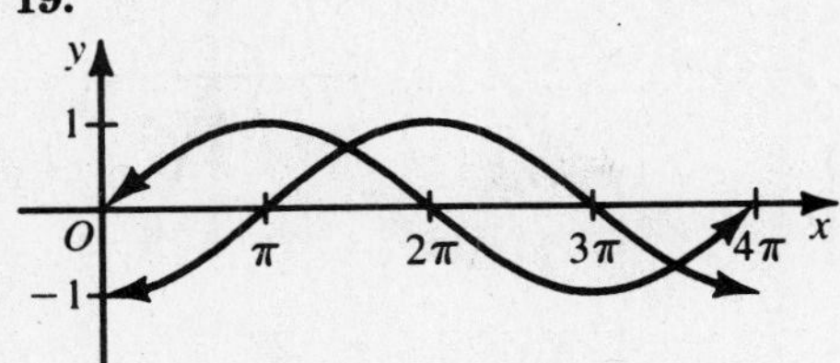

20.

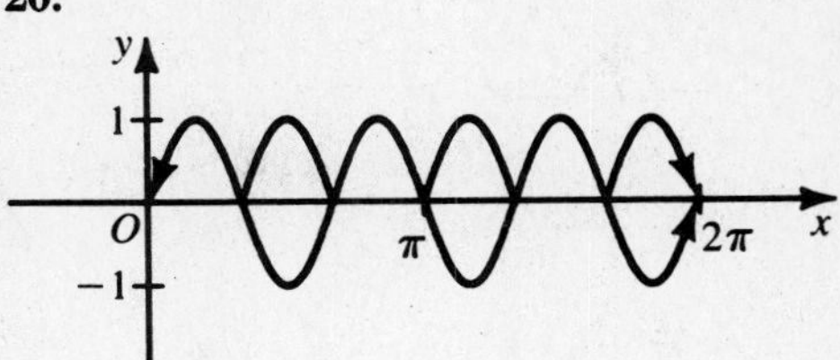

Page 601 • SELF-TEST 2

1. $r = \sqrt{(-4)^2 + (4\sqrt{3})^2} = \sqrt{16 + 48} = \sqrt{64} = 8$; $\sin \alpha = \frac{4\sqrt{3}}{8} = \frac{\sqrt{3}}{2}$;

   $\cos \alpha = -\frac{4}{8} = -\frac{1}{2}$

2. $\sin^2\alpha + \cos^2\alpha = 1$; $\left(-\frac{15}{17}\right)^2 + \cos^2\alpha = 1$; $\cos^2\alpha = 1 - \frac{225}{289} = \frac{64}{289}$; $\cos \alpha = \pm\frac{8}{17}$;

   since α is in Quad. III, $\cos \alpha = -\frac{8}{17}$.

3. $\sin 83.7° = 0.9940$
4. $\cos 242° = -\cos 62° = -0.4695$
5. $\sin 1.3^R = 0.9636$
6. $\cos(-7.8^R) = \cos 1.52^R = 0.0508$

7.

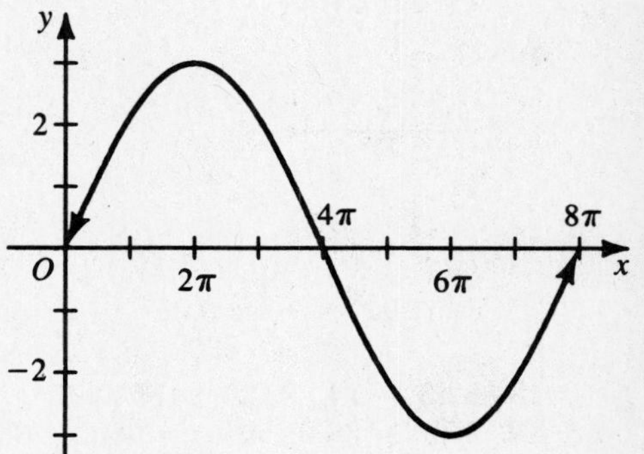

Pages 605–606 • WRITTEN EXERCISES

A 1.

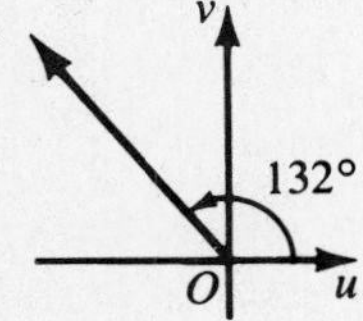

$\tan 132° = -\tan 48°$
$= -1.111$

2.

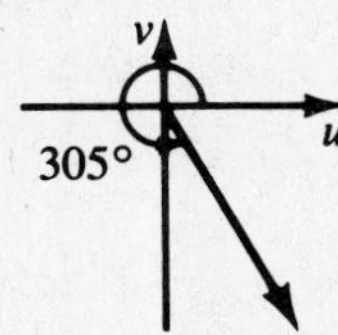

$\sec 305° = \sec 55°$
$= 1.743$

3.

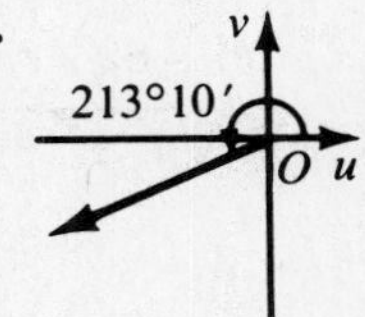

$\csc 213°10'$
$= -\csc 33°10'$
$= -1.828$

4.

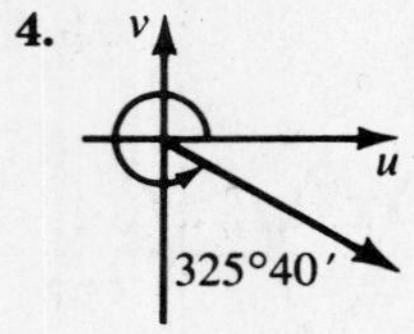

tan 325°40′
= −tan 34°20′
= −0.6830

5.

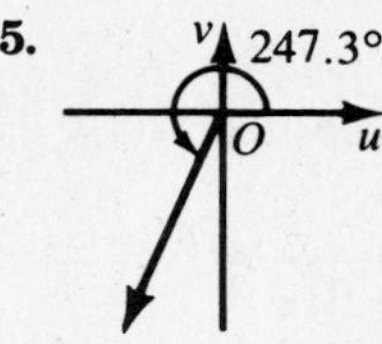

cot 247.3° = cot 67.3°
= 0.4183

6.

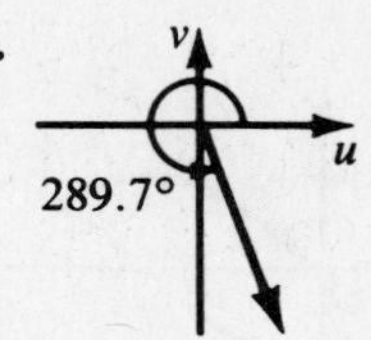

csc 289.7°
= −csc 70.3°
= −1.062

7.

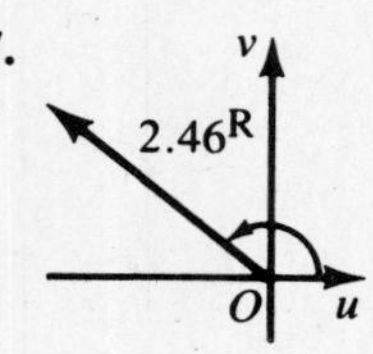

sec 2.46^R
$\approx -\sec 0.68^R$
$= -1.286$

8.

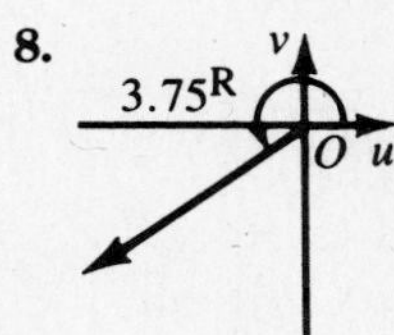

cot $3.75^R \approx$ cot 0.61^R
= 1.431

9.

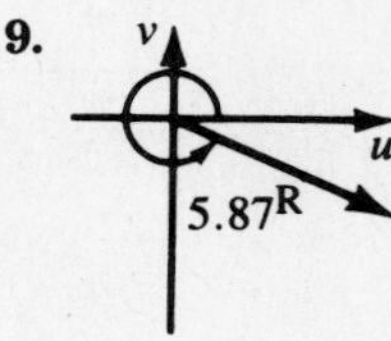

tan 5.87^R
$\approx -\tan 0.41^R$
$= -0.4346$

10.

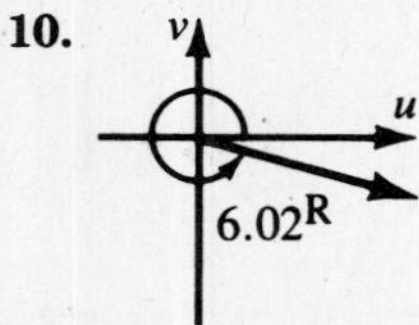

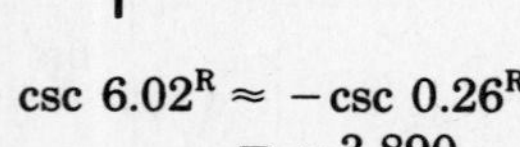

csc $6.02^R \approx -\csc 0.26^R$
$= -3.890$

11.

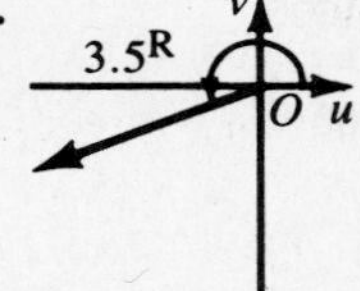

cot $3.5^R \approx$ cot 0.36^R
= 2.657

12.

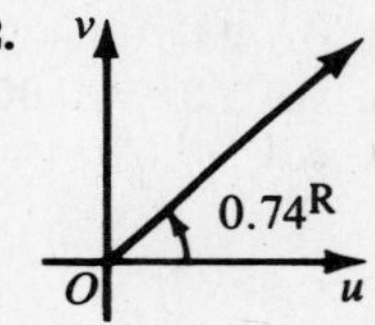

sec $0.74^R \approx 1.354$

13. 62.3° **14.** 22.2° **15.** 49.6° **16.** 78.7° **17.** 38.1° **18.** 12.9°

19. $m(\theta) = 36.0°$; $m(\alpha) = 360° - 36° = 324°$

20. $m(\theta) = 53.5°$; $m(\alpha) = 180° - 53.5° = 126.5°$

21. $m(\theta) = 25.5°$; $m(\alpha) = 180° + 25.5° = 205.5°$

22. $m(\theta) = 68.7°$; $m(\alpha) = 180° - 68.7° = 111.3°$

23. $m(\theta) \approx 22.9°$; $m(\alpha) \approx 360° - 22.9° = 337.1°$

24. $m(\theta) \approx 82.9°$; $m(\alpha) \approx 180° + 82.9° = 262.9°$

25. $r = \sqrt{6^2 + (-8)^2} = \sqrt{100} = 10$; $\sin \alpha = -\frac{4}{5}$; $\cos \alpha = \frac{3}{5}$; $\tan \alpha = -\frac{4}{3}$;

$\cot \alpha = -\frac{3}{4}$; $\sec \alpha = \frac{5}{3}$; $\csc \alpha = -\frac{5}{4}$

26. $r = \sqrt{(-5)^2 + 12^2} = \sqrt{169} = 13$; $\sin \alpha = \frac{12}{13}$; $\cos \alpha = -\frac{5}{13}$; $\tan \alpha = -\frac{12}{5}$;

$\cot \alpha = -\frac{5}{12}$; $\sec \alpha = -\frac{13}{5}$; $\csc \alpha = \frac{13}{12}$

27. $r = \sqrt{0^2 + (-3)^2} = 3$; $\sin \alpha = -\frac{3}{3} = -1$; $\cos \alpha = \frac{0}{3} = 0$; $\tan \alpha$ is not defined;

$\cot \alpha = 0$; $\sec \alpha$ is not defined; $\csc \alpha = -1$

28. $r = \sqrt{(-7)^2 + (-24)^2} = \sqrt{625} = 25$; $\sin \alpha = -\frac{24}{25}$; $\cos \alpha = -\frac{7}{25}$; $\tan \alpha = \frac{24}{7}$;

$\cot \alpha = \frac{7}{24}$; $\sec \alpha = -\frac{25}{7}$; $\csc \alpha = -\frac{25}{24}$

29. $r = \sqrt{(-2)^2 + 0^2} = 2$; $\sin\alpha = \frac{0}{2} = 0$; $\cos\alpha = \frac{-2}{2} = -1$; $\tan\alpha = \frac{0}{-1} = 0$; $\cot\alpha$ is not defined; $\sec\alpha = -1$; $\csc\alpha$ is not defined.

30. $r = \sqrt{3^2 + 6^2} = \sqrt{45} = 3\sqrt{5}$; $\sin\alpha = \frac{6}{3\sqrt{5}} = \frac{2}{\sqrt{5}}$; $\cos\alpha = \frac{3}{3\sqrt{5}} = \frac{1}{\sqrt{5}}$;

$\tan\alpha = \frac{6}{3} = 2$; $\cot\alpha = \frac{1}{2}$; $\sec\alpha = \sqrt{5}$; $\csc\alpha = \frac{\sqrt{5}}{2}$

31. $r = \sqrt{(2\sqrt{6})^2 + (-5)^2} = \sqrt{49} = 7$; $\sin\alpha = -\frac{5}{7}$; $\cos\alpha = \frac{2\sqrt{6}}{7}$; $\tan\alpha = -\frac{5}{2\sqrt{6}}$;

$\cot\alpha = -\frac{2\sqrt{6}}{5}$; $\sec\alpha = \frac{7}{2\sqrt{6}}$; $\csc\alpha = -\frac{7}{5}$

32. $r = \sqrt{(-4)^2 + (2\sqrt{5})^2} = \sqrt{36} = 6$; $\sin\alpha = \frac{2\sqrt{5}}{6} = \frac{\sqrt{5}}{3}$; $\cos\alpha = -\frac{4}{6} = -\frac{2}{3}$;

$\tan\alpha = \frac{2\sqrt{5}}{-4} = -\frac{\sqrt{5}}{2}$; $\cot\alpha = -\frac{2}{\sqrt{5}}$; $\sec\alpha = -\frac{3}{2}$; $\csc\alpha = \frac{3}{\sqrt{5}}$

33. $\sin\alpha = \frac{-\sqrt{3}}{2} = \frac{v}{r}$; $2^2 = (-\sqrt{3})^2 + u^2$; $u^2 = 4 - 3 = 1$; $u = \pm 1$; since α is in

Quad. IV, $u = 1$; $\cos\alpha = \frac{1}{2}$; $\tan\alpha = \frac{-\sqrt{3}}{1} = -\sqrt{3}$; $\cot\alpha = -\frac{1}{\sqrt{3}}$; $\sec\alpha = 2$;

$\csc\alpha = -\frac{2}{\sqrt{3}}$

34. $\sin\alpha = \frac{-1}{\sqrt{2}} = \frac{v}{r}$; $(\sqrt{2})^2 = (-1)^2 + u^2$; $u^2 = 2 - 1 = 1$; $u = \pm 1$; since α is in

Quad. III, $u = -1$; $\cos\alpha = -\frac{1}{\sqrt{2}}$; $\tan\alpha = 1$; $\cot\alpha = 1$; $\sec\alpha = -\sqrt{2}$;

$\csc\alpha = -\sqrt{2}$

35. $\sin\alpha = \frac{\sqrt{5}}{3} = \frac{v}{r}$; $3^2 = (\sqrt{5})^2 + u^2$; $u^2 = 9 - 5 = 4$; $u = \pm 2$; since α is in

Quad. II, $u = -2$; $\cos\alpha = -\frac{2}{3}$; $\tan\alpha = \frac{\sqrt{5}}{-2} = -\frac{\sqrt{5}}{2}$; $\cot\alpha = -\frac{2}{\sqrt{5}}$;

$\sec\alpha = -\frac{3}{2}$; $\csc\alpha = \frac{3}{\sqrt{5}}$

36. $\sec\alpha = \frac{3}{-2\sqrt{2}} = \frac{r}{u}$; $3^2 = (-2\sqrt{2})^2 + v^2$; $v^2 = 9 - 8 = 1$; $v = \pm 1$; since α is in

Quad. III, $v = -1$; $\sin\alpha = -\frac{1}{3}$; $\cos\alpha = -\frac{2\sqrt{2}}{3}$; $\tan\alpha = \frac{-1}{-2\sqrt{2}} = \frac{1}{2\sqrt{2}}$;

$\cot\alpha = 2\sqrt{2}$; $\csc\alpha = -3$

37. $\csc\alpha = \frac{7}{\sqrt{15}} = \frac{r}{v}$; $7^2 = (\sqrt{15})^2 + u^2$; $u^2 = 49 - 15 = 34$; $u = \pm\sqrt{34}$; since α is in Quad. II, $u = -\sqrt{34}$; $\sin\alpha = \frac{\sqrt{15}}{7}$; $\cos\alpha = -\frac{\sqrt{34}}{7}$; $\tan\alpha = \frac{\sqrt{15}}{-\sqrt{34}} = -\frac{\sqrt{510}}{34}$; $\cot\alpha = \frac{-\sqrt{34}}{\sqrt{15}} = -\frac{\sqrt{510}}{15}$; $\sec\alpha = -\frac{7}{\sqrt{34}}$

38. $\cos\alpha = \frac{2}{\sqrt{13}} = \frac{u}{r}$; $(\sqrt{13})^2 = 2^2 + v^2$; $v^2 = 13 - 4 = 9$; $v = \pm 3$; since α is in Quad. IV, $v = -3$; $\sin\alpha = -\frac{3}{\sqrt{13}}$; $\tan\alpha = \frac{-3}{2} = -\frac{3}{2}$; $\cot\alpha = -\frac{2}{3}$; $\sec\alpha = \frac{\sqrt{13}}{2}$; $\csc\alpha = -\frac{\sqrt{13}}{3}$

B 39. $\tan\alpha = \frac{1}{5} = \frac{u}{v}$, where $u = \sin\alpha$ and $v = \cos\alpha$; $v = 5u$ and $u^2 + v^2 = 1$; $(5u)^2 + u^2 = 1$; $26u^2 = 1$; $u^2 = \frac{1}{26}$, $u = \pm\frac{1}{\sqrt{26}}$; since α is in Quad. III, $u = -\frac{1}{\sqrt{26}}$ and $v = -\frac{5}{\sqrt{26}}$; $\sin\alpha = -\frac{1}{\sqrt{26}}$; $\cos\alpha = -\frac{5}{\sqrt{26}}$; $\cot\alpha = 5$; $\sec\alpha = -\frac{\sqrt{26}}{5}$; $\csc\alpha = -\sqrt{26}$

40. $\cot\alpha = -\frac{3\sqrt{5}}{2} = \frac{v}{u}$, where $u = \sin\alpha$ and $v = \cos\alpha$; $v = -\frac{3\sqrt{5}}{2}u$ and $u^2 + v^2 = 1$; $\left(-\frac{3\sqrt{5}}{2}u\right)^2 + u^2 = 1$; $45u^2 + 4u^2 = 4$; $u^2 = \frac{4}{49}$, $u = \pm\frac{2}{7}$ since α is in Quad. II, $u = \frac{2}{7}$ and $v = -\frac{3\sqrt{5}}{7}$; $\sin\alpha = \frac{2}{7}$; $\cos\alpha = -\frac{3\sqrt{5}}{7}$; $\tan\alpha = -\frac{2}{3\sqrt{5}}$; $\sec\alpha = -\frac{7}{3\sqrt{5}}$; $\csc\alpha = \frac{7}{2}$

41. $1 + \tan^2\alpha = 1 + \left(\frac{v}{u}\right)^2 = \frac{u^2 + v^2}{u^2} = \frac{r^2}{u^2} = \left(\frac{r}{u}\right)^2 = \sec^2\alpha$

42. $1 + \cot^2\alpha = 1 + \left(\frac{u}{v}\right)^2 = \frac{v^2 + u^2}{v^2} = \frac{r^2}{v^2} = \left(\frac{r}{v}\right)^2 = \csc^2\alpha$

43.

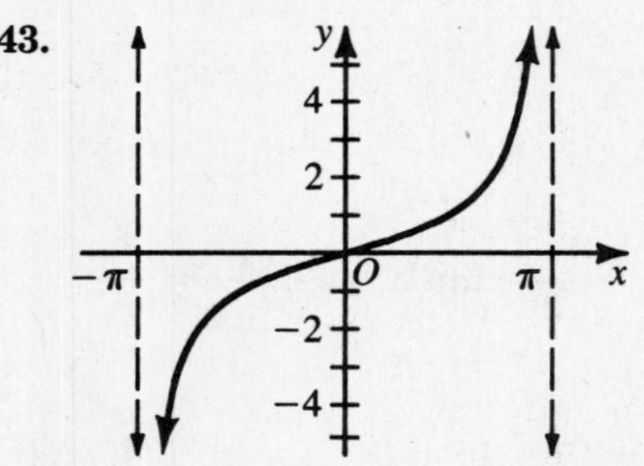

44.

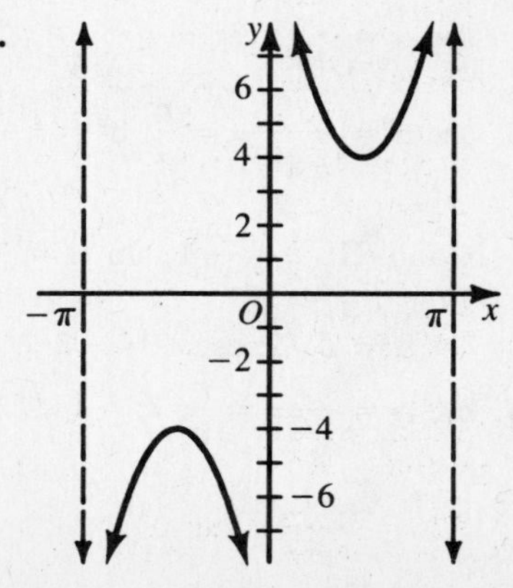

45.

46.

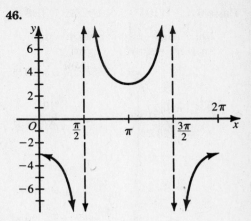

47.

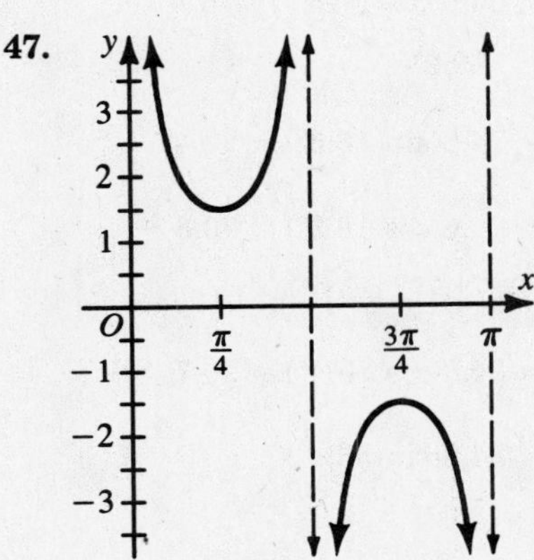

48.

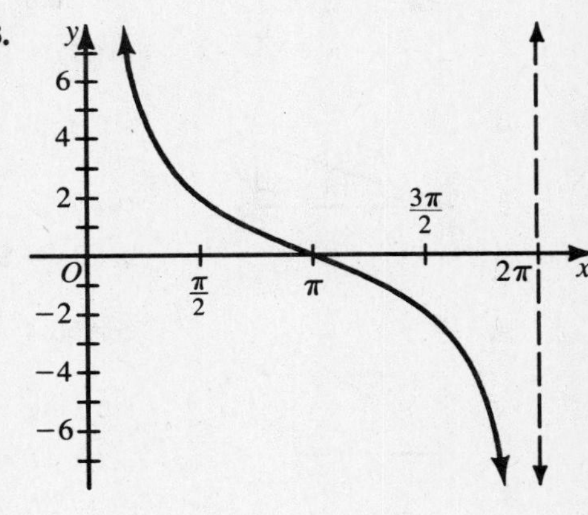

Note: For Exs. 49–52, $u = \cos \alpha$ and $v = \sin \alpha$.

C **49.** $\angle AOB \cong \angle AOB$; $m\angle DEO = 90° = m\angle ABO$; $\triangle ABO \sim \triangle PEO$ by AA

$$\frac{AB}{v} = \frac{OB}{u}$$

$$\frac{AB}{\sin \alpha} = \frac{1}{\cos \alpha}$$

$$AB = \frac{\sin \alpha}{\cos \alpha} = \tan \alpha$$

50. $\triangle ABO \sim \triangle PEO$ (See Ex. 49.)

$$\frac{AO}{PO} = \frac{OB}{u}$$

$$\frac{AO}{1} = \frac{1}{\cos \alpha}$$

$$AO = \sec \alpha$$

51. Since $\overline{DC}$ is tangent to the circle at C, $m\angle DCO = 90° = m\angle PEO$. Since $\overline{DC}$ and $\overline{EO}$ are both perpendicular to $\overline{CO}$, $\overline{DC} \parallel \overline{EO}$ and $m\angle CDO = m\angle EOP$. Thus $\triangle CDO \sim \triangle EOP$ by AA.

$$\frac{CD}{u} = \frac{CO}{v}$$

$$\frac{CD}{\cos \alpha} = \frac{1}{\sin \alpha}$$

$$CD = \frac{\cos \alpha}{\sin \alpha} = \cot \alpha$$

52. $\triangle CDO \sim \triangle EOP$ (See Ex. 51.)

$$\frac{OD}{PO} = \frac{CO}{v}$$

$$\frac{OD}{1} = \frac{1}{\sin \alpha}$$

$$OD = \csc \alpha$$

Page 610 • WRITTEN EXERCISES

Note: Answers given were obtained using a calculator. If a table is used, some answers may vary slightly.

A 1.

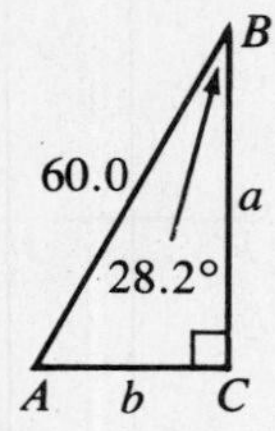

$\cos 28.2° = \frac{a}{60.0}$; $a = 60.0(\cos 28.2°) \approx 52.9$

$\sin 28.2° = \frac{b}{60.0}$; $b = 60.0(\sin 28.2°) \approx 28.4$

$m(A) = 90° - 28.2° = 61.8°$

2.

$\sin 20.8° = \frac{a}{50.0}$; $a = 50.0(\sin 20.8°) \approx 17.8$

$\cos 20.8° = \frac{b}{50.0}$; $b = 50.0(\cos 20.8°) \approx 46.7$

$m(B) = 90° - 20.8° = 69.2°$

3.

$\sin 18.3° = \frac{a}{74.6}$; $a = 74.6(\sin 18.3°) \approx 23.4$

$\cos 18.3° = \frac{b}{74.6}$; $b = 74.6(\cos 18.3°) \approx 70.8$

$m(B) = 90° - 18.3° = 71.7°$

4.

$\cos 53.4° = \frac{a}{34.8}$; $a = 34.8(\cos 53.4°) \approx 20.7$

$\sin 53.4° = \frac{b}{34.8}$; $b = 34.8(\sin 53.4°) \approx 27.9$

$m(A) = 90° - 53.4° = 36.6°$

5.

$\tan 32.7° = \frac{a}{25.0}$; $a = 25.0(\tan 32.7°) \approx 16.0$

$\cos 32.7° = \frac{25.0}{c}$; $c = \frac{25.0}{\cos 32.7°} \approx 29.7$

$m(B) = 90° - 32.7° = 57.3°$

6.

$\tan 71.7° = \frac{320}{a}$; $a = \frac{320}{\tan 71.7°} \approx 106$

$\sin 71.7° = \frac{320}{c}$; $c = \frac{320}{\sin 71.7°} \approx 337$

$m(A) = 90° - 71.7° = 18.3°$

7.

$\tan 56.1° = \frac{48.0}{b}$; $b = \frac{48.0}{\tan 56.1°} \approx 32.3$

$\sin 56.1° = \frac{48.0}{c}$; $c = \frac{48.0}{\sin 56.1°} \approx 57.8$

$m(B) = 90° - 56.1° = 33.9°$

8.

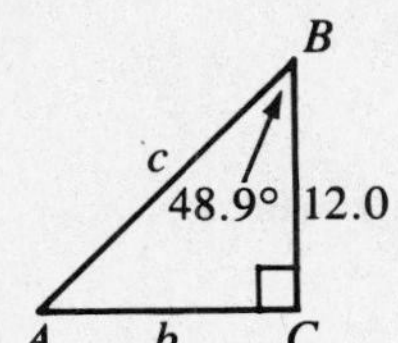

$\tan 48.9° = \dfrac{b}{12.0}$; $b = 12.0(\tan 48.9°) \approx 13.8$

$\cos 48.9° = \dfrac{12.0}{c}$; $c = \dfrac{12.0}{\cos 48.9°} \approx 18.3$

$m(A) = 90° - 48.9° = 41.1°$

9.

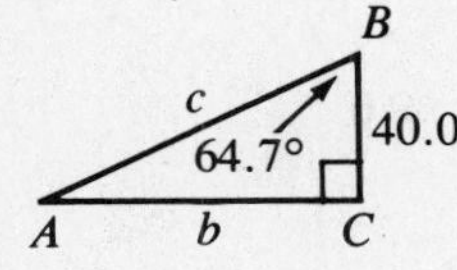

$\tan 64.7° = \dfrac{b}{40.0}$; $b = 40.0(\tan 64.7°) \approx 84.6$

$\cos 64.7° = \dfrac{40.0}{c}$; $c = \dfrac{40.0}{\cos 64.7°} \approx 93.6$

$m(A) = 90° - 64.7° = 25.3°$

10.

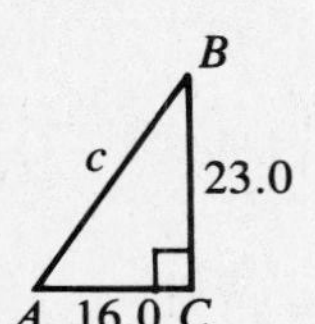

$\tan 39.2° = \dfrac{a}{225}$; $a = 225(\tan 39.2°) \approx 184$

$\cos 39.2° = \dfrac{225}{c}$; $c = \dfrac{225}{\cos 39.2°} \approx 290$

$m(B) = 90° - 39.2° = 50.8°$

11.

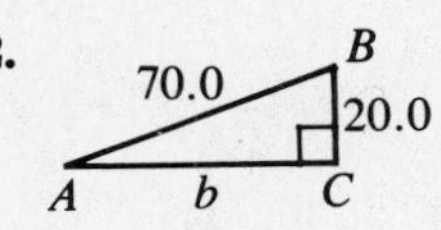

$\tan A = \dfrac{23.0}{16.0} = 1.4375$; $m(A) \approx 55.2°$

$m(B) \approx 90° - 55.2° = 44.8°$

$\sin 55.2° \approx \dfrac{23.0}{c}$; $c \approx \dfrac{23.0}{\sin 55.2°} \approx 28.0$

12.

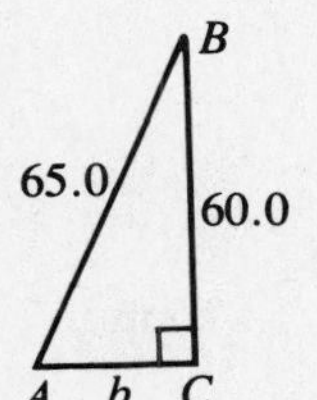

$\sin A = \dfrac{20.0}{70.0} \approx 0.2857$; $m(A) \approx 16.6°$

$m(B) \approx 90° - 16.6° = 73.4°$

$\tan 16.6° \approx \dfrac{20.0}{b}$; $b \approx \dfrac{20.0}{\tan 16.6°} \approx 67.1$

13.

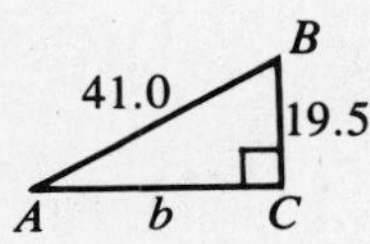

$\sin A = \dfrac{60.0}{65.0} \approx 0.9231$; $m(A) \approx 67.4°$

$m(B) \approx 90° - 67.4° = 22.6°$

$\tan 67.4° \approx \dfrac{60.0}{b}$; $b \approx \dfrac{60.0}{\tan 67.4°} \approx 25.0$

14.

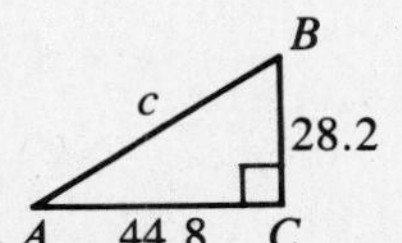

$\sin A = \dfrac{19.5}{41.0} \approx 0.4756$; $m(A) \approx 28.4°$

$m(B) \approx 90° - 28.4° = 61.6°$

$\cos 28.4° \approx \dfrac{b}{41.0}$; $b \approx 41.0(\cos 28.4°) \approx 36.1$

15.

B, c, 28.2, A, 44.8, C

$\tan A = \dfrac{28.2}{44.8} \approx 0.6295$; $m(A) \approx 32.2°$

$m(B) \approx 90° - 32.2° = 57.8°$

$\sin 32.2° \approx \dfrac{28.2}{c}$; $c \approx \dfrac{28.2}{\sin 32.2°} \approx 52.9$

16. $\cos A = \dfrac{11.4}{61.5} \approx 0.1854;\ m(A) \approx 79.3°$

$m(B) \approx 90° - 79.3° = 10.7°$

$\sin 79.3° \approx \dfrac{a}{61.5};\ a \approx 61.5(\sin 79.3°) \approx 60.4$

17. $\sin A = \dfrac{12.4}{25.1} \approx 0.4940;\ m(A) \approx 29.6°$

$m(B) \approx 90° - 29.6° = 60.4°$

$\cos 29.6° \approx \dfrac{b}{25.1};\ b \approx 25.1(\cos 29.6°) \approx 21.8$

18. $\tan A = \dfrac{103}{496} \approx 0.2077;\ m(A) \approx 11.7°$

$m(B) \approx 90° - 11.7° = 78.3°$

$\sin 11.7° \approx \dfrac{103}{c};\ c \approx \dfrac{103}{\sin 11.7°} \approx 508$

19. $\cos A = \dfrac{17.5}{29.3} \approx 0.5973;\ m(A) \approx 53.3°$

$m(B) \approx 90° - 53.3° = 36.7°$

$\sin 53.3° \approx \dfrac{a}{29.3};\ a \approx 29.3(\sin 53.3°) \approx 23.5$

Pages 610–612 • PROBLEMS

A **1.**

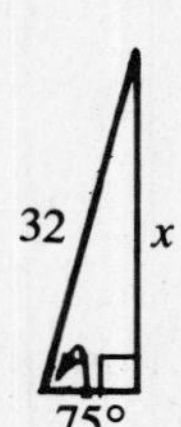

$\sin 75° = \dfrac{x}{32};\ x = 32(\sin 75°) \approx 30.9$; 30.9 ft

2. $\sin 22.5° = \dfrac{8.3}{x};\ x = \dfrac{8.3}{\sin 22.5°} \approx 21.7$; 21.7 km

3. $\tan A = \dfrac{17}{11} \approx 1.5455;\ m(A) \approx 57.1°$

4. $\tan 10.9° = \dfrac{156}{x};\ x = \dfrac{156}{\tan 10.9°} \approx 810$; 810 m

5.

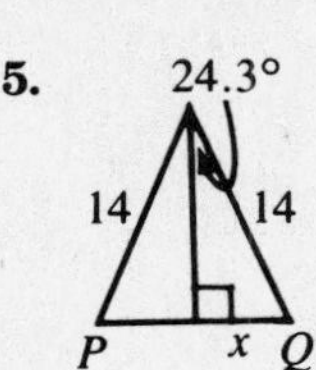

$\sin 24.3° = \frac{x}{14}$; $x = 14(\sin 24.3°) \approx 5.76$; $PQ = 2x \approx 11.5$

6.

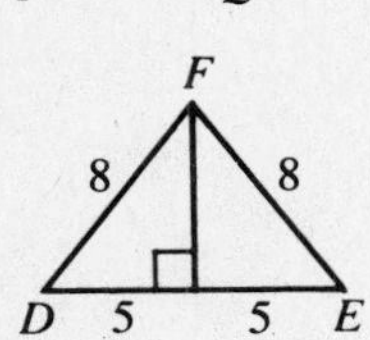

$\cos D = \frac{5}{8} = 0.625$; $m(D) \approx 51.3°$

7. $\sin 69° = \frac{13.1}{x}$; $x = \frac{13.1}{\sin 69°} \approx 14.0$; 14.0 m

B **8.**

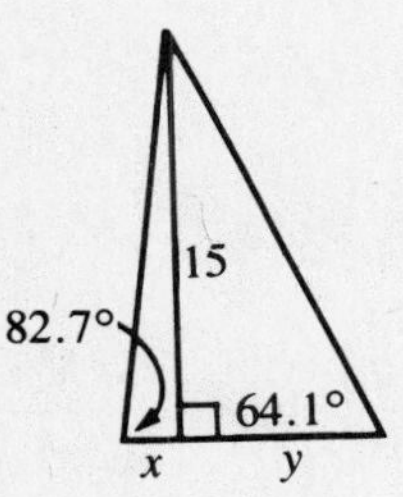

$\tan 82.7° = \frac{15}{x}$; $x = \frac{15}{\tan 82.7°} \approx 1.92$

$\tan 64.1° = \frac{15}{y}$; $y = \frac{15}{\tan 64.1°} \approx 7.28$

$x + y \approx 1.92 + 7.28 = 9.20$; 9.20 km

9.

$\sin 54.3° = \frac{y}{25}$; $y = 25(\sin 54.3°) \approx 20.3$

$\cos 54.3° = \frac{x}{25}$; $x = 25(\cos 54.3°) \approx 14.6$

$\tan 60.7° = \frac{z}{x}$; $z \approx 14.6(\tan 60.7°) \approx 26.0$

$z + y \approx 26.0 + 20.3 = 46.3$; 46.3 km

10.

$\tan 63.7° = \frac{x}{12.1}$; $x = 12.1(\tan 63.7°) \approx 24.5$

$\tan A = \frac{x}{8.5} \approx \frac{24.5}{8.5} \approx 2.882$; $m(A) \approx 70.9°$

11.

$\tan 11.3° = \frac{160}{BC}$; $BC = \frac{160}{\tan 11.3°} \approx 800.7$

$\tan 10.2° = \frac{160}{AC}$; $AC = \frac{160}{\tan 10.2°} \approx 889.2$

$AB = AC - BC \approx 889.2 - 800.7 \approx 89$; 89 m

C **12.**

$\tan 11.3° = \frac{y}{x}$; $0.1998x \approx y$

$\tan 10.2° \approx \frac{y}{15 + x}$; $0.1799(15 + x) \approx y$

$0.1998x \approx 2.6985 + 0.1799x$; $0.0199x \approx 2.6985$;

$x \approx 135.6$; $y \approx 0.1998(135.6) \approx 27.1$; 27.1 m

Page 612 • SELF-TEST 3

1. $r = \sqrt{(-3)^2 + (2\sqrt{10})^2} = \sqrt{49} = 7$; $\sin \alpha = \frac{2\sqrt{10}}{7}$; $\cos \alpha = -\frac{3}{7}$;
$\tan \alpha = -\frac{2\sqrt{10}}{3}$; $\cot \alpha = -\frac{3}{2\sqrt{10}}$; $\sec \alpha = -\frac{7}{3}$; $\csc \alpha = \frac{7}{2\sqrt{10}}$
2. $\tan 37.2° = \frac{7}{b}$; $b = \frac{7}{\tan 37.2°} \approx 9.22$
$\sin 37.2° = \frac{7}{c}$; $c = \frac{7}{\sin 37.2°} \approx 11.6$
$m(B) = 90° - 37.2° = 52.8°$

Pages 613–614 • CHAPTER REVIEW

1. $C \approx 2(3.2)(3.14) = 20.096$; $(20.096)(2.3) = 46.2$; a
2. $\frac{5\pi}{4} \cdot \frac{180}{\pi} = 225$; c
3. $-330 \cdot \frac{\pi}{180} = -\frac{11\pi}{6}$; b
4. $r = \sqrt{3^2 + (-4)^2} = \sqrt{25} = 5$; $\cos \alpha = \frac{3}{5}$; d
5. $\sin^2\alpha + \cos^2\alpha = 1$; $\sin^2\alpha + \left(\frac{3}{4}\right)^2 = 1$; $\sin^2\alpha = 1 - \frac{9}{16} = \frac{7}{16}$; $\sin \alpha = \pm\frac{\sqrt{7}}{4}$;
since $\sin \alpha > 0$, $\sin \alpha = \frac{\sqrt{7}}{4}$; d
6. $\sin \frac{19\pi^R}{6} = \sin\left(\frac{7\pi^R}{6} + 2\pi^R\right) = \sin\left(\frac{7\pi^R}{6}\right) = -\frac{1}{2}$; c
7. $r = \sqrt{(-\sqrt{6})^2 + (-\sqrt{6})^2} = \sqrt{12} = 2\sqrt{3}$; $\sin \alpha = \frac{-\sqrt{6}}{2\sqrt{3}} = -\frac{\sqrt{2}}{2}$; since α is in Quad. III, $m(\alpha) = 225° + k \cdot 360°$; d
8. $\cos 143.7° = -\cos 36.3° = -0.8059$; b
9. $m(\alpha) = 1.2^R$; a
10. amplitude is $|-3|$, or 3; c
11. fundamental period is $\frac{2\pi}{|5|}$, or $\frac{2\pi}{5}$; c
12. $\sin^2\alpha + \cos^2\alpha = 1$; $\sin^2\alpha + \left(-\frac{\sqrt{5}}{3}\right)^2 = 1$; $\sin^2\alpha = \frac{4}{9}$; $\sin \alpha = \pm\frac{2}{3}$; since $\sin \alpha > 0$,
$\sin \alpha = \frac{2}{3}$; $\tan \alpha = \frac{\frac{2}{3}}{-\frac{\sqrt{5}}{3}} = -\frac{2}{\sqrt{5}}$; a
13. $\tan B = \frac{AC}{BC}$; $\tan 42° = \frac{10}{BC}$; $BC = \frac{10}{\tan 42°} \approx 11.1$; b

Pages 614–615 • CHAPTER TEST

1. $\frac{4.2}{2.4} = 1.75 = C$; $1.75 \approx d(3.14)$; $d \approx 0.56$; 0.56 m

2. $C \approx 2 \cdot 25 \cdot 3.14 = 157$; $\frac{\frac{2\pi}{5}}{2\pi} = \frac{1}{5}$; $\frac{1}{5} \cdot 157 = 31.4$; 31.4 cm

3. $108 \cdot \frac{\pi}{180} = \frac{3\pi}{5}$; $\frac{3\pi^R}{5}$

4. $\frac{4\pi}{5} \cdot \frac{180}{\pi} = 144$; $144°$

5. $\sin^2\alpha + \cos^2\alpha = 1$; $\left(-\frac{15}{17}\right)^2 + \cos^2\alpha = 1$; $\cos^2\alpha = \frac{64}{289}$; $\cos\alpha = \pm\frac{8}{17}$; since $\cos\alpha > 0$, $\cos\alpha = \frac{8}{17}$.

6. Since α is in Quad. II, $m(\alpha) = \frac{5\pi^R}{6}$.

7. $r = \sqrt{(2\sqrt{3})^2 + (-2)^2} = \sqrt{16} = 4$; $\sin\alpha = -\frac{1}{2}$; since α is in Quad. IV, $m(\alpha) = 330° + k \cdot 360°$, or $\frac{11\pi^R}{6} + 2k\pi^R$

8. $\cos 672° = \cos 48° = 0.6691$

9. $\sin(-4.2^R) \approx \sin(4.2^R - 3.14^R) = \sin 1.06^R = 0.8724$

10. 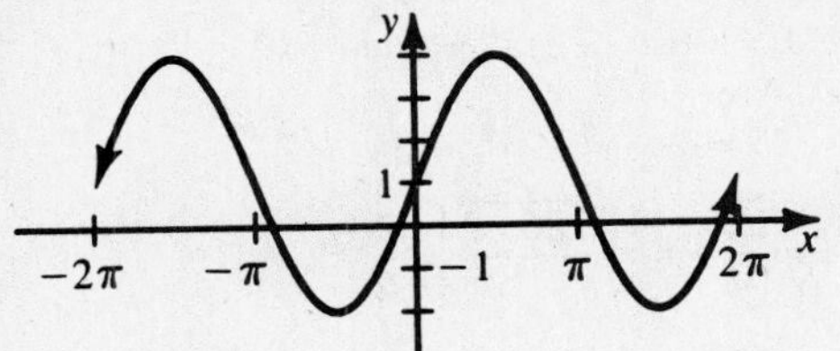

11. 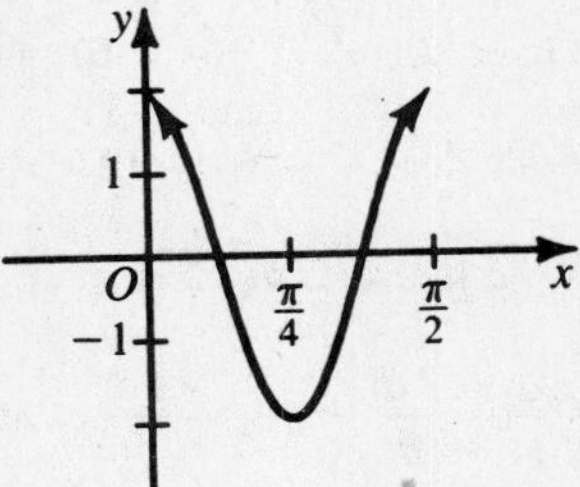

12. $\sin^2\alpha + \cos^2\alpha = 1$; $\left(\frac{1}{2}\right)^2 + \cos^2\alpha = 1$; $\cos^2\alpha = \frac{3}{4}$; $\cos\alpha = \pm\frac{\sqrt{3}}{2}$; since $\cos\alpha < 0$, $\cos\alpha = -\frac{\sqrt{3}}{2}$; $\tan\alpha = \frac{\frac{1}{2}}{-\frac{\sqrt{3}}{2}} = -\frac{1}{\sqrt{3}}$; $\cot\alpha = -\sqrt{3}$; $\sec\alpha = -\frac{2}{\sqrt{3}}$; $\csc\alpha = 2$

13. $\sin B = \frac{AC}{AB}$; $AC = 121(\sin 37.2°) \approx 73.2$

Page 615 • CONTEST PROBLEMS

1. Let x be the distance from point Q to where the girl should head in order to try to catch the boat, and let y be the distance to Q from the boat's position when the girl leaves her house. Let r be the girl's rate of walking; then $2r$ is the rate of the boat. The girl will travel $(\sqrt{1 + x^2})$ mi while the boat travels $(x + y)$ mi.

Since their times are the same,

$$\frac{\sqrt{1 + x^2}}{r} = \frac{x + y}{2r}, \quad \text{or} \quad 2\sqrt{1 + x^2} = x + y.$$

$$4(1 + x^2) = (x + y)^2$$
$$4 + 4x^2 = x^2 + 2xy + y^2$$
$$3x^2 - 2xy + 4 - y^2 = 0$$
$$x = \frac{2y \pm \sqrt{4y^2 - 12(4 - y^2)}}{6}$$
$$= \frac{y \pm 2\sqrt{y^2 - 3}}{3}$$

The discriminant must be nonnegative (if $y < \sqrt{3}$, it is impossible for the girl to catch the boat), so $y \geq \sqrt{3}$. If $y = \sqrt{3}$, then $x = \frac{\sqrt{3}}{3}$, or about 0.58. Thus, to maximize her chances of catching the boat, the girl should head for a point on the canal about 0.58 mi down from point Q. (If $y > \sqrt{3}$, then the boat was further upstream, and she will definitely catch the boat by heading to the point 0.58 mi from Q.)

Pages 615–616 • MIXED REVIEW

1. $\left(-\frac{1}{4}c + 7c\right) \div \left(\frac{-3}{4}\right) = \frac{27c}{4} \cdot \frac{4}{-3} = -9c$ 2. $\frac{-3x^6y^3}{x^2y^4} = -3x^4y^{-1}\left(\text{or } -\frac{3x^4}{y}\right)$
3. $(w - 3)(w^2 - 2w + 6) = w^3 - 2w^2 + 6w - 3w^2 + 6w - 18 =$ $w^3 - 5w^2 + 12w - 18$
4. $(-3y^3 + y^2 - 8) - (2y^2 + 3y - 15) = -3y^3 - y^2 - 3y + 7$
5. $\sqrt[3]{(-8)^4} = -8\sqrt[3]{-8} = 16$ 6. $\sqrt[3]{\frac{125}{4}} = \sqrt[3]{\frac{125 \cdot 2}{4 \cdot 2}} = \frac{5}{2}\sqrt[3]{2}$
7. $\frac{3d}{5} - \frac{7d}{5} = 16$; $-\frac{4d}{5} = 16$; $d = -20$; $\{-20\}$
8. $x^2 - 13x = 40$; $x^2 - 13x - 40 = 0$; $x = \frac{13 \pm \sqrt{169 + 160}}{2} = \frac{13 \pm \sqrt{329}}{2}$; $\left\{\frac{13 + \sqrt{329}}{2}, \frac{13 - \sqrt{329}}{2}\right\}$
9. $2y^2 + 5y - 3 < 0$; the related equation $2y^2 + 5y - 3 = 0$, or $(2y - 1)(y + 3) = 0$, has roots $\frac{1}{2}$ and -3. The inequality has solution set $\left\{y\colon -3 < y < \frac{1}{2}\right\}$.
10. $y - \sqrt{y + 5} = 7$; $y - 7 = \sqrt{y + 5}$; $y^2 - 14y + 49 = y + 5$; $y^2 - 15y + 44 = 0$; $(y - 11)(y - 4) = 0$; $y = 11$ or $y = 4$ *Check:* $11 - \sqrt{11 + 5} \stackrel{?}{=} 7$; $11 - 4 = 7$; $4 - \sqrt{4 + 5} \stackrel{?}{=} 7$; $4 - 3 \neq 7$; $\{11\}$
11. $y^2 - 6x^2 = 4$, $y = 3x^2 - 2$; $y^2 - 2(y + 2) = 4$; $y^2 - 2y - 4 = 4$; $y^2 - 2y - 8 = 0$; $(y - 4)(y + 2) = 0$; $y = 4$ or $y = -2$; if $y = 4$, then $4 = 3x^2 - 2$, or $x^2 = 2$, or $x = \pm\sqrt{2}$; if $y = -2$, then $-2 = 3x^2 - 2$, or $x^2 = 0$, or $x = 0$; $\{(\sqrt{2}, 4), (-\sqrt{2}, 4), (0, -2)\}$

12.
$$\begin{aligned} 3x + 5y + z &= 3 \\ -2x - y + 3z &= 7 \\ x + 4y - 2z &= -6 \end{aligned}$$

$$\begin{array}{r} 9x + 15y + 3z = 9 \\ -(-2x - y + 3z = 7) \\ \hline 11x + 16y = 2 \end{array} \qquad \begin{array}{r} 6x + 10y + 2z = 6 \\ x + 4y - 2z = -6 \\ \hline 7x + 14y = 0 \end{array}$$

$$\begin{array}{r} 77x + 112y = 14 \\ -(77x + 154y = 0) \\ \hline -42y = 14 \\ y = -\frac{1}{3} \end{array} \qquad \begin{aligned} 7x + 14\left(-\frac{1}{3}\right) &= 0 \\ 7x &= \frac{14}{3} \\ x &= \frac{2}{3} \end{aligned} \qquad \begin{aligned} \left(\frac{2}{3}\right) + 4\left(-\frac{1}{3}\right) - 2z &= -6 \\ -2z &= -5\frac{1}{3} \\ z &= \frac{8}{3} \end{aligned}$$

$$\left\{\left(\frac{2}{3}, -\frac{1}{3}, \frac{8}{3}\right)\right\}$$

13. $|3x + 2| > 1$

$$\begin{array}{rcr} 3x + 2 > 1 & \text{or} & -(3x + 2) > 1 \\ 3x > -1 & \text{or} & -3x > 3 \\ x > -\frac{1}{3} & \text{or} & x < -1 \end{array}$$

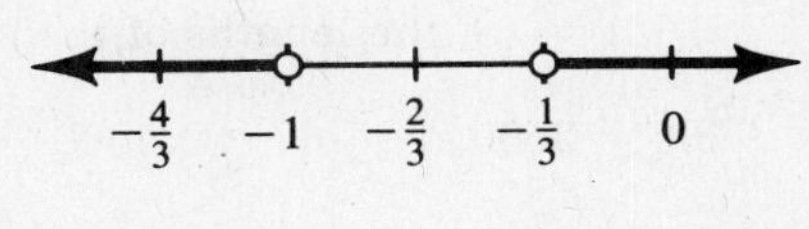

14.

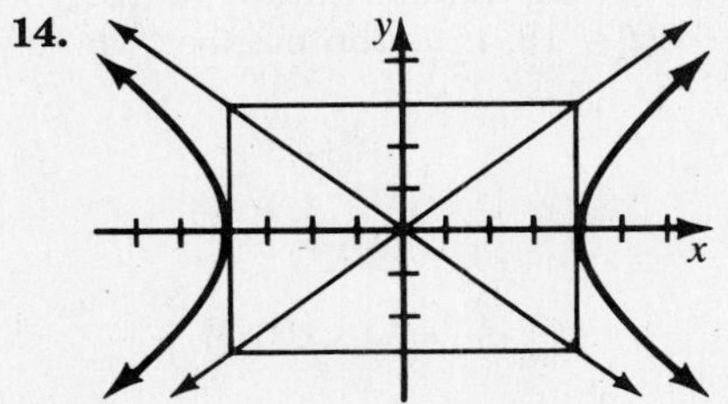

15.

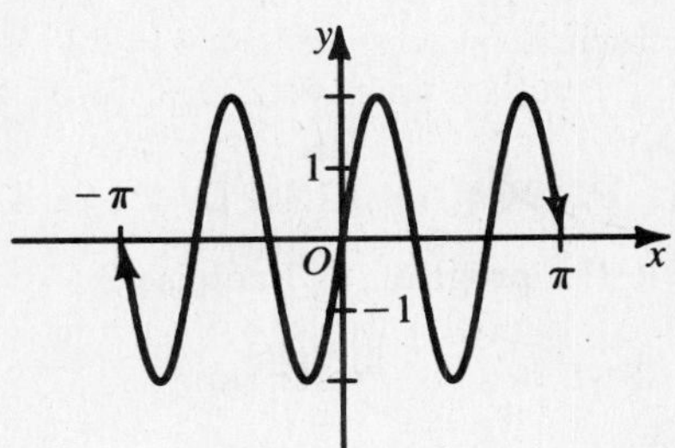

16. $m = \frac{5 + 1}{2 - 4} = \frac{6}{-2} = -3$; $-3 = \frac{y + 1}{x - 4}$; $-3x + 12 = y + 1$; $y = -3x + 11$

17. $x = -\frac{b}{2a} = \frac{4}{-2} = -2$; $x = -2$

18. This is an arithmetic series whose first term, a_1, is -1, common difference, d, is 2, and whose 9th term, a_9, is 15, so $S_9 = \frac{9}{2}(a_1 + a_9) = \frac{9}{2}(-1 + 15) = 63$.

19. This is a geometric series whose first term, a_1, is $\frac{1}{3}$, whose last term, a_8, is $\frac{128}{3}$, and whose common ratio, r, is 2, so $S_8 = \frac{\frac{1}{3} - 2 \cdot \frac{128}{3}}{1 - 2} = \frac{\frac{1}{3} - \frac{256}{3}}{-1} = \frac{255}{3} = 85$.

20. $\frac{y_1}{x_1 z_1} = \frac{y_2}{x_2 z_2}$; $\frac{30}{3 \cdot 15} = \frac{y_2}{5 \cdot 2}$; $y_2 = \frac{30 \cdot 5 \cdot 2}{3 \cdot 15} = \frac{20}{3}$

21. $\frac{5 \cdot 4}{1 \cdot 2}(2x)^3(3y)^2 = 10(8x^3)(9y^2) = 720x^3y^2$

22. $\cos \frac{13\pi^R}{3} = \cos\left(\frac{13\pi^R}{3} - 4\pi^R\right) = \cos \frac{\pi^R}{3} = \frac{1}{2}$

23. $\sin A = \frac{a}{c}$; $\frac{3}{4} = \frac{12}{c}$; $3c = 48$; $c = 16$; $b^2 = c^2 - a^2 = (16)^2 - (12)^2 = 256 - 144 = 112$; $b = \sqrt{112} \approx 10.6$

24. $\begin{bmatrix} 8 \\ -3 \end{bmatrix} = \begin{bmatrix} x \\ y \end{bmatrix} + \begin{bmatrix} 9 \\ -3 \end{bmatrix}$; $\begin{bmatrix} x \\ y \end{bmatrix} = \begin{bmatrix} 8 \\ -3 \end{bmatrix} - \begin{bmatrix} 9 \\ -3 \end{bmatrix} = \begin{bmatrix} -1 \\ 0 \end{bmatrix}$; $(-1, 0)$

25. $A + B = \begin{bmatrix} 1 & -2 \\ -2 & 4 \end{bmatrix}$, $A - B = \begin{bmatrix} 3 & 4 \\ -6 & -8 \end{bmatrix}$; $(A + B)(A - B) =$
$\begin{bmatrix} 1 & -2 \\ -2 & 4 \end{bmatrix}\begin{bmatrix} 3 & 4 \\ -6 & -8 \end{bmatrix} = \begin{bmatrix} 15 & 20 \\ -30 & -40 \end{bmatrix}$

26. Let x be the length of a side of one square and $2x - 1$ be the length of a side of the other square. $(2x - 1)^2 - x^2 = 385$; $4x^2 - 4x + 1 - x^2 = 385$;
$3x^2 - 4x - 384 = 0$; $(3x + 32)(x - 12) = 0$; $x = -\frac{32}{3}$ (reject) or $x = 12$;
$2x - 1 = 23$; the lengths of the sides of the squares are 12 and 23.

27. $7000 = 2500\left(1 + \frac{0.08}{4}\right)^{4t}$; $7000 = 2500(1.02)^{4t}$; $2.8 = (1.02)^{4t}$;
$4t \log 1.02 = \log 2.8$; $t = \frac{\log 2.8}{4 \log 1.02} = \frac{0.4472}{4(0.0086)} \approx 13$; 13 years

28. Let $t =$ the number of people who own both. By the fundamental counting principle on page 483, $400 = 230 + 185 - t$; $t = 15$; thus the number who own a car but no truck is $230 - 15$, or 215.

Page 617 • PROGRAMMING IN PASCAL

1. See the program in Exercise 2.
2.

```
PROGRAM solve_rt_triangle (INPUT, OUTPUT);

CONST
  pi = 3.141593;

VAR
  angle_a, angle_b, side_a, side_b, side_c : real;
  option : integer;

( ************************************************************ )
PROCEDURE display_menu;

BEGIN
  writeln('  WHAT IS YOUR OPTION');
  writeln('<1>  measure angle A and length of side a');
  writeln('<2>  measure angle A and length of side b');
  writeln('<3>  measure angle A and length of side c');
  writeln('<4>  measure angle B and length of side a');
  writeln('<5>  measure angle B and length of side b');
  writeln('<6>  measure angle B and length of side c');
  writeln('<7>  lengths of side a and side c');
  writeln('<8>  lengths of side a and side b');
  writeln('<9>  lengths of side b and side c');
  write('ENTER: ');
  readln(option);
  writeln;
END;
```

```
(*****************************************************************)
PROCEDURE enter_angle_and_side (VAR angle_1, side, angle_2 : real);

BEGIN
  write('Enter the measure of the angle: ');
  readln(angle_1);
  angle_2 := 90 - angle_1;
  write('Enter the length of the side: ');
  readln(side);
END;

(*****************************************************************)
PROCEDURE enter_side_and_hyp (VAR side_1, hyp, side_2 : real);

BEGIN
  write('Enter the length of the side: ');
  readln(side_1);
  side_2 := sqrt(hyp * hyp - side_1 * side_1);
  write('Enter the length of the hypotenuse: ');
  readln(hyp);
END;

(*****************************************************************)
PROCEDURE find_opp_and_hyp (ang_1, opp_1, ang_2 : real; VAR opp_2,
                                                        hyp : real);

BEGIN
  opp_2 := (opp_1 * sin((ang_2 * pi)/180))/(sin((ang_1 * pi)/180));
  hyp := sqrt(opp_1 * opp_1 + opp_2 * opp_2);
END;

(*****************************************************************)
PROCEDURE find_adj_sides (ang_1, hyp, ang_2 : real; VAR opp_1,
                                                     opp_2 : real);

BEGIN
  opp_1 := hyp * sin((ang_1 * pi)/180);
  opp_2 := hyp * sin((ang_2 * pi)/180);
END;

(*****************************************************************)
PROCEDURE find_acute_angles (side_1, side_2 : real; VAR ang_1,
                                                     ang_2 : real);

BEGIN
  ang_1 := 180 * arctan(side_1/side_2)/pi;
  ang_2 := 90 - ang_1;
END;
```

[*Program continued on next page*]

```
(**************************************************************)
BEGIN (* main *)
  display_menu;
    CASE option OF
         1 : BEGIN
               enter_angle_and_side(angle_a, side_a, angle_b);
               find_opp_and_hyp(angle_a, side_a, angle_b, side_b,
                                                          side_c);
             END;
         2 : BEGIN
               enter_angle_and_side(angle_a, side_b, angle_b);
               find_opp_and_hyp(angle_b, side_b, angle_a, side_a,
                                                          side_c);
             END;
         3 : BEGIN
               enter_angle_and_side(angle_a, side_c, angle_b);
               find_adj_sides(angle_a, side_c, angle_b, side_a,
                                                          side_b);
             END;
         4 : BEGIN
               enter_angle_and_side(angle_b, side_a, angle_a);
               find_opp_and_hyp(angle_a, side_a, angle_b, side_b,
                                                          side_c);
             END;
         5 : BEGIN
               enter_angle_and_side(angle_b, side_b, angle_a);
               find_opp_and_hyp(angle_b, side_b, angle_a, side_a,
                                                          side_c);
             END;
         6 : BEGIN
               enter_angle_and_side(angle_b, side_c, angle_a);
               find_adj_sides(angle_b, side_c, angle_a, side_b,
                                                          side_a);
             END;
         7 : BEGIN
               enter_side_and_hyp(side_a, side_c, side_b);
               find_acute_angles(side_a, side_b, angle_a, angle_b);
             END;
         8 : BEGIN
               write('Enter side a: ');
               readln(side_a);
               write('Enter side b: ');
               readln(side_b);
               side_c := sqrt(side_a * side_a + side_b * side_b);
               angle_a := 180 * arctan(side_a/side_b)/pi;
               angle_b := 90 - angle_a;
             END;
```

```
          9 : BEGIN
                enter_side_and_hyp(side_b, side_c, side_a);
                find_acute_angles(side_b, side_a, angle_b, angle_a);
              END;
    END; (* case *)
  writeln;
  writeln('ANGLE A = ', angle_a:1:3, '     SIDE A = ' , side_a:1:3);
  writeln('ANGLE B = ', angle_b:1:3, '     SIDE B = ' , side_b:1:3);
  writeln('ANGLE C = ', 90, '          SIDE C = ', side_c:1:3);
END.
```

See pages 610–611 for the answers to Written Exercises 1–19 in Section 14-10.

3.
```
PROGRAM sin_x_over_x (INPUT, OUTPUT);

VAR
  count, term : integer;
  x : real;

(**************************************************************)
BEGIN (* main *)
  write('Enter how many times (sin x)/x is to be evaluated: ');
  readln(count);
  writeln;
  x := 10;
  writeln('x':8, '(sin x)/x':16);
  writeln;
  FOR term := 1 TO count DO
      BEGIN
        x := 0.1 * x;
        writeln(x:1:11, '  ', (sin(x)/x):1:11);
      END;
END.
```

x	$\frac{\sin x}{x}$	x	$\frac{\sin x}{x}$
1	0.841470984	−1	0.841470984
0.1	0.998334166	−0.1	0.998334166
0.01	0.999983333	−0.01	0.999983333
0.001	0.999999833	−0.001	0.999999833
0.0001	0.999999998	−0.0001	0.999999998

It is unnecessary to compute $\frac{\sin x}{x}$ for the negative values because $\frac{\sin x}{x} = \frac{\sin(-x)}{-x}$.

4. a. See the program in Exercise 4b.

b.
```
PROGRAM taylor (INPUT, OUTPUT);

VAR
  x, sum : real;
  count, n : integer;
```

[*Program continued on next page*]

```
(***********************************************************)
FUNCTION power (x : real; n : integer) : real;

BEGIN
  IF n = 0
     THEN power := 1
     ELSE power := power(x, n - 1) * x;
END;

(***********************************************************)
FUNCTION factorial (n : integer) : real;

VAR
  factor : integer;
  prod : real;

BEGIN
  prod := 1;
  FOR factor := n DOWNTO 1 DO
      prod := prod * factor;
  factorial := prod;
END;

(***********************************************************)
BEGIN (* main *)
  write('Enter the value of x to be used: ');
  readln(x);
  write('Enter the number of terms to be evaluated: ');
  readln(count);
  writeln;
  sum := 0;
  FOR n := 1 TO count DO
      sum := sum - power(-1, n) * power(x, 2 * n - 1)/
                                           factorial(2 * n - 1);
  writeln('sin x = ', sin(x):1:11);
  write('x - x^3/3! + x^5/5! - x^7/7! + x^9/9! - ... = ');
  writeln(sum:1:11);
  writeln;
  sum := 0;
  FOR n := 1 TO count DO
      sum := sum - power(-1, n) * power(x, 2 * n - 2)/
                                           factorial(2 * n - 2);
  writeln('cos x = ', cos(x):1:11);
  write('1 - x^2/2! + x^4/4! - x^6/6! + x^8/8! - ... = ');
  writeln(sum:1:11);
END.
```

CHAPTER 15 • Trigonometric Identities

Pages 621–622 • WRITTEN EXERCISES

A 1. $\cos^2\alpha \tan^2\alpha = \cos^2\alpha \cdot \frac{\sin^2\alpha}{\cos^2\alpha} = \sin^2\alpha$; $\cos\alpha \neq 0$

2. $(1 - \sin^2\alpha)\cot\alpha = \cos^2\alpha \cdot \frac{\cos\alpha}{\sin\alpha} = \frac{\cos^3\alpha}{\sin\alpha}$; $\sin\alpha \neq 0$

3. $\tan^2\alpha \csc^2\alpha \cos\alpha = \frac{\sin^2\alpha}{\cos^2\alpha} \cdot \frac{1}{\sin^2\alpha} \cdot \cos\alpha = \frac{1}{\cos\alpha}$; $\sin\alpha \neq 0$, $\cos\alpha \neq 0$

4. $\tan\alpha(\cot\alpha + \tan\alpha) = \tan\alpha\cot\alpha + \tan^2\alpha = 1 + \tan^2\alpha = \sec^2\alpha = \frac{1}{\cos^2\alpha}$; $\sin\alpha \neq 0$, $\cos\alpha \neq 0$

5. $\sin\alpha\tan\alpha + \cos\alpha = \frac{\sin^2\alpha}{\cos\alpha} + \cos\alpha = \frac{\sin^2\alpha + \cos^2\alpha}{\cos\alpha} = \frac{1}{\cos\alpha}$; $\cos\alpha \neq 0$

6. $\cot\alpha(\sec^2\alpha - 1) = \cot\alpha\tan^2\alpha = \tan\alpha = \frac{\sin\alpha}{\cos\alpha}$; $\sin\alpha \neq 0$, $\cos\alpha \neq 0$

7. $\sec\alpha(\cos\alpha + \sin\alpha\tan\alpha) = \frac{1}{\cos\alpha}\left(\cos\alpha + \frac{\sin^2\alpha}{\cos\alpha}\right) = 1 + \frac{\sin^2\alpha}{\cos^2\alpha} =$
$1 + \tan^2\alpha = \sec^2\alpha = \frac{1}{\cos^2\alpha}$; $\cos\alpha \neq 0$

8. $(\csc\alpha - \sin\alpha)\sec^2\alpha = \left(\frac{1}{\sin\alpha} - \sin\alpha\right)\frac{1}{\cos^2\alpha} = \left(\frac{1 - \sin^2\alpha}{\sin\alpha}\right)\frac{1}{\cos^2\alpha} =$
$\frac{\cos^2\alpha}{\sin\alpha} \cdot \frac{1}{\cos^2\alpha} = \frac{1}{\sin\alpha}$; $\sin\alpha \neq 0$, $\cos\alpha \neq 0$

9. $\frac{1 + \tan^2\alpha}{\tan^2\alpha} = \frac{1}{\tan^2\alpha} + \frac{\tan^2\alpha}{\tan^2\alpha} = \cot^2\alpha + 1 = \csc^2\alpha = \frac{1}{\sin^2\alpha}$; $\sin\alpha \neq 0$, $\cos\alpha \neq 0$

10. $\frac{1 - \sin^2\alpha}{\cot\alpha} = \frac{\cos^2\alpha}{\frac{\cos\alpha}{\sin\alpha}} = \cos\alpha\sin\alpha$; $\sin\alpha \neq 0$, $\cos\alpha \neq 0$

11. $\frac{\sec^2 x - 1}{1 - \cos^2 x} = \frac{\tan^2 x}{\sin^2 x} = \frac{\sin^2 x}{\cos^2 x} \cdot \frac{1}{\sin^2 x} = \frac{1}{\cos^2 x}$; $\sin x \neq 0$, $\cos x \neq 0$

12. $\frac{1 + \cot^2 x}{\cot x \csc x} = \frac{\csc^2 x}{\cot x \csc x} = \frac{\csc x}{\cot x} = \frac{1}{\sin x} \cdot \frac{\sin x}{\cos x} = \frac{1}{\cos x}$; $\sin x \neq 0$, $\cos x \neq 0$

13. $\cos^2 x \csc x + \sin x = \frac{\cos^2 x}{\sin x} + \sin x = \frac{\cos^2 x + \sin^2 x}{\sin x} = \frac{1}{\sin x}$; $\sin x \neq 0$

14. $\frac{\sec^2 x - \tan^2 x}{\tan x \csc x} = \frac{1}{\tan x \csc x} = \frac{1}{\frac{\sin x}{\cos x} \cdot \frac{1}{\sin x}} = \cos x$; $\sin x \neq 0$, $\cos x \neq 0$

15. $(\sec x - \cos x)\cot^2 x = \left(\frac{1}{\cos x} - \cos x\right)\cot^2 x = \left(\frac{1 - \cos^2 x}{\cos x}\right)\cot^2 x =$
$\left(\frac{\sin^2 x}{\cos x}\right) \cdot \frac{\cos^2 x}{\sin^2 x} = \cos x$; $\sin x \neq 0$, $\cos x \neq 0$

16. $\cos x(\sin x + \cos x \cot x) = \cos x\left(\frac{\sin^2 x}{\sin x} + \frac{\cos^2 x}{\sin x}\right) = \cos x\left(\frac{1}{\sin x}\right) = \frac{\cos x}{\sin x}$; $\sin x \neq 0$

17. $\frac{\cos x + \sin x \tan x}{1 + \tan^2 x} = \frac{\cos x + \frac{\sin^2 x}{\cos x}}{\sec^2 x} = \frac{\frac{\cos^2 x + \sin^2 x}{\cos x}}{\frac{1}{\cos^2 x}} = \cos x$; $\cos x \neq 0$

18. $\tan x(1 - \sin x)(1 + \sin x) = \tan x(1 - \sin^2 x) = \frac{\sin x}{\cos x} \cdot \cos^2 x = \sin x \cos x;$
$\cos x \neq 0$

B 19. $\frac{\sin x}{1 + \cos x} - \frac{\sin x}{1 - \cos x} = \frac{\sin x(1 - \cos x) - \sin x(1 + \cos x)}{(1 - \cos^2 x)} =$
$\frac{\sin x - \sin x \cos x - \sin x \cos x - \sin x}{\sin^2 x} = \frac{-2 \sin x \cos x}{\sin^2 x} = \frac{-2 \cos x}{\sin x} = -2 \cot x;$
$\sin x \neq 0$

20. $\frac{1}{\sin x \cos x} - \tan x = \frac{1}{\sin x \cos x} - \frac{\sin^2 x}{\sin x \cos x} = \frac{\cos^2 x}{\sin x \cos x} = \frac{\cos x}{\sin x} = \cot x;$
$\sin x \neq 0, \cos x \neq 0$

21. $\frac{\sec^2\alpha - \tan^2\alpha}{1 - \sin^2\alpha} = \frac{1}{\cos^2\alpha} = \sec^2\alpha; \cos \alpha \neq 0$

22. $\frac{\csc^2\alpha - \cot^2\alpha}{1 + \tan^2\alpha} = \frac{1}{\sec^2\alpha} = \cos^2\alpha; \sin \alpha \neq 0, \cos \alpha \neq 0$

23. $\sec x \csc x - \cot x = \frac{1}{\sin x \cos x} - \frac{\cos^2 x}{\sin x \cos x} = \frac{\sin^2 x}{\sin x \cos x} = \frac{\sin x}{\cos x} = \tan x;$
$\sin x \neq 0, \cos x \neq 0$

24. $\sin x(\tan x + \cot x) = \sin x\left(\frac{\sin x}{\cos x} + \frac{\cos x}{\sin x}\right) = \sin x\left(\frac{\sin^2 x + \cos^2 x}{\cos x \sin x}\right) =$
$\frac{1}{\cos x} = \sec x; \sin x \neq 0, \cos x \neq 0$

25. $\frac{\tan x}{1 - \cos x} - \frac{\tan x}{1 + \cos x} = \tan x\left(\frac{(1 + \cos x) - (1 - \cos x)}{1 - \cos^2 x}\right) = \frac{\sin x}{\cos x}\left(\frac{2 \cos x}{\sin^2 x}\right) =$
$\frac{2}{\sin x} = 2 \csc x; \sin x \neq 0, \cos x \neq 0$

26. $\frac{\cos \alpha}{\sec \alpha + 1} + \frac{\cos \alpha}{\sec \alpha - 1} = \cos \alpha\left(\frac{(\sec \alpha - 1) + (\sec \alpha + 1)}{\sec^2\alpha - 1}\right) = \cos \alpha\left(\frac{2 \sec \alpha}{\tan^2\alpha}\right) =$
$\frac{2}{\tan^2\alpha} = 2 \cot^2\alpha; \sin \alpha \neq 0, \cos \alpha \neq 0$

27. $\frac{(\tan \alpha + 1)^2}{\sec \alpha} - \sec \alpha = \frac{\tan^2\alpha + 2 \tan \alpha + 1 - \sec^2\alpha}{\sec \alpha} =$
$\frac{\sec^2\alpha + 2 \tan \alpha - \sec^2\alpha}{\sec \alpha} = \frac{2 \tan \alpha}{\sec \alpha} = \frac{2\frac{\sin \alpha}{\cos \alpha}}{\frac{1}{\cos \alpha}} = 2 \sin \alpha; \cos \alpha \neq 0$

28. $\frac{1 + \tan x}{\sin x + \cos x} = \frac{\left(1 + \frac{\sin x}{\cos x}\right)\cos x}{(\sin x + \cos x)\cos x} = \frac{\cos x + \sin x}{(\sin x + \cos x)\cos x} = \frac{1}{\cos x} = \sec x;$
$\cos x \neq 0, \sin x \neq -\cos x$

29. $\frac{\sin \alpha}{1 + \cos \alpha} + \cot \alpha = \frac{\sin \alpha}{1 + \cos \alpha} + \frac{\cos \alpha}{\sin \alpha} = \frac{\sin^2\alpha + \cos \alpha(1 + \cos \alpha)}{\sin \alpha(1 + \cos \alpha)} =$
$\frac{\sin^2\alpha + \cos \alpha + \cos^2\alpha}{\sin \alpha(1 + \cos \alpha)} = \frac{1 + \cos \alpha}{\sin \alpha(1 + \cos \alpha)} = \frac{1}{\sin \alpha} = \csc \alpha; \sin \alpha \neq 0$

30. $(\sec x + 1)(\csc x - \cot x) = \left(\frac{1}{\cos x} + 1\right)\left(\frac{1}{\sin x} - \frac{\cos x}{\sin x}\right) = \left(\frac{1 + \cos x}{\cos x}\right)\left(\frac{1 - \cos x}{\sin x}\right) =$
$\frac{1 - \cos^2 x}{\cos x \sin x} = \frac{\sin^2 x}{\cos x \sin x} = \frac{\sin x}{\cos x} = \tan x; \sin x \neq 0, \cos x \neq 0$

31. $\dfrac{\cos x}{1 + \sin x} + \dfrac{1 + \sin x}{\cos x} = \dfrac{\cos^2 x + (1 + \sin x)^2}{\cos x(1 + \sin x)} = \dfrac{\cos^2 x + 1 + 2 \sin x + \sin^2 x}{\cos x(1 + \sin x)} =$
$\dfrac{2 + 2 \sin x}{\cos x(1 + \sin x)} = \dfrac{2(1 + \sin x)}{\cos x(1 + \sin x)} = \dfrac{2}{\cos x} = 2 \sec x$; $\cos x \neq 0$

32. $\dfrac{\tan \alpha \csc \alpha}{\cos \alpha} - \dfrac{\sec \alpha - \cos \alpha}{\sin \alpha \tan \alpha} = \dfrac{\frac{\sin \alpha}{\cos \alpha} \cdot \frac{1}{\sin \alpha}}{\cos \alpha} - \dfrac{\frac{1}{\cos \alpha} - \cos \alpha}{\frac{\sin^2 \alpha}{\cos \alpha}} =$

$\dfrac{1}{\cos^2 \alpha} - \dfrac{1 - \cos^2 \alpha}{\sin^2 \alpha} = \sec^2 \alpha - \dfrac{\sin^2 \alpha}{\sin^2 \alpha} = \sec^2 \alpha - 1 = \tan^2 \alpha$; $\sin \alpha \neq 0$, $\cos \alpha \neq 0$

C 33. $\cos \alpha = \dfrac{1}{\sec \alpha}$ and $\sec \alpha = -\sqrt{\tan^2 \alpha + 1}$ in Quadrant II; $\cos \alpha = -\dfrac{1}{\sqrt{\tan^2 \alpha + 1}}$

34. $\csc^2 \alpha = 1 + \cot^2 \alpha = 1 + \dfrac{1}{\tan^2 \alpha} = \dfrac{\tan^2 \alpha + 1}{\tan^2 \alpha}$; since $\csc \alpha$ is negative and $\tan \alpha$ is positive in Quadrant III, $\csc \alpha = -\sqrt{\dfrac{\tan^2 \alpha + 1}{\tan^2 \alpha}} = -\dfrac{\sqrt{\tan^2 \alpha + 1}}{\tan \alpha}$.

35. $\cos \alpha = \sqrt{1 - \sin^2 \alpha} = \sqrt{1 - \left(\dfrac{1}{r}\right)^2} = \dfrac{\sqrt{r^2 - 1}}{r}$; $\tan \alpha = \dfrac{\sin \alpha}{\cos \alpha} = \dfrac{1}{r} \div \dfrac{\sqrt{r^2 - 1}}{r} =$

$\dfrac{1}{r} \cdot \dfrac{r}{\sqrt{r^2 - 1}} = \dfrac{1}{\sqrt{r^2 - 1}}$

Pages 623–624 • WRITTEN EXERCISES

A 1. $\cos \alpha(\sec \alpha - \cos \alpha) = \cos \alpha\left(\dfrac{1}{\cos \alpha} - \cos \alpha\right) = \dfrac{\cos \alpha}{\cos \alpha} - \cos^2 \alpha = 1 - \cos^2 \alpha =$
$\sin^2 \alpha$; $\cos \alpha \neq 0$

2. $\csc \alpha - \sin \alpha = \dfrac{1}{\sin \alpha} - \sin \alpha\left(\dfrac{\sin \alpha}{\sin \alpha}\right) = \dfrac{1 - \sin^2 \alpha}{\sin \alpha} = \dfrac{\cos^2 \alpha}{\sin \alpha} = \cos \alpha \cdot \dfrac{\cos \alpha}{\sin \alpha} =$
$\cos \alpha \cot \alpha$; $\sin \alpha \neq 0$

3. $\tan \alpha + \cot \alpha = \dfrac{\sin \alpha}{\cos \alpha}\left(\dfrac{\sin \alpha}{\sin \alpha}\right) + \dfrac{\cos \alpha}{\sin \alpha}\left(\dfrac{\cos \alpha}{\cos \alpha}\right) = \dfrac{\sin^2 \alpha + \cos^2 \alpha}{\sin \alpha \cos \alpha} = \dfrac{1}{\cos \alpha \sin \alpha} =$
$\dfrac{1}{\cos \alpha} \cdot \dfrac{1}{\sin \alpha} = \sec \alpha \csc \alpha$; $\cos \alpha \neq 0$, $\sin \alpha \neq 0$

4. $\cos x + \sin x \tan x = \cos x\left(\dfrac{\cos x}{\cos x}\right) + \sin x \cdot \dfrac{\sin x}{\cos x} = \dfrac{\cos^2 x}{\cos x} + \dfrac{\sin^2 x}{\cos x} =$
$\dfrac{\cos^2 x + \sin^2 x}{\cos x} = \dfrac{1}{\cos x} = \sec x$; $\cos x \neq 0$

5. $\cos x(\sec x + \cos x \csc^2 x) = \cos x\left(\dfrac{1}{\cos x} + \cos x \cdot \dfrac{1}{\sin^2 x}\right) = 1 + \dfrac{\cos^2 x}{\sin^2 x} =$
$1 + \cot^2 x = \csc^2 x$; $\cos x \neq 0$, $\sin x \neq 0$

6. $\dfrac{1 + \tan^2 x}{1 + \cot^2 x} = \dfrac{\sec^2 x}{\csc^2 x} = \dfrac{\frac{1}{\cos^2 x}}{\frac{1}{\sin^2 x}} = \dfrac{\sin^2 x}{\cos^2 x} = \tan^2 x$; $\cos x \neq 0$, $\sin x \neq 0$

7. $\dfrac{(\cos\alpha - \sin\alpha)^2}{\sin\alpha} = \dfrac{\cos^2\alpha - 2\cos\alpha\sin\alpha + \sin^2\alpha}{\sin\alpha} = \dfrac{1 - 2\cos\alpha\sin\alpha}{\sin\alpha} =$
$\dfrac{1}{\sin\alpha} - 2\cos\alpha = \csc\alpha - 2\cos\alpha$; $\sin\alpha \neq 0$

8. $\dfrac{\cos x}{1 - \cos x} - \dfrac{\cos x}{1 + \cos x} = \dfrac{\cos x(1 + \cos x)}{(1 - \cos x)(1 + \cos x)} - \dfrac{\cos x(1 - \cos x)}{(1 + \cos x)(1 - \cos x)} =$
$\dfrac{2\cos x\cos x}{1 - \cos^2 x} = \dfrac{2\cos^2 x}{\sin^2 x} = \dfrac{2\cos^2 x}{\sin^2 x} = 2\cot^2 x$; $\sin x \neq 0$

9. $\dfrac{\sin\alpha}{\sec\alpha - 1} + \dfrac{\sin\alpha}{\sec\alpha + 1} = \dfrac{\sin\alpha(\sec\alpha + 1)}{(\sec\alpha - 1)(\sec\alpha + 1)} + \dfrac{\sin\alpha(\sec\alpha - 1)}{(\sec\alpha + 1)(\sec\alpha - 1)} =$
$\dfrac{2\sin\alpha\sec\alpha}{\sec^2\alpha - 1} = \dfrac{2\sin\alpha\cdot\dfrac{1}{\cos\alpha}}{\tan^2\alpha} = \dfrac{2\dfrac{\sin\alpha}{\cos\alpha}}{\tan^2\alpha} = \dfrac{2\tan\alpha}{\tan^2\alpha} = \dfrac{2}{\tan\alpha} = 2\cot\alpha$; $\cos\alpha \neq 0$;
$\sin\alpha \neq 0$

10. $\sec\alpha(\csc\alpha - \cot\alpha\cos\alpha) = \dfrac{1}{\cos\alpha}\cdot\dfrac{1}{\sin\alpha} - \dfrac{1}{\cos\alpha}\cdot\cot\alpha\cos\alpha = \dfrac{1}{\cos\alpha\sin\alpha} -$
$\cot\alpha = \dfrac{1}{\cos\alpha\sin\alpha} - \dfrac{\cos\alpha(\cos\alpha)}{\sin\alpha(\cos\alpha)} = \dfrac{1 - \cos^2\alpha}{\cos\alpha\sin\alpha} = \dfrac{\sin^2\alpha}{\cos\alpha\sin\alpha} = \dfrac{\sin\alpha}{\cos\alpha} = \tan\alpha$;
$\cos\alpha \neq 0$, $\sin\alpha \neq 0$

11. $\sec^2 x + \csc^2 x = \dfrac{1}{\cos^2 x} + \dfrac{1}{\sin^2 x} = \dfrac{\sin^2 x + \cos^2 x}{\cos^2 x\cdot\sin^2 x} = \dfrac{1}{\cos^2 x\sin^2 x} = \dfrac{1}{\cos^2 x}\cdot\dfrac{1}{\sin^2 x} =$
$\sec^2 x\csc^2 x$; $\sin x \neq 0$, $\cos x \neq 0$

12. $\sec x(\csc x - \sin x) = \dfrac{1}{\cos x}\left(\dfrac{1}{\sin x} - \sin x\right) = \dfrac{1}{\cos x}\left(\dfrac{1 - \sin^2 x}{\sin x}\right) =$
$\dfrac{1}{\cos x}\left(\dfrac{\cos^2 x}{\sin x}\right) = \dfrac{\cos x}{\sin x} = \cot x$; $\sin x \neq 0$, $\cos x \neq 0$

13. $\dfrac{1 + \cot\alpha}{\sin\alpha + \cos\alpha} = \dfrac{1 + \dfrac{\cos\alpha}{\sin\alpha}}{\sin\alpha + \cos\alpha} = \dfrac{\dfrac{\sin\alpha + \cos\alpha}{\sin\alpha}}{\sin\alpha + \cos\alpha} = \dfrac{1}{\sin\alpha} = \csc\alpha$; $\sin\alpha \neq 0$,
$\sin\alpha \neq -\cos\alpha$

14. $\dfrac{\sec\alpha + \csc\alpha}{1 + \cot\alpha} = \dfrac{\dfrac{1}{\cos\alpha}\left(\dfrac{\sin\alpha}{\sin\alpha}\right) + \dfrac{1}{\sin\alpha}\left(\dfrac{\cos\alpha}{\cos\alpha}\right)}{\dfrac{\sin\alpha}{\sin\alpha} + \dfrac{\cos\alpha}{\sin\alpha}} = \dfrac{\dfrac{\sin\alpha + \cos\alpha}{\cos\alpha\sin\alpha}}{\dfrac{\cos\alpha + \sin\alpha}{\sin\alpha}} =$
$\dfrac{\sin\alpha + \cos\alpha}{\cos\alpha\sin\alpha}\cdot\dfrac{\sin\alpha}{\cos\alpha + \sin\alpha} = \dfrac{1}{\cos\alpha} = \sec\alpha$; $\sin\alpha \neq 0$, $\cos\alpha \neq 0$,
$\sin\alpha \neq -\cos\alpha$

15. $(\sec x + 1)(\csc x - \cot x) = \left(\dfrac{1}{\cos x} + 1\right)\left(\dfrac{1}{\sin x} - \dfrac{\cos x}{\sin x}\right) =$
$\left(\dfrac{1 + \cos x}{\cos x}\right)\left(\dfrac{1 - \cos x}{\sin x}\right) = \dfrac{1 - \cos^2 x}{\cos x\sin x} = \dfrac{\sin^2 x}{\cos x\sin x} = \dfrac{\sin x}{\cos x} = \tan x$; $\sin x \neq 0$,
$\cos x \neq 0$

16. $(\tan x + \cos x)(\cot x + \tan x) = \left(\dfrac{\sin x}{\cos x} + \cos x\right)\left(\dfrac{\cos x}{\sin x} + \dfrac{\sin x}{\cos x}\right) =$
$\left(\dfrac{\sin x}{\cos x} + \cos x\right)\left(\dfrac{\cos^2 x + \sin^2 x}{\sin x\cos x}\right) = \left(\dfrac{\sin x}{\cos x} + \cos x\right)\left(\dfrac{1}{\sin x\cos x}\right) =$
$\dfrac{\sin x}{\sin x\cos^2 x} + \dfrac{\cos x}{\sin x\cos x} = \dfrac{1}{\cos^2 x} + \dfrac{1}{\sin x} = \sec^2 x + \csc x$; $\sin x \neq 0$, $\cos x \neq 0$

B **17.** $(\tan\alpha + \cot\alpha)^2 = \tan^2\alpha + 1 + 1 + \cot^2\alpha = \sec^2\alpha + \csc^2\alpha = \dfrac{1}{\cos^2\alpha} + \dfrac{1}{\sin^2\alpha} =$
$\dfrac{\sin^2\alpha + \cos^2\alpha}{\cos^2\alpha\sin^2\alpha} = \dfrac{1}{\cos^2\alpha\sin^2\alpha} = \sec^2\alpha\csc^2\alpha$; $\sin\alpha \neq 0$, $\cos\alpha \neq 0$

18. $(\csc\alpha - \sin\alpha)^2 = \csc^2\alpha - 1 - 1 + \sin^2\alpha = \cot^2\alpha - (1 - \sin^2\alpha) =$
$\cot^2\alpha - \cos^2\alpha$; $\sin\alpha \neq 0$

19. $\dfrac{1}{1-\cos\alpha} - \dfrac{1}{1+\sec\alpha} = \dfrac{1}{1-\cos\alpha} - \dfrac{1}{1+\dfrac{1}{\cos\alpha}} = \dfrac{1}{1-\cos\alpha} - \dfrac{\cos\alpha}{\cos\alpha+1} =$
$\dfrac{1+\cos\alpha-\cos\alpha(1-\cos\alpha)}{(1-\cos\alpha)(1+\cos\alpha)} = \dfrac{1+\cos\alpha-\cos\alpha+\cos^2\alpha}{1-\cos^2\alpha} = \dfrac{1+\cos^2\alpha}{\sin^2\alpha} =$
$\dfrac{1}{\sin^2\alpha} + \dfrac{\cos^2\alpha}{\sin^2\alpha} = \csc^2\alpha + \cot^2\alpha$; $\sin\alpha \neq 0$

20. $\dfrac{\sec\alpha}{\sec\alpha - 1} - \dfrac{\sec\alpha+1}{\tan^2\alpha} = \dfrac{\sec\alpha(\sec\alpha+1)}{(\sec\alpha-1)(\sec\alpha+1)} - \dfrac{\sec\alpha+1}{\tan^2\alpha} = \dfrac{\sec^2\alpha+\sec\alpha}{\sec^2\alpha-1} -$
$\dfrac{\sec\alpha+1}{\tan^2\alpha} = \dfrac{(\sec^2\alpha+\sec\alpha)-(\sec\alpha+1)}{\tan^2\alpha} = \dfrac{\tan^2\alpha}{\tan^2\alpha} = 1$; $\sin\alpha \neq 0$, $\cos\alpha \neq 0$

21. $\sec^4\alpha - \tan^4\alpha = (\sec^2\alpha - \tan^2\alpha)(\sec^2\alpha + \tan^2\alpha) = 1(\sec^2\alpha + \tan^2\alpha) =$
$\sec^2\alpha + \tan^2\alpha$; $\cos\alpha \neq 0$

22. $\dfrac{\sec x - \tan x}{\cos x} - \dfrac{\cos x}{\sec x + \tan x} = \dfrac{(\sec x - \tan x)(\sec x + \tan x) - \cos^2 x}{\cos x(\sec x + \tan x)} =$
$\dfrac{\sec^2 x - \tan^2 x - \cos^2 x}{\cos x(\sec x + \tan x)} = \dfrac{1-\cos^2 x}{\cos x\left(\dfrac{1}{\cos x} + \dfrac{\sin x}{\cos x}\right)} = \dfrac{\sin^2 x}{1+\sin x}$; $\cos x \neq 0$

23. $\dfrac{\cos\alpha}{1-\sin\alpha} + \dfrac{1-\sin\alpha}{\cos\alpha} = \dfrac{\cos^2\alpha + (1-\sin\alpha)^2}{(1-\sin\alpha)\cos\alpha} = \dfrac{\cos^2\alpha + 1 - 2\sin\alpha + \sin^2\alpha}{\cos\alpha(1-\sin\alpha)} =$
$\dfrac{2-2\sin\alpha}{\cos\alpha(1-\sin\alpha)} = \dfrac{2(1-\sin\alpha)}{\cos\alpha(1-\sin\alpha)} = \dfrac{2}{\cos\alpha} = 2\sec\alpha$; $\cos\alpha \neq 0$

24. $\dfrac{\sec x - 1}{\tan x} + \dfrac{\tan x}{\sec x + 1} = \dfrac{(\sec x - 1)(\sec x + 1) + \tan^2 x}{\tan x(\sec x + 1)} = \dfrac{\sec^2 x - 1 + \tan^2 x}{\tan x(\sec x + 1)} =$
$\dfrac{2\tan^2 x}{\tan x(\sec x + 1)} = \dfrac{2\tan x}{\sec x + 1} = \dfrac{\left(\dfrac{2\sin x}{\cos x}\right)\cos x}{\left(\dfrac{1}{\cos x} + 1\right)\cos x} = \dfrac{2\sin x}{1+\cos x}$; $\sin x \neq 0$, $\cos x \neq 0$

25. $\dfrac{\tan x + 1}{\tan x} - \dfrac{\sec x\csc x + 1}{\tan x + 1} = \dfrac{(\tan x + 1)^2 - \tan x(\sec x\csc x + 1)}{\tan x(\tan x + 1)} =$
$\dfrac{\tan^2 x + 2\tan x + 1 - \sec^2 x - \tan x}{\tan x(\tan x + 1)} = \dfrac{\tan x}{\tan x(\tan x + 1)} = \dfrac{1}{\tan x + 1} =$
$\dfrac{1\cdot\cos x}{\left(\dfrac{\sin x}{\cos x} + 1\right)\cos x} = \dfrac{\cos x}{\sin x + \cos x}$; $\sin x \neq 0$, $\cos x \neq 0$, $\sin x \neq -\cos x$

26. $\dfrac{1-\sin^4 x}{\sec^2 x + \tan^2 x} = \dfrac{(1-\sin^2 x)(1+\sin^2 x)}{\dfrac{1}{\cos^2 x} + \dfrac{\sin^2 x}{\cos^2 x}} = \dfrac{\cos^2 x(1+\sin^2 x)}{\dfrac{1+\sin^2 x}{\cos^2 x}} = \cos^4 x$; $\cos x \neq 0$

C **27.** $\dfrac{\sec\alpha - \tan\alpha}{\sec\alpha + \tan\alpha} = \dfrac{(\sec\alpha - \tan\alpha)(\sec\alpha - \tan\alpha)}{(\sec\alpha + \tan\alpha)(\sec\alpha - \tan\alpha)} = \dfrac{(\sec\alpha - \tan\alpha)^2}{\sec^2\alpha - \tan^2\alpha} =$
$\dfrac{(\sec\alpha - \tan\alpha)^2}{1} = \left(\dfrac{1}{\cos\alpha} - \dfrac{\sin\alpha}{\cos\alpha}\right)^2 = \left(\dfrac{1-\sin\alpha}{\cos\alpha}\right)^2$; $\cos\alpha \neq 0$

28. $\frac{\tan x + \sin x}{\tan x - \sin x} = \frac{\left(\frac{\sin x}{\cos x} + \sin x\right)\frac{\cos x}{\sin x}}{\left(\frac{\sin x}{\cos x} - \sin x\right)\frac{\cos x}{\sin x}} = \frac{(1 + \cos x)(1 + \cos x)}{(1 - \cos x)(1 + \cos x)} = \frac{(1 + \cos x)^2}{1 - \cos^2 x} =$
$\frac{(1 + \cos x)^2}{\sin^2 x} = \left(\frac{1 + \cos x}{\sin x}\right)^2$; $\sin x \neq 0$, $\cos x \neq 0$

Page 624 • SELF-TEST 1

1. $\sec \alpha(\cos \alpha + \sin \alpha \tan \alpha) = \frac{1}{\cos \alpha}\left(\cos \alpha + \sin \alpha \frac{\sin \alpha}{\cos \alpha}\right) = \frac{\cos \alpha}{\cos \alpha} + \frac{\sin^2 \alpha}{\cos^2 \alpha} =$
$\frac{\cos^2 \alpha + \sin^2 \alpha}{\cos^2 \alpha} = \frac{1}{\cos^2 \alpha} = \sec^2 \alpha$; $\cos \alpha \neq 0$

2. $\frac{\cos x}{\sec x} + \frac{\sin x}{\csc x} = \frac{\cos x}{\frac{1}{\cos x}} + \frac{\sin x}{\frac{1}{\sin x}} = \cos^2 x + \sin^2 x = 1$; $\sin x \neq 0$, $\cos x \neq 0$

3. $\frac{\sin^2 \alpha}{1 - \cos \alpha} = \frac{1 - \cos^2 \alpha}{1 - \cos \alpha} = \frac{(1 - \cos \alpha)(1 + \cos \alpha)}{1 - \cos \alpha} = 1 + \cos \alpha$; $\cos \alpha \neq 1$

4. $\sec \alpha \csc \alpha - \tan \alpha = \frac{1}{\cos \alpha} \cdot \frac{1}{\sin \alpha} - \frac{\sin \alpha}{\cos \alpha} = \frac{1 - \sin^2 \alpha}{\cos \alpha \sin \alpha} = \frac{\cos^2 \alpha}{\cos \alpha \sin \alpha} = \frac{\cos \alpha}{\sin \alpha} =$
$\cot \alpha$; $\sin \alpha \neq 0$, $\cos \alpha \neq 0$

Pages 630–631 • WRITTEN EXERCISES

A **1.** $\cos \frac{7\pi}{12} \cos \frac{\pi}{12} - \sin \frac{7\pi}{12} \sin \frac{\pi}{12} = \cos\left(\frac{7\pi}{12} + \frac{\pi}{12}\right) = \cos \frac{8\pi}{12} = \cos \frac{2\pi}{3} = -\frac{1}{2}$

2. $\cos \frac{\pi}{8} \cos \frac{7\pi}{8} + \sin \frac{\pi}{8} \sin \frac{7\pi}{8} = \cos\left(\frac{\pi}{8} - \frac{7\pi}{8}\right) = \cos\left(-\frac{6\pi}{8}\right) = \cos\left(-\frac{3\pi}{4}\right) =$
$\cos \frac{3\pi}{4} = -\frac{1}{\sqrt{2}}$

3. $\cos \frac{5\pi}{9} \cos \frac{2\pi}{9} + \sin \frac{5\pi}{9} \sin \frac{2\pi}{9} = \cos\left(\frac{5\pi}{9} - \frac{2\pi}{9}\right) = \cos \frac{\pi}{3} = \frac{1}{2}$

4. $\cos 175° \cos 25° + \sin 175° \sin 25° = \cos(175° - 25°) = \cos 150° = -\frac{\sqrt{3}}{2}$

5. $\cos 1.2\pi \cos 0.3\pi - \sin 1.2\pi \sin 0.3\pi = \cos(1.2\pi + 0.3\pi) = \cos 1.5\pi = 0$

6. $\cos 185° \cos 40° - \sin 185° \sin 40° = \cos(185° + 40°) = \cos 225° = -\frac{1}{\sqrt{2}}$

7. $\cos 135° \cos(-15°) + \sin 135° \sin(-15°) = \cos(135° - (-15°)) = \cos 150° = -\frac{\sqrt{3}}{2}$

8. $\cos(-5°)\cos 245° - \sin(-5°)\sin 245° = \cos(-5° + 245°) = \cos 240° = -\frac{1}{2}$

9. $\cos 15° = \cos(45° - 30°) = \cos 45° \cos 30° + \sin 45° \sin 30° =$
$\frac{1}{\sqrt{2}} \cdot \frac{\sqrt{3}}{2} + \frac{1}{\sqrt{2}} \cdot \frac{1}{2} = \frac{\sqrt{3}}{2\sqrt{2}} + \frac{1}{2\sqrt{2}} = \frac{(\sqrt{3} + 1)\sqrt{2}}{2\sqrt{2} \cdot \sqrt{2}} = \frac{\sqrt{6} + \sqrt{2}}{4}$

10. $\cos 75° = \cos(45° + 30°) = \cos 45° \cos 30° - \sin 45° \sin 30° =$
$\frac{1}{\sqrt{2}} \cdot \frac{\sqrt{3}}{2} - \frac{1}{\sqrt{2}} \cdot \frac{1}{2} = \frac{(\sqrt{3} - 1)\sqrt{2}}{2\sqrt{2} \cdot \sqrt{2}} = \frac{\sqrt{6} - \sqrt{2}}{4}$

11. $\cos 165° = \cos(120° + 45°) = \cos 120° \cos 45° - \sin 120° \sin 45° =$
$\frac{1}{\sqrt{2}}\left(-\frac{1}{2}\right) - \frac{1}{\sqrt{2}} \cdot \frac{\sqrt{3}}{2} = \frac{(-1 - \sqrt{3})\sqrt{2}}{2\sqrt{2} \cdot \sqrt{2}} = \frac{-\sqrt{2} - \sqrt{6}}{4}$

12. $\cos 195° = \cos(150° + 45°) = \cos 150° \cos 45° - \sin 150° \sin 45° =$
$\left(-\frac{\sqrt{3}}{2}\right)\frac{1}{\sqrt{2}} - \frac{1}{2} \cdot \frac{1}{\sqrt{2}} = \frac{(-\sqrt{3} - 1)\sqrt{2}}{2\sqrt{2} \cdot \sqrt{2}} = \frac{-\sqrt{6} - \sqrt{2}}{4}$

13. $\cos 345° = \cos(300° + 45°) = \cos 300° \cos 45° - \sin 300° \sin 45° =$
$\frac{1}{2} \cdot \frac{1}{\sqrt{2}} - \left(-\frac{\sqrt{3}}{2}\right)\frac{1}{\sqrt{2}} = \frac{(1 + \sqrt{3})\sqrt{2}}{2\sqrt{2} \cdot \sqrt{2}} = \frac{\sqrt{2} + \sqrt{6}}{4}$

14. $\cos 285° = \cos(240° + 45°) = \cos 240° \cos 45° - \sin 240° \sin 45° =$
$\left(-\frac{1}{2}\right)\frac{1}{\sqrt{2}} - \left(-\frac{\sqrt{3}}{2}\right)\frac{1}{\sqrt{2}} = \frac{(-1 + \sqrt{3})\sqrt{2}}{2\sqrt{2} \cdot \sqrt{2}} = \frac{-\sqrt{2} + \sqrt{6}}{4}$

15. $\cos(-15°) = \cos(30° - 45°) = \cos 30° \cos 45° + \sin 30° \sin 45° =$
$\frac{\sqrt{3}}{2} \cdot \frac{1}{\sqrt{2}} + \frac{1}{2} \cdot \frac{1}{\sqrt{2}} = \frac{(\sqrt{3} + 1)\sqrt{2}}{2\sqrt{2} \cdot \sqrt{2}} = \frac{\sqrt{6} + \sqrt{2}}{4}$

16. $\cos(-105°) = \cos(45° - 150°) = \cos 45° \cos 150° + \sin 45° \sin 150° =$
$\frac{1}{\sqrt{2}}\left(-\frac{\sqrt{3}}{2}\right) + \frac{1}{\sqrt{2}} \cdot \frac{1}{2} = \frac{(-\sqrt{3} + 1)\sqrt{2}}{2\sqrt{2} \cdot \sqrt{2}} = \frac{-\sqrt{6} + \sqrt{2}}{4}$

17. $\cos\frac{\pi}{12} = \cos\left(\frac{\pi}{3} - \frac{\pi}{4}\right) = \cos\frac{\pi}{3}\cos\frac{\pi}{4} + \sin\frac{\pi}{3}\sin\frac{\pi}{4} = \frac{1}{2} \cdot \frac{1}{\sqrt{2}} + \frac{\sqrt{3}}{2} \cdot \frac{1}{\sqrt{2}} =$
$\frac{(1 + \sqrt{3})\sqrt{2}}{2\sqrt{2} \cdot \sqrt{2}} = \frac{\sqrt{2} + \sqrt{6}}{4}$

18. $\cos\frac{13\pi}{12} = \cos\left(\frac{5\pi}{6} + \frac{\pi}{4}\right) = \cos\frac{5\pi}{6}\cos\frac{\pi}{4} - \sin\frac{5\pi}{6}\sin\frac{\pi}{4} = \left(-\frac{\sqrt{3}}{2}\right)\frac{1}{\sqrt{2}} - \frac{1}{2} \cdot \frac{1}{\sqrt{2}} =$
$\frac{(-\sqrt{3} - 1)\sqrt{2}}{2\sqrt{2} \cdot \sqrt{2}} = \frac{-\sqrt{6} - \sqrt{2}}{4}$

19. $\cos\frac{17\pi}{12} = \cos\left(\frac{5\pi}{4} + \frac{\pi}{6}\right) = \cos\frac{5\pi}{4}\cos\frac{\pi}{6} - \sin\frac{5\pi}{4}\sin\frac{\pi}{6} =$
$\left(-\frac{1}{\sqrt{2}}\right)\frac{\sqrt{3}}{2} - \left(-\frac{1}{\sqrt{2}}\right)\frac{1}{2} = \frac{(-\sqrt{3} + 1)\sqrt{2}}{2\sqrt{2} \cdot \sqrt{2}} = \frac{-\sqrt{6} + \sqrt{2}}{4}$

20. $\cos\left(-\frac{5\pi}{12}\right) = \cos\left(\frac{\pi}{3} - \frac{3\pi}{4}\right) = \cos\frac{\pi}{3}\cos\frac{3\pi}{4} + \sin\frac{\pi}{3}\sin\frac{3\pi}{4} =$
$\frac{1}{2}\left(-\frac{1}{\sqrt{2}}\right) + \frac{\sqrt{3}}{2} \cdot \frac{1}{\sqrt{2}} = \frac{(-1 + \sqrt{3})\sqrt{2}}{2\sqrt{2} \cdot \sqrt{2}} = \frac{-\sqrt{2} + \sqrt{6}}{4}$

21. $\cos\left(x - \frac{\pi}{2}\right) = \cos x \cos\frac{\pi}{2} + \sin x \sin\frac{\pi}{2} = \cos x(0) + \sin x(1) = \sin x$

22. $\cos(\pi + x) = \cos \pi \cos x - \sin \pi \sin x = (-1) \cos x - (0) \sin x = -\cos x$

23. $\cos\left(\frac{3\pi}{2} + x\right) = \cos\frac{3\pi}{2}\cos x - \sin\frac{3\pi}{2}\sin x = (0) \cos x - (-1) \sin x = \sin x$

24. $\cos\left(\frac{3\pi}{2} - x\right) = \cos\frac{3\pi}{2}\cos x + \sin\frac{3\pi}{2}\sin x = (0) \cos x + (-1) \sin x = -\sin x$

25. $\cos\left(\frac{\pi}{2} + x\right) = \cos\frac{\pi}{2}\cos x - \sin\frac{\pi}{2}\sin x = (0)\cos x - (1)\sin x = -\sin x$

26. $\cos\left(x - \frac{\pi}{6}\right) = \cos x \cos\frac{\pi}{6} + \sin x \sin\frac{\pi}{6} = \cos x\left(\frac{\sqrt{3}}{2}\right) + \sin x\left(\frac{1}{2}\right) =$
$\frac{\sqrt{3}\cos x + \sin x}{2}$

27. $\cos\left(\frac{\pi}{3} - x\right) = \cos\frac{\pi}{3}\cos x + \sin\frac{\pi}{3}\sin x = \left(\frac{1}{2}\right)\cos x + \left(\frac{\sqrt{3}}{2}\right)\sin x =$
$\frac{\cos x + \sqrt{3}\sin x}{2}$

28. $\cos\left(\frac{3\pi}{4} + x\right) = \cos\frac{3\pi}{4}\cos x - \sin\frac{3\pi}{4}\sin x = \left(-\frac{1}{\sqrt{2}}\right)\cos x - \left(\frac{1}{\sqrt{2}}\right)\sin x =$
$-\frac{1}{\sqrt{2}}(\cos x + \sin x)$

B **29.** In Quad. I, $\sin x = \sqrt{1 - \left(\frac{4}{5}\right)^2} = \frac{3}{5}$; in Quad. I, $\sin y = \sqrt{1 - \left(\frac{12}{13}\right)^2} = \frac{5}{13}$;
$\cos(x - y) = \cos x \cos y + \sin x \sin y = \frac{4}{5}\cdot\frac{12}{13} + \frac{3}{5}\cdot\frac{5}{13} = \frac{48}{65} + \frac{15}{65} = \frac{63}{65}$

30. In Quad. II, $\sin x = \sqrt{1 - \left(-\frac{3}{5}\right)^2} = \frac{4}{5}$; in Quad. I, $\sin y = \sqrt{1 - \left(\frac{5}{13}\right)^2} = \frac{12}{13}$;
$\cos(x - y) = \left(-\frac{3}{5}\right)\frac{5}{13} + \frac{4}{5}\cdot\frac{12}{13} = \frac{-15 + 48}{65} = \frac{33}{65}$

31. In Quad. III, $\sin x = -\sqrt{1 - \left(-\frac{7}{25}\right)^2} = -\frac{24}{25}$; in Quad. II, $\sin y =$
$\sqrt{1 - \left(-\frac{4}{5}\right)^2} = \frac{3}{5}$; $\cos(x - y) = \left(-\frac{7}{25}\right)\left(-\frac{4}{5}\right) + \left(-\frac{24}{25}\right)\frac{3}{5} = \frac{28 - 72}{125} = -\frac{44}{125}$

32. In Quad. IV, $\sin x = -\sqrt{1 - \left(\frac{15}{17}\right)^2} = -\frac{8}{17}$; in Quad. III,
$\sin y = -\sqrt{1 - \left(-\frac{3}{5}\right)^2} = -\frac{4}{5}$; $\cos(x - y) = \frac{15}{17}\left(-\frac{3}{5}\right) + \left(-\frac{8}{17}\right)\left(-\frac{4}{5}\right) =$
$\frac{-45 + 32}{85} = -\frac{13}{85}$

33. In Quad. I, $\sin x = \sqrt{1 - \left(\frac{2}{3}\right)^2} = \frac{5}{9}$; in Quad. IV, $\sin y = -\sqrt{1 - \left(\frac{\sqrt{3}}{2}\right)^2} = -\frac{1}{2}$;
$\cos(x + y) = \cos x \cos y - \sin x \sin y = \frac{2}{3}\cdot\frac{\sqrt{3}}{2} - \frac{\sqrt{5}}{3}\left(-\frac{1}{2}\right) = \frac{2\sqrt{6} + \sqrt{5}}{6}$

34. In Quad. III, $\sin x = -\sqrt{1 - \left(-\frac{3}{4}\right)^2} = -\frac{\sqrt{7}}{4}$; in Quad. I,

$\sin y = \sqrt{1 - \left(\frac{2\sqrt{2}}{3}\right)^2} = \frac{1}{3}$; $\cos(x + y) = \left(-\frac{3}{4}\right)\frac{2\sqrt{2}}{3} - \left(-\frac{\sqrt{7}}{4}\right)\frac{1}{3} = \frac{-6\sqrt{2} + \sqrt{7}}{12}$

35. In Quad. II, $\sin x = \sqrt{1 - \left(-\frac{\sqrt{15}}{4}\right)^2} = \frac{1}{4}$; in Quad. IV, $\sin y =$

$-\sqrt{1 - \left(\frac{\sqrt{5}}{3}\right)^2} = -\frac{2}{3}$; $\cos(x + y) = \left(-\frac{\sqrt{15}}{4}\right)\frac{\sqrt{5}}{3} - \frac{1}{4}\left(-\frac{2}{3}\right) = \frac{-5\sqrt{3} + 2}{12}$

C **36.** In Quad. I, $\sin x = \sqrt{1 - a^2}$; in Quad. I, $\sin y = \sqrt{1 - (\sqrt{1 - b^2})^2} =$
$\sqrt{1 - (1 - b^2)} = \sqrt{b^2} = |b|$; $\cos(x + y) = a\sqrt{1 - b^2} - |b|\sqrt{1 - a^2}$

Pages 635–637 • WRITTEN EXERCISES

A **1.** $\sin 80° \cos 20° - \cos 80° \sin 20° = \sin(80° - 20°) = \sin 60° = \frac{\sqrt{3}}{2}$

2. $\sin 70° \cos 170° + \cos 70° \sin 170° = \sin(70° + 170°) = \sin 240° = -\frac{\sqrt{3}}{2}$

3. $\sin 275° \cos 55° + \cos 275° \sin 55° = \sin(275° + 55°) = \sin 330° = -\frac{1}{2}$

4. $\sin 335° \cos 110° - \cos 335° \sin 110° = \sin(335° - 110°) = \sin 225° = -\frac{1}{\sqrt{2}}$

5. $\frac{\tan 95° + \tan 25°}{1 - \tan 95° \tan 25°} = \tan(95° + 25°) = \tan 120° = -\sqrt{3}$

6. $\frac{\tan 155° - \tan 20°}{1 + \tan 155° \tan 20°} = \tan(155° - 20°) = \tan 135° = -1$

7. $2 \sin 112.5° \cos 112.5° = \sin(112.5° + 112.5°) = \sin 225° = -\frac{1}{\sqrt{2}}$

8. $2 \sin 165° \cos 165° = \sin(165° + 165°) = \sin 330° = -\frac{1}{2}$

9. $\frac{2 \tan 105°}{1 - \tan^2 105°} = \tan(105° + 105°) = \tan 210° = \frac{1}{\sqrt{3}}$

10. $\frac{2 \tan 67.5°}{1 - \tan^2 67.5°} = \tan(67.5° + 67.5°) = \tan 135° = -1$

11. $\sin 15° = \sin(45° - 30°) = \sin 45° \cos 30° - \cos 45° \sin 30° =$
$\frac{1}{\sqrt{2}} \cdot \frac{\sqrt{3}}{2} - \frac{1}{\sqrt{2}} \cdot \frac{1}{2} = \frac{(\sqrt{3} - 1)\sqrt{2}}{2\sqrt{2} \cdot \sqrt{2}} = \frac{\sqrt{6} - \sqrt{2}}{4}$

12. $\sin 105° = \sin(60° + 45°) = \sin 60° \cos 45° + \cos 60° \sin 45° =$
$\frac{\sqrt{3}}{2} \cdot \frac{1}{\sqrt{2}} + \frac{1}{2} \cdot \frac{1}{\sqrt{2}} = \frac{\sqrt{6} + \sqrt{2}}{4}$

13. $\sin\frac{11\pi}{12} = \sin\left(\frac{2\pi}{3} + \frac{\pi}{4}\right) = \sin\frac{2\pi}{3}\cos\frac{\pi}{4} + \cos\frac{2\pi}{3}\sin\frac{\pi}{4} =$

$\frac{\sqrt{3}}{2}\cdot\frac{1}{\sqrt{2}} + \left(-\frac{1}{2}\right)\frac{1}{\sqrt{2}} = \frac{\sqrt{6} - \sqrt{2}}{4}$

14. $\sin\frac{17\pi}{12} = \sin\left(\frac{\pi}{4} + \frac{7\pi}{6}\right) = \sin\frac{\pi}{4}\cos\frac{7\pi}{6} + \cos\frac{\pi}{4}\sin\frac{7\pi}{6} =$

$\frac{1}{\sqrt{2}}\left(-\frac{\sqrt{3}}{2}\right) + \frac{1}{\sqrt{2}}\left(-\frac{1}{2}\right) = \frac{-\sqrt{6} - \sqrt{2}}{4}$

15. $\sin(-75°) = \sin(45° - 120°) = \sin 45° \cos 120° - \cos 45° \sin 120° =$

$\frac{1}{\sqrt{2}}\left(-\frac{1}{2}\right) - \frac{1}{\sqrt{2}}\cdot\frac{\sqrt{3}}{2} = \frac{-\sqrt{2} - \sqrt{6}}{4}$

16. $\sin 165° = \sin(120° + 45°) = \sin 120° \cos 45° + \cos 120° \sin 45° =$

$\frac{\sqrt{3}}{2}\cdot\frac{1}{\sqrt{2}} + \left(-\frac{1}{2}\right)\frac{1}{\sqrt{2}} = \frac{\sqrt{6} - \sqrt{2}}{4}$

17. $\sin\left(-\frac{7\pi}{12}\right) = \sin\left(\frac{\pi}{4} - \frac{5\pi}{6}\right) = \sin\frac{\pi}{4}\cos\frac{5\pi}{6} - \cos\frac{\pi}{4}\sin\frac{5\pi}{6} =$

$\frac{1}{\sqrt{2}}\left(-\frac{\sqrt{3}}{2}\right) - \frac{1}{\sqrt{2}}\cdot\frac{1}{2} = \frac{-\sqrt{6} - \sqrt{2}}{4}$

18. $\sin\frac{13\pi}{12} = \sin\left(\frac{5\pi}{6} + \frac{\pi}{4}\right) = \sin\frac{5\pi}{6}\cos\frac{\pi}{4} + \cos\frac{5\pi}{6}\sin\frac{\pi}{4} =$

$\frac{1}{2}\cdot\frac{1}{\sqrt{2}} + \left(-\frac{\sqrt{3}}{2}\right)\frac{1}{\sqrt{2}} = \frac{\sqrt{2} - \sqrt{6}}{4}$

19. $\tan 75° = \tan(45° + 30°) = \frac{\tan 45° + \tan 30°}{1 - \tan 45° \tan 30°} = \frac{1 + \frac{1}{\sqrt{3}}}{1 - 1\cdot\frac{1}{\sqrt{3}}} =$

$\frac{\sqrt{3} + 1}{\sqrt{3} - 1}\cdot\frac{\sqrt{3} + 1}{\sqrt{3} + 1} = \frac{3 + 2\sqrt{3} + 1}{3 - 1} = \frac{4 + 2\sqrt{3}}{2} = 2 + \sqrt{3}$

20. $\tan 165° = \tan(45° + 120°) = \frac{\tan 45° + \tan 120°}{1 - \tan 45° \tan 120°} = \frac{1 + (-\sqrt{3})}{1 - 1(-\sqrt{3})} =$

$\frac{1 - \sqrt{3}}{1 + \sqrt{3}}\cdot\frac{1 - \sqrt{3}}{1 - \sqrt{3}} = \frac{4 - 2\sqrt{3}}{-2} = -2 + \sqrt{3}$

21. $\tan 195° = \tan(45° + 150°) = \frac{\tan 45° + \tan 150°}{1 - \tan 45° \tan 150°} = \frac{1 + \left(-\frac{1}{\sqrt{3}}\right)}{1 - 1\left(-\frac{1}{\sqrt{3}}\right)} =$

$\frac{\sqrt{3} - 1}{\sqrt{3} + 1}\cdot\frac{\sqrt{3} - 1}{\sqrt{3} - 1} = \frac{4 - 2\sqrt{3}}{2} = 2 - \sqrt{3}$

22. $\tan(-15°) = \tan(30° - 45°) = \dfrac{\tan 30° - \tan 45°}{1 + \tan 30° \tan 45°} = \dfrac{\frac{1}{\sqrt{3}} - 1}{1 + \frac{1}{\sqrt{3}} \cdot 1} =$

$\dfrac{1 - \sqrt{3}}{\sqrt{3} + 1} \cdot \dfrac{\sqrt{3} - 1}{\sqrt{3} - 1} = \dfrac{-4 + 2\sqrt{3}}{2} = -2 + \sqrt{3}$

23. $\tan \dfrac{7\pi}{12} = \tan\left(\dfrac{\pi}{4} + \dfrac{\pi}{3}\right) = \dfrac{\tan \frac{\pi}{4} + \tan \frac{\pi}{3}}{1 - \tan \frac{\pi}{4} \tan \frac{\pi}{3}} = \dfrac{1 + \sqrt{3}}{1 - \sqrt{3}} \cdot \dfrac{1 + \sqrt{3}}{1 + \sqrt{3}} =$

$\dfrac{4 + 2\sqrt{3}}{-2} = -2 - \sqrt{3}$

24. $\tan\left(-\dfrac{5\pi}{12}\right) = \tan\left(\dfrac{\pi}{4} - \dfrac{2\pi}{3}\right) = \dfrac{\tan \frac{\pi}{4} - \tan \frac{2\pi}{3}}{1 + \tan \frac{\pi}{4} \tan \frac{2\pi}{3}} = \dfrac{1 - (-\sqrt{3})}{1 + 1(-\sqrt{3})} =$

$\dfrac{1 + \sqrt{3}}{1 - \sqrt{3}} \cdot \dfrac{1 + \sqrt{3}}{1 + \sqrt{3}} = \dfrac{4 + 2\sqrt{3}}{-2} = -2 - \sqrt{3}$

25. $\tan\left(-\dfrac{11\pi}{12}\right) = \tan\left(\dfrac{\pi}{4} - \dfrac{7\pi}{6}\right) = \dfrac{\tan \frac{\pi}{4} - \tan \frac{7\pi}{6}}{1 + \tan \frac{\pi}{4} \tan \frac{7\pi}{6}} = \dfrac{1 - \frac{1}{\sqrt{3}}}{1 + \frac{1}{\sqrt{3}}} =$

$\dfrac{\sqrt{3} - 1}{\sqrt{3} + 1} \cdot \dfrac{\sqrt{3} - 1}{\sqrt{3} - 1} = \dfrac{4 - 2\sqrt{3}}{2} = 2 - \sqrt{3}$

26. $\tan\left(\dfrac{17\pi}{12}\right) = \tan\left(\dfrac{\pi}{4} + \dfrac{7\pi}{6}\right) = \dfrac{\tan \frac{\pi}{4} + \tan \frac{7\pi}{6}}{1 - \tan \frac{\pi}{4} \tan \frac{7\pi}{6}} = \dfrac{1 + \frac{1}{\sqrt{3}}}{1 - \frac{1}{\sqrt{3}}} =$

$\dfrac{\sqrt{3} + 1}{\sqrt{3} - 1} \cdot \dfrac{\sqrt{3} + 1}{\sqrt{3} + 1} = \dfrac{4 + 2\sqrt{3}}{2} = 2 + \sqrt{3}$

B **27.** $\sin\left(\dfrac{\pi}{2} + x\right) = \sin \dfrac{\pi}{2} \cos x + \cos \dfrac{\pi}{2} \sin x = (1) \cos x + (0) \sin x = \cos x$

28. $\sin(\pi + x) = \sin \pi \cos x + \cos \pi \sin x = (0) \cos x + (-1) \sin x = -\sin x$

29. $\sin(\pi - x) = \sin \pi \cos x - \cos \pi \sin x = (0) \cos x - (-1) \sin x = \sin x$

30. $\sin\left(\dfrac{3\pi}{2} + x\right) = \sin \dfrac{3\pi}{2} \cos x + \cos \dfrac{3\pi}{2} \sin x = (-1) \cos x + (0) \sin x = -\cos x$

31. $\sin\left(\dfrac{3\pi}{2} - x\right) = \sin \dfrac{3\pi}{2} \cos x - \cos \dfrac{3\pi}{2} \sin x = (-1) \cos x - (0) \sin x = -\cos x$

32. $\sin\left(x - \dfrac{\pi}{2}\right) = \sin x \cos \dfrac{\pi}{2} - \cos x \sin \dfrac{\pi}{2} = \sin x(0) - \cos x(1) = -\cos x$

33. $\sin(x - \pi) = \sin x \cos \pi - \cos x \sin \pi = \sin x(-1) - \cos x(0) = -\sin x$

34. $\sin\left(x - \dfrac{3\pi}{2}\right) = \sin x \cos \dfrac{3\pi}{2} - \cos x \sin \dfrac{3\pi}{2} = \sin x(0) - \cos x(-1) = \cos x$

35. $\tan\left(\frac{\pi}{2} - x\right) = \frac{\sin\left(\frac{\pi}{2} - x\right)}{\cos\left(\frac{\pi}{2} - x\right)} = \frac{\sin\frac{\pi}{2}\cos x - \cos\frac{\pi}{2}\sin x}{\cos\frac{\pi}{2}\cos x + \sin\frac{\pi}{2}\sin x} = \frac{(1)\cos x - (0)\sin x}{(0)\cos x + (1)\sin x} =$

$\frac{\cos x}{\sin x} = \cot x$

36. $\tan\left(\frac{3\pi}{2} + x\right) = \frac{\sin\left(\frac{3\pi}{2} + x\right)}{\cos\left(\frac{3\pi}{2} + x\right)} = \frac{\sin\frac{3\pi}{2}\cos x + \cos\frac{3\pi}{2}\sin x}{\cos\frac{3\pi}{2}\cos x - \sin\frac{3\pi}{2}\sin x} = \frac{-\cos x}{\sin x} = -\cot x$

37. $\tan\left(\frac{3\pi}{2} - x\right) = \frac{\sin\left(\frac{3\pi}{2} - x\right)}{\cos\left(\frac{3\pi}{2} - x\right)} = \frac{\sin\frac{3\pi}{2}\cos x - \cos\frac{3\pi}{2}\sin x}{\cos\frac{3\pi}{2}\cos x + \sin\frac{3\pi}{2}\sin x} = \frac{-\cos x}{-\sin x} = \cot x$

38. $\tan\left(x - \frac{\pi}{2}\right) = \frac{\sin\left(x - \frac{\pi}{2}\right)}{\cos\left(x - \frac{\pi}{2}\right)} = \frac{\sin x\cos\frac{\pi}{2} - \cos x\sin\frac{\pi}{2}}{\cos x\cos\frac{\pi}{2} + \sin x\sin\frac{\pi}{2}} = \frac{-\cos x}{\sin x} = -\cot x$

39. $\sec 105° = \frac{1}{\cos 105°} = \frac{1}{\cos(45° + 60°)} = \frac{1}{\cos 45°\cos 60° - \sin 45°\sin 60°} =$

$\frac{1}{\frac{1}{\sqrt{2}}\cdot\frac{1}{2} - \frac{1}{\sqrt{2}}\cdot\frac{\sqrt{3}}{2}} = \frac{2\sqrt{2}}{1 - \sqrt{3}}\cdot\frac{1 + \sqrt{3}}{1 + \sqrt{3}} = \frac{2\sqrt{2} + 2\sqrt{6}}{-2} = -\sqrt{2} - \sqrt{6}$

40. $\csc 195° = \frac{1}{\sin(150° + 45°)} = \frac{1}{\sin 150°\cos 45° + \cos 150°\sin 45°} =$

$\frac{1}{\frac{1}{2}\cdot\frac{1}{\sqrt{2}} + \left(-\frac{\sqrt{3}}{2}\right)\frac{1}{\sqrt{2}}} = \frac{2\sqrt{2}}{1 - \sqrt{3}}\cdot\frac{1 + \sqrt{3}}{1 + \sqrt{3}} = \frac{2\sqrt{2} + 2\sqrt{6}}{-2} = -\sqrt{2} - \sqrt{6}$

41. $\cot 75° = \frac{1}{\tan(45° + 30°)} = \frac{1 - \tan 45°\tan 30°}{\tan 45° + \tan 30°} = \frac{1 - 1\cdot\frac{1}{\sqrt{3}}}{1 + \frac{1}{\sqrt{3}}} =$

$\frac{\sqrt{3} - 1}{\sqrt{3} + 1}\cdot\frac{\sqrt{3} - 1}{\sqrt{3} - 1} = \frac{3 - 2\sqrt{3} + 1}{2} = 2 - \sqrt{3}$

42. $\csc 285° = \frac{1}{\sin(45° + 240°)} = \frac{1}{\sin 45°\cos 240° + \cos 45°\sin 240°} =$

$\frac{1}{\frac{1}{\sqrt{2}}\left(-\frac{1}{2}\right) + \frac{1}{\sqrt{2}}\left(-\frac{\sqrt{3}}{2}\right)} = \frac{2\sqrt{2}}{-1 - \sqrt{3}}\cdot\frac{-1 + \sqrt{3}}{-1 + \sqrt{3}} = \frac{-2\sqrt{2} + 2\sqrt{6}}{-2} = \sqrt{2} - \sqrt{6}$

43. $\cot \frac{11\pi}{12} = \frac{1}{\tan\left(\frac{2\pi}{3} + \frac{\pi}{4}\right)} = \frac{1 - \tan \frac{2\pi}{3} \tan \frac{\pi}{4}}{\tan \frac{2\pi}{3} + \tan \frac{\pi}{4}} = \frac{1 - (-\sqrt{3})1}{(-\sqrt{3}) + 1} =$

$\frac{1 + \sqrt{3}}{-\sqrt{3} + 1} \cdot \frac{1 + \sqrt{3}}{1 + \sqrt{3}} = \frac{4 + 2\sqrt{3}}{-2} = -2 - \sqrt{3}$

44. $\sec \frac{17\pi}{12} = \frac{1}{\cos\left(\frac{\pi}{4} + \frac{7\pi}{6}\right)} = \frac{1}{\cos \frac{\pi}{4} \cos \frac{7\pi}{6} - \sin \frac{\pi}{4} \sin \frac{7\pi}{6}} =$

$\frac{1}{\frac{1}{\sqrt{2}}\left(-\frac{\sqrt{3}}{2}\right) - \frac{1}{\sqrt{2}}\left(-\frac{1}{2}\right)} = \frac{2\sqrt{2}}{-\sqrt{3} + 1} \cdot \frac{1 + \sqrt{3}}{1 + \sqrt{3}} = \frac{2\sqrt{2} + 2\sqrt{6}}{-2} = -\sqrt{2} - \sqrt{6}$

45. $\csc \frac{7\pi}{12} = \frac{1}{\sin\left(\frac{\pi}{4} + \frac{\pi}{3}\right)} = \frac{1}{\sin \frac{\pi}{4} \cos \frac{\pi}{3} + \cos \frac{\pi}{4} \sin \frac{\pi}{3}} = \frac{1}{\frac{1}{\sqrt{2}} \cdot \frac{1}{2} + \frac{1}{\sqrt{2}} \cdot \frac{\sqrt{3}}{2}} =$

$\frac{2\sqrt{2}}{1 + \sqrt{3}} \cdot \frac{1 - \sqrt{3}}{1 - \sqrt{3}} = \frac{2\sqrt{2} - 2\sqrt{6}}{-2} = \sqrt{6} - \sqrt{2}$

46. $\cot \frac{19\pi}{12} = \frac{1}{\tan\left(\frac{\pi}{4} + \frac{4\pi}{3}\right)} = \frac{1 - \tan \frac{\pi}{4} \tan \frac{4\pi}{3}}{\tan \frac{\pi}{4} + \tan \frac{4\pi}{3}} = \frac{1 - 1 \cdot \sqrt{3}}{1 + \sqrt{3}} = \frac{1 - \sqrt{3}}{1 + \sqrt{3}} \cdot \frac{1 - \sqrt{3}}{1 - \sqrt{3}} =$

$\frac{4 - 2\sqrt{3}}{-2} = -2 + \sqrt{3}$

47. $\cos x = \sqrt{1 - \sin^2 x} = \sqrt{1 - \frac{16}{25}} = \frac{3}{5}$; $\cos y = \sqrt{1 - \sin^2 y} = \sqrt{1 - \frac{144}{169}} = \frac{5}{13}$;

$\sin(x + y) = \sin x \cos y + \cos x \sin y = \frac{4}{5} \cdot \frac{5}{13} + \frac{3}{5} \cdot \frac{12}{13} = \frac{20 + 36}{65} = \frac{56}{65}$

48. $\cos x = -\sqrt{1 - \sin^2 x} = -\sqrt{1 - \frac{25}{169}} = -\frac{12}{13}$; $\sin y = -\sqrt{1 - \cos^2 y} =$

$-\sqrt{1 - \frac{9}{25}} = -\frac{4}{5}$; $\sin(x + y) = \sin x \cos y + \cos x \sin y =$

$\frac{5}{13}\left(-\frac{3}{5}\right) + \left(-\frac{12}{13}\right)\left(-\frac{4}{5}\right) = \frac{-15 + 48}{65} = \frac{33}{65}$

49. $\sin x = \sqrt{1 - \cos^2 x} = \sqrt{1 - \left(\frac{15}{17}\right)^2} = \frac{8}{17}$; $\cos y = \sqrt{1 - \sin^2 y} =$

$\sqrt{1 - \left(-\frac{4}{5}\right)^2} = \frac{3}{5}$; $\sin(x + y) = \sin x \cos y + \cos x \sin y =$

$\frac{8}{17} \cdot \frac{3}{5} + \left(-\frac{15}{17}\right)\left(-\frac{4}{5}\right) = \frac{24 + 60}{85} = \frac{84}{85}$

50. $\cos x = -\sqrt{1 - \sin^2 x} = -\sqrt{1 - \left(-\frac{1}{2}\right)^2} = -\frac{\sqrt{3}}{2}$; $\sin y = \sqrt{1 - \cos^2 y} = \sqrt{1 - \left(\frac{2\sqrt{2}}{3}\right)^2} = \frac{1}{3}$; $\sin(x + y) = \sin x \cos y + \cos x \sin y = \left(-\frac{1}{2}\right)\left(-\frac{2\sqrt{2}}{3}\right) + \left(-\frac{\sqrt{3}}{2}\right)\frac{1}{3} = \frac{2\sqrt{2} - \sqrt{3}}{6}$

51. $\cos x = \sqrt{1 - \sin^2 x} = \sqrt{1 - \left(\frac{24}{25}\right)^2} = \frac{7}{25}$; $\sin y = \sqrt{1 - \cos^2 y} = \sqrt{1 - \left(\frac{4}{5}\right)^2} = \frac{3}{5}$; $\sin(x - y) = \sin x \cos y - \cos x \sin y = \frac{24}{25} \cdot \frac{4}{5} - \frac{7}{25} \cdot \frac{3}{5} = \frac{75}{125} = \frac{3}{5}$

52. $\cos x = -\sqrt{1 - \sin^2 x} = -\sqrt{1 - \left(\frac{2}{3}\right)^2} = -\frac{\sqrt{5}}{3}$; $\cos y = \sqrt{1 - \sin^2 y} = \sqrt{1 - \left(\frac{\sqrt{7}}{4}\right)^2} = \frac{3}{4}$; $\sin(x - y) = \sin x \cos y - \cos x \sin y = \frac{2}{3} \cdot \frac{3}{4} - \left(-\frac{\sqrt{5}}{3}\right)\frac{\sqrt{7}}{4} = \frac{6 + \sqrt{35}}{12}$

53. $\sin x = -\sqrt{1 - \cos^2 x} = -\sqrt{1 - \left(-\frac{1}{4}\right)^2} = -\frac{\sqrt{15}}{4}$; $\sin y = \sqrt{1 - \cos^2 y} = \sqrt{1 - \left(-\frac{3}{5}\right)^2} = \frac{4}{5}$; $\sin(x - y) = \sin x \cos y - \cos x \sin y = \left(-\frac{\sqrt{15}}{4}\right)\left(-\frac{3}{5}\right) - \left(-\frac{1}{4}\right)\frac{4}{5} = \frac{3\sqrt{15} + 4}{20}$

54. $\sin x = -\sqrt{1 - \cos^2 x} = -\sqrt{1 - \left(\frac{\sqrt{5}}{3}\right)^2} = -\frac{2}{3}$; $\cos y = -\sqrt{1 - \sin^2 y} = -\sqrt{1 - \left(-\frac{1}{3}\right)^2} = -\frac{2\sqrt{2}}{3}$; $\sin(x - y) = \sin x \cos y - \cos x \sin y = \left(-\frac{2}{3}\right)\left(-\frac{2\sqrt{2}}{3}\right) - \frac{\sqrt{5}}{3}\left(-\frac{1}{3}\right) = \frac{4\sqrt{2} + \sqrt{5}}{9}$

C **55.** $\cot(x + y) = \frac{\cos(x + y)}{\sin(x + y)} = \frac{\cos x \cos y - \sin x \sin y}{\sin x \cos y + \cos x \sin y} = \frac{\frac{\cos x \cos y}{\sin x \sin y} - \frac{\sin x \sin y}{\sin x \sin y}}{\frac{\sin x \cos y}{\sin x \sin y} + \frac{\cos x \sin y}{\sin x \sin y}} = \frac{\cot x \cot y - 1}{\cot y + \cot x} = \frac{\cot x \cot y - 1}{\cot x + \cot y}$; $\sin(x + y) \neq 0$, $\sin x \neq 0$, $\sin y \neq 0$

56. $\sec(x + y) = \dfrac{1}{\cos(x + y)} = \dfrac{1}{\cos x \cos y - \sin x \sin y} = \dfrac{\dfrac{1}{\cos x \cos y}}{\dfrac{\cos x \cos y}{\cos x \cos y} - \dfrac{\sin x \sin y}{\cos x \cos y}} =$

$\dfrac{\sec x \sec y}{1 - \tan x \tan y}$; $\cos(x + y) \neq 0, \cos x \neq 0, \cos y \neq 0$

57. $\csc(x + y) = \dfrac{1}{\sin(x + y)} = \dfrac{1}{\sin x \cos y + \cos x \sin y} = \dfrac{\dfrac{1}{\sin x \sin y}}{\dfrac{\sin x \cos y}{\sin x \sin y} + \dfrac{\cos x \sin y}{\sin x \sin y}} =$

$\dfrac{\csc x \csc y}{\cot y + \cot x} = \dfrac{\csc x \csc y}{\cot x + \cot y}$; $\sin(x + y) \neq 0, \sin x \neq 0, \sin y \neq 0$

58. $\dfrac{\sin(x - y)}{\sin y} + \dfrac{\cos(x - y)}{\cos y} = \dfrac{\sin x \cos y - \cos x \sin y}{\sin y} + \dfrac{\cos x \cos y + \sin x \sin y}{\cos y} =$

$\dfrac{\sin x \cos y}{\sin y} - \cos x + \cos x + \dfrac{\sin x \sin y}{\cos y} = \sin x\left(\dfrac{\cos y}{\sin y} + \dfrac{\sin y}{\cos y}\right) =$

$\sin x\left(\dfrac{\cos^2 y + \sin^2 y}{\sin y \cos y}\right) = \dfrac{\sin x}{\sin y \cos y} = \sin x \sec y \csc y$; $\sin y \neq 0, \cos y \neq 0$

59. $\dfrac{\cos(x + y)}{\sin y} + \dfrac{\sin(x + y)}{\cos y} = \dfrac{\cos x \cos y - \sin x \sin y}{\sin y} + \dfrac{\sin x \cos y + \cos x \sin y}{\cos y} =$

$\dfrac{\cos x \cos y}{\sin y} - \sin x + \sin x + \dfrac{\cos x \sin y}{\cos y} = \cos x\left(\dfrac{\cos y}{\sin y} + \dfrac{\sin y}{\cos y}\right) =$

$\cos x\left(\dfrac{\cos^2 y + \sin^2 y}{\sin y \cos y}\right) = \cos x\left(\dfrac{1}{\sin y \cos y}\right) = \cos x \sec y \csc y$; $\sin y \neq 0, \cos y \neq 0$

Pages 640–641 • WRITTEN EXERCISES

A **1.** $\cos \alpha = \dfrac{3}{5}$; $\sin 2\alpha = 2 \sin \alpha \cos \alpha = 2 \cdot \dfrac{4}{5} \cdot \dfrac{3}{5} = \dfrac{24}{25}$

2. $\cos \alpha = -\dfrac{4}{5}$; $\sin 2\alpha = 2\left(-\dfrac{3}{5}\right)\left(-\dfrac{4}{5}\right) = \dfrac{24}{25}$

3. $\sin \alpha = \dfrac{12}{13}$; $\sin 2\alpha = 2\left(\dfrac{12}{13}\right)\left(-\dfrac{5}{13}\right) = -\dfrac{120}{169}$

4. $\sin \alpha = -\dfrac{\sqrt{5}}{3}$; $\sin 2\alpha = 2\left(-\dfrac{\sqrt{5}}{3}\right)\left(\dfrac{2}{3}\right) = -\dfrac{4\sqrt{5}}{9}$

5. $\cos \alpha = \dfrac{3}{\sqrt{13}}$; $\sin 2\alpha = 2\left(-\dfrac{2}{\sqrt{13}}\right)\left(\dfrac{3}{\sqrt{13}}\right) = -\dfrac{12}{13}$

6. $\tan \alpha = \dfrac{2\sqrt{10}}{3}$; $\sec \alpha = \sqrt{1 + \tan^2 \alpha} = \sqrt{1 + \left(\dfrac{2\sqrt{10}}{3}\right)^2} = \dfrac{7}{3}$; $\cos \alpha = \dfrac{1}{\sec \alpha} = \dfrac{3}{7}$

and $\sin \alpha = \dfrac{2\sqrt{10}}{7}$; $\sin 2\alpha = 2\left(\dfrac{2\sqrt{10}}{7}\right)\left(\dfrac{3}{7}\right) = \dfrac{12\sqrt{10}}{49}$

7. $\cos 2\alpha = 2 \cos^2 \alpha - 1 = 2\left(\dfrac{5}{13}\right)^2 - 1 = -\dfrac{119}{169}$

8. $\cos 2\alpha = 1 - 2\sin^2\alpha = 1 - 2\left(\frac{2}{\sqrt{5}}\right)^2 = -\frac{3}{5}$

9. $\cos 2\alpha = 1 - 2\sin^2\alpha = 1 - 2\left(-\frac{\sqrt{7}}{4}\right)^2 = \frac{1}{8}$

10. $\cos 2\alpha = 2\cos^2\alpha - 1 = 2\left(-\frac{3}{\sqrt{10}}\right)^2 - 1 = \frac{4}{5}$

11. $\cos 2\alpha = 2\cos^2\alpha - 1 = 2\left(\frac{\sqrt{5}}{3}\right)^2 - 1 = \frac{1}{9}$

12. $\sec\alpha = -\sqrt{1 + \tan^2\alpha} = -\sqrt{1 + \left(\frac{2\sqrt{6}}{5}\right)^2} = -\frac{7}{5}$; $\cos\alpha = -\frac{5}{7}$;

$\cos 2\alpha = 2\cos^2\alpha - 1 = 2\left(\frac{5}{7}\right)^2 - 1 = \frac{1}{49}$

13. $\cos x = -\sqrt{1 - \left(\frac{2}{3}\right)^2} = -\frac{\sqrt{5}}{3}$; $\tan x = \frac{\sin x}{\cos x} = \frac{\frac{2}{3}}{-\frac{\sqrt{5}}{3}} = -\frac{2}{\sqrt{5}}$;

$\tan 2x = \frac{2\tan x}{1 - \tan^2 x} = \frac{2\left(-\frac{2}{\sqrt{5}}\right)}{1 - \left(-\frac{2}{\sqrt{5}}\right)^2} = \frac{-\frac{4}{\sqrt{5}}}{\frac{1}{5}} = \frac{-20}{\sqrt{5}} = -4\sqrt{5}$

14. $\tan 2x = \frac{2\tan x}{1 - \tan^2 x} = \frac{2\left(\frac{1}{5}\right)}{1 - \left(\frac{1}{5}\right)^2} = \frac{\frac{2}{5}}{\frac{24}{25}} = \frac{5}{12}$; $\tan 4x = \frac{2\tan 2x}{1 - \tan^2 2x} = \frac{2\left(\frac{5}{12}\right)}{1 - \frac{25}{144}} =$

$\frac{\frac{10}{12}}{\frac{119}{144}} = \frac{120}{119}$

15. $\cos 75° = \cos\frac{150°}{2} = \sqrt{\frac{1 + \cos 150°}{2}} = \sqrt{\frac{1 + \left(-\frac{\sqrt{3}}{2}\right)}{2}} =$

$\sqrt{\frac{2 - \sqrt{3}}{4}} = \frac{\sqrt{2 - \sqrt{3}}}{2}$

16. $\sin 22.5° = \sin\frac{45°}{2} = \sqrt{\frac{1 - \cos 45°}{2}} = \sqrt{\frac{1 - \frac{\sqrt{2}}{2}}{2}} = \frac{\sqrt{2 - \sqrt{2}}}{2}$

17. $\sin 105° = \sin\frac{210°}{2} = \sqrt{\frac{1 - \cos 210°}{2}} = \sqrt{\frac{1 - \left(-\frac{\sqrt{3}}{2}\right)}{2}} = \frac{\sqrt{2 + \sqrt{3}}}{2}$

18. $\tan 165^\circ = \tan \frac{330^\circ}{2} = \frac{\sin 330^\circ}{1 + \cos 330^\circ} = \frac{\frac{1}{2}}{1 + \frac{\sqrt{3}}{2}} = \frac{1}{2 + \sqrt{3}} = \frac{2 - \sqrt{3}}{1} = 2 - \sqrt{3}$

19. $\tan 112.5^\circ = \tan \frac{225^\circ}{2} = \frac{\sin 225^\circ}{1 + \cos 225^\circ} = \frac{-\frac{\sqrt{2}}{2}}{1 - \frac{\sqrt{2}}{2}} = \frac{-\sqrt{2}}{2 - \sqrt{2}} \cdot \frac{2 + \sqrt{2}}{2 + \sqrt{2}} =$

$\frac{-2\sqrt{2} - 2}{2} = -\sqrt{2} - 1$

20. $\cos 15^\circ = \cos \frac{30^\circ}{2} = \sqrt{\frac{1 + \cos 30^\circ}{2}} = \sqrt{\frac{1 + \frac{\sqrt{3}}{2}}{2}} = \sqrt{\frac{2 + \sqrt{3}}{4}} = \frac{\sqrt{2 + \sqrt{3}}}{2}$

21. $\sin \frac{3\pi}{8} = \sin \frac{\frac{3\pi}{4}}{2} = \sqrt{\frac{1 - \cos \frac{3\pi}{4}}{2}} = \sqrt{\frac{1 - \left(-\frac{\sqrt{2}}{2}\right)}{2}} = \frac{\sqrt{2 + \sqrt{2}}}{2}$

22. $\cos \frac{\pi}{8} = \cos \frac{\frac{\pi}{4}}{2} = \sqrt{\frac{1 + \cos \frac{\pi}{4}}{2}} = \sqrt{\frac{1 + \frac{\sqrt{2}}{2}}{2}} = \frac{\sqrt{2 + \sqrt{2}}}{2}$

23. $\tan \frac{\pi}{8} = \tan \frac{\frac{\pi}{4}}{2} = \frac{\sin \frac{\pi}{4}}{1 + \cos \frac{\pi}{4}} = \frac{\frac{\sqrt{2}}{2}}{1 + \frac{\sqrt{2}}{2}} = \frac{\sqrt{2}}{2 + \sqrt{2}} \cdot \frac{2 - \sqrt{2}}{2 - \sqrt{2}} = \frac{2\sqrt{2} - 2}{2} = \sqrt{2} - 1$

24. $\sin \frac{5\pi}{8} = \sin \frac{\frac{5\pi}{4}}{2} = \sqrt{\frac{1 - \cos \frac{5\pi}{4}}{2}} = \sqrt{\frac{1 - \left(-\frac{\sqrt{2}}{2}\right)}{2}} = \frac{\sqrt{2 + \sqrt{2}}}{2}$

25. $\cos \frac{7\pi}{12} = \cos \frac{\frac{7\pi}{6}}{2} = -\sqrt{\frac{1 + \cos \frac{7\pi}{6}}{2}} = -\sqrt{\frac{1 - \frac{\sqrt{3}}{2}}{2}} = -\sqrt{\frac{2 - \sqrt{3}}{4}} =$

$-\frac{\sqrt{2 - \sqrt{3}}}{2}$

26. $\tan \frac{11\pi}{12} = \tan \frac{\frac{11\pi}{6}}{2} = \frac{\sin \frac{11\pi}{6}}{1 + \cos \frac{11\pi}{6}} = \frac{-\frac{1}{2}}{1 + \frac{\sqrt{3}}{2}} = \frac{-1}{2 + \sqrt{3}} \cdot \frac{2 - \sqrt{3}}{2 - \sqrt{3}} =$

$\frac{-2 + \sqrt{3}}{4 - 3} = -2 + \sqrt{3}$

27. $\sin\frac{\alpha}{2} = \sqrt{\frac{1-\frac{7}{9}}{2}} = \sqrt{\frac{1}{9}} = \frac{1}{3}$

$\cos\frac{\alpha}{2} = \sqrt{\frac{1+\frac{7}{9}}{2}} = \sqrt{\frac{8}{9}} = \frac{2\sqrt{2}}{3}$

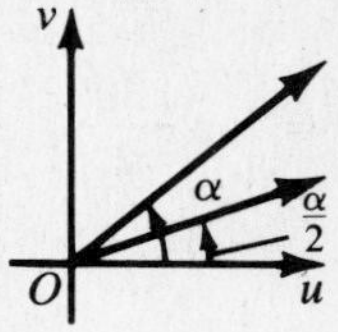

28. $\sin\frac{\alpha}{2} = \sqrt{\frac{1-\left(-\frac{7}{9}\right)}{2}} = \sqrt{\frac{9+7}{18}} = \frac{2\sqrt{2}}{3}$

$\cos\frac{\alpha}{2} = \sqrt{\frac{1-\frac{7}{9}}{2}} = \sqrt{\frac{9-7}{18}} = \frac{1}{3}$

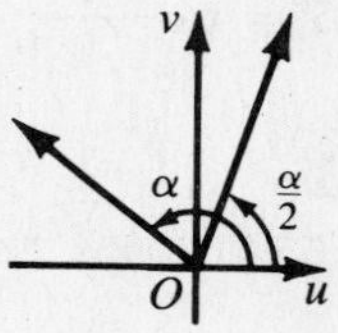

29. $\cos\alpha = \sqrt{1-\left(\frac{24}{25}\right)^2} = \frac{7}{25}$

$\sin\frac{\alpha}{2} = \sqrt{\frac{1-\frac{7}{25}}{2}} = \sqrt{\frac{9}{25}} = \frac{3}{5}$

$\cos\frac{\alpha}{2} = \sqrt{1-\sin^2\frac{\alpha}{2}} = \frac{4}{5}$

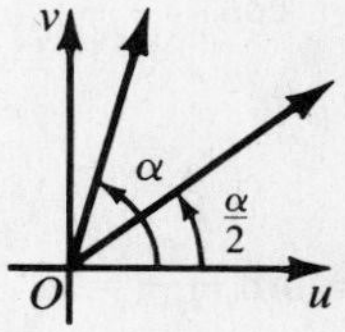

30. $\cos\alpha = -\sqrt{1-\left(\frac{24}{25}\right)^2} = -\frac{7}{25}$

$\sin\frac{\alpha}{2} = \sqrt{\frac{1-\left(-\frac{7}{25}\right)}{2}} = \sqrt{\frac{16}{25}} = \frac{4}{5}$

$\cos\frac{\alpha}{2} = \sqrt{1-\left(\frac{4}{5}\right)^2} = \frac{3}{5}$

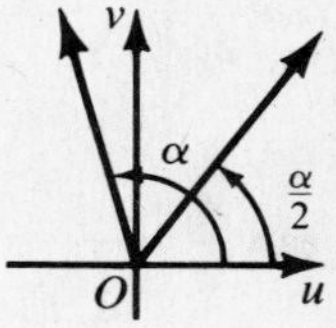

31. $\sin\frac{\alpha}{2} = \sqrt{\frac{1-\left(-\frac{7}{8}\right)}{2}} = \sqrt{\frac{8+7}{16}} = \frac{\sqrt{15}}{4}$

$\cos\frac{\alpha}{2} = -\sqrt{\frac{1+\left(-\frac{7}{8}\right)}{2}} = -\sqrt{\frac{8-7}{16}} = -\frac{1}{4}$

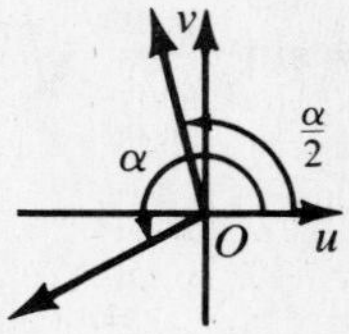

32. $\cos\alpha = \sqrt{1-\left(-\frac{7}{8}\right)^2} = \frac{\sqrt{15}}{8}$

$\sin\frac{\alpha}{2} = \sqrt{\frac{1-\frac{\sqrt{15}}{8}}{2}} = \sqrt{\frac{8-\sqrt{15}}{16}} = \frac{\sqrt{8-\sqrt{15}}}{4}$

$\cos\frac{\alpha}{2} = -\sqrt{\frac{1+\frac{\sqrt{15}}{8}}{2}} = -\frac{\sqrt{8+\sqrt{15}}}{4}$

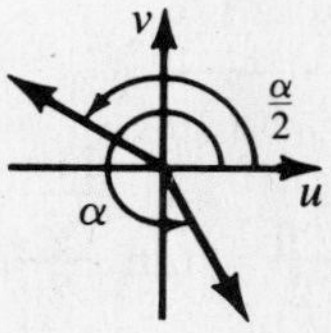

B **33.** $\sin\left[2\left(\frac{x}{2}\right)\right] = 2\sin\frac{x}{2}\cos\frac{x}{2}$

Case 1: $0 < \frac{x}{2} \le \frac{\pi}{2}$ $\quad \sin\left[2\left(\frac{x}{2}\right)\right] = 2\sqrt{\frac{1-\cos x}{2}}\sqrt{\frac{1+\cos x}{2}} = 2\sqrt{\frac{1-\cos^2 x}{4}} =$

$\sqrt{1-\cos^2 x} = \sin x$

Case 2: $\frac{\pi}{2} < \frac{x}{2} \le \pi$ $\quad \sin\left[2\left(\frac{x}{2}\right)\right] = 2\sqrt{\frac{1-\cos x}{2}}\left(-\sqrt{\frac{1+\cos x}{2}}\right) =$

$-\sqrt{1-\cos^2 x} = \sin x$ (*Note:* Since $\pi < x \le 2\pi$, $\sin x$ is negative.)

Case 3: $\pi < \frac{x}{2} \le \frac{3\pi}{2}$ $\quad \sin\left[2\left(\frac{x}{2}\right)\right] = 2\left(-\sqrt{\frac{1-\cos x}{2}}\right)\left(-\sqrt{\frac{1+\cos x}{2}}\right) =$

$\sqrt{1-\cos^2 x} = \sin x$

Case 4: $\frac{3\pi}{2} < \frac{x}{2} \le 2\pi$ $\quad \sin\left[2\left(\frac{x}{2}\right)\right] = 2\left(-\sqrt{\frac{1-\cos x}{2}}\right)\left(\sqrt{\frac{1+\cos x}{2}}\right) =$

$-\sqrt{1-\cos^2 x} = \sin x$ (*Note:* Since $3\pi < x \le 4\pi$, $\sin x$ is negative.)

34. $\cos\left[2\left(\frac{x}{2}\right)\right] = 2\cos^2\left(\frac{x}{2}\right) - 1 = 2\left(\pm\sqrt{\frac{1+\cos x}{2}}\right)^2 - 1 = 2\left(\frac{1+\cos x}{2}\right) - 1 =$

$1 + \cos x - 1 = \cos x$

35. $\sin^2\frac{x}{2} + \cos^2\frac{x}{2} = \left(\pm\sqrt{\frac{1-\cos x}{2}}\right)^2 + \left(\pm\sqrt{\frac{1+\cos x}{2}}\right)^2 =$

$\frac{1-\cos x}{2} + \frac{1+\cos x}{2} = 1$

36. $\sin^2 2x + \cos^2 2x = (2\sin x\cos x)^2 + (\cos^2 x - \sin^2 x)^2 =$
$4\sin^2 x\cos^2 x + \cos^4 x - 2\sin^2 x\cos^2 x + \sin^4 x =$
$\cos^4 x + 2\sin^2 x\cos^2 x + \sin^4 x = (\cos^2 x + \sin^2 x)^2 = 1$

37. $1 + \tan^2\frac{x}{2} = 1 + \left(\pm\sqrt{\frac{1-\cos x}{1+\cos x}}\right)^2 = 1 + \frac{1-\cos x}{1+\cos x} = \frac{2}{1+\cos x}$; $\sec^2\frac{x}{2} =$

$\left(\frac{1}{\cos\frac{x}{2}}\right)^2 = \left(\frac{1}{\pm\sqrt{\frac{1+\cos x}{2}}}\right)^2 = \frac{1}{\frac{1+\cos x}{2}} = \frac{2}{1+\cos x}$; $\therefore\ 1 + \tan^2\frac{x}{2} = \sec^2\frac{x}{2}$.

C **38.** Since α and β are in Quadrant I,
$\sin\beta = \sqrt{1-\cos^2\beta} = \sqrt{1-\sin^2\alpha} = \cos\alpha$;
$\sin 2\alpha = 2\sin\alpha\cos\alpha = 2\cos\beta\sin\beta =$
$\sin 2\beta$

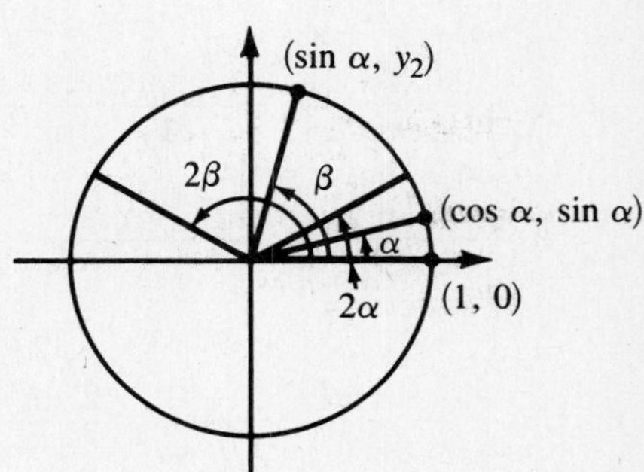

Page 641 • COMPUTER EXERCISES

1.
```
10  LET PI = 3.141593
20  INPUT A
30  LET A2 = A/2
40  LET A = A * PI/180
50  LET X = COS(A)
60  IF ABS(X - INT(X + .05)) < .001 THEN LET X = INT(X + .05)
70  LET SN = SQR((1 - X)/2)
80  LET SN = INT(SN * 10000 + .5)/10000
90  IF A2 > 360 THEN LET A2 = A2 - 360 : GOTO 90
100  IF A2 > 180 AND A2 < 360 THEN LET S$ = "-" ELSE S$ = ""
110  PRINT S$;SN
120  END
```

2. 0.5 **3.** −0.5 **4.** −0.5 **5.** −0.8660

Page 643 • WRITTEN EXERCISES

A **1.** $\dfrac{\cos 2\alpha}{\sin^2 \alpha} = \dfrac{\cos^2 \alpha - \sin^2 \alpha}{\sin^2 \alpha} = \dfrac{\cos^2 \alpha}{\sin^2 \alpha} - \dfrac{\sin^2 \alpha}{\sin^2 \alpha} = \cot^2 \alpha - 1; \sin \alpha \neq 0$

2. $\tan \alpha(\cos 2\alpha + 1) = \dfrac{\sin \alpha}{\cos \alpha}(2 \cos^2 \alpha - 1 + 1) = \dfrac{\sin \alpha}{\cos \alpha}(2 \cos^2 \alpha) = 2 \sin \alpha \cos \alpha =$ $\sin 2\alpha; \cos \alpha \neq 0$

3. $(\sin \alpha - \cos \alpha)^2 = \sin^2 \alpha - 2 \sin \alpha \cos \alpha + \cos^2 \alpha =$ $\sin^2 \alpha + \cos^2 \alpha - 2 \sin \alpha \cos \alpha = 1 - \sin 2\alpha$

4. $\cos 2\alpha + 2 \sin^2 \alpha = \cos^2 \alpha - \sin^2 \alpha + 2 \sin^2 \alpha = \cos^2 \alpha + \sin^2 \alpha = 1$

5. $1 - \sin 2\alpha \tan \alpha = 1 - (2 \sin \alpha \cos \alpha)\left(\dfrac{\sin \alpha}{\cos \alpha}\right) = 1 - 2 \sin^2 \alpha = \cos 2\alpha; \cos \alpha \neq 0$

6. $\cot \alpha \cos \alpha - \sin \alpha = \dfrac{\cos \alpha}{\sin \alpha} \cdot \cos \alpha - \dfrac{\sin^2 \alpha}{\sin \alpha} = \dfrac{\cos^2 \alpha - \sin^2 \alpha}{\sin \alpha} = \dfrac{\cos 2\alpha}{\sin \alpha} =$ $\cos 2\alpha \cdot \dfrac{1}{\sin \alpha} = \cos 2\alpha \csc \alpha; \sin \alpha \neq 0$

7. $\dfrac{\cos 2\alpha}{\cos \alpha + \sin \alpha} = \dfrac{\cos^2 \alpha - \sin^2 \alpha}{\cos \alpha + \sin \alpha} = \dfrac{(\cos \alpha - \sin \alpha)(\cos \alpha + \sin \alpha)}{(\cos \alpha + \sin \alpha)} = \cos \alpha - \sin \alpha;$ $\cos \alpha \neq -\sin \alpha$

8. $\dfrac{\sin 2\alpha}{1 - \cos 2\alpha} = \dfrac{2 \sin \alpha \cos \alpha}{1 - (1 - 2 \sin^2 \alpha)} = \dfrac{2 \sin \alpha \cos \alpha}{1 - 1 + 2 \sin^2 \alpha} = \dfrac{2 \sin \alpha \cos \alpha}{2 \sin^2 \alpha} = \dfrac{\cos \alpha}{\sin \alpha} =$ $\cot \alpha; \sin \alpha \neq 0$

9. $\dfrac{1 - \sin 2x}{\cos^2 x} = \dfrac{\sin^2 x + \cos^2 x - 2 \sin x \cos x}{\cos^2 x} = \dfrac{(\sin x - \cos x)^2}{\cos^2 x} =$ $\left(\dfrac{\cos x - \sin x}{\cos x}\right)^2 = (1 - \tan x)^2; \cos x \neq 0$

10. $\dfrac{1 - \cos 2\alpha}{1 + \cos 2\alpha} = \dfrac{1 - (1 - 2 \sin^2 \alpha)}{1 + (2 \cos^2 \alpha - 1)} = \dfrac{1 - 1 + 2 \sin^2 \alpha}{1 + 2 \cos^2 \alpha - 1} = \dfrac{2 \sin^2 \alpha}{2 \cos^2 \alpha} = \tan^2 \alpha; \cos \alpha \neq 0$

11. $\dfrac{1 - \tan^2 \alpha}{1 + \tan^2 \alpha} = \dfrac{1 - \dfrac{\sin^2 \alpha}{\cos^2 \alpha}}{\sec^2 \alpha} = \dfrac{\dfrac{\cos^2 \alpha - \sin^2 \alpha}{\cos^2 \alpha}}{\dfrac{1}{\cos^2 \alpha}} = \cos^2 \alpha - \sin^2 \alpha = \cos 2\alpha; \cos \alpha \neq 0$

12. $\cot 2\alpha = \dfrac{\cos 2\alpha}{\sin 2\alpha} = \dfrac{\cos^2\alpha - \sin^2\alpha}{2\sin\alpha\cos\alpha} = \dfrac{\cos^2\alpha}{2\sin\alpha\cos\alpha} - \dfrac{\sin^2\alpha}{2\sin\alpha\cos\alpha} =$
$\dfrac{1}{2}\left(\dfrac{\cos\alpha}{\sin\alpha} - \dfrac{\sin\alpha}{\cos\alpha}\right) = \dfrac{1}{2}(\cot\alpha - \tan\alpha)$; $\sin\alpha, \cos\alpha \neq 0$

B **13.** $\cos^4\alpha - \sin^4\alpha = (\cos^2\alpha + \sin^2\alpha)(\cos^2\alpha - \sin^2\alpha) = (1)(\cos 2\alpha) = \cos 2\alpha$

14. $1 - \tan^4 x = (1 + \tan^2 x)(1 - \tan^2 x) = \sec^2 x\left(1 - \dfrac{\sin^2 x}{\cos^2 x}\right) =$
$\dfrac{1}{\cos^2 x}\left(\dfrac{\cos^2 x - \sin^2 x}{\cos^2 x}\right) = \dfrac{\cos 2x}{\cos^4 x}$; $\cos x \neq 0$

15. $\dfrac{1 - \cos x}{\sin x} = \dfrac{(1 - \cos x)(1 + \cos x)}{\sin x(1 + \cos x)} = \dfrac{1 - \cos^2 x}{\sin x(1 + \cos x)} = \dfrac{\sin^2 x}{\sin x(1 + \cos x)} =$
$\dfrac{\sin x}{1 + \cos x} = \tan\dfrac{x}{2}$; $\sin x \neq 0$

16. $\tan\dfrac{\alpha}{2}(\cos 2\alpha - \cos^2\alpha) = \dfrac{\sin\alpha}{1 + \cos\alpha}(2\cos^2\alpha - 1 - \cos^2\alpha) =$
$\dfrac{\sin\alpha(\cos\alpha + 1)(\cos\alpha - 1)}{1 + \cos\alpha} = \sin\alpha(\cos\alpha - 1)$; $\dfrac{\sin 2\alpha}{2} - \sin\alpha =$
$\dfrac{2\sin\alpha\cos\alpha}{2} - \sin\alpha = \sin\alpha(\cos\alpha - 1)$; thus $\tan\dfrac{\alpha}{2}(\cos 2\alpha - \cos^2\alpha) =$
$\dfrac{\sin 2\alpha}{2} - \sin\alpha$; $\cos\alpha \neq -1$

17. $\sin 3\alpha = \sin(2\alpha + \alpha) = \sin 2\alpha\cos\alpha + \cos 2\alpha\sin\alpha = (2\sin\alpha\cos\alpha)\cos\alpha +$
$(1 - 2\sin^2\alpha)\sin\alpha = 2\sin\alpha\cos^2\alpha + \sin\alpha - 2\sin^3\alpha = 2\sin\alpha(1 - \sin^2\alpha) +$
$\sin\alpha - 2\sin^3\alpha = 2\sin\alpha - 2\sin^3\alpha + \sin\alpha - 2\sin^3\alpha = 3\sin\alpha - 4\sin^3\alpha$

18. $\cos 3\alpha = \cos(2\alpha + \alpha) = \cos 2\alpha\cos\alpha - \sin 2\alpha\sin\alpha = (2\cos^2\alpha - 1)(\cos\alpha) -$
$(2\sin\alpha\cos\alpha)(\sin\alpha) = 2\cos^3\alpha - \cos\alpha - 2\sin^2\alpha\cos\alpha = 2\cos^3\alpha -$
$\cos\alpha - 2(1 - \cos^2\alpha)\cos\alpha = 2\cos^3\alpha - \cos\alpha - 2\cos\alpha - 2\cos^3\alpha =$
$4\cos^3\alpha - 3\cos\alpha$

19. $\cos 4\alpha = \cos^2 2\alpha - \sin^2 2\alpha = (\cos^2\alpha - \sin^2\alpha)^2 - (2\sin\alpha\cos\alpha)^2 = \cos^4\alpha -$
$2\sin^2\alpha\cos^2\alpha + \sin^4\alpha - 4\sin^2\alpha\cos^2\alpha = \cos^4\alpha + \sin^4\alpha - 6\sin^2\alpha\cos^2\alpha$

20. $\sin 4\alpha = 2\sin 2\alpha\cos 2\alpha = 2(2\sin\alpha\cos\alpha)(1 - 2\sin^2\alpha) =$
$4\sin\alpha\cos\alpha - 8\sin^3\alpha\cos\alpha$

21. $\dfrac{\cos\alpha}{\cos\alpha - \sin\alpha} - \dfrac{\cos\alpha}{\cos\alpha + \sin\alpha} = \dfrac{\cos\alpha(\cos\alpha + \sin\alpha) - \cos\alpha(\cos\alpha - \sin\alpha)}{(\cos\alpha - \sin\alpha)(\cos\alpha + \sin\alpha)} =$
$\dfrac{\cos^2\alpha + \cos\alpha\sin\alpha - \cos^2\alpha + \cos\alpha\sin\alpha}{\cos^2\alpha - \sin^2\alpha} = \dfrac{2\cos\alpha\sin\alpha}{\cos 2\alpha} = \dfrac{\sin 2\alpha}{\cos 2\alpha} = \tan 2\alpha$;
$\cos\alpha \neq \pm\sin\alpha$

22. $\dfrac{\cos\alpha}{\cos\alpha + \sin\alpha} - \dfrac{\sin\alpha}{\cos\alpha - \sin\alpha} = \dfrac{\cos\alpha(\cos\alpha - \sin\alpha) - \sin\alpha(\cos\alpha + \sin\alpha)}{(\cos\alpha + \sin\alpha)(\cos\alpha - \sin\alpha)} =$
$\dfrac{\cos^2\alpha - \cos\alpha\sin\alpha - \sin\alpha\cos\alpha - \sin^2\alpha}{\cos^2\alpha - \sin^2\alpha} = \dfrac{\cos^2\alpha - \sin^2\alpha}{\cos^2\alpha - \sin^2\alpha} - \dfrac{2\sin\alpha\cos\alpha}{\cos^2\alpha - \sin^2\alpha} =$
$1 - \dfrac{\sin 2\alpha}{\cos 2\alpha} = 1 - \tan 2\alpha$; $\cos\alpha \neq \pm\sin\alpha$

23. $\sin(\alpha + \beta)\sin(\alpha - \beta) = (\sin\alpha\cos\beta + \cos\alpha\sin\beta)(\sin\alpha\cos\beta - \cos\alpha\sin\beta) =$
$\sin^2\alpha\cos^2\beta - \cos^2\alpha\sin^2\beta = \sin^2\alpha(1 - \sin^2\beta) - (1 - \sin^2\alpha)\sin^2\beta =$
$\sin^2\alpha - \sin^2\alpha\sin^2\beta - \sin^2\beta + \sin^2\alpha\sin^2\beta = \sin^2\alpha - \sin^2\beta$

24. $\cos(a + \beta)\cos(\alpha - \beta) = (\cos\alpha\cos\beta - \sin\alpha\sin\beta)(\cos\alpha\cos\beta + \sin\alpha\sin\beta) =$
$\cos^2\alpha\cos^2\beta - \sin^2\alpha\sin^2\beta = \cos^2\alpha(1 - \sin^2\beta) - (1 - \cos^2\alpha)\sin^2\beta =$
$\cos^2\alpha - \cos^2\alpha\sin^2\beta - \sin^2\beta + \cos^2\alpha\sin^2\beta = \cos^2\alpha - \sin^2\beta$

25. $\sin(a + \beta) + \sin(\alpha - \beta) = \sin\alpha\cos\beta + \cos\alpha\sin\beta + \sin\alpha\cos\beta - \cos\alpha\sin\beta =$
$2\sin\alpha\cos\beta$

26. $\sin(\alpha + \beta) - \sin(\alpha - \beta) = \sin\alpha\cos\beta + \cos\alpha\sin\beta - (\sin\alpha\cos\beta - \cos\alpha\sin\beta) =$
$2\cos\alpha\sin\beta$

27. $\cos(\alpha + \beta) + \cos(\alpha - \beta) = \cos\alpha\cos\beta - \sin\alpha\sin\beta + \cos\alpha\cos\beta + \sin\alpha\sin\beta =$
$2\cos\alpha\cos\beta$

28. $\cos(\alpha + \beta) - \cos(\alpha - \beta) = \cos\alpha\cos\beta - \sin\alpha\sin\beta - (\cos\alpha\cos\beta + \sin\alpha\sin\beta) =$
$-2\sin\alpha\sin\beta$

C 29. By Exercise 25, $2\sin\frac{A + B}{2}\cos\frac{A - B}{2} = \sin\left(\frac{A + B}{2} + \frac{A - B}{2}\right) +$
$\sin\left(\frac{A + B}{2} - \frac{A - B}{2}\right) = \sin A + \sin B.$

30. By Exercise 26, $2\cos\frac{A + B}{2}\sin\frac{A - B}{2} = \sin\left(\frac{A + B}{2} + \frac{A - B}{2}\right) -$
$\sin\left(\frac{A + B}{2} - \frac{A - B}{2}\right) = \sin A - \sin B.$

31. By Exercise 27, $2\cos\frac{A + B}{2}\sin\frac{A - B}{2} = \cos\left(\frac{A + B}{2} + \frac{A - B}{2}\right) +$
$\cos\left(\frac{A + B}{2} - \frac{A - B}{2}\right) = \cos A + \cos B.$

32. By Exercise 28, $-2\sin\frac{A + B}{2}\sin\frac{A - B}{2} = \cos\left(\frac{A + B}{2} + \frac{A - B}{2}\right) -$
$\cos\left(\frac{A + B}{2} - \frac{A - B}{2}\right) = \cos A - \cos B.$

Page 644 • SELF-TEST 2

1. $\frac{\tan 165° - \tan 15°}{1 + \tan 165° \tan 15°} = \tan(165° - 15°) = \tan 150° = -\frac{1}{\sqrt{3}}$

2. $\cos\frac{\pi}{12} = \cos\left(\frac{\pi}{3} - \frac{\pi}{4}\right) = \cos\frac{\pi}{3}\cos\frac{\pi}{4} + \sin\frac{\pi}{3}\sin\frac{\pi}{4} = \frac{1}{2}\cdot\frac{1}{\sqrt{2}} + \frac{\sqrt{3}}{2}\cdot\frac{1}{\sqrt{2}} =$
$\frac{1 + \sqrt{3}}{2\sqrt{2}} = \frac{\sqrt{2} + \sqrt{6}}{4}$

3. $\sin 345° = \sin(300° + 45°) = \sin 300° \cos 45° + \cos 300° \sin 45° =$
$-\frac{\sqrt{3}}{2}\cdot\frac{1}{\sqrt{2}} + \frac{1}{2}\cdot\frac{1}{\sqrt{2}} = \frac{-\sqrt{3} + 1}{2\sqrt{2}} = \frac{-\sqrt{6} + \sqrt{2}}{4}$

4. $\cos 2\alpha = 1 - 2\sin^2\alpha = 1 - 2\left(\frac{2}{3}\right)^2 = \frac{1}{9}$

5. $\sin 112.5° = \sin\frac{225°}{2} = \sqrt{\frac{1 - \cos 225°}{2}} = \sqrt{\frac{1 - \left(-\frac{\sqrt{2}}{2}\right)}{2}} =$
$\sqrt{\frac{2 + \sqrt{2}}{4}} = \frac{\sqrt{2 + \sqrt{2}}}{2}$

6. $\csc 2\alpha = \frac{1}{\sin 2\alpha} = \frac{1}{2\sin\alpha\cos\alpha} = \frac{1}{\sin\alpha}\cdot\frac{1}{2\cos\alpha} = \frac{\csc\alpha}{2\cos\alpha}$; $\sin\alpha \neq 0$, $\cos\alpha \neq 0$

Page 646 • WRITTEN EXERCISES

A 1. $c^2 = a^2 + b^2 - 2ab\cos C = 5^2 + 8^2 - 2 \cdot 5 \cdot 8 \cdot \cos 60° = 25 + 64 - 80\left(\frac{1}{2}\right) = 49;$
$c = \sqrt{49} = 7.0$

2. $b^2 = a^2 + c^2 - 2 \cdot a \cdot c \cdot \cos B = 14^2 + 16^2 - 2 \cdot 14 \cdot 16 \cdot \cos 120° =$
$196 + 256 - 448\left(-\frac{1}{2}\right) = 676;\ b = \sqrt{676} = 26.0$

3. $a^2 = (6.5)^2 + (4\sqrt{3})^2 - 2(6.5)(4\sqrt{3})\cos 30° = 42.25 + 48 - 52\sqrt{3} \cdot \frac{\sqrt{3}}{2} =$
$12.25;\ a = \sqrt{12.25} = 3.5$

4. $c^2 = 4^2 + (7.5)^2 - 2(4)(7.5)\cos 60° = 16 + 56.25 - 60\left(\frac{1}{2}\right) = 42.25;$
$c = \sqrt{42.25} = 6.5$

5. $b^2 = 15^2 + 11^2 - 2 \cdot 15 \cdot 11 \cdot \cos 26° = 225 + 121 - 330(0.8988) = 49.396;$
$b \approx 7.0$

6. $a^2 = 35^2 + 9^2 - 2 \cdot 35 \cdot 9 \cdot \cos 75° = 1225 + 81 - 630(0.2588) = 1142.956;$
$a \approx 33.8$

7. $\cos C = \frac{7^2 + 15^2 - 13^2}{2 \cdot 7 \cdot 15} = \frac{49 + 225 - 169}{210} = 0.5;\ m(C) = 60.0°$

8. $\cos B = \frac{7^2 + 14^2 - 10^2}{2 \cdot 7 \cdot 14} = \frac{145}{196} \approx 0.7398;\ m(B) \approx 42.3°$

9. $\cos A = \frac{22^2 + 30^2 - 19^2}{2(22)(30)} = \frac{1023}{1320} = 0.775;\ m(A) \approx 39.2°$

10. $\cos C = \frac{8^2 + 9^2 - 13^2}{2 \cdot 8 \cdot 9} = \frac{-24}{144} \approx -0.1667;\ m(C) \approx 180° - 80.4° = 99.6°$

Note: Answers may vary in the following exercises.

B 11. $c^2 = 4^2 + 10^2 - 2 \cdot 4 \cdot 10 \cos 59° = 16 + 100 - 80(0.5150) = 74.8;\ c \approx 8.65 \approx 9;$
$\cos A \approx \frac{10^2 + 74.8 - 4^2}{2(10)(8.65)} \approx 0.9179;\ m(A) \approx 23°;\ m(B) = 180° - (23° + 59°) = 98°$

12. $b^2 = 25^2 + 5^2 - 2 \cdot 5 \cdot 25 \cos 38° = 625 + 25 - 250(0.7880) = 453;$
$b \approx 21.28 \approx 21;\ \cos A \approx \frac{5^2 + 453 - 25^2}{2(5)(21.28)} \approx -0.6908;\ m(A) \approx 180° - 46° = 134°;$
$m(C) \approx 180° - (134° + 38°) = 8°$

13. $a^2 = 8^2 + 9^2 - 2 \cdot 8 \cdot 9 \cos 60° = 64 + 81 - 144\left(\frac{1}{2}\right) = 73;\ a \approx 8.54 \approx 9;$
$\cos B \approx \frac{9^2 + 73 - 8^2}{2(9)(8.54)} \approx 0.5855;\ m(B) \approx 54°;\ m(C) \approx 180° - (60° + 54°) = 66°$

14. $c^2 = 15^2 + 8^2 - 2 \cdot 8 \cdot 15 \cos 63° = 225 + 64 - 240(0.4540) = 180.04;$
$c \approx 13.42 \approx 13;\ \cos A \approx \frac{8^2 + 180.04 - 15^2}{2(8)(13.42)} \approx 0.0887;\ m(A) \approx 85°;$
$m(B) \approx 180° - (85° + 63°) = 32°$

15. $\cos A = \frac{7^2 + 8^2 - 4^2}{2(7)(8)} \approx 0.8661;\ m(A) \approx 30°;$
$\cos B = \frac{4^2 + 8^2 - 7^2}{2(4)(8)} \approx 0.4844;\ m(B) \approx 61°;\ m(C) \approx 180° - (30° + 61°) = 89°$

16. $\cos A = \frac{7^2 + 9^2 - 12^2}{2(7)(9)} \approx -0.1111;\ m(A) \approx 96°;$
$\cos B = \frac{12^2 + 9^2 - 7^2}{2(12)(9)} \approx 0.8148;\ m(B) \approx 35°;\ m(C) \approx 180° - (96° + 35°) = 49°$

17. 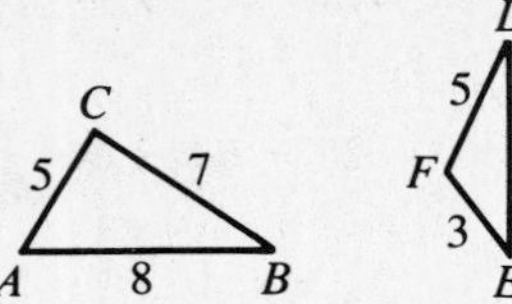

$\cos B = \dfrac{8^2 + 7^2 - 5^2}{2(8)(7)} = \dfrac{88}{112} = \dfrac{11}{14}$

$\cos E = \dfrac{7^2 + 3^2 - 5^2}{2(7)(3)} = \dfrac{33}{42} = \dfrac{11}{14}$

Since $\angle B$ and $\angle E$ are both acute angles, $m(B) = m(E)$.

Page 647 • PROBLEMS

A

1. The angle between the hands is 60°. Let x = distance between the outer ends of the hands. Then $x^2 = 8^2 + 15^2 - 2(8)(15)\cos 60° = 64 + 225 - 240\left(\frac{1}{2}\right) = 169$; $x = 13$ cm

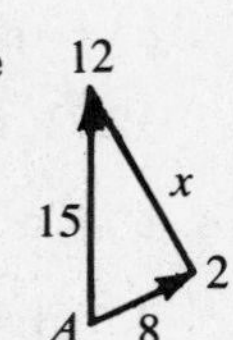

2. Let x = distance between the ships after 1 h. $x^2 = 12^2 + 17^2 - 2(12)(17)\cos 33° = 144 + 289 - 408(0.8387) = 90.81$; $x \approx 9.53$ km

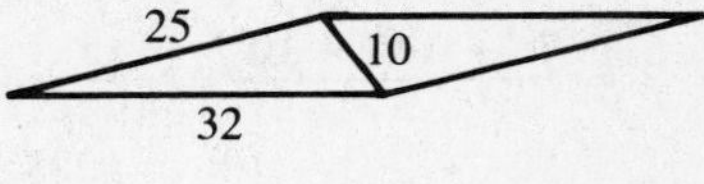

3. $\cos A = \dfrac{400^2 + 450^2 - 350^2}{2(400)(450)} \approx 0.6667$; $m(A) \approx 48.2°$

4. Let $\angle A$ be the acute angle. $\cos A = \dfrac{25^2 + 32^2 - 10^2}{2(25)(32)} \approx 0.9681$; $m(A) \approx 14.5°$

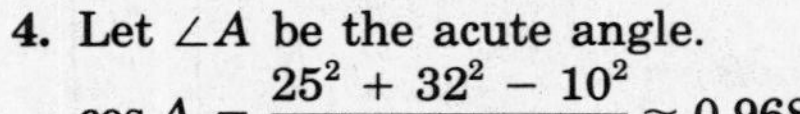

5. Let x = the distance between the outermost corners of the area swept out. $x^2 = 31^2 + 31^2 - 2(31)(31)\cos 120° = 31^2 + 31^2 - 2(31^2)\left(-\frac{1}{2}\right) = 2883$; $x \approx 53.7$ cm

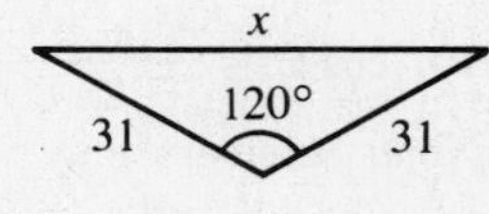

6. Let x = distance from A to B. $x^2 = 300^2 + 280^2 - 2(300)(280)\cos 26° = 90{,}000 + 78{,}400 - 168{,}000(0.8988) = 17{,}401.6$; $x \approx 132$ m

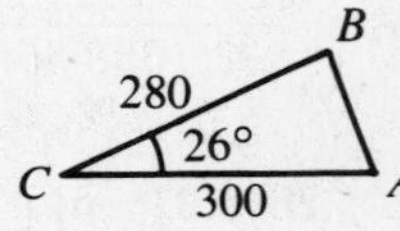

B

7. $(AC)^2 = 10^2 + 42^2 - 2(10)(42)\cos 60° = 100 + 1764 - 840\left(\frac{1}{2}\right) = 1444$; $AC = 38$; $m(\angle CAB) = m(\angle DCA) = 14°$ since $\overline{AB} \parallel \overline{DC}$; $(BC)^2 = 10^2 + 38^2 - 2(10)(38)\cos 14° = 100 + 1444 - 760(0.9703) = 806.57$; $BC \approx 28.4$

8. The measure of each angle of a regular pentagon is 108°, so the measure of each base angle of a triangle formed by a diagonal and two sides of the pentagon is $\frac{1}{2}(180° - 108°)$, or 36°, and thus the vertex angle of the triangle formed by two diagonals and a side is $108° - 2(36°)$, or 36°. $x^2 = 50^2 + 50^2 - 2(50)(50)\cos 36° = 2(50^2) - 2(50^2)(0.8090) = 954.9$; $x \approx 30.9$ cm

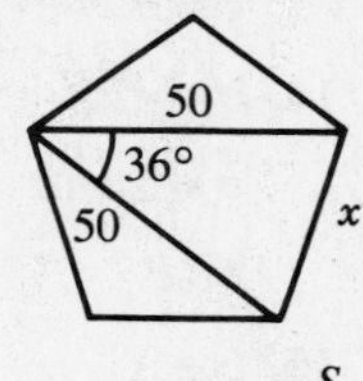

C

9. Let x = distance from Venus to the sun. $x^2 = (1.5 \times 10^8)^2 + (5.4 \times 10^7)^2 - 2(1.5 \times 10^8)(5.4 \times 10^7)\cos 39° = 2.25 \times 10^{16} + 2.916 \times 10^{15} - 1.62 \times 10^{16}(0.7771) = 1.2827 \times 10^{16}$; $x \approx 1.13 \times 10^8$ km

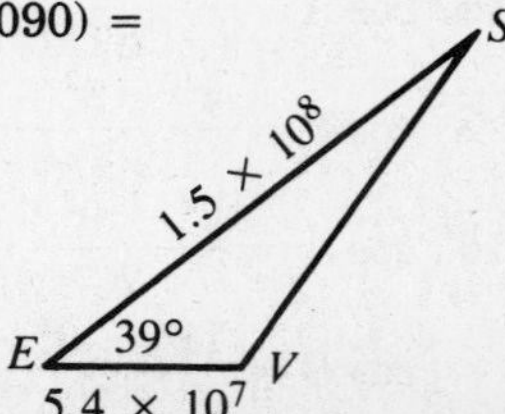

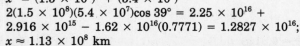

10. In $\triangle PSQ$, $x^2 = u^2 + v^2 - 2uv\cos\alpha$. In $\triangle QSR$, $y^2 = v^2 + u^2 - 2uv\cos(180 - \alpha)$. Since $\cos(180 - \alpha) = -\cos\alpha$, $y^2 = v^2 + u^2 + 2uv\cos\alpha$. $\therefore x^2 + y^2 = 2u^2 + 2v^2 = 2(u^2 + v^2)$.

Pages 651–652 • WRITTEN EXERCISES

A **1.** $\dfrac{a}{\sin 30°} = \dfrac{b}{\sin 54°}$; $\dfrac{12}{0.5} = \dfrac{b}{0.8090}$; $b \approx 19.4$; $m(C) = 180° - (30° + 54°) = 96°$; $\dfrac{a}{\sin 30°} = \dfrac{c}{\sin 96°}$; $\dfrac{12}{0.5} = \dfrac{c}{0.9945}$; $c \approx 23.9$

2. $\dfrac{20}{\sin 30°} = \dfrac{c}{\sin 67°}$; $\dfrac{20}{0.5} = \dfrac{c}{0.9205}$; $c \approx 36.8$; $m(A) = 180° - (30° + 67°) = 83°$; $\dfrac{20}{\sin 30°} = \dfrac{a}{\sin 83°}$; $\dfrac{20}{0.5} = \dfrac{a}{0.9925}$; $a \approx 39.7$

3. $\dfrac{40}{\sin 120°} = \dfrac{20}{\sin B}$; $40\sin B = 20(0.8660)$; $\sin B = 0.4330$; $m(B) \approx 25.7°$; $m(C) \approx 180° - (120° + 25.7°) = 34.3°$; $\dfrac{c}{\sin 34.3°} = \dfrac{40}{\sin 120°}$; $c(0.8660) = 40(0.5635)$; $c \approx 26.0$

4. Since $8 < 25\left(\dfrac{1}{2}\right)$, there is no solution.

5. $\dfrac{\sin 14°}{75} = \dfrac{\sin 128°}{b}$; $75(0.7880) = b(0.2419)$; $b \approx 244.3$; $m(C) = 180° - (14° + 128°) = 38°$; $\dfrac{\sin 14°}{75} = \dfrac{\sin 38°}{c}$; $c(0.2419) = 75(0.6157)$; $c \approx 190.9$

6. $c\sin A = 34\left(\dfrac{1}{2}\right) = 17$. Since $c\sin A < a < c$, there are two solutions; $\dfrac{\sin 30°}{20} = \dfrac{\sin C}{34}$; $\sin C = \dfrac{(0.5)(34)}{20} \approx 0.85$; $m(C) \approx 58.2°$ or $121.8°$. If $m(C) \approx 58.2°$, $m(B) \approx 180° - (30° + 58.2°) = 91.8°$ and $\dfrac{\sin 91.8°}{b} \approx \dfrac{\sin 30°}{20}$, or $b \approx \dfrac{(0.9995)(20)}{0.5} \approx 40.0$. If $m(C) \approx 121.8°$, then $m(B) \approx 180° - (30° + 121.8°) = 28.2°$ and $b \approx \dfrac{(\sin 28.2°)(20)}{\sin 30°} \approx \dfrac{(0.4726)(20)}{0.5} \approx 18.9$

7. $m(A) = 180° - (33.3° + 137°) = 9.7°$; $\dfrac{\sin 9.7°}{48} = \dfrac{\sin 33.3°}{b}$; $b \approx \dfrac{48(0.5490)}{0.1685} \approx 156.4$; $\dfrac{\sin 9.7°}{48} = \dfrac{\sin 137°}{c}$; $c \approx \dfrac{48(0.6820)}{0.1685} \approx 194.3$

8. $b\sin A = 35\sin 61.7° = 35(0.8805) \approx 30.8$. Since $a < b\sin A$, there is no solution.

9. $a\sin B = 50\sin 17° = 50(0.2924) \approx 14.6$. Since $a\sin B < b < a$, there are two solutions; $\dfrac{\sin 17°}{35} = \dfrac{\sin A}{50}$; $\sin A = \dfrac{50(0.2924)}{35} \approx 0.4177$; $m(A) \approx 24.7°$ or $155.3°$. If $m(A) \approx 24.7°$, $m(C) \approx 180° - (24.7° + 17°) = 138.3°$; $\dfrac{\sin 138.3°}{c} \approx \dfrac{\sin 17°}{35}$;

$c \approx \dfrac{35(0.6652)}{0.2924} \approx 79.6.$ If $m(A) \approx 155.3°$, $m(C) \approx 180° - (155.3° + 17°) = 7.7°$;

$\dfrac{\sin 7.7°}{c} = \dfrac{\sin 17°}{35}$; $c \approx \dfrac{(0.1340)(35)}{0.2924} \approx 16.0$

10. $\dfrac{\sin 25°}{55} = \dfrac{\sin A}{27}$; $\sin A = \dfrac{27(0.4226)}{55} = 0.2075$; $m(A) \approx 12.0°$;

$m(B) \approx 180° - (12.0° + 25°) = 143.0°$; $\dfrac{\sin 143°}{b} \approx \dfrac{\sin 25°}{55}$; $b \approx \dfrac{55(0.6018)}{0.4226} \approx 78.3$

11. $\frac{1}{2}ab \sin C = \frac{1}{2}(8)(9)(0.5) = 18$

12. $\frac{1}{2}(24)(17)\sin 150° = \frac{1}{2}(24)(17)(0.5) = 102$

13. $\frac{1}{2}(15)(16)\sin 63° = \frac{1}{2}(15)(16)(0.8910) \approx 107$

14. $\frac{1}{2}(28)(18)\sin 73° = \frac{1}{2}(28)(18)(0.9563) \approx 241$

15. $\dfrac{\sin A}{50} = \dfrac{\sin 30°}{40}$; $\sin A = \dfrac{(0.5)(50)}{40} = 0.6250$; $m(A) \approx 38.7°$ or $141.3°$.

If $m(A) \approx 38.7°$, $m(C) \approx 111.3°$. If $m(A) \approx 141.3°$, $m(C) \approx 8.7°$.

Area $= \frac{1}{2}(50)(40)\sin 111.3° \approx \frac{1}{2}(50)(40)(0.9317) \approx 932$ or

Area $= \frac{1}{2}(50)(40)\sin 8.7° \approx \frac{1}{2}(50)(40)(0.1513) \approx 151$

16. $\dfrac{\sin A}{25} = \dfrac{\sin 21°}{17}$; $\sin A = \dfrac{25(0.3584)}{17} \approx 0.5270$; $m(A) \approx 31.8°$ or $148.2°$;

$m(C) \approx 127.2°$ or $10.8°$; Area $= \frac{1}{2}(25)(17)\sin 127.2° \approx \frac{1}{2}(25)(17)(0.7965) \approx 169$ or

Area $= \frac{1}{2}(25)(17)\sin 10.8° \approx \frac{1}{2}(25)(17)(0.1874) \approx 40$

B **17.** Since $\angle D$ is a right angle, $\sin A = \dfrac{CD}{b}$; $CD = b \sin A$

18. $\sin B = \dfrac{CD}{a}$; $CD = a \sin B$

19.

1. $CD = b \sin A$, $CD = a \sin B$ — Result of Exs. 17 and 18
2. $a \sin B = b \sin A$ — Substitution principle
3. $\dfrac{\sin B}{b} = \dfrac{\sin A}{a}$ — Div. prop. of =

20.

1. $\sin B = \dfrac{AD}{c}$ — Def. of sine
2. $c \sin B = AD$ — Mult. prop. of =
3. $\sin(180° - C) = \dfrac{AD}{b}$ — Def. of sine
4. $\sin(180° - C) = \sin C$ — Supplements have = sine values
5. $\sin C = \dfrac{AD}{b}$ — Substitution prin.
6. $b \sin C = AD$ — Mult. prop. of =
7. $c \sin B = b \sin C$ — Trans. prop. of =
8. $\dfrac{\sin B}{b} = \dfrac{\sin C}{c}$ — Div. prop. of =

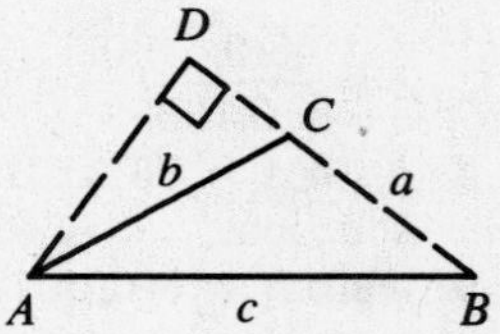

21. Since $\frac{\sin B}{b} = \frac{\sin C}{c}$, $c = \frac{b \sin C}{\sin B}$. Area $= \frac{1}{2}bc \sin A =$
$\frac{1}{2}b\left(\frac{b \sin C}{\sin B}\right)\sin A = \frac{b^2 \sin C \sin A}{2 \sin B}$

22. $m(A) = 180° - [m(B) + m(C)]$; $\sin A = \sin(B + C)$, since supplements have equal sine values. Since $\frac{\sin A}{a} = \frac{\sin B}{b}$, $\frac{\sin(B + C)}{a} = \frac{\sin B}{b}$.

C **23. a.** Both inscribed angles intercept the same arc, so their measures must be equal.

b. $m(\angle PBC) = 90°$ since the angle intercepts a semicircle; therefore, $\sin P = \frac{a}{PC}$ and $PC = \frac{a}{\sin P}$. Since $m(P) = m(A)$, $\sin P = \sin A$. Thus $PC = \frac{a}{\sin A}$.

c.

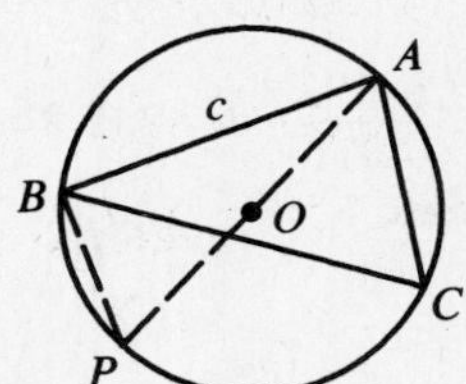

$m\angle PAC = 90°$	$m\angle PBA = 90°$
$\sin P = \frac{b}{PC}$	$\sin P = \frac{c}{PA}$
$PC = \frac{b}{\sin P}$	$PA = \frac{c}{\sin P}$
$m(P) = m(B)$	$m(P) = m(C)$
$PC = \frac{b}{\sin B}$	$PA = \frac{c}{\sin C}$

Pages 652–653 • PROBLEMS

A **1.** Area $= (20)(15)(\sin 117°) = 20(15)(0.8910) = 267.3$; 267.3 cm^2

2. The area is the sum of the areas of 20 congruent isosceles triangles, each with legs 12 cm long and vertex angle measuring $\frac{360°}{20}$, or 18°.

Area $= 20\left[\frac{1}{2}(12)(12)\sin 18°\right] = 20(72)(0.3090) \approx 445.0$; 445.0 cm^2

3. Let $x =$ distance from second observer to plane.

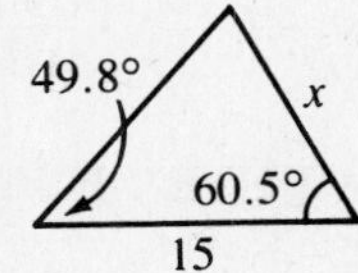

$$\frac{\sin 49.8°}{x} = \frac{\sin(180° - (49.8° + 60.5°))}{15}$$
$$15(0.7638) = x \sin 69.7°$$
$$11.457 = x(0.9379)$$
$$x \approx 12.2 \text{ km}$$

4.

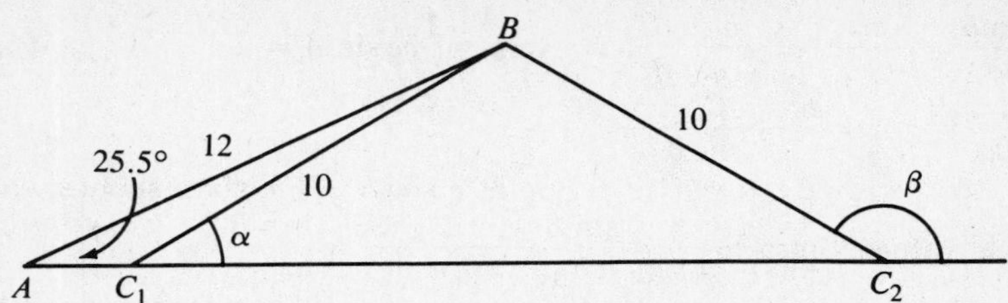

$\frac{\sin 25.5°}{10} = \frac{\sin C}{12}$; $\sin C = \frac{12(0.4305)}{10} \approx 0.5166$. Since $12(\sin 25.5°) < 10 < 12$, there are two solutions; $m(\angle AC_2B) \approx 31.1°$ or $m(\angle AC_1B) \approx 148.9°$; depending on how far Otis Washington has already walked, he should turn through $180° - 148.9°$, or $31.1°$ (α in the diagram), or through $180° - 31.1°$, or $148.9°$ (β in the diagram).

5. $\frac{\sin 30°}{x} = \frac{\sin 45°}{10.2}$; $x = \frac{(0.5)(10.2)}{0.7071} \approx 7.2$; 7.2 m

6. Let x = length of the shaft.

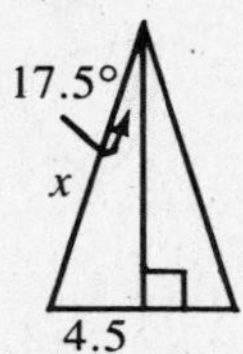

$$\frac{\sin 72.5°}{x} = \frac{\sin 35°}{9}$$

$$x = \frac{9(0.9537)}{0.5736} \approx 15.0$$

The length is about 15.0 cm.

B 7. Let x, y, and z be as shown in the diagram.

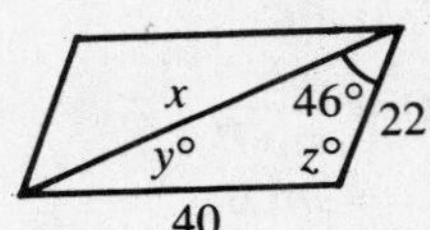

$$\frac{\sin 46°}{40} = \frac{\sin y°}{22} = \frac{\sin z°}{x}$$

$$\sin y° = \frac{22(0.7193)}{40} \approx 0.3956$$

$y° \approx 23.3°$

$z° \approx 180° - (46° + 23.3°) = 110.7°$

$$\frac{\sin 46°}{40} = \frac{\sin 110.7°}{x}$$

$$x = \frac{40(0.9354)}{0.7193} \approx 52.0$$

The diagonal is about 52.0 cm long.

8. $m(\angle DAC) = 180° - (60° + 54°) = 66°$; $m(\angle CAB) = 54°$, since $\overline{AB} \parallel \overline{DC}$; $m(B) = 180° - (54° + 48°) = 78°$; $\frac{\sin 66°}{75} = \frac{\sin 60°}{AC}$; $AC = \frac{(0.8660)75}{0.9135} \approx 71.099$; $\frac{\sin 48°}{AB} = \frac{\sin 78°}{AC}$; $\frac{0.7431}{AB} = \frac{0.9781}{71.099}$; $AB \approx 54.0$ m

Page 653 • COMPUTER EXERCISES

1.
```
10 LET PI = 3.141593
20 INPUT A,B,MA
30 LET AR = MA * PI/180
40 IF MA >= 90 AND MA < 180 AND A <= B THEN PRINT
                                          "NO TRIANGLE":END
50 IF MA < 90 AND A < B * SIN(AR) THEN PRINT "NO TRIANGLE":END
```

```
60  LET X = B * SIN(AR)/A
70  LET BR = ATN(X/SQR(-X^2 + 1))
80  LET MB = BR * 180/PI
90  PRINT MB
100  IF B * SIN(AR) < A AND A < B THEN PRINT " OR ";180 - MB
110  END
```

2. 43.98° or 136.02° **3.** no triangle **4.** 31.28°

Page 653 • SELF-TEST 3

1. $a^2 = 5^2 + 9^2 - 2(5)(9)\cos 49° = 25 + 81 - 90(0.6561) = 46.95;\ a \approx 6.9$

2. $\dfrac{\sin 31°}{a} = \dfrac{\sin 57°}{95};\ a = \dfrac{95(0.5150)}{0.8387} \approx 58.3$

3. $\text{Area} = \dfrac{1}{2}(12)(20)\sin 49° = 120(0.7547) \approx 91$

4. The area of the polygon is the sum of the areas of 9 isosceles triangles, each of which has legs 50 cm long and vertex angles measuring $\dfrac{360°}{9}$, or 40°. Let x = the length of a side of the polygon. Then
$x^2 = 50^2 + 50^2 - 2(50)(50)\cos 40° = 1169.78$, and $x \approx 34.2$ cm.

5. Let x = distance from ship to the second station.

$$\frac{\sin 43°}{x} = \frac{\sin(180° - (43° + 107°))}{125}$$

$$x = \frac{125(0.6820)}{0.5} = 170.5$$

The ship is 170.5 m from the second station.

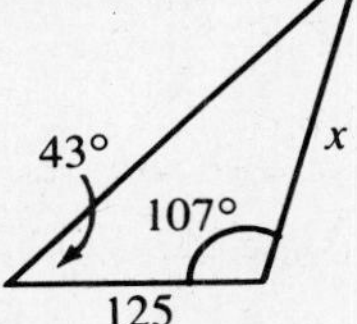

Pages 654–655 • CHAPTER REVIEW

1. $\dfrac{1}{\csc^2\alpha} = \sin^2\alpha$; c

2. $\dfrac{1}{\sec\alpha}(\tan\alpha + \cot\alpha) = \cos\alpha\left(\dfrac{\sin\alpha}{\cos\alpha} + \dfrac{\cos\alpha}{\sin\alpha}\right) = \cos\alpha\left(\dfrac{\sin^2\alpha + \cos^2\alpha}{\sin\alpha\cos\alpha}\right) =$

$\cos\alpha\left(\dfrac{1}{\cos\alpha\sin\alpha}\right) = \dfrac{1}{\sin\alpha} = \csc\alpha$; a

3. $\dfrac{(\sin x - \cos x)^2}{\cos x} = \dfrac{\sin^2 x - 2\sin x\cos x + \cos^2 x}{\cos x} = \dfrac{1 - 2\sin x\cos x}{\cos x} =$

$\dfrac{1}{\cos x} - 2\sin x = \sec x - 2\sin x$; b

4. $\dfrac{1}{\sec\alpha - 1} + \dfrac{1}{\sec\alpha + 1} = \dfrac{\sec\alpha - 1 + \sec\alpha + 1}{\sec^2\alpha - 1} = \dfrac{2\sec\alpha}{\tan^2\alpha}$

$$= \frac{\dfrac{2}{\cos\alpha}}{\dfrac{\sin^2\alpha}{\cos^2\alpha}} = \frac{2\cos\alpha}{\sin^2\alpha}$$

$$= 2\cdot\frac{\cos\alpha}{\sin\alpha}\cdot\frac{1}{\sin\alpha} = 2\cot\alpha\csc\alpha$$

$$= 2\cos\alpha\cdot\frac{1}{\sin^2\alpha} = 2\cos\alpha\csc^2\alpha \neq 2\cos\alpha\sec\alpha;\ \text{d}$$

5. $\cos\frac{7\pi}{8}\cos\frac{\pi}{8} - \sin\frac{7\pi}{8}\sin\frac{\pi}{8} = \cos\left(\frac{7\pi}{8} + \frac{\pi}{8}\right) = \cos\pi = -1$; a

6. $\cos x = \sqrt{1 - \sin^2 x} = \sqrt{1 - \left(\frac{4}{5}\right)^2} = \frac{3}{5}$; $\sin y = \sqrt{1 - \cos^2 y} = \sqrt{1 - \left(-\frac{12}{13}\right)^2} = \frac{5}{13}$; $\sin(x - y) = \sin x\cos y - \cos x\sin y = \frac{4}{5}\left(-\frac{12}{13}\right) - \frac{3}{5}\cdot\frac{5}{13} = -\frac{63}{65}$; a

7. c 8. b 9. d

10. $\frac{\cos 2\alpha}{\cos\alpha} = \frac{1 - 2\sin^2\alpha}{\cos\alpha} = \sec\alpha - \frac{2\sin^2\alpha}{\cos\alpha} = \sec\alpha - 2\sin\alpha\tan\alpha \neq \sec\alpha - 2\sin^2\alpha\tan\alpha$; d

11. b 12. $\frac{\sin A}{a} = \frac{\sin B}{b}$; $\frac{0.8}{4} = \frac{0.2}{b}$; $0.8b = 0.8$; $b = 1$; c

Pages 655–656 • CHAPTER TEST

1. $\cot^2\alpha = \frac{1}{\tan^2\alpha}$; $\tan\alpha \neq 0$, $\cot\alpha \neq 0$

2. $(1 - \sin\alpha)\sec\alpha = (1 - \sin\alpha)\frac{1}{\cos\alpha} = \frac{(1 - \sin\alpha)}{\cos\alpha} = \frac{(1 - \sin\alpha)(1 + \sin\alpha)}{\cos\alpha(1 + \sin\alpha)} = \frac{1 - \sin^2\alpha}{\cos\alpha(1 + \sin\alpha)} = \frac{\cos^2\alpha}{\cos\alpha(1 + \sin\alpha)} = \frac{\cos\alpha}{1 + \sin\alpha}$, $\cos\alpha \neq 0$

3. $\sin x = -\sqrt{1 - \cos^2 x} = -\frac{3}{5}$; $\sin y = -\sqrt{1 - \cos^2 y} = -\sqrt{1 - \left(-\frac{24}{25}\right)^2} = -\frac{7}{25}$; $\cos(x - y) = \cos x\cos y + \sin x\sin y = \frac{4}{5}\left(-\frac{24}{25}\right) + \left(-\frac{3}{5}\right)\left(-\frac{7}{25}\right) = -\frac{75}{125} = -\frac{3}{5}$

4. $\sin\frac{7\pi}{12} = \sin\left(\frac{\pi}{4} + \frac{\pi}{3}\right) = \sin\frac{\pi}{4}\cos\frac{\pi}{3} + \cos\frac{\pi}{4}\sin\frac{\pi}{3} = \frac{1}{\sqrt{2}}\cdot\frac{1}{2} + \frac{1}{\sqrt{2}}\cdot\frac{\sqrt{3}}{2} = \frac{1 + \sqrt{3}}{2\sqrt{2}} = \frac{\sqrt{2} + \sqrt{6}}{4}$

5. $\tan 2x = \frac{2\tan x}{1 - \tan^2 x} = \frac{2(0.6)}{1 - (0.6)^2} = \frac{1.2}{0.64} = 1.875$

6. $\cos x = -\sqrt{1 - (-0.6)^2} = -0.8$; $\sin\frac{x}{2} = \pm\sqrt{\frac{1 - (-0.8)}{2}} = \pm\sqrt{0.9} = \pm\sqrt{\frac{9}{10}} = \pm\frac{3}{\sqrt{10}}$; since $\frac{\pi}{2} < \frac{x}{2} < \frac{3\pi}{4}$, $\sin\frac{x}{2} = \frac{3}{\sqrt{10}}$.

7. $\frac{2}{\sin 2\alpha} = \frac{2}{2\sin\alpha\cos\alpha} = \frac{1}{\sin\alpha\cos\alpha} = \csc\alpha\sec\alpha$; $\sin\alpha \neq 0$, $\cos\alpha \neq 0$

8. $c^2 = 15^2 + 30^2 - 2(15)(30)\cos 60° = 225 + 900 - 900(0.5) = 675$; $c \approx 26.0$

9. $\frac{\sin 45°}{30} = \frac{\sin 60°}{b}$; $b = \frac{30\sin 60°}{\sin 45°} = \frac{30(0.8660)}{(0.7071)} \approx 36.7$

Page 657 • APPLICATION

1. **a.** 13 h
   b. $x = 22$; $y = 3.0 \sin \frac{2\pi}{365}(22) + 12.2 \approx 3.0 \sin(0.3787) + 12.2 \approx 13.3$; 13.3 h

2. $A = \frac{1}{2}(16 - 8) = 4.0$ **3.** $A = \frac{1}{2}(24 - 0) = 12.0$

Page 658 • PREPARING FOR COLLEGE ENTRANCE EXAMS

1. $\frac{13\pi}{6} \cdot \frac{180}{\pi} = 390$; C
2. $s = r \cdot m(\alpha)$; $\frac{8\pi}{5} = r \cdot \frac{3\pi}{2}$; $r = \frac{8\pi}{5} \cdot \frac{2}{3\pi} = \frac{16}{15}$; B
3. E
4. E
5. $3(5^n) = 27$; $5^n = 9$; $n \log 5 = \log 9$; $n = \frac{\log 9}{\log 5}$; D
6. $\frac{1}{\csc \alpha}(\tan \alpha + \cot \alpha) = \sin \alpha\left(\frac{\sin \alpha}{\cos \alpha} + \frac{\cos \alpha}{\sin \alpha}\right) = \sin \alpha\left(\frac{\sin^2\alpha + \cos^2\alpha}{\sin \alpha \cos \alpha}\right) =$
   $\frac{\sin \alpha(1)}{\sin \alpha \cos \alpha} = \frac{1}{\cos \alpha} = \sec \alpha$; B
7. $\sin 2\alpha = 2 \sin \alpha \cos \alpha = 2 \sin \alpha\left(-\frac{12}{13}\right) = 2\left(-\frac{12}{13}\right)\left(\sqrt{1 - \left(-\frac{12}{13}\right)^2}\right) =$
   $2\left(-\frac{12}{13}\right)\left(\frac{5}{13}\right) = -\frac{120}{169}$; A
8. $2a^{-2}(30a^5b - a^2) = 60a^3b - 2$; B

Page 659 • PROGRAMMING IN PASCAL

1. **a.** Use the trigonometric identities to compute the remaining functions. Since the other circular functions can be written in terms of sin and cos, only sin and cos need to be included in the implementation.
   b.

```
(******************************************************************)
FUNCTION tan (x : real) : real;

BEGIN
  tan := sin(x)/cos(x);
END;

(******************************************************************)
FUNCTION cot (x : real) : real;

BEGIN
  cot := cos(x)/sin(x);
END;

(******************************************************************)
FUNCTION sec (x : real) : real;

BEGIN
  sec := 1/cos(x);
END;
```

[*Program continued on next page*]

```
(************************************************************)
FUNCTION csc (x : real) : real;

BEGIN
  csc := 1/sin(x);
END;
```

2.

```
PROGRAM solve_triangle_abc (INPUT, OUTPUT);

TYPE
  flag = 1..2;

VAR
  a, b, c, b2, c2 : real;
  angle_a, angle_b, angle_c, angle_b2, angle_c2 : real;
  pi : real;
  option : integer;

(**************************************************************)
PROCEDURE display_menu (VAR option : integer);

BEGIN
  writeln('WHAT IS YOUR OPTION');
  writeln('<1>   2 sides and included angle');
  writeln('<2>   2 angles and included side');
  writeln('<3>   3 sides');
  write('ENTER: ');
  readln(option);
  writeln;
END;

(**************************************************************)
PROCEDURE get_data (option : integer);

BEGIN
  CASE option OF
       1 : BEGIN
             write('Enter the length of a side: ');
             readln(a);
             write('Enter the length of the other side: ');
             readln(b);
             write('Enter the measure of the included angle: ');
             readln(angle_c);
           END;
```

```
      2 : BEGIN
            write('Enter the measure of an angle: ');
            readln(angle_a);
            write('Enter the measure of the other angle: ');
            readln(angle_b);
            angle_c := 180 - (angle_a + angle_b);
            write('Enter the length of the included side: ');
            readln(c);
          END;
      3 : BEGIN
            write('Enter the length of a side: ');
            readln(a);
            write('Enter the length of another side: ');
            readln(b);
            write('Enter the length of the third side: ');
            readln(c);
          END;
  END; (* case *)
  writeln;
END;

(* The flag tells the procedure whether to solve for an angle *)
(* or a side using the law of cosines. *)
(*************************************************************)
PROCEDURE law_of_cos (VAR c, ang_c : real; a, b : real;
                                               choice : flag);
VAR
  cosine : real;

BEGIN
  IF choice = 1
     THEN c := sqrt(a * a + b * b - 2 * a * b * cos(pi *
                                                  ang_c/180))
     ELSE BEGIN
            cosine := (c * c - a * a - b * b)/(-2 * a * b);
            IF cosine > 0
               THEN BEGIN
                      ang_c := arctan(sqrt(1 - cosine * cosine)/
                                                      cosine);
                      ang_c := 180 * ang_c/pi;
                    END;
```

[*Program continued on next page*]

```
            IF cosine < 0
               THEN BEGIN
                      cosine := -cosine;
                      ang_c := arctan(sqrt(1 - cosine * cosine)/
                                                           cosine);
                      ang_c := 180 - 180 * ang_c/pi;
                    END;
            IF cosine = 0
               THEN ang_c := 90;
          END;
END;

(* The flag tells the procedure whether to solve for an angle *)
(* or a side using the law of sines. *)
(*************************************************************)
PROCEDURE law_of_sine (VAR a, ang_a : real; b, ang_b : real;
                                                 choice : flag);
VAR
  sine : real;

BEGIN
  IF choice = 2
     THEN BEGIN
            sine := a * sin(pi * ang_b/180)/b;
            IF sine > 0
               THEN BEGIN
                      ang_a := arctan(sine/sqrt(1 - sine * sine));
                      ang_a := 180 * ang_a/pi;
                    END;
            IF sine < 0
               THEN BEGIN
                      sine := -sine;
                      ang_a := arctan(sine/sqrt(1 - sine * sine));
                      ang_a := 180 - 180 * ang_a/pi;
                    END;
          END
     ELSE a := b * sin(pi * ang_a/180)/(sin(pi * ang_b/180));
END;

(*************************************************************)
PROCEDURE display_results;

BEGIN
  write('The lengths of the sides are: ');
  writeln(a:1:3, ' ', b:1:3, ' ', c:1:3);
  write('The measures of the angles are: ');
  writeln(angle_a:1:3, ' ', angle_b:1:3, ' ', angle_c:1:3);
  writeln;
```

```
  IF second_sol
     THEN BEGIN
            writeln('OR');
            writeln;
            write('The lengths of the sides are: ');
            writeln(a:1:3, ' ', b2:1:3, ' ', c2:1:3);
            write('The measures of the angles are: ');
            write(angle_a:1:3, ' ', angle_b2:1:3, ' ');
            writeln(angle_c2:1:3);
          END;
END;

(**************************************************************)
BEGIN (* main *)
  display_menu(option);
  get_data(option);
  pi := 4 * arctan(1);
  CASE option OF
       1 : BEGIN
             law_of_cos(c, angle_c, a, b, 1);
             law_of_sine(a, angle_a, c, angle_c, 2);
             angle_b := 180 - (angle_a + angle_c);
           END;
       2 : BEGIN
             law_of_sine(a, angle_a, c, angle_c, 1);
             law_of_sine(b, angle_b, c, angle_c, 1);
           END;
       3 : BEGIN
             law_of_cos(c, angle_c, a, b, 2);
             law_of_sine(a, angle_a, c, angle_c, 2);
             angle_b := 180 - (angle_a + angle_c);
           END;
  END; (* case *)
  display_results;
END.
```

3. Modify the program in Exercise 2.
Insert the following line in the VAR declaration section.

```
second_sol : boolean;
```

Insert the following line in the procedure *display_menu*.

```
writeln('<4>   2 sides and angle opposite one of them');
```

Insert the following lines in the procedure *get_data*.

```
4 : BEGIN
      write('Enter the measure of the angle: ');
      readln(angle_a);
      write('Enter length of the side opposite the angle: ');
      readln(a);
      write('Enter the length of the other side: ');
      readln(b);
    END;
```

Insert the following line in the main body of the program. It should be the next line after the assignment statement for *pi*.

```
second_sol := FALSE;
```

Insert the following lines in the main body of the program.

```
4 : BEGIN
      IF angle_a < 90
        THEN BEGIN
               IF a < b * sin(pi * angle_a/180)
                 THEN writeln('NO SOLUTION')
                 ELSE BEGIN
                        law_of_sine(b, angle_b, a, angle_a, 2);
                        angle_c := 180 - (angle_a + angle_b);
                        law_of_sine(c, angle_c, a, angle_a, 1);
                      END;
               IF (b * sin(pi * angle_a/180) < a) AND (a < b)
                 THEN BEGIN
                        angle_b2 := 180 - angle_b;
                        angle_c2 := 180 - (angle_a + angle_b2);
                        law_of_sine(b2, angle_b2, a, angle_a, 1);
                        law_of_sine(c2, angle_c2, a, angle_a, 1);
                        second_sol := TRUE
                      END;
             END;
      IF (90 <= angle_a) AND (angle_a < 180)
        THEN IF a <= b
               THEN writeln('NO SOLUTION')
               ELSE BEGIN
                      law_of_sine(b, angle_b, a, angle_a, 2);
                      angle_c := 180 - (angle_a + angle_b);
                      law_of_sine(c, angle_c, a, angle_a, 1);
                    END;
    END;
```

Pages 666–667 • WRITTEN EXERCISES

A **1.** $m(\alpha) = 0° + k \cdot 180°$ **2.** $m(\alpha) = 0° + k \cdot 360°$
3. $m(\alpha) = 60° + k \cdot 360°$ and $m(\alpha) = 300° + k \cdot 360°$
4. $m(\alpha) = 210° + k \cdot 360°$ and $m(\alpha) = 330° + k \cdot 360°$ **5.** $m(\alpha) = 0° + k \cdot 180°$
6. $m(\alpha) = 30° + k \cdot 180°$ **7.** $x = \frac{\pi}{4} + 2k\pi$ and $x = \frac{7\pi}{4} + 2k\pi$ **8.** $x = \frac{\pi}{2} + 2k\pi$
9. $x = \frac{4\pi}{3} + 2k\pi$ and $x = \frac{5\pi}{3} + 2k\pi$ **10.** $x = \frac{2\pi}{3} + 2k\pi$ and $x = \frac{4\pi}{3} + 2k\pi$
11. $x = \frac{5\pi}{6} + k\pi$ **12.** $x = \frac{\pi}{3} + k\pi$ **13.** $\frac{\pi}{3}$, 60° **14.** $\frac{3\pi}{4}$, 135° **15.** $-\frac{\pi}{2}$, −90°
16. $\frac{\pi}{6}$, 30° **17.** $-\frac{\pi}{3}$, −60° **18.** $\frac{\pi}{6}$, 30° **19.** 0.42, 24°
20. 0.63, 36° **21.** −1.43, −81.8°
22. $\cos 0.47 = 0.8910$; $\text{Cos}^{-1}(-0.8910) = 3.14 - 0.14 = 2.67$; $180° - 27° = 153°$
23. 0.16, 9° **24.** −1.26, −72°

B **25.** $\text{Sin}^{-1}\left(\cos \frac{\pi}{6}\right) = \text{Sin}^{-1} \frac{\sqrt{3}}{2} = \frac{\pi}{3}$ **26.** $\text{Sin}^{-1}(\cos 180°) = \text{Sin}^{-1}(-1) = -90°$
27. $\text{Cos}^{-1}(\sin 270°) = \text{Cos}^{-1}(-1) = 180°$ **28.** $\text{Cos}^{-1}\left(\cos \frac{5\pi}{4}\right) = \text{Cos}^{-1}\left(-\frac{1}{\sqrt{2}}\right) = \frac{3\pi}{4}$
29. $\text{Sin}^{-1}\left(\sin \frac{5\pi}{4}\right) = \text{Sin}^{-1}\left(-\frac{1}{\sqrt{2}}\right) = -\frac{\pi}{4}$ **30.** $\text{Sin}^{-1}\left(\cos \frac{5\pi}{3}\right) = \text{Sin}^{-1} \frac{1}{2} = \frac{\pi}{6}$
31. $\text{Cos}^{-1}\left[\sin\left(-\frac{\pi}{6}\right)\right] = \text{Cos}^{-1}\left(-\frac{1}{2}\right) = \frac{2\pi}{3}$
32. $\text{Tan}^{-1}(\tan 135°) = \text{Tan}^{-1}(-1) = -45°$
33. $\text{Tan}^{-1}\left(\tan \frac{5\pi}{6}\right) = \text{Tan}^{-1}\left(-\frac{1}{\sqrt{3}}\right) = -\frac{\pi}{6}$

34.

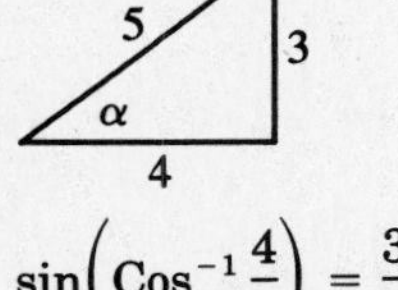

$\sin\left(\text{Cos}^{-1} \frac{4}{5}\right) = \frac{3}{5}$

35.

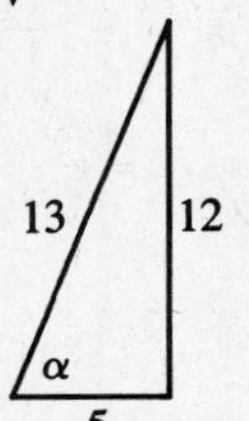

$\cos\left(\text{Sin}^{-1} \frac{12}{13}\right) = \frac{5}{13}$

36.

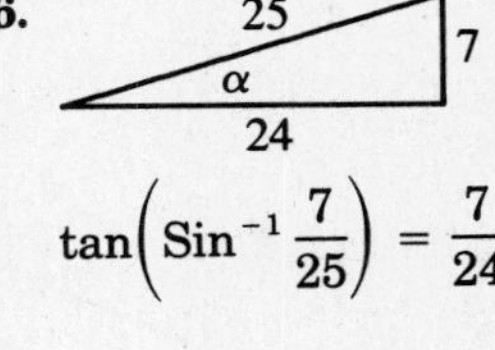

$\tan\left(\text{Sin}^{-1} \frac{7}{25}\right) = \frac{7}{24}$

37.

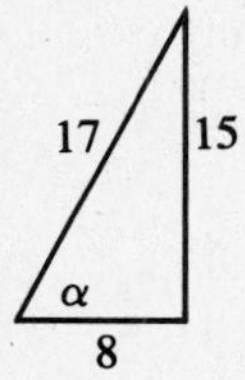

$\sin\left(\text{Tan}^{-1} \frac{15}{8}\right) = \frac{15}{17}$

38.

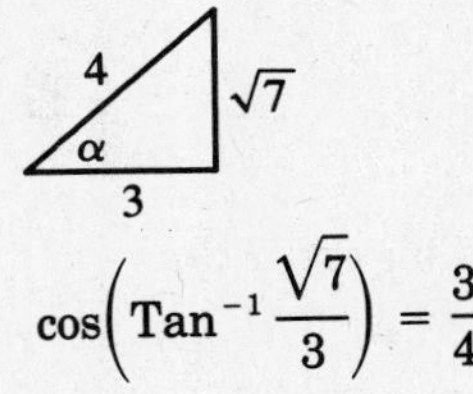

$\cos\left(\text{Tan}^{-1} \frac{\sqrt{7}}{3}\right) = \frac{3}{4}$

39.

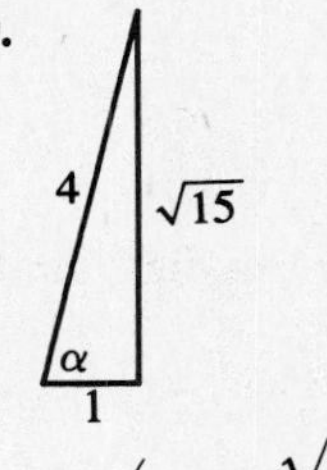

$\tan\left(\text{Sin}^{-1} \frac{\sqrt{15}}{4}\right) = \sqrt{15}$

40. 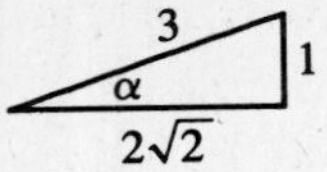

$\sin\left(\text{Cos}^{-1}\dfrac{2\sqrt{2}}{3}\right) = \dfrac{1}{3}$

41. 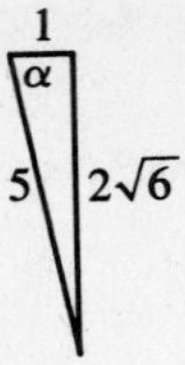

$\cos\left[\text{Sin}^{-1}\left(-\dfrac{2\sqrt{6}}{5}\right)\right] = \dfrac{1}{5}$

42. See diagrams for Exs. 34 and 36; $\cos\left(\text{Sin}^{-1}\dfrac{3}{5} + \text{Cos}^{-1}\dfrac{24}{25}\right) =$

$\left[\cos\left(\text{Sin}^{-1}\dfrac{3}{5}\right)\right]\left[\cos\left(\text{Cos}^{-1}\dfrac{24}{25}\right)\right] - \left[\sin\left(\text{Sin}^{-1}\dfrac{3}{5}\right)\right]\left[\sin\left(\text{Cos}^{-1}\dfrac{24}{25}\right)\right] =$

$\dfrac{4}{5}\cdot\dfrac{24}{25} - \dfrac{3}{5}\cdot\dfrac{7}{25} = \dfrac{96}{125} - \dfrac{21}{125} = \dfrac{3}{5}$

43. 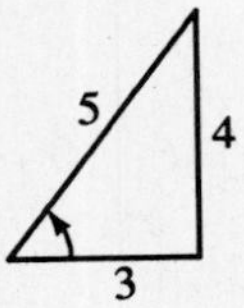

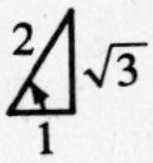

$\sin\left(\text{Sin}^{-1}\dfrac{4}{5} + \text{Cos}^{-1}\dfrac{1}{2}\right) =$

$\left[\sin\left(\text{Sin}^{-1}\dfrac{4}{5}\right)\right]\left[\cos\left(\text{Cos}^{-1}\dfrac{1}{2}\right)\right] +$

$\left[\cos\left(\text{Sin}^{-1}\dfrac{4}{5}\right)\right]\left[\sin\left(\text{Cos}^{-1}\dfrac{1}{2}\right)\right] = \dfrac{4}{5}\cdot\dfrac{1}{2} + \dfrac{3}{5}\cdot\dfrac{\sqrt{3}}{2} =$

$\dfrac{4}{10} + \dfrac{3\sqrt{3}}{10} = \dfrac{4 + 3\sqrt{3}}{10}$

44. 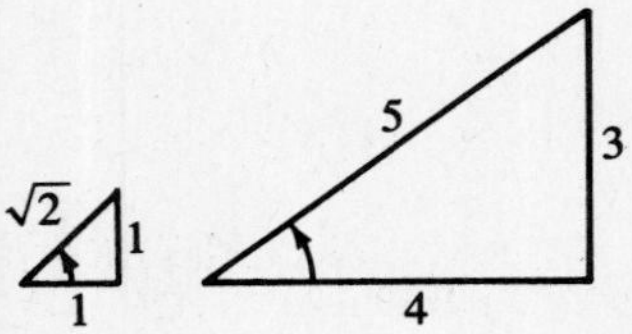

$\sin\left(\text{Sin}^{-1}\dfrac{1}{\sqrt{2}} + \text{Tan}^{-1}\dfrac{3}{4}\right) =$

$\left[\sin\left(\text{Sin}^{-1}\dfrac{1}{\sqrt{2}}\right)\right]\left[\cos\left(\text{Tan}^{-1}\dfrac{3}{4}\right)\right] +$

$\left[\cos\left(\text{Sin}^{-1}\dfrac{1}{\sqrt{2}}\right)\right]\left[\sin\left(\text{Tan}^{-1}\dfrac{3}{4}\right)\right] = \dfrac{1}{\sqrt{2}}\cdot\dfrac{4}{5} +$

$\dfrac{1}{\sqrt{2}}\cdot\dfrac{3}{5} = \dfrac{4}{5\sqrt{2}} + \dfrac{3}{5\sqrt{2}} = \dfrac{7}{5\sqrt{2}}$

45. 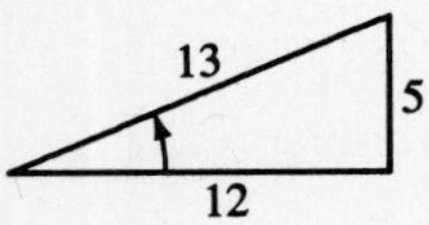

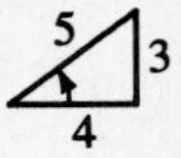

$\cos\left(\text{Sin}^{-1}\dfrac{5}{13} - \text{Cos}^{-1}\dfrac{4}{5}\right) =$

$\left[\cos\left(\text{Sin}^{-1}\dfrac{5}{13}\right)\right]\left[\cos\left(\text{Cos}^{-1}\dfrac{4}{5}\right)\right] +$

$\left[\sin\left(\text{Sin}^{-1}\dfrac{5}{13}\right)\right]\left[\sin\left(\text{Cos}^{-1}\dfrac{4}{5}\right)\right] = \dfrac{12}{13}\cdot\dfrac{4}{5} +$

$\dfrac{5}{13}\cdot\dfrac{3}{5} = \dfrac{48}{65} + \dfrac{15}{65} = \dfrac{63}{65}$

46. See the diagram for Ex. 36; $\cos\left(2\,\text{Sin}^{-1}\dfrac{7}{25}\right) = \cos^2\left(\text{Sin}^{-1}\dfrac{7}{25}\right) -$

$\sin^2\left(\text{Sin}^{-1}\dfrac{7}{25}\right) = \left(\dfrac{24}{25}\right)^2 - \left(\dfrac{7}{25}\right)^2 = \dfrac{576}{625} - \dfrac{49}{625} = \dfrac{527}{625}$

47. $\sin\left(2\,\text{Cos}^{-1}\dfrac{\sqrt{2}}{2}\right) = \sin(2\cdot 45°) = \sin 90° = 1$

C **48.** **a.** Yes; the graph of $y = \mathrm{Sin}^{-1} x$ is symmetric with respect to the origin.
b. $\cos(\mathrm{Sin}^{-1}(-x)) = \cos(-\mathrm{Sin}^{-1} x) = \cos(\mathrm{Sin}^{-1} x)$

49. **a.** Yes; the graph of $y = \mathrm{Cos}^{-1} x$ is symmetric with respect to the point $\left(0, \frac{\pi}{2}\right)$.
b. $\sin(\mathrm{Cos}^{-1}(-x)) = \sin(\pi - \mathrm{Cos}^{-1} x) = \sin \pi \cos(\mathrm{Cos}^{-1} x) - \cos \pi \sin(\mathrm{Cos}^{-1} x) = 0 - (-1) \sin(\mathrm{Cos}^{-1} x) = \sin(\mathrm{Cos}^{-1} x)$

50. Let $x = \sin y = \cos\left(\frac{\pi}{2} - y\right)$, $0 \le y \le \frac{\pi}{2}$. Then $\mathrm{Sin}^{-1} x + \mathrm{Cos}^{-1} x =$
$\mathrm{Sin}^{-1}(\sin y) + \mathrm{Cos}^{-1}\left(\cos\left(\frac{\pi}{2} - y\right)\right) = y + \left(\frac{\pi}{2} - y\right) = \frac{\pi}{2}$.

51. Let $x = \tan y = \cot\left(\frac{\pi}{2} - y\right)$, $0 \le y \le \frac{\pi}{2}$. Then $\mathrm{Tan}^{-1} x + \mathrm{Cot}^{-1} x =$
$\mathrm{Tan}^{-1}(\tan y) + \mathrm{Cot}^{-1}\left(\cot\left(\frac{\pi}{2} - y\right)\right) = y + \left(\frac{\pi}{2} - y\right) = \frac{\pi}{2}$.

Pages 669–670 • WRITTEN EXERCISES

A **1.** **a.** $\sin \alpha = \frac{\sqrt{3}}{2}$; $\{60° + k \cdot 360°, 120° + k \cdot 360°\}$ **b.** $\{60°, 120°\}$

2. **a.** $\cos \alpha = -\frac{1}{2}$; $\{120° + k \cdot 360°, 240° + k \cdot 360°\}$ **b.** $\{120°, 240°\}$

3. **a.** $\cos \alpha = -\frac{1}{\sqrt{2}}$; $\{135° + k \cdot 360°, 225° + k \cdot 360°\}$ **b.** $\{135°, 225°\}$

4. **a.** $\{45° + k \cdot 180°\}$ **b.** $\{45°, 225°\}$

5. **a.** $\sec \alpha = -2$; $\cos \alpha = -\frac{1}{2}$; $\{120° + k \cdot 360°, 240° + k \cdot 360°\}$ **b.** $\{120°, 240°\}$

6. **a.** $\sin^2 \alpha = \frac{1}{2}$; $\sin \alpha = \pm\frac{1}{\sqrt{2}}$; $\{45° + k \cdot 90°\}$ **b.** $\{45°, 135°, 225°, 315°\}$

7. **a.** $\cos^2 \alpha = \frac{3}{4}$; $\cos \alpha = \pm\frac{\sqrt{3}}{2}$; $\{30° + k \cdot 180°, 150° + k \cdot 180°\}$
b. $\{30°, 150°, 210°, 330°\}$

8. **a.** $\tan^2 \alpha = \frac{1}{3}$; $\tan \alpha = \pm\frac{1}{\sqrt{3}}$; $\{30° + k \cdot 180°, 150° + k \cdot 180°\}$
b. $\{30°, 150°, 210°, 330°\}$

9. **a.** $\sin x = -\cos x$; $\left\{\frac{3\pi}{4} + k\pi\right\}$ **b.** $\left\{\frac{3\pi}{4}, \frac{7\pi}{4}\right\}$

10. **a.** $3 \sin^2 x = 1 - \sin^2 x$; $4 \sin^2 x = 1$; $\sin^2 x = \frac{1}{4}$; $\sin x = \pm\frac{1}{2}$;
$\left\{\frac{\pi}{6} + k\pi, \frac{5\pi}{6} + k\pi\right\}$
b. $\left\{\frac{\pi}{6}, \frac{5\pi}{6}, \frac{7\pi}{6}, \frac{11\pi}{6}\right\}$

11. **a.** $\sin x = 2 \sin x \cos x$; $2 \sin x \cos x - \sin x = 0$; $\sin x(2 \cos x - 1) = 0$;
$\sin x = 0$ or $\cos x = \frac{1}{2}$; $\left\{k\pi, \frac{\pi}{3} + 2k\pi, \frac{5\pi}{3} + 2k\pi\right\}$
b. $\left\{0, \frac{\pi}{3}, \pi, \frac{5\pi}{3}\right\}$

12. a. $\cos x = -2\sin x\cos x$; $\cos x + 2\sin x\cos x = 0$; $\cos x(1 + 2\sin x) = 0$;
$\cos x = 0$ or $\sin x = -\frac{1}{2}$; $\left\{\frac{\pi}{2} + k\pi, \frac{7\pi}{6} + 2k\pi, \frac{11\pi}{6} + 2k\pi\right\}$

b. $\left\{\frac{\pi}{2}, \frac{7\pi}{6}, \frac{3\pi}{2}, \frac{11\pi}{6}\right\}$

13. a. $\cos 2x = -1$; $2\cos^2 x - 1 = -1$; $2\cos^2 x = 0$; $\cos x = 0$; $\left\{\frac{\pi}{2} + k\pi\right\}$

b. $\left\{\frac{\pi}{2}, \frac{3\pi}{2}\right\}$

14. a. $1 - 2\sin^2 x = 2\sin^2 x - 1$; $4\sin^2 x = 2$; $\sin x = \pm\frac{1}{\sqrt{2}}$; $\left\{\frac{\pi}{4} + \frac{k\pi}{2}\right\}$

b. $\left\{\frac{\pi}{4}, \frac{3\pi}{4}, \frac{5\pi}{4}, \frac{7\pi}{4}\right\}$

15. $2\sin^2\alpha + \sin\alpha - 1 = 0$; $(2\sin\alpha - 1)(\sin\alpha + 1) = 0$; $\sin\alpha = \frac{1}{2}$ or $\sin\alpha = -1$;
$\{30°, 150°, 270°\}$

16. $\tan^2\alpha + \tan\alpha = 0$; $\tan\alpha(\tan\alpha + 1) = 0$; $\tan\alpha = 0$ or $\tan\alpha = -1$;
$\{0°, 180°, 135°, 315°\}$

17. $3\cos\alpha = 2(1 - \cos^2\alpha)$; $3\cos\alpha = 2 - 2\cos^2\alpha$; $2\cos^2\alpha + 3\cos\alpha - 2 = 0$;
$(2\cos\alpha - 1)(\cos\alpha + 2) = 0$; $\cos\alpha = \frac{1}{2}$ or $\cos\alpha = -2$; reject $\cos\alpha = -2$;
$\{60°, 300°\}$

18. $2\cos^2\alpha - 1 = -\cos\alpha$; $2\cos^2\alpha + \cos\alpha - 1 = 0$; $(2\cos\alpha - 1)(\cos\alpha + 1) = 0$;
$\cos\alpha = \frac{1}{2}$ or $\cos\alpha = -1$; $\{60°, 180°, 300°\}$

19. $\sin^2 2\alpha - \sin 2\alpha = 0$; $\sin 2\alpha(\sin 2\alpha - 1) = 0$; $\sin 2\alpha = 0$ or $\sin 2\alpha = 1$;
$m(2\alpha) = 0°, 90°, 180°, 360°, 450°, 540°$; $\{0°, 45°, 90°, 180°, 225°, 270°\}$

20. $\cos^2 2\alpha = \frac{1}{2}$; $\cos 2\alpha = \pm\frac{1}{\sqrt{2}}$; $m(2\alpha) = 45°, 135°, 225°, 315°, 405°, 495°, 585°, 675°$;
$\{22.5°, 67.5°, 112.5°, 157.5°, 202.5°, 247.5°, 292.5°, 337.5°\}$, or
$\{22°30', 67°30', 112°30', 157°30', 202°30', 247°30', 292°30', 337°30'\}$

21. $8\sin^2 x - 6\sin x + 1 = 0$; $(4\sin x - 1)(2\sin x - 1) = 0$; $\sin x = \frac{1}{4}$ or $\sin x = \frac{1}{2}$;
$\left\{0.25, \frac{\pi}{6}, \frac{5\pi}{6}, 2.89\right\}$

22. $\sin x + \frac{1}{\sin x} = 2$; $\sin^2 x + 1 = 2\sin x$; $\sin^2 x - 2\sin x + 1 = 0$; $(\sin x - 1)^2 = 0$;
$\sin x = 1$; $\left\{\frac{\pi}{2}\right\}$

23. $3(2\cos^2 x - 1) = 2\cos^2 x$; $6\cos^2 x - 3 = 2\cos^2 x$; $4\cos^2 x = 3$; $\cos^2 x = \frac{3}{4}$;
$\cos x = \pm\frac{\sqrt{3}}{2}$; $\left\{\frac{\pi}{6}, \frac{5\pi}{6}, \frac{7\pi}{6}, \frac{11\pi}{6}\right\}$

24. $2\sin x(2\sin x\cos x) = 3\cos x$; $4\sin^2 x\cos x = 3\cos x$;
$4\sin^2 x\cos x - 3\cos x = 0$; $\cos x(4\sin^2 x - 3) = 0$; $\cos x = 0$ or $\sin x = \pm\frac{\sqrt{3}}{2}$;
$\left\{\frac{\pi}{3}, \frac{\pi}{2}, \frac{2\pi}{3}, \frac{4\pi}{3}, \frac{3\pi}{2}, \frac{5\pi}{3}\right\}$

25. $\tan^2 x = 3 - (\tan^2 x + 1)$; $2\tan^2 x = 2$; $\tan x = \pm 1$; $\left\{\frac{\pi}{4}, \frac{3\pi}{4}, \frac{5\pi}{4}, \frac{7\pi}{4}\right\}$

26. $2(1 - \sin^2 \alpha) - \sin\alpha = 1$; $1 - \sin\alpha - 2\sin^2\alpha = 0$; $(1 - 2\sin\alpha)(1 + \sin\alpha) = 0$; $\sin\alpha = \frac{1}{2}$ or $\sin\alpha = -1$; $\{30°, 150°, 270°\}$

B **27.** $\sin^2 x - 2\sin x\cos x + \cos^2 x = \frac{1}{2}$; $1 - 2\sin x\cos x = \frac{1}{2}$; $2\sin x\cos x = \frac{1}{2}$; $\sin 2x = \frac{1}{2}$; $2x = \frac{\pi}{6}, \frac{5\pi}{6}, \frac{13\pi}{6}, \frac{17\pi}{6}$; $\left\{\frac{\pi}{12}, \frac{5\pi}{12}, \frac{13\pi}{12}, \frac{17\pi}{12}\right\}$

28. $(\sec^2 x - 1) - 1 = \sec x$; $\sec^2 x - \sec x - 2 = 0$; $(\sec x - 2)(\sec x + 1) = 0$; $\sec x = 2$ or $\sec x = -1$; $\left\{\frac{\pi}{3}, \pi, \frac{5\pi}{3}\right\}$

29. $3\sin 2x = 2(1 - \sin^2 2x)$; $2\sin^2 2x + 3\sin 2x - 2 = 0$; $(2\sin 2x - 1)(\sin 2x + 2) = 0$; $\sin 2x = \frac{1}{2}$ or $\sin 2x = -2$ (impossible); $2x = \frac{\pi}{6}, \frac{5\pi}{6}, \frac{13\pi}{6}, \frac{17\pi}{6}$; $\left\{\frac{\pi}{12}, \frac{5\pi}{12}, \frac{13\pi}{12}, \frac{17\pi}{12}\right\}$

30. $2(1 - \sin^2 2x) = 5\sin 2x - 1$; $2\sin^2 2x + 5\sin 2x - 3 = 0$; $(2\sin 2x - 1)(\sin 2x + 3) = 0$; $\sin 2x = \frac{1}{2}$ or $\sin 2x = -3$ (impossible); $2x = \frac{\pi}{6}, \frac{5\pi}{6}, \frac{13\pi}{6}, \frac{17\pi}{6}$; $\left\{\frac{\pi}{12}, \frac{5\pi}{12}, \frac{13\pi}{12}, \frac{17\pi}{12}\right\}$

31. $\sin^2\alpha + 2\sin\alpha\cos\alpha + \cos^2\alpha = 2\sin 2\alpha$; $1 + \sin 2\alpha = 2\sin 2\alpha$; $\sin 2\alpha = 1$; $m(2\alpha) = 90°, 450°$; $\{45°, 225°\}$

32. $1 - 2\sin^2 x - 3\sin x - 2 = 0$; $2\sin^2 x + 3\sin x + 1 = 0$; $(2\sin x + 1)(\sin x + 1) = 0$; $\sin x = -\frac{1}{2}$ or $\sin x = -1$; $\left\{\frac{7\pi}{6}, \frac{3\pi}{2}, \frac{11\pi}{6}\right\}$

33. $2\cos^2 x - 1 = \sqrt{3}\cos x - 1$; $2\cos^2 x - \sqrt{3}\cos x = 0$; $\cos x(2\cos x - \sqrt{3}) = 0$; $\cos x = 0$ or $\cos x = \frac{\sqrt{3}}{2}$; $\left\{\frac{\pi}{6}, \frac{\pi}{2}, \frac{3\pi}{2}, \frac{11\pi}{6}\right\}$

34. $8\sin^4\alpha - 10\sin^2\alpha + 3 = 0$; $(4\sin^2\alpha - 3)(2\sin^2\alpha - 1) = 0$; $\sin\alpha = \pm\frac{\sqrt{3}}{2}$ or $\sin\alpha = \pm\frac{1}{\sqrt{2}}$; $\{45°, 60°, 120°, 135°, 225°, 240°, 300°, 315°\}$

C **35.** $\sqrt{2 - 2\cos\alpha} = 2\cos\frac{\alpha}{2}$; $\sqrt{4\left(\frac{1 - \cos\alpha}{2}\right)} = 2\cos\frac{\alpha}{2}$; $2\sin\frac{\alpha}{2} = 2\cos\frac{\alpha}{2}$; $\frac{\sin\frac{\alpha}{2}}{\cos\frac{\alpha}{2}} = 1$; $\tan\frac{\alpha}{2} = 1$; $\{90°\}$

36. $\cot\alpha - \tan\alpha = 2$; $\frac{1}{\tan\alpha} - \tan\alpha = 2$; $\tan^2\alpha + 2\tan\alpha - 1 = 0$; $\tan\alpha = -1 + \sqrt{2}$ or $\tan\alpha = -1 - \sqrt{2}$; $\{22.5°, 112.5°, 202.5°, 292.5°\}$

37. $\sqrt{3}\cos^2\alpha - 2\sin\alpha\cos\alpha = \sqrt{3}\sin^2\alpha$; $\sqrt{3}(\cos^2\alpha - \sin^2\alpha) - 2\sin\alpha\cos\alpha = 0$; $\sqrt{3}\cos 2\alpha - \sin 2\alpha = 0$; $\frac{\sin 2\alpha}{\cos 2\alpha} = \sqrt{3}$; $\tan 2\alpha = \sqrt{3}$; $m(2\alpha) = 60°, 240°, 420°, 600°$; $\{30°, 120°, 210°, 300°\}$

38. $\frac{\cos 3x}{\sin x} + \frac{\sin 3x}{\cos x} = 2\sqrt{3}$; $\cos 3x\cos x + \sin 3x\sin x = 2\sqrt{3}\sin x\cos x$;

$\cos(3x - x) = \sqrt{3}\sin 2x$; $\frac{\sin 2x}{\cos 2x} = \frac{1}{\sqrt{3}}$; $\tan 2x = \frac{1}{\sqrt{3}}$; $2x = \frac{\pi}{6}, \frac{7\pi}{6}, \frac{13\pi}{6}, \frac{19\pi}{6}$;
$\left\{\frac{\pi}{12}, \frac{7\pi}{12}, \frac{13\pi}{12}, \frac{19\pi}{12}\right\}$

Page 670 • SELF-TEST 1

1. $y = \frac{5\pi}{4} + 2k\pi$ and $y = \frac{7\pi}{4} + 2k\pi$ **2.** $\text{Cos}^{-1}\, 0.8678 = 0.52$ **3.** $-60°$

4. $\cos 2x + 1 = \cos x$; $(2\cos^2 x - 1) + 1 = \cos x$; $2\cos^2 x - \cos x = 0$;
$\cos x(2\cos x - 1) = 0$; $\cos x = 0$ or $\cos x = \frac{1}{2}$;
$\left\{\frac{\pi}{2} + 2k\pi, \frac{\pi}{3} + 2k\pi, \frac{3\pi}{2} + 2k\pi, \frac{5\pi}{3} + 2k\pi\right\}$

5. $\sin^2 \alpha = \frac{3}{4}$; $\sin \alpha = \pm\frac{\sqrt{3}}{2}$; $\{60°, 120°\}$

Pages 675–676 • WRITTEN EXERCISES

A 1. $r = \sqrt{6^2 + 6^2} = 6\sqrt{2}$; $\cos\theta = \frac{6}{\sqrt{6^2 + 6^2}} = \frac{6}{6\sqrt{2}} = \frac{1}{\sqrt{2}}$;
$\sin\theta = \cos\theta = \frac{6}{\sqrt{6^2 + 6^2}} = \frac{6}{6\sqrt{2}} = \frac{1}{\sqrt{2}}$; $m(\theta) = 45°$; $(6\sqrt{2}, 45°)$
Alternatively, by choosing $r = -6\sqrt{2}$, you obtain $(-6\sqrt{2}, -135°)$.

2.

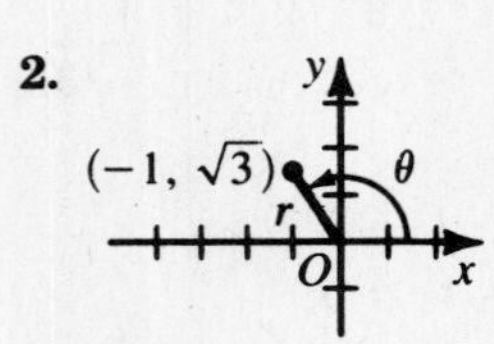

$r = \sqrt{(-1)^2 + (\sqrt{3})^2} = 2$; $\cos\theta = \frac{-1}{\sqrt{(-1)^2 + (\sqrt{3})^2}} = -\frac{1}{2}$;
$\sin\theta = \frac{\sqrt{3}}{\sqrt{(-1)^2 + (\sqrt{3})^2}} = \frac{\sqrt{3}}{2}$; $m(\theta) = 120°$; $(2, 120°)$ or, alternatively, $(-2, -60°)$

3.

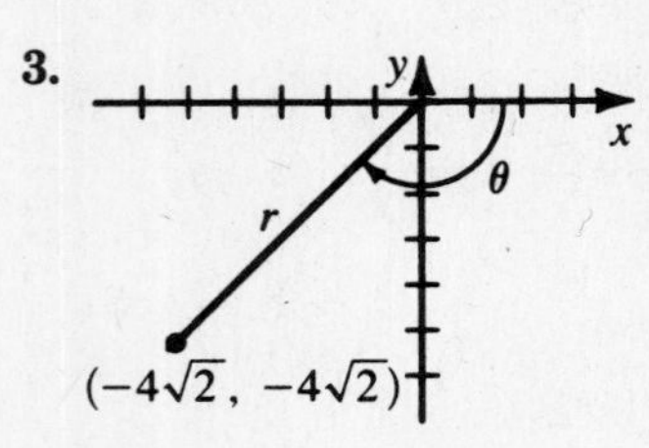

$r = \sqrt{(-4\sqrt{2})^2 + (-4\sqrt{2})^2} = 8$;
$\cos\theta = \frac{-4\sqrt{2}}{\sqrt{(-4\sqrt{2})^2 + (-4\sqrt{2})^2}} = \frac{-4\sqrt{2}}{8} = -\frac{\sqrt{2}}{2}$;
$\sin\theta = \frac{-4\sqrt{2}}{\sqrt{(-4\sqrt{2})^2 + (-4\sqrt{2})^2}} = -\frac{\sqrt{2}}{2}$;
$m(\theta) = -135°$; $(8, -135°)$ or, alternatively, $(-8, 45°)$

4. $r = \sqrt{(2)^2 + (-2\sqrt{3})^2} = 4$; $\cos\theta = \frac{2}{\sqrt{2^2 + (-2\sqrt{3})^2}} = \frac{2}{4} = \frac{1}{2}$;
$\sin\theta = \frac{-2\sqrt{3}}{\sqrt{2^2 + (-2\sqrt{3})^2}} = \frac{-2\sqrt{3}}{4} = -\frac{\sqrt{3}}{2}$; $m(\theta) = -60°$;
$(4, -60°)$ or, alternatively, $(-4, 120°)$

5. $\left(\frac{-5\sqrt{2}}{2}, \frac{5\sqrt{2}}{2}\right)$

$$r = \sqrt{\left(\frac{-5\sqrt{2}}{2}\right)^2 + \left(\frac{5\sqrt{2}}{2}\right)^2} = \sqrt{\frac{25}{2} + \frac{25}{2}} = 5;$$

$$\cos\theta = \frac{\frac{-5\sqrt{2}}{2}}{5} = -\frac{\sqrt{2}}{2};$$

$$\sin\theta = \frac{\frac{5\sqrt{2}}{2}}{5} = \frac{\sqrt{2}}{2};$$

$m(\theta) = 135°$; $(5, 135°)$ or, alternatively, $(-5, -45°)$

6. $(\sqrt{6}, -\sqrt{2})$

$$r = \sqrt{(\sqrt{6})^2 + (-\sqrt{2})^2} = \sqrt{8} = 2\sqrt{2};$$

$$\cos\theta = \frac{\sqrt{6}}{2\sqrt{2}} = \frac{\sqrt{3}}{2};$$

$$\sin\theta = \frac{-\sqrt{2}}{2\sqrt{2}} = -\frac{1}{2};\ m(\theta) = -30°;$$

$(2\sqrt{2}, -30°)$ or, alternatively, $(-2\sqrt{2}, 150°)$

7.

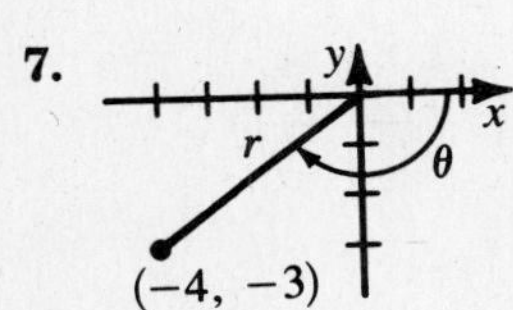

$$r = \sqrt{(-4)^2 + (-3)^2} = \sqrt{25} = 5;$$

$$\cos\theta = -\frac{4}{5};\ \sin\theta = -\frac{3}{5};$$

$m(\theta) \approx -143.1°$; $(5, -143.1°)$, or, alternatively, $(-5, 36.9°)$

8.

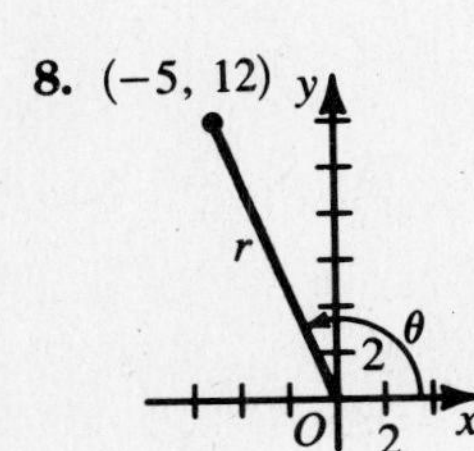

$$r = \sqrt{(-5)^2 + (12)^2} = \sqrt{169} = 13;\ \cos\theta = -\frac{5}{13};$$

$$\sin\theta = \frac{12}{13};\ m(\theta) \approx 112.6°;\ (13, 112.6°)$$

or, alternatively, $(-13, -67.4°)$

9.

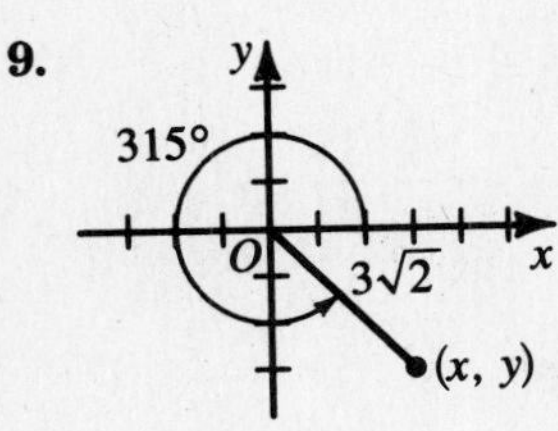

$$x = 3\sqrt{2}\cos 315° = 3\sqrt{2}\cdot\frac{1}{\sqrt{2}} = 3;$$

$$y = 3\sqrt{2}\sin 315° = 3\sqrt{2}\left(-\frac{1}{\sqrt{2}}\right) = -3;\ (3, -3)$$

10. 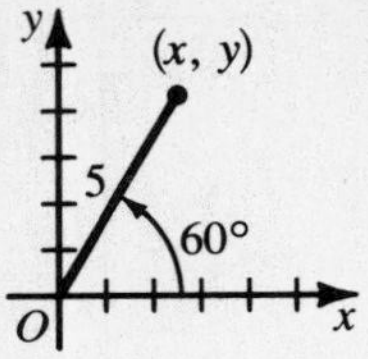 $x = 5\cos 60° = 5 \cdot \frac{1}{2} = \frac{5}{2}$;
$y = 5\sin 60° = 5 \cdot \frac{\sqrt{3}}{2} = \frac{5\sqrt{3}}{2}$; $\left(\frac{5}{2}, \frac{5\sqrt{3}}{2}\right)$

11. 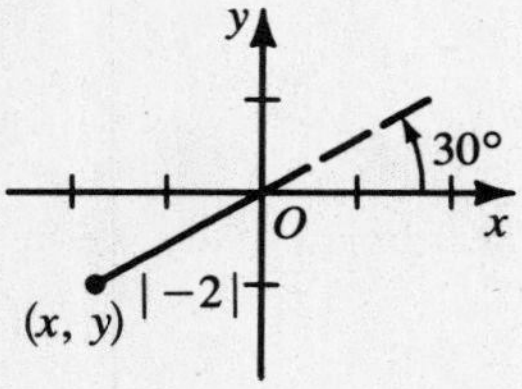 $x = -2\cos 30° = -2 \cdot \frac{\sqrt{3}}{2} = -\sqrt{3}$;
$y = -2\sin 30° = -2 \cdot \frac{1}{2} = -1$;
$(-\sqrt{3}, -1)$

12. 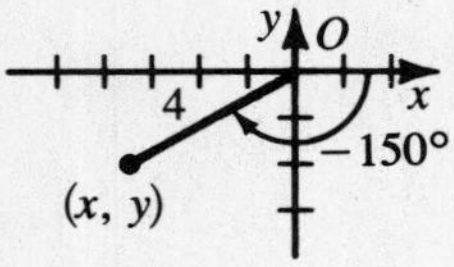 $x = 4\cos(-150°) = 4\left(-\frac{\sqrt{3}}{2}\right) = -2\sqrt{3}$;
$y = 4\sin(-150°) = 4\left(-\frac{1}{2}\right) = -2$; $(-2\sqrt{3}, -2)$

13. 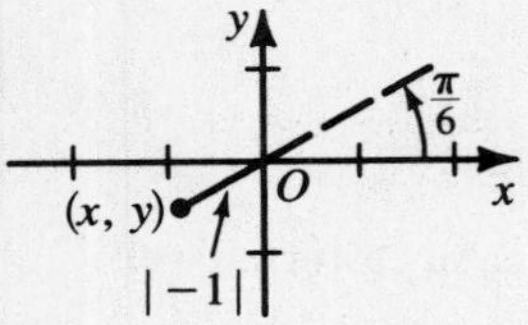 $x = -\cos\frac{\pi}{6} = -1 \cdot \frac{\sqrt{3}}{2} = -\frac{\sqrt{3}}{2}$;
$y = -\sin\frac{\pi}{6} = -1 \cdot \frac{1}{2} = -\frac{1}{2}$; $\left(-\frac{\sqrt{3}}{2}, -\frac{1}{2}\right)$

14. 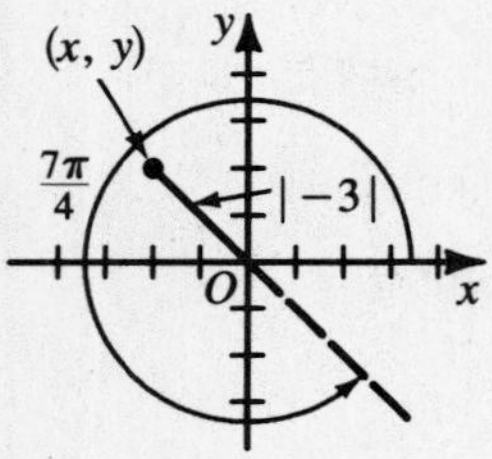 $x = -3\cos\frac{7\pi}{4} = -3 \cdot \frac{1}{\sqrt{2}} = -\frac{3}{\sqrt{2}}$;
$y = -3\sin\frac{7\pi}{4} = -3\left(-\frac{1}{\sqrt{2}}\right) = \frac{3}{\sqrt{2}}$;
$\left(-\frac{3}{\sqrt{2}}, \frac{3}{\sqrt{2}}\right)$

15. 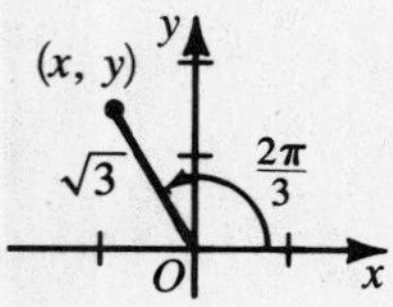 $x = \sqrt{3}\cos\frac{2\pi}{3} = \sqrt{3}\left(-\frac{1}{2}\right) = -\frac{\sqrt{3}}{2}$;
$y = \sqrt{3}\sin\frac{2\pi}{3} = \sqrt{3} \cdot \frac{\sqrt{3}}{2} = \frac{3}{2}$; $\left(-\frac{\sqrt{3}}{2}, \frac{3}{2}\right)$

16. 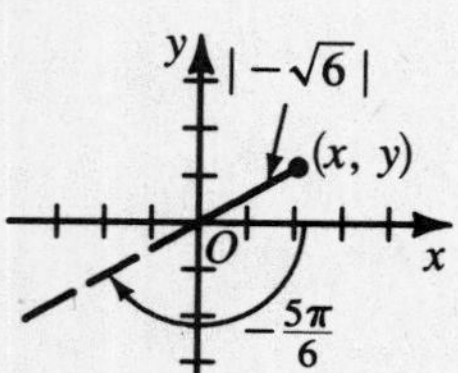 $x = -\sqrt{6}\cos\left(-\frac{5\pi}{6}\right) = -\sqrt{6}\left(-\frac{\sqrt{3}}{2}\right) = \frac{3\sqrt{2}}{2}$;
$y = -\sqrt{6}\sin\left(-\frac{5\pi}{6}\right) = -\sqrt{6}\left(-\frac{1}{2}\right) = \frac{\sqrt{6}}{2}$;
$\left(\frac{3\sqrt{2}}{2}, \frac{\sqrt{6}}{2}\right)$

17. 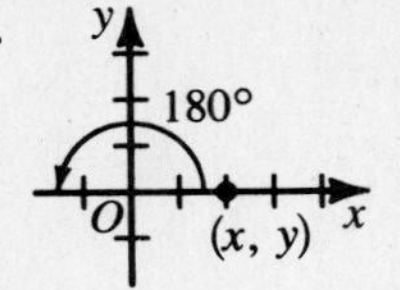

$x = -2 \cos 180° = -2(-1) = 2;$
$y = -2 \sin 180° = -2 \cdot 0 = 0;\ (2, 0)$

18. $r \cos \theta = 3;\ r = 3 \sec \theta$

19. $r \sin \theta = -2;\ r = -2 \csc \theta$

20. $r \cos \theta + r \sin \theta - 3 = 0;\ r(\cos \theta + \sin \theta) = 3;\ r = \dfrac{3}{\cos \theta + \sin \theta}$

21. $(r \cos \theta)^2 + (r \sin \theta)^2 = 9;\ r^2 \cos^2 \theta + r^2 \sin^2 \theta = 9;\ r^2(\cos^2 \theta + \sin^2 \theta) = 9;\ r^2 = 9$

22. $(r \cos \theta)^2 - (r \sin \theta)^2 = 25;\ r^2 \cos^2 \theta - r^2 \sin^2 \theta = 25;\ r^2(\cos^2 \theta - \sin^2 \theta) = 25;$
$r^2 \cos 2\theta = 25;\ r^2 = \dfrac{25}{\cos 2\theta}$

23. $(r \cos \theta)^2 + (r \sin \theta)^2 - 8(r \cos \theta) = 0;\ r^2 \cos^2 \theta + r^2 \sin^2 \theta - 8r \cos \theta = 0;$
$r^2(\cos^2 \theta + \sin^2 \theta) - 8r \cos \theta = 0;\ r^2 - 8r \cos \theta = 0$

B 24. a. 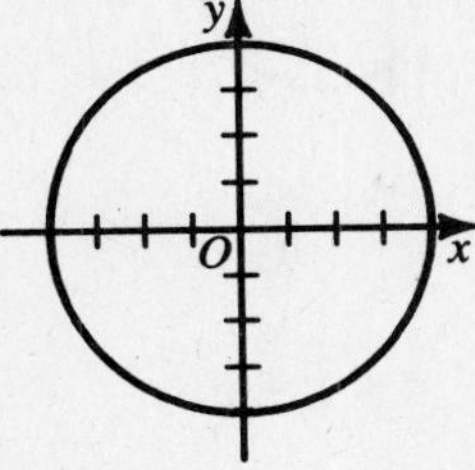

b. $\sqrt{x^2 + y^2} = 4$
$x^2 + y^2 = 16$

25. a. 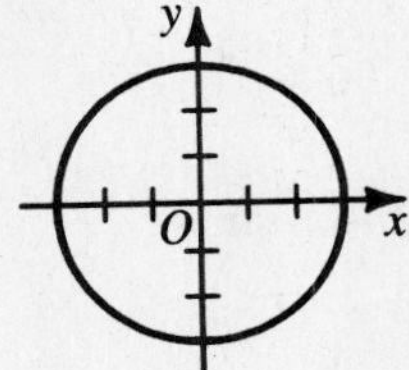

b. $-\sqrt{x^2 + y^2} = -3$
$x^2 + y^2 = 9$

26. a. 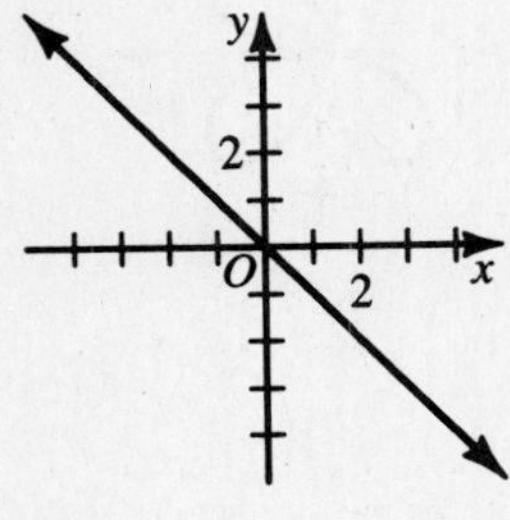

b. $x = r \cos(-45°) = \dfrac{r}{\sqrt{2}}$

$y = r \sin(-45°) = -\dfrac{r}{\sqrt{2}}$

$x = -y$

27. a. 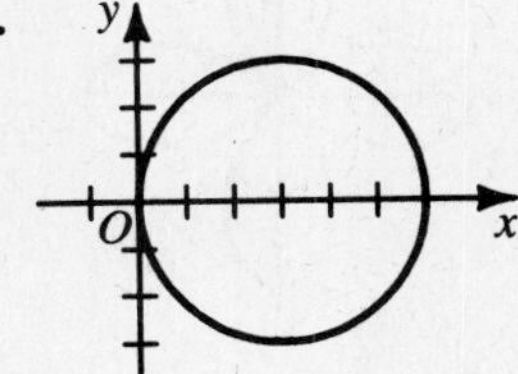

b. $r = 6 \cos \theta$
$r = 6\left(\dfrac{x}{r}\right)$
$r^2 = 6x$
$x^2 + y^2 = 6x$

28. a.

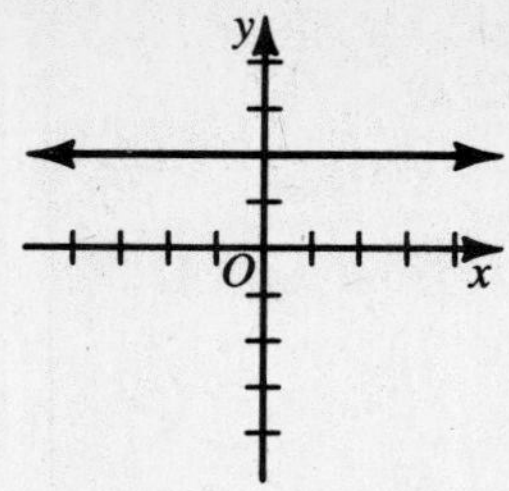

b. $r = 2 \csc \theta$
$r \sin \theta = 2$
$y = 2$

29. a.

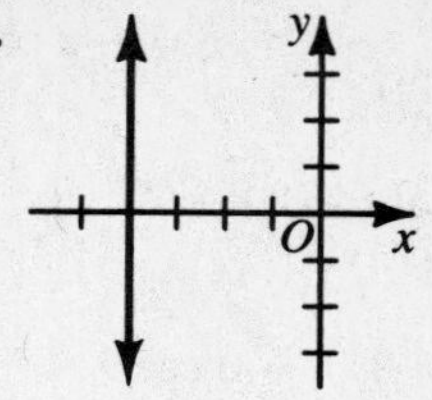

b. $r = -4 \sec \theta$
$r \cos \theta = -4$
$x = -4$

30. a.

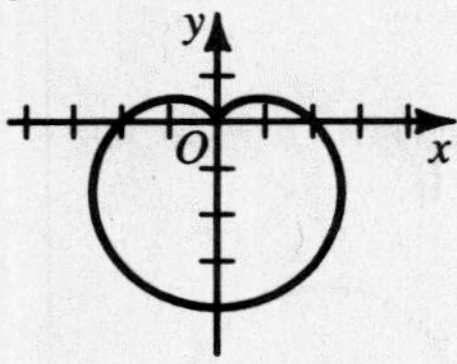

b. $r = 2 - 2 \sin \theta$
$r = 2 - 2\left(\frac{y}{r}\right)$
$r^2 = 2r - 2y$
$(r^2 + 2y)^2 = (2r)^2$
$(x^2 + y^2 + 2y)^2 = 4(x^2 + y^2)$

31. a.

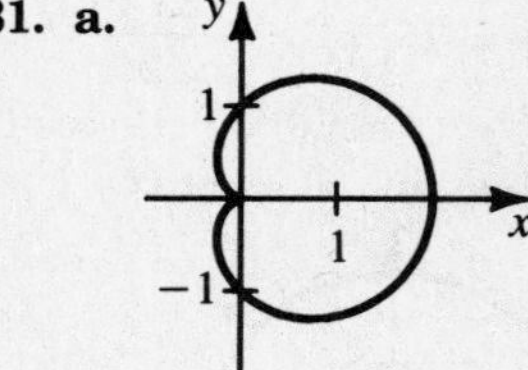

b. $r = 1 + \cos \theta$
$r = 1 + \left(\frac{x}{r}\right)$
$r^2 = r + x$
$(r^2 - x)^2 = r^2$
$(x^2 + y^2 - x)^2 = x^2 + y^2$

32. a.

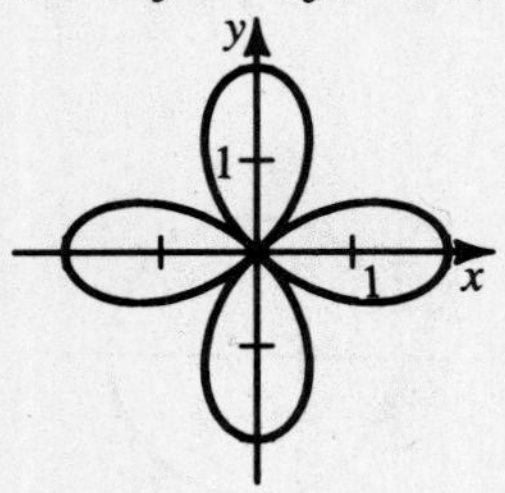

b. $r = 2 \cos 2\theta$
$r = 2(2 \cos^2 \theta - 1)$
$r = 2\left(\frac{2x^2}{r^2} - 1\right)$
$(r^3)^2 = (4x^2 - 2r^2)^2$
$(x^2 + y^2)^3 = 4(x^2 - y^2)^2$

33. a.

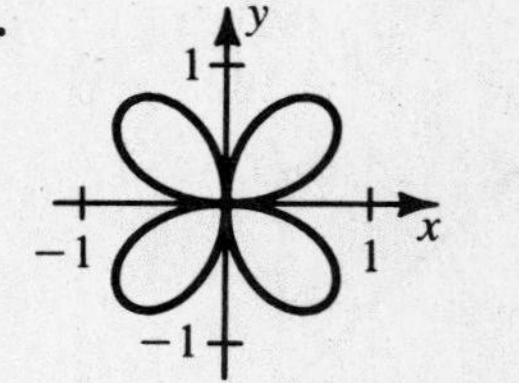

b. $r = \sin 2\theta$
$r = 2 \sin \theta \cos \theta$
$r = 2\left(\frac{y}{r}\right)\left(\frac{x}{r}\right)$
$(r^3)^2 = (2yx)^2$
$(x^2 + y^2)^3 = 4x^2y^2$

34. a.

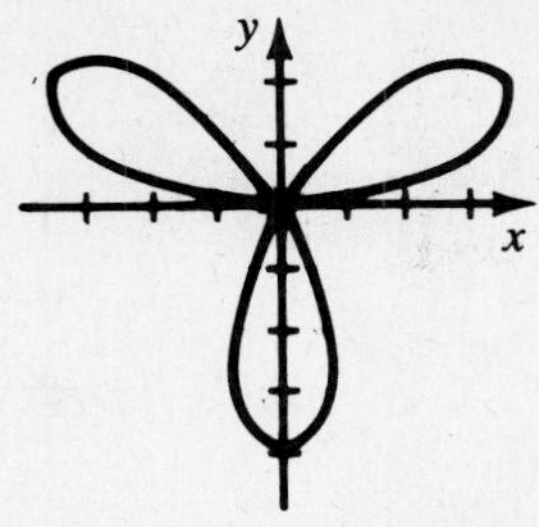

b. $r = 4 \sin 3\theta = 4 \sin(2\theta + \theta)$
$r = 4(\sin 2\theta \cos \theta + \cos 2\theta \sin \theta)$
$r = 4(2 \sin \theta \cos^2\theta + (2 \cos^2\theta - 1)\sin \theta)$
$r = 4(4 \sin \theta \cos^2\theta - \sin \theta)$
$r = 4\left(\frac{4yx^2}{r^3} - \frac{y}{r}\right)$
$r^4 = 16x^2y - 4r^2y$
$(x^2 + y^2)^2 = 16x^2y - 4(x^2 + y^2)y$
$(x^2 + y^2)^2 = 12x^2y - 4y^3$

35. a.

b. $r = 1 - 2 \sin \theta$
$r = 1 - 2\left(\frac{y}{r}\right)$
$r^2 = r - 2y$
$(r^2 + 2y)^2 = r^2$
$(x^2 + y^2 + 2y)^2 = x^2 + y^2$

C **36. a.**

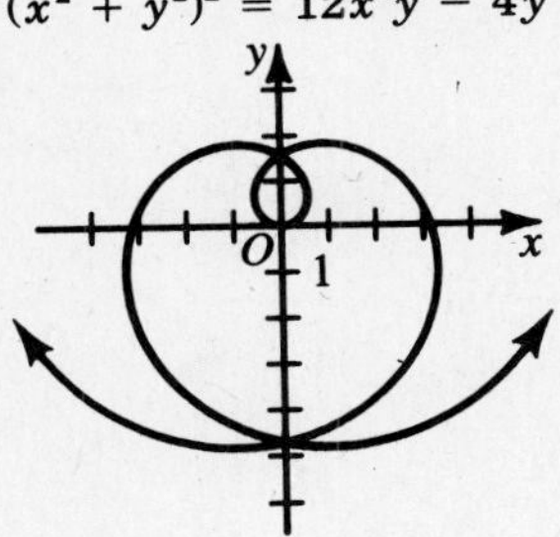

b. $r = \pm\sqrt{x^2 + y^2}$,
$\theta = \text{Cos}^{-1}\left(\frac{x}{\pm\sqrt{x^2 + y^2}}\right) + 2k\pi$;
$\pm\sqrt{x^2 + y^2} = \text{Cos}^{-1}\left(\frac{x}{\pm\sqrt{x^2 + y^2}}\right) + 2k\pi$

37. a.

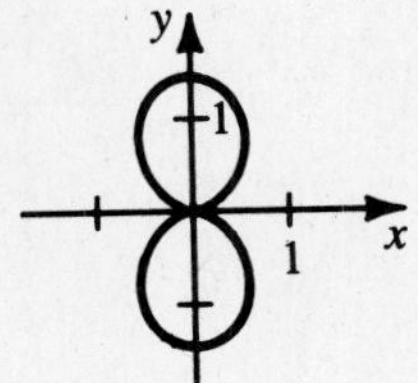

b. $r^2 = 2 \sin \theta$
$r^2 = \frac{2y}{r}$
$(r^3)^2 = (2y)^2$
$(x^2 + y^2)^3 = 4y^2$

38. a.

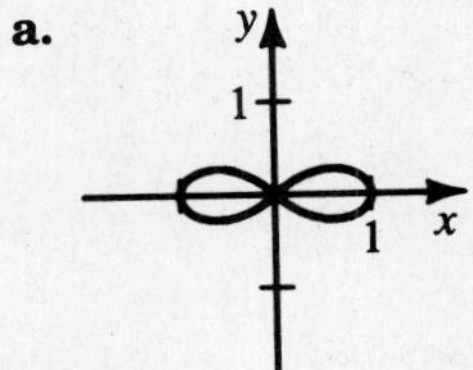

b. $r^2 = \cos 2\theta$
$r^2 = 2 \cos^2\theta - 1$
$r^2 = \frac{2x^2}{r^2} - 1$
$r^4 = 2x^2 - r^2$
$(x^2 + y^2)^2 = x^2 - y^2$

Pages 678–679 • WRITTEN EXERCISES

A **1.** $|z| = \sqrt{3^2 + 3^2} = \sqrt{18} = 3\sqrt{2}$; $\cos\theta = \frac{3}{3\sqrt{2}} = \frac{1}{\sqrt{2}}$;

$\sin\theta = \frac{3}{3\sqrt{2}} = \frac{1}{\sqrt{2}}$; $m(\theta) = 45°$;

$3 + 3i = 3\sqrt{2}(\cos 45° + i \sin 45°)$

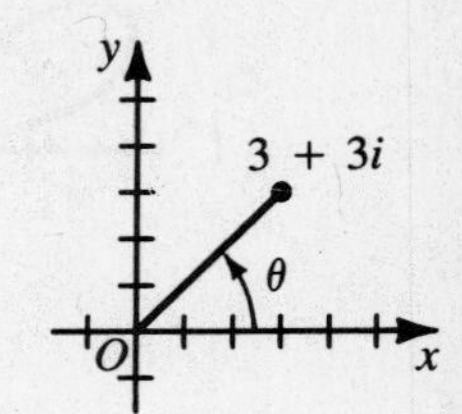

2. $|z| = \sqrt{0^2 + (-4)^2} = \sqrt{16} = 4$; $\cos\theta = \frac{0}{4} = 0$; $\sin\theta = \frac{-4}{4} = -1$;
$m(\theta) = 270°$; $-4i = 4(\cos 270° + i \sin 270°)$

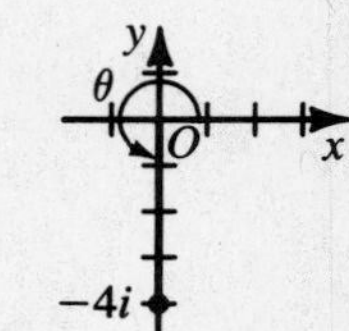

3. $|z| = \sqrt{(-7)^2 + 0^2} = 7$; $\cos\theta = \frac{-7}{7} = -1$; $\sin\theta = \frac{0}{7} = 0$;
$m(\theta) = 180°$; $-7 = 7(\cos 180° + i \sin 180°)$

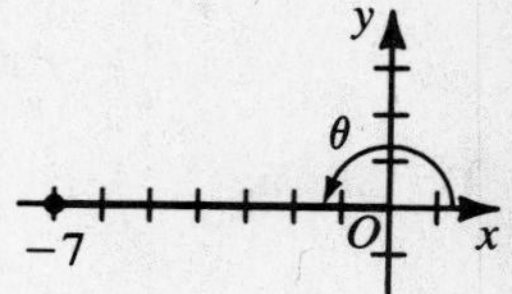

4. $|z| = \sqrt{(-1)^2 + (\sqrt{3})^2} = \sqrt{4} = 2$; $\cos\theta = -\frac{1}{2}$; $\sin\theta = \frac{\sqrt{3}}{2}$;
$m(\theta) = 120°$; $-1 + i\sqrt{3} = 2(\cos 120° + i \sin 120°)$

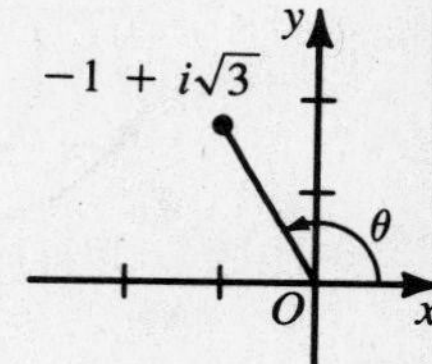

5. $|z| = \sqrt{(2\sqrt{2})^2 + (-2\sqrt{2})^2} = \sqrt{16} = 4$; $\cos\theta = \frac{2\sqrt{2}}{4} = \frac{\sqrt{2}}{2}$;
$\sin\theta = \frac{-2\sqrt{2}}{4} = -\frac{\sqrt{2}}{2}$; $m(\theta) = 315°$; $2\sqrt{2} - 2i\sqrt{2} =$
$4(\cos 315° + i \sin 315°)$

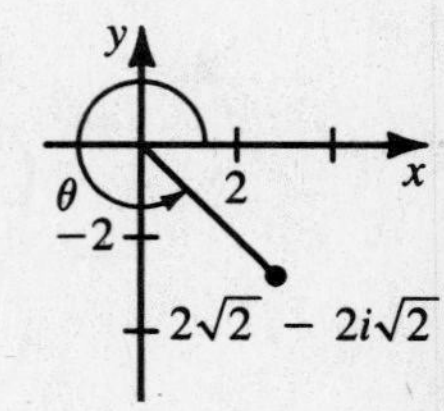

6. $|z| = \sqrt{\left(-\frac{1}{2}\right)^2 + \left(-\frac{\sqrt{3}}{2}\right)^2} = \sqrt{1} = 1$; $\cos\theta = \frac{-\frac{1}{2}}{1} = -\frac{1}{2}$;

$\sin\theta = \frac{-\frac{\sqrt{3}}{2}}{1} = -\frac{\sqrt{3}}{2}$; $m(\theta) = 240°$; $-\frac{1}{2} - \frac{\sqrt{3}}{2}i = 1(\cos 240° + i \sin 240°)$

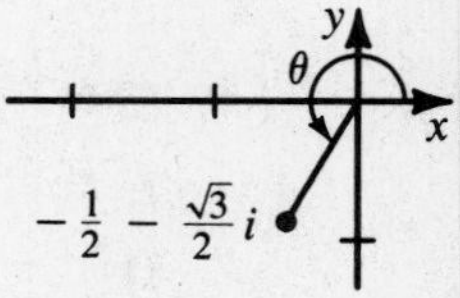

7. $|z| = \sqrt{(-6)^2 + (6)^2} = \sqrt{72} = 6\sqrt{2}$; $\cos\theta = \frac{-6}{6\sqrt{2}} = -\frac{1}{\sqrt{2}}$;

$\sin\theta = \frac{6}{6\sqrt{2}} = \frac{1}{\sqrt{2}}$; $m(\theta) = 135°$; $-6 + 6i =$

$6\sqrt{2}(\cos 135° + i \sin 135°)$

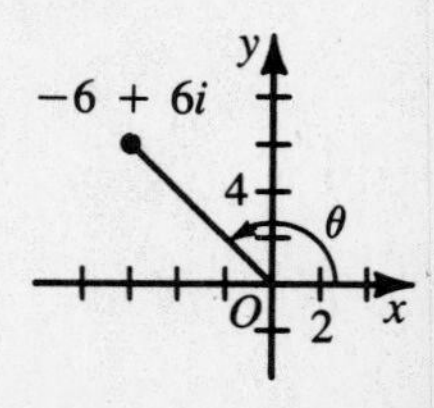

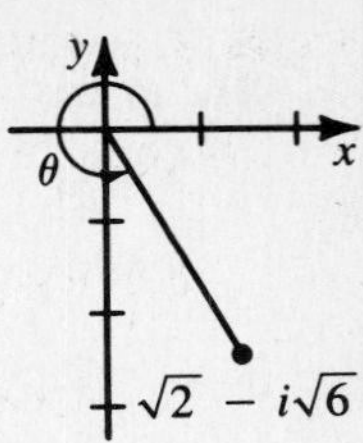

8. $|z| = \sqrt{(\sqrt{2})^2 + (-\sqrt{6})^2} = \sqrt{8} = 2\sqrt{2}$; $\cos\theta = \dfrac{\sqrt{2}}{2\sqrt{2}} = \dfrac{1}{2}$;

$\sin\theta = \dfrac{-\sqrt{6}}{2\sqrt{2}} = -\dfrac{\sqrt{3}}{2}$; $m(\theta) = 300°$; $\sqrt{2} - i\sqrt{6} =$

$2\sqrt{2}(\cos 300° + i\sin 300°)$

9. $4\sqrt{2}\left(-\dfrac{1}{\sqrt{2}} + i\cdot\dfrac{1}{\sqrt{2}}\right) = -4 + 4i$

10. $10\left(\dfrac{\sqrt{3}}{2} + i\cdot\dfrac{1}{2}\right) = 5\sqrt{3} + 5i$

11. $2\left(-\dfrac{1}{2} + i\cdot\dfrac{\sqrt{3}}{2}\right) = -1 + i\sqrt{3}$

12. $12\left(\dfrac{\sqrt{3}}{2} + i\left(-\dfrac{1}{2}\right)\right) = 6\sqrt{3} - 6i$

13. $6\left(-\dfrac{1}{2} + i\left(-\dfrac{\sqrt{3}}{2}\right)\right) = -3 - 3i\sqrt{3}$

14. $\dfrac{1}{3}\left(-\dfrac{\sqrt{3}}{2} + i\cdot\dfrac{1}{2}\right) = -\dfrac{\sqrt{3}}{6} + \dfrac{1}{6}i$

15. $\sqrt{6}\left(\dfrac{1}{\sqrt{2}} + i\left(-\dfrac{1}{\sqrt{2}}\right)\right) = \sqrt{3} - i\sqrt{3}$

16. $8\left(\dfrac{\sqrt{3}}{2} + i\left(-\dfrac{1}{2}\right)\right) = 4\sqrt{3} - 4i$

17. a. $z_1z_2 = 2\cdot 5[\cos(300° + 60°) + i\sin(300° + 60°)] = 10(\cos 360° + i\sin 360°) =$
$10(1 + 0i) = 10$

b. $\dfrac{z_1}{z_2} = \dfrac{2}{5}[\cos(300° - 60°) + i\sin(300° - 60°)] = \dfrac{2}{5}(\cos 240° + i\sin 240°) =$

$\dfrac{2}{5}\left(-\dfrac{1}{2} - \dfrac{\sqrt{3}}{2}i\right) = -\dfrac{1}{5} - \dfrac{\sqrt{3}}{5}i$

18. a. $z_1z_2 = 2\sqrt{2}\cdot\sqrt{2}[\cos(45° + 225°) + i\sin(45° + 225°)] =$
$4(\cos 270° + i\sin 270°) = 4(0 + -i) = -4i$

b. $\dfrac{z_1}{z_2} = \dfrac{2\sqrt{2}}{\sqrt{2}}\left[\cos(45° - 225°) + i\sin(45° - 225°)\right] = 2[\cos(-180°) + i\sin(-180°)] =$
$2(-1 + 0i) = -2$

19. a. $z_1z_2 = 6\cdot 4[\cos(135° + 105°) + i\sin(135° + 105°)] = 24(\cos 240° + i\sin 240°) =$

$24\left(-\dfrac{1}{2} - \dfrac{\sqrt{3}}{2}i\right) = -12 - 12i\sqrt{3}$

b. $\dfrac{z_1}{z_2} = \dfrac{6}{4}[\cos(135° - 105°) + i\sin(135° - 105°)] = \dfrac{3}{2}(\cos 30° + i\sin 30°) =$

$\dfrac{3}{2}\left(\dfrac{\sqrt{3}}{2} + \dfrac{1}{2}i\right) = \dfrac{3\sqrt{3}}{4} + \dfrac{3}{4}i$

20. a. $z_1z_2 = 8\cdot\dfrac{1}{2}[\cos(255° + 75°) + i\sin(255° + 75°)] = 4(\cos 330° + i\sin 330°) =$

$4\left(\dfrac{\sqrt{3}}{2} - \dfrac{1}{2}i\right) = 2\sqrt{3} - 2i$

b. $\dfrac{z_1}{z_2} = \dfrac{8}{\frac{1}{2}}[\cos(255° - 75°) + i\sin(255° - 75°)] = 16(\cos 180° + i\sin 180°) =$
$16(-1 + 0i) = -16$

21. a. $z_1z_2 = \frac{3}{2} \cdot 4[\cos(105° + 15°) + i\sin(105° + 15°)] = 6(\cos 120° + i\sin 120°) = 6\left(-\frac{1}{2} + \frac{\sqrt{3}}{2}i\right) = -3 + 3i\sqrt{3}$

b. $\frac{z_1}{z_2} = \frac{\frac{3}{2}}{4}[\cos(105° - 15°) + i\sin(105° - 15°)] = \frac{3}{8}(\cos 90° + i\sin 90°) = \frac{3}{8}(0 + i) = \frac{3}{8}i$

22. a. $z_1z_2 = \sqrt{2} \cdot 3\sqrt{2}[\cos(225° + 165°) + i\sin(225° + 165°)] = 6(\cos 390° + i\sin 390°) = 6(\cos 30° + i\sin 30°) = 6\left(\frac{\sqrt{3}}{2} + \frac{1}{2}i\right) = 3\sqrt{3} + 3i$

b. $\frac{z_1}{z_2} = \frac{\sqrt{2}}{3\sqrt{2}}[\cos(225° - 165°) + i\sin(225° - 165°)] = \frac{1}{3}(\cos 60° + i\sin 60°) = \frac{1}{3}\left(\frac{1}{2} + \frac{\sqrt{3}}{2}i\right) = \frac{1}{6} + \frac{1}{6}i\sqrt{3}$

B 23. a. $z_1z_2 = \left(\frac{1}{2} \cdot \frac{\sqrt{2}}{2} - \frac{\sqrt{3}}{2} \cdot \frac{\sqrt{2}}{2}\right) + \left(\frac{1}{2} \cdot \frac{\sqrt{2}}{2} + \frac{\sqrt{3}}{2} \cdot \frac{\sqrt{2}}{2}\right)i = \left(\frac{\sqrt{2}}{4} - \frac{\sqrt{6}}{4}\right) + \left(\frac{\sqrt{2}}{4} + \frac{\sqrt{6}}{4}\right)i = \frac{\sqrt{2} - \sqrt{6}}{4} + \frac{\sqrt{2} + \sqrt{6}}{4}i$

b. $|z_1| = \sqrt{\left(\frac{1}{2}\right)^2 + \left(\frac{\sqrt{3}}{2}\right)^2} = \sqrt{1} = 1$; $\cos\theta_1 = \frac{\frac{1}{2}}{1} = \frac{1}{2}$; $\sin\theta_1 = \frac{\frac{\sqrt{3}}{2}}{1} = \frac{\sqrt{3}}{2}$; $m(\theta_1) = 60°$; $\frac{1}{2} + \frac{\sqrt{3}}{2}i = 1(\cos 60° + i\sin 60°)$; $|z_2| = \sqrt{\left(\frac{\sqrt{2}}{2}\right)^2 + \left(\frac{\sqrt{2}}{2}\right)^2} = \sqrt{1} = 1$; $\cos\theta_2 = \frac{\frac{\sqrt{2}}{2}}{1} = \frac{\sqrt{2}}{2}$; $\sin\theta_2 = \frac{\frac{\sqrt{2}}{2}}{1} = \frac{\sqrt{2}}{2}$; $m(\theta_2) = 45°$; $\frac{\sqrt{2}}{2} + \frac{\sqrt{2}}{2}i = 1(\cos 45° + i\sin 45°)$; $z_1z_2 = 1 \cdot 1[\cos(60° + 45°) + i\sin(60° + 45°)] = \cos 105° + i\sin 105°$

c. $\cos 105° = \frac{\sqrt{2} - \sqrt{6}}{4} \approx -0.2588$; $\sin 105° = \frac{\sqrt{2} + \sqrt{6}}{4} \approx 0.9659$

d. $\cos 105° = \cos(60° + 45°) = \cos 60° \cos 45° - \sin 60° \sin 45° = \frac{1}{2} \cdot \frac{1}{\sqrt{2}} - \frac{\sqrt{3}}{2} \cdot \frac{1}{\sqrt{2}} = \frac{\sqrt{2} - \sqrt{6}}{4}$; $\sin 105° = \sin(60° + 45°) = \sin 60° \cos 45° + \cos 60° \sin 45° = \frac{\sqrt{3}}{2} \cdot \frac{1}{\sqrt{2}} + \frac{1}{2} \cdot \frac{1}{\sqrt{2}} = \frac{\sqrt{6} + \sqrt{2}}{4}$

C **24.** $|\bar{z}| = \sqrt{a^2 + (-b)^2} = \sqrt{a^2 + b^2} = r$; $\cos \bar{\theta} = \dfrac{a}{\sqrt{a^2 + b^2}}$; $\sin \bar{\theta} = \dfrac{-b}{\sqrt{a^2 + b^2}}$;

$m(\bar{\theta}) = -\theta$; $\bar{z} = r(\cos(-\theta) + i \sin(-\theta))$

a. $z\bar{z} = r \cdot r[\cos(\theta + (-\theta)) + i \sin(\theta + (-\theta))] = r^2(\cos 0° + i \sin 0°)$

b. $\dfrac{z}{\bar{z}} = \dfrac{r}{r}[\cos(\theta - (-\theta)) + i \sin(\theta - (-\theta))] = \cos 2\theta + i \sin 2\theta$

Pages 679–680 • COMPUTER EXERCISES

1.
```
10  LET PI = 3.141593
20  PRINT "ENTER THE VALUES OF X AND Y";
30  INPUT X,Y
40  LET R = SQR(X^2 + Y^2)
50  PRINT "ABSOLUTE VALUE OF Z IS: ";R
60  IF X = 0 AND Y = 0 THEN PRINT "ARGUMENT IS ANY VALUE" : END
70  IF X = 0 THEN LET A = 90 : GOTO 110
80  LET A = ATN(ABS(Y/X))
90  LET A = A * 180/PI
100  IF X < 0 AND Y > 0 THEN LET A = -A + 180
110  IF X <= 0 AND Y <= 0 THEN LET A = A + 180
120  IF X > 0 AND Y < 0 THEN LET A = -A + 360
130  PRINT "ARGUMENT IS: ";A;" DEGREES"
140  END
```

2. 4.243; 315° **3.** 2.828; 135° **4.** 5; 233° **5.** 7; 270° **6.** 9.43; 58° **7.** 6; 0°

8.
```
10  DIM X(2),Y(2),R(2),A(2)
20  LET PI = 3.141593
30  FOR I = 1 TO 2
40  PRINT "ENTER THE VALUES OF X AND Y FOR Z(";I;")";
50  INPUT X,Y
60  LET R = SQR(X^2 + Y^2)
70  IF X = 0 AND Y = 0 THEN LET A = 0 : GOTO 130
80  IF X = 0 THEN LET A = PI/2 : GOTO 110
90  LET A = ATN(ABS(Y/X))
100  IF X < 0 AND Y > 0 THEN LET A = -A + PI
110  IF X <= 0 AND Y <= 0 THEN LET A = A + PI
120  IF X > 0 AND Y < 0 THEN LET A = -A + 2 * PI
130  LET R(I) = R : LET A(I) = A
140  NEXT I
150  LET A = R(1) * R(2) * COS(A(1) + A(2))
160  LET B = R(1) * R(2) * SIN(A(1) + A(2))
170  LET A = INT(A + .5) : LET B = INT(B + .5)
180  PRINT A;" + ";B;"I"
```

9. $11 + 17i$ **10.** $-3 - 6i$ **11.** $2 - 14i$

Pages 683–684 • WRITTEN EXERCISES

A **1.** $r = \sqrt{\left(\frac{\sqrt{3}}{2}\right)^2 + \left(\frac{1}{2}\right)^2} = 1$; $\cos\theta = \frac{\sqrt{3}}{2}$; $\sin\theta = \frac{1}{2}$;

$m(\theta) = 30°$; $\frac{\sqrt{3}}{2} + \frac{1}{2}i = 1(\cos 30° + i\sin 30°)$;

$\left(\frac{\sqrt{3}}{2} + \frac{1}{2}i\right)^3 = 1^3(\cos 3 \cdot 30° + i\sin 3 \cdot 30°) = \cos 90° + i\sin 90° = 0 + i = i$

2. $r = \sqrt{(-1)^2 + (\sqrt{3})^2} = 2$; $\cos\theta = -\frac{1}{2}$; $\sin\theta = \frac{\sqrt{3}}{2}$; $m(\theta) = 120°$; $-1 + i\sqrt{3} = 2(\cos 120° + i\sin 120°)$; $(-1 + i\sqrt{3})^3 = 2^3(\cos 3 \cdot 120° + i\sin 3 \cdot 120°) = 8(\cos 360° + i\sin 360°) = 8(1 + 0i) = 8$

3. $r = \sqrt{(-1)^2 + (1)^2} = \sqrt{2}$; $\cos\theta = -\frac{1}{\sqrt{2}}$; $\sin\theta = \frac{1}{\sqrt{2}}$; $m(\theta) = 135°$; $-1 + i = \sqrt{2}(\cos 135° + i\sin 135°)$; $(-1 + i)^4 = (\sqrt{2})^4(\cos 4 \cdot 135° + i\sin 4 \cdot 135°) = 4(\cos 540° + i\sin 540°) = 4(\cos 180° + i\sin 180°) = 4(-1 + 0i) = -4$

4. $r = \sqrt{(\sqrt{2})^2 + (-\sqrt{2})^2} = 2$; $\cos\theta = \frac{\sqrt{2}}{2}$; $\sin\theta = -\frac{\sqrt{2}}{2}$;
$m(\theta) = 315°$; $\sqrt{2} - i\sqrt{2} = 2(\cos 315° + i\sin 315°)$;
$(\sqrt{2} - i\sqrt{2})^4 = 2^4(\cos 4 \cdot 315° + i\sin 4 \cdot 315°) = 16(\cos 1260° + i\sin 1260°) = 16(\cos 180° + i\sin 180°) = 16(-1 + 0i) = -16$

5. $r = \sqrt{(\sqrt{3})^2 + (-1)^2} = 2$; $\cos\theta = \frac{\sqrt{3}}{2}$; $\sin\theta = -\frac{1}{2}$; $m(\theta) = 330°$;

$\sqrt{3} - i = 2(\cos 330° + i\sin 330°)$; $(\sqrt{3} - i)^5 = 2^5(\cos 5 \cdot 330° + i\sin 5 \cdot 330°) =$

$32(\cos 1650° + i\sin 1650°) = 32(\cos 210° + i\sin 210°) = 32\left(-\frac{\sqrt{3}}{2} - \frac{1}{2}i\right) = -16\sqrt{3} - 16i$

6. $r = \sqrt{(-\sqrt{3})^2 + 1^2} = 2$; $\cos\theta = -\frac{\sqrt{3}}{2}$; $\sin\theta = \frac{1}{2}$; $m(\theta) = 150°$;
$-\sqrt{3} + i = 2(\cos 150° + i\sin 150°)$;
$(-\sqrt{3} + i)^4 = 2^4(\cos 4 \cdot 150° + i\sin 4 \cdot 150°) = 16(\cos 600° + i\sin 600°) =$

$16(\cos 240° + i\sin 240°) = 16\left(-\frac{1}{2} - \frac{\sqrt{3}}{2}i\right) = -8 - 8i\sqrt{3}$

7. $r = \sqrt{\left(\frac{\sqrt{3}}{2}\right)^2 + \left(-\frac{1}{2}\right)^2} = 1$; $\cos\theta = \frac{\sqrt{3}}{2}$; $\sin\theta = -\frac{1}{2}$; $m(\theta) = 330°$;

$\frac{\sqrt{3}}{2} - \frac{1}{2}i = 1(\cos 330° + i\sin 330°)$; $\left(\frac{\sqrt{3}}{2} - \frac{1}{2}i\right)^{10} =$
$1^{10}(\cos 10 \cdot 330° + i\sin 10 \cdot 330°) = \cos 3300° + i\sin 3300° =$

$\cos 60° + i\sin 60° = \frac{1}{2} + \frac{\sqrt{3}}{2}i$

8. $r = \sqrt{(-2)^2 + (-2\sqrt{3})^2} = 4$; $\cos\theta = -\frac{1}{2}$; $\sin\theta = -\frac{\sqrt{3}}{2}$; $m(\theta) = 240°$;
$-2 - 2i\sqrt{3} = 4(\cos 240° + i\sin 240°)$; $(-2 - 2i\sqrt{3})^3 =$
$4^3(\cos 3\cdot 240° + i\sin 3\cdot 240°) = 64(\cos 720° + i\sin 720°) =$
$64(\cos 0° + i\sin 0°) = 64(1 + 0i) = 64$

9. $r = \sqrt{(1)^2 + (-1)^2} = \sqrt{2}$; $\cos\theta = \frac{1}{\sqrt{2}}$; $\sin\theta = \frac{-1}{\sqrt{2}}$; $m(\theta) = 315°$; $1 - i =$
$\sqrt{2}(\cos 315° + i\sin 315°)$; $(1 - i)^{-3} = (\sqrt{2})^{-3}[\cos(-3\cdot 315°) + i\sin(-3\cdot 315°)] =$
$\frac{1}{2\sqrt{2}}[\cos(-945°) + i\sin(-945°)] = \frac{1}{2\sqrt{2}}(\cos 135° + i\sin 135°) =$
$\frac{1}{2\sqrt{2}}\left(-\frac{1}{\sqrt{2}} + \frac{1}{\sqrt{2}}i\right) = -\frac{1}{4} + \frac{1}{4}i$

10. $r = \sqrt{(\sqrt{2})^2 + (\sqrt{2})^2} = 2$; $\cos\theta = \frac{\sqrt{2}}{2}$; $\sin\theta = \frac{\sqrt{2}}{2}$; $m(\theta) = 45°$; $\sqrt{2} + i\sqrt{2} =$
$2(\cos 45° + i\sin 45°)$; $(\sqrt{2} + i\sqrt{2})^{-4} = 2^{-4}[\cos(-4\cdot 45°) + i\sin(-4\cdot 45°)] =$
$\frac{1}{16}[\cos(-180°) + i\sin(-180°)] = \frac{1}{16}(\cos 180° + i\sin 180°) = \frac{1}{16}(-1 + 0i) = -\frac{1}{16}$

11. $r = \sqrt{1^2 + (\sqrt{3})^2} = 2$; $\cos\theta = \frac{1}{2}$; $\sin\theta = \frac{\sqrt{3}}{2}$; $m(\theta) = 60°$; $1 + i\sqrt{3} =$
$2(\cos 60° + i\sin 60°)$; $(1 + i\sqrt{3})^{-4} = 2^{-4}[\cos(-4\cdot 60°) + i\sin(-4\cdot 60°)] =$
$\frac{1}{16}[\cos(-240°) + i\sin(-240°)] = \frac{1}{16}(\cos 120° + i\sin 120°) = \frac{1}{16}\left(-\frac{1}{2} + \frac{\sqrt{3}}{2}i\right) =$
$-\frac{1}{32} + \frac{\sqrt{3}}{32}i$

12. $r = \sqrt{\left(\frac{3\sqrt{3}}{2}\right)^2 + \left(-\frac{3}{2}\right)^2} = 3$; $\cos\theta = \frac{\sqrt{3}}{2}$; $\sin\theta = -\frac{1}{2}$;
$m(\theta) = 330°$; $\frac{3\sqrt{3}}{2} - \frac{3}{2}i = 3(\cos 330° + i\sin 330°)$;
$\left(\frac{3\sqrt{3}}{2} - \frac{3}{2}i\right)^{-3} = 3^{-3}[\cos(-3\cdot 330°) + i\sin(-3\cdot 330°)] =$
$\frac{1}{27}[\cos(-990°) + i\sin(-990°)] = \frac{1}{27}(\cos 90° + i\sin 90°) = \frac{1}{27}(0 + i) = \frac{1}{27}i$

13. $r = \sqrt{(0.2588)^2 + (0.9659)^2} \approx 1$; $\cos\theta = 0.2588$; $\sin\theta = 0.9659$;
$m(\theta) = 75°$; $0.2588 + 0.9659i = 1(\cos 75° + i\sin 75°)$; $(0.2588 + 0.9659i)^2 =$
$1^2(\cos 2\cdot 75° + i\sin 2\cdot 75°) = \cos 150° + i\sin 150° = -\frac{\sqrt{3}}{2} + \frac{1}{2}i$

14. $r = \sqrt{(-0.2588)^2 + (0.9659)^2} \approx 1$; $\cos\theta = -0.2588$; $\sin\theta = 0.9659$;
$m(\theta) = 105°$; $-0.2588 + 0.9659i = 1(\cos 105° + i\sin 105°)$;
$[2(-0.2588 + 0.9659i)]^3 = 2^3(-0.2588 + 0.9659i)^3 =$
$8(\cos 3\cdot 105° + i\sin 3\cdot 105°) = 8(\cos 315° + i\sin 315°) = \frac{8}{\sqrt{2}} - \frac{8}{\sqrt{2}}i$

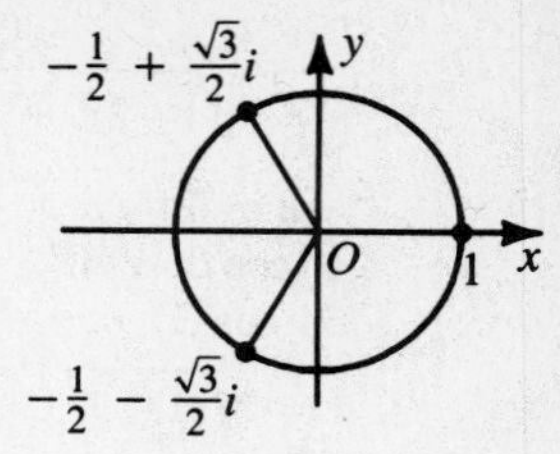

B **15.** $1^{1/3} = 1^{1/3}(\cos 0° + i \sin 0°)^{1/3} = \cos \frac{0° + k \cdot 360°}{3} + i \sin \frac{0° + k \cdot 360°}{3}$; $\cos 0° + i \sin 0° = 1$, $\cos 120° + i \sin 120° = -\frac{1}{2} + \frac{\sqrt{3}}{2}i$, $\cos 240° + i \sin 240° = -\frac{1}{2} - \frac{\sqrt{3}}{2}i$

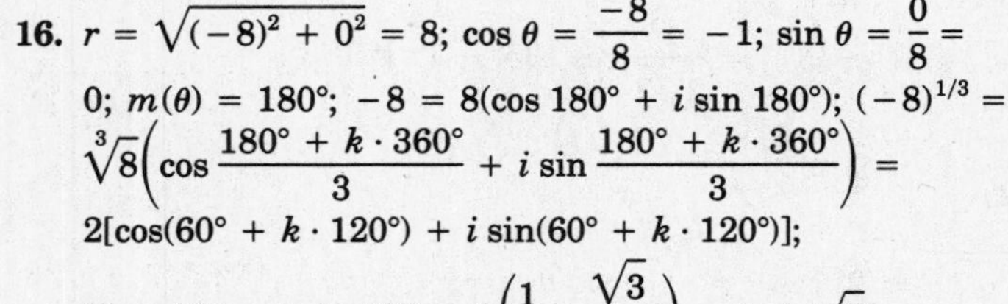

16. $r = \sqrt{(-8)^2 + 0^2} = 8$; $\cos \theta = \frac{-8}{8} = -1$; $\sin \theta = \frac{0}{8} = 0$; $m(\theta) = 180°$; $-8 = 8(\cos 180° + i \sin 180°)$; $(-8)^{1/3} = \sqrt[3]{8}\left(\cos \frac{180° + k \cdot 360°}{3} + i \sin \frac{180° + k \cdot 360°}{3}\right) =$ $2[\cos(60° + k \cdot 120°) + i \sin(60° + k \cdot 120°)]$;

$2(\cos 60° + i \sin 60°) = 2\left(\frac{1}{2} + \frac{\sqrt{3}}{2}i\right) = 1 + \sqrt{3}i$,

$2(\cos 180° + i \sin 180°) = 2(-1 + 0i) = -2$,

$2(\cos 300° + i \sin 300°) = 2\left(\frac{1}{2} - \frac{\sqrt{3}}{2}i\right) = 1 - \sqrt{3}i$

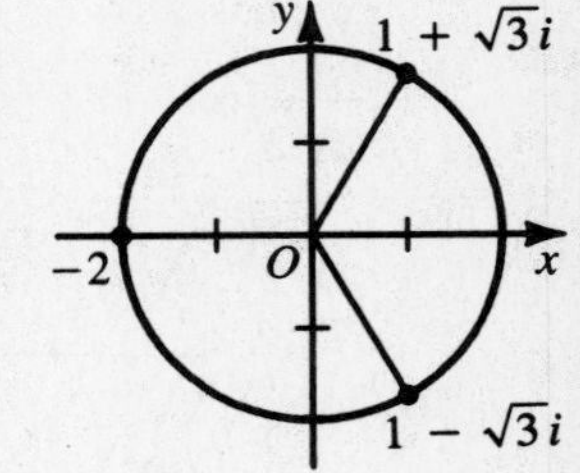

17. $r = \sqrt{(-16)^2 + 0^2} = 16$; $\cos \theta = \frac{-16}{16} = -1$; $\sin \theta = \frac{0}{16} = 0$; $m(\theta) = 180°$; $-16 = 16(\cos 180° + i \sin 180°)$; $(-16)^{1/4} = \sqrt[4]{16}\left(\cos \frac{180° + k \cdot 360°}{4} + i \sin \frac{180° + k \cdot 360°}{4}\right) =$ $2[\cos(45° + k \cdot 90°) + i \sin(45° + k \cdot 90°)]$;

$2(\cos 45° + i \sin 45°) = 2\left(\frac{1}{\sqrt{2}} + \frac{1}{\sqrt{2}}i\right) = \sqrt{2} + i\sqrt{2}$,

$2(\cos 135° + i \sin 135°) = 2\left(-\frac{1}{\sqrt{2}} + \frac{1}{\sqrt{2}}i\right) = -\sqrt{2} + i\sqrt{2}$,

$2(\cos 225° + i \sin 225°) = 2\left(-\frac{1}{\sqrt{2}} - \frac{1}{\sqrt{2}}i\right) = -\sqrt{2} - i\sqrt{2}$,

$2(\cos 315° + i \sin 315°) = 2\left(\frac{1}{\sqrt{2}} - \frac{1}{\sqrt{2}}i\right) = \sqrt{2} - i\sqrt{2}$

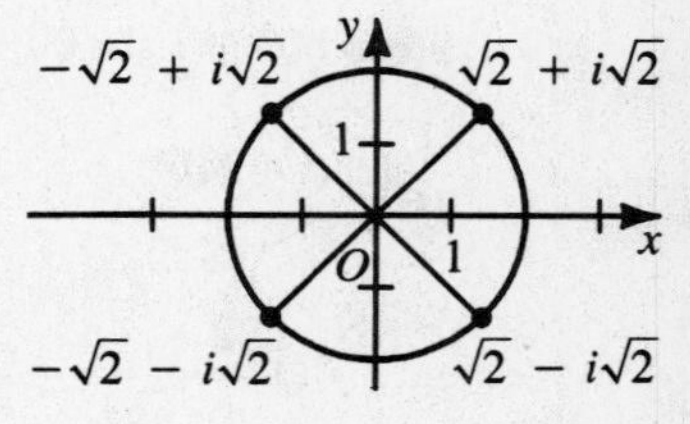

18. $r = \sqrt{(-8)^2 + (8\sqrt{3})^2} = \sqrt{256} = 16$; $\cos \theta = \frac{-8}{16} = -\frac{1}{2}$; $\sin \theta = \frac{8\sqrt{3}}{16} = \frac{\sqrt{3}}{2}$; $m(\theta) = 120°$; $-8 + 8i\sqrt{3} = 16(\cos 120° + i \sin 120°)$; $(-8 + 8i\sqrt{3})^{1/4} = \sqrt[4]{16}\left(\cos \frac{120° + k \cdot 360°}{4} + i \sin \frac{120° + k \cdot 360°}{4}\right) =$ $2[\cos(30° + k \cdot 90°) + i \sin(30° + k \cdot 90°)]$;

$-1 + i\sqrt{3}$ y $\sqrt{3} + i$ x $-\sqrt{3} - i$ $1 - i\sqrt{3}$

$2(\cos 30° + i \sin 30°) = 2\left(\frac{\sqrt{3}}{2} + \frac{1}{2}i\right) = \sqrt{3} + i$,

$2(\cos 120° + i \sin 120°) = 2\left(-\frac{1}{2} + \frac{\sqrt{3}}{2}i\right) = -1 + i\sqrt{3}$, $2(\cos 210° + i \sin 210°) =$ $2\left(-\frac{\sqrt{3}}{2} - \frac{1}{2}i\right) = -\sqrt{3} - i$, $2(\cos 300° + i \sin 300°) = 2\left(\frac{1}{2} - \frac{\sqrt{3}}{2}i\right) = 1 - i\sqrt{3}$

19. $r = \sqrt{(-128)^2 + (-128\sqrt{3})^2} = 256$; $\cos\theta =$ $\frac{-128}{251} = -\frac{1}{2}$; $\sin\theta = \frac{-128\sqrt{3}}{256} = -\frac{\sqrt{3}}{2}$;
$m(\theta) = 240°$; $-128 - 128i\sqrt{3} =$
$256(\cos 240° + i \sin 240°)$; $(-128 - 128i\sqrt{3})^{1/4} =$
$\sqrt[4]{256}\left(\cos\frac{240° + k \cdot 360°}{4} + i \sin\frac{240° + k \cdot 360°}{4}\right) =$
$4[\cos(60° + k \cdot 90°) + i \sin(60° + k \cdot 90°)]$;

$4(\cos 60° + i \sin 60°) = 4\left(\frac{1}{2} + \frac{\sqrt{3}}{2}i\right) = 2 + 2i\sqrt{3}$, $4(\cos 150° + i \sin 150°) =$ $4\left(-\frac{\sqrt{3}}{2} + \frac{1}{2}i\right) = -2\sqrt{3} + 2i$, $4(\cos 240° + i \sin 240°) = 4\left(-\frac{1}{2} - \frac{\sqrt{3}}{2}i\right) =$ $-2 - 2i\sqrt{3}$, $4(\cos 330° + i \sin 330°) = 4\left(\frac{\sqrt{3}}{2} - \frac{1}{2}i\right) = 2\sqrt{3} - 2i$

20. $r = \sqrt{0^2 + (32)^2} = 32$; $\cos\theta = \frac{0}{32} = 0$; $\sin\theta = \frac{32}{32} = 1$;
$m(\theta) = 90°$; $32i = 32(\cos 90° + i \sin 90°)$; $(32i)^{1/5} =$
$\sqrt[5]{32}\left(\cos\frac{90° + k \cdot 360°}{5} + i \sin\frac{90° + k \cdot 360°}{5}\right) =$
$2[\cos(18° + k \cdot 72°) + i \sin(18° + k \cdot 72°)]$;
$2(\cos 18° + i \sin 18°)$, $2(\cos 90° + i \sin 90°)$,
$2(\cos 162° + i \sin 162°)$, $2(\cos 234° + i \sin 234°)$,
$2(\cos 306° + i \sin 306°)$

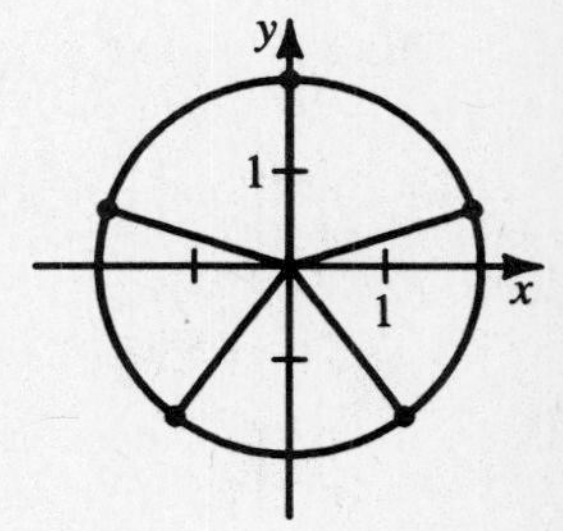

21. $z = a + bi = r(\cos\theta + i \sin\theta)$; $\frac{1}{z} = \frac{1}{r(\cos\theta + i\sin\theta)} =$
$\frac{1}{r} \cdot \frac{1}{\cos\theta + i\sin\theta} \cdot \frac{\cos\theta - i\sin\theta}{\cos\theta - i\sin\theta} = \frac{1}{r} \cdot \frac{\cos\theta - i\sin\theta}{\cos^2\theta + \sin^2\theta} = \frac{1}{r}(\cos\theta - i\sin\theta) =$
$\frac{1}{r}(\cos(-\theta) + i\sin(-\theta))$

22. $z = a + bi = r(\cos\theta + i\sin\theta)$; $z^{-n} = (z^{-1})^n = \left[\frac{1}{r}(\cos(-\theta) + i\sin(-\theta))\right]^n =$
$\left(\frac{1}{r}\right)^n(\cos(-n\theta) + i\sin(-n\theta)) = r^{-n}(\cos(-n\theta) + i\sin(-n\theta))$

23. $r = \sqrt{1^2 + (-1)^2} = \sqrt{2}$; $\cos\theta = \frac{1}{\sqrt{2}}$; $\sin\theta = -\frac{1}{\sqrt{2}}$;
$m(\theta) = 315°$; $1 - i = \sqrt{2}(\cos 315° + i\sin 315°)$;
$(1 - i)^{1/3} = (\sqrt{2})^{1/3}\left(\cos\frac{315° + k \cdot 360°}{3} + i\sin\frac{315° + k \cdot 360°}{3}\right) =$
$2^{1/6}[\cos(105° + k \cdot 120°) + i\sin(105° + k \cdot 120°)]$; $2^{1/6}(\cos 105° + i\sin 105°)$,

$2^{1/6}(\cos 225° + i \sin 225°)$, $2^{1/6}(\cos 345° + i \sin 345°)$; the product of the three cube roots equals $(2^{1/6})(2^{1/6})(2^{1/6})[\cos(105° + 225° + 345°) + i \sin(105° + 225° + 345°)] = \sqrt{2}[\cos 675° + i \sin 675°] = \sqrt{2}[\cos 315° + i \sin 315°] = \sqrt{2}\left[\frac{1}{\sqrt{2}} - \frac{1}{\sqrt{2}}i\right] = 1 - i$

24. $r = \sqrt{(-1)^2 + 0^2} = 1$; $\cos\theta = \frac{-1}{1} = -1$; $\sin\theta = \frac{0}{1} = 0$; $m(\theta) = 180°$;
$-1 = \cos 180° + i \sin 180°$; $(-1)^{1/4} = \cos\frac{180° + k \cdot 360°}{4} + i \sin\frac{180° + k \cdot 360°}{4} =$
$\cos(45° + k \cdot 90°) + i \sin(45° + k \cdot 90°)$; $\cos 45° + i \sin 45°$, $\cos 135° + i \sin 135°$, $\cos 225° + i \sin 225°$, $\cos 315° + i \sin 315°$; the product of the four fourth roots equals $\cos(45° + 135° + 225° + 315°) + i \sin(45° + 135° + 225° + 315°) = \cos 720° + i \sin 720° = \cos 0° + i \sin 0° = 1 + 0i = 1$

C **25.** $z = r(\cos\theta + i \sin\theta)$; $z^{1/2} = r^{1/2}\left(\cos\frac{\theta + k \cdot 360°}{2} + i \sin\frac{\theta + k \cdot 360°}{2}\right)$;
the roots are $r^{1/2}\left(\cos\frac{\theta}{2} + i \sin\frac{\theta}{2}\right)$ and $r^{1/2}\left[\cos\left(\frac{\theta}{2} + 180°\right) + i \sin\left(\frac{\theta}{2} + 180°\right)\right]$;
the product of the two square roots equals
$(r^{1/2})(r^{1/2})\left[\cos\left(\frac{\theta}{2} + \frac{\theta}{2} + 180°\right) + i \sin\left(\frac{\theta}{2} + \frac{\theta}{2} + 180°\right)\right] =$
$r[\cos(\theta + 180°) + i \sin(\theta + 180°)] = r(-\cos\theta - i \sin\theta) = -r(\cos\theta + i \sin\theta) = -z$

26. $z = r(\cos\theta + i \sin\theta)$; $z^{1/3} = r^{1/3}\left(\cos\frac{\theta + k \cdot 360°}{3} + i \sin\frac{\theta + k \cdot 360°}{3}\right) =$
$r^{1/3}\left[\cos\left(\frac{\theta}{3} + k \cdot 120°\right) + i \sin\left(\frac{\theta}{3} + k \cdot 120°\right)\right]$; the roots are $r^{1/3}\left(\cos\frac{\theta}{3} + i \sin\frac{\theta}{3}\right)$,
$r^{1/3}\left[\cos\left(\frac{\theta}{3} + 120°\right) + i \sin\left(\frac{\theta}{3} + 120°\right)\right]$, and
$r^{1/3}\left[\cos\left(\frac{\theta}{3} + 240°\right) + i \sin\left(\frac{\theta}{3} + 240°\right)\right]$; the product of the three cube roots
equals $(r^{1/3})(r^{1/3})(r^{1/3})\left[\cos\left(\frac{\theta}{3} + \frac{\theta}{3} + 120° + \frac{\theta}{3} + 240°\right) +\right.$
$\left. i \sin\left(\frac{\theta}{3} + \frac{\theta}{3} + 120° + \frac{\theta}{3} + 240°\right)\right] = r[\cos(\theta + 360°) + i \sin(\theta + 360°)] =$
$r[\cos\theta + i \sin\theta] = z$

27. $z = a + bi = r(\cos\theta + i \sin\theta)$; $\bar{z} = a - bi = r[\cos(-\theta) + i \sin(-\theta)]$; $(\bar{z})^n = r^n[\cos n(-\theta) + i \sin n(-\theta)] = r^n[\cos(-n\theta) + i \sin(-n\theta)] = \overline{z^n}$

28. $1 = 1(\cos 0° + i \sin 0°)$; $1^{1/5} = 1\left(\cos\frac{0° + k \cdot 360°}{5} + i \sin\frac{0° + k \cdot 360°}{5}\right) =$
$1[\cos(k \cdot 72°) + i \sin(k \cdot 72°)]$; the fifth roots are 1, $\cos 72° + i \sin 72°$, $\cos 144° + i \sin 144°$, $\cos 216° + i \sin 216°$, and $\cos 288° + i \sin 288°$. 1 is not a primitive fifth root of unity because the other fifth roots are not integral powers of 1. Each root has itself and 1 as integral powers, so for each root we need to check whether the other three roots are integral powers of the root.
$(\cos 72° + i \sin 72°)^2 = \cos 144° + i \sin 144°$
$(\cos 72° + i \sin 72°)^3 = \cos 216° + i \sin 216°$
$(\cos 72° + i \sin 72°)^4 = \cos 288° + i \sin 288°$

$(\cos 144° + i \sin 144°)^2 = \cos 288° + i \sin 288°$
$(\cos 144° + i \sin 144°)^3 = \cos 432° + i \sin 432° = \cos 72° + i \sin 72°$
$(\cos 144° + i \sin 144°)^4 = \cos 576° + i \sin 576° = \cos 216° + i \sin 216°$

$(\cos 216° + i \sin 216°)^2 = \cos 432° + i \sin 432° = \cos 72° + i \sin 72°$
$(\cos 216° + i \sin 216°)^3 = \cos 648° + i \sin 648° = \cos 288° + i \sin 288°$
$(\cos 216° + i \sin 216°)^4 = \cos 864° + i \sin 864° = \cos 144° + i \sin 144°$

$(\cos 288° + i \sin 288°)^2 = \cos 576° + i \sin 576° = \cos 216° + i \sin 216°$
$(\cos 288° + i \sin 288°)^3 = \cos 864° + i \sin 864° = \cos 144° + i \sin 144°$
$(\cos 288° + i \sin 288°)^4 = \cos 1152° + i \sin 1152° = \cos 72° + i \sin 72°$

Thus the primitive fifth roots of unity are $\cos 72° + i \sin 72°$, $\cos 144° + i \sin 144°$, $\cos 216° + i \sin 216°$, and $\cos 288° + i \sin 288°$.

Page 684 • SELF-TEST 2

1. $r = \sqrt{(-2)^2 + (2\sqrt{3})^2} = \sqrt{4 + 12} = 4$; $\cos \theta = \frac{-2}{4} = -\frac{1}{2}$; $\sin \theta = \frac{2\sqrt{3}}{4} = \frac{\sqrt{3}}{2}$; $m(\theta) = 120°$; $(4, 120°)$, $(-4, -60°)$

2. $x = 2 \cos(-45°) = 2\left(\frac{1}{\sqrt{2}}\right) = \sqrt{2}$; $y = 2 \sin(-45°) = 2\left(-\frac{1}{\sqrt{2}}\right) = -\sqrt{2}$; $(\sqrt{2}, -\sqrt{2})$

3. a. $z_1 z_2 = 3 \cdot 2[\cos(330° + 30°) + i \sin(330° + 30°)] = 6(\cos 360° + i \sin 360°) = 6(1 + 0i) = 6$

b. $\frac{z_1}{z_2} = \frac{3}{2}[\cos(330° - 30°) + i \sin(330° - 30°)] = \frac{3}{2}(\cos 300° + i \sin 300°) = \frac{3}{2}\left(\frac{1}{2} - \frac{\sqrt{3}}{2}i\right) = \frac{3}{4} - \frac{3\sqrt{3}}{4}i$

c. $(z_1)^3 = [3(\cos 330° + i \sin 330°)]^3 = 3^3[\cos(3 \cdot 330°) + i \sin(3 \cdot 330°)] = 27(\cos 990° + i \sin 990°) = 27(\cos 270° + i \sin 270°) = 27(0 - i) = -27i$

4. $r = \sqrt{1^2 + 1^2} = \sqrt{2}$; $\cos \theta = \frac{1}{\sqrt{2}}$; $\sin \theta = \frac{1}{\sqrt{2}}$; $m(\theta) = 45°$;

$1 + i = \sqrt{2}(\cos 45° + i \sin 45°)$; $(1 + i)^{1/3} = (\sqrt{2})^{1/3}\left(\cos \frac{45° + k \cdot 360°}{3} + i \sin \frac{45° + k \cdot 360°}{3}\right) =$

$2^{1/6}[\cos(15° + k \cdot 120°) + i \sin(15° + k \cdot 120°)]$; $\sqrt[6]{2}(\cos 15° + i \sin 15°)$, $\sqrt[6]{2}(\cos 135° + i \sin 135°)$, $\sqrt[6]{2}(\cos 255° + i \sin 255°)$

Page 685 • READING ALGEBRA

Note: Explanations will vary.

1. Nine times a minus two; nine times the quantity a minus two
2. The quantity two times x plus three, all divided by seven times y; two times x plus three divided by seven times y
3. The cosine of three times x plus one; the cosine of the quantity three times x plus 1
4. The tangent of one-half times x; one-half times the tangent of x
5. The square root of the quantity one minus x squared; one minus x
6. Nine to the two times x minus 1 power; nine to the two times x power minus 1

7. The logarithm base five of the quantity y minus four; the logarithm base five of y minus four
8. Eight times b minus three squared; the quantity eight times b minus three, all squared
9. The principal inverse sine of one; the inverse sine of one
10. The quantity secant of x, all squared; the secant of x squared
11. Four times x squared; the quantity 4 times x, all squared
12. Two times y minus three times x plus four; two times y minus the quantity three times y plus four

Pages 690–691 • WRITTEN EXERCISES

A 1.

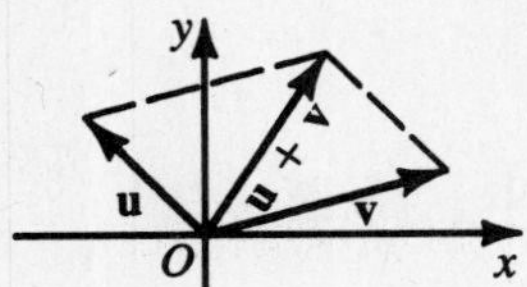

$180° - (135° - 15°) = 60°$

$\|\mathbf{u} + \mathbf{v}\|^2 = 5^2 + 8^2 - 2(5)(8)\cos 60°$

$= 89 - 80\left(\frac{1}{2}\right) = 49$

$\|\mathbf{u} + \mathbf{v}\| = 7$

2.

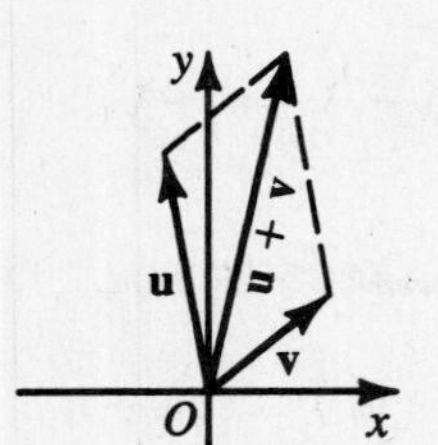

$180° - (100° - 40°) = 120°$

$\|\mathbf{u} + \mathbf{v}\|^2 = 10^2 + 6^2 - 2(10)(6)\cos 120°$

$= 136 - 120\left(-\frac{1}{2}\right) = 196$

$\|\mathbf{u} + \mathbf{v}\| = 14$

3.

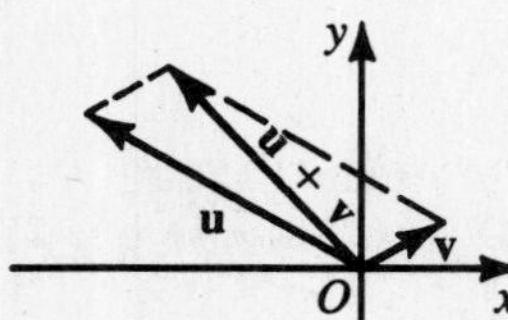

$180° - (150° - 30°) = 60°$

$\|\mathbf{u} + \mathbf{v}\|^2 = 21^2 + 5^2 - 2(21)(5)\cos 60°$

$= 466 - 210\left(\frac{1}{2}\right) = 361$

$\|\mathbf{u} + \mathbf{v}\| = 19$

4.

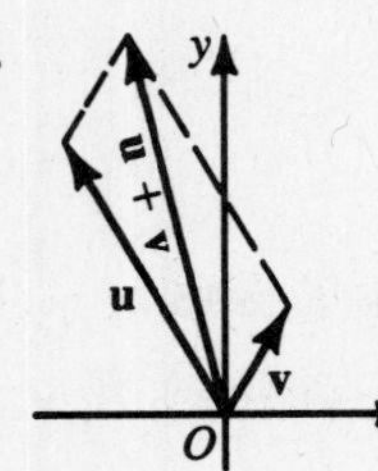

$180° - (120° - 60°) = 120°$

$\|\mathbf{u} + \mathbf{v}\|^2 = 16^2 + 5^2 - 2(16)(5)\cos 120°$

$= 281 - 160\left(-\frac{1}{2}\right) = 361$

$\|\mathbf{u} + \mathbf{v}\| = 19$

5.

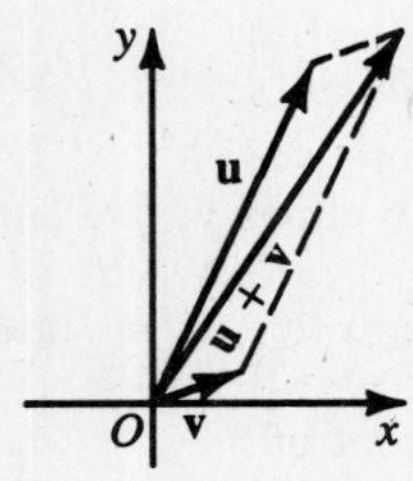

$180° - (65° - 20°) = 135°$

$\|\mathbf{u} + \mathbf{v}\|^2 = 6^2 + (\sqrt{2})^2 - 2(6)(\sqrt{2})\cos 135°$

$= 38 - 12\sqrt{2}\left(-\frac{1}{\sqrt{2}}\right) = 50$

$\|\mathbf{u} + \mathbf{v}\| = 5\sqrt{2}$

6. 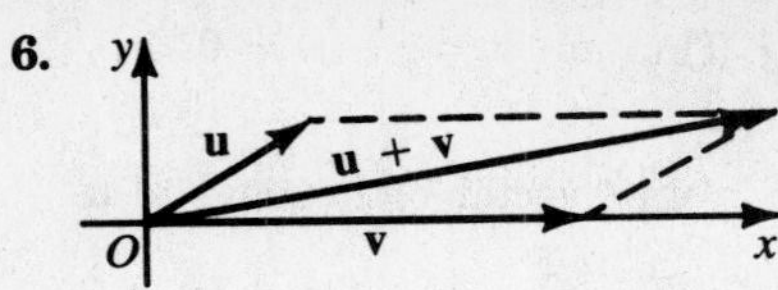

$180° - 30° = 150°$

$\|\mathbf{u} + \mathbf{v}\|^2 = 2^2 + (3\sqrt{3})^2 - 2(2)(3\sqrt{3})\cos 150°$

$= 31 - 12\sqrt{3}\left(-\frac{\sqrt{3}}{2}\right) = 49$

$\|\mathbf{u} + \mathbf{v}\| = 7$

7. 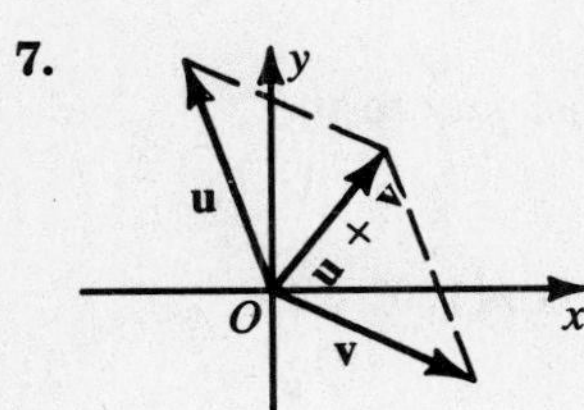

$180° - (110° + 25°) = 45°$

$\|\mathbf{u} + \mathbf{v}\|^2 = 12^2 + (8\sqrt{2})^2 - 2(12)(8\sqrt{2})\cos 45°$

$= 272 - 192\sqrt{2}\left(\frac{1}{\sqrt{2}}\right) = 80$

$\|\mathbf{u} + \mathbf{v}\| = 4\sqrt{5}$

8. 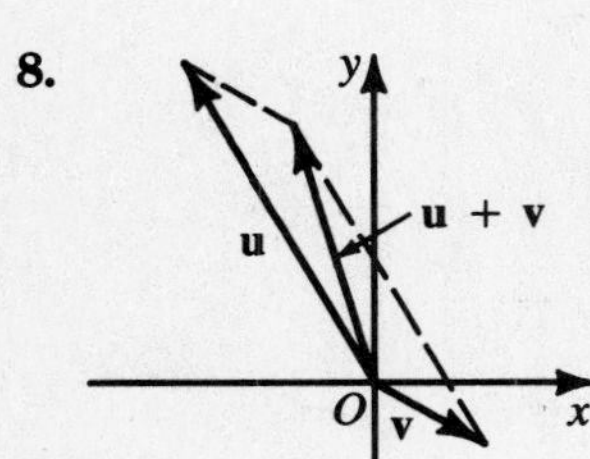

$180° - (120° + 30°) = 30°$

$\|\mathbf{u} + \mathbf{v}\|^2 = (5\sqrt{3})^2 + 2^2 - 2(5\sqrt{3})(2)\cos 30°$

$= 79 - 20\sqrt{3}\left(\frac{\sqrt{3}}{2}\right) = 49$

$\|\mathbf{u} + \mathbf{v}\| = 7$

9. $a = -3$, $b = 4$, $c = 7$, $d = 24$; $\|\mathbf{u}\| = 5$ and $\|\mathbf{v}\| = 25$;
$\mathbf{u} \cdot \mathbf{v} = (-3)(7) + (4)(24) = -21 + 96 = 75$; $(5)(25)\cos\theta = 75$;
$\cos\theta = \frac{75}{125} = \frac{3}{5} = 0.6$; $m(\theta) \approx 53.1°$

10. $a = 5$, $b = -12$, $c = -2$, $d = 1$; $\|\mathbf{u}\| = 13$ and $\|\mathbf{v}\| = \sqrt{5}$;
$\mathbf{u} \cdot \mathbf{v} = (5)(-2) + (-12)(1) = -22$; $(13)(\sqrt{5})\cos\theta = -22$;
$\cos\theta = \frac{-22}{13\sqrt{5}} \approx -0.7568$; $m(\theta) \approx 139.2°$

11. $a = 15$, $b = 8$, $c = -1$, $d = -13$; $\|\mathbf{u}\| = 17$ and $\|\mathbf{v}\| = \sqrt{170}$;
$\mathbf{u} \cdot \mathbf{v} = (15)(-1) + (8)(-13) = -119$; $(17)(\sqrt{170})\cos\theta = -119$;
$\cos\theta = \frac{-119}{17\sqrt{170}} = \frac{-7}{\sqrt{170}} \approx -0.5367$; $m(\theta) \approx 122.5°$

12. $a = -3$, $b = 3$, $c = 6$, $d = -9$; $\|\mathbf{u}\| = \sqrt{18} = 3\sqrt{2}$ and $\|\mathbf{v}\| = \sqrt{117} = 3\sqrt{13}$;
$\mathbf{u} \cdot \mathbf{v} = (-3)(6) + (3)(-9) = -45$; $(3\sqrt{2})(3\sqrt{13})\cos\theta = -45$;
$\cos\theta = \frac{-45}{9\sqrt{26}} = \frac{-5}{\sqrt{26}} \approx -0.9806$; $m(\theta) \approx 168.7°$

Note: Answers for Exs. 13–32 may vary.

13. Let $\mathbf{v} = x\mathbf{i} + y\mathbf{j}$ be orthogonal to $\mathbf{u} = 3\mathbf{i} + 4\mathbf{j}$. Thus $\mathbf{u} \cdot \mathbf{v} = 3x + 4y = 0$.
One solution is $x = 4$ and $y = -3$, or $\mathbf{v} = 4\mathbf{i} - 3\mathbf{j}$.
$\frac{\mathbf{v}}{\|\mathbf{v}\|} = \frac{4\mathbf{i} - 3\mathbf{j}}{5} = \frac{4}{5}\mathbf{i} - \frac{3}{5}\mathbf{j}$ is a unit vector orthogonal to **u**.

14. Let $\mathbf{v} = x\mathbf{i} + y\mathbf{j}$ be orthogonal to $\mathbf{u} = 5\mathbf{i} - 12\mathbf{j}$. Thus $\mathbf{u} \cdot \mathbf{v} = 5x - 12y = 0$.
One solution is $x = 12$ and $y = 5$, or $\mathbf{v} = 12\mathbf{i} + 5\mathbf{j}$.
$\frac{\mathbf{v}}{\|\mathbf{v}\|} = \frac{12\mathbf{i} + 5\mathbf{j}}{13} = \frac{12}{13}\mathbf{i} + \frac{5}{13}\mathbf{j}$ is a unit vector orthogonal to **u**.

15. Let $\mathbf{v} = x\mathbf{i} + y\mathbf{j}$ be orthogonal to $\mathbf{u} = \mathbf{i} - \mathbf{j}$. Thus $\mathbf{u} \cdot \mathbf{v} = x - y = 0$.
One solution is $x = 1$ and $y = 1$, or $\mathbf{v} = \mathbf{i} + \mathbf{j}$.
$\dfrac{\mathbf{v}}{\|\mathbf{v}\|} = \dfrac{\mathbf{i} + \mathbf{j}}{\sqrt{2}} = \dfrac{1}{\sqrt{2}}\mathbf{i} + \dfrac{1}{\sqrt{2}}\mathbf{j} = \dfrac{1}{\sqrt{2}}\mathbf{i} + \dfrac{1}{\sqrt{2}}\mathbf{j}$ is a unit vector orthogonal to $\mathbf{u}$.

16. Let $\mathbf{v} = x\mathbf{i} + y\mathbf{j}$ be orthogonal to $\mathbf{u} = 5\mathbf{i} - 7\mathbf{j}$. Thus $\mathbf{u} \cdot \mathbf{v} = 5x - 7y = 0$.
One solution is $x = 7$ and $y = 5$, or $\mathbf{v} = 7\mathbf{i} + 5\mathbf{j}$.
$\dfrac{\mathbf{v}}{\|\mathbf{v}\|} = \dfrac{7\mathbf{i} + 5\mathbf{j}}{\sqrt{74}} = \dfrac{7}{\sqrt{74}}\mathbf{i} + \dfrac{5}{\sqrt{74}}\mathbf{j}$ is a unit vector orthogonal to $\mathbf{u}$.

B 17.

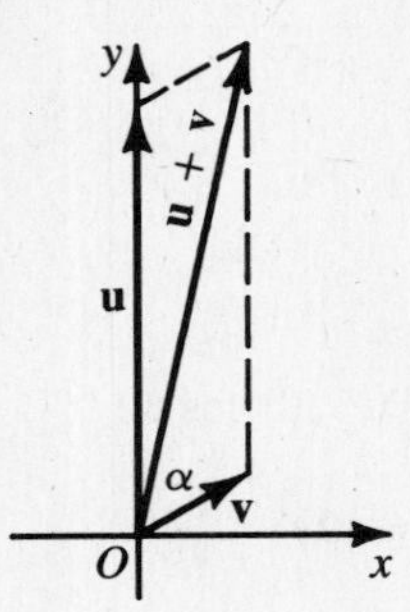

$180° - (90° - 30°) = 120°$

$\|\mathbf{u} + \mathbf{v}\|^2 = 32^2 + 10^2 - 2(32)(10)\cos 120°$

$= 1124 - 640\left(-\dfrac{1}{2}\right) = 1444$

$\|\mathbf{u} + \mathbf{v}\| = \sqrt{1444} = 38$

$\dfrac{\sin \alpha}{32} = \dfrac{\sin 120°}{38}$

$\sin \alpha \approx \dfrac{32(0.8660)}{38} \approx 0.7293$

$m(\alpha) \approx 47°$; $m(\theta) \approx 47° + 30° = 77°$

18.

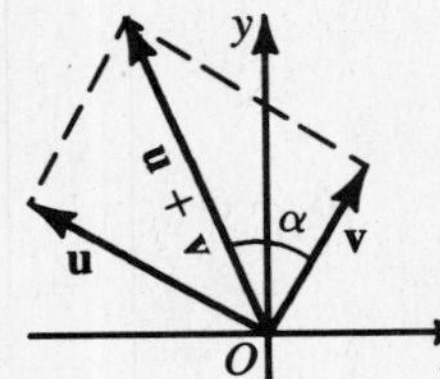

$180° - (150° - 60°) = 90°$

$\|\mathbf{u} + \mathbf{v}\|^2 = 8^2 + 6^2 = 100$

$\|\mathbf{u} + \mathbf{v}\| = 10$

$\sin \alpha = \dfrac{8}{10} = 0.8$

$m(\alpha) \approx 53°$; $m(\theta) \approx 53° + 60° = 113°$

19.

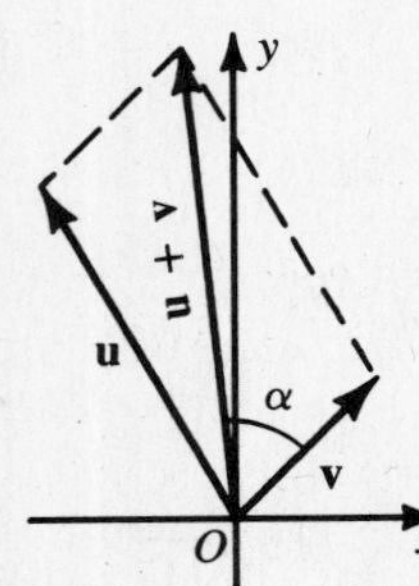

$180° - (120° - 45°) = 105°$

$\|\mathbf{u} + \mathbf{v}\|^2 = 20^2 + 10^2 - 2(20)(10)\cos 105°$

$\approx 500 - 400(-0.2588) = 603.52$

$\|\mathbf{u} + \mathbf{v}\| \approx \sqrt{603.52} \approx 24.6$

$\dfrac{\sin \alpha}{20} \approx \dfrac{\sin 105°}{24.6}$

$\sin \alpha \approx \dfrac{20(0.9659)}{24.6} \approx 0.7853$

$m(\alpha) \approx 52°$; $m(\theta) \approx 52° + 45° = 97°$

20.

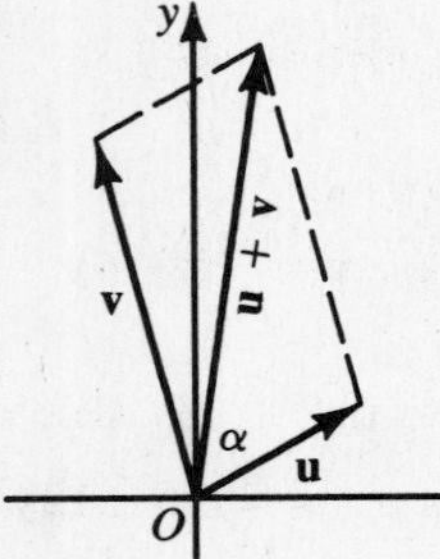

$180° - (105° - 30°) = 105°$

$\|\mathbf{u} + \mathbf{v}\|^2 = 10^2 + 20^2 - 2(10)(20)\cos 105°$

$\approx 500 - 400(-0.2588) = 603.52$

$\|\mathbf{u} + \mathbf{v}\| = \sqrt{603.52} \approx 24.6$

$\dfrac{\sin \alpha}{20} \approx \dfrac{\sin 105°}{24.6}$

$\sin \alpha \approx \dfrac{20(0.9659)}{24.6} \approx 0.7852$

$m(\alpha) \approx 52°$; $m(\theta) \approx 52° + 30° = 82°$

21.

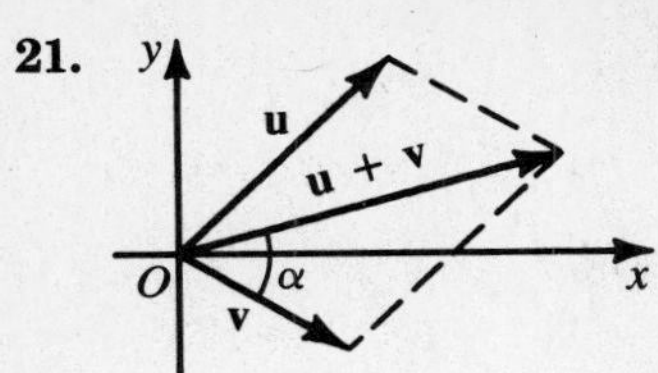

$180° - (45° + 30°) = 105°$

$\|\mathbf{u} + \mathbf{v}\|^2 = 12^2 + 8^2 - 2(12)(8)\cos 105°$

$\approx 208 - 192(-0.2588) = 257.6896$

$\|\mathbf{u} + \mathbf{v}\| \approx \sqrt{257.6896} \approx 16.1$

$\dfrac{\sin \alpha}{12} \approx \dfrac{\sin 105°}{16.1}$; $\sin \alpha \approx \dfrac{12(0.9659)}{16.1} \approx 0.7199$

$m(\alpha) \approx 46°$; $m(\theta) \approx 46° - 30° = 16°$

22.

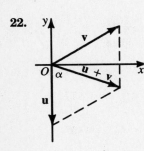

$180° - (30° + 90°) = 60°$

$\|\mathbf{u} + \mathbf{v}\|^2 = 16^2 + 20^2 - 2(16)(20)\cos 60°$

$= 656 - 640\left(\dfrac{1}{2}\right) = 336$

$\|\mathbf{u} + \mathbf{v}\| \approx 18.3$

$\dfrac{\sin \alpha}{20} \approx \dfrac{\sin 60°}{18.3}$

$\sin \alpha \approx \dfrac{20(0.8660)}{18.3} \approx 0.9464$

$m(\alpha) \approx 71°$; $m(\theta) \approx 71° - 90° = -19°$

23.

$180° - (80° - 50°) = 150°$

$\|\mathbf{u} + \mathbf{v}\|^2 = 10^2 + 24^2 - 2(10)(24)\cos 150°$

$\approx 676 - 480(-0.8660) \approx 1091.68$

$\|\mathbf{u} + \mathbf{v}\| \approx 33.0$

$\dfrac{\sin \alpha}{10} \approx \dfrac{\sin 150°}{33}$

$\sin \alpha \approx \dfrac{10(0.5)}{33} \approx 0.1515$

$m(\alpha) \approx 9°$; $m(\theta) \approx 9° + 50° = 59°$

24.

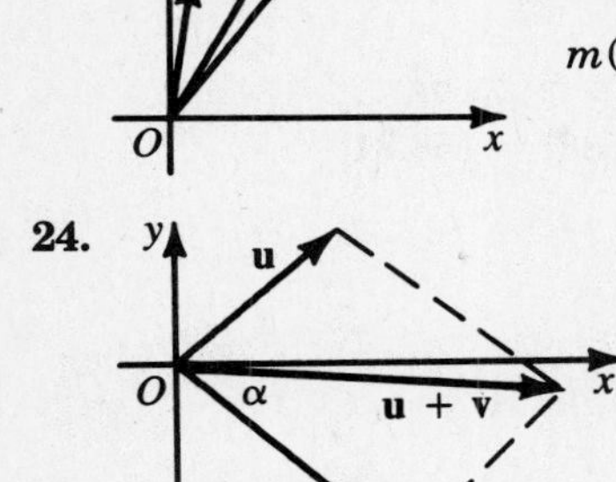

$180° - (40° + 40°) = 100°$

$\|\mathbf{u} + \mathbf{v}\|^2 = 8^2 + 10^2 - 2(8)(10)\cos 100°$

$\approx 164 - 160(-0.1736) \approx 191.776$

$\|\mathbf{u} + \mathbf{v}\| \approx 13.8$

$\dfrac{\sin \alpha}{8} \approx \dfrac{\sin 100°}{13.8}$

$\sin \alpha \approx \dfrac{8(0.9848)}{13.8} \approx 0.5709$

$m(\alpha) \approx 35°$; $m(\theta) \approx 35° - 40° = -5°$

25.

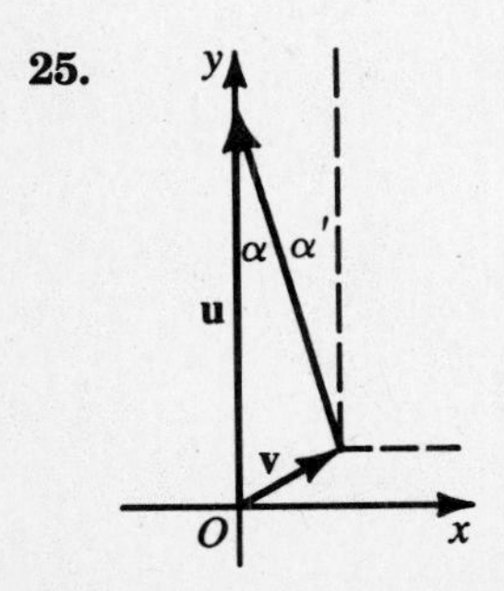

$90° - 30° = 60°$

$\|\mathbf{u} - \mathbf{v}\|^2 = 32^2 + 10^2 - 2(32)(10)\cos 60°$

$= 1124 - 640\left(\dfrac{1}{2}\right) = 804$

$\|\mathbf{u} - \mathbf{v}\| \approx 28.4$

$\dfrac{\sin \alpha}{10} \approx \dfrac{\sin 60°}{28.4}$

$\sin \alpha \approx \dfrac{10(0.8660)}{28.4} \approx 0.3049$

$m(\alpha) \approx 18°$

$m(\theta) = m(\alpha') + 90° = m(\alpha) + 90° \approx 18° + 90° = 108°$

26.

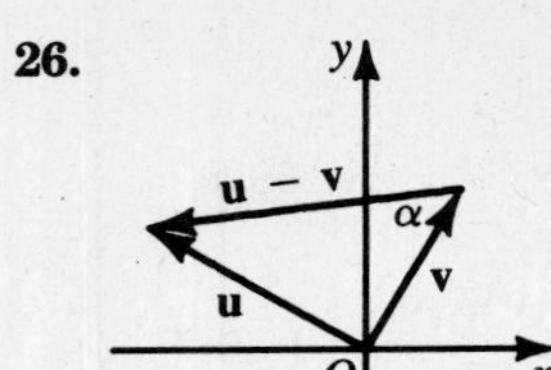

$150° - 60° = 90°$

$\|\mathbf{u} - \mathbf{v}\|^2 = 8^2 + 6^2 = 100$

$\|\mathbf{u} - \mathbf{v}\| = 10$

$\sin \alpha = \frac{8}{10} = 0.8$

$m(\alpha) \approx 53°;\ m(\theta) \approx -(180° - 60°) - 53° = -173°$

27.

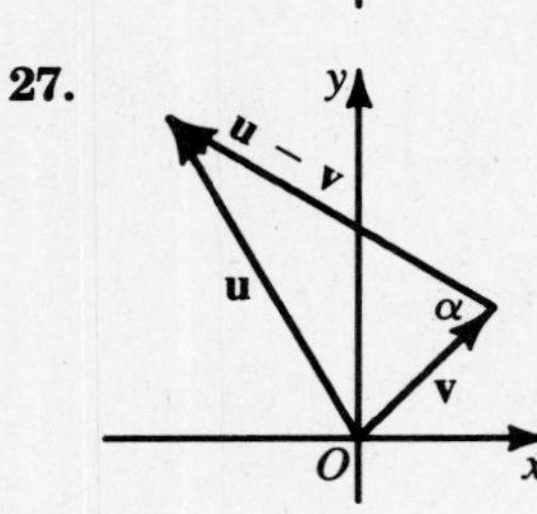

$120° - 45° = 75°$

$\|\mathbf{u} - \mathbf{v}\|^2 = 20^2 + 10^2 - 2(20)(10)\cos 75°$

$\approx 500 - 400(0.2588) \approx 396.48$

$\|\mathbf{u} - \mathbf{v}\| \approx 19.9$

$\frac{\sin \alpha}{20} \approx \frac{\sin 75°}{19.9}$

$\sin \alpha \approx \frac{20(0.9659)}{19.9} \approx 0.9708$

$m(\alpha) \approx 76°;\ m(\theta) \approx (180° - 76°) + 45° = 149°$

28.

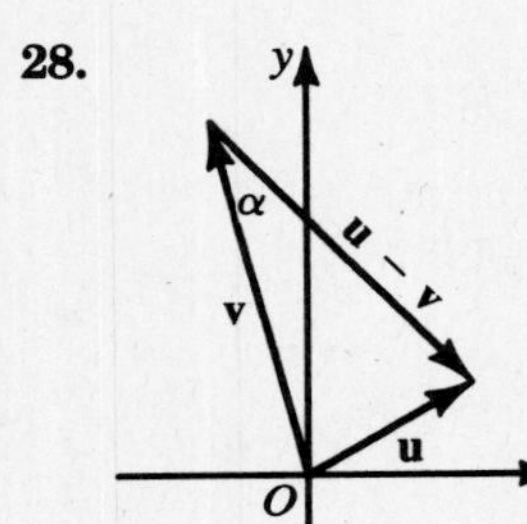

$105° - 30° = 75°$

$\|\mathbf{u} - \mathbf{v}\|^2 = 10^2 + 20^2 - 2(10)(20)\cos 75°$

$\approx 500 - 400(0.2588) \approx 396.48$

$\|\mathbf{u} - \mathbf{v}\| \approx 19.9$

$\frac{\sin \alpha}{10} = \frac{\sin 75°}{19.9}$

$\sin \alpha \approx \frac{10(0.9659)}{19.9} \approx 0.4854$

$m(\alpha) \approx 29°;\ m(\theta) \approx 29° - 75° = -46°$

29.

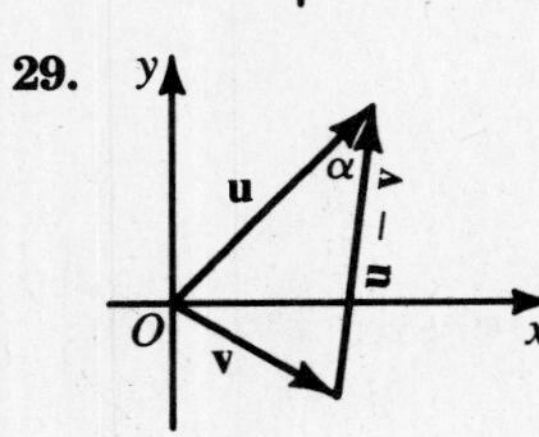

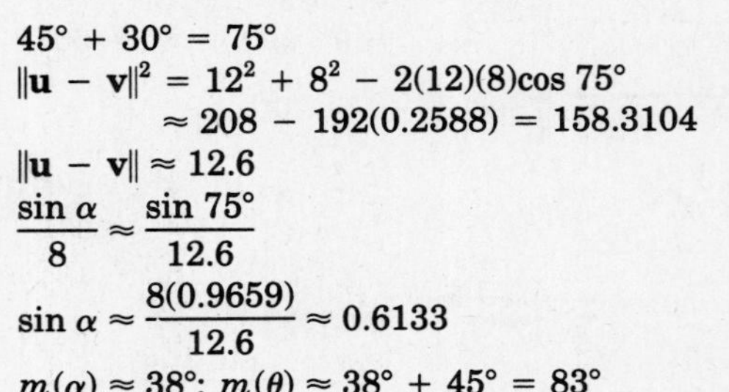

$45° + 30° = 75°$

$\|\mathbf{u} - \mathbf{v}\|^2 = 12^2 + 8^2 - 2(12)(8)\cos 75°$

$\approx 208 - 192(0.2588) = 158.3104$

$\|\mathbf{u} - \mathbf{v}\| \approx 12.6$

$\frac{\sin \alpha}{8} \approx \frac{\sin 75°}{12.6}$

$\sin \alpha \approx \frac{8(0.9659)}{12.6} \approx 0.6133$

$m(\alpha) \approx 38°;\ m(\theta) \approx 38° + 45° = 83°$

30.

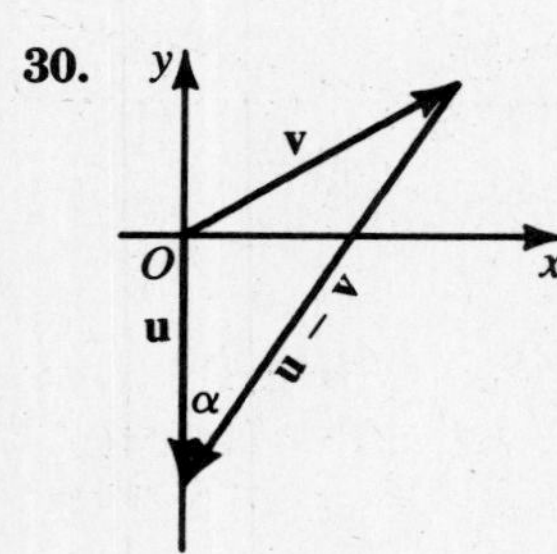

$90° + 30° = 120°$

$\|\mathbf{u} - \mathbf{v}\|^2 = 16^2 + 20^2 - 2(16)(20)\cos 120°$

$= 656 - 640\left(-\frac{1}{2}\right) = 976$

$\|\mathbf{u} - \mathbf{v}\| \approx 31.2$

$\frac{\sin \alpha}{20} \approx \frac{\sin 120°}{31.2}$

$\sin \alpha \approx \frac{20(0.8660)}{31.2} \approx 0.5551$

$m(\alpha) \approx 34°;\ m(\theta) \approx -(34° + 90°) = -124°$

31. 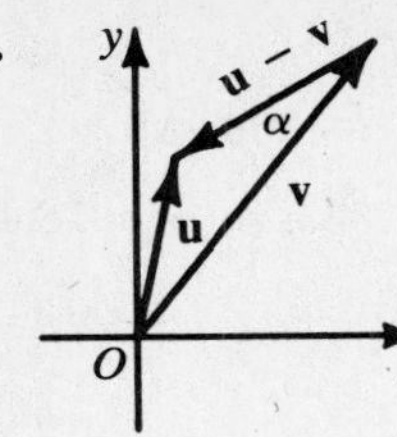

$80° - 50° = 30°$

$\|\mathbf{u} - \mathbf{v}\|^2 = 10^2 + 24^2 - 2(10)(24)\cos 30°$

$\approx 676 - 480(0.8660) \approx 260.32$

$\|\mathbf{u} - \mathbf{v}\| \approx 16.1$

$\frac{\sin \alpha}{10} \approx \frac{\sin 30°}{16.1}$

$\sin \alpha \approx \frac{10(0.5)}{16.1} \approx 0.3106$

$m(\alpha) \approx 18°;\ m(\theta) \approx -[(180° - 50°) + 18°] = -148°$

32. 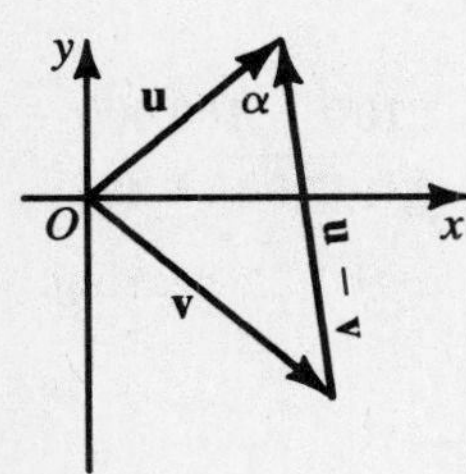

$40° + 40° = 80°$

$\|\mathbf{u} - \mathbf{v}\|^2 = 8^2 + 10^2 - 2(8)(10)\cos 80°$

$\approx 164 - 160(0.1736) \approx 136.224$

$\|\mathbf{u} - \mathbf{v}\| \approx 11.7$

$\frac{\sin \alpha}{10} \approx \frac{\sin 80°}{11.7}$

$\sin \alpha \approx \frac{10(0.9848)}{11.7} \approx 0.8417$

$m(\alpha) \approx 57°;\ m(\theta) \approx 40° + 57° = 97°$

C 33. 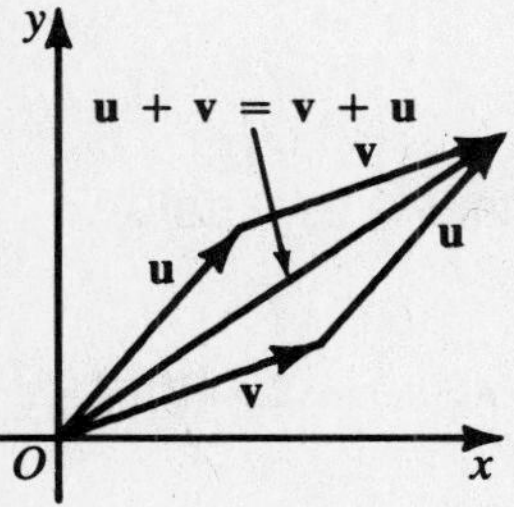

34. 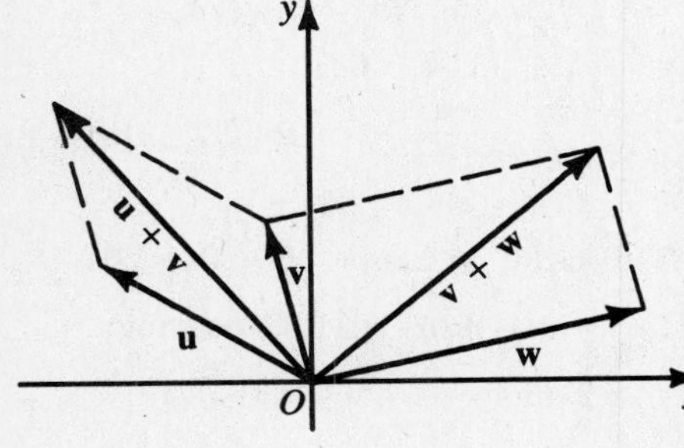

35. $(PQ)^2 = \|\mathbf{u}\|^2 + \|\mathbf{v}\|^2 - 2\|\mathbf{u}\|\,\|\mathbf{v}\|\cos\theta$. Since $\mathbf{u} \cdot \mathbf{v} = \|\mathbf{u}\|\,\|\mathbf{v}\|\cos\theta$, we can substitute into the first equation to get $(PQ)^2 = \|\mathbf{u}\|^2 + \|\mathbf{v}\|^2 - 2\mathbf{u} \cdot \mathbf{v}$. Using the distance formula, $(PQ)^2 = (a - c)^2 + (b - d)^2 = (a^2 + b^2) + (c^2 + d^2) - 2\mathbf{u} \cdot \mathbf{v}$. Expanding and simplifying gives $ac + bd = \mathbf{u} \cdot \mathbf{v}$.

Pages 693–694 • PROBLEMS

A 1. 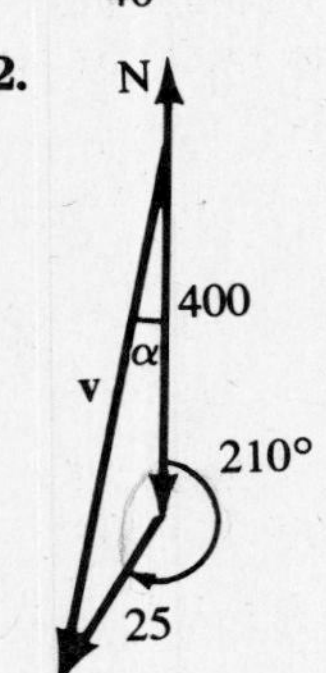

$\|\mathbf{v}\|^2 = 120^2 + 40^2 - 2(120)(40)\cos 90° = 14{,}400 + 1600 - 0 = 16{,}000$; $\|\mathbf{v}\| = \sqrt{16{,}000} \approx 126$; $\tan\alpha = \frac{40}{120} \approx 0.3333$; $m(\alpha) \approx 18.4°$; bearing: $180° + 18.4° = 198.4°$; 126 km/h

2.

$\|\mathbf{v}\|^2 = 400^2 + 25^2 - 2(400)(25)\cos(360° - 210°) = 160{,}000 + 625 - 20{,}000(-0.8660) = 177{,}945$; $\|\mathbf{v}\| = \sqrt{177{,}945} \approx 422$; $\cos\alpha = \frac{400^2 + 422^2 - 25^2}{2(400)(422)} \approx 0.99958$; $m(\alpha) \approx 1.7°$; bearing: $180° + 1.7° = 181.7°$; 422 mi/h

3. $\|\mathbf{F}\| = 55 \times 9.8 = 5.39 \times 10^2$ N; $W = (5.39 \times 10^2)(60) \approx 3.23 \times 10^4$ J; $\frac{3.23 \times 10^4}{3.6 \times 10^6} \approx 0.00897$ kW · h

4. $\|\mathbf{F}\| = 2500 \times 9.8 = 2.45 \times 10^4$ N; $W = (2.45 \times 10^4)(50) \approx 1.2 \times 10^6$ J

5. 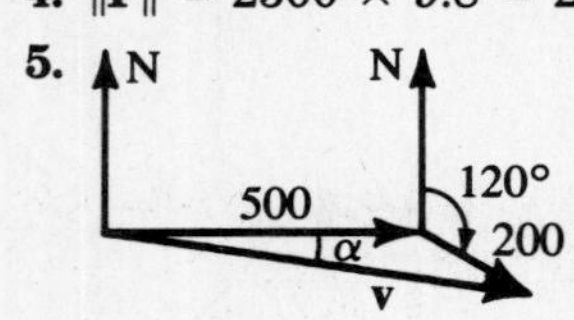

$360° - 120° - 90° = 150°$
$\|\mathbf{v}\| = 500^2 + 200^2 - 2(500)(200)\cos 150° = 250{,}000 + 40{,}000 - 200{,}000(-0.8660) = 463{,}200$; $\|\mathbf{v}\| = \sqrt{463{,}200} \approx 681$;
$\cos\alpha = \frac{500^2 + 681^2 - 200^2}{2(500)(681)} \approx 0.98937$; $m(\alpha) \approx 8.4°$;
bearing: $90° + 8.4° = 98.4°$; 681 mi

6. 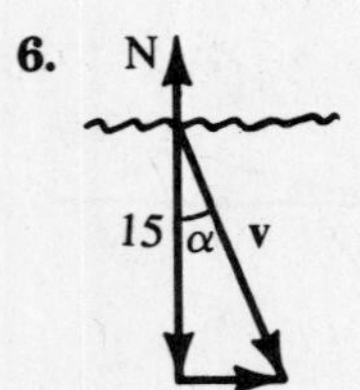

$\|\mathbf{v}\|^2 = 15^2 + 6^2 - 2(15)(6)\cos 90° = 225 + 36 - 0 = 261$; $\|\mathbf{v}\| = \sqrt{261} \approx 16.2$; $\tan\alpha = \frac{6}{15} = 0.4$; $m(\alpha) \approx 21.8°$; bearing: $180° - 21.8° = 158.2°$; 16.2 km/h

7. 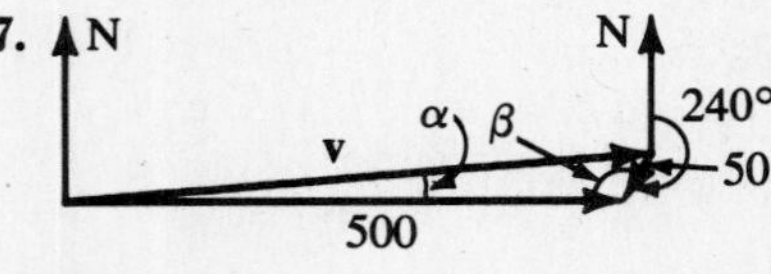

Find β, the angle opposite **v**. $240° - 180° = 60°$. Since that angle is 60°, the right triangle with 50 as its hypotenuse has acute angles of 60° and 30°. The 30° angle is the supplement of β. Thus $m(\beta) = 150°$.

$\|\mathbf{v}\|^2 = 500^2 + 50^2 - 2(500)(50)\cos 150° = 250{,}000 + 2500 - 50{,}000(-0.8660) = 295{,}800$; $\|\mathbf{v}\| = \sqrt{295{,}800} \approx 544$; $\cos\alpha = \frac{500^2 + 544^2 - 50^2}{2(500)(544)} \approx 0.99896$; $m(\alpha) \approx 2.6°$; bearing: $90° - 2.6° = 87.4°$; 544 km/h

8. The current in the river flows parallel to the river's edge. $\sin 25° = \frac{\|\mathbf{v}\|}{12}$; $\|\mathbf{v}\| = 12(0.4226) \approx 5.07$; 5.07 km/h

9. $\|\mathbf{F}\| = 30(1800) \times 9.8 = 5.292 \times 10^5$ N; $W = (5.292 \times 10^5)(40) \approx 2.12 \times 10^7$ J; $\frac{2.12 \times 10^7}{3.6 \times 10^6} \approx 5.89$ kW · h

B 10. For each hour, the boat travels 15 km across the river but is forced 5 km downstream. Since 45 m = 0.045 km, the boat takes 0.003 h to cross the river and is forced 5(0.003), or 0.015 km downstream. 0.015 km = 15 m

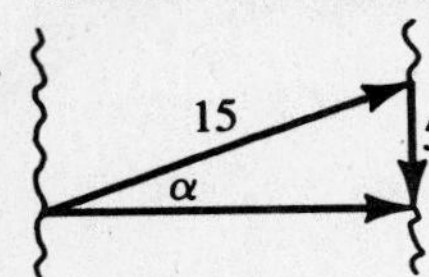

$\sin \alpha = \frac{5}{15} \approx 0.3333$; $m(\alpha) \approx 19.5°$; 70.5°

11.

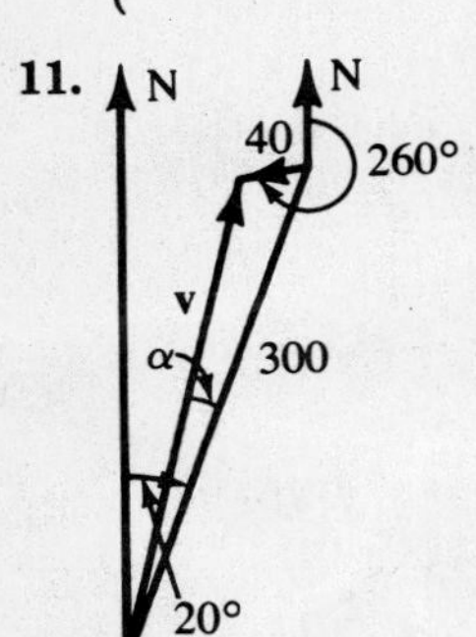

The angle opposite **v** measures 260° − 180° − 20°, or 60°. $\|\mathbf{v}\|^2 = 300^2 + 40^2 - 2(300)(40)\cos 60° = 90{,}000 + 1600 - 24{,}000(0.5) = 79{,}600$; $\|\mathbf{v}\| = \sqrt{79{,}600} \approx 282$; $\cos \alpha = \frac{300^2 + 282^2 - 40^2}{2(300)(282)} \approx 0.9925$; $m(\alpha) \approx 7.0°$; bearing: 20° − 7.0° = 13.0°; 282 km/h

12.

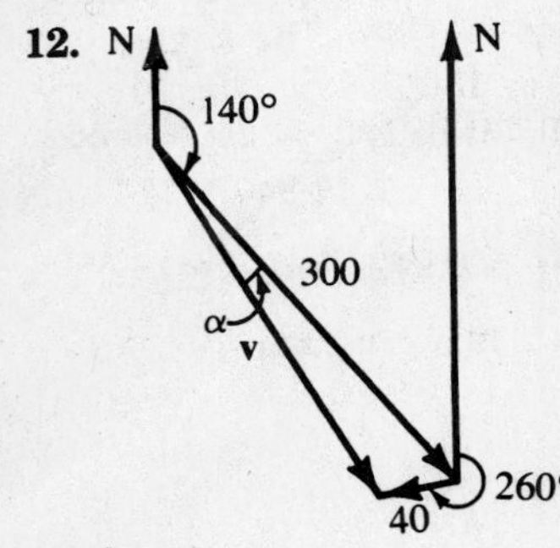

The angle opposite **v** measures 360° − 260° − 40°, or 60°. $\|\mathbf{v}\|^2 = 300^2 + 40^2 - 2(300)(40)\cos 60° = 90{,}000 + 1600 - 24{,}000(0.5) = 79{,}600$; $\|\mathbf{v}\| = \sqrt{79{,}600} \approx 282$; $\cos \alpha = \frac{300^2 + 282^2 - 40^2}{2(300)(282)} \approx 0.9925$; $m(\alpha) \approx 7.0°$; bearing: 140° + 7.0° = 147.0°; 282 km/h

C 13. 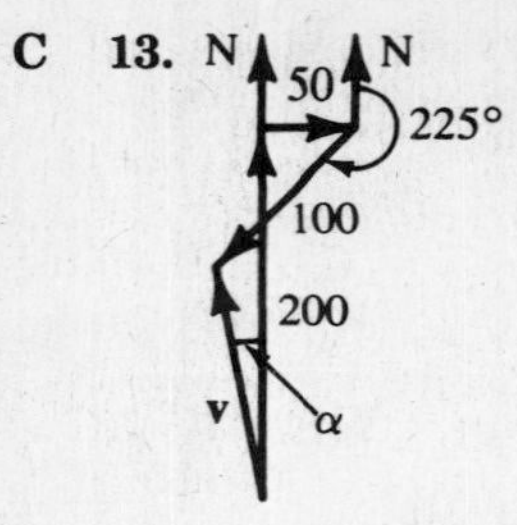

The angle between the vectors 50 and 100 measures $360° - 225° - 90°$, or 45°. The other acute angle in the small triangle measures 45°. Therefore, by the theorem that vertical angles are congruent, the angle opposite **v** measures 45°. Since the small triangle is a 45°–45°–90° triangle, the other leg is 50 and the hypotenuse is $50\sqrt{2}$, or approx. 70.7. Thus, in the triangle containing **v**, the longest side is $200 - 50$, or 150, and the shortest side is $100 - 70.7$, or 29.3.
$\|\mathbf{v}\|^2 = 150^2 + 29.3^2 - 2(150)(29.3)\cos 45° = 22{,}500 + 858.49 - 8790(0.7071) = 17{,}143.081$; $\|\mathbf{v}\| = \sqrt{17{,}143.081} \approx 131$; $\cos\alpha = \dfrac{150^2 + 131^2 - 29.3^2}{2(150)(131)} \approx 0.9873$; $m(\alpha) \approx 9.1°$; bearing: $360° - 9.1° = 350.9°$; 131 mi

Page 695 • SELF-TEST 3

1. $180° - (105° - 45°) = 120°$

$$\|\mathbf{u} + \mathbf{v}\|^2 = 6^2 + 8^2 - 2(6)(8)\cos 120° = 100 + 48 = 148$$

$\|\mathbf{u} + \mathbf{v}\| = \sqrt{148} \approx 12.2$

$\dfrac{\sin\alpha}{8} \approx \dfrac{\sin 120°}{12.2}$; $\sin\alpha \approx \dfrac{8(0.8660)}{12.2} \approx 0.5679$

$m(\alpha) \approx 35°$; $m(\theta) \approx 35° + 45° = 80°$

2. $a = 3$, $b = -4$, $c = 5$, $d = 12$; $\|\mathbf{u}\| = 5$ and $\|\mathbf{v}\| = 13$; $\mathbf{u} \cdot \mathbf{v} = (3)(5) + (-4)(12) = -33$; $(5)(13)\cos\theta = -33$; $\cos\theta = -\dfrac{33}{65} \approx -0.5077$; $m(\theta) \approx 120.5°$

3. 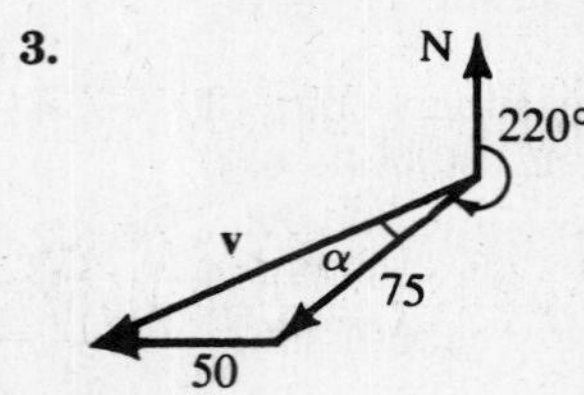

The supplement to the angle opposite **v** measures $180° - 90° - (220° - 180°)$, or 50°. The angle opposite **v** measures $180° - 50°$, or 130°.
$\|\mathbf{v}\|^2 = 50^2 + 75^2 - 2(50)(75)\cos 130° = 2500 + 5625 - 7500(-0.6428) = 12{,}946$; $\|\mathbf{v}\| = \sqrt{12{,}946} \approx 114$;
$\cos\alpha = \dfrac{75^2 + 114^2 - 50^2}{2(75)(114)} \approx 0.9427$; $m(\alpha) \approx 19.5°$;
bearing: $220° + 19.5° = 239.5°$; 114 km

Pages 696–697 • CHAPTER REVIEW

1. $\text{Cos}^{-1}\dfrac{1}{\sqrt{2}} = \dfrac{\pi}{4}$; c 2. $\text{Cos}^{-1}\left(\cos\dfrac{5\pi}{3}\right) = \text{Cos}^{-1}\left(\dfrac{1}{2}\right) = \dfrac{\pi}{3}$; c

3. $4\cos^2 x - 8\cos x + 7 = 0$; $\cos x = \dfrac{8 \pm \sqrt{64 - 4(28)}}{8} = \dfrac{8 \pm \sqrt{-48}}{8}$; c

4. $x = 5\cos 330° = 5\left(\dfrac{\sqrt{3}}{2}\right) = \dfrac{5\sqrt{3}}{2}$; $y = 5\sin 330° = 5\left(-\dfrac{1}{2}\right) = -\dfrac{5}{2}$; $\left(\dfrac{5\sqrt{3}}{2}, -\dfrac{5}{2}\right)$; b

5. $r = \sqrt{(3\sqrt{2})^2 + (-3\sqrt{2})^2} = \sqrt{36} = 6$; $\cos\theta = \dfrac{3\sqrt{2}}{6} = \dfrac{\sqrt{2}}{2}$;
$\sin\theta = \dfrac{-3\sqrt{2}}{6} = \dfrac{-\sqrt{2}}{2}$; $m(\theta) = 315°$; $(6, 315°)$; a

6. $|z| = \sqrt{(\sqrt{3})^2 + 1^2} = \sqrt{4} = 2$; $\cos\theta = \frac{\sqrt{3}}{2}$; $\sin\theta = \frac{1}{2}$; $m(\theta) = 30°$;
$\sqrt{3} + i = 2(\cos 30° + i\sin 30°)$; b

7. $\frac{z_1}{z_2} = \frac{9}{3}[\cos(180° - 60°) + i\sin(180° - 60°)] = 3(\cos 120° + i\sin 120°) =$
$3\left(-\frac{1}{2} + i\frac{\sqrt{3}}{2}\right)$; $-\frac{3}{2} + \frac{3\sqrt{3}}{2}i$; c

8. $(z_2)^3 = 3^3(\cos 3\cdot 60° + i\sin 3\cdot 60°) = 27(\cos 180° + i\sin 180°) = 27(-1 + 0i) =$
-27; b

9. 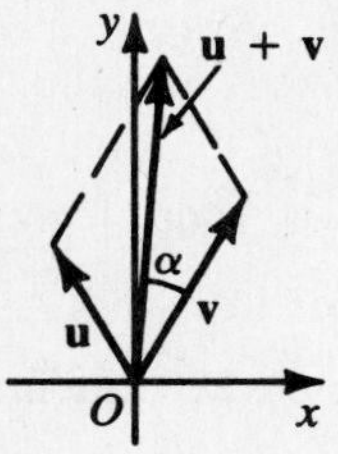 $180° - (120° - 60°) = 120°$
$\|\mathbf{u} + \mathbf{v}\|^2 = 5^2 + 7^2 - 2(5)(7)\cos 120° = 74 - 70(-0.5) = 109$;
$\|\mathbf{u} + \mathbf{v}\| = \sqrt{109}$; b

10. Let $\mathbf{u} = 4\mathbf{i} + 3\mathbf{j}$; $\mathbf{v} = \mathbf{i} + 2\mathbf{j}$. $a = 4$, $b = 3$, $c = 1$, $d = 2$; $\|\mathbf{u}\| = 5$ and $\|\mathbf{v}\| = \sqrt{5}$;
$\mathbf{u}\cdot\mathbf{v} = (4)(1) + (3)(2) = 10$; $5\sqrt{5}\cos\theta = 10$; $\cos\theta = \frac{10}{5\sqrt{5}} = \frac{2}{\sqrt{5}} \approx 0.8944$;
$m(\theta) \approx 27°$; c

11. 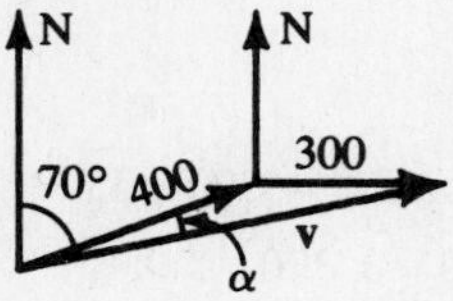 The angle opposite $\mathbf{v}$ measures $360° - 90° - (180° - 70°)$, or $160°$. $\|\mathbf{v}\|^2 = 400^2 + 300^2 - 2(400)(300)\cos 160° =$
$160{,}000 + 90{,}000 - 240{,}000(-0.9397) = 475{,}528$; $\|\mathbf{v}\| =$
$\sqrt{475{,}528} \approx 690$; a

Page 697 • CHAPTER TEST

1. $\text{Cos}^{-1}\left(\sin\frac{5\pi}{6}\right) = \text{Cos}^{-1}\left(\frac{1}{2}\right) = \frac{\pi}{3}$

2. $\tan x\sec x = \tan x$; $\tan x\sec x - \tan x = 0$; $\tan x(\sec x - 1) = 0$;
$\tan x = 0$ or $\sec x = 1$; $\tan x = 0$ or $\cos x = 1$; $\{k\pi\}$

3. $x^2 + y^2 = 4x$; $(r\cos\theta)^2 + (r\sin\theta)^2 = 4r\cos\theta$; $r^2\cos^2\theta + r^2\sin^2\theta = 4r\cos\theta$;
$r^2(\cos^2\theta + \sin^2\theta) = 4r\cos\theta$; $r^2 = 4r\cos\theta$; $r^2 - 4r\cos\theta = 0$

4. $z_1z_2 = 5\cdot 2[\cos(135° + 15°) + i\sin(135° + 15°)] = 10(\cos 150° + i\sin 150°) =$
$10\left(-\frac{\sqrt{3}}{2} + \frac{1}{2}i\right) = -5\sqrt{3} + 5i$

5. $\frac{z_1}{z_2} = \frac{5}{2}[\cos(135° - 15°) + i\sin(135° - 15°)] = \frac{5}{2}(\cos 120° + i\sin 120°) =$
$\frac{5}{2}\left(-\frac{1}{2} + \frac{\sqrt{3}}{2}i\right) = -\frac{5}{4} + \frac{5\sqrt{3}}{4}i$

6. $(z_1)^3 = 5^3(\cos 3\cdot 135° + i\sin 3\cdot 135°) = 125(\cos 405° + i\sin 405°) =$
$125(\cos 45° + i\sin 45°) = 125\left(\frac{1}{\sqrt{2}} + \frac{1}{\sqrt{2}}i\right) = \frac{125}{\sqrt{2}} + \frac{125}{\sqrt{2}}i$

7. $r = \sqrt{1^2 + (-1)^2} = \sqrt{2}$; $\cos\theta = \frac{1}{\sqrt{2}}$; $\sin\theta = -\frac{1}{\sqrt{2}}$;

$m(\theta) = 315°$; $1 - i = \sqrt{2}(\cos 315° + i \sin 315°)$; $(1 - i)^{1/3} =$
$(\sqrt{2})^{1/3}\left(\cos \dfrac{315° + k \cdot 360°}{3} + i \sin \dfrac{315° + k \cdot 360°}{3}\right) =$
$2^{1/6}[\cos(105° + k \cdot 120°) + i \sin(105° + k \cdot 120°)]$; $\sqrt[6]{2}(\cos 105° + i \sin 105°)$,
$\sqrt[6]{2}(\cos 225° + i \sin 225°)$, $\sqrt[6]{2}(\cos 345° + i \sin 345°)$

8. 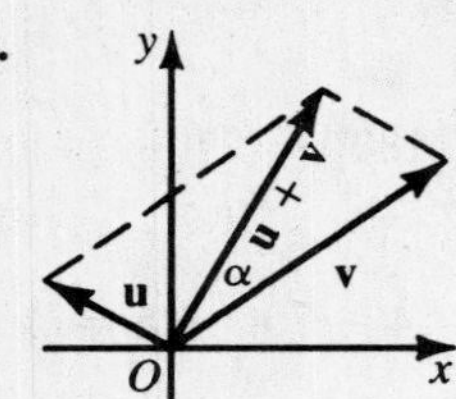

$180° - (150° - 30°) = 60°$
$\|\mathbf{u} + \mathbf{v}\|^2 = 3^2 + 8^2 - 2(3)(8)\cos 60° = 73 - 48(0.5) = 49$;
$\|\mathbf{u} + \mathbf{v}\| = 7.0$; $\dfrac{\sin \alpha}{3} \approx \dfrac{\sin 60°}{7}$; $\sin \alpha \approx \dfrac{3(0.8660)}{7} \approx 0.3711$;
$m(\alpha) \approx 22°$; $m(\theta) \approx 22° + 30° = 52°$

9. $\|\mathbf{v}\|^2 = 300^2 + 45^2 - 2(300)(45)\cos(180° - 120°) = 90{,}000 + 2025 -$
$27{,}000(0.5) = 78{,}525$; $\|\mathbf{v}\| = \sqrt{78{,}525} \approx 280$; $\cos \alpha =$
$\dfrac{300^2 + 280^2 - 45^2}{2(300)(280)} \approx 0.9903$; $m(\alpha) \approx 8°$; bearing: 8°; 280 km/h

Page 699 • APPLICATION

1. $90° - 60° = 30°$; $x = 600 \cdot \cos 30° = 300\sqrt{3}$;
$y = 600 \cdot \sin 30° = 300$; $(300\sqrt{3}, 300)$

2. $r = \sqrt{(-156)^2 + (-90)^2} \approx 180$; $\sin \theta = \dfrac{-90}{180} = -0.5$; $\cos \theta = \dfrac{-156}{180} \approx -0.8667$;
$m(\theta) = 210°$; 180, 210°

Pages 699–701 • CUMULATIVE REVIEW

1. $X - \begin{bmatrix} 6 & -5 \\ 2 & 4 \end{bmatrix} + \begin{bmatrix} 6 & -5 \\ 2 & 4 \end{bmatrix} = \begin{bmatrix} 2 & 5 \\ -2 & -7 \end{bmatrix} + \begin{bmatrix} 6 & -5 \\ 2 & 4 \end{bmatrix}$;
$X + \begin{bmatrix} 0 & 0 \\ 0 & 0 \end{bmatrix} = \begin{bmatrix} 8 & 0 \\ 0 & -3 \end{bmatrix}$; $X = \begin{bmatrix} 8 & 0 \\ 0 & -3 \end{bmatrix}$

2. $X + \begin{bmatrix} 3 & 8 \\ -2 & 4 \end{bmatrix} - \begin{bmatrix} 3 & 8 \\ -2 & 4 \end{bmatrix} = \begin{bmatrix} 1 & -3 \\ 5 & -6 \end{bmatrix} - \begin{bmatrix} 3 & 8 \\ -2 & 4 \end{bmatrix}$;
$X + \begin{bmatrix} 0 & 0 \\ 0 & 0 \end{bmatrix} = \begin{bmatrix} -2 & -11 \\ 7 & -10 \end{bmatrix}$; $X = \begin{bmatrix} -2 & -11 \\ 7 & -10 \end{bmatrix}$

3. $\begin{bmatrix} 3 & 0 \\ 0 & 3 \end{bmatrix} - 6X = \begin{bmatrix} -3 & 0 \\ 0 & -3 \end{bmatrix}$; $\begin{bmatrix} 3 & 0 \\ 0 & 3 \end{bmatrix} - \begin{bmatrix} 3 & 0 \\ 0 & 3 \end{bmatrix} - 6X = \begin{bmatrix} -3 & 0 \\ 0 & -3 \end{bmatrix} - \begin{bmatrix} 3 & 0 \\ 0 & 3 \end{bmatrix}$;
$-6X = \begin{bmatrix} -6 & 0 \\ 0 & -6 \end{bmatrix}$; $X = -\frac{1}{6}\begin{bmatrix} -6 & 0 \\ 0 & -6 \end{bmatrix}$; $X = \begin{bmatrix} 1 & 0 \\ 0 & 1 \end{bmatrix}$

4. $\begin{bmatrix} 1 & 2 \\ -1 & 3 \end{bmatrix}^{-1}\begin{bmatrix} 1 & 2 \\ -1 & 3 \end{bmatrix}X = \begin{bmatrix} 1 & 2 \\ -1 & 3 \end{bmatrix}^{-1}\begin{bmatrix} 3 & -1 \\ 2 & 1 \end{bmatrix}$; $X = \frac{1}{5}\begin{bmatrix} 3 & -2 \\ 1 & 1 \end{bmatrix}\begin{bmatrix} 3 & -1 \\ 2 & 1 \end{bmatrix}$;
$X = \begin{bmatrix} 1 & -1 \\ 1 & 0 \end{bmatrix}$

5. $\begin{bmatrix} 2 & 8 \\ 4 & 1 \end{bmatrix}$ **6.** $\begin{bmatrix} 3 + 4 + 0 \\ 4 - 2 - 15 \end{bmatrix} = \begin{bmatrix} 7 \\ -13 \end{bmatrix}$

7. $\begin{bmatrix} 1 & 3 \\ 0 & 0 \end{bmatrix}\begin{bmatrix} 1 & 3 \\ 0 & 0 \end{bmatrix} = \begin{bmatrix} 1+0 & 3+0 \\ 0+0 & 0+0 \end{bmatrix} = \begin{bmatrix} 1 & 3 \\ 0 & 0 \end{bmatrix}$

8. $\begin{bmatrix} -2 & \frac{7}{2} \\ 1 & 2 \end{bmatrix}^{-1} = \frac{2}{-15}\begin{bmatrix} 2 & -\frac{7}{2} \\ -1 & -2 \end{bmatrix} = \begin{bmatrix} -\frac{4}{15} & \frac{7}{15} \\ \frac{2}{15} & \frac{4}{15} \end{bmatrix}$

9. $\begin{bmatrix} 1 & 2 \\ 3 & 2 \end{bmatrix}^{-1} = \frac{1}{-4}\begin{bmatrix} 2 & -2 \\ -3 & 1 \end{bmatrix} = \begin{bmatrix} -\frac{1}{2} & \frac{1}{2} \\ \frac{3}{4} & -\frac{1}{4} \end{bmatrix}$; $\begin{bmatrix} x \\ y \end{bmatrix} = \begin{bmatrix} -\frac{1}{2} & \frac{1}{2} \\ \frac{3}{4} & -\frac{1}{4} \end{bmatrix}\begin{bmatrix} -1 \\ 1 \end{bmatrix} = \begin{bmatrix} 1 \\ -1 \end{bmatrix}$;
$x = 1, y = -1$

10. $\begin{bmatrix} x' \\ y' \end{bmatrix} = \begin{bmatrix} 3 & 7 \\ 9 & 5 \end{bmatrix}\begin{bmatrix} 1 \\ -1 \end{bmatrix} = \begin{bmatrix} -4 \\ 4 \end{bmatrix}$; $(-4, 4)$

11. $\begin{bmatrix} 6 \\ -4 \end{bmatrix} = \begin{bmatrix} 0 & 2 \\ 2 & 0 \end{bmatrix}\begin{bmatrix} x \\ y \end{bmatrix}$; $\begin{bmatrix} 0 & 2 \\ 2 & 0 \end{bmatrix}^{-1} = \frac{1}{-4}\begin{bmatrix} 0 & -2 \\ -2 & 0 \end{bmatrix} = \begin{bmatrix} 0 & \frac{1}{2} \\ \frac{1}{2} & 0 \end{bmatrix}$; $\begin{bmatrix} 0 & \frac{1}{2} \\ \frac{1}{2} & 0 \end{bmatrix}\begin{bmatrix} 6 \\ -4 \end{bmatrix} =$
$\begin{bmatrix} x \\ y \end{bmatrix}$; $\begin{bmatrix} -2 \\ 3 \end{bmatrix} = \begin{bmatrix} x \\ y \end{bmatrix}$; $(-2, 3)$

12. $\frac{\pi}{180} \cdot 270 = \frac{3\pi}{2}$; $\frac{3\pi}{2}^{R}$ **13.** $\frac{180}{\pi}\left(-\frac{4\pi}{5}\right) = -144$; $-144°$

14. $\left(-\frac{8}{17}\right)^2 + \cos^2\alpha = 1$; $\cos^2\alpha = 1 - \frac{64}{289} = \frac{225}{289}$; $\cos\alpha = \pm\frac{15}{17}$;
since α is in Quadrant IV, $\cos\alpha = \frac{15}{17}$.

15. $r = \sqrt{12^2 + 5^2} = \sqrt{169} = 13$; $\sin\alpha = \frac{5}{13} \approx 0.3846$; $\alpha = 22.6° + k \cdot 360°$

16. $-\frac{1}{2}$ **17.** $\cos 1014° = \cos 294° = \cos 66° = 0.4067$ **18.** 2; $\frac{2\pi}{4} = \frac{\pi}{2}$

19. $\sec\alpha = \frac{3}{-1} = \frac{r}{u}$; $3^2 = (-1)^2 + v^2$; $v^2 = 9 - 1$; $v = \pm\sqrt{8} = \pm 2\sqrt{2}$;
since α is in Quadrant III, $v = -2\sqrt{2}$; $\sin\alpha = \frac{-2\sqrt{2}}{3}$, $\cos\alpha = \frac{-1}{3}$,
$\tan\alpha = 2\sqrt{2}$, $\csc\alpha = \frac{3}{-2\sqrt{2}}$, $\cot\alpha = \frac{1}{2\sqrt{2}}$

20. A, C, B; 19.6; 19.7

$\tan B = \frac{19.6}{19.7} \approx 0.9949$; $m(B) \approx 44.9°$

21. B, C, A; 36.8°; 40

$\tan 36.8° = \frac{a}{40}$; $a = 40(\tan 36.8°)$; $a \approx 29.9$

22. $\frac{\sec^2 x - 1}{\cos^2 x - 1} = \frac{\tan^2 x}{-\sin^2 x} = \frac{\frac{\sin^2 x}{\cos^2 x}}{-\sin^2 x} = \frac{\sin^2 x}{\cos^2 x} \cdot \frac{1}{-\sin^2 x} = \frac{1}{-\cos^2 x}$, $\cos x \neq 0, \pm 1$

23. $(\tan x + \sin x)(1 - \cos x) = \tan x - \cos x \tan x + \sin x - \sin x \cos x = \frac{\sin x}{\cos x} - \cos x\left(\frac{\sin x}{\cos x}\right) + \sin x - \sin x \cos x = \frac{\sin x}{\cos x} - \sin x \cos x = \frac{\sin x - \sin x \cos^2 x}{\cos x} = \frac{\sin x(1 - \cos^2 x)}{\cos x} = \frac{\sin^3 x}{\cos x}$, $\cos x \neq 0$

24. $\cos \frac{17\pi}{12} = \cos\left(\frac{5\pi}{4} + \frac{\pi}{6}\right) = \cos \frac{5\pi}{4} \cos \frac{\pi}{6} - \sin \frac{5\pi}{4} \sin \frac{\pi}{6} = -\frac{1}{\sqrt{2}} \cdot \frac{\sqrt{3}}{2} - \left(-\frac{1}{\sqrt{2}}\right)\frac{1}{2} = \frac{-\sqrt{6} + \sqrt{2}}{4}$

25. $\cos 175° \cos 25° + \sin 175° \sin 25° = \cos 150° = -\frac{\sqrt{3}}{2}$

26. $\sin 225° \cos 165° - \cos 225° \sin 165° = \sin 60° = \frac{\sqrt{3}}{2}$

27. $\sin \frac{5\pi}{8} = \sin \frac{\frac{5\pi}{4}}{2} = \sqrt{\frac{1 - \cos \frac{5\pi}{4}}{2}} = \sqrt{\frac{1 - \left(-\frac{\sqrt{2}}{2}\right)}{2}} = \frac{1}{2}\sqrt{2 + \sqrt{2}}$

28. $\cos B = \frac{a^2 + c^2 - b^2}{2ac} = \frac{81 + 25 - 100}{2(9)(5)} = \frac{6}{90} \approx 0.0667$; $m(B) \approx 86°$

29. $\frac{20}{\sin 30°} = \frac{b}{\sin 45°}$; $\frac{20}{0.5} = \frac{b}{0.7071}$; $b \approx 28.3$

30. $\frac{3\pi}{2} + 2k\pi$; $270° + k \cdot 360°$

31. $\text{Sin}^{-1}\left(-\frac{1}{\sqrt{2}}\right) = -\frac{\pi}{4}$

32. $\sec^2 x - 2 = 0$; $\sec^2 x = 2$; $\sec x = \sqrt{2}$, $\sec x = -\sqrt{2}$; $\cos x = \frac{1}{\sqrt{2}}$, $\cos x = -\frac{1}{\sqrt{2}}$; $\frac{\pi}{4} + k \cdot \frac{\pi}{2}$; $\left\{\frac{\pi}{4} + k \cdot \frac{\pi}{2}\right\}$

33. $1 - \cos^2 x = 3 \cos^2 x$; $1 = 4 \cos^2 x$; $\frac{1}{4} = \cos^2 x$; $\cos x = \pm\frac{1}{2}$; $\left\{\frac{\pi}{3}, \frac{2\pi}{3}, \frac{4\pi}{3}, \frac{5\pi}{3}\right\}$

34. $r = \sqrt{(-2\sqrt{3})^2 + 2^2} = 4$; since $\cos\theta = \frac{-2\sqrt{3}}{4} = \frac{-\sqrt{3}}{2}$ and $\sin\theta = \frac{2}{4} = \frac{1}{2}$, $\theta = 150°$; $(4, 150°)$ or, alternatively $(-4, -30°)$

35. $x = -3\cos 150° = \frac{3\sqrt{3}}{2}$, $y = -3\sin 150° = -\frac{3}{2}$; $\left(\frac{3\sqrt{3}}{2}, -\frac{3}{2}\right)$

36. $4\left(-\frac{1}{2} + i\left(-\frac{\sqrt{3}}{2}\right)\right) = -2 - 2i\sqrt{3}$

37. $r = \sqrt{1^2 + 1^2} = \sqrt{2}$; $\cos\theta = \frac{1}{\sqrt{2}}$; $\sin\theta = \frac{1}{\sqrt{2}}$; $m(\theta) = 45°$; $(1 + i)^5 =$
$[\sqrt{2}(\cos 45° + i\sin 45°)]^5 = (\sqrt{2})^5(\cos 5 \cdot 45° + i\sin 5 \cdot 45°) =$
$4\sqrt{2}(\cos 225° + i\sin 225°) = 4\sqrt{2}\left(-\frac{\sqrt{2}}{2} - i\frac{\sqrt{2}}{2}\right) = -4 - 4i$

38.

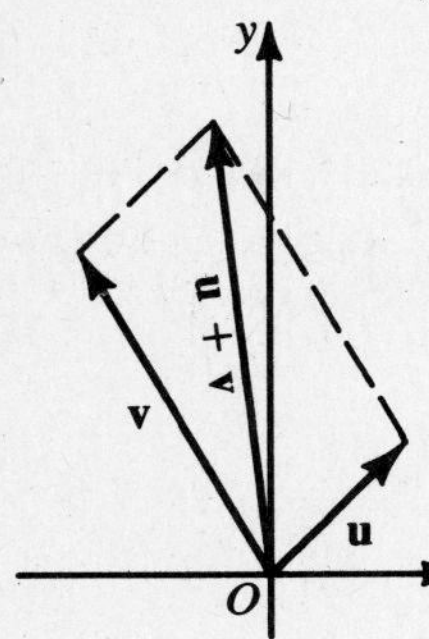

$180° - (45° + 30°) = 105°$;
$\|\mathbf{u} + \mathbf{v}\|^2 = 10^2 + 20^2 - 2(10)(20)\cos 105°$
$= 100 + 400 - (400)(-0.2588)$
$= 603.52$;
$\|\mathbf{u} + \mathbf{v}\| \approx 24.6$

Page 701 • PROGRAMMING IN PASCAL

1. a. See the procedure *get_polar_coords* in Exercise 2b.
   b. See the procedure *get_cartesian_coords* in Exercise 2b.
2. a. See the procedure *recip* in Exercise 2b.
   b.

```
PROGRAM de_moivre (INPUT, OUTPUT);

TYPE
  degree = real;
  complex = RECORD
              a : real;
              b : real;
            END;

VAR
  r, pi, n : real;
  angle : degree;
  z : complex;
```

[*Program continued on next page*]

```
(**********************************************************)
PROCEDURE get_polar_coords (x, y : real; VAR r : real;
                                         VAR angle : degree);

BEGIN
  pi := 4 * arctan(1);
  IF x > 0
     THEN angle := 180 * arctan(y/x)/pi
     ELSE IF x < 0
             THEN angle := 180 + (180 * arctan(y/x)/pi)
             ELSE IF y > 0
                     THEN angle := 90
                     ELSE angle := 270;
  r := sqrt(x * x + y * y);
END;

(**********************************************************)
PROCEDURE get_cartesian_coords (r : real; angle : degree;
                                            VAR x, y : real);

BEGIN
  x := r * cos((pi * angle)/180);
  y := r * sin((pi * angle)/180);
END;

(**********************************************************)
PROCEDURE recip (VAR z : complex);

VAR
  r : real;

BEGIN
  r := z.a * z.a + z.b * z.b;
  z.a := z.a/r;
  z.b := -z.b/r;
END;

(**********************************************************)
BEGIN (* main *)
  write('Enter real coordinate of the complex number: ');
  readln(z.a);
  write('Enter imaginary coordinate of complex number: ');
  readln(z.b);
  writeln;
  write('Enter power to which complex number is raised: ');
  readln(n);
```

```
    IF n < 1
       THEN BEGIN
              n := -n;
              recip(z);
            END;
    get_polar_coords(z.a, z.b, r, angle);
    r := exp(n * ln(r));
    angle := n * angle;
    get_cartesian_coords(r, angle, z.a, z.b);
    writeln;
    IF abs(z.a - 0) > 0
       THEN write(z.a:3:3);
    IF abs(z.b - 0) > 0.0005
       THEN IF z.b < 0
               THEN writeln(' - ', -z.b:3:3, 'i')
               ELSE IF abs(z.a - 0) > 0.0005
                       THEN writeln(' + ', z.b:3:3, 'i')
                       ELSE writeln(z.b:3:3, 'i');
  END.
```

3. See the procedure *get_components* in Exercise 4.
4. PROGRAM resultant (INPUT, OUTPUT);

```
TYPE
  degree = real;

VAR
  u_norm, u_x, u_y : real;
  v_norm, v_x, v_y : real;
  result_norm, result_x, result_y : real;
  pi : real;
  u_angle, v_angle, result_angle : degree;

(************************************************************)
PROCEDURE get_components (norm : real; angle : degree;
                                          VAR x, y : real);

BEGIN
  x := norm * cos((angle * pi)/180);
  y := norm * sin((angle * pi)/180);
END;

(************************************************************)
BEGIN (* main *)
  write('Enter the norm of the first vector: ');
  readln(u_norm);
  write('Enter the direction angle of the first vector: ');
  readln(u_angle);
  writeln;
```

[*Program continued on next page*]

```
  write('Enter the norm of the second vector: ');
  readln(v_norm);
  write('Enter the direction angle of the second vector: ');
  readln(v_angle);
  writeln;
  pi := 4 * arctan(1);
  get_components(u_norm, u_angle, u_x, u_y);
  get_components(v_norm, v_angle, v_x, v_y);
  result_x := (u_x + v_x);
  result_y := (u_y + v_y);
  result_norm := sqrt(result_x*result_x + result_y*result_y);
  IF result_x = 0
     THEN IF result_y > 0
             THEN result_angle := 90
             ELSE result_angle := 270
     ELSE result_angle := (180 * arctan(result_y/result_x))/pi;
  writeln('Norm of the resultant is: ', result_norm:3:3);
  writeln('Direction angle is: ', result_angle:3:3, ' degrees');
END.
```

Pages 708–710 • WRITTEN EXERCISES

A **1.** 70, 76, 79, 82, 82, 82, 88, 92, 92, 93: sum of data points = 836, mean = $\frac{836}{10}$ = 83.6; median = 82; mode = 82

2. 2, 3, 3, 4, 4, 4, 5, 6, 7, 8, 9: sum of data points = 55, mean = $\frac{55}{11}$ = 5; median = 4; mode = 4

3. 1, 2, 3, 5, 8, 9, 10, 12, 13: sum of data points = 63, mean = $\frac{63}{9}$ = 7; median = 8; no mode

4. 0, 0, 1, 2, 3, 4, 5, 6, 8, 12: sum of data points = 41, mean = $\frac{41}{10}$ = 4.1; median = $\frac{3 + 4}{2}$ = 3.5; mode = 0

5. 51, 53, 55, 60, 60, 68, 72, 75, 83, 85, 88, 91: sum of data points = 841, mean = $\frac{841}{12} = 70.08\overline{3} \approx 70$; median = $\frac{68 + 72}{2}$ = 70; mode = 60

6. 30, 32, 34, 34, 36, 36, 36, 36, 38, 40, 40, 42, 44, 44, 46: sum of data points = 568, mean = $\frac{568}{15} = 37.8\overline{6} \approx 38$; median = 36; mode = 36. The mode is of most interest to the manager; the mode tells which size was sold most often.

7.

Score	Freq.
70	1
76	1
79	1
82	3
88	1
92	2
93	1

8.

Size	Freq.
30	1
32	1
34	2
36	4
38	1
40	2
42	1
44	2
46	1

9.

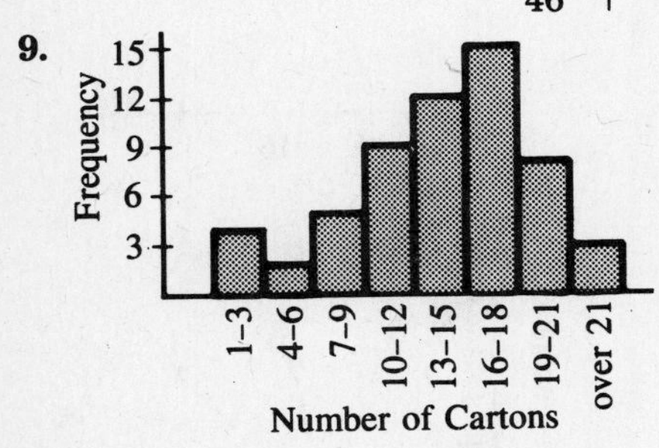

10.

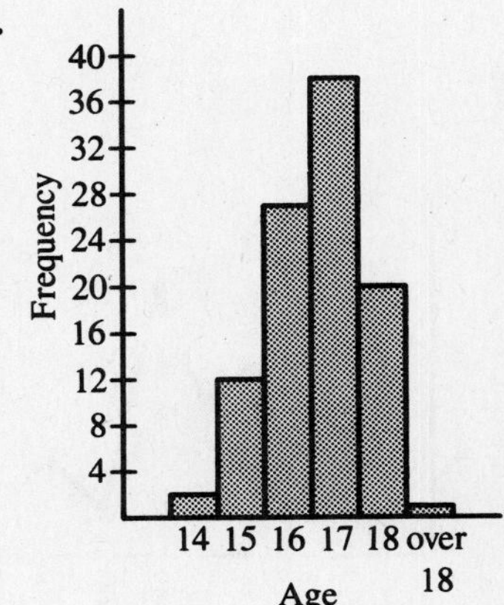

11.

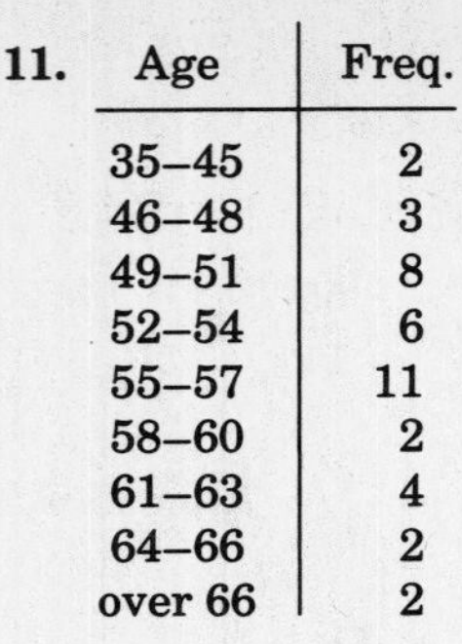

Age	Freq.
35–45	2
46–48	3
49–51	8
52–54	6
55–57	11
58–60	2
61–63	4
64–66	2
over 66	2

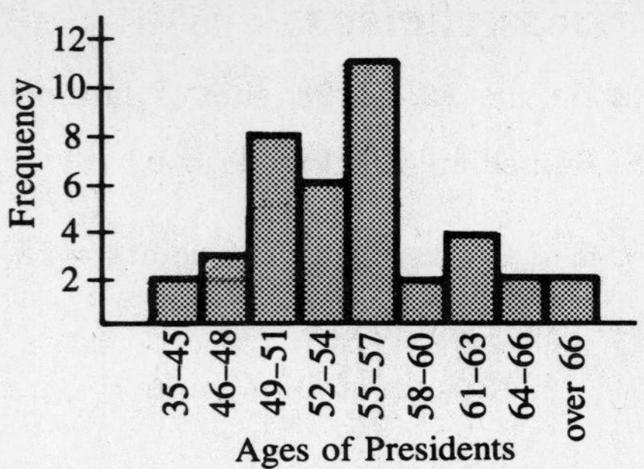

12.

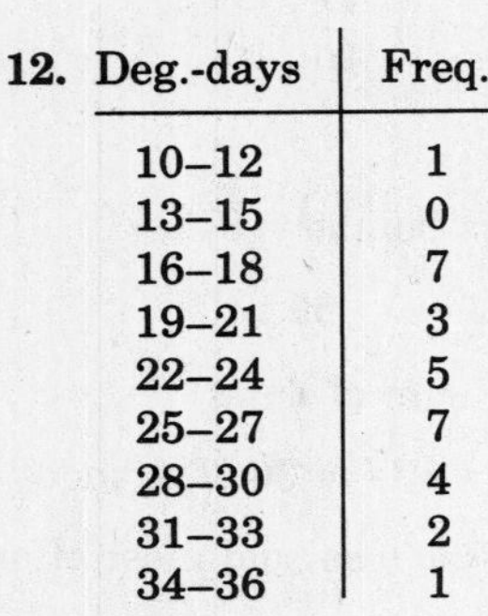

Deg.-days	Freq.
10–12	1
13–15	0
16–18	7
19–21	3
22–24	5
25–27	7
28–30	4
31–33	2
34–36	1

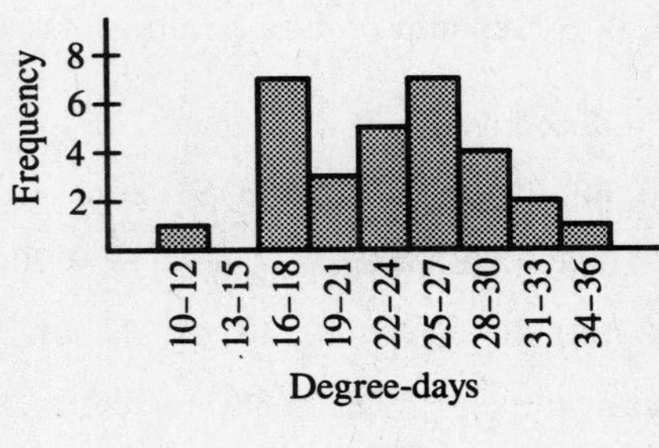

13.

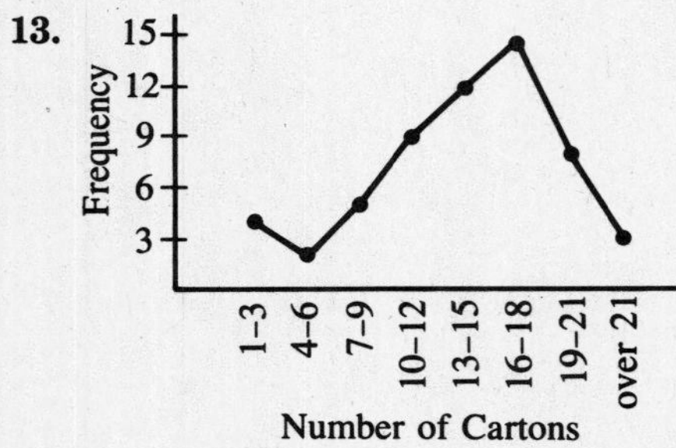

14.

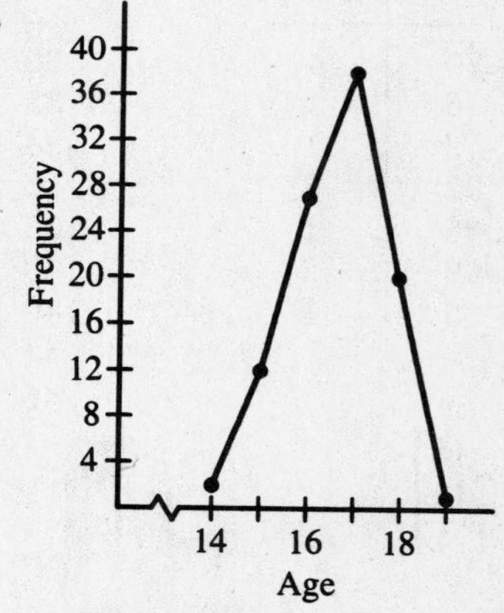

15.

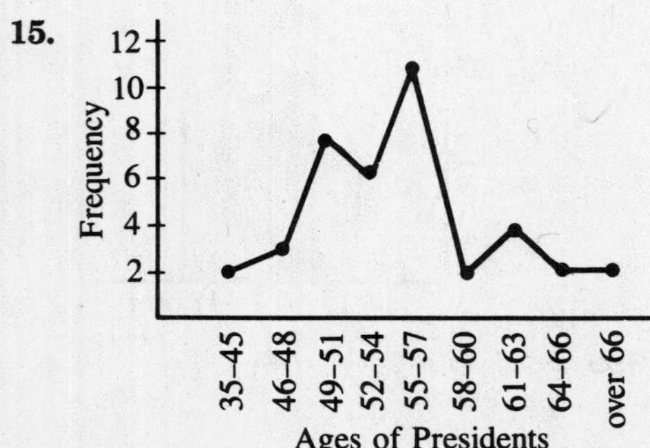

16.

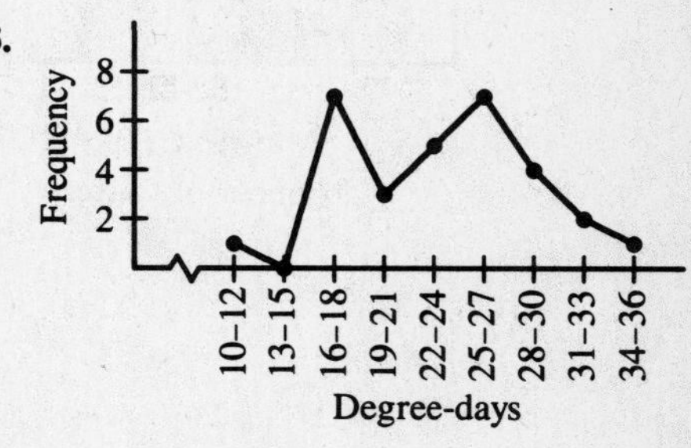

17. Sum of cartons $= 2 \cdot 4 + 5 \cdot 2 + 8 \cdot 5 + 11 \cdot 9 + 14 \cdot 12 + 17 \cdot 15 + 20 \cdot 8 + 22 \cdot 3 = 806$; mean $= \frac{806}{58} \approx 13.9$

18. Sum of ages $= 14 \cdot 2 + 15 \cdot 12 + 16 \cdot 27 + 17 \cdot 38 + 18 \cdot 20 + 19 \cdot 1 = 1665$; mean $= \frac{1665}{100} \approx 16.7$

19. Sum of ages $= 2193$; mean $= \frac{2193}{40} \approx 54.8$

20. Sum of degree-days $= 700$; mean $= \frac{700}{30} \approx 23.3$

B 21. Mean $= \frac{38(76.5) + 22(85.0) + 35(78.8)}{38 + 22 + 35} = \frac{7535}{95} \approx 79.3$

22. **a.** The mean of Brand X is $\frac{50}{10}$, or 5. The mode of Brand Y is 5. The median of Brand Z is 5. **b.** Since over half of the Brand Z sets last at least 5 years, Brand Z seems most likely to last 5 or more years.

C 23. percentile rank of 32: $\frac{87}{93} + \frac{1}{2}\left(\frac{1}{93}\right) \approx 93.55\% + \frac{1}{2}(1.075\%) = 94.0875\%$; 94

24. percentile rank of 28: $\frac{80}{93} + \frac{1}{2}\left(\frac{3}{93}\right) \approx 86.02\% + \frac{1}{2}(3.226\%) = 87.633\%$; 88

25. percentile rank of 21: $\frac{45}{93} + \frac{1}{2}\left(\frac{12}{93}\right) \approx 48.39\% + \frac{1}{2}(12.903\%) = 54.8415\%$; 55

Page 710 • COMPUTER EXERCISES

1.
```
5 PRINT "THE PROGRAM IS READING THE NECESSARY DATA."
6 PRINT "PLEASE WAIT." : PRINT
10 READ N
20 DIM D(N),F(N),T(N)
30 FOR I = 1 TO N
40 READ D(I)
50 NEXT I
60 FOR J = 1 TO N - 1
70 FOR I = 1 TO N - J
80 IF D(I) <= D(I + 1) THEN GOTO 100
90 LET T = D(I) : LET D(I) = D(I + 1) : LET D(I + 1) = T
100 NEXT I
110 NEXT J
120 FOR I = 1 TO N
130 LET F(I) = 1
140 NEXT I
150 FOR I = 1 TO N
160 FOR J = I + 1 TO N - 1
170 IF D(I) <> D(J) THEN GOTO 220
180 IF D(I) = 999 OR D(J) = 999 THEN GOTO 220
190 LET D(J) = 999
200 LET F(I) = F(I) + 1
210 NEXT J
```

[Program continued on next page]

```
220  NEXT I
230  PRINT "FUEL ECONOMY","FREQUENCY","RELATIVE"
240  PRINT "   RATING"," ","FREQUENCY"
250  PRINT
260  LET X = 12 : LET Y = 14 : LET I = 1
270  FOR J = 1 TO N
280  IF D(J) = 999 THEN GOTO 330
290  IF D(J) >= X AND D(J) <= Y THEN LET T(I) =
                                  T(I) + F(J) : GOTO 330
300  LET J = J - 1 : LET I = I + 1
310  LET X = X + 3 : IF X = 39 THEN LET Y = Y + 5 : GOTO 330
320  LET Y = Y + 3
330  NEXT J
340  LET X = 12 : LET Y = 14
350  FOR I = 1 TO 9
360  PRINT X;" - ";Y,T(I),INT(T(I)/N * 1000 + .5)/1000
370  LET X = X + 3 : LET Y = Y + 3
380  NEXT I
390  PRINT "OVER 38",T(10),INT(T(10)/N * 1000 + .5)/1000
900  DATA 93
910  DATA 20,19,20,26,19,25,18,16,16,21,18,18,19,20,19
920  DATA 21,21,20,19,26,21,19,15,19,19,18,18,17,21,36
930  DATA 26,32,27,23,25,17,21,28,27,18,18,17,28,24,22
940  DATA 26,30,20,19,16,20,16,27,35,14,30,16,20,16,30
950  DATA 16,20,21,19,19,17,23,17,37,16,25,25,21,19,19
960  DATA 21,21,21,20,16,24,28,25,21,27,25,22,39,29,42
970  DATA 25,25,25
980  END
```

2. Insert the following lines:

```
45  LET S = S + D(I)
400  PRINT : PRINT : PRINT "ACTUAL MEAN: ";S/N
```

Actual mean: 22.172043

3. Insert the following lines:

```
410  LET C = 13
420  FOR I = 1 TO 10
430  LET S2 = S2 + C * T(I)
440  LET C = C + 3
450  NEXT I
460  PRINT : PRINT "ESTIMATED MEAN: ";S2/N
```

Estimated mean: 22.225806

4. Insert the following lines:

```
111  LET A = INT((N + 1)/2)
112  LET B = INT((N + 2)/2)
113  LET MD = (D(A) + D(B))/2
281  IF MD >= X AND MD <= Y THEN LET MP = X + 1
470  PRINT : PRINT "MEDIAN: ";MD
480  PRINT "MIDPOINT OF INTERVAL CONTAINING MEDIAN: ";MP
```

Median: 21; midpoint: 22

Pages 714–715 • WRITTEN EXERCISES

A **1.** $\bar{x} = \frac{134}{6} \approx 22.33$; $\sigma^2 \approx \frac{(-10.33)^2 + (-5.33)^2 + (2.67)^2 \cdot 3 + (7.67)^2}{6} = \frac{215.3334}{6} \approx$ 35.89; $\sigma = \sqrt{35.89} \approx 5.99$

2. $\bar{x} = \frac{14.5}{5} = 2.9$; $\sigma^2 = \frac{(-0.3)^2 + (-0.2)^2 + 0^2 + (0.2)^2 + (0.3)^2}{5} = \frac{0.26}{5} = 0.052$; $\sigma = \sqrt{0.052} \approx 0.23$

3. $\bar{x} = \frac{56}{7} = 8$, $\sigma^2 = \frac{(-4)^2 + 0^2 + 4^2 + 2^2 + (-1)^2 + 1^2 + (-2)^2}{7} = \frac{42}{7} = 6$; $\sigma = \sqrt{6} \approx 2.45$

4. $\bar{x} = \frac{103}{7} \approx 14.71$; $\sigma^2 \approx$ $\frac{(5.59)^2 + (2.29)^2 + (2.09)^2 + (-2.21)^2 + (-2.51)^2 \cdot 2 + (-2.71)^2}{7} = \frac{65.6887}{7} \approx$ 9.38; $\sigma = \sqrt{9.38} \approx 3.06$

5. $\bar{x} = \frac{168}{12} = 14$; $\sigma^2 = \frac{6^2 \cdot 2 + 3^2 + 2^2 \cdot 2 + 0^2 + (-2)^2 + (-4)^2 \cdot 2 + (-3)^2 \cdot 3}{12} =$ $\frac{152}{12} \approx 12.67$; $\sigma = \sqrt{12.67} \approx 3.56$

6. $\bar{x} = \frac{86}{10} = 8.6$; $\sigma^2 =$ $\frac{(0.6)^2 + (0.2)^2 + (-0.2)^2 + (-0.5)^2 + (-0.3)^2 + (0.1)^2 \cdot 2 + 0^2 \cdot 3}{10} = \frac{0.8}{10} = 0.08$; $\sigma = \sqrt{0.08} \approx 0.28$

7. $z = \frac{56 - 52}{20} = 0.2$ **8.** $z = \frac{82 - 52}{20} = 1.5$ **9.** $z = \frac{32 - 52}{20} = -1$

10. $z = \frac{40 - 52}{20} = -0.6$ **11.** $1 = \frac{x_i - 48}{15}$; $15 = x_i - 48$; $x_i = 63$

12. $-2 = \frac{x_i - 48}{15}$; $-30 = x_i - 48$; $x_i = 18$ **13.** $0.8 = \frac{x_i - 48}{15}$; $12 = x_i - 48$; $x_i = 60$

14. $-1.5 = \frac{x_i - 48}{15}$; $-22.5 = x_i - 48$; $x_i = 25.5$

15. mean of set A = mean of set B = 50 (See Oral Ex. 2.);
for set A: $\frac{(30 - 50) + (52 - 50) + (68 - 50)}{3} = \frac{-20 + 2 + 18}{3} = 0$;
for set B: $\frac{(48 - 50) + (49 - 50) + (53 - 50)}{3} = \frac{-2 + (-1) + 3}{3} = 0$;
this expression is always equal to zero.

16–19. The table below shows the z-score for each test score.

student	A	B	C	D
law	1.5	3.17	2.17	2.33
medicine	−1.75	0	0.25	2.25
engineering	2.5	−3	1	−1

On the basis of the test, each student should choose the field in which he/she earned the highest z-score.

16. engineering **17.** law **18.** law **19.** law

B **20.** Herb: $1.6 = \frac{86 - \bar{x}}{\sigma}$; $1.6\sigma = 86 - \bar{x}$; $\bar{x} = 86 - 1.6\sigma$. Maria: $2.4 = \frac{90 - \bar{x}}{\sigma}$; $2.4\sigma = 90 - \bar{x}$; $\bar{x} = 90 - 2.4\sigma$. Then $86 - 1.6\sigma = 90 - 2.4\sigma$, $0.8\sigma = 4$, and $\sigma = 5$. $\bar{x} = 86 - 1.6(5) = 78$

21. If every data point is increased by 5, the mean is $\frac{\sum_{i=1}^{n}(x_i + 5)}{n} = \frac{\sum_{i=1}^{n} x_i + 5n}{n} = \frac{\sum_{i=1}^{n} x_i}{n} + \frac{5n}{n} = \bar{x} + 5$. The mean is increased by 5. The variance becomes $\frac{\sum_{i=1}^{n}[(x_i + 5) - (\bar{x} + 5)]^2}{n} = \frac{\sum_{i=1}^{n}(x_i - \bar{x})^2}{n} = \sigma^2$, the original variance. The variance and standard deviation are unchanged.

22. If every data point is doubled, the mean is $\frac{\sum_{i=1}^{n} 2x_i}{n} = \frac{2\sum_{i=1}^{n} x_i}{n} = 2\bar{x}$. The mean is doubled. The variance becomes $\frac{\sum_{i=1}^{n}(2x_i - 2\bar{x})^2}{n} = \frac{\sum_{i=1}^{n} 2^2(x_i - \bar{x})^2}{n} = \frac{4\sum_{i=1}^{n}(x_i - \bar{x})^2}{n} = 4\sigma^2$. The variance is quadrupled. The standard deviation is $\sqrt{4\sigma^2} = 2\sigma$; the standard deviation is doubled.

23. If the frequency of each data point is doubled, the mean is $\frac{\sum_{i=1}^{r} x_i \cdot 2f_i}{\sum_{i=1}^{r} 2f_i} = \frac{2\sum_{i=1}^{r} x_i f_i}{2\sum_{i=1}^{r} f_i} = \bar{x}$. The mean is unchanged. The variance becomes $\frac{\sum_{i=1}^{r}(x_i - \bar{x})^2 \cdot 2f_i}{\sum_{i=1}^{r} 2f_i} = \frac{2\sum_{i=1}^{r}(x_i - \bar{x})^2 f_i}{2\sum_{i=1}^{r} f_i} = \sigma^2$. The variance and standard deviation are unchanged.

24. The mean becomes $\frac{\sum_{i=1}^{n}\frac{x_i + \bar{x}}{2}}{n} = \frac{\frac{1}{2}\sum_{i=1}^{n}(x_i + \bar{x})}{n} = \frac{1}{2}\left(\frac{\sum_{i=1}^{n} x_i}{n}\right) + \frac{1}{2}\left(\frac{n\bar{x}}{n}\right) = \frac{1}{2}\bar{x} + \frac{1}{2}\bar{x} = \bar{x}$. The mean is unchanged. The variance becomes $\frac{\sum_{i=1}^{n}\left(\frac{x_i + \bar{x}}{2} - \bar{x}\right)^2}{n} = \frac{\sum_{i=1}^{n}\left(\frac{x_i - \bar{x}}{2}\right)^2}{n} = \frac{\sum_{i=1}^{n}\frac{1}{4}(x_i - \bar{x})^2}{n} = \frac{\frac{1}{4}\sum_{i=1}^{n}(x_i - \bar{x})^2}{n} = \frac{1}{4}\sigma^2$. The variance is multiplied by $\frac{1}{4}$. The standard deviation is $\sqrt{\frac{1}{4}\sigma^2} = \frac{1}{2}\sigma$. The standard deviation is halved.

C **25.** $\bar{y} = \dfrac{\sum_{i=1}^{n} y_i}{n} = \dfrac{\sum_{i=1}^{n} mx_i}{n} = \dfrac{m\sum_{i=1}^{n} x_i}{n} = m\dfrac{\sum_{i=1}^{n} x_i}{n} = m\bar{x}$. $\sigma_y{}^2 = \dfrac{\sum_{i=1}^{n}(y_i - \bar{y})^2}{n} = \dfrac{\sum_{i=1}^{n}(mx_i - m\bar{x})^2}{n} = \dfrac{\sum_{i=1}^{n}(m(x_i - \bar{x}))^2}{n} = m^2\dfrac{\sum_{i=1}^{n}(x_i - \bar{x})^2}{n} = m^2\sigma_x{}^2$. $\therefore \sigma_y = \sqrt{m^2\sigma_x{}^2} = |m|\sigma_x$.

26. $\bar{x} = \dfrac{x_1 + x_2}{2}$; $\sigma^2 = \dfrac{\sum_{i=1}^{2}(x_i - \bar{x})^2}{2} = \dfrac{(x_1 - \bar{x})^2 + (x_2 - \bar{x})^2}{2} = \dfrac{\left(x_1 - \frac{x_1 + x_2}{2}\right)^2 + \left(x_2 - \frac{x_1 + x_2}{2}\right)^2}{2} = \dfrac{\left(\frac{x_1}{2} - \frac{x_2}{2}\right)^2 + \left(\frac{x_2}{2} - \frac{x_1}{2}\right)^2}{2} = \dfrac{2\left(\frac{x_2}{2} - \frac{x_1}{2}\right)^2}{2} = \left(\dfrac{x_2}{2} - \dfrac{x_1}{2}\right)^2 = \dfrac{1}{4}(x_2 - x_1)^2$; $\therefore \sigma = \sqrt{\dfrac{(x_2 - x_1)^2}{4}} = \dfrac{|x_2 - x_1|}{2}$

27. $\sigma^2 = \dfrac{\sum_{i=1}^{n}(x_i - \bar{x})^2}{n} = \dfrac{\sum_{i=1}^{n}(x_i^2 - 2\bar{x}x_i + \bar{x}^2)}{n} = \dfrac{\sum_{i=1}^{n} x_i^2 - 2\bar{x}\sum_{i=1}^{n} x_i + \sum_{i=1}^{n}\bar{x}^2}{n}$. Since $\sum_{i=1}^{n} x_i = n\bar{x}$ and $\sum_{i=1}^{n}\bar{x}^2 = n\bar{x}^2$, $\sigma^2 = \dfrac{\sum_{i=1}^{n} x_i^2 - 2\bar{x}n\bar{x} + n\bar{x}^2}{n} = \dfrac{\sum_{i=1}^{n} x_i^2 - n\bar{x}^2}{n}$. Since $\bar{x} = \dfrac{\sum_{i=1}^{n} x_i}{n}$, $\sigma^2 = \dfrac{\sum_{i=1}^{n} x_i^2 - n\left(\frac{1}{n}\right)^2\left(\sum_{i=1}^{n} x_i\right)^2}{n} = \dfrac{n\sum_{i=1}^{n} x_i^2 - \left(\sum_{i=1}^{n} x_i\right)^2}{n^2}$.

Page 715 • COMPUTER EXERCISES

1.
```
5 PRINT "THE PROGRAM IS READING THE NECESSARY DATA."
6 PRINT "PLEASE WAIT." : PRINT
10 READ N
20 DIM D(N),F(N)
30 FOR I = 1 TO N
40 READ D(I)
50 NEXT I
60 FOR J = 1 TO N - 1
70 FOR I = 1 TO N - J
80 IF D(I) <= D(I + 1) THEN GOTO 100
90 LET T = D(I) : LET D(I) = D(I + 1) : LET D(I + 1) = T
100 NEXT I
110 NEXT J
120 FOR I = 1 TO N
130 LET F(I) = 1
140 NEXT I
150 FOR I = 1 TO N
```

[Program continued on next page]

```
160 FOR J = I + 1 TO N - 1
170 IF D(I) <> D(J) THEN GOTO 220
180 IF D(I) = 999 OR D(J) = 999 THEN GOTO 220
190 LET D(J) = 999
200 LET F(I) = F(I) + 1
210 NEXT J
220 NEXT I
230 FOR I = 1 TO N
240 IF D(I) = 999 THEN GOTO 270
250 LET S = S + D(I) * F(I)
260 LET FR = FR + F(I)
270 NEXT I
280 LET MN = S/FR
290 FOR I = 1 TO N
300 IF D(I) = 999 THEN GOTO 320
310 LET V = V + (D(I) - MN)^2 * F(I)
320 NEXT I
330 LET VR = V/FR
340 LET SD = SQR(VR)
350 PRINT "MEAN: ";INT(MN * 100 + .5)/100
360 PRINT "VARIANCE: ";INT(VR * 100 + .5)/100
370 PRINT "STANDARD VARIATION: ";INT(SD * 100 + .5)/100
900 DATA 48
910 DATA 65,70,70,70,70,70,70
920 DATA 75,75,75,75,75,75,75,75,75,75,75,75,75,75
930 DATA 80,80,80,80,80,80,80,80,80,80,80,80
940 DATA 85,85,85,85,85,85,85,85
950 DATA 90,90,90,90,95,95,95
960 END
```

2. 79.58; 50.87; 7.13

3. Change the DATA lines 900–950 for each exercise.
1. 22.33; 35.89; 5.99 2. 2.9; 0.05; 0.23 3. 8; 6; 2.45 4. 14.71; 9.38; 3.06
5. 14; 12.67; 3.56 6. 8.6; 0.08; 0.28

4. Change the following lines in the program. Delete lines 320–370.

```
270 LET SX = SX + D(I)^2 * F(I)
280 NEXT I
290 LET VR = (FR * SX - S^2)/FR^2
300 LET VR = INT(VR * 100 + .5)/100
310 PRINT "VARIANCE: ";VR
```

5. 50.87 **6.** Use 19 for the value in the interval "over 18". 1.05

Page 716 • SELF-TEST 1

1.

In.	Freq.
8–10	4
11–13	5
14–16	4
17–19	5
20–22	3
23–25	8
over 25	1

2. 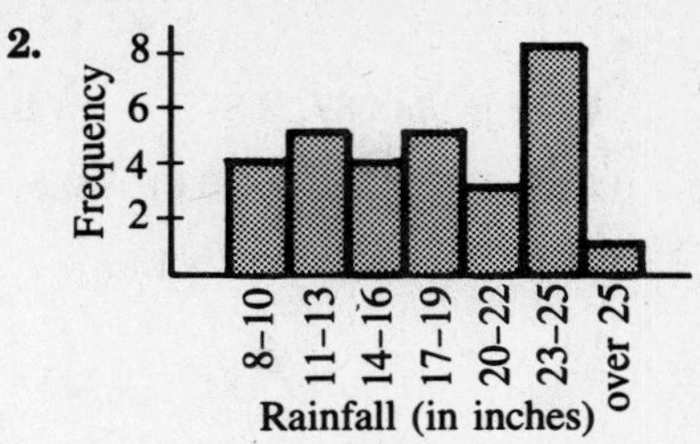

3. 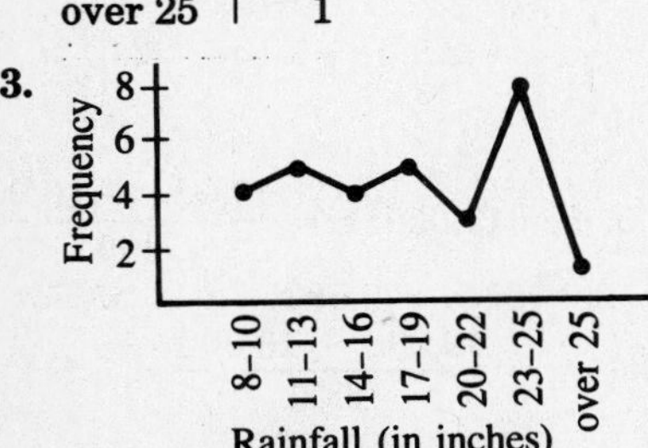

4. 0, 1, 4, 10, 18, 26, 37, 37, 74, 89, 98, 107; sum of diaries = 501, mean = $\frac{501}{12} = 41.75$; median = $\frac{26 + 37}{2} = 31.5$; mode = 37

5. $\bar{x} = 41.75$; $\sigma^2 = [(56.25)^2 + (32.25)^2 + (-4.75)^2 \cdot 2 + (-15.75)^2 + (-31.75)^2 + (-41.75)^2 + (-37.75)^2 + (-40.75)^2 + (-23.75)^2 + (47.25)^2 + (65.25)^2]/12 \approx \frac{17{,}388.25}{12} \approx 1449.02$; $\sigma = \sqrt{1449.02} \approx 38.07$

6. Class 1: $\bar{x} = \frac{350}{5} = 70$; $\sigma^2 = \frac{(-11)^2 + (-8)^2 \cdot 2 + 13^2 + 14^2}{5} = \frac{614}{5} = 122.8$; $\sigma = \sqrt{122.8} \approx 11.08$; $z = \frac{83 - 70}{11.08} \approx 1.17$; class 2: $\bar{x} = \frac{350}{5} = 70$; $\sigma^2 = \frac{(-3)^2 + (-22)^2 + (-17)^2 + 13^2 + 29^2}{5} = \frac{1792}{5} = 358.4$; $\sigma = \sqrt{358.4} \approx 18.93$; $z = \frac{83 - 70}{18.93} \approx 0.69$; Class 1

Pages 724–725 • WRITTEN EXERCISES

A 1. a. The z-score for 30 is $z = \frac{30 - 42}{15} = -0.8$. The z-score for 60 is $z = \frac{60 - 42}{15} = 1.2$. $P(-0.8 < z < 1.2) = A(0.8) + A(1.2) = 0.2881 + 0.3849 = 0.673$; 67.3% **b.** The z-score for 36 is $z = \frac{36 - 42}{15} = -0.4$. $P(z < -0.4) = P(z < 0) - A(0.4) = 0.5 - 0.1554 = 0.3446$; 34.46% **c.** The z-score for 54 is $z = \frac{54 - 42}{15} = 0.8$. $P(z > 0.8) = P(z > 0) - A(0.8) = 0.5 - 0.2881 = 0.2119$; 21.19% **d.** $P(z < -0.8) + P(z > 1.2) = [P(z < 0) - A(0.8)] + [P(z > 0) - A(1.2)] = (0.5 - 0.2881) + (0.5 - 0.3849) = 0.2119 + 0.1151 = 0.327$; 32.7%

2. **a.** The z-score for 2 is $z = \dfrac{2 - 9.6}{4} = -1.9$. $P(z < -1.9) = P(z < 0) - A(1.9) =$ $0.5 - 0.4713 = 0.0287$; 2.87% **b.** The z-score of 12 is $z = \dfrac{12 - 9.6}{4} = 0.6$. $P(z > 0.6) = P(z > 0) - A(0.6) = 0.5 - 0.2257 = 0.2743$; 27.43%

3. **a.** $z = \dfrac{15 - 12}{2} = 1.5$; $P(z > 1.5) = P(z > 0) - A(1.5) = 0.5 - 0.4332 = 0.0668$

b. $z = \dfrac{10 - 12}{2} = -1$; $P(z < -1) = P(z < 0) - P(1) = 0.5 - 0.3413 = 0.1587$

c. For 8: $z = \dfrac{8 - 12}{2} = -2$. For 17: $z = \dfrac{17 - 12}{2} = 2.5$. $P(-2 < z < 2.5) = A(2) +$ $A(2.5) = 0.4772 + 0.4938 = 0.9710$

4. **a.** For 15,000: $z = \dfrac{15{,}000 - 16{,}000}{2000} = -0.5$. For 17,000: $z = \dfrac{17{,}000 - 16{,}000}{2000} =$ 0.5. $P(-0.5 < z < 0.5) = A(0.5) + A(0.5) = 2(0.1915) = 0.3830$ **b.** For 18,000: $z = \dfrac{18{,}000 - 16{,}000}{2000} = 1$. For 19,000: $z = \dfrac{19{,}000 - 16{,}000}{2000} = 1.5$. $P(1 < z < 1.5) = A(1.5) - A(1) = 0.4332 - 0.3413 = 0.0919$ **c.** $z =$ $\dfrac{10{,}000 - 16{,}000}{2000} = -3$; $P(z < -3) = P(z < 0) - A(3) = 0.5 - 0.4987 = 0.0013$

d. $z = \dfrac{14{,}000 - 16{,}000}{2000} = -1$; $P(z > -1) = A(1) + P(z > 0) =$ $0.3413 + 0.5 = 0.8413$

5. $z = \dfrac{95 - 84}{5} = 2.2$; $P(z > 2.2) = P(z > 0) - A(2.2) = 0.5 - 0.4861 = 0.0139$; $1.39\% \approx 1.4\%$

6. **a.** $z = \dfrac{40 - 48}{5} = -1.6$; $P(z < -1.6) = P(z < 0) - A(1.6) = 0.5 - 0.4452 =$ 0.0548; 5.48% **b.** For 50: $z = \dfrac{50 - 48}{5} = 0.4$. $P(-1.6 < z < 0.4) = A(1.6) +$ $A(0.4) = 0.4452 + 0.1554 = 0.6006$; 60.06%

B **7.** $8\% = 0.08 = 0.5 - 0.42$; the z-score corresponding to the area 0.42 is 1.4. Let x = the cutoff grade that results in $P(z > 1.4)$. $1.4 = \dfrac{x - 76}{5}$; $7 = x - 76$; $83 = x$; 83

8. The probability of winning a prize is $\dfrac{40}{40{,}000} = 0.001 = 0.5 - 0.499$. The z-score corresponding to the area 0.499 is 3.1. Let x = the minimum score that results in $P(z > 3.1)$. $3.1 = \dfrac{x - 72}{10}$; $31 = x - 72$; $103 = x$; 103

C **9.** $1\% = 0.01 = 0.5 - 0.49$; the z-score corresponding to the area 0.49 is 2.3. Let x = length of guarantee that results in $P(z < -2.3)$. $-2.3 = \dfrac{x - 5}{1.5}$; $-3.45 = x - 5$; $1.55 = x$; 1.55 yr

10. 80% represents an area of 0.80. To minimize the adjustments, center the area around the mean. That divides the area 0.80 into two equal areas of 0.40 each. The z-score corresponding to the area 0.4 is 1.3. Let x = minimum height and y = maximum height that result in $P(-1.3 < z < 1.3)$. $-1.3 = \dfrac{x - 170}{20}$,

$-26 = x - 170$, $144 = x$; $1.3 = \dfrac{y - 170}{20}$, $26 = y - 170$, $196 = y$; 144 cm to 196 cm

Pages 730–731 • WRITTEN EXERCISES

A 1. $p = \frac{4}{5} = 0.8$; $\overline{p} = \frac{70}{100} = 0.7$; $\sigma_{\overline{p}} = \sqrt{\frac{0.8(1 - 0.8)}{100}} = 0.04$; $0.7 - 2(0.04) < p < 0.7 + 2(0.04)$, $0.62 < p < 0.78$; since p does not fall within this interval, the hypothesis is rejected.

2. $p = 0.1$; $\overline{p} = \frac{6}{25} = 0.24$; $\sigma_{\overline{p}} = \sqrt{\frac{0.1(1 - 0.1)}{25}} = 0.06$; $0.24 - 2(0.06) < p < 0.24 + 2(0.06)$, $0.12 < p < 0.36$; since p does not fall within this interval, the hypothesis is rejected.

3. $\overline{p} = \frac{47}{96} \approx 0.490$; $\sigma_{\overline{p}} \approx \sqrt{\frac{(0.490)(0.510)}{96}} \approx 0.051$; $0.490 - 2(0.051) < p < 0.490 + 2(0.051)$, $0.388 < p < 0.592$

4. $\overline{p} = \frac{120}{400} = 0.3$; $\sigma_{\overline{p}} \approx \sqrt{\frac{(0.3)(0.7)}{400}} \approx 0.023$; $0.3 - 2(0.023) < p < 0.3 + 2(0.023)$, $0.254 < p < 0.346$

5. $\overline{p} = \frac{3}{12} = 0.25$; $\sigma_{\overline{p}} \approx \sqrt{\frac{(0.25)(0.75)}{12}} = 0.125$; $0.25 - 2(0.125) < p < 0.25 + 2(0.125)$, $0 < p < 0.5$

6. $\overline{p} = \frac{90}{150} = 0.6$; $\sigma_{\overline{p}} \approx \sqrt{\frac{(0.6)(0.4)}{150}} = 0.04$; $0.6 - 2(0.04) < p < 0.6 + 2(0.04)$, $0.52 < p < 0.68$

7. $\overline{p} = \frac{90}{240} = 0.375$; $\sigma_{\overline{p}} \approx \sqrt{\frac{(0.375)(0.625)}{240}} \approx 0.031$; $0.375 - 2(0.031) < p < 0.375 + 2(0.031)$, $0.313 < p < 0.437$

8. $\overline{p} = \frac{8}{72} \approx 0.111$; $\sigma_{\overline{p}} \approx \sqrt{\frac{(0.111)(0.889)}{72}} \approx 0.037$; $0.111 - 2(0.037) < p < 0.111 + 2(0.037)$, $0.037 < p < 0.185$; the fraction of the parts that are defective falls between 0.037 and 0.185. Multiply by 270 to find the number of defective parts. $270(0.037) = 9.95 \approx 10$; $270(0.185) = 49.95 \approx 50$; 10 to 50 defective parts

9. $p = 2\% = 0.02$; $\overline{p} = \frac{16}{400} = 0.04$; $\sigma_{\overline{p}} = \sqrt{\frac{(0.02)(0.98)}{400}} = 0.007$; $0.04 - 2(0.007) < p < 0.04 + 2(0.007)$, $0.026 < p < 0.054$; since p does not fall within this interval, the results of this year's test are inconsistent with previous results. Answers may vary for the explanation. One possible explanation is that the test was unusually easy.

B 10. a.

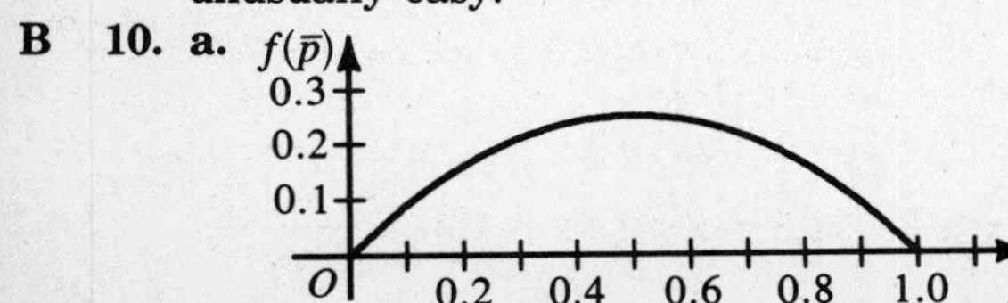

b. $f(\bar{p}) = \bar{p}(1 - \bar{p}) = -\bar{p}^2 + \bar{p}$; maximum point occurs at $\left(-\frac{b}{2a}, -\frac{b^2 - 4ac}{4a}\right)$;
maximum value of function $= -\frac{b^2 - 4ac}{4a} = -\frac{1^2 - 4(-1)(0)}{4(-1)} = -\frac{1}{-4} = \frac{1}{4} = 0.25$

c. Since maximum value of function $= \frac{1}{4}$, $\bar{p}(1 - \bar{p}) \le \frac{1}{4}$. Thus
$\sigma_{\bar{p}} = \sqrt{\frac{\bar{p}(1 - \bar{p})}{n}} \le \sqrt{\frac{1}{4n}}$.

11. To find the widest interval, let $\sigma_{\bar{p}} = \sqrt{\frac{1}{4n}}$. $\bar{p} - 2\left(\sqrt{\frac{1}{4n}}\right) < p < \bar{p} + 2\left(\sqrt{\frac{1}{4n}}\right)$, $\bar{p} - 2\left(\frac{1}{2}\sqrt{\frac{1}{n}}\right) < p < \bar{p} + 2\left(\frac{1}{2}\sqrt{\frac{1}{n}}\right)$, $\bar{p} - \sqrt{\frac{1}{n}} < p < \bar{p} + \sqrt{\frac{1}{n}}$

12. From Exercise 11, the widest possible 95% confidence interval is $\bar{p} - \sqrt{\frac{1}{n}} < p < \bar{p} + \sqrt{\frac{1}{n}}$. Thus $|p - \bar{p}| < \sqrt{\frac{1}{n}} \le 0.02$. $\sqrt{\frac{1}{n}} \le 0.02$, $\frac{1}{n} \le 0.04$, $n \ge \frac{1}{0.04}$, $n \ge 25$; a sample of 25 or more

13. $p - 2\sigma_{\bar{p}} < \bar{p} < p + 2\sigma_{\bar{p}}$; $-2\sigma_{\bar{p}} < \bar{p} - p < 2\sigma_{\bar{p}}$; $-\bar{p} - 2\sigma_{\bar{p}} < -p < -\bar{p} + 2\sigma_{\bar{p}}$; $\bar{p} + 2\sigma_{\bar{p}} > p > \bar{p} - 2\sigma_{\bar{p}}$; $\bar{p} - 2\sigma_{\bar{p}} < p < \bar{p} + 2\sigma_{\bar{p}}$

C **14.** $\bar{p} = 36\% = 0.36$; $\sigma_{\bar{p}} = \sqrt{\frac{(0.36)(0.64)}{n}} = \frac{0.48}{\sqrt{n}}$; $0.36 - 2\left(\frac{0.48}{\sqrt{n}}\right) < p < 0.36 + 2\left(\frac{0.48}{\sqrt{n}}\right)$, $0.36 - \frac{0.96}{\sqrt{n}} < p < 0.36 + \frac{0.96}{\sqrt{n}}$; $\frac{0.96}{\sqrt{n}} \le 0.04$, $24 \le \sqrt{n}$, $576 \le n$; minimum sample of 576

Page 731 • SELF-TEST 2

1. The z-score for 3 is $z = \frac{3 - 4.2}{0.8} = -1.5$. The z-score for 5 is $z = \frac{5 - 4.2}{0.8} = 1$.
$P(-1.5 < z < 1) = A(1.5) + A(1) = 0.4332 + 0.3413 = 0.7745$; 77.45%

2. The z-score for 10 is $z = \frac{10 - 15}{3.5} \approx -1.4$. $P(z < -1.4) = P(z < 0) - A(1.4) = 0.5 - 0.4192 = 0.0808$

3. $\bar{p} = \frac{4}{100} = 0.04$; $\sigma_{\bar{p}} = \sqrt{\frac{(0.04)(0.96)}{100}} \approx 0.020$; $0.04 - 2(0.020) < p < 0.04 + 2(0.020)$, $0 < p < 0.08$

Pages 738–739 • WRITTEN EXERCISES

A **1.**

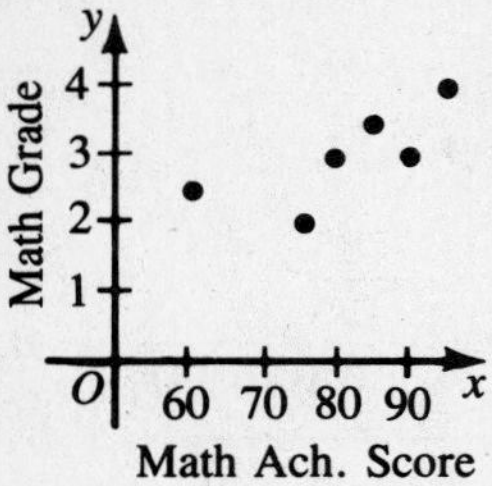
y
4
3
2
1
O
60 70 80 90
x
Math Grade
Math Ach. Score

2.

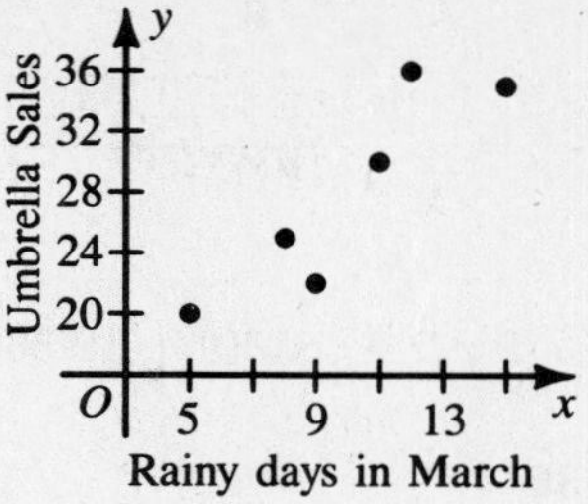
y
36
32
28
24
20
O
5 9 13
x
Umbrella Sales
Rainy days in March

3.

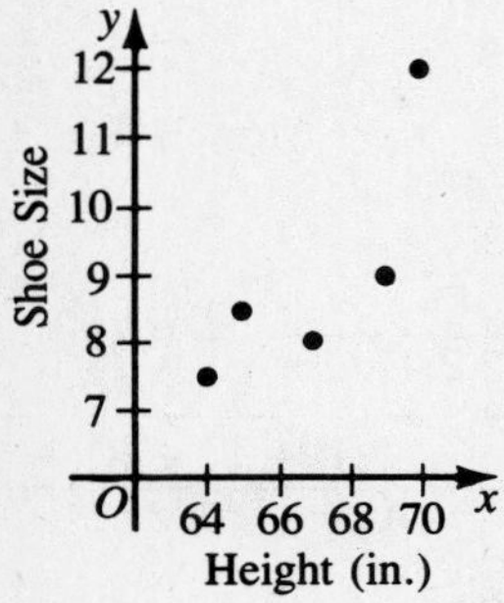
y
12
11
10
9
8
7
O
64 66 68 70
x
Shoe Size
Height (in.)

4.

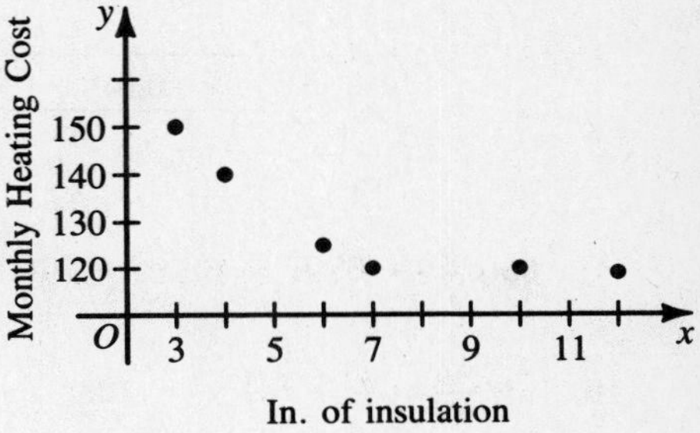
y
150
140
130
120
O
3 5 7 9 11
x
Monthly Heating Cost
In. of insulation

5.

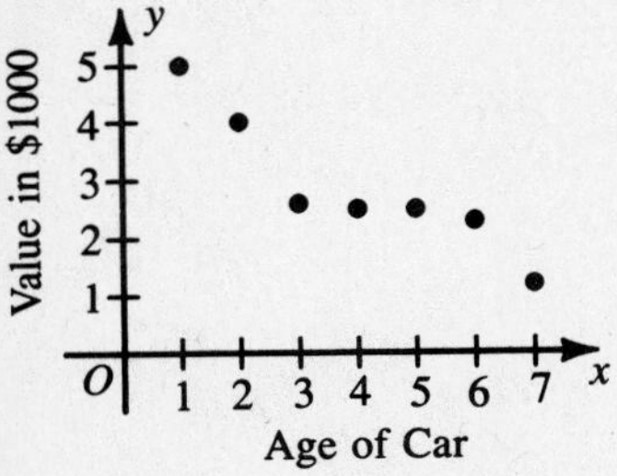
y
5
4
3
2
1
O
1 2 3 4 5 6 7
x
Value in $1000
Age of Car

6.

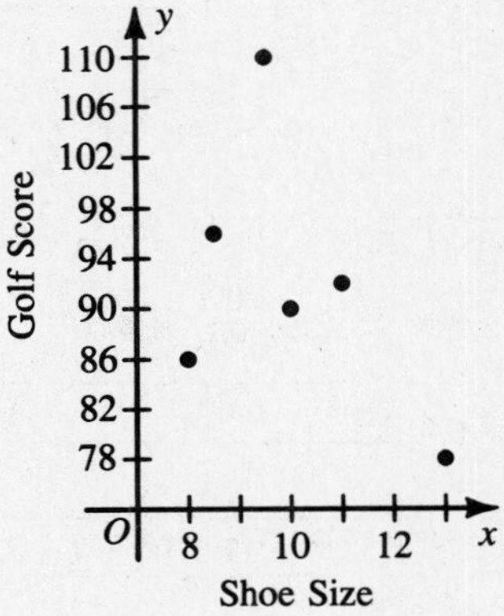
y
110
106
102
98
94
90
86
82
78
O
8 10 12
x
Golf Score
Shoe Size

7. $\bar{x} = \dfrac{485}{6} \approx 80.83$; $\bar{y} = \dfrac{18}{6} = 3$; $\sigma_x =$

$$\sqrt{\frac{(-20.83)^2 + (-5.83)^2 + (-0.83)^2 + (4.17)^2 + (9.17)^2 + (14.17)^2}{6}} \approx 11.33; \ \sigma_y =$$

$$\sqrt{\frac{(-0.5)^2 + (-1)^2 + 0^2 + (0.5)^2 + 0^2 + 1^2}{6}} \approx 0.65; \ \sum_{i=1}^{n} x_i \cdot y_i = 60(2.5) + 75(2) +$$

$$80(3) + 85(3.5) + 90(3) + 95(4) = 1487.5; \ r = \frac{\frac{1487.5}{6} - 80.83(3)}{11.33(0.65)} \approx 0.74$$

8. $\bar{x} = \frac{60}{6} = 10$; $\bar{y} = \frac{168}{6} = 28$; $\sigma_x = \sqrt{\frac{(-5)^2 + (-2)^2 + (-1)^2 + 1^2 + 2^2 + 5^2}{6}} \approx$ 3.16; $\sigma_y = \sqrt{\frac{(-8)^2 + (-3)^2 + (-6)^2 + 2^2 + 8^2 + 7^2}{6}} \approx 6.14$; $\sum_{i=1}^{n} x_i \cdot y_i = 5(20) +$ $8(25) + 9(22) + 11(30) + 12(36) + 15(35) = 1785$; $r = \frac{\frac{1785}{6} - 10(28)}{3.16(6.14)} \approx 0.90$

9. $\bar{x} = \frac{335}{5} = 67$; $\bar{y} = \frac{45}{5} = 9$; $\sigma_x = \sqrt{\frac{(-3)^2 + (-2)^2 + 0^2 + 2^2 + 3^2}{5}} \approx 2.28$; $\sigma_y = \sqrt{\frac{(-1.5)^2 + (-0.5)^2 + (-1)^2 + 0^2 + 3^2}{5}} \approx 1.58$; $\sum_{i=1}^{n} x_i \cdot y_i = 64(7.5) +$ $65(8.5) + 67(8) + 69(9) + 70(12) = 3029.5$; $r = \frac{\frac{3029.5}{5} - 67(9)}{2.28(1.58)} \approx 0.81$

10. $\bar{x} = \frac{42}{6} = 7$; $\bar{y} = \frac{774}{6} = 129$; $\sigma_x = \sqrt{\frac{(-4)^2 + (-3)^2 + (-1)^2 + 0^2 + 3^2 + 5^2}{6}} \approx$ 3.16; $\sigma_y = \sqrt{\frac{21^2 + 11^2 + (-4)^2 + (-9)^2 \cdot 2 + (-10)^2}{6}} \approx 11.83$; $\sum_{i=1}^{n} x_i \cdot y_i =$ $3(150) + 4(140) + 6(125) + 7(120) + 10(120) + 12(119) = 5228$; $r =$ $\frac{\frac{5228}{6} - 7(129)}{3.16(11.83)} \approx -0.85$

11. $\bar{x} = \frac{28}{7} = 4$; $\bar{y} = \frac{20.1}{7} \approx 2.87$;

$$\sigma_x = \sqrt{\frac{(-3)^2 + (-2)^2 + (-1)^2 + 0^2 + 1^2 + 2^2 + 3^2}{7}} = 2;$$

$$\sigma_y = \sqrt{\frac{(2.13)^2 + (1.13)^2 + (-0.27)^2 + (-0.37)^2 \cdot 2 + (-0.57)^2 + (-1.67)^2}{7}} \approx 1.15;$$

$$\sum_{i=1}^{n} x_i \cdot y_i = 1(5) + 2(4) + 3(2.6) + 4(2.5) + 5(2.5) + 6(2.3) + 7(1.2) = 65.5;$$

$$r = \frac{\frac{65.5}{7} - 4(2.87)}{2(1.15)} \approx -0.92$$

12. $\bar{x} = \frac{60}{6} = 10$; $\bar{y} = \frac{552}{6} = 92$;

$$\sigma_x = \sqrt{\frac{(-2)^2 + (-1.5)^2 + (-0.5)^2 + 0^2 + 1^2 + 3^2}{6}} \approx 1.66;$$

$\sigma_y = \sqrt{\frac{(-6)^2 + 4^2 + 18^2 + (-2)^2 + 0^2 + (-14)^2}{6}} \approx 9.80$; $\sum_{i=1}^{n} x_i \cdot y_i = 8(86) +$ $8.5(96) + 9.5(110) + 10(90) + 11(92) + 13(78) = 5475$;

$$r = \frac{\frac{5475}{6} - 10(92)}{1.66(9.80)} \approx -0.46$$

B **13. a.** $\bar{x} = \frac{a+b}{2}$; $\bar{y} = \frac{c+d}{2}$; $\sigma_x = \frac{b-a}{2}$; $\sigma_y = \frac{d-c}{2}$

b. $$r = \frac{\frac{\sum_{i=1}^{2} x_i y_i}{n} - \bar{x}\cdot\bar{y}}{\sigma_x \sigma_y} = \frac{\frac{ac+bd}{2} - \frac{(a+b)(c+d)}{4}}{\frac{(b-a)(d-c)}{4}} = \frac{ac+bd-bc-ad}{bd-bc-ad+ac} = 1$$

14. $\bar{x} = \frac{a+b}{2}$; $\bar{y} = \frac{c+d}{2}$; since $a < b$ and $c < d$, then $\sigma_x = \frac{b-a}{2}$ and $\sigma_y = \frac{c-d}{2}$.

$$\therefore\ r = \frac{\frac{ac+bd}{2} - \frac{(a+b)(c+d)}{4}}{\frac{(b-a)(c-d)}{4}} = \frac{ac+bd-ad-bc}{bc-bd-ac+ad} = -1$$

15. By Exercise 25 on page 715, $\bar{y} = m\bar{x}$ and $\sigma_y = |m| \cdot \sigma_x = m\sigma_x$.

$$\therefore\ r = \frac{\frac{\sum_{i=1}^{n} x_i(mx_i)}{n} - \bar{x}(m\bar{x})}{\sigma_x(m\sigma_x)} = \frac{\frac{\sum_{i=1}^{n} mx_i^2}{n} - m\bar{x}^2}{m\sigma_x^2} = \frac{m\left(\frac{\sum_{i=1}^{n} x_i^2}{n} - \bar{x}^2\right)}{m\sigma_x^2} =$$

$$\frac{\frac{\sum_{i=1}^{n} x_i^2}{n} - \left(\frac{\sum_{i=1}^{n} x_i}{n}\right)^2}{\sigma_x^2} = \frac{\frac{1}{n^2}\left(n\sum_{i=1}^{n} x_i^2 - \left(\sum_{i=1}^{n} x_i\right)^2\right)}{\sigma_x^2} = \frac{\sigma_x^2}{\sigma_x^2} = 1$$

16. Since $m < 0$, $\sigma_y = |m|\sigma_x = -m\sigma_x$. Proceed as in Ex. 15. $r = \frac{\frac{\sum_{i=1}^{n} mx_i^2}{n} - m\bar{x}^2}{-m\sigma_x^2} =$

$$\frac{\sigma_x^2}{-\sigma_x^2} = -1$$

C **17. a.** $$z_{x_i} \cdot z_{y_i} = \frac{(x_i - \bar{x})(y_i - \bar{y})}{\sigma_x \cdot \sigma_y} = \frac{x_i y_i - \bar{x}y_i - x_i\bar{y} + \bar{x}\cdot\bar{y}}{\sigma_x \sigma_y}$$

b. Since $\frac{1}{n}\sum_{i=1}^{n} x_i = \bar{x}$, $\sum_{i=1}^{n} x_i = n\bar{x}$. Similarly, $\sum_{i=1}^{n} y_i = n\bar{y}$.

$$\therefore\ \sum_{i=1}^{n}(x_i y_i - \bar{x}y_i - x_i\bar{y} + \bar{x}\cdot\bar{y}) = \sum_{i=1}^{n} x_i y_i - \sum_{i=1}^{n}\bar{x}y_i - \sum_{i=1}^{n} x_i\bar{y} + \sum_{i=1}^{n}\bar{x}\cdot\bar{y} =$$

$$\sum_{i=1}^{n} x_i y_i - \bar{x}\sum_{i=1}^{n} y_i - \bar{y}\sum_{i=1}^{n} x_i + \sum_{i=1}^{n}\bar{x}\cdot\bar{y} = \sum_{i=1}^{n} x_i y_i - n\bar{x}\cdot\bar{y} - n\bar{x}\cdot\bar{y} + n\bar{x}\cdot\bar{y} =$$

$$\sum_{i=1}^{n} x_i y_i - n\bar{x}\cdot\bar{y}$$

c. $\dfrac{\sum_{i=1}^{n} z_{x_i} \cdot z_{y_i}}{n} = \dfrac{\sum_{i=1}^{n} \left(\dfrac{x_i y_i - \overline{x} y_i - x_i \overline{y} + \overline{x} \cdot \overline{y}}{\sigma_x \sigma_y} \right)}{n} = \dfrac{1}{n\sigma_x \sigma_y}\left(\sum_{i=1}^{n} x_i y_i - n\overline{x} \cdot \overline{y} \right) =$

$$\frac{\dfrac{\sum_{i=1}^{n} x_i y_i}{n} - \overline{x} \cdot \overline{y}}{\sigma_x \sigma_y}$$

Page 739 • COMPUTER EXERCISES

1.

```
10  PRINT "THE PROGRAM IS READING THE NECESSARY DATA."
20  PRINT "PLEASE WAIT." : PRINT
30  READ N
40  FOR I = 1 TO N
50  READ X(I),Y(I)
60  LET PR = PR + X(I) * Y(I)
70  LET SX = SX + X(I)
80  LET SY = SY + Y(I)
90  NEXT I
100  LET MX = SX/N
110  LET MY = SY/N
120  FOR I = 1 TO N
130  LET NX = NX + (X(I) - MX)^2
140  LET NY = NY + (Y(I) - MY)^2
150  NEXT I
160  LET DX = SQR(NX/N)
170  LET DY = SQR(NY/N)
180  LET CC = (PR/N - MX * MY)/(DX * DY)
190  LET CC = INT(CC * 100 + .5)/100
200  PRINT "CORRELATION COEFFICIENT: ";CC
900  DATA 5
910  DATA 1,14,3,11,4,7,5,8,7,5
920  END
```

Correlation coefficient: −0.95

2. 0.74 **3.** 0.90 **4.** 0.80 **5.** −0.85 **6.** −0.92 **7.** −0.46

Page 739 • SELF-TEST 3

1.

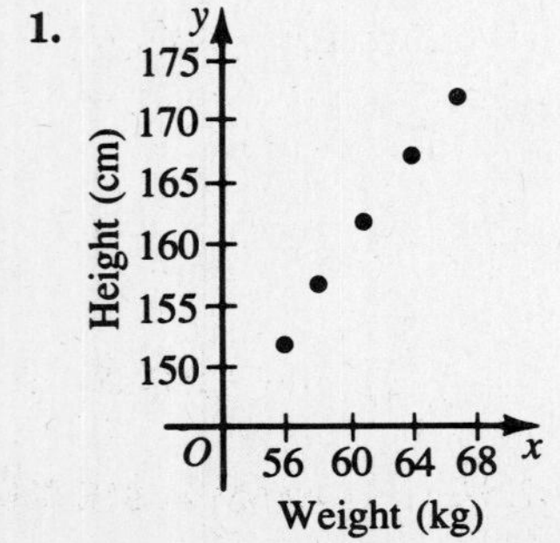

2.

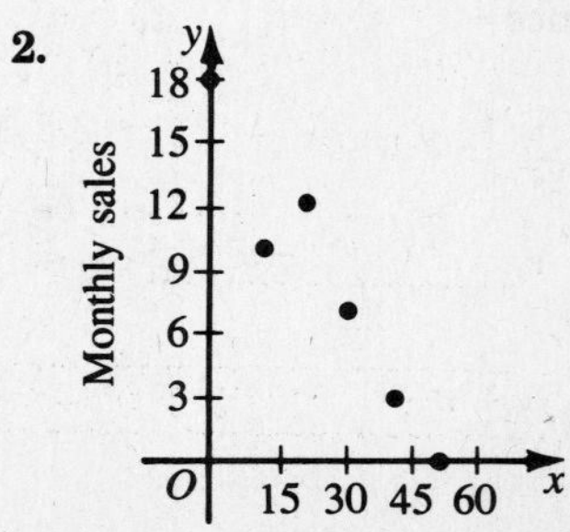

3. $\bar{x} = \frac{306}{5} = 61.2$; $\bar{y} = \frac{810}{5} = 162$;

$$\sigma_x = \sqrt{\frac{(-5.2)^2 + (-3.2)^2 + (-0.2)^2 + (2.8)^2 + (5.8)^2}{5}} \approx 3.97;$$

$$\sigma_y = \sqrt{\frac{(-10)^2 + (-5)^2 + 0^2 + 5^2 + 10^2}{5}} \approx 7.07; \sum_{i=1}^{n} x_i \cdot y_i = 56(152) +$$

$$58(157) + 61(162) + 64(167) + 67(172) = 49{,}712; \; r = \frac{\frac{49{,}712}{5} - 61.2(162)}{3.97(7.07)} \approx 0.998$$

4. $\bar{x} = \frac{150}{6} = 25$; $\bar{y} = \frac{50}{6} \approx 8.33$;

$$\sigma_x = \sqrt{\frac{(-25)^2 + (-15)^2 + (-5)^2 + 5^2 + 15^2 + 25^2}{6}} \approx 17.08;$$

$$\sigma_y = \sqrt{\frac{(9.67)^2 + (1.67)^2 + (3.67)^2 + (-1.33)^2 + (-5.33)^2 + (-8.33)^2}{6}} \approx 5.91;$$

$$\sum_{i=1}^{n} x_i \cdot y_i = 0(18) + 10(10) + 20(12) + 30(7) + 40(3) + 50(0) = 670;$$

$$r = \frac{\frac{670}{6} - 25(8.33)}{17.08(5.91)} \approx -0.96$$

Pages 741–742 • CHAPTER REVIEW

1. mean $= \frac{2150}{10} = 215$; b

2. median $= \frac{195 + 209}{2} = 202$; a

3. $\sigma^2 = [(-68)^2 + (-61)^2 + (-43)^2 + (-20)^2 \cdot 2 + (-6)^2 + 3^2 + 26^2 + 68^2 + 121^2]/10 = \frac{30{,}980}{10} = 3098$; $\sigma = \sqrt{3098} \approx 55.7$; a

4. $z = \frac{195 - 215}{55.7} \approx -0.36$; c

5. $P(z > 2) = P(z > 0) - A(2) = 0.5 - 0.4772 = 0.0228$; b

6. For 14: $z = \frac{14 - 16}{4} = -0.5$. For 18: $z = \frac{18 - 16}{4} = 0.5$. $P(-0.5 < z < 0.5) = A(0.5) + A(0.5) = 2(0.1915) = 0.383 \approx 0.38$; d

7. $p = 60\% = 0.6$; $\sigma_{\bar{p}} = \sqrt{\frac{(0.6)(0.4)}{25}} \approx 0.098$; c

8. $0.6 - 2(0.098) < \bar{p} < 0.6 + 2(0.098)$, $0.404 < \bar{p} < 0.796$, $0.40 < \bar{p} < 0.80$; b

9. $\bar{x} = \frac{13{,}776}{7} = 1968$; $\bar{y} = \frac{413}{7} = 59$;

$$\sigma_x = \sqrt{\frac{(-12)^2 + (-8)^2 + (-4)^2 + 0^2 + 4^2 + 8^2 + 12^2}{7}} = 8;$$

$$\sigma_y = \sqrt{\frac{3^2 + 2^2 + 1^2 \cdot 2 + 0^2 + (-3)^2 + (-4)^2}{7}} \approx 2.39; \sum_{i=1}^{n} x_i \cdot y_i = 1956(62) +$$

$$1960(61) + 1964(60) + 1968(60) + 1972(59) + 1976(56) + 1980(55) = 812{,}656;$$

$$r = \frac{\frac{812{,}656}{7} - 1968(59)}{8(2.39)} \approx -0.96; \text{ a}$$

10. $\bar{x} = \frac{356}{6} = 59.33$; $\bar{y} = \frac{1936}{6} \approx 322.67$;

$\sigma_x = \sqrt{\frac{(-46.33)^2 + (-12.33)^2 + (-11.33)^2 + (3.67)^2 + (17.67)^2 + (48.67)^2}{6}} \approx$

29.22; $\sigma_y = \sqrt{\frac{(-55.67)^2 + (-51.67)^2 + (81.33)^2 + (96.33)^2 + (-72.67)^2 + (2.33)^2}{6}}$

≈ 67.02; $\sum_{i=1}^{n} x_i \cdot y_i = 13(267) + 47(271) + 48(404) + 63(419) + 77(250) +$

$108(325) = 116{,}347$; $r = \frac{\frac{116{,}347}{6} - 59.33(322.67)}{29.22(67.02)} \approx 0.13$; b

Page 742 • CHAPTER TEST

1. 0, 1, 3, 4, 6, 7, 7, 8, 9, 10; mean $= \frac{55}{10} = 5.5$; median $= \frac{6 + 7}{2} = 6.5$; mode $= 7$

2. $\bar{x} = \frac{371}{5} = 74.2$; $\sigma^2 = \frac{(-4.2)^2 + (-3.2)^2 + (0.8)^2 + (2.8)^2 + (3.8)^2}{5} = \frac{50.8}{5} = 10.16$;

$\sigma = \sqrt{10.16} \approx 3.19$

3. $P(-1 < z < 2) = A(1) + A(2) = 0.3413 + 0.4772 = 0.8185$

4. For 7: $z = \frac{7 - 12}{2} = -2.5$. $P(z > -2.5) = A(2.5) + P(z > 0) = 0.4938 + 0.5 =$ 0.9938; 99.38%

5. $p = 80\% = 0.8$; $\bar{p} = \frac{7}{16} = 0.4375$; $\sigma_{\bar{p}} = \sqrt{\frac{(0.8)(0.2)}{16}} = 0.1$; $0.4375 - 2(0.1) < p <$

$0.4375 + 2(0.1)$, $0.2375 < p < 0.6375$; since p does not fall within this interval, the flower shop's claim is rejected.

6. $\bar{p} = \frac{8}{100} = 0.08$; $\sigma_{\bar{p}} = \sqrt{\frac{(0.08)(0.92)}{100}} \approx 0.027$; $0.08 - 2(0.027) < p < 0.08 +$

$2(0.027)$, $0.026 < p < 0.134$. Multiply by 100 to find the number of defective light bulbs. $100(0.026) = 2.6 \approx 3$; $100(0.134) = 13.4 \approx 13$; 3 to 13 defective light bulbs

7.

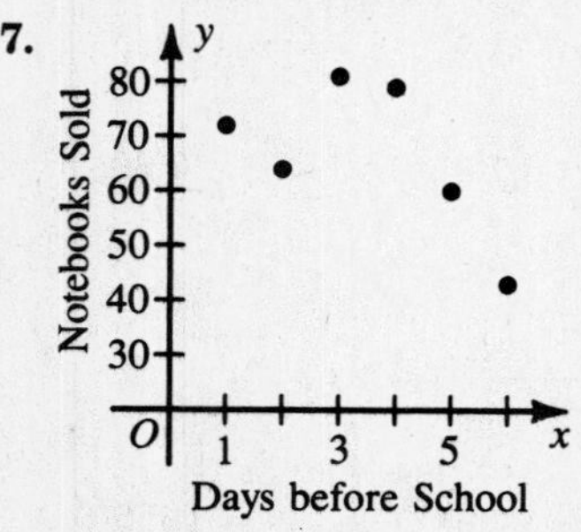

8. $\bar{x} = \dfrac{21}{6} = 3.5$; $\bar{y} = \dfrac{399}{6} = 66.5$;

$$\sigma_x = \sqrt{\frac{(2.5)^2 + (1.5)^2 + (0.5)^2 + (-0.5)^2 + (-1.5)^2 + (-2.5)^2}{6}} \approx 1.71;$$

$$\sigma_y = \sqrt{\frac{(-23.5)^2 + (-6.5)^2 + (12.5)^2 + (14.5)^2 + (-2.5)^2 + (5.5)^2}{6}} \approx 12.89;$$

$$\sum_{i=1}^{n} x_i \cdot y_i = 6(43) + 5(60) + 4(79) + 3(81) + 2(64) + 1(72) = 1317;$$

$$r = \frac{\frac{1317}{6} - 3.5(66.5)}{1.71(12.89)} \approx -0.60$$

Page 743 • PREPARING FOR COLLEGE ENTRANCE EXAMS

1. $\sin \dfrac{10\pi}{3} = -\dfrac{\sqrt{3}}{2}$; since $0 < \dfrac{3}{10\pi} < \dfrac{\pi}{2}$, $\cot \dfrac{3}{10\pi} > 0$; $\therefore \cot \dfrac{3}{10\pi} > \sin \dfrac{10\pi}{3}$; B
2. If x is a multiple of π, $\tan x = \tan(-x)$. If $\tan x > 0$, $\tan x > \tan(-x)$. If $\tan x < 0$, $\tan x < \tan(-x)$. D
3. If $x = \dfrac{\pi}{4}$, $\sin x + \cos x = \dfrac{1}{\sqrt{2}} + \dfrac{1}{\sqrt{2}} = \sqrt{2}$ and $\cot x + \tan x = 1 + 1 = 2$. Thus $(\sin x + \cos x) < (\cot x + \tan x)$. If $x = -\dfrac{\pi}{4}$, $\sin x + \cos x = -\dfrac{1}{\sqrt{2}} + \dfrac{1}{\sqrt{2}} = 0$ and $\cot x + \tan x = -1 + (-1) = -2$. Thus $(\sin x + \cos x) > (\cot x + \tan x)$. D
4. $1^{2n} = (1^2)^n = 1^n$; C
5. $(\sqrt{2} + \sqrt{5})^2 = 7 + 2\sqrt{10}$; $(\sqrt{7})^2 = 7$; since $(\sqrt{2} + \sqrt{5})^2 > (\sqrt{7})^2$, $\sqrt{2} + \sqrt{5} > \sqrt{7}$; A
6. If $a = b = c = 0$, $a + b + c = abc$. If $a = b = 1$ and $c = 0$, $a + b + c > abc$. If $a = 1$, $b = 3$, and $c = 4$, $a + b + c < abc$. D
7. Square: $A = 2^2 = 4$; circle: $A = \pi 2^2 = 4\pi$; B
8. $\dfrac{1}{y^2} = \dfrac{1^2}{y^2} = \left(\dfrac{1}{y}\right)^2$; C
9. $1 - \dfrac{a}{c} = \dfrac{c - a}{c}$; $\dfrac{-1}{\frac{c}{c-a}} = -\left(\dfrac{c - a}{c}\right)$; the expressions are opposites. If $c > a$, value of expression A is greater. If $c < a$, value of expression B is greater. D
10. $\text{Cos}^{-1}(\sin 180°) = \text{Cos}^{-1} 0 = 90°$; B
11. $r = \sqrt{1^2 + (\sqrt{3})^2} = 2$; $\cos\theta = \dfrac{1}{2}$, $m(\theta) = 60°$; $1 + i\sqrt{3} = 2(\cos 60° + i \sin 60°)$; E
12. $\|\mathbf{v}\| = \sqrt{6^2 + 8^2} = 10$; C
13. The terms of the sequence are 2, 3, 5, 10, 20, 40, . . . ; B

Page 744 • CONTEST PROBLEMS

1.

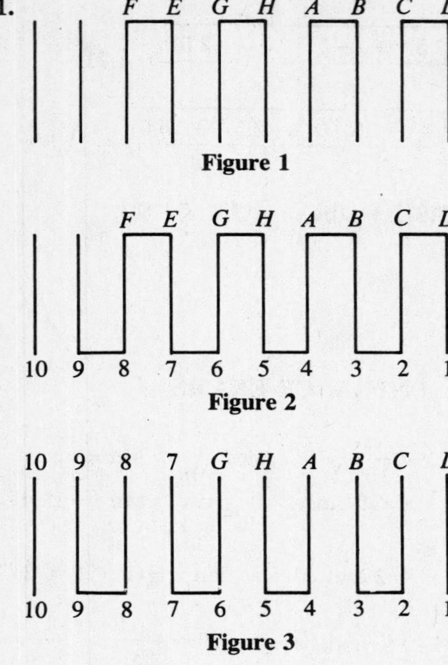

Figure 1

Figure 2

Figure 3

Connect four pairs of wires on one side. Label the connected wires in pairs — A and B, C and D, E and F, G and H. See Figure 1.

Go to the other side. By using the continuity tester, determine which pairs of wires have a closed circuit (i.e., which pairs had been connected on the first side). Label the wires in pairs — 1 and 2, 3 and 4, 5 and 6, 7 and 8. Connect 2 to 3, 4 to 5, 6 to 7. Connect 8 to one of other two wires; label the wire 9. The remaining wire is 10. See Figure 2.

Return to the first side. Determine which of the original two nonpaired wires is not part of the circuit; label that wire 10. The other wire is 9. Remove the wire connectors from this side. Only one wire will now form a closed circuit with 9; label that wire 8. The wire that was originally paired with 8 becomes 7 (i.e., if 8 was F, then E becomes 7). See Figure 3.

Only one wire will now form a closed circuit with 7; label that wire 6. The wire that was originally paired with 6 becomes 5. Only one wire will now form a closed circuit with 5; label that wire 4. The wire that was originally paired with 4 becomes 3. And so on.

Page 745 • PROGRAMMING IN PASCAL

1. **a.** See the procedure *range_of_data* in Exercise 4.
   b. See the procedure *mean_and_sd* in Exercise 4.
   c. See the procedure *mean_and_sd* in Exercise 4.
2. **a.** Run the program in Exercise 4. See the results under "FOR THE STANDARDIZED DATA".
   b. Run the program in Exercise 4. See the results under "FOR THE STANDARDIZED DATA".
   c. See the procedures *standardize* and *display_data* in Exercise 4.
3. **a.** Run the program in Exercise 4. Compare the results for the original data with those for the translated data.
   b. Run the program in Exercise 4. Compare the results for the original data with those for the translated data.
   c. Determine the mean of the test grades. Find the difference between that mean and the desired mean. Add that difference to each of the test grades in the class.

4.
```
PROGRAM stats (INPUT, OUTPUT);

CONST
  max = 20;    (* maximum number of items *)

TYPE
  values = ARRAY[1..max] OF real;

VAR
  data, z_data, trans_data : values;
  i, counter : integer;
  mean, sd, k, new_mean, old_mean : real;

(*************************************************************)
PROCEDURE obtain_data;

BEGIN
  counter := 0;
  writeln('Enter -1 to stop.');  (* impossible data value *)
  REPEAT
    counter := counter + 1;
    write('Enter number: ');
    readln(data[counter]);
  UNTIL (data[counter] = -1) OR (counter = max);
  IF data[counter] = -1
     THEN counter := counter - 1;
END;

(*************************************************************)
PROCEDURE range_of_data;

VAR
  max, min : real;
  i : integer;

BEGIN
  max := data[1];
  min := data[1];
  FOR i := 2 TO counter DO
      BEGIN
        IF data[i] > max
           THEN max := data[i];
        IF data[i] < min
           THEN min := data[i]
      END;
```

[*Program continued on next page*]

```
  writeln;
  writeln('Maximum: ', max:5:3);
  writeln('Minimum: ', min:5:3);
  writeln('Range: ', (max - min):5:3);
  writeln;
END;

(**************************************************************)
PROCEDURE mean_and_sd (score : values);

VAR
  sum, sqrs_summed : real;
  i : integer;

BEGIN
  sum := 0;
  sqrs_summed := 0;
  FOR i := 1 TO counter DO
      BEGIN
        sum := sum + score[i];
        sqrs_summed := sqrs_summed + score[i] * score[i];
      END;
  mean := sum/counter;
  sd := sqrt(sqrs_summed/counter - mean * mean);
END;

(**************************************************************)
PROCEDURE standardize;

VAR
  i : integer;

BEGIN
  FOR i := 1 TO counter DO
      z_data[i] := (data[i] - mean)/sd;
END;

(**************************************************************)
PROCEDURE display_data (score : values);

VAR
  i, position : integer;
```

```
BEGIN
  position := 1;
  FOR i := 1 TO counter DO
      BEGIN
        position := position + 1;
        write(score[i]:5:3, ' ':4);
        IF position MOD 8 = 0
           THEN writeln;
      END;
  IF position MOD 8 <> 0
     THEN writeln;
END;

(*************************************************************)
BEGIN (* main *)
  obtain_data;
  IF counter > 0
     THEN BEGIN
            range_of_data;
            mean_and_sd(data);
            writeln('Mean: ', mean:5:3);
            writeln('Standard Deviation: ', sd:5:3);
            old_mean := mean;
            writeln;
            standardize;
            mean_and_sd(z_data);
            writeln('FOR THE STANDARDIZED DATA');
            writeln('Mean: ', mean:5:3);
            writeln('Standard Deviation: ', sd:5:3);
            writeln;
            writeln('The standardized data is: ');
            display_data(z_data);
            writeln;
            write('Enter the amount by which you want to ');
            write('increase the data: ');
            readln(k);
            writeln;
            FOR i := 1 TO counter DO
                trans_data[i] := data[i] + k;
            mean_and_sd(trans_data);
            writeln('FOR THE TRANSLATED DATA');
            writeln('Mean: ', mean:5:3);
            writeln('Standard Deviation: ', sd:5:3);
            writeln;
```

[*Program continued on next page*]

```
            write('Enter the value for the new mean: ');
            readln(new_mean);
            writeln;
            k := new_mean - old_mean;
            FOR i := 1 TO counter DO
                data[i] := data[i] + k;
            mean_and_sd(data);
            write('FOR THE ADJUSTED DATA HAVING ');
            writeln('MEAN ', mean:5:3);
            writeln('The data is: ');
            display_data(data);
          END
     ELSE writeln('There isn''t any data to analyze.');
END.
```

Page 776 • EXERCISES

1. Valid
2. Invalid; the first character may not be a numeral.
3. Invalid; the hyphen, a special character, may not be used.
4. Valid 5. Valid
6. Invalid; the word "program" should begin the line.
7. Invalid; the hyphen, a special character, may not be used and a semicolon should end the line.
8. Invalid; the word "program" should begin the line, the special character "&" may not be used, and a semicolon should end the line.
9. False 10. True 11. True 12. False

Page 777 • EXERCISES

1. Valid
2. Invalid; the variable must be of type real to be assigned the result of a "/" division.
3. Valid 4. Valid
5. Invalid; the variable must be of type boolean to be assigned the value "true".
6. Invalid; only one variable must appear on the left side of an assignment statement, and the variables must be of the same type.
7. Valid 8. Valid 9. Valid
10. Invalid; only one variable must appear on the left side of an assignment statement.
11. Invalid; a comma must not appear in a number.
12. Valid 13. Valid
14. Invalid; the variables must be of type integer to be used with a "DIV" division.
15. Valid

Pages 781–782 • EXERCISES

Note: You may need to adjust the read and readln commands in order to accommodate the specifics of the version of Pascal that you are using.

1. 1)
```
writeln('PLEASE ENTER YOUR CODE');
```
   2)
```
write('PLEASE ENTER ');
writeln('YOUR CODE');
```
2. 1)
```
writeln('CLIENT NUMBER');
writeln('      10');
writeln('      11');
writeln('      12');
writeln('      13');
```
   2)
```
write('CLIENT ');
writeln('NUMBER');
writeln('      10');
writeln('      11');
writeln('      12');
writeln('      13');
```

3. 1)
```
writeln('TODAY''S DATE ____________');
writeln('TIME ____________');
```
2)
```
write('TODAY''S ');
writeln('DATE ____________');
writeln('TIME ____________');
```
4.
```
The even numbers between 1 and 11 are
2, 4, 6, 8, 10
```
5.
```
The batter's average is
0.365
```
6. Cab **7.** Can

8. **a.** Assume the user enters 10. The program displays the following:
```
Enter a number; press <return> : 10
Your number is : 10
```
b. The program now displays:
```
Enter a number : 10 Your number is : 10
```
The display is now on one line. The change results from the difference between *read* and *readln.*

c. The program now displays:
```
Enter a number :
10 Your number is :
10
```
The display is now on three lines. The change results from the difference between *write* and *writeln.*

d. (i.) *read:* The user may press the spacebar or the carriage return in order to have the input read. The cursor remains on the same line, ready to read the next input value or to execute the next command. *readln:* The user must press the carriage return in order to have the input read. The cursor then moves down to the next line.

(ii.) *write:* After the text of the *write* statement is printed, the cursor remains on the same line, ready to print the next output or to execute the next command. *writeln:* After the text of the *writeln* statement is printed, the cursor moves down to the next line.

9.
```
PROGRAM echo_and_square (input,output);

VAR number : integer;

BEGIN
  write('Enter a number; press <return> ');
  readln(number);
  writeln('Your number is ', number);
  writeln('Its square is ', number * number);
END.
```

Pages 786–787 • EXERCISES

1. False **2.** False **3.** True **4.** True

5.
```
PROGRAM pass_or_fail (input,output);

VAR grade : char;
    score : integer;
```

```
BEGIN
write('Enter the percent average for the course: ');
readln(score);
IF score < 60
   THEN grade := 'F'
   ELSE grade := 'P';
writeln(grade);
END.
```

6.

```
PROGRAM postage (input, output);

VAR weight, cost : real;

BEGIN
write('Enter weight of letter: ');
readln(weight);
IF weight <= 1
 THEN cost := 0.22
 ELSE IF weight <= 2
       THEN cost := 0.39
       ELSE IF weight <= 3
             THEN cost := 0.56
             ELSE IF weight <= 4
                   THEN cost := 0.73
                   ELSE IF weight <= 5
                         THEN cost := 0.90
                         ELSE IF weight <= 6
                               THEN cost := 1.07
                               ELSE IF weight <= 7
                                     THEN cost := 1.24
                                     ELSE IF weight <= 8
                                           THEN cost := 1.41
                                           ELSE IF weight <= 9
                                                 THEN cost := 1.58
                                                 ELSE cost := 1.75;
writeln('Cost is $',cost);
END.
```

7.

```
PROGRAM award_grade_and_comment (input,output);

  VAR grade : 'A'..'F';
      score : integer;

  BEGIN
  writeln('Enter integer score between 0 and 100 inclusive');
  readln(score);
```

[*Program continued on next page*]

```
IF score < 60
   THEN grade := 'F'
   ELSE IF score < 70
           THEN grade := 'D'
           ELSE IF score < 80
                   THEN grade := 'C'
                   ELSE IF score < 90
                           THEN grade := 'B'
                           ELSE grade := 'A';
IF grade = 'F'
   THEN writeln(grade, ' Unsatisfactory')
   ELSE IF grade = 'D'
           THEN writeln(grade, ' Needs improvement')
           ELSE IF grade = 'C'
                   THEN writeln(grade, ' Average')
                   ELSE IF grade = 'B'
                           THEN writeln(grade, ' Superior')
                           ELSE writeln(grade, ' Excellent');
END.
```

8.

```
PROGRAM postage_using_case (input, output);

VAR reply : integer;
    cost : real;

BEGIN
  writeln('Enter the number for the weight of the letter.');
  writeln('  (1)   weight <= 1 ounce');
  writeln('  (2)   1 ounce < weight <= 2 ounces');
  writeln('  (3)   2 ounces < weight <= 3 ounces');
  writeln('  (4)   3 ounces < weight <= 4 ounces');
  writeln('  (5)   4 ounces < weight <= 5 ounces');
  writeln('  (6)   5 ounces < weight <= 6 ounces');
  writeln('  (7)   6 ounces < weight <= 7 ounces');
  writeln('  (8)   7 ounces < weight <= 8 ounces');
  writeln('  (9)   8 ounces < weight <= 9 ounces');
  writeln(' (10)   9 ounces < weight <= 10 ounces');
  readln(reply);
  CASE reply OF
    1: cost := 0.22;
    2: cost := 0.39;
    3: cost := 0.56;
    4: cost := 0.73;
    5: cost := 0.90;
    6: cost := 1.07;
    7: cost := 1.24;
    8: cost := 1.41;
    9: cost := 1.58;
   10: cost := 1.75;
```

```
  END; (* end of case *)
  writeln('Cost is $',cost);
END.
```

Page 791 • EXERCISES

1. **a.** Answers will vary depending on the number chosen.
   b. See the function *cube* in Exercise 1c.
   c.

```
PROGRAM echo_and_square_and_cube (input, output);

VAR number : real;

(**********************************************************)
PROCEDURE get_num;

BEGIN
  write('Enter a number; press (return): ');
  readln(number);
END;

(**********************************************************)
FUNCTION square (x : real) : real;

BEGIN
  square := x * x;
END;

(**********************************************************)
FUNCTION cube (x : real) : real;

BEGIN
  cube := x * x * x;
END;

(**********************************************************)
PROCEDURE display_num_and_sqr_and_cube;

BEGIN
  writeln('Your number is : ', number);
  writeln('Its square is : ', square(number));
  writeln('Its cube is : ', cube(number));
END;

(**********************************************************)
BEGIN   (* Main Program *)
  get_num;
  display_num_and_sqr_and_cube
END.
```

2. a.
```
PROGRAM function_values (input, output);

VAR number : real;

(*********************************************************)
PROCEDURE get_num;

BEGIN
  write('Enter a value for x; press (return): ');
  readln(number);
END;

(*********************************************************)
FUNCTION f_of_x (x : real) : real;

BEGIN
  f_of_x := 2 * x * x - 3 * x - 5;
END;

(*********************************************************)
PROCEDURE display_x_and_f_of_x;

BEGIN
  writeln('The value of x is : ', number);
  writeln('The value of f(x) is : ', f_of_x(number));
END;

(*********************************************************)
BEGIN   (* Main Program *)
  get_num;
  display_x_and_f_of_x
END.
```

b. Replace the function in Exercise 2a with the following function.
```
(*********************************************************)
FUNCTION f_of_x (x : real) : real;

BEGIN
  f_of_x := x * x + x;
END;
```

c. Replace the function in Exercise 2a with the following function.
```
(*********************************************************)
FUNCTION f_of_x (x : real) : real;

BEGIN
  f_of_x := x * x - 1;
END;
```

3. a. The output should appear as shown at the top of page 791.

b.
```
PROCEDURE change (VAR num2, num3 : integer);
```

Page 794 • EXERCISES

1. No errors; 20
2. The initial value 20 is greater than the final value 10 so the reserved word DOWNTO must be used.
3. No errors; 5
4. No errors; 13
5. No errors; 11
6. The initial value 1 is less than the final value 6 so the reserved word TO must be used.
7. The complete output is:

```
This is loop number 6 and product = 6
This is loop number 5 and product = 30
This is loop number 4 and product = 120
This is loop number 3 and product = 360
This is loop number 2 and product = 720
This is loop number 1 and product = 720
The final product is 720
```

8. See the function *square* in Exercise 10.
9. See the procedure *display_nums_and_sqrs* in Exercise 10.
10.

```
PROGRAM echo_and_square (input, output);

VAR numbers : ARRAY[1..10] OF real;

(************************************************************)
PROCEDURE get_nums;

VAR index : integer;

BEGIN
  FOR index := 1 TO 10 DO
    BEGIN
      write('Enter a number; press <return>: ');
      readln(numbers[index])
    END;
END;

(************************************************************)
FUNCTION square (x : real) : real;

BEGIN
  square := x * x;
END;
```

[*Program continued on next page*]

```
(*****************************************************************)
PROCEDURE display_nums_and_sqrs;

VAR index : integer;

BEGIN
  FOR index := 1 TO 10 DO
    BEGIN
      writeln('Number ',index,' is : ',numbers[index]);
      writeln('Its square is : ',square(numbers[index]));
    END;
END;

(*****************************************************************)
BEGIN   (* Main Program *)
  get_nums;
  display_nums_and_sqrs
END.
```

Page 796 • EXERCISES

1. a.
```
x := 0;
REPEAT
  writeln(x);
  x := x + 2;
UNTIL x > 100;
```
b.
```
FOR x := 2 TO 100 DO
  IF x MOD 2 = 0 THEN writeln(x);
```
2. a.
```
x := 1;
sum := 0;
REPEAT
  sum := sum + x;
  x := x + 2;
UNTIL x > 99;
```
b.
```
sum := 0;
FOR x := 1 TO 99 DO
  IF x MOD 2 = 1 THEN sum := sum + x;
```
3. a.
```
number := 0;
terms := -1;
sum := 0;
REPEAT
  sum := sum + number;
  terms := terms + 1;
  write('Enter a number: ');
  readln(number);
UNTIL number = -1;
average := sum/terms;
```

b. No. In the REPEAT-UNTIL loop, the user controls termination of the loop by signaling the end of the input. In a FOR-NEXT loop, the number of loop executions must be known before the user enters any values.

Page 797 • EXERCISES

1.
```
x := 0;
WHILE x <= 100 DO
  BEGIN
    writeln(x);
    x := x + 2;
  END;
```
2.
```
x := 1;
sum := 0;
WHILE x <= 99 DO
  BEGIN
    sum := sum + x;
    x := x + 2;
  END;
```
3.
```
number := 0;
terms := -1;
sum := 0;
WHILE number <> -1 DO
  BEGIN
    sum := sum + number;
    terms := terms + 1;
    write('Enter a number: ');
    readln(number);
  END;
average := sum/terms;
```

Page 802 • EXERCISES

1. (1) For $n = 1$, $2 \cdot 1 - 1 = 1 = 1^2$. (2) If $1 + 3 + 5 + \ldots + (2k - 1) = k^2$, then $1 + 3 + 5 + \ldots + (2k - 1) + [2(k + 1) - 1] = k^2 + [2(k + 1) - 1] = k^2 + 2k + 1 = (k + 1)^2$.
2. (1) For $n = 1$, $2 \cdot 1 = 2 = 1^2 + 1$. (2) If $2 + 4 + 6 + \ldots + 2k = k^2 + k$, then $2 + 4 + 6 + \ldots + 2k + 2(k + 1) = k^2 + k + 2(k + 1) = k^2 + k + 2k + 2 = (k^2 + 2k + 1) + (k + 1) = (k + 1)^2 + (k + 1)$.
3. (1) For $n = 1$, $ar^{1-1} = a = \dfrac{a(1 - r^1)}{1 - r}$. (2) If $a + ar + ar^2 + \ldots + ar^{k-1} = \dfrac{a(1 - r^k)}{1 - r}$, then $a + ar + ar^2 + \ldots + ar^{k-1} + ar^k = \dfrac{a(1 - r^k)}{1 - r} + ar^k = \dfrac{a(1 - r^k)}{1 - r} + \dfrac{ar^k(1 - r)}{1 - r} = \dfrac{a - ar^k + ar^k - ar^{k+1}}{1 - r} = \dfrac{a - ar^{k+1}}{1 - r} = \dfrac{a(1 - r^{k+1})}{1 - r}$.
4. (1) For $n = 1$, $\dfrac{1}{1(1 + 1)} = \dfrac{1}{2} = \dfrac{1}{1 + 1}$. (2) If $\dfrac{1}{1 \cdot 2} + \dfrac{1}{2 \cdot 3} + \dfrac{1}{3 \cdot 4} + \ldots + \dfrac{1}{k(k + 1)} = \dfrac{k}{k + 1}$, then $\dfrac{1}{1 \cdot 2} + \dfrac{1}{2 \cdot 3} + \dfrac{1}{3 \cdot 4} + \ldots + \dfrac{1}{k(k + 1)} + \dfrac{1}{(k + 1)[(k + 1) + 1]} = \dfrac{k}{k + 1} + \dfrac{1}{(k + 1)[(k + 1) + 1]} = \dfrac{k(k + 2)}{(k + 1)(k + 2)} + \dfrac{1}{(k + 1)(k + 2)} = \dfrac{k^2 + 2k + 1}{(k + 1)(k + 2)} = \dfrac{(k + 1)^2}{(k + 1)(k + 2)} = \dfrac{k + 1}{k + 2} = \dfrac{k + 1}{(k + 1) + 1}$.

Page 804 • EXERCISES

1. $70 \div 42 = 1$ (remainder 28); $42 \div 28 = 1$ (remainder 14); $28 \div 14 = 2$ (remainder 0); GCF = 14.
2. $624 \div 456 = 1$ (remainder 168); $456 \div 168 = 2$ (remainder 120); $168 \div 120 = 1$ (remainder 48); $120 \div 48 = 2$ (remainder 24); $48 \div 24 = 2$ (remainder 0); GCF = 24.
3. $1436 \div 465 = 3$ (remainder 41); $465 \div 41 = 11$ (remainder 14); $41 \div 14 = 2$ (remainder 13); $14 \div 13 = 1$ (remainder 1); $13 \div 1 = 13$ (remainder 0); GCF = 1.
4. $1122 \div 374 = 3$ (remainder 0); GCF = 374.
5. Because n is a positive integer, exponentiation can be treated as repeated multiplication: $x^1 = x$, $x^2 = x \cdot x$, $x^3 = x \cdot x^2$, and so on. Therefore, $x^n = \begin{cases} x \cdot x^{n-1} & \text{if } n \geq 2 \\ x & \text{if } n = 1 \end{cases}$
6. Consider the population sequence: $P(0) = P_0$, $P(1) = 2 \cdot P_0 = 2 \cdot P(0)$, $P(2) = 2 \cdot 2 \cdot P_0 = 2 \cdot P(1)$, and so on. Therefore, $P(t) = \begin{cases} 2 \cdot P(t - 1) & \text{if } t \geq 1 \\ P_0 & \text{if } t = 0 \end{cases}$

Pages 804–805 • EXERCISES

1. a. Starting at A, the other path is $A - B - C - D - F - E$; starting at B, the paths are $B - A - C - D - E - F$ and $B - A - C - D - F - E$; starting at E, $E - F - D - C - B - A$ and $E - F - D - C - A - B$; starting at F, $F - E - D - C - B - A$ and $F - E - D - C - A - B$.

b. Starting at either C or D would require the tourist to pass through the starting point a second time in order to reach all the towns.

2. a.

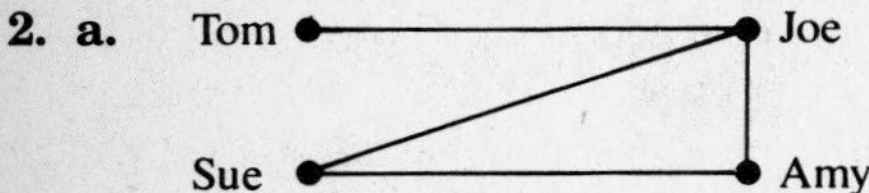

b. Tom does not share a class with either Amy or Sue.

Page 806 • EXERCISES

1. Answers may vary. For example: $A_1^* = \{no, on\}$; $A_2^* = \{ono, noo, nono, noon\}$; $A_3^* = \{o, oo, ooo, oooo, ooooo\}$
2. Answers may vary. For example: "A student greeted the principal."; "A teacher helped a student."; "The principal praised the teacher."
3.

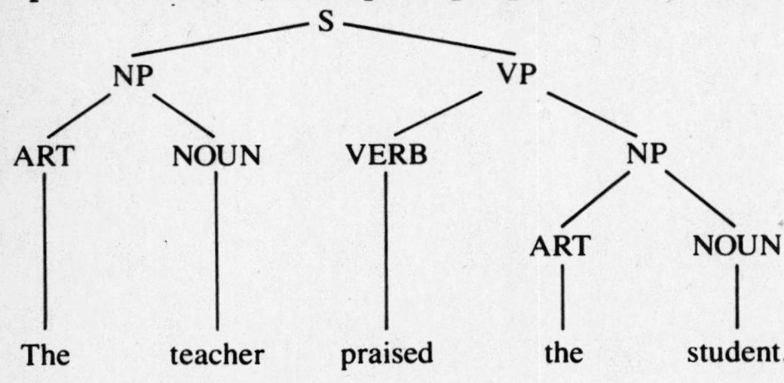